Optical Particle Sizing
Theory and Practice

Optical Particle Sizing

Theory and Practice

Edited by

Gérard Gouesbet and Gérard Gréhan

INSA de Rouen
CNRS Associated Laboratory 230
Mont-Saint-Aignan, France

Plenum Press • New York and London

Library of Congress Cataloging in Publication Data

Optical particle sizing: theory and practice / edited by Gérard Gouesbet and Gérard Gréhan.
p.   cm.
"Proceedings of an international symposium on Optical Particle Sizing: Theory and Practice, held May 12–15, 1987, in Rouen, France" — T.p. verso.
Includes bibliographies and index.
ISBN 0-306-42781-8
1. Particle size determination — Congresses. 2. Particles — Optical properties — Congresses. I. Gouesbet, Gérard. II. Gréhan, Gérard. TA418.8.076 1987
620′.43 — dc19   87-29148
                                                                                                            CIP

Proceedings of an international symposium on Optical Particle Sizing:
Theory and Practice, held May 12-15, 1987, in Rouen, France

© 1988 Plenum Press, New York
A Division of Plenum Publishing Corporation
233 Spring Street, New York, N.Y. 10013

Printed in the United States of America

PREFACE

     Optical particle sizing is undoubtedly a fascinating field of
research of the utmost practical importance. In the Universe fluids are
nearly everywhere, and when they occur they almost invariably contain
particles. Inside our bodies we can take the example of blood
transporting a vital procession of red and white cells. Around us, we
can find various particles in the air we breathe, bubbles in the
champagne or the soda we drink, or natural and artificial (polluting!)
particles in the lakes we swim in. Industrial processes and systems are
also concerned with particles, from pulverized coal flames to fluidized
beds, in a range of applications involving rocket exhausts, pneumatic
transport and more generally the infinite realm of multiphase
situations. Such an obviously vast field would require a whole volume
like this one merely to attempt to describe it superficially. To be
sure, we would need a scientific Prevert to catalogue such an endless
inventory. Finally, even outside our terrestrial spaceship particles can
be detected in alien atmospheres or between stars.

     Theorists will enjoy analyzing the richness of light/particle
interaction, a subject which is very far from being exhausted.
Experimental researchers will love designing and studying various
probing instruments with a laser source at the input and a computer at
the output, two requisites of today's technological revolution. Others
will be fond of using these instruments to learn and to understand the
behavior of various systems in industrial processes or geophysical flows
or to assimilate the theories needed to probe remote astrophysical
situations. The organization of the Rouen Conference adhered to the
above filiation, from theory to applications, via measurement
techniques.

     The aim was to bring together for four days experts on the theory
of light scattering and the optical sizing of particles in order to
emphasize the most significant advances, spread new knowledge and
outline future developments. This symposium was motivated by the
increasing interest in the field, thus requiring the sharing of mutual
experience between scientists working in various areas of physics.
Plenary lectures and formal presentations were given, and posters
displayed during the whole conference. This volume contains the papers
which were presented.

     As an editor of these proceedings, I was given the privilege (and
the responsibility) of writing this preface. However, my main
contribution to the conference had been simply to throw out the initial
idea of holding it. It would therefore not have been possible without
the help of numerous organizations and individuals whom I warmly thank.

v

Financial support and scientific sponsorship came from :

Conseil Régional de Haute-Normandie
Direction des Recherches Etudes et Techniques
Centre National de la Recherche Scientifique
Institut National des Sciences Appliquées de Rouen
Université de Haute-Normandie
Chambre de Commerce et de l'Industrie
GAMS/COFERA
Optical Society of America
Société Française d'Optique

Each of the numerous proposed abstracts was first reviewed by two members of the Advisory Committee, consisting of :

Dr. L.P. Bayvel
Dr. N. Chigier
Prof. D.F.G. Durao
Prof. S.E. Elgobashi
Vice Pres. W.M. Farmer
Prof. E.D. Hirleman
Dr. Ing. R. Kleine
Dr. J. Lahaye
Prof. M. Ledoux

Prof. S.S. Lee
Dr. Ing. B. Rück
Prof. C. Sorensen
Prof. D.E. Stock
Prof. J. Swithenbank
Dr. H.C. Van De Hulst
Prof. J.H. Whitelaw
Dr. M.L. Yeoman

Then, relying on these reviews, the Scientific Committee members

Prof. C.F. Bohren
Prof. F. Durst
Prof. G. Gouesbet
Prof. C. Imbert

Dr. A.R. Jones
Prof. M. Kerker
Dr. G. Madelaine
Prof. M. Maeda

were responsible for the final decision of acceptance. Because of the large number of abstracts received, and their overall high quality, selection was a difficult task. I particularly thank all the committee members who undertook to fulfill this task. A large number of papers were unfortunately not accepted, not because they were found to be poor, but merely because of the limited time available for presentations. I must apologize for these rejections and acknowledge that the authors of the rejected papers had their share in the success of this conference.

Special thanks should also be extended to the authors of the plenary lectures (M. Kerker, F. Durst, S. Galant) who introduced the first three days of the conference. A.R. Jones should be warmly commended for having performed the formidable task of compiling all the papers, and getting the gist of them to provide us with a concluding plenary lecture.

Besides the assistance of the Scientific Committee and the Advisory Committee members, the organization of the symposium was made possible thanks to F. Aubry, P. Bied, A. Fouquer, M. Grébonval, G. Gréhan, T. Lecordier, M. Ledoux, M.O. Monnet and A. Souillard. I am particularly indebted to my collaborator G. Gréhan (CNRS) and to P. Bied (INSCIR student), who carried out a lot of the work of organizing the scientific program and the social events, and helped me with the large amount of mailing which was necessary in preparing for the conference.

I would also like to express my friendly gratitude to P. Valentin, head of the Rouen INSA, who helped us solve the numerous problems that anyone trying to organize a conference is inevitably faced with in those unexpected encounters of the third kind.

    Finally, while gathering the advisory committee, I was shocked to
learn that Diran Deirmendjian died suddenly on the 15th june 1986. He is
well known by all people working in the field of light scattering. His
famous book entitled 'Electromagnetic Scattering on Spherical
Polydispersions', published in 1969 by American Elsevier, New York,
deserves a privileged place in the bokshelf of anyone working in that
field. In agreement with his friends and colleagues, this symposium is
dedicated to his memory.

Prof. G. GOUESBET

CONTENTS

LIGHT SCATTERING THEORY: A PROGRESS REPORT

Milton Kerker

Clarkson University
Potsdam, N.Y. 13676

INTRODUCTION

Having reached an age when egoism overwhelmes common sense, I have
decided to survey the general direction of recent research on light scattering by particles rather than to lecture on some of my research. And
so I must warn at the outset that the subject in hand is mainly biblio-
graphic, and to a lesser extent historical and philosophical. It should
establish my credentials as an elder if not as a statesman.

Of course one could go to that marvelous book by van de Hulst,[1] or
to one by myself,[2] or to the more recent one by Bohren and Huffman,[3] for
a bibliographical survey but even the latter book is at least five or six
years out-of-date. Progress continues unabated in this old field as is
evident from a meeting such as this.

It was about a year ago that I was invited to assemble a collection
of papers for a "Milestone Volume on Light Scattering."[4] This is to be
part of a reprint series devoted to various topics in optical sciences
and engineering. The thought was that there is no substitute for reading
the original papers and the further notion was that having these available
in a single volume would be a great convenience. The weeding and selection from the repository of the literature by a presumed expert would
make matters quite convenient, especially for someone new to the field.
The pitfall of course is that the bias exhibited in that selection will
offset the convenience.

I accepted; the task is complete and this lecture is a report, of
sorts.

It did not take long to appreciate that many areas would have to be
excluded in order to keep the size of the collection within prescribed
limits. The literature of scattering by molecular and macromolecular
systems is vast and that topic has been omitted. A reprint volume covering the early years was published some time ago.[5] Unfortunately this
has not been supplemented to include the tremendous outpouring of recent
work, including that dealing with quasi-elastic scattering. Yet, molecular scattering does stray from the main area of interest of this meeting
and of my own recent interests. Moreover, most of the physical optics is
constrained to simple Rayleigh dipolar scattering theory. This theory is

then used to elucidate the statistical thermodynamics, the molecular architecture, the molecular dynamics, the intermolecular and intramolecular interactions, etc., whether in the gaseous, the liquid or the solid state.

My excuse for omitting multiple scattering is similar. The present focus is on the complexities of the physical optics of single scattering.

Even the remaining corpus was still unmanageable and it became necessary to make still tougher decisions. Purely computational results were eliminated unless they illustrated new or unusual physical effects. Lest this omission be of great concern, be assured that most papers containing new theoretical conbributions or physical insights also include more than enough illustrative numerical results to satisfy even a voracious appetite for peering at tables and graphs.

I must be wary in how I express accounting for the avoidance to a large extent of experimental techniques, descriptions of instruments and experimental papers, particularly since the subject of this meeting is mainly experimental. The principal theme in our survey is the elucidation of the physical phenomenon of light scattering itself rather than its applications. Of course there is no intention to exclude experiments or numerical studies leading to the <u>discovery</u> of phenomena which may be accounted for later by theoretical analysis. But that does not seem to have been the most usual pattern.

It was the demands of space that necessitated both the omission of first-order Rayleigh-Debye-Born approximations and the great classical papers by Rayleigh, Love, Lorenz, Mie, Debye, Bromwich, Schwarzchild, etc. These are so detailed that they would require a separate volume. Instead there is included a perspicuous survey of early scattering studies written by Logan.

Because of this omission, I am impelled to offer some comments on Lord Rayleigh's contributions. Consider, for example, that the eponym Rayleigh scattering is a constant source of confusion because of the varied range of phenomena with which it can be associated. He wrote on the subject throughout his career from "On the light from the sky, its polarization and colour" in 1871 to "On the dispersal of light by a dielectric cylinder" in 1918. Rayleigh's most inventive paper on light scattering was published in 1881. By this time he evinced sufficient confidence in Maxwell's equations to utilize electromagnetic theory in deriving the volume integral for the scattered field as obtained from Green's theorem. This integral can be represented as an expansion in $\Delta\varepsilon$, the difference between the dielectric constant of the particle and the medium. The solution is perfectly general with no need to maintain any restriction either on the particle size or shape or $\Delta\varepsilon$, provided the indicated integrations which are far from trivial can be performed.

The restriction on size and shape could be relaxed when $\Delta\varepsilon$ is not too large, permitting Rayleigh to derive closed form expressions for angular scattering by spheres and cylinders, an approximation that masquerades under some combination of the names Rayleigh, Gans, Debye, Born—with Rayleigh's name frequently omitted. It was many years later that Rayleigh integrated the expression for a sphere to obtain the scattering cross section in closed form. Still later he extended the results to a sphere with a radially varying refractive index, including the case of a spherical shell. The model can hardly be stated more cogently than in Rayleigh's words: "each element of a homogeneous obstacle acts independently as a center of distribution and (that) the aggregate effect in any direction depends upon the phases of the elementary second disturbances as affected by the situation of the element along the paths of the primary

and of the secondary light.... It is evident that the formulae are applicable only when the whole retardation of the primary light in traversing
the obstacles can be neglected in comparison with the wavelength."

Next Rayleigh examines the squared term in the expansion of $\Delta\varepsilon$ for
the spherical case and notes that whereas the leading terms in the expansion does not contribute to the 90° scattering polarized parallel to the
scattering plane, this second-order term gives a contribution whose intensity varies as λ^{-8}. He proposed that this corresponds to Tyndall's
observation of such "residual blue" as particles become larger. One
recognizes this inverse eighth-power scatter as due to the quadrupole
terms in the multipolar expansion later discussed by G. Mie.

Next Rayleigh considers the case that the sphere is small relative to
the wavelength but for which the restriction on $\Delta\varepsilon$ is relaxed. This leads
to the famous Rayleigh formula for spheres, equivalent to that obtained in
1871, but now in terms of the dielectric constant rather than the density.

Up to this point Rayleigh has used the Green's theorem method to
express the scattered field as a volume integral over the field within
the particle. This is a completely general formulation, but its solution
by expansion into a power series in $\Delta\varepsilon$ is fraught with great computational
difficulties except for the leading term which gives the Rayleigh-Gans-
Debye-Born approximation. In the last part of this great 1881 paper,
Rayleigh returns to scattering by a cylinder which he now formulates as a
boundary value problem. The incident, internal, and scattered waves are
expanded in a Fourier series, the latter two with unknown coefficients,
so as to permit satisfying the boundary conditions. Rayleigh writes out
the general expressions for the scattering coefficients and discusses
some of the limiting cases. The method can be applied to objects whose
surface conforms to a coordinate surface in a coordinate system in which
the wave equation is separable. There are no limitations on particle size,
dielectric constant, or magnetic permeability. Later derivations by others
ignored Rayleigh's work, an oversight to which he called attention when
later carrying out some detailed calculations.

The first application of this boundary value method to spheres was
by L. Lorenz in 1890, and is now generically termed Mie scattering following the detailed discussion of its applicability to the colors of gold sols
by G. Mie. We have discussed elsewhere[2] the large number of more or less
independent boundary value treatments of the sphere of arbitrary size
including those by Lorenz (1890), Schwarzchild (1901), Thomson (1893),
Love (1899), Walker (1900), Mie (1908), Debye (1909), Bateman (1918), and
Bromwich (1919). It is surprising that Rayleigh did not trouble to treat
the sphere. The physical statement of the boundary value problem is identical whether a cylinder or sphere, and the mathematical techniques were
quite similar to those which he had already used to describe the scattering of sound waves. Indeed, in 1910 he comments on Love's treatment
of the sphere (1899) by the boundary value method: "Prof. Love, in a
valuable paper on the 'Scattering of Electric Waves by a Dielectric
Sphere,' has treated this problem by a different method, limited on the
one hand to the sphere, but on the other applicable whatever may be the
value of (K - 1), and (so far as the general analytical expressions are
concerned) whatever may be the size of the sphere. In the 1881 paper I
had treated in this way the problem of a dielectric cylinder." Rayleigh
then compares the leading terms in Love's solution with higher order
terms in his own treatment based on the Green's function integral treatment. He notes a discrepancy whereupon he locates an error in Love's
work thereby obtaining perfect concordance between the integral and
boundary value solutions. Rayleigh then carries out detailed numerical
calculations comparable to those in Mie's 1908 paper except that these

are for refractive index 1.5, whereas Mie dealt with the complex refractive index corresponding to gold.

Rayleigh maintained an interest in light scattering through his long scientific career. He opened up and established virtually all of the main lines of investigation of elastic scattering. Although the work is spread over many years throughout several papers, the great paper of 1881 is so rich that it must stand as both a seminal and a definitive classic of light scattering.

The papers selected for the Milestone volume are tabulated at the end of this article. The remainder of this talk will survey these in a cursory manner.

<u>References</u>
1. H.C. van de Hulst, "Light Scattering by Small Particles," John Wiley, New York, 1957.
2. M. Kerker, "The Scattering of Light and other Electromagnetic Radiation," Academic Press, New York, 1969.
3. C.F. Bohren and D.R. Huffman, "Absorption and Scattering of Light by Small Particles," Wiley-Interscience, New York, 1983.
4. M. Kerker (Editor), "Selected Papers on Light Scattering," SPIE Milestone Volume (1988).
5. M.B. Huglun (Editor), "Light Scattering from Polymer Solutions," Academic Press, New York, 1972.

BOUNDARY VALUE SOLUTIONS (References cited from here are found in the bibliography at the end of this narrative).

A. <u>Historical</u>

The initial paper by Logan[1] surveys the literature on the scattering of plane waves by a sphere from an 1861 memoir by A. Clebsch to work just prior to World War II. It includes the great papers by Rayleigh, Lorenz, J.J. Thomson, Mie, Debye and others of somewhat lesser fame. Indeed it sets an appropriate perspective for altering the eponym Mie scattering to an only somewhat more appropriate term Lorenz–Mie scattering.

B. <u>Spheres</u>

1. <u>New and alternate solutions</u>
This group of papers provides new or alternate boundary value solutions for spheres. Several authors have treated a somewhat more general formulation of the optical constants. Bohren[2] has derived the scattering matrix for media exhibiting optical rotation and circular dichroism for which the optical constitutive relations are

$$\vec{D} = \varepsilon\vec{E} + \alpha\varepsilon\vec{\nabla}\mathsf{x}\vec{E}, \ \vec{B} = \mu\vec{H} + \beta\mu\vec{\nabla}\mathsf{x}\vec{H} \tag{1}$$

Ford and Werner[3] have treated gyroelectric media for which the most general form of the dielectric tensor consistent with axial symmetry is given by

$$\overset{\leftrightarrow}{\varepsilon} = \begin{pmatrix} \varepsilon_{xx} & \varepsilon_{xy} & 0 \\ -\varepsilon_{xy} & \varepsilon_{xx} & 0 \\ 0 & 0 & \varepsilon_{zz} \end{pmatrix} \tag{2}$$

This solution can be extended to gyromagnetic media such as ferromagnetic materials for which the magnetic permeability tensor is of the same form. Ruppin has considered the case which allows for the excitation of longitudinal polarization waves within the medium of which the particle is composed.[4] He has also considered the effect of selecting several alternate boundary conditions.[5] Dasgupta and Fuchs[6] have provided a unified approach for treating these so-called nonlocal effects which may be due to surface effects on the dielectric properties of small particles.

De Zutter[7] has provided a relativistic analysis for scattering by a rotating sphere, and Twersky[8] and Michielsen et al.[9] considered the scattering by a moving sphere in the light of special relativity.

Interest in inelastic scattering by molecules embedded within or at the outer surface of spheres has required the boundary value solution for radiation by a radiating electric[10] or magnetic dipole[11] embedded with a sphere or by an electric dipole radiating at the outer surface.[12,13] This work was stimulated by an interest in the recently discovered phenomenon of surface enhanced Raman scattering. Hua and Gersten[14] have provided a general treatment of non-linear second-harmonic generation by a sphere illuminated by a plane wave.

A group of studies has extended the Lorenz-Mie boundary value solution to spheres which are inhomogeneous but spherically symmetric. Aden and Kerker[15] and Guttler[16] dealt with a spherical core surrounded by a concentric spherical shell. Fikioris and Uzunoglu have extended this to the case that the spherical inclusion need not be at the center.[17] Wait[18] and Bhandari[19] have generalized this to a sphere with an arbitrary number of homogeneous layers. Wyatt[20] has extended the model to include a continuously variable but radially symmetric refractive index and Levine and Kerker[21] have treated the case where the refractive index of the outer layer of a layered sphere has a power law dependence.

2. Computations

Two papers which discuss the details of computational routines are included[22,23] in order to provide touchstones against which to measure both accuracy and speed.

The Bessel and Legendre functions on which the Lorenz-Mie scattering coefficients depend are usually computed by recurrence. Verner[24] has shown that the denominators of the coefficients can be generated recurrently and Bohren[25] has proceeded to demonstrate that since these can be formulated as the multipolar expansion of the electromagnetic field, the coefficients themselves must satisfy recurrence relations. Chew[26] has derived an interesting relation between successive coefficients for perfectly conducting spheres.

3. Optical resonances

It became apparent as soon as detailed computations became available that the scattering cross section, as a function of either size parameter or refractive index, exhibited a very complicated fine structure which could be ascribed to resonances in particular multipole modes. These have been discussed numerically by Mevel[27] and by Rosasco and Bennett.[28] Fuchs and Kliewer[29] have ascribed them to polarization excitation resulting from the coupling of optical phonons with photons. Ashkin and Djeidic[30] were able to observe such exceedingly sharp resonance by measuring the power needed to support a particle at a given height in a laser beam by radiation pressure as the wavelength is continuously varied. These measured results agreed precisely with those calculated by Lorenz-Mie theory, by Chylek, Kiehl and Ko.[31]

4. Shaped beams

It has been useful to consider scattering by a sphere in an incident beam which is no longer a plane wave. Chew et al. have treated

diverging,[32] converging[33] and evanescent beams.[34] The first may occur
in multiple scattering and surface-enhanced Raman scattering; the second
beams in focussed down to observed single particles; the third in procedures which utilize attenuated total reflection. A particle may be
illuminated by a laser beam with a Gaussian profile. The particle need
not be on the beam axis. This problem has been treated by Tau and
Corriveau[35] and by Gouesbet et al.[36] There are fundamental difficulties
involved in the rigorous description of the laser beam. The order of
approximation involved because of this has been discussed by Gouesbet
et al.[37]

A great convenience in the Lorenz-Mie scattering formalism is that
scattering coefficients depend only upon the physical parameters viz.
the optical size and the optical constants of the sphere. These are then
modulated by Legendre functions which are purely geometric in that they
depend only upon the spherical coordinates of the observer. With a
shaped beam there is a further modulation which depends upon the particular shape of the incident beam. However the coefficients are identical
with those obtained for the incident plain wave.

5. Interpretations and applications

We now consider various scattering phenomena brought to light either
by further analysis, numerical evaluation or experiments. The spate of
newly recognized effects attests to the amazing richness of the interaction of light with small particles and of the underlying theory in
which these phenomena were implicit.

Sinclair[38] and Brillouin[39] have discussed the finding that extinction cross section for large particles is just double the geometrical
cross section and Sinclair demonstrated the effect experimentally.
Nussenzweig and Wiscombe[40] have derived asymptotic expressions for the
extinction, absorption and radiation pressure cross sections based on
complex-angular-momentum theory. These resemble geometrical-optics
results plus corrections from the edge domain. Bohren[41] has provided a
cogent physical discussion of the very large enhancement of the scattering and absorption cross sections due to excitation of surface plasmons
and phonons.

Ambiquities which arise when a particle is embedded within an
absorbing medium have been discussed by Bohren and Gilra[42] who specify
an appropriate operational procedure for defining and measuring the
extinction.

The radiation torque on a sphere caused by a circularly-polarized
plane wave has been derived by Marston and Crichton[43] and that exerted by
a circularly polarized Gaussian laser beam has been obtained by Chang and
Lee.[44]

The electric field within an irradiated sphere is quite nonuniform
and this gives rise to a correspondingly nonuniform distribution of heat
sources whenever the sphere is absorptive. The thermal effect due to the
interaction of ambient gas with the nonuniformly heated surface, termed
photophoresis, has been analyzed by Yalamov et al.[45] in the limit of low
values of the Knudsen number (ratio of mean free path of gas molecules to
particle radius). Pluchino and Arnold[46] have incorporated the effect of
finite values of the Knudsen number and found agreement between calculations and experiments. Sitarski and Kerker[47] have treated the free
molecular regime (high values of Knudsen number), employing a Monte Carlo
scheme for calculating the heat conduction within the particle including
the case of a solid absorbing core coating with a nonabsorbing volatile
shell.

Wagner and Lynch[48] have shown that a scatterer with axis of symmetry oriented along the incident direction of a plane wave has a zero back scatter cross section whenever the relative permittivity and relative permeability are equal. Kerker et al.[49] have noted additionally that the polarization of the scattered radiation in any direction is the same as that of the incident radiation including the case of incident elliptical polarization. Furthermore, Kerker has noted that for small spheres having a nonzero value of μ, the scattering radiation may be completely polarized for incident unpolarized radiation at scattering angles other than 90°, that there may be either preferential back scatter or forward scatter and that because the refractive index, $m = (\mu\varepsilon)^{\frac{1}{2}}$, there may be significant and interesting scattering effects even when $m = 1$, provided $\varepsilon \neq \mu$.

Kerker[50] has noted that the scattering cross section for a coated sphere with small size parameter may become vanishingly small when the relative dielectric constant of one region $\varepsilon_1 < 1$ and that of the other region $\varepsilon_2 > 1$. Alexopoulos and Uzunoglu[51] have noted when the medium is active, i.e. for negative values of the absorption index, that the extinction cross section may become zero and they termed these invisible scatterers. However, Kerker[52] pointed out that in such cases the zero extinction arises from cancellation between the scattering cross section and the negative absorption cross section which denotes emission. He has discussed the physical implications for stimulated emission as well as for amplification of the incident wave. Tzeng et al.[53] have observed such stimulated emission.

Cooney and Gross[54] have noted that three wave mixing to produce a coherent anti-Stokes Raman scattering (CARS) dipole within a particle must utilize the local Lorenz-Mie fields and that the subsequent emission by the dipole with the sphere can be described by the model utilized earlier by Chew et al.[10] Qian, Snow and Chang[55] have reported observations of stimulated Raman scattering, CARS and coherent Raman mixing from micron size droplets. Chew, Wang and Kerker[56] have extended the model of Cooney and Gross[54] to the case where the CARS molecules are located at the outer surface of the sphere by utilizing the expressions derived earlier by Chew, Kerker and Cooke[32] for scattering of a radiation from a dipole source located outside a spherical particle. They have predicted strong enhancement for frequencies at which surface plasmons are excited.

C. Cylinders

Lord Rayleigh had solved the boundary value problem of scattering by a homogeneous cylinder at normal incidence in 1881 and this was extended to oblique incidence only in 1955 by Wait[57] who also gave the limiting expressions for small values of the size parameter.[58] Cooke and Kerker[59] reported observations of the manner in which the scattered radiation propagates along conical surfaces as the cylinder is tilted.

Wait[60] has treated scattering by a magnetized plasmon cylinder at perpendicular incidence and this has been extended to oblique incidence by Wilhelmsson[61] and by Samaddar.[62] Yeh and Wang[63] have formulated the problem of the interaction of obliquely incident radiation with a radially inhomogeneous cylinder and have specifically discussed numerical results for a Luneberg profile.

Owen, Chang and Barber[64] have utilized cylindrical geometry to discuss morphology-dependent resonances as these are manifested in elastic scattering, fluorescence emission and Raman scattering. The experimental observations have been compared with computed results for each of these three phenomena. Kerker[65] has also utilized cylindrical geometry to

illustrate the resonances that occur with media characterized by negative
absorption indexes. These resonances exhibited amazingly high Q values.
For cylinders with small radii the unusual simultaneous occurrence in two
different modes was noted.

D. Spheroids

Oguchi's treatment of a spheroid mimics that for a sphere.[66] The
boundary conditions are expressed in terms of the r, θ, ϕ components of
the field vectors, which in turn are expanded in spherical vector wave
functions. The r coordinate at the surface is expanded in powers of the
eccentricity and only the linear term is retained. This permits computation of scattering functions for spheroids of arbitrary size and optical
constants, provided the eccentricity is not too large. Morrison and
Cross[67] follow a similar procedure but obtain numerical results by a
point matching technique.

Asano and Yamamoto[68] have provided a complete classical solution by
separating the vector wave equations in spheroidal coordinates and expanding the field vectors in spheroidal wave functions. The unknown
coefficients in the expansions are determined by the boundary conditions.
Asano and Sato[69] have developed a scheme for integrating the solution of
Asano and Yamamoto over all particle orientations so that the light scattering properties of an assembly of randomly oriented particles can be
obtained.

E. Clusters of Spheres

When sufficiently separated, signals from arrays of particles add
coherently or incoherently depending upon whether or not the particles
are positioned randomly. In many cases multiple scattering must be considered. Whenever the particles are sufficiently close, they interact
electromagnetically so that the boundary conditions must be satisfied coherently at each surface. Bruning and Lo[70] have treated the case of two
spheres of arbitrary optical properties and unequal radii separated by
an arbitrary distance by the multipole expansion method with the aid of a
translation addition theorem which expresses the vector spherical wave
functions about a displaced origin in terms of those about another origin.
A recursive relation which simplifies calculation of the translation coefficients was crucial for practical computations. Kattawar and Dean[71] have
corrected several errors in the work of Bruning and Lo and compared numerical results with measurements.

Borghese et al.[72] have generalized the problem to an arbitrary cluster
of spheres with no restrictions on radii, optical constants or locations
of the various spheres in the cluster. The field scattered by the cluster
is expanded in a multicentered series of multipoles. Computational simplifications are obtained by using group theory to exploit any symmetry
properties of the cluster.[73] The result for 2-, 3- and 4-sphere clusters
shows that optical behavior of aggregates differs substantially from
that of the constituting spheres and is sensitive to the particular
structure of the cluster.[74]

F. Arbitrary Shape

Erma has proposed a boundary-perturbation technique which is claimed
to be valid to all orders in the perturbation parameter which is imposed
on a radius. The boundary condition is expanded in a Taylor series which
in effect provides a succession of boundary conditions at the unperturbed
boundary. The problem is discussed in three phases, viz. for a cylin-
drically, symmetric, though otherwise arbitrarily shaped conductor;[75] for

8

a conductor of entirely arbitrary shape;[76] for an arbitrarily shaped
object with arbitrary but homogeneous optical properties.[77] There are no
computed results.

INTEGRAL EQUATION SOLUTIONS

A. <u>General</u>

Saxon[78] has discussed an integral equation formulation of the scattering problem which automatically incorporates all of the boundary conditions. The essential feature is the appearance of an integral which requires a knowledge of the internal field. The treatments discussed in
this section will utilize in some form such an integral equation formulation. Strom[79] has discussed interrelations among several volume and
surface integral expressions for the scattered field that arise in these
treatments.

B. <u>Extended Boundary Condition Method</u>

The extended boundary condition method was first formulated by
Waterman[80] for perfectly conducting objects. The basic physical notion
was that the boundary condition could be satisfied by induced currents
on the surface sufficient to cancel the incident field throughout the
interior. The scattered radiation in turn could be expressed as an
integral over the surface currents. Arbitrary optical constants were
included in a later treatment.[81] Conceptually, there were now additional
sources outside the object which were annulled by surface currents which
in turn gave rise to the appropriate internal field. Peterson and
Strom[82] formulated the method for multilayered scatterers, for clusters
of scatterers and for a scatterer consisting of the two non-enclosing
parts.

Barber and Yeh[83] proposed an alternate albeit conceptually similar
derivation using Schelkunoff's equivalence theorem. Wang and Barber[84]
extended the treatment to layered spheroids. Wang, Kerker and Chew[85]
considered inelastic scattering by molecules embedded in dielectric
spheroids. In analogy with the spherical case,[11] this required establishment of appropriate surface currents for both the scattered and internal
fields due to a source dipole within the particle rather than for an
incident plane wave. Yeh, Colak and Barber[86] treated the case of a
sharply focussed incident beam by expanding the beam in terms of its
plane wave spectrum.

Waterman[87] rederived the method in a way that eliminated conceptual
recourse to fictitious fields.

C. <u>Iterative Procedures</u>

Stevenson[88] has expressed the scattered field and the field inside
an arbitrarily shaped scatterer as a power series in the ratio of a
dimension of the scatterer to the wavelength and he has formulated the
partial field as a surface integral. Calculation of successive terms
requires only solution of Laplace's equation in the appropriate coordinate system.

Purcell and Pennypacker[89] have modeled the scattering object by an
array of polarizable dipoles located on a cubic lattice. Each of the
oscillating dipoles is driven by the field of the incident wave and the
fields of all the other oscillators. The amplitudes are determined by a
self-consistent iteration. Yung[90] has suggested numerical techniques for
making the computations more viable.

Andreasen[91] expands the incident wave in cylindrical modes which
induced corresponding current distributions on the surface of the body.
Each scattered mode is related to a current distribution by two complex
integral equations. These are reduced to linear complex equations for
solution on a computer. Holt, Uzunoglu and Evans[92] utilize an integral
equation equivalent to that given by Saxon.[78] The essence of their method
is to employ a transform which leaves the integral equation with a nonsingular kernel rather than the singular kernel present initially.
Ohtaka and Inoue[93] utilize a self-consistent iterative technique to determine the scattered field from the integral formulation of the internal
fields. Inoue and Ohtaka[94] utilize the same technique to solve for the
scattered fields of a cluster of two spheres. This solution is then used
to discuss surface-enhanced Raman scattering for such a configuration.

GENERAL

Van de Hulst[95] has derived a relation between the extinction of a
plane wave by an arbitrary obstacle and the complex amplitude of the
forward scattered wave. Zimm and Dandliker[96] have utilized this approach,
calculated the retardation of the transmitted light due to interference
between the incident and forward scattered beam. This retardation can be
identified in this sense with the refractive index of a dispersion of scattering particles. Also in the same sense and by utilizing the Kramers-
Kronig relations, McKeller, Box and Bohrens[97] have shown that the total
volume fraction of a dispersion of arbitrarily shaped scatterers is proportional to the integral of the attentuation coefficient over wavelength.

Bibliography

BOUNDARY VALUE SOLUTIONS

A. Historical
1. Survey of some early studies of the scattering of plane waves by a
    sphere, Nelson A. Logan, Proceedings of the IEEE, 773-785, August
    1965.

B. Spheres
    1. New and alternate solutions
2. Light scattering by an optically active sphere, Craig F. Bohren, Chem.
    Phys. Lett. $\underline{29}$, 458-462 (1974).
3. Scattering and absorption of electromagnetic waves by a gyrotropic
    sphere, G.W. Ford and S.A. Werner, Phys. Rev. B $\underline{18}$, 6752-6769 (1978).
4. Optical properties of small metal spheres, R. Ruppin, Phys. Rev. B $\underline{11}$,
    2871-2876 (1975).
5. Optical properties of spatially dispersive dielectric spheres, R.
    Ruppin, J. Opt. Soc. Am. $\underline{71}$, 755-758 (1981).
6. Polarizability of a small sphere including nonlocal effects, Basab
    B. Dasgupta and Ronald Fuchs, Phys. Rev. B $\underline{24}$, 554-561 (1981).
7. Scattering by a rotating dielectric sphere, Daniel De Zutter, IEEE
    Transactions on Antennas and Propagation, $\underline{AP-28}$, 643-651 (1980).
8. Relativistic scattering of electromagnetic waves by moving obstacles,
    Victor Twersky, J. of Mathematical Phys. $\underline{12}$, 2328-2341 (1971).
9. Three-dimensional relativistic scattering of electromagnetic waves
    by an object in uniform translational motion, B.L. Michielsen, G.C.
    Herman, A.T. de Hoop and D. De Zutter, J. Math. Phys. $\underline{22}$, 2716-2722
    (1981).
10. Model for Raman and fluorescent scattering by molecules embedded in
    small particles, H. Chew, P.J. McNulty and M. Kerker, Phys. Rev. A
    $\underline{13}$, 396-404 (1976).
11. Fluorescent and Raman scattering by molecules embedded in small
    particles: Magnetic dipole transitions, H. Chew, Phys. Rev. A $\underline{19}$,
    2137-2138 (1979).

12. Surface enhanced Raman scattering (SERS) by molecules adsorbed at
     spherical particles: errata, Milton Kerker, Dau-Sing Wang and H.
     Chew, Appl. Opt. 19, 4159-4174 (1980).
13. Electromagnetic aspects of the enhanced Raman scattering by a
     molecule adsorbed on a polarizable sphere, K. Ohtaka and M. Inoue,
     J. Phys. C 15, 6463-6480 (1982).
14. Theory of second harmonic generation by small metal spheres, Xias
     Ming Hua and Joel L. Gersten, Phys. Rev. B 33, 3756-3764 (1986).
15. Scattering of electromagnetic waves from two concentric spheres,
     Arthur L. Aden and Milton Kerker, J. of Appl. Phys. 22, 1242-1246
     (1951).
16. Die Miesche Theorie der Beugung durch dielektrische Kugeln mit
     absorbierendem Kern und ihre Bedeutung fur probleme der interstel-
     laren materie und des atmospharischen aerosols, A. Guttler, Annalen
     der Physik. 6, 65-78 (1952).
17. Scattering from an eccentrically stratified dielectric sphere, J.
     G. Fikioris and N.K. Uzunoglu, J. Opt. Soc. Am. 69, 1359-1366 (1979).
18. Electromagnetic scattering from a radially inhomogeneous spheres,
     James R. Wait, Appl. Sci. Res. 10, 441-450 (1963).
19. Scattering coefficients for a multilayered sphere: analytic expres-
     sions and algorithms, Ramesh Bhandari, Appl. Opt. 24, 1060-1967
     (1985).
20. Scattering of electromagnetic plane waves from inhomogeneous spher-
     ically symmetric objects, Philip J. Wyatt, Phys. Rev. 127, 1837-1843
     (1962). Erratum, ibid, 134 (1964).
21. Scattering of electromagnetic waves from two concentric spheres, when
     outer shell has a variable refractive index, S. Levine and M. Kerker,
     I.C.E.S. 37-46 (1963).

  2. Computations
22. Scattering of electromagnetic radiation by a large, absorbing sphere,
     J.V. Dave, IBM J. Res. Develop. 302-313 (1969).
23. Improved Mie scattering algorithms, W.J. Wiscombe, Appl. Opt. 19,
     1505-1509 (1980).
24. Note on the recurrence between Mie's coefficients, B. Verner, J. Opt.
     Soc. of Am. 66, 1424-1425 (1976).
25. Recurrence relations for Mie scattering coefficients, C.F. Bohren,
     J. Opt. Soc. of Am. A 4, 612 (1987).
26. Relation between contiguous Mie coefficients for perfectly conduct-
     ing spheres, H. Chew, Phys. Lett. A 115, 191-192 (1986).

  3. Optical Resonances
27. Etude de la structure detaillee des courses de diffusion des ondes
     electromagnetiques par les spheres dielectriques, Jean Mevel,
     Journal de Physique et le Radium 19, 630-636 (1958).
28. Internal field resonance structure: Implications for optical absorp-
     tion and scattering by microscopic particles, Gregory J. Rosasco
     and Herbert S. Bennett, J. Opt. Soc. Am. 68, 1242-1250 (1978).
29. Optical modes of vibration in an ionic crystal sphere, Ronald Fuchs
     and K.L. Kliewer, J. of the Opt. Soc. of Am. 58, 319-330 (1968).
30. Observation of optical resonances of dielectric spheres by light
     scattering, A. Ashkin and J.M. Dziedzic, Appl. Opt. 20, 1803-1814
     (1981).
31. Optical levitation and partial-wave resonances, P. Chylek, J. Kiehl
     and M.K.W. Ko, Phys. Rev. A 18, 2229-2233 (1978).

  4. Shaped beams
32. Electromagnetic scattering by a dielectric sphere in a diverging
     radiation field, H. Chew, M. Kerker and D.D. Cooke, Phys. Rev. A 16,
     320-323 (1977).

33. Light scattering in converging beams, Herman Chew, Milton Kerker and Derry D. Cooke, Opt. Lett. 1, 138-140 (1977).
34. Elastic scattering of evanescent electromagnetic waves, Herman Chew, Dau-Sing Wang and Milton Kerker, Appl. Opt. 18, 2679-2687 (1979).
35. Scattering of electromagnetic beams by spherical objects, W.G. Tam and Robert Corriveau, J. Opt. Soc. Am. 68, 763-767 (1978).
36. The order of approximation in a theory of the scattering of a Gaussian beam by a Mie scatter center, G. Gouesbet, B. Maheu and G. Grehan, J. Optics 16, 239-247 (1985).
37. Scattering of a Gaussian beam by a Mie scatter center using a Bromwich formalism, G. Gouesbet, G. Grehan and B. Maheu, J. Optics 16, 83-93 (1985).

5. <u>Interpretations and applications</u>
38. Light scattering by spherical particles, David Sinclair, J. of the Opt. Soc. of Am. 37, 475-480 (1947).
39. The scattering cross section of spheres for electromagnetic waves, L. Brillouin, J. of Appl. Phys. 20, 1110-1125 (1949).
40. Efficiency factors in Mie scattering, H.M. Nussenzveig and W.J. Wiscombe, Phys. Rev. Lett. 45, 1490-1494 (1980).
41. How can a particle absorb more than the light incident on it? Craig F. Bohren, Am. J. Phys. 51, 323-327 (1983).
42. Extinction by a spherical particle in an absorbing medium, Craig F. Bohren and Daya P. Gilra, J. Colloid and Interf. Sci. 72, 215-221 (1979).
43. Radiation torque on a sphere caused by a circularly-polarized electromagnetic wave, Philip L. Marston and James H. Crichton, Phys. Rev. A 30, 2508-2516 (1984).
44. Optical torque exerted on a homogeneous sphere levitated in the circularly polarized fundamental-mode laser beam, Soo Chang and Sand Soo Lee, J. Opt. Soc. Am. B 2, 1853-1860 (1985).
45. Theory of the photophoretic motion of the large-size volatile aerosol particle, Yu. I. Yalamov, V.B. Kutukov and E.R. Shchukin, J. Colloid Interface Sci. 57, 564-571 (1976).
46. Comprehensive model of the photophoretic force on a spherical microparticle, A.B. Pluchino and S. Arnold, Opt. Lett. 10, 261-263 (1985).
47. Monte Carlo simulation of photophoresis of submicron aerosol particles, Marek Sitarski and Milton Kerker, J. of the Atm. Sciences 41, 2250-2262 (1984).
48. Theorem on electromagnetic backscatter, R.J. Wagner and P.J. Lynch, Phys. Rev. 131, 21-23 (1963).
49. Electromagnetic scattering by magnetic spheres, M. Kerker, D.-S. Wang and C.L. Giles, J. Opt. Soc. Am. 73, 765-767 (1983).
50. Invisible bodies, Milton Kerker, J. of the Opt. Soc. of Am. 65, 376-379 (1975).
51. Electromagnetic scattering from active objects: invisible scatterers, N.G. Alexopoulos and N.K. Uzunoglu, Appl. Opt. 17, 235-239 (1978).
52. Electromagnetic scattering from active objects, Milton Kerker, Appl. Opt. 17, 3337-3339 (1978).
53. Laser emission from individual droplets at wavelengths corresponding to morphology-dependent resonances, H.-M. Tzeng, K.F. Wall, M.B. Long and R.K. Chang, Opt. Lett. 9, 499-501 (1984).
54. Coherent anti-Stokes Raman scattering by droplets in the Mie size range, John Cooney and Abraham Cross, Opt. Lett. 7, 218-220 (1982).
55. Coherent Raman mixing and coherent anti-Stokes Raman scattering from individual micrometer-size droplets, Shi-Xiong Qian, Judith B. Snow, and Richard K.Chang, Opt. Lett. 10, 499-501 (1985).
56. Surface enhancement of coherent anti-Stokes Raman scattering by colloidal spheres, H. Chew, D.-S. Wang and M. Kerker, J. of Opt. Soc. of Am. 1, 56-66 (1984).

C. Cylinders

57. Scattering of a plane wave from a circular dielectric cylinder at oblique incidence, James R. Wait, Canadian J. of Phys. __33__, 189-195 (1955).
58. The long wavelength limit in scattering from a dielectric cylinder at oblique incidence, James R. Wait, Canadian J. of Phys. __43__, 2212-2215 (1965).
59. Light scattering from long thin glass cylinders at oblique incidence, D.D. Cooke and M. Kerker, J. Opt. Soc. Am. __59__, 43-48 (1969).
60. Some boundary value problems involving plasma media, James R. Wait, J. of Res. of the Nat. Bureau of Standards __65B__, 137-150 (1961).
61. Interaction between an obliquely incident plane electromagnetic wave and an electron beam in the presence of a static magnetic field of arbitrary strength, K.H.B. Wilhelmsson, J. of Res. of the Nat. Bureau of Standards __66D__, 439-451 (1962).
62. Scattering of plane waves from an infinitely long cylinder of anisotropic materials at oblique incidence with an application to an electronic scanning antenna, S.N. Samaddar, Appl. Sci. Res. __10__, 385-411 (1963).
63. Scattering of obliquely incident waves by inhomogeneous fibers, C. Yeh and P.K.C. Wang, J. Appl. Phys. __43__, 3999-4006 (1972).
64. Morphology-dependent resonances in Raman scattering, fluorescence emission, and elastic scattering from microparticles, J.F. Owen, R. K. Chang and P.W. Barber, Aerosol Sci. and Tech. __1__, 293-302 (1982).
65. Resonances in electromagnetic scattering by objects with negative absorption, M. Kerker, Appl. Opt. __18__, 1184-1189 (1979).

D. Spheroids

66. Attenuation of electromagnetic wave due to rain with distorted raindrops, J. Oguchi, J. of Radio Res. Labs. __7__, 467-485 (1960).
67. Scattering of a plane electromagnetic wave by axisymmetric raindrops, J.A. Morrison and M.-J. Cross, Bell system Tech. J. __53__, 955-978, 1008-1019 (1974).
68. Light scattering by a spheroidal particle, S. Asano and G. Yamamoto, Appl. Opt. __14__, 29-48 (1975). Errata, ibid. __15__, 2028 (1976).
69. Light scattering by randomly oriented spheroidal particles, S. Asano and M. Sato, Appl. Opt. __19__, 962-974 (1980).

E. Clusters of Spheres

70. Multiple scattering of EM waves by spheres, Part I-Multipole expansion and ray-optical solutions, John H. Bruning and Yuen T. Lo, IEEE Transactions on Antennas and Propagation __AP-19__, 378-390 (1971).
71. Electromagnetic scattering from two dielectric spheres: Comparison between theory and experiment, George W. Kattawar and Cleon E. Dean, Opt. Lett. __8__, 48-50 (1983).
72. Electromagnetic scattering by a cluster of spheres, F. Borghese, P. Denti, G. Toscano and O.I. Sindoni, Appl. Opt. __18__, 116-120 (1979).
73. Use of group theory for the description of electromagnetic scattering from molecular systems, F. Borghese, P. Denti, R. Saija, G. Toscano and O.I. Sindoni, J. Opt. Soc. Am. A __1__, 183-188 (1984).
74. Effects of aggregation on the electromagnetic resonance scattering of dielectric spherical objects, F. Borghese, P. Denti, R. Saija and G. Toscano, Il Nuovo Cimento __6__, 545-557 (1985).

F. Arbitrary Shape

75. An exact solution for the scattering of electromagnetic waves from conductors of arbitrary shape. I. Case of cylindrical symmetry, Victorr A. Erma, The Phys. Rev. __173__, 1243-1257 (1968).
76. Exact solution for the scattering of electromagnetic waves fron conductors of arbitrary shape. II. General case, Victorr A. Erma, Phys. Rev. __176__, 1544-1553 (1968).

77. Exact solution for the scattering of electromagnetic waves from
     bodies of arbitrary shape. III. Obstacles with arbitrary electro-
     magnetic properties, Victorr A. Erma, Phys. Rev. 179, 1238-1246
     (1969).

INTEGRAL EQUATION SOLUTIONS
A. General
78. Tensor scattering matrix for the electromagnetic field, David S.
     Saxon, Phys. Rev. 100, 1771-1775 (1955).
79. On the integral equations for electromagnetic scattering, Staffan
     Strom, Am. J. of Phys. 43, 1060-1069 (1975).

B. Extended Boundary Condition Method
80. Matrix formulation of electromagnetic scattering, P.C. Waterman,
     Proceedings of the IEEE, 805-812, August 1965.
81. Symmetry, unitarity, and geometry in electromagnetic scattering, P.
     C. Waterman, Phys. Rev. D 3, 825-839 (1971).
82. T-matrix formulation of electromagnetic scattering from multilayered
     scatterers,Bo Peterson and Staffan Strom, Phys. Rev. D 10, 2670-
     2684 (1974).
83. Scattering of electromagnetic waves by arbitrarily shaped dielectric
     bodies, P. Barber and C. Yeh, Appl. Opt. 14, 2864-2872 (1975).
84. Scattering by inhomogeneous nonspherical objects, Dau-Sing Wang and
     Peter W. Barber, Appl. Opt. 18, 1190-1197 (1979).
85. Raman and fluorescent scattering by molecules embedded in dielectric
     spheroids, Dau-Sing Wang, Milton Kerker and Herman W. Chew, Appl.
     Opt. 19, 2135-2328 (1980).
86. Scattering of sharply focused beams by arbitrarily shaped dielectric
     particles: An exact solution, C. Yeh, S. Colak and P. Barber, Appl.
     Opt. 21, 4426-4433 (1982).
87. Matrix methods in potential theory and electromagnetic scattering,
     P.C. Waterman, J. Appl. Phys. 50, 4550-4566 (1979).

C. Iterative Procedures
88. Solution of electromagnetic scattering problems as power series in
     the ratio (dimension of scatterer)/wavelength, A.F. Stevenson, J. of
     Appl. Phys. 24, 1134-1142 (1953).
89. Scattering and absorption of light by nonspherical dielectric grains,
     Edward M. Purcell and Carlton R. Pennypacker, The Astrophysical J.
     186, 705-714 (1973).
90. Variational principle for scattering of light by dielectric particles,
     Yuk L. Yung, Appl. Opt. 17, 3707-3709 (1978).
91. Scattering from bodies of revolution, Mogens G. Andreasen, IEEE Trans.
     on Antennas and Propagation, AP-13, 303-310, March 1965. Erratum,
     ibid. AP-14, 659 (1966).
92. An integral equation solution to the scattering of electromagnetic
     radiation by dielectric spheroids and ellipsoids, A.R. Hold, N.K.
     Uzunoglu, B.G. Evans, IEEE Transactions on Antennas and Propagation,
     AP-26, 706-712, September 1978.
93. Light scattering from macroscopic spherical bodies. I. Integrated
     density of states of transverse electromagnetic fields, K. Ohtaka
     M. Inoue, Phys. Rev. B 25, 677-688 (1982).
94. Surface enhanced Raman scattering by metal spheres. I. Cluster
     effect, Masahiro Inoue and Kazuo Ohtaka, J. of the Phys. Society
     of Japan 52, 3853-3864 (1983).

GENERAL

95. On the attenuation of plane waves by obstacles of arbitrary size and
    form, H.C. van de Hulst, Physica XV. No. 8-9, 740-746 (1949).

96. Theory of light scattering and refractive index of solutions of large
    colloidal particles, B.H. Zimm and W.B. Dandliker, J. Phys. Chem. $\underline{58}$,
    644 (1954).

97. Sum rules for optical scattering amplitudes, Bruce H.J. McKellar,
    Michael A. Box and Craig F. Bohren, J. Opt. Soc. Am. $\underline{72}$, 535-538
    (1982).   Erratum.

MIE SCATTERING NEAR THE CRITICAL ANGLE

N. Fiedler-Ferrari

Laboratório de Física de Plasmas
Instituto de Física, Universidade de São Paulo
01498 - São Paulo, SP, Brazil

H.M. Nussenzveig[*]

Departamento de Física, Pontifícia Universidade Católica
22453 - Rio de Janeiro, RJ, Brazil

Abstract - Complex angular momentum theory is applied to the problem of
high-frequency critical light scattering by a spherical cavity near the
critical angle. The main contributions to the scattering arise from a critical domain close to critical incidence. The results are in good agreement
with the exact Mie solution.

1. INTRODUCTION

The complex angular momentum (CAM) theory of Mie scattering and its
application to the problems of the rainbow and the glory have been reviewed
elsewhere[1].

We present here the latest application of CAM theory: the treatment of
critical scattering[2]. This is a new diffraction effect found in the transition region around the critical scattering angle for refractive index N
relative to the external medium < 1 (e.g., for an air bubble in water). The
assumptions are $(ka)^{1/3} \gg 1$ and $(1-N)^{1/2}(ka)^{1/3} \gg 1$, where k is the
wavenumber in the external medium and $\underline{a}$ is the radius of the cavity. The

____

[*]Work partially supported by FINEP, CNPq and CAPES.

main contributions arise from a "critical domain" close to critical incidence, and they lead to a new kind of diffraction integral.

In Mie scattering for $N < 1$, the critically incident ray is reflected at a critical scattering angle

$$\Theta_t = \pi - 2\Theta_c = \pi - 2\sin^{-1}N \quad . \tag{1}$$

According to ray optics, total reflection takes place for angles of incidence beyond Θ_c, i.e., for $\Theta < \Theta_t$.

In the geometrical optics approximation[3], the angular distribution of the scattered intensity goes through a <u>cusp</u> at $\Theta = \Theta_t$. This singularity arises from the abrupt approach of the Fresnel reflectivities to unity at the critical angle[4].

Exact Mie calculations[5] show an oscillatory behavior of the intensity in the total reflection region near Θ_t ($\Theta \lesssim \Theta_t$). These diffraction fringes have also been observed experimentally[6].

A "physical optics approximation" along the lines of classical diffraction theory has been proposed by Marston[6]. The contribution from surface reflection is treated by a procedure similar to Airy's theory of the rainbow: a Kirchhoff-type approximation is applied to the amplitude distribution along a virtual reflected wavefront. In view of their steep approach to total reflection, the reflectivities are approximated by step functions. This "reflectivity edge" gives rise to an angular distribution of scattered intensity similar to a Fresnel straight-edge pattern, which would account for the diffraction fringes.

The actual angular distribution[5] differs from the Fresnel one: the oscillation amplitude increases as one goes farther away from Θ_t. This reinforcement was explained[7] through interference with directly transmitted rays due to below-critical incidence.

Superimposed on the "slow" oscillations just discussed, the Mie patterns[5] show fine structure, represented by rapid oscillations of relatively smaller amplitude. This arises from interference with "far-side" contributions (in nuclear scattering terminology[8]), mainly from rays that have undergone one internal reflection. The fine structure is unrelated with critical scattering, so that it should be subtracted out or averaged over in order to isolate pure critical scattering effects.

The physical optics approximation is in reasonable agreement with the general features of the slow oscillations; however, in the neighborhood of Θ_t, the quantitative agreement is poor, specially for $\Theta > \Theta_t$.

The CAM theory applied for $N < 1$ corrects the deviations shows by the physical optics approximation near Θ_t and its results are in good agreement with the exact Mie solution.

The critical domain and the dominant contributions to critical scattering are discussed in Section 2. An outline of the method and the main results are presented in Section 3. Finally, Section 4 lists the relevant conclusions.

2. THE CRITICAL DOMAIN

To discuss critical scattering in terms of CAM theory it is convenient to employ the well-known analogy[9] with Schrödinger scattering of particles with energy $E = k^2$ (in units $h = m = 1$, $k =$ wave number) by a square potential

$$V(r) = V_0 (0 \leqslant r < a), = 0 (r > a) \quad . \tag{2}$$

The associated refractive index is

$$N^2 = 1 - (V_0/E) \quad , \tag{3}$$

so that $N < 1$ corresponds to a square barrier $(V_0 > 0)$. The effective potential for radial motion is

$$V_{eff}^{\lambda}(r) = V(r) + \lambda^2/r^2 \quad , \tag{4}$$

where λ is the complex angular momentum variable, with physical values $\lambda = \ell + 1/2$ $(\ell = 1, 2, 3, \ldots)$ associated with the partial wave terms, and the last term in (4) represents the centrifugal barrier. Therefore, $V_{eff}^{\lambda}(r)$ represents a cusped potential step. The <u>critical angular momentum</u> $\lambda_c = N\beta = \alpha$ $(\beta =$ size parameter $= ka$ $(a =$ sphere radius$))$ associated with critical incidence corresponds to an energy level E at the top of the step. There is a <u>critical domain</u> (analogue to the <u>edge domain</u>[1,9] in the $N > 1$ case)

$$\alpha - O(\alpha^{1/3}) \;\lesssim\; \lambda \;\lesssim\; \alpha + O(\alpha^{1/3}) \tag{5}$$

For incident rays in the <u>lower critical domain</u> $0 \leqslant \lambda_c - \lambda \lesssim \alpha^{1/3}$, the radial turning point within the sphere lies very close to the surface, corresponding to rays in a boundary layer that undergo near-total internal reflection. In the <u>upper critical domain</u> $0 \leqslant \lambda - \lambda_c \lesssim \alpha^{1/3}$, the penetration depth for tunnelling into the sphere is still much larger than the wavelength; correspondingly, the evanescent waves generated by total reflection become <u>inner surface waves</u>, travelling internally along the surface (<u>whispering gallery modes</u>).

In the λ plane, whispering gallery modes are associated with Regge poles[9,10] of $S(\lambda,\beta)$ near $\lambda = \lambda_c = \alpha$. At $\Theta = \Theta_t$, λ_c is an accumulation point of saddle points associated with different terms of the Debye multiple reflection expansion[1,9]. For $\Theta < \Theta_t$ and sufficiently far from Θ_t, Ludwig[10] proposed including $O(\beta^{1/4})$ saddle point contributions and $O(\beta^{1/4})$ Regge pole contributions; such a representation would be very difficult to evaluate in practice. The critical scattering region was excluded from his treatment.

The dominant contributions to critical scattering in the CAM theory[2] arise from the critical domain (5). The dominant terms from the lower critical domain are the <u>direct reflection</u> and <u>direct transmission</u> Debye terms. The main new effects are contained in the <u>above-critical total reflection term</u>, arising from the upper critical domain. The far-side once internally reflected contribution, which is mainly responsible for the fine-structure oscillations, is given by the WKB approximation[9,11] and it need not be considered any further.

3. RESULTS

A. <u>Preliminary Considerations</u>

The <u>critical region</u>, where our solution is supposed to be valid, is defined as:

$$\varepsilon > 0 \; : \; \varepsilon = O(\beta^{-1/3})$$

$$\varepsilon < 0 \; : \; \varepsilon = O(\beta^{-1/2}) \qquad\qquad (6)$$

$$\varepsilon = \frac{1}{2}\,(\Theta_t - \Theta) \quad ,$$

where Θ is the scattering angle.

The Debye expansion is used for below-critical incidence. For above-critical incidence we use expressions without making this expansion. After deforming judiciously[2] the paths in the λ-plane, the dominant contributions to the critical scattering are obtained as integrals on the real axis of the CAM plane. In the below-critical (above-critical) terms λ_c is taken as the upper (lower) limit in the integrals, as a consequence, only the part of the range of the saddle point which is in the lower (upper)-critical domain is considered in each term. The dominant contributions are obtained in lowest order approximation.

B. The Dominant Contributions

The <u>below-critical direct transmission term</u> is the interference term included in the physical optics approximation[7], where it was evaluated by the stationary phase (WKB) method. Since the critical scattering region is a Fock transition region between 1-ray and 0-ray domains for this term, the WKB approximation is not valid: the evaluation leads[9,11] to

$$S^{<}_{j1}(\beta,\Theta) \approx e^{-\frac{7i\pi}{12}} \left(\frac{\beta}{2\pi \sin\Theta}\right) \exp\left\{-\,2i\beta\,(M - \varepsilon N)\right\} \frac{N^{3/2}\,\eta_j}{\pi M}$$

$$\times \int_0^\infty \frac{\exp\left[2\,e^{-i\frac{\pi}{6}}\,\frac{\varepsilon}{\gamma'}\,x\right]}{\left[\mathrm{Ai}\left(e^{\frac{2i\pi}{3}}\,x\right)\right]^2}\; dx \qquad (j = 1,2) \quad , \qquad (7)$$

where $M = (1 - N^2)^{1/2}$, $\gamma' = (2/\alpha)^{1/3}$, $\eta_1 = 1$, $\eta_2 = N^{-2}$, Ai is the Airy function, and $j = 1$ $(j = 2)$ is associated with perpendicular (parallel) polarization. The contribution (7) is given by an incomplete generalized Fock function, containing only part of the range of the direct transmission saddle point.

The <u>below critical reflection term</u> was not taken into account in the physical optics approximation. It is given[2,9,11] by

$$
S_{j0}^{<}(\beta,\Theta) \approx e^{\frac{3i\pi}{4}} \left[\frac{N \sin\left(\frac{\Theta}{2}\right)}{2\pi \sin\Theta} \right]^{1/2} \beta \exp\left\{ -2i\beta \sin\left(\frac{\Theta}{2}\right) \right\}
$$

$$
\times \int_{-\infty}^{Z=0} \left[\frac{1+\Delta}{1-\Delta} \right] \exp(-iv^2)\, dv \quad , \quad (j=1,2) \quad , \tag{8}
$$

where

$$
Z = \gamma' \left[\left(\beta \sin\left(\frac{\Theta}{2}\right) \right)^{1/2} v + \beta \cos\left(\frac{\Theta}{2}\right) - \alpha \right] \quad ,
$$

$$
v = \left[\frac{2}{\beta \sin(\Theta/2)} \right]^{1/2} \left(\lambda - \beta \cos\left(\frac{\Theta}{2}\right) \right) \quad , \tag{9}
$$

$$
\Delta = e^{-i\frac{\pi}{6}} \frac{N\gamma' \eta_j}{M} \ell n'\, Ai\left(e^{-\frac{2i\pi}{3}} Z \right)
$$

($\ell n'$ denotes the logarithmic derivative). The contribution (8) is given by an incomplete Fresnel–Fock integral, containing only part of the range of the direct reflection saddle point.

The <u>above critical total reflection term</u>, containing the new diffraction effects associated with critical scattering, is given by[2]

$$
S_j^{>}(\beta,\Theta) \approx e^{-i\frac{\pi}{4}} \left[\frac{MN}{2\pi \sin\Theta} \right]^{1/2} \beta \exp\left[-2i\beta(M-\epsilon N) \right] P_F(x,y) \quad , \tag{10}
$$

$$
(j=1,2) \quad ,
$$

where

$$
P_F(x,y) = \int_0^{\infty} \exp\left\{ -i\left[u^2 - xu - \left(\frac{y}{\sqrt{2}} \right)^{4/3} \right. \right.
$$

$$
\left. \left. \times \eta_j\, \ell n'\, Ai\left(2^{4/3}\, \frac{u}{y^{2/3}} \right) \right] \right\} du \quad , \tag{11}
$$

is a new type of diffraction integral, y depends only on N and β, and x is proportional to $\Theta - \Theta_t$.

If we neglect the variation of the last term in the exponent, replacing it by a constant, we get a Fresnel integral, as in the physical optics approximation[6]. Since $y \ \beta^{-1/4}$, the argument of the Airy function is $\gg 1$ for $\beta \gg 1$, except near $u = 0$, so that one may employ the asymptotic approximation

$$\ell n' \ Ai(z) \ \approx \ - \sqrt{z} \quad , \quad z \gg 1 \quad . \tag{12}$$

This corresponds to the "plane surface limit", in which the effects spherical curvature are neglected. In this limit, setting $u = t^2$, (11) with $j = 1$ becomes

$$P_F(x,y) \ \approx \ 2 \int_0^\infty \exp\left[- i(t^4 - x t^2 + y t)\right] t \, dt \quad . \tag{13}$$

Pearcey's integral[12], associated with the cusp diffraction catastrophe[13], is given by

$$P(x,y) \ = \ \int_{-\infty}^\infty \exp\left[i \ (t^4 + x t^2 + y t)\right] dt \quad . \tag{14}$$

Thus, $\partial P/\partial y$ is related with $P_F(x,y)$ given by (13).

In this plane surface limit, the y term in the exponent gives rise to a shifted Fresnel-like pattern. For each above-critical ray, this shift corresponds to the <u>Goos-Hänchen lateral shift</u>[14].

In the present case, we have the spherical analogue of this shift which is a <u>Goos-Hänchen angular displacement</u> $\Delta\vartheta$. A ray with angle of incidence above Θ_c tunnels along the surface thorugh an extra angle $\Delta\vartheta$ as an inner surface wave (evanescent wave) before reemerging at the angle of reflection.

To obtain $\Delta\vartheta$, one may employ the concept of <u>angular displacement in a scattering process</u>[15], which is analogous to the Wigner time delay[16,17], applied to the conjugate pair angular momentum and angle. For an "angular momentum wave packet" centered around λ_0 , the angular displacement ϑ is given by

$$\vartheta = 2\, d\eta(k,\lambda_0)/d\lambda \quad , \tag{15}$$

where $\eta(k,\lambda)$ is the scattering phase shift as a function of the (continuous) angular momentum λ. The Goss-Hänchen effect appears as an additional angular displacement arising from the last term in (11).

We propose to call the new diffraction integral (11) the <u>Pearcey-Fock half-range integral</u>, because of its connection both with generalized Fock functions and with Pearcey's integral.

4. CONCLUSIONS

The combined effect of the dominant CAM terms (below-critical direct reflection and transmission terms and above-critical total reflection term) was compared with the exact Mie solution within the critical scattering region, for $\beta = 10^3$ and $\beta = 10^4$. The results[2] are in good agreement with the "slow" component of the Mie solution, which, as explained before, represents the critical scattering effects (fine structure arises from the far-side contribution). We conclude that CAM theory also accounts for critical scattering.

ACKNOWLEDGEMENTS

One of us (N.F.F.) is grateful to the Brazilian agency FAPESP, for a fellowship.

REFERENCES

1. H.M. Nussenzveig, <u>J. Opt. Soc. Am.</u> 69: 1068 (1979).
2. N. Fiedler-Ferrari Jr., Ph.D. thesis, submitted to the University of São Paulo (1983); N. Fiedler-Ferrari Jr. and H.M. Nussenzveig, to be published.
3. G.E. Davis, <u>J. Opt. Soc. Am.</u> 45: 572 (1955).
4. A. Sommerfeld, "Optics", Sect. 5, Academic Press, New York (1954).
5. D.L. Kingsbury and P.L. Marston, <u>J. Opt. Soc. Am.</u> 71: 358 (1981).

6. P.L. Marston, J. Opt. Soc. Am. 69: 1205 (1979).

7. P.L. Marston and D.L. Kingsbury, J. Opt. Soc. Am. 71: 192 (1981).

8. W.E. Frahn, in "Heavy-Ion Science", vol. 1, D.A. Bromley, ed., Plenum Press, New York (1982).

9. H.M. Nussenzveig, J. Math. Phys. 10: 82, 125 (1969).

10. D. Ludwig, J. Math. Phys. 11: 1617 (1970).

11. V. Khare, Ph.D. thesis, University of Rochester (1975).

12. T. Pearcey, Phil. Mag. 37: 311 (1946).

13. M.V. Berry and C. Upstill, in "Progress in Optics", vol. 18, E. Wolf, ed., North-Holland, Amsterdam (1980).

14. F. Goos and H. Hänchen, Ann. Phys. Lpz. (6) 1: 333 (1947); H.K.V. Lotsch, Optik 32: 116 (1970).

15. N. Fiedler-Ferrari Jr. and H.M. Nussenzveig, in Proc. II Brazilian Meeting on Particles and Fields", p. 73, Braz. Phys. Soc., São Paulo (1981).

16. E.P. Wigner, Phys. Rev. 98: 145 (1955).

17. H.M. Nussenzveig, Phys. Rev. D 6: 1535 (1972).

SCATTERING OF A GAUSSIAN BEAM BY A SPHERE USING A BROMWICH FORMULATION :

CASE OF AN ARBITRARY LOCATION

Gérard Gouesbet, Bruno Maheu, and Gérard Gréhan

Laboratoire d'Energétique des Systèmes et Procédés
UA CNRS 230 - INSA de Rouen - BP 08
76130 Mont-Saint-Aignan - France

I - INTRODUCTION

The present paper is devoted to the generalization[1] of the Mie scattering theory for a sphere illuminated by a plane wave[1] to the case when the scatter center is illuminated by a Gaussian beam. Such a fundamental theory may lead, in other steps, to important applications in optical sizing, by enabling the researchers to design rigorous approaches to the principles of a few optical sizing methods (the visibility or the phase Doppler techniques, for instance).

The problem of Mie theory generalization has been considered by several authors ([2,3,4,5], among others), presenting theoretical approaches and numerical results of various extents. The work carried out in this field by our team traces back to 1980[6]. Formal works have been published in a series of papers in the Journal of Optics[7,8,9] and numerical results appeared in Applied Optics[10,11], including the design of a so-called localized approximation (see ref 12 in the present symposium). However, these works were restricted to the case when the scatter center is located on the axis of the incident beam. The present work is devoted to a final generalization, the location of the scatter center being arbitrary. The formulation is not given for the cross-sections and pressure radiation for lack of room, although it has been established.

II - THE BROMWICH FORMULATION

To solve the Maxwell equations, taking into account boundary conditions, in a spherical coordinate system (r,θ,φ), fig 1, the Bromwich formulation[13,14,7,8] writes the solution as the sum of two special solutions : the Tranversal Magnetic (TM) wave for which $H_r = o$ and the Transversal Electric (TE) wave for which $E_r = o$. The TM and TE-fields are deduced from Bromwich Scalar Potentials (BSP), U_{TM} and U_{TE}, respectively. Any BSP,U, complies with the following equation :

$$\frac{\partial^2 U}{\partial r^2} + k^2 U + \frac{1}{r^2 \sin\theta} \frac{\partial}{\partial\theta} \left(\sin\theta \frac{\partial U}{\partial\theta}\right) + \frac{1}{r^2 \sin^2\theta} \frac{\partial^2 U}{\partial\varphi^2} = 0 \qquad (1)$$

in which k is the wave-number :

$$k = \omega \, (\mu\epsilon)^{1/2} = M \, \frac{\omega}{c} \tag{2}$$

where ω is the angular frequency of the electromagnetic sinusoidal wave varying in time like $\exp(i\omega t)$, μ and ϵ are respectively the permeability and the permittivity of the medium, M its complex refractive index and c the speed of light.

The following BSP :

$$U = \sum_{n=1}^{\infty} c_n \, r\Psi_n^1 (kr) \, P_n^m (\cos\theta) \begin{Bmatrix} \sin \\ \cos \end{Bmatrix} (m\varphi) \tag{3}$$

are solutions of (1). The $\psi_n^1 (kr)$ are the spherical Bessel functions [7,8] and P_n^m the associated Legendre polynomials [7,8]. The following BSP :

$$U = \sum_{n=1}^{\infty} c_n \, \xi_n (kr) \, P_n^m (\cos\theta) \begin{Bmatrix} \sin \\ \cos \end{Bmatrix} (m\varphi) \tag{4}$$

are also solutions of (1). The $\xi_n (kr)$ are given by :

$$\xi_n (kr) = \Psi_n (kr) + i \, (-1)^n \sqrt{\frac{\pi \, kr}{2}} \, J_{-n-1/2} (kr) \tag{5}$$

in which $\psi_n (kr)$ are the Ricatti-Bessel functions, equal to $kr\psi_n^1 (kr)$, and the $J_{-n-1/2} (kr)$ are the Bessel functions of (negative) half-integer order. $\psi_n (kr)$ remains finite for $r = 0$ while $\xi_n (kr)$ tends to a spherical wave for $r \to \infty$.

From the definition of the TM- and TE-waves :

$$H_{r,TM} = E_{r,TE} = 0 \tag{6}$$

When the BSP are determined, the other field components are obtained from the following relations :

$$E_{r,TM} = \frac{\partial^2 U_{TM}}{\partial r^2} + k^2 \, U_{TM} \tag{7}$$

$$E_{\theta,TM} = \frac{1}{r} \frac{\partial^2 U_{TM}}{\partial r \, \partial\theta} \tag{8}$$

$$E_{\varphi,TM} = \frac{1}{r \, \sin\theta} \frac{\partial^2 U_{TM}}{\partial r \, \partial\varphi} \tag{9}$$

$$H_{\theta,TM} = \frac{i\omega\epsilon}{r \, \sin\theta} \frac{\partial U_{TM}}{\partial\varphi} \tag{10}$$

$$H_{\varphi,TM} = - \frac{i\omega\epsilon}{r} \frac{\partial U_{TM}}{\partial\theta} \tag{11}$$

$$E_{\theta,TE} = - \frac{i\omega\mu}{r \, \sin\theta} \frac{\partial U_{TE}}{\partial\varphi} \tag{12}$$

$$E_{\varphi,TE} = \frac{i\omega\mu}{r} \frac{\partial U_{TE}}{\partial\theta} \tag{13}$$

$$H_{r,TE} = \frac{\partial^2 U_{TE}}{\partial r^2} + k^2 U_{TE} \tag{14}$$

$$H_{\theta,TE} = \frac{1}{r} \frac{\partial^2 U_{TE}}{\partial r \, \partial \theta} \tag{15}$$

$$H_{\varphi,TE} = \frac{1}{r \sin\theta} \frac{\partial^2 U_{TE}}{\partial r \, \partial \varphi} \tag{16}$$

III – THE SCATTERING PROBLEM (FIG 1)

The center of the scatterer (diameter d, complex refractive index M relatively to the surrounding non-absorbing medium) is located at the point O_p of a Cartesian coordinate system $(O_p xyz)$. It is illuminated by a Gaussian beam, the middle of the waist being located at point O_G. An accessory Cartesian coordinate system $(O_G uvw)$ will be used, with $O_G u$ parallel to $O_p x$, and similarly for the other axes. The incident wave propagates from the negative w to the positive w, with Cartesian components E_u, H_v, E_w, H_w, other components being zero, in the system (uvw). We set : $O_p O_G = (x_o, y_o, z_o)$.

.The aim is to compute the properties of the scattered light observed at point $P(r,\theta,\varphi)$ and some associated quantities.

Fig. 1. The Scattering Problem
(V stands for E, elec-
tric field, or H,
magnetic field)

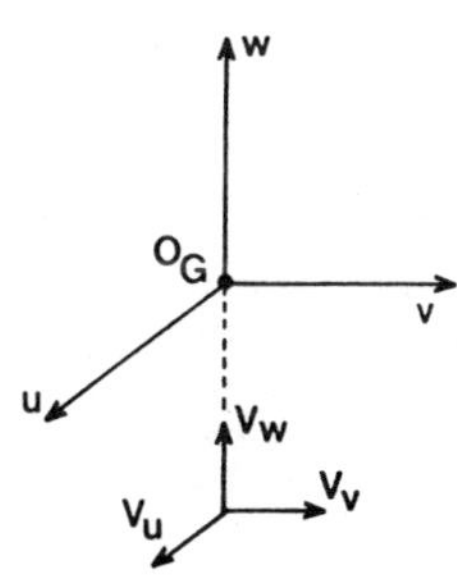

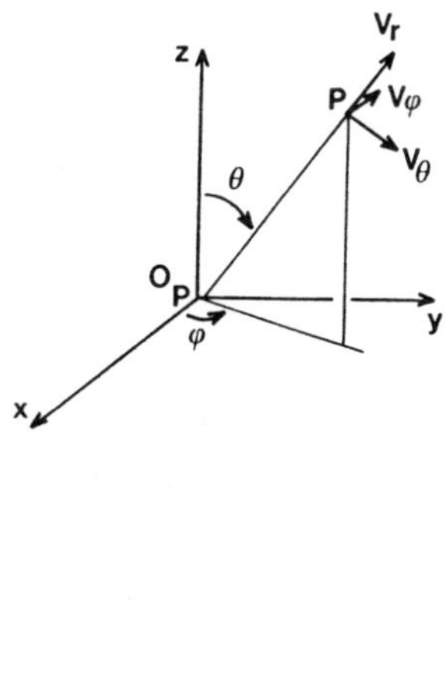

IV – CARTESIAN DESCRIPTION OF THE INCIDENT WAVE

The incident wave is described according to Lax et al [15] who designed a paraxial approximation and produced a procedure to derive higher order corrections and to Davis [16] who presented the same theory in a simpler and more appealing way. For a different description of the wave, or even for another kind of wave, we expect that mainly all the formulae of our scattering theory would remain unmodified, except for the expression of two infinite sequences of coefficients, $g_{n,TM}^m$ and $g_{n,TE}^m$ later discussed.

We introduce a small parameter s :

$$s = w_o/l = 1/(kw_o) \tag{17}$$

in which w_o is the waist radius, and l the spreading length. Following Davis, the vector potential describing the beam may be expanded in power series of s. At the lowest-order (order L), we neglect in Davis' formulation all terms in the field components with powers of s higher than 1. We then obtain the Cartesian field components in the system (u,v,w) given also in ref 8, from which we deduce in a second step the field components in the particle system (x,y,z) :

$$E_y = H_x = 0 \tag{18}$$

$$E_x = E_0 \Psi_0 \exp\left[- ik (z-z_0)\right] \tag{19}$$

$$E_z = - \frac{2Q}{l} (x-x_0) E_x \tag{20}$$

$$H_y = H_0 \Psi_0 \exp\left[- ik (z-z_0)\right] \tag{21}$$

$$H_z = - \frac{2Q}{l} (y-y_0) H_y \tag{22}$$

in which :

$$\Psi_0 = \exp\left[- i \left(P + Qh_+^2\right)\right] \tag{23}$$

$$iP = - Ln\ iQ \tag{24}$$

$$Q = \frac{1}{i + \frac{2}{l} (z-z_0)} \tag{25}$$

$$h_+^2 = \frac{1}{w_0^2} \left[(x-x_0)^2 + (y-y_0)^2\right] \tag{26}$$

The set (18)-(22) does not rigorously comply with Maxwell's equations. Consequences have been extensively discussed in ref 9. The conclusion is that the relative inconsistencies introduced in the theory is $O(s^2)$ for the field components. For typical values (λ = 0.5 μm, w_o = 50 μm), we have $s^2 \sim 10^{-6}$, a very small value indeed. For practical applications, errors introduced in the theory could be only detected in very exotic situations.

V - THE RADIAL COMPONENTS E_r AND H_r.

The radial components E_r and H_r are the only ones required to compute the BSP associated with the incident wave.

From the set (18)-(22), they are found to be :

$$E_r = E_0 \Psi_0 \left[\cos\varphi\ \sin\theta \left(1 - \frac{2Q}{l} r \cos\theta\right) + \frac{2Q}{l} x_0 \cos\theta\right] \exp(K) \tag{27}$$

$$H_r = H_0 \Psi_0 \left[\sin\varphi\ \sin\theta \left(1 - \frac{2Q}{l} r \cos\theta\right) + \frac{2Q}{l} y_0 \cos\theta\right] \exp(K) \tag{28}$$

in which :

$$K = -ik(r\cos\theta - z_0) \tag{29}$$

For further processing, we need identify where the argument φ appears and express this dependence on φ in terms of the plain trigonometric functions sin and cos involved in the general expressions for the BSP 3,4. By tedious algebraic manipulations based on various expansions, we find :

$$E_r = E_r^0 + E_r^\varphi \tag{30}$$

$$H_r = H_r^0 + H_r^\varphi \tag{31}$$

where all the terms which do not depend on φ have been grouped together in E_r^0 and H_r^0. We have :

$$E_r^0 = E_0\,\frac{F}{2}\left[\sum_{j_+=0}^{jplt}\Psi_{jplt} + \sum_{j_-=0}^{jplt}\Psi_{jplt}\right] + E_0\,x_0\,G\sum_{j_0=0}^{jplt}\Psi_{jplt} \tag{32}$$

$$E_r^\varphi = E_0\,\frac{F}{2}\left[\sum_{j_+\neq0}^{jplt}\Psi_{jplt}\exp(ij_+\varphi) + \sum_{j_-\neq0}^{jplt}\Psi_{jplt}\exp(ij_-\varphi)\right]$$

$$+ E_0\,x_0\,G\sum_{j_0\neq0}^{jplt}\Psi_{jplt}\exp(ij_0\varphi) \tag{33}$$

$$H_r^0 = H_0\,\frac{F}{2i}\left[\sum_{j_+=0}^{jplt}\Psi_{jplt} - \sum_{j_-=0}^{jplt}\Psi_{jplt}\right] + H_0\,y_0\,G\sum_{j_0=0}^{jplt}\Psi_{jplt} \tag{34}$$

$$H_r^\varphi = H_0\,\frac{F}{2i}\left[\sum_{j_+\neq0}^{jplt}\Psi_{jplt}\exp(ij_+\varphi) - \sum_{j_-\neq0}^{jplt}\Psi_{jplt}\exp(ij_-\varphi)\right]$$

$$+ H_0\,y_0\,G\sum_{j_0\neq0}^{jplt}\Psi_{jplt}\exp(ij_0\varphi) \tag{35}$$

in which the functions F and G are :

$$F = \Psi_0^0\,\sin\theta\left(1 - \frac{2Q}{l}\,r\cos\theta\right)\exp(ikz_0)\,\exp(-ikr\cos\theta) \tag{36}$$

$$G = \Psi_0^0\,\frac{2Q}{l}\cos\theta\,\exp(ikz_0)\,\exp(-ikr\cos\theta) \tag{37}$$

with :

$$\Psi_0^0 = \exp(-iP)\,\exp\left(-iQ\,\frac{r^2\sin^2\theta}{w_0^2}\right)\exp\left(-iQ\,\frac{x_0^2+y_0^2}{w_0^2}\right) \tag{38}$$

We also have :

$$\sum_{jplt} = \sum_{j=0}^{\infty} \sum_{p=0}^{j} \sum_{l=0}^{j-p} \sum_{t=0}^{p} \qquad (39)$$

The sign $\displaystyle\sum_{c}^{jplt}$ designates the summation $\displaystyle\sum^{jplt}$ restricted to the terms for which the condition C is satisfied.

The ψ_{jplt} are given by :

$$\Psi_{jplt} = \left(\frac{iQ\ r\sin\theta}{w_0^2}\right)^j \frac{x_0^{j-p}\ y_0^p\ (-1)^t}{i^p\ 1!\ (j-p-1)!\ t!\ (p-t)!} \qquad (40)$$

Finally, we have :

$$j_+ = j + 1 - 21 - 2t \qquad (41)$$

$$j_- = j - 1 - 21 - 2t \qquad (42)$$

$$j_0 = j - 21 - 2t \qquad (43)$$

VI - THE BSP'S FOR THE INCIDENT WAVE

From the definition of the TE-wave, $E_{r,TE} = o$. Then, using also (7), the radial component of the electric field is simply :

$$E_r = E_{r,TM} = \frac{\partial^2 U_{TM}^i}{\partial r^2} + k^2 U_{TM}^i \qquad (44)$$

in which U_{TM}^i is the TM-BSP for the incident wave.

The equations being linear, and E_r being a sum of terms, U_{TM}^i is researched as a sum of corresponding terms. Let U_{TM}^{io} be the TM-BSP associated to the field E_r^o. Using the general form (3), and not (4) for boundary conditions to appear later, we set, with $m = o$ since E_r^o does not depend on φ :

$$U_{TM}^{io} = \sum_{n=1}^{\infty} \frac{E_0}{k}\ i^{n-1}\ (-1)^n\ \frac{2n+1}{n(n+1)}\ g_{n,TM}^0\ r\Psi_n^1(kr)\ P_n(\cos\theta) \qquad (45)$$

where P_n is P_n^o (Legendre polynomials), the form given to the coefficients being chosen for further simplification (see the last section, for instance).

The determination of the unknown coefficients $g_{n,TM}^o$ is now required. To the purpose, we inject (45) into (44), and equal the result to E_r^o given by (32), leading to an equation for the $g_{n,TM}^o$'s. To solve it, we use orthogonality relations :

$$\int_0^{\pi} P_n^m(\cos\theta)\ P_l^m(\cos\theta)\ \sin\theta\ d\theta = \frac{2}{2n+1}\ \frac{(n+m)!}{(n-m)!}\ \delta_{nl} \qquad (46)$$

$$\int_0^{\infty} \Psi_n^1(kr)\ \Psi_m^1(kr)\ d(kr) = \frac{\pi}{2(2n+1)}\ \delta_{nm} \qquad (47)$$

where m must be taken equal to 0 in relation (46). We multiply the equation to solve by two integral operators in order to use (46), (47), and isolate the unknown coefficients.

We obtain :

$$g^0_{n,TM} = \frac{k(2n+1)}{\pi \, i^{n-1} \, (-1)^n} \int_0^\pi \int_0^\infty \left\{ \frac{F}{2} \left(\sum_{j_+=0}^{jplt} \Psi_{jplt} + \sum_{j_-=0}^{jplt} \Psi_{jplt} \right) \right.$$

$$\left. + x_0 G \sum_{j_0=0}^{jplt} \Psi_{jplt} \right\} . r\Psi^1_n(kr) \, P_n(\cos\theta) \, \sin\theta \, d\theta \, d(kr) \qquad (48)$$

Let U^{iB}_{TM}, U^{iC}_{TM} and U^{iD}_{TM} be the TM-BSP's associated respectively with each of the three terms appearing in E^φ_r. Working similarly as before, we obtain :

$$\left(U^{iB}_{TM} \, , \, U^{iC}_{TM} \, , \, U^{iD}_{TM} \right) = \sum_{n=1}^\infty \sum_{|m|=1}^n \frac{E_0}{k} \left(C^{B,m}_n \, , \, C^{C,m}_n \, , \, x_0 C^{D,m}_n \right)$$

$$r\Psi^1_n(kr) \, P^{|m|}_n(\cos\theta) \, \exp(im\varphi) \qquad (49)$$

where $C^{B,m}_n$ corresponds to U^{iB}_{TM}, and similarly for the two other terms. The coefficients are given by :

$$C^{B,m}_n = \frac{k(2n+1)^2}{\pi n(n+1)} \frac{(n-|m|)!}{(n+|m|)!} \int_0^\pi \int_0^\infty \frac{F}{2} \left[\sum_{j_+=m\neq0}^{jplt} \Psi_{jplt} \right]$$

$$r\Psi^1_n(kr) \, P^{|m|}_n(\cos\theta) \, \sin\theta \, d\theta \, d(kr) \qquad (50)$$

$$C^{C,m}_n = \frac{k(2n+1)^2}{\pi n(n+1)} \frac{(n-|m|)!}{(n+|m|)!} \int_0^\pi \int_0^\infty \frac{F}{2} \left[\sum_{j_-=m\neq0}^{jplt} \Psi_{jplt} \right]$$

$$r\Psi^1_n(kr) \, P^{|m|}_n(\cos\theta) \, \sin\theta \, d\theta \, d(kr) \qquad (51)$$

$$C^{D,m}_n = \frac{k(2n+1)^2}{\pi n(n+1)} \frac{(n-|m|)!}{(n+|m|)!} \int_0^\pi \int_0^\infty G \left[\sum_{j_0=m\neq0}^{jplt} \Psi_{jplt} \right]$$

$$r\Psi^1_n(kr) \, P^{|m|}_n(\cos\theta) \, \sin\theta \, d\theta \, d(kr) \qquad (52)$$

We set :

$$g^m_{n,TM} = \frac{1}{kc^{pw}_n} \left[C^{B,m}_n + C^{C,m}_n + x_0 C^{D,m}_n \right] \qquad (53)$$

in which the superscript 'pw' is reminiscent for 'plane wave', with :

$$kc^{pw}_n = i^{n-1} (-1)^n \frac{2n+1}{n(n+1)} \qquad (54)$$

and observe that the case m = o in relation (53) leads to (48) if we extend the relations (50)-(52) to this case. Then, we add the four BSP's, U^{io}_{TM}, U^{iB}_{TM}, U^{iC}_{TM} and U^{iD}_{TM}, and simply find :

$$U^i_{TM} = E_0 \sum_{n=1}^{\infty} \sum_{m=-n}^{+n} c_n^{pw} \, g^m_{n,TM} \; r\Psi^i_n(kr) \, P_n^{|m|}(\cos\theta) \, \exp(im\varphi) \qquad (55)$$

In terms of Ricatti-Bessel functions to be now preferred, (55) becomes :

$$U^i_{TM} = \frac{E_0}{k} \sum_{n=1}^{\infty} \sum_{m=-n}^{+n} c_n^{pw} \, g^m_{n,TM} \; \Psi_n(kr) \, P_n^{|m|}(\cos\theta) \, \exp(im\varphi) \qquad (56)$$

Working similarly with $H_r = H_{r,TE}$ instead of $E_r = E_{r,TM}$, we obtain the TE-BSP for the incident wave :

$$U^i_{TE} = \frac{H_0}{k} \sum_{n=1}^{\infty} \sum_{m=-n}^{+n} c_n^{pw} \, g^m_{n,TE} \; \Psi_n(kr) \, P_n^{|m|}(\cos\theta) \, \exp(im\varphi) \qquad (57)$$

in which :

$$g^m_{n,TE} = \frac{1}{kc_n^{pw}} \left[- i \, C_n^{B,m} + i \, C_n^{C,m} + y_0 \, C_n^{D,m} \right] \qquad (58)$$

VII - THE BSP'S FOR THE EXTERNAL AND SPHERE WAVES

We call external wave the wave scattered by the particle and sphere wave the wave inside the particle. The BSP's for the external (U^e_{TM}, U^e_{TE}) and sphere (U^{sp}_{TM}, U^{sp}_{TE}) waves are set to be :

$$U^e_{TM} = \frac{-E_0}{k} \sum_{n=1}^{\infty} \sum_{m=-n}^{+n} c_n^{pw} \, A_n^m \, \xi_n(kr) \, P_n^{|m|}(\cos\theta) \, \exp(im\varphi) \qquad (59)$$

$$U^e_{TE} = \frac{-H_0}{k} \sum_{n=1}^{\infty} \sum_{m=-n}^{+n} c_n^{pw} \, B_n^m \, \xi_n(kr) \, P_n^{|m|}(\cos\theta) \, \exp(im\varphi) \qquad (60)$$

$$U^{sp}_{TM} = \frac{kE_0}{k^2_{sp}} \sum_{n=1}^{\infty} \sum_{m=-n}^{+n} c_n^{pw} \, E_n^m \, \Psi_n(k_{sp}r) \, P_n^{|m|}(\cos\theta) \, \exp(im\varphi) \qquad (61)$$

$$U^{sp}_{TE} = \frac{kH_0}{k^2_{sp}} \sum_{n=1}^{\infty} \sum_{m=-n}^{+n} c_n^{pw} \, D_n^m \, \Psi_n(k_{sp}r) \, P_n^{|m|}(\cos\theta) \, \exp(im\varphi) \qquad (62)$$

k_{sp} is the wavenumber in the sphere material. The functions $\xi_n(kr)$ are used in (59)-(60) to satisfy the boundary conditions in the limit $r \to \infty$, i.e. to give field expressions which tend to a spherical wave in this limit.

VIII - DETERMINATION OF THE EXTERNAL WAVE

The scattering coefficients A_n^m and B_n^m of the external wave are determined by writing the tangential continuity of the electric and

magnetic fields at the surface of the sphere ($r = d/2$), the field
components being obtained from the set (7)-(16), using (56)-(57) and
(59)-(62). We write :

$$V^i_{\theta,x} + V^e_{\theta,x} = V^{sp}_{\theta,x} \tag{63}$$

where V stands for E or H, and X for TM or TE. These are four boundary
conditions which lead to :

$$M \left[g^m_{n,TM} \, \Psi'_n(\alpha) - A^m_n \, \xi'_n(\alpha) \right] = E^m_n \, \Psi'_n(\beta) \tag{64}$$

$$M^2 \left[g^m_{n,TE} \, \Psi_n(\alpha) - B^m_n \, \xi_n(\alpha) \right] = D^m_n \, \Psi_n(\beta) \tag{65}$$

$$\left[g^m_{n,TM} \, \Psi_n(\alpha) - A^m_n \, \xi_n(\alpha) \right] = E^m_n \, \Psi_n(\beta) \tag{66}$$

$$M \left[g^m_{n,TE} \, \Psi'_n(\alpha) - B^m_n \, \xi'_n(\alpha) \right] = D^m_n \, \Psi'_n(\beta) \tag{67}$$

where M is the complex refractive index of the particle relatively to
the surrounding medium, α the size parameter $\pi d/\lambda$ (λ being the
wavelength in the surrounding medium) and β is $M\alpha$. The "prime" means
derivative of the function for the value of the argument indicated
between parentheses. To establish these equations, we have assumed a
non-magnetic particle ($\mu_{sp}/\mu = 1$) and used the accessory relation, valid
for a non-magnetic particle :

$$M = k_{sp}/k = \sqrt{\epsilon_{sp}/\epsilon} \tag{68}$$

The same set (64)-(67) is obtained if we write the relation (63)
with φ instead of θ.

The set (64)-(67) is easily solved to find :

$$A^m_n = a_n \, g^m_{n,TM} \tag{69}$$

$$B^m_n = b_n \, g^m_{n,TE} \tag{70}$$

where a_n and b_n are the usual scattering coefficients of the Mie
theory :

$$a_n = \frac{\Psi_n(\alpha) \, \Psi'_n(\beta) - M \, \Psi'_n(\alpha) \, \Psi_n(\beta)}{\xi_n(\alpha) \, \Psi'_n(\beta) - M \, \xi'_n(\alpha) \, \Psi_n(\beta)} \tag{71}$$

$$b_n = \frac{M \, \Psi_n(\alpha) \, \Psi'_n(\beta) - \Psi'_n(\alpha) \, \Psi_n(\beta)}{M \, \xi_n(\alpha) \, \Psi'_n(\beta) - \xi'_n(\alpha) \, \Psi_n(\beta)} \tag{72}$$

IX - THE SCATTERED FIELD COMPONENTS

From the external BSP's now determined, the field components of the
scattered wave are obtained using the set (7)-(16). We find :

$$E_r = -kE_0 \sum_{n=1}^{\infty} \sum_{m=-n}^{+n} c_n^{pw} \, a_n g_{n,TM}^m \left[\xi_n''(kr) \right. \tag{73}$$

$$\left. + \xi_n(kr) \right] P_n^{|m|}(\cos\theta) \, \exp(im\varphi)$$

$$E_\theta = -\frac{E_0}{r} \sum_{n=1}^{\infty} \sum_{m=-n}^{+n} c_n^{pw} \left[a_n g_{n,TM}^m \, \xi_n'(kr) \, \tau_n^{|m|}(\cos\theta) \right. \tag{74}$$

$$\left. + m b_n g_{n,TE}^m \, \xi_n(kr) \, \Pi_n^{|m|}(\cos\theta) \right] \exp(im\varphi)$$

$$E_\varphi = -\frac{iE_0}{r} \sum_{n=1}^{\infty} \sum_{m=-n}^{+n} c_n^{pw} \left[m a_n g_{n,TM}^m \, \xi_n'(kr) \, \Pi_n^{|m|}(\cos\theta) \right. \tag{75}$$

$$\left. + b_n g_{n,TE}^m \, \xi_n(kr) \, \tau_n^{|m|}(\cos\theta) \right] \exp(im\varphi)$$

$$H_r = -kH_0 \sum_{n=1}^{\infty} \sum_{m=-n}^{+n} c_n^{pw} b_n g_{n,TE}^m \left[\xi_n''(kr) \right. \tag{76}$$

$$\left. + \xi_n(kr) \right] P_n^{|m|}(\cos\theta) \, \exp(im\varphi)$$

$$H_\theta = \frac{H_0}{r} \sum_{n=1}^{\infty} \sum_{m=-n}^{+n} c_n^{pw} \left[m a_n g_{n,TM}^m \, \xi_n(kr) \, \Pi_n^{|m|}(\cos\theta) \right. \tag{77}$$

$$\left. - b_n g_{n,TE}^m \, \xi_n'(kr) \, \tau_n^{|m|}(\cos\theta) \right] \exp(im\varphi)$$

$$H_\varphi = \frac{iH_0}{r} \sum_{n=1}^{\infty} \sum_{m=-n}^{+n} c_n^{pw} \left[a_n g_{n,TM}^m \, \xi_n(kr) \, \tau_n^{|m|}(\cos\theta) \right. \tag{78}$$

$$\left. - m b_n g_{n,TE}^m \, \xi_n'(kr) \, \Pi_n^{|m|}(\cos\theta) \right] \exp(im\varphi)$$

where we have used :

$$k = \omega\mu \frac{H_0}{E_0} = \omega\epsilon \frac{E_0}{H_0} \tag{79}$$

We have defined generalized Legendre functions according to :

$$\tau_n^k(\cos\theta) = \frac{d}{d\theta} P_n^k(\cos\theta) \tag{80}$$

$$\Pi_n^k(\cos\theta) = \frac{P_n^k(\cos\theta)}{\sin\theta} \tag{81}$$

36

For k = 1, these are the usual Legendre functions appearing in the Mie theory, $\tau_n(\cos\theta)$ and $\pi_n(\cos\theta)$. Note the following correspondence with classical notations : $\tau_n^1 = \tau_n$, $\pi_n^1 = \pi_n$ but $P_n^0 = P_n$.

X - SCATTERED FIELD COMPONENTS IN THE FAR-FIELD

The set (73)-(78) can be of interest in some applications. An example of near-field computations, although limited to Mie theory, is given in ref 17. However, in most cases, interest is limited to the so-called far field, defined by the inequality $r \gg \lambda$. The exact meaning of this inequality is not a trivial matter [17]. Nevertheless, it leads to an asymptotic expression for the functions $\xi_n(kr)$ which is [18] :

$$\xi_n(kr) \to i^{n+1} \exp(-ikr) \qquad (82)$$

Then :

$$\left[\xi_n''(kr) + \xi_n(kr)\right] \to 0 \qquad (83)$$

leading to :

$$E_r = H_r \to 0 \qquad (84)$$

In this limit, the non-zero field components simplify to :

$$E_\theta = \frac{iE_0}{kr} \exp(-ikr) \sum_{n=1}^{\infty} \sum_{m=-n}^{+n} \frac{2n+1}{n(n+1)} \left[a_n g_{n,TM}^m \tau_n^{|m|}(\cos\theta) \right.$$
$$\left. + imb_n g_{n,TE}^m \Pi_n^{|m|}(\cos\theta) \right] \exp(im\varphi) \qquad (85)$$

$$E_\varphi = \frac{-E_0}{kr} \exp(-ikr) \sum_{n=1}^{\infty} \sum_{m=-n}^{+n} \frac{2n+1}{n(n+1)} \left[ma_n g_{n,TM}^m \Pi_n^{|m|}(\cos\theta) \right.$$
$$\left. + ib_n g_{n,TE}^m \tau_n^{|m|}(\cos\theta) \right] \exp(im\varphi) \qquad (86)$$

$$H_\varphi = \frac{H_0}{E_0} E_\theta \qquad (87)$$

$$H_\theta = - \frac{H_0}{E_0} E_\varphi \qquad (88)$$

The scattered wave has become a transverse wave.

XI - THE SCATTERED INTENSITIES

The scattered intensities are then computed with the aid of the Poynting theorem :

$$S = \frac{1}{2} \text{Re} \left(E_\theta H_\varphi^* - E_\varphi H_\theta^* \right) \qquad (89)$$

Let I_θ and I_φ be the intensities associated with the electric fields E_θ and E_φ respectively, reduced with the normalizing assumption :

$$\frac{1}{2} \sqrt{\frac{\varepsilon}{\mu}} \; E_0^2 = 1 \tag{90}$$

Injecting the set (85)-(88) in (89), and separating the contributions I_θ and I_φ to S, we obtain :

$$\begin{pmatrix} I_\theta \\ I_\varphi \end{pmatrix} = \frac{\lambda^2}{4\pi^2 r^2} \begin{pmatrix} |S_2|^2 \\ |S_1|^2 \end{pmatrix} \tag{91}$$

where S_2 and S_1 are defined by :

$$E_\theta = \frac{iE_0}{kr} \exp(-ikr) \; S_2 \tag{92}$$

$$E_\varphi = - \frac{E_0}{kr} \exp(-ikr) \; S_1 \tag{93}$$

XII - THE PHASE ANGLE

The phase angle δ between the components E_θ and E_φ is equal to the phase angle between the components E'_θ and E'_φ given by :

$$E'_\theta = iS_2 = A'_\theta \; e^{i\varphi_2} \tag{94}$$

$$E'_\varphi = - S_1 = A'_\varphi \; e^{i\varphi_1} \tag{95}$$

where A'_θ and A'_φ are real numbers. We readily obtain :

$$\tan\delta = \tan(\varphi_2 - \varphi_1) = \frac{\mathrm{Re}(S_1) \; \mathrm{Re}(S_2) + \mathrm{Im}(S_1) \; \mathrm{Im}(S_2)}{\mathrm{Im}(S_1) \; \mathrm{Re}(S_2) - \mathrm{Re}(S_1) \; \mathrm{Im}(S_2)} \tag{96}$$

The expressions for the cross sections (respectively for absorption, scattering, extinction and radiation pressure) should here follow the results for the scattered intensities and for the phase angle. But, as it has been explained in the introduction, the lack of room led us to report these relations into another paper. The end of the present paper will focus on the relations between our previous work and the present one.

XIII - ANOTHER FORMULATION

To put a bridge between the present formulation and the results obtained previously in refs 7,8,9, another formulation is required.

Let S_2 and S_1 be the amplitude functions given by [7,8,9] :

$$S_2 = \sum_{n=1}^{\infty} \frac{2n+1}{n(n+1)} \; g_n \; (a_n \tau_n(\cos\theta) + b_n \Pi_n(\cos\theta)) \tag{97}$$

$$S_1 = \sum_{n=1}^{\infty} \frac{2n+1}{n(n+1)} \; g_n \; (a_n \Pi_n(\cos\theta) + b_n \tau_n(\cos\theta)) \tag{98}$$

where the coefficients g_n are given by :

$$g_n = \frac{k(2n+1)}{i^{n-1}(-1)^n \pi n(n+1)} \int_0^\pi \int_0^\infty F\, r\, \Psi_n^1(kr)\, P_n^1(\cos\theta)\, \sin\theta\, d\theta\, d(kr)$$

(99)

These coefficients appeared in the previous versions of our generalization [7,8,9].

Then we establish that :

$$\mathcal{S}_2 = \cos\varphi\, S_2 + \mathcal{S}_2'$$

(100)

$$\mathcal{S}_1 = i\sin\varphi\, S_1 + \mathcal{S}_1'$$

(101)

with :

$$\mathcal{S}_2' = \sum_{\substack{n=1 \\ }}^{\infty} \sum_{\substack{m=-n \\ |m|\neq 1}}^{+n} \frac{2n+1}{n(n+1)} \left[a_n g_{n,TM}^m\, \tau_n^{|m|}(\cos\theta) \right.$$

$$\left. + imb_n g_{n,TE}^m\, \Pi_n^{|m|}(\cos\theta) \right] \exp(im\varphi)$$

$$+ \sum_{n=1}^{\infty} \frac{2n+1}{n(n+1)} \left[a_n \tau_n(\cos\theta) \left(\cos\varphi\, G_{n,TM}^+ + i\sin\varphi\, G_{n,TM}^- \right) \right.$$

$$\left. + b_n \Pi_n(\cos\theta) \left(i\cos\varphi\, G_{n,TE}^- - \sin\varphi\, G_{n,TE}^+ \right) \right]$$

(102)

and :

$$\mathcal{S}_1' = \sum_{\substack{n=1 \\ }}^{\infty} \sum_{\substack{m=-n \\ |m|\neq 1}}^{+n} \frac{2n+1}{n(n+1)} \left[ma_n g_{n,TM}^m\, \Pi_n^{|m|}(\cos\theta) \right.$$

$$\left. + ib_n g_{n,TE}^m\, \tau_n^{|m|}(\cos\theta) \right] \exp(im\varphi)$$

$$+ \sum_{n=1}^{\infty} \frac{2n+1}{n(n+1)} \left[a_n \Pi_n(\cos\theta) \left(\cos\varphi\, G_{n,TM}^- + i\sin\varphi\, G_{n,TM}^+ \right) \right.$$

$$\left. + b_n \tau_n(\cos\theta) \left(i\cos\varphi\, G_{n,TE}^+ - \sin\varphi\, G_{n,TE}^- \right) \right]$$

(103)

in which :

$$G_{n,TM}^+ = g_{n,TM}^1 + g_{n,TM}^{-1} - g_n$$

(104)

$$G_{n,TE}^+ = g_{n,TE}^1 + g_{n,TE}^{-1}$$

(105)

$$G_{n,TM}^- = g_{n,TM}^1 - g_{n,TM}^{-1}$$

(106)

$$G_{n,TE}^- = g_{n,TE}^1 - g_{n,TE}^{-1} + ig_n$$

(107)

Using the relations (100)-(101), the expressions for the scattered intensities become :

$$\begin{pmatrix} I_\theta \\ I_\varphi \end{pmatrix} = \begin{pmatrix} I_\theta^L \\ I_\varphi^L \end{pmatrix} + \begin{pmatrix} I_\theta^C \\ I_\varphi^C \end{pmatrix} + \begin{pmatrix} I_\theta^S \\ I_\varphi^S \end{pmatrix} \tag{108}$$

with :

$$\begin{pmatrix} I_\theta^L \\ I_\varphi^L \end{pmatrix} = \frac{\lambda^2}{4\pi^2 r^2} \begin{pmatrix} i_2 \cos^2\varphi \\ i_1 \sin^2\varphi \end{pmatrix} \tag{109}$$

where the intensity functions i_j are $|S_j|^2$ and :

$$\begin{pmatrix} I_\theta^C \\ I_\varphi^C \end{pmatrix} = \frac{2\lambda^2}{4\pi^2 r^2} \begin{pmatrix} \cos\varphi \; \mathrm{Re}\left(S_2 S_2'^{\;*}\right) \\ \sin\varphi \; \mathrm{Re}\left(iS_1 S_1'^{\;*}\right) \end{pmatrix} \tag{110}$$

$$\begin{pmatrix} I_\theta^S \\ I_\varphi^S \end{pmatrix} = \frac{\lambda^2}{4\pi^2 r^2} \begin{pmatrix} \left|S_2'\right|^2 \\ \left|S_1'\right|^2 \end{pmatrix} \tag{111}$$

The first term (superscript L) has the structure of the one appearing in the Mie theory and will be called the Leader term. The last one contains S_2' and S_1' which must be added to the Mie theory terms to produce S_2 and S_1 (relations 100-101), and will be called the Secondary term (superscript S). The second term involves a coupling between the Leader and Secondary terms and will be called the Cross-term (superscript C).

For the phase angle δ, we put :

$$\tan\delta_0 = \frac{\mathrm{Re}(S_1)\;\mathrm{Im}(S_2) - \mathrm{Re}(S_2)\;\mathrm{Im}(S_1)}{\mathrm{Re}(S_1)\;\mathrm{Re}(S_2) + \mathrm{Im}(S_1)\;\mathrm{Im}(S_2)} \tag{112}$$

where the r.h.s is formally identical to the expression of the phase angle tangent in the pure Mie theory. Injecting (100)-(101) in (96), we then find the link between δ and δ_0 as follows :

$$\tan\delta = \tan(\delta_0 + \delta_1) + \tan\delta_2 \tag{113}$$

where the expressions for δ_1 and δ_2 are not given , being not essential in the present discussion.

XIV - SPECIAL CASES

We now specify the formulation to some special cases of interest and first consider the case when the center of the scattering sphere is located on the axis of the incident beam ($x_0 = y_0 = 0$). Then the formulation simplifies dramatically because :

$$\Psi_{jplt} = \delta_j^0 \tag{114}$$

where δ_j^0 is zero, except for $j = o$ where it is 1 (Kronecker symbol). Consequently, all the $g_{n,TM}^m$ and $g_{n,TE}^m$ become equal to 0 except for $|m| = 1$. For these coefficients, with $|m| = 1$, we find :

$$g_{n,TM}^1 = g_{n,TM}^{-1} = \frac{1}{2} g_n \tag{115}$$

$$g_{n,TE}^1 = - g_{n,TE}^{-1} = - \frac{1}{2} g_n \tag{116}$$

leading to (relations 104-107) :

$$G_{n,TM}^+ = G_{n,TE}^+ = G_{n,TM}^- = G_{n,TE}^- = 0 \tag{117}$$

Hence :

$$S_2' = S_1' = 0 \tag{118}$$

The scattered intensities are given by the leader terms, any cross- or secondary term being now identical to zero. For the phase angle, we find $\delta_1 = \delta_2 = 0$, and $\tan \delta$ is given by the classical relation (112). The formulation becomes identical to the one of the classical Mie theory except for the appearance of the g_n's in the amplitude functions (97,98). This simpler formulation, given and discussed in ref 8,9, corresponded to the more special case when $z_o = o$ which merely simplifies a bit the expression for F (relation 36). In ref 7, we considered a simpler case when the incident field components E_w and H_w are neglected. This approximation has been called the order L- of approximation and its status has been discussed in ref 9. It only produces another simpler expression for the function F.

Finally, assume $w_o \to \infty$. The incident Gaussian wave becomes a plane wave. Using orthogonality relations for the $\psi_n(kr)$ and $P_n^1(\cos\theta)$, we prove, using (99), that all the g_n's become equal to 1 (refs 7,8). Then, the formulation becomes fully identical with the Mie theory formulation just as given by Kerker[18]. The Mie theory has become a special case of a more general theory, as it should for correctness.

XV - CONCLUSION

We have built a theoretical formulation for the scattering of a laser beam by a spherical scatterer wherever it is located with respect to the beam. The general features of the theory and of its formulation are related to the Lorenz-Mie theory and, therefore, it is called Generalized Lorenz-Mie Theory. In the above paper the main steps of the GLMT have been outlined and the formulae have been given for the scattered intensities and for the phase angle in the case when the beam is a Gaussian one. Other scattering parameters will be given in further papers with more details. Other beam descriptions could be used instead of a Gaussian beam without changing the method, namely the Bromwich formulation. The present formulation of the GLMT represents the theoretical achievement of a work which began years ago. In the next step of our work, we will numerically compute the scattering parameters of an arbitrarily located scatterer as this has yet been done for centered scatterers.

REFERENCES

1. G. Mie, "Beiträge zur Optik Trüber Medien, speziell Kolloidaler Metalösungen", Ann. der Phys., 25, 377-452 (1908).
2. W.G. Tam and R. Corriveau, "Scattering of Electromagnetic beams by Spherical Objects", J. Opt. Soc. Am., 68, 6, 763-767 (1978).
3. N. Morita, T. Takenaka, T. Yamasaki and Y. Nakanishi, "Scattering of a Beam Wave by a Spherical Object", IEEE Trans. Antennas Propag., AP 16, 6, 724-727 (1968).
4. W.C. Tsai and R.J. Pogorzelski, "Eigenfunction Solution of the Scattering of Beam Radiation Fields by Spherical Objects", J. Opt. Soc. Am., 65, 12, 1457-1463 (1975).
5. J.S. Kim and S.S. Lee, "Scattering of Laser Beams and the Optical Potential Well for a Homogeneous Sphere", J. Opt. Soc. Am., 73, 3, 303-312 (1983).
6. G. Gréhan, "Nouveaux progrès en théorie de Lorenz-Mie. Application à la mesure de diamètres de particules dans des écoulements", Thèse, Rouen (1980).
7. G. Gouesbet and G. Gréhan, "Sur la généralisation de la théorie de Lorenz-Mie", J. Opt., 13, 2, 97-103 (1982).
8. G. Gouesbet, G. Gréhan and B. Maheu, "Scattering of a Gaussian Beam by a Mie Scatter Center Using a Bromwich Formalism", J. Opt., 16, 2, 83-93 (1985).
9. G. Gouesbet, B. Maheu and G. Gréhan, "The Order of Approximation in a Theory of the Scattering of a Gaussian Beam by a Mie Scatter Center", J. Opt., 16, 5, 239-247 (1985).
10. G. Gréhan, B. Maheu and G. Gouesbet, "Scattering of Laser Beams by Mie Scatter Centers : Numerical Results Using a Localized Approximation", Appl. Opt., 25, 19, 3539-3548 (1986).
11. B. Maheu, G. Gréhan and G. Gouesbet, "Generalized Lorenz-Mie Theory : First Exact Values and Comparisons with the Localized Approximation", Appl. Opt., 26, 1, 23-25 (1987).
12. B. Maheu, G. Gréhan and G. Gouesbet, "Theory of laser beam scattering by individual spherical particles : numerical results and applications to optical sizing", symposium on Optical Sizing, 12-15 May 1987, Rouen, France (1987).
13. T.J. Bromwich, "Electromagnetic Waves", Phil. Mag., S.6., 38, 223, 143-164 (1919).
14. P. Poincelot, Précis d'électromagnétisme théorique, Masson, Paris (1963).
15. M. Lax, W.H. Louisell and W.B. Mc Knight, "From Maxwell to Paraxial Wave Optics", Phys. Rev. A, 11, 4, 1365-1370 (1975).
16. L.W. Davis, "Theory of Electromagnetic Beams", Phys. Rev. A, 19, 3, 1177-1179 (1979).
17. F. Slimani, G. Gréhan, G. Gouesbet and D. Allano, "Near-field Lorenz-Mie Theory and its application to microholography", Appl. Opt., 23, 22, 4140-4148 (1984).
18. M. Kerker, "The Scattering of Light and Other Electromagnetic Radiations", Academic Press, New-York and London (1969).

EFFECTS OF PARTICLE NONSPHERICITY ON LIGHT-SCATTERING

Peter W. Barber and Steven C. Hill

Electrical and Computer Engineering Department
Clarkson University
Potsdam, New York 13676

INTRODUCTION

Characterizing particles by analyzing their light scattering
properties is of interest in many diverse fields. The most common
measurement is particle size. Measurements of particle refractive index
and shape are also of interest. Most optical sizing methods are based
on the assumption that the particles are spherical.

It is important to study the light scattering characteristics of
nonspherical particles: 1) to understand the effects of nonsphericity on
optical sizing methods and to quantify the errors that may be introduced
by the assumption of sphericity; 2) to develop techniques for measuring
the size of nonspherical particles; and 3) to design new optical
measurement systems which may detect, for example, the nonsphericity of
particles or the alignment of nonspherical particles in a flow system.

The purpose of this paper is to show the effects of particle
nonsphericity on light scattering and to indicate how these effects
might be exploited to develop new measurement systems. Two modalities
are considered, the scattering matrix (angular scattering) and spectral
scattering. The results shown here are all derived from theoretical
solutions of the electromagnetic scattering problem. Mie theory[1] is
used for spherical particles and the T-matrix method[2] for nonspherical
particles.

SCATTERING MATRIX

The polarization state of a beam of light can be completely defined
by the Stoke's vector, a column vector containing the Stoke's parameters
I,Q,U, and V.[1] The polarization state of the incident beam can be
changed upon interaction with a scatterer. The Stoke's parameters of
the incident and scattered beams are related by a 4 x 4 matrix, the
scattering matrix, given by

$$
\begin{bmatrix} I_s \\ Q_s \\ U_s \\ V_s \end{bmatrix} = \frac{1}{k^2 r^2} \begin{bmatrix} S_{11} & S_{12} & S_{13} & S_{14} \\ S_{21} & S_{22} & S_{23} & S_{24} \\ S_{31} & S_{32} & S_{33} & S_{34} \\ S_{41} & S_{42} & S_{43} & S_{44} \end{bmatrix} \begin{bmatrix} I_i \\ Q_i \\ U_i \\ V_i \end{bmatrix}
$$

where r is the distance from the particle to the detector, $k = 2\pi/\lambda$, and λ is the wavelength of the incident light.

Each of the elements of the scattering matrix describes the angular variation of some feature of the scattered light. For example, S_{11} is the scattered irradiance when the particle is illuminated by an unpolarized incident beam. While all sixteen elements of the scattering matrix are present in the most general case, the matrix takes on special forms for specific particulate systems. For example, the system of identical randomly oriented spheroidal particles which will be illustrated here has the form

$$
\begin{bmatrix} S_{11} & S_{12} & 0 & 0 \\ S_{12} & S_{22} & 0 & 0 \\ 0 & 0 & S_{33} & -S_{43} \\ 0 & 0 & S_{43} & S_{44} \end{bmatrix}
$$

Spherical particles have an even simpler form with $S_{22} = S_{11}$ and $S_{44} = S_{33}$.

In our work, we use the normalized scattering matrix P of Asano and Sato.[3] With the normalization condition

$$
\frac{1}{4\pi} \int_{4\pi} P_{11} \, d\Omega = 1
$$

defining the phase function P_{11}, the normalized scattering matrix P is related to the scattering matrix S by

$$
\frac{k^2 \overline{C_{sca}}}{4\pi} P_{ij} = S_{ij} \qquad (i,j = 1,2,3,4)
$$

where $\overline{C_{sca}}$ is the scattering cross section averaged over all orientations. The final form is then

$$
\begin{bmatrix} I_s \\ Q_s \\ U_s \\ V_s \end{bmatrix} = \frac{C_{sca}}{4\pi r^2} \begin{bmatrix} P_{11} & P_{12} & 0 & 0 \\ P_{12} & P_{22} & 0 & 0 \\ 0 & 0 & P_{33} & -P_{43} \\ 0 & 0 & P_{43} & P_{44} \end{bmatrix} \begin{bmatrix} I_i \\ Q_i \\ U_i \\ V_i \end{bmatrix}
$$

For purposes of illustration, the six independent scattering matrix elements are given in Fig. 1 for a 1.5:1 prolate spheroid with m = 1.68 + i0.0001, which is a typical refractive index for silicates, and size parameter ka of 3.276.[4] The calculated quantities are the phase function P_{11} (on a logarithmic scale) and the other five matrix elements normalized to P_{11}. Of particular interest is the quantity $1- P_{22}/P_{11}$. P_{22}/P_{11} is the ratio of the depolarized light to the total scattered light. Since $P_{22} = P_{11}$ for spherical scatterers, it has been suggested that $1- P_{22}/P_{11}$ is a good measure of nonsphericity since the quantity will be zero at all angles for a sphere, but nonzero for nonspheres.[1,3,5] Similarly, since $P_{44} = P_{33}$ for a sphere, differences between P_{33}/P_{11} and P_{44}/P_{11} indicate nonsphericity. It is clear from this example that where a suspension of particles is concerned, some of the features of the polarized scattered light may be used to detect nonsphericity.

Another potential application pertains to detecting the alignment of nonspherical particles in a flow system. For illumination along the axis of symmetry, a prolate spheroidal particle has a scattering matrix with eight zero elements and with eight nonzero elements symmetrical about 180 degrees as shown in Fig. 2(a). When the spheroid tilts from perfect alignment with the incident beam, the scattering matrix changes dramatically as shown in Fig. 2(b). The matrix elements are all nonzero and nonsymmetric about 180 degrees. Measurement of a subset of the scattering matrix can provide an indication of alignment.

SPECTRAL SCATTERING

Resonance peaks corresponding to the natural modes of oscillation[6,7] have been observed in elastic scattering spectra[8-10] and in Raman[11] and fluorescence[12,13] emission spectra of microspheres. Since the resonance frequencies are sensitive functions of the size, shape[14] and refractive index, such resonances have been termed structural resonances or morphology dependent resonances (MDR's).

The locations and shapes of such MDR's have been used to accurately determine the sizes and refractive indices of spheres. Chylek et. al.[10] described a method for determining both the size and refractive index by comparing the shapes and locations of measured and computed resonance peaks. Using this technique the relative uncertainties in size were reduced from 0.01 to 5×10^{-5}, and the relative uncertainties in

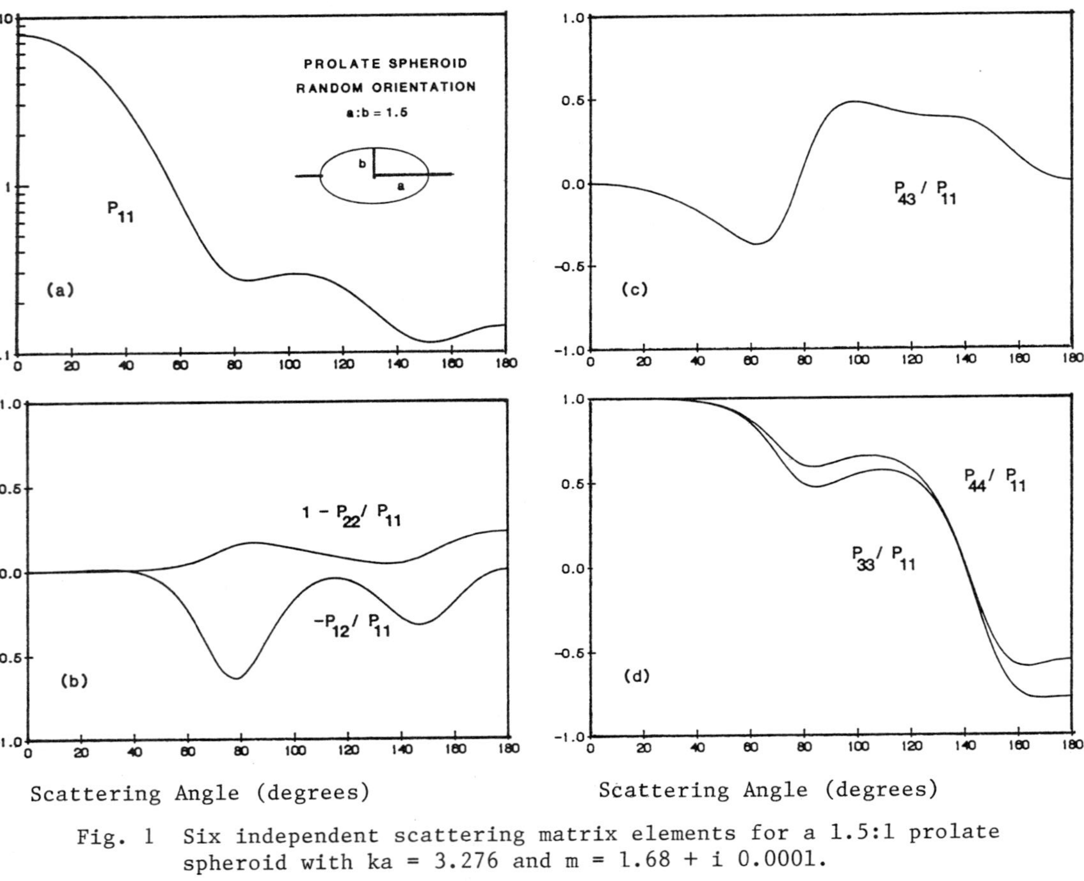

Fig. 1 Six independent scattering matrix elements for a 1.5:1 prolate spheroid with ka = 3.276 and m = 1.68 + i 0.0001.

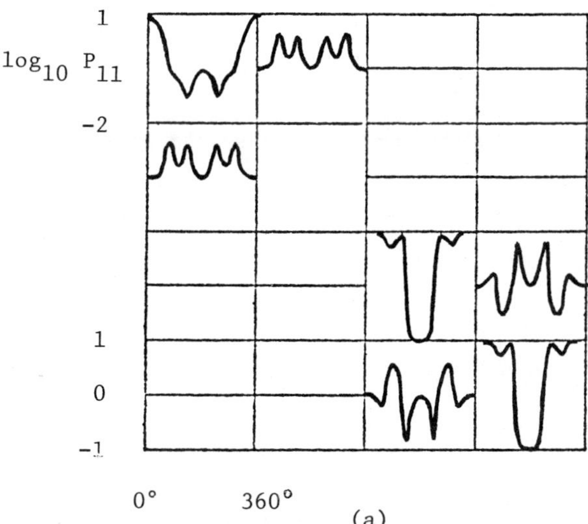

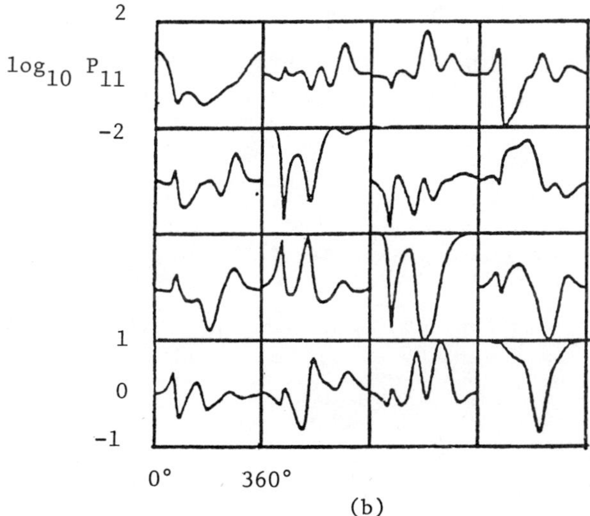

Fig. 2 Scattering matrix elements for a 2:1 oblate
spheroid with ka = 1.575 (a = semiminor axis)
and m = 1.68 + i 0.0001. (a) alignment with
the incident beam, (b) nonalignment.

refractive index were reduced from 5×10^{-3} to 5×10^{-5}. Algorithms for sizing that use only resonance locations and are more suitable for automated sizing have also been described.[15,16] With these algorithms the refractive index of the spheres is assumed to be known beforehand. Advantages of spectral methods for sizing are that the spectra can be determined relatively easily, and the sizes can be determined extremely accurately. The relative uncertainties obtained with this method are much smaller than the reported uncertainties associated with the determination of size from angular scattering data. The method is sensitive enough to determine, for example, the evaporation rates of falling droplets. If fluorescence emission or Raman emission spectra of microspheres are used, the spectra may be obtained while the spheres are resting on substrates.[9] This is useful because many particles to be sized are collected on filters.

Although it has been shown that resonances occur in spectra of oriented or randomly oriented spheroids,[14] there have been no analyses of the effects of particle nonsphericity on sizing using structural resonances. Fig. 3(a) shows the computed resonance spectrum of a sphere having a refractive index of 2.0. The size parameter $(2\pi a/\lambda)$ varies from 7.0 to 9.0, which corresponds to a range of wavelengths of 0.5 um to 0.643 um if the diameter of the sphere is 1.29 um. Figure 3(b) shows the computed resonance spectrum of a 1.05:1 prolate spheroid having a volume equal to the sphere of Fig. 3(a). The orientation of the spheroid in the incident field is indicated on the figure. Both the broad and narrow resonances differ markedly from the resonances of the sphere. Several new sharp resonances also appear. One implication of the spectra of Fig. 3 is that the algorithms developed for sizing spheres using resonance spectra would not be directly applicable to the sizing of spheroids with aspect ratios even as small as 1.05. However, the spectra of Fig. 3 do suggest that spectral data might be used to determine aspect ratios and/or orientations of spheroids.

T-MATRIX STUDIES

The calculated results shown in Fig. 3 were obtained using the T-matrix method. The T-matrix method is ideally suited for studying the spectral scattering characteristics of slightly deformed spherical particles.

In the T-matrix solution, the incident field is given by

$$\overline{E}_i(r) = \sum_{\nu=1}^{\infty} D_\nu [a_\nu \overline{M}_\nu^1(k\overline{r}) + b_\nu \overline{N}_\nu^1(k\overline{r})]$$

where $\overline{M}$ and $\overline{N}$ are the vector spherical wave functions, ν is a combined index incorporating the spherical indices, D is a normalization constant, and the expansion coefficients a snd b are assumed to be known for a specified incident field. The superscript 1 on $\overline{M}$ and $\overline{N}$ indicates that these functions have a Bessel function radial dependence.

The scattered field has a similar form

$$\overline{E}_s(r) = \sum_{\nu=1}^{\infty} D_\nu [f_\nu \overline{M}_\nu^3(k\overline{r}) + g_\nu \overline{N}_\nu^3(k\overline{r})]$$

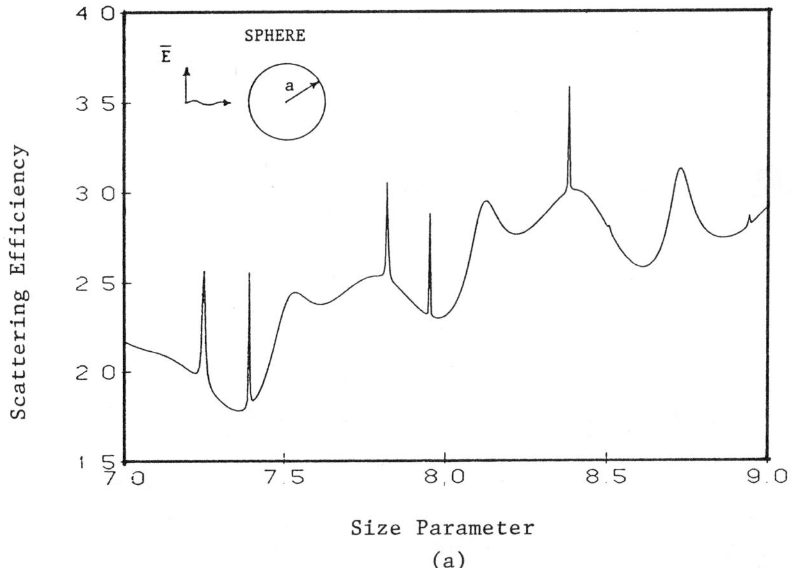

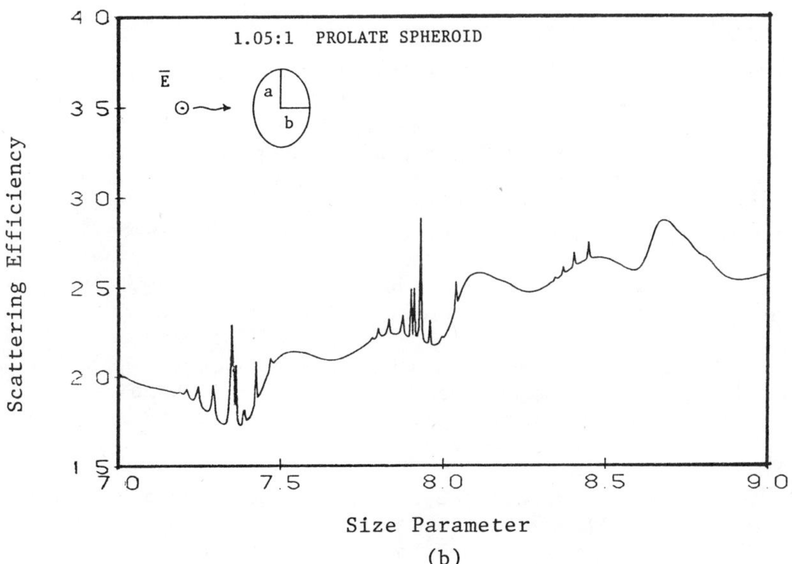

Fig. 3 Spectral scattering (scattering efficiency vs size parameter) by a sphere and 1.05:1 prolate spheroid with m = 2 + i 0.0.

where the f and g coefficients are unknown and the superscript 3 on $\overline{M}$ and $\overline{N}$ indicates that these functions are of the type suitable for radiation fields (Hankel function radial dependence).

Truncating the expansions results in a linear system of equations relating the unknown scattered field coefficients to the known incident field coefficients.

$$\begin{bmatrix} f_\nu \\ \\ g_\nu \end{bmatrix} = - [T] \begin{bmatrix} a_\nu \\ \\ b_\nu \end{bmatrix}$$

The [T] matrix transforms the incident field coefficients to the scattered field coefficients and hence is called the transition matrix or T-matrix. The matrix contains all of the information about the scatterer, such as its size, shape and index of refraction. A similar form relates the internal field coefficients to the incident field coefficients. Because the T-matrix method is based on spherical harmonic expansions of the field quantities, the form of the [T] matrix explicitly shows the light scattering effects of particle nonsphericity.

Figure 4 shows the absolute value of the elements of the [T] matrix for a spherical particle of size parameter 30 and index of refraction of 1.33. We note that the matrix is diagonal. In fact, the T-matrix method for a spherical particle is identical to the Mie Theory. The important feature here is that there is a one-to-one relationship between incident and scattered field coefficients. In other words, a particular mode of the incident field is coupled to the same mode of the scattered field.

The character of the [T] matrix changes dramatically when a spherical particle is slightly deformed. Figure 5 shows the matrix for a 1.1:1 prolate spheroid with the same volume as the sphere in Figure 4. The index of refraction is still 1.33. The off-diagonal terms indicate that the particle is causing mode coupling to occur such that a particular mode of the incident field can couple to more than one — in fact can couple to many modes of the scattered field. We believe that this mode coupling may provide a mechanism for the generation of additional resonance peaks which are observed in the scattering cross section, although this has not yet been shown. One other feature of the matrix which is important from a computational standpoint is that while the matrix is full for a general nonspherical particle, the matrix is sparse for slightly deformed spheres. We plan to use this characteristic of the matrix to improve the efficiency of spectral calculations for slightly deformed spheres.

We have explored the resonance behavior in more detail by making a series of calculations for a sphere as it is deformed to slightly oblate and prolate shapes. Figure 6 shows the scattering efficiency for equatorial incidence as a particle of constant volume changes from a 0.985:1 oblate spheroid to a 1.015:1 prolate spheroid in steps of 0.005. The index of refraction is 2.0. The curves have been displaced vertically to better show the resonance behavior. The scattering efficiency for a sphere is the middle curve in each plot. Focusing on the right hand set of peaks we note that additional peaks occur on the right hand side of the central peak for oblate spheroids and on the left hand side of the central peak for prolate spheroids. However,

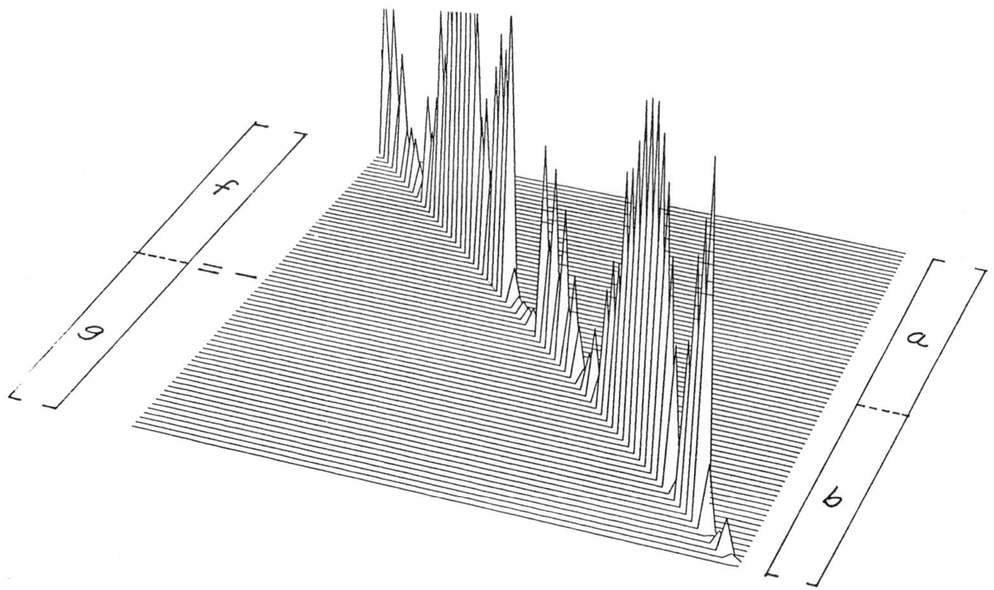

Fig. 4   The T-matrix for a sphere with size parameter = 30 and
an index of refraction of m = 1.33.

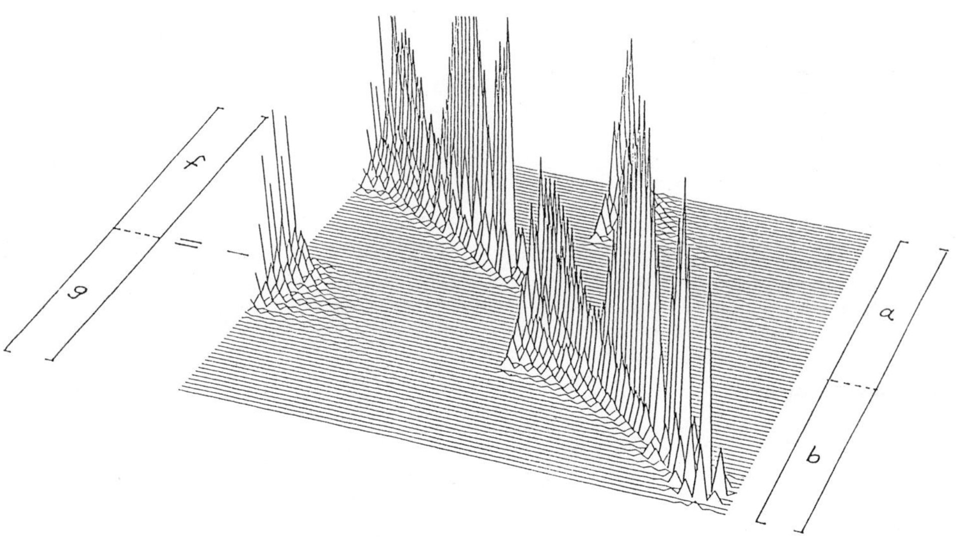

Fig. 5   The T-matrix for a 1.1:1 prolate spheroid with the same
volume as the sphere in Fig. 4 and an index of refraction
of m = 1.33.

additional calculated results show that the peaks switch when the
incident polarization is rotated by 90 degrees. This is not unexpected
as the resonance behavior is believed to be related to trapped waves
circumnavigating the particle, and we should expect similar effects to
occur when both the particle shape and the incident polarization
change. However, the complete reasons for this behavior need to be
explained further. Another feature noted is that the amplitude of the
additional peaks increases as the particle deviates further from a
spherical shape.

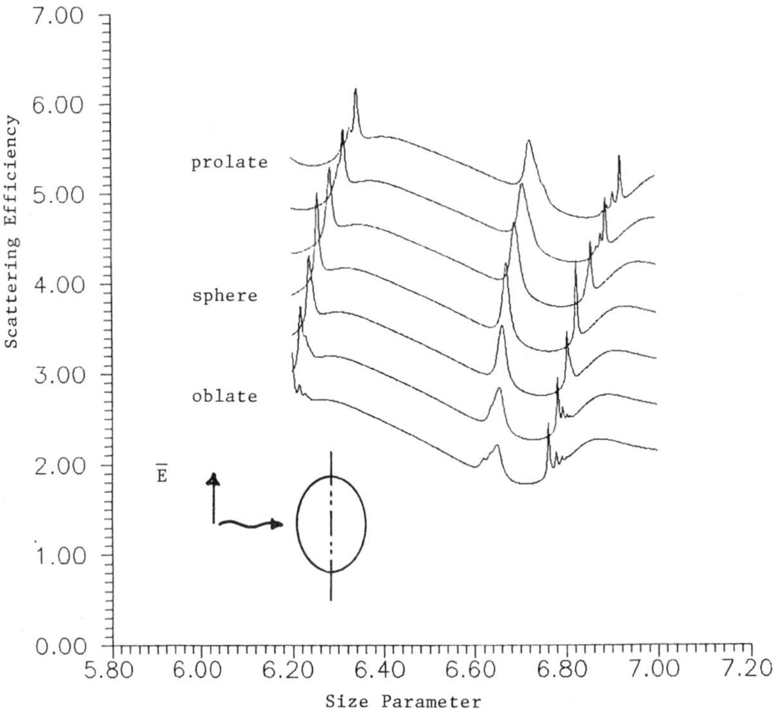

Fig. 6 Scattering efficiency as a function of particle
shape for an index of refraction of m = 2.0.

CONCLUSION

The angular and spectral scattering characteristics of nonspherical
particles differ markedly from those for spherical particles. The
scattering matrix explicitly shows the effects of nonsphericity on the
angular scattering. No studies of the amount of information about shape
and orientation that might be retrieved from scattering spectra have
been reported. Investigations of such spectra may have important
applications in determining the size, shape, refractive index and
orientation of particles.

52

REFERENCES

1. C. F. Bohren and D. R. Huffman, ''Absorption and Scattering of Light by Small Particles,'' Wiley, New York, (1983).

2. P. Barber and C. Yeh, ''Scattering of Electromagnetic Waves by Arbitrarily Shaped Dielectric Bodies,'' Appl. Opt. 14: 2864 (1975).

3. S. Asano and M. Sato, ''Light Scattering by Randomly Oriented Spheroidal Particles,'' Appl. Opt. 19: 962 (1980).

4. P. E. Geller, T. G. Tsuei, and P. W. Barber, ''Information Content of the Scattering Matrix for Spheroidal Particles,'' Appl. Opt. 24: 2391 (1985).

5. R. W. Schaefer, ''Calculations of the Light Scattered by Randomly Oriented Ensembles of Spheroids of Size Comparable to the Wavelength,'' Ph.D. Dissertation, State University of New York at Albany (1980).

6. J. A. Stratton, ''Electromagnetic Theory,'' McGraw-Hill, New York (1941), pp. 554-557.

7. R. Fuchs and K. L. Kliewer, ''Optical Modes of Vibration in an Ionic Crystal Sphere,'' J. Opt. Soc. Am. 58: 319 (1968).

8. P. Chylek, J. T. Kiehl, and M. K. W. Ko, ''Optical Levitation and Partial-Wave Resonances,'' Phys. Rev. A 18: 2229 (1978).

9. A. Ashkin and J. M. Dziedzic, ''Observations of Optical Resonances of Dielectric Spheres by Light Scattering,'' Appl. Opt. 20: 1803 (1981).

10. P. Chylek, V. Ramaswamy, A. Ashkin, and J. M. Dziedzic, ''Simultaneous Determination of Refractive Index and Size of Spherical Dielectric Particles from Light Scattering Data,'' Appl. Opt. 22: 2302 (1983).

11. R. Thurn and W. Kiefer, ''Raman-Microsampling Technique Applying Optical Levitation of Radiation Pressure,'' Appl. Spectrosc. 38: 78 (1984).

12. R. E. Benner, P. W. Barber, J. F. Owen, and R. K. Chang, ''Observation of Structure Resonances in the Fluorescence Spectra from Microspheres,'' Phys Rev. Lett. 44: 475 (1980).

13. S. C. Hill, R. E. Benner, C. K. Rushforth, and P. R. Conwell, ''Structural Resonances Observed in the Fluorscence Emission from Small Spheres on Substrates,'' Appl. Opt. 23: 1680 (1984).

14. P. W. Barber, J. F. Owen, and R. K. Chang, ''Resonant Scattering for Characterization of Axisymmetric Dielectric Objects,'' IEEE Trans Antennas Propag. AP-30: 168 (1982).

15. P. R. Conwell, C. K. Rushforth, R. E. Benner, and S. C. Hill, ''An Efficient Automated Algorithm for the Sizing of Dielectric Microspheres Using the Resonance Spectrum,'' J. Opt. Soc. Am. A 1: 1181 (1984).

16. S. C. Hill, C. K. Rushforth, R. E. Benner, and P. R. Conwell, ''Sizing Dielectric Spheres and Cylinders by Aligning Measured and Computed Resonance Locations: Algorithm for Multiple Orders,'' Appl. Opt. 24: 2380 (1985).

PARTICLE SIZING BY INVERSION OF EXTINCTION DATA

M. Bertero [o], C. De Mol [+] and E. R. Pike [*]

(o) Dipartimento di Fisica dell'Università di Genova and
Istituto Nazionale di Fisica Nucleare, Genova, Italy
(+) Département de Mathématique, Université Libre de Bruxelles
Bruxelles, Belgium
(*) Department of Physics, King's College, London and RSRE
Great Malvern, England

INTRODUCTION

We consider the problem of inverting light scattering data, namely
extinction data, in order to retrieve information about particle size distributions. In the so-called extinction methods,[1] one measures the spectral
turbidity of the sample or, in other words, the extinction coefficient $\tau(k)$,
for different values of the wavenumber k of the incident light. More precisely, k represents a reduced wave number which takes into account the refractive index characterizing the suspension.

In the case of dilute scatterers and in the anomalous diffraction
approximation,[1] the extinction coefficient and the distribution of the sizes
of spherical particles are related by the following Fredholm integral equation of the first kind:

$$g(k) = \int_0^\infty K(kr) \, f(r) \, dr \tag{1}$$

The data function $g(k)$ is given by

$$g(k) = \tau(k)/k \tag{2}$$

and $f(r)$ is the volume distribution

$$f(r) = (4/3) \, \pi \, r^3 \, N(r) \tag{3}$$

where $N(r)$ is the density of particles of radius r. Using ρ for the product
kr, the integral kernel in equation (1) is given by

$$K(\rho) = 3(2\rho^2 + 4 - 4\rho\sin\rho - 4\cos\rho)/4\rho^3$$

$$= 3(\pi/2)^{1/2} \, \rho^{-3/2} \, \mathbb{H}_{3/2}(\rho) \tag{4}$$

where $\mathbb{H}_\nu(\rho)$ is the Struve function of order ν.

The inverse problem we consider consists in solving (1), i. e. in recovering the function f(r) from continuous or sampled values of the data function g(k). There is extensive previous literature on this subject, including various exact inversion formulae, dating back to Shifrin and Perelman.[2] In general, such exact inversion formulae require on the one hand complete data and on the other hand a priori knowledge of some parameters of the size distributions which are not readily available in most cases. Therefore, we have tried to develop a method which can be used in practical cases, namely for incomplete and sampled data, with arbitrary sampling schemes. The basic tool we used for analysing the inverse problem and estimating the size distribution is the so-called singular system of the linear operator to be inverted. We summarize here briefly the main features of the reconstruction method and a few properties of the relevant singular systems.[3]

INVERSION FROM COMPLETE DATA

For further reference, let us first consider the hypothetical case where the data function g(k) is known over the complete semi-axis $0 < k < \infty$ and where we reconstruct solutions over the entire semi-axis $0 < r < \infty$. We assume that g(k) and f(r) belong to the space $L^2(0, \infty)$ of real square integrable functions over the positive semi-axis. We must then invert the integral operator A which is defined by

$$(Af)(k) = \int_0^\infty K(kr) \, f(r) \, dr \tag{5}$$

and is a bounded self-adjoint operator in $L^2(0, \infty)$. Since the integral kernel K(kr) depends only on the product of the two variables k and r, the spectral representation of A can be found by means of the Mellin transform, as in the case of the Laplace transform inversion.[4] If $\hat{f}(\omega)$ denotes the Mellin transform of the square integrable function f(r)

$$\hat{f}(\omega) = \int_0^\infty f(r) \, r^{-\frac{1}{2} + i\omega} \, dr \tag{6}$$

then the integral equation becomes a simple multiplicative relation:

$$\hat{g}(\omega) = \hat{K}(\omega) \, \hat{f}(-\omega) \tag{7}$$

where $\hat{g}(\omega)$ and $\hat{K}(\omega)$ are the Mellin transforms of g and K, respectively. Hence from (6) and (7) we obtain an exact inversion formula by means of an inverse Mellin transform:

$$f(r) = \int_{-\infty}^{+\infty} \frac{\hat{g}(-\omega)}{\hat{K}(-\omega)} \, r^{-\frac{1}{2} - i\omega} \, \frac{d\omega}{2\pi} \tag{8}$$

For the kernel (4), we have found the following expression for the Mellin transform of $K(\rho)$ and for its modulus:[3]

$$\hat{K}(\omega) = 3 \, \left(\frac{\pi}{2}\right)^{1/2} 2^{-2+i\omega} \, \frac{\Gamma\left(\frac{1}{4} + i\frac{\omega}{2}\right)}{\Gamma\left(\frac{5}{2} - \frac{1}{4} - i\frac{\omega}{2}\right)} \, \mathrm{tg}\left(\frac{\pi}{4} + i\pi\frac{\omega}{2}\right) \tag{9}$$

$$|\hat{K}(\omega)| = 3 \, \left(\frac{\pi}{2}\right)^{1/2} \left[\left(\omega^2 + \frac{25}{4}\right) \left(\omega^2 + \frac{1}{4}\right) \right]^{-1/2} \tag{10}$$

Since $|\hat{K}(\omega)|$ tends to zero like ω^{-2} when ω tends to infinity, inversion by means of formula (8) will be unstable when the data are affected by noise. This is a manifestation of the well-known ill-posedness inherent to Fredholm integral equations of the first kind. In order to avoid uncontrolled amplification of the data noise affecting the reconstructed solutions, an ill-posed problem has to be regularized by means of some regularization method.[5] One of the simplest standard procedures for doing this is the so-called "optimum filtering" technique which consists here in eliminating in the integral of formula (8) all frequencies ω which are greater than some cutoff frequency ω_o. The corresponding solution $\tilde{f}_o$, obtained from (8) by integrating only over the interval $[-\omega_o, +\omega_o]$, will be numerically stable if the cutoff frequency ω_o is properly chosen in function of the signal-to-noise ratio. More precisely, if $g(k)$ is corrupted by additive white noise with zero mean and power spectrum ε^2 and if it is assumed that $f(r)$ is also a white noise process with zero mean and power spectrum E^2, then ω_o is determined from the following condition:

$$|\hat{K}(\omega_o)|^2 = (\varepsilon/E)^2 \tag{11}$$

Note that more sophisticated regularization techniques are available but will not be considered here.

The suppression of the highest frequency components in the Mellin transform of f implies limitations on the resolution with which the solution is recovered. As in the case of the Laplace transform inversion,[4] one can introduce a characteristic resolution ratio δ_o given by

$$\delta_o = \exp(\pi/\omega_o) \tag{12}$$

which represents the ratio of the abscissas of closest resolvable elements. Because of the dependence of the integral kernel on the product kr, it is clear indeed that fine details of $f(r)$ can be more easily retrieved near the origin than far away from it. For the spectrum (10) the resolution ratio is found to be $\delta_o = 1.177$ for $E/\varepsilon = 10^{+2}$ and $\delta_o = 1.05$ for $E/\varepsilon = 10^{+3}$.

INVERSION FROM LIMITED DATA

In practical situations, the data function can be measured only for a restricted range of values of the wavenumber k. Let us assume that $g(k)$ is known on the finite interval $[k_A, k_B]$. We can try to compensate the loss of information due to limited data by introducing a priori knowledge about the support of the unknown function $f(r)$. This can be done by means of a "profile function" $P(r)$, which should resemble roughly the expected solution $f(r)$. In other words, $P(r)$ should be localized mainly where $f(r)$ is expected to take significant values and small where $f(r)$ is negligible. If a first approximation of the actual shape of the size distribution is also available, $P(r)$ can convey this information as well. We can then try to invert the data by looking for solutions that belong to a weighted L^2-space, say $L_P^2(0, \infty)$, with scalar product

$$(f, h) = \int_0^\infty \frac{f(r)\, h(r)}{P^2(r)}\, dr \tag{13}$$

Regularization is in general equivalent to some control on the magnitude of the norm of the solution $f(r)$, i.e. here of the following quantity

$$||f|| = \left[\int_0^\infty \frac{f^2(r)}{P^2(r)}\, dr \right]^{1/2} \tag{14}$$

Hence we see that regularized solutions in such a weighted L^2-space are forced to be small where $P(r)$ is small and to vanish on the intervals where $P(r)$ is identically zero. It is assumed that $P(r)$ is a bounded square integrable function over the positive semi-axis.

The integral operator to be inverted, as defined by formula (5), is now an operator from the weighted space $L_P^2(0,\infty)$ into the data space $L^2(k_A,k_B)$. It is a compact operator, so that we can introduce its singular system $\{\alpha_k;u_k,v_k\}$. We recall that the singular values α_k (which are positive by definition) and the singular functions u_k and v_k are the solutions of the following coupled equations

$$Au_k = \alpha_k v_k \qquad A^* v_k = \alpha_k u_k \tag{15}$$

where A^* denotes the adjoint operator given by

$$(A^*g)(r) = P^2(r) \int_{k_A}^{k_B} K(kr)\, g(k)\, dk \tag{16}$$

We have investigated the properties of the singular system of A and established a series of properties of the distribution of the singular values, mainly for the case where $P(r)$ is the characteristic function of a finite interval (i. e. $P(r)$ is equal to one inside this interval and to zero elsewhere) and where $P(r)$ is a gamma distribution.[3] The method of optimal filter provides also in the present case regularized solutions of the equation $Af = g$. These solutions can be expressed in terms of the singular system of A.[4]

INVERSION FROM SAMPLED DATA

We consider now the real-life situation where the data are no longer continuous but discrete. We assume that $g(k)$ has been measured only in N sampling points, say k_1, k_2, ..., k_N. The data space is then a N-dimensional vector space G_N, in which we define the following weighted euclidean scalar product

$$(g,h) = \sum_{n=1}^{N} w_n\, g_n\, h_n \tag{17}$$

The inverse problem we must solve now is to estimate the size distribution $f(r)$ from the N values

$$g_n = (A_N f)_n \qquad n = 1,\ldots,N \tag{18}$$

where $g_n = g(k_n)$ and the operator A_N given by

$$(A_N f)_n = \int_0^\infty K(k_n r)\, f(r)\, dr \qquad n = 1,\ldots,N \tag{19}$$

applies the functions of $L_P^2(0,\infty)$ onto vectors of G_N.

The choice of the sampling points is free. For example, one could consider a set of N equidistant points in the interval $[k_A,k_B]$:

$$k_n = k_A + (n-1)\, d \qquad n = 1,\ldots,N \tag{20}$$

with equidistance $d = (k_B-k_A)/(N-1)$. However, the dependence of the integral kernel on the product kr suggests that sets of points forming geometric progressions might be more efficient, namely

$$k_n = k_A\, \Delta^{n-1} \qquad n=1,\ldots,N \tag{21}$$

where the dilation factor Δ is chosen in such a way that $k_N = k_B$.

For the weights in the scalar product (17), we take then $w_n = d$ in the case of the sampling scheme (20) and $w_n = k_n \ln(\Delta)$ in the case of the distribution (21).

We denote by $\{\alpha_{N,k}; u_{N,k}, v_{N,k}\}$ the singular system of the operator A_N defined by (19) and we assume that the singular values are always numbered by decreasing order of magnitude. A multiple singular value is counted as many times as required by its multiplicity. When the number N of data points in the distributions (20) or (21) tends to infinity, it follows from perturbation lemmas for compact operators that the singular value $\alpha_{N,k}$ converges to the corresponding singular value α_k of the operator A considered in the previous section, in the case of continuous data on the same interval. The singular function $u_{N,k}$ also converges, in the norm of $L_P^2(0,\infty)$, to the corresponding singular function u_k of A, while the N-dimensional singular vector $v_{N,k}$ converges to sampled values of the function v_k.

It is interesting to remark that the singular vectors $v_{N,k}$ can be computed by means of standard methods since they are the eigenvectors, with eigenvalues $\alpha_{N,k}^2$, of the operator $A_N A_N^*$, as can be easily devised from formula (15). This operator is represented by a $N \times N$ matrix of elements

$$a_{mn} = w_n \int_0^\infty K(k_m r)\ P^2(r)\ K(k_n r)\ dr \tag{22}$$

The latter expression follows immediately from the explicit form of the adjoint operator A_N^* which transforms the vectors g of G_N into functions of $L_P^2(0,\infty)$:

$$(A_N^* g)(r) = P^2(r) \sum_{n=1}^{N} w_n\ g_n\ K(k_n r) \tag{23}$$

From the second of the coupled equations (15), we see that $u_{N,k}$ can be determined from the singular vector $v_{N,k}$ by applying the operator A_N^* :

$$u_{N,k}(r) = \frac{1}{\alpha_{N,k}} P^2(r) \sum_{n=1}^{N} w_n\ (v_{N,k})_n\ K(k_n r) \tag{24}$$

where $(v_{N,k})_n$ denotes the nth component of the vector $v_{N,k}$.

The solution of equation (18) is not unique: there are many possible size distributions which correspond to zero data or in other words, which belong to the null-space of the operator A_N. Such distributions can be called "transparent" or "invisible" distributions since they cannot be recovered from the available data. A usual procedure is then to assume a priori that the invisible part of the solution $f(r)$ is just zero or equivalently to look for the solution that has minimal norm. This solution, called the normal solution, belongs to the subspace spanned by the functions $u_{N,k}$ and admits the following representation in terms of the singular system:

$$f^\dagger(r) = \sum_{k=1}^{N} \frac{(g, v_{N,k})}{\alpha_{N,k}}\ u_{N,k}(r) \tag{25}$$

When the condition number $\alpha_{N,1}/\alpha_{N,N}$, i. e. the ratio between the largest and the smallest singular value is very large, the problem of computing $f^\dagger$ is ill-conditioned. The normal solution is then unstable in the presence of noise, so that some kind of regularization is needed in order to get stable meaningful solutions. The method of optimal filtering is equivalent here in keeping in the sum (25) only those terms for which $\alpha_{N,k} \geq \epsilon/E$, where $(E/\epsilon)^2$ is the signal-to-noise ratio. We call "number of degrees of freedom" the number P of terms appearing in the stable solution $\tilde{f}$ obtained in this way.

The inversion of extinction data in practical situations is a problem characterized by a rather small number of degrees of freedom. It appears moreover
from numerical experience that the number N of data values required in order
that the first and relevant components of the singular system $\{\alpha_k; u_k, v_k\}$
corresponding to continuous data should be well approximated by the corresponding components of $\{\alpha_{N,k}; u_{N,k}, v_{N,k}\}$ is not too high. This number can in
fact be lowered roughly to the number of degrees of freedom, provided that
the sampling points are "optimally" placed. In particular, as observed in the
case of the Laplace transform,[4] geometric distributions like (21) perform
better in that respect than equidistant points.

We have computed singular systems corresponding to various sampling
schemes and to different choices of the profile function. Some typical
examples are reported elsewhere.[3] Let us just mention here that when the
profile function is the characteristic function of a given interval, the
values of the singular functions $u_k(r)$ can be very large at the edges of
this interval. This is the source of undesirable edge effects in the reconstructed solutions. These edge effects can be eliminated by choosing a profile function which vanishes smoothly outside of the support of the solution.
The singular functions, being modulated by $P^2(r)$ as seen from (24), are then
also gently behaving. Let us still mention that the number of zeroes of $u_{N,k}$
is regularly increasing with the index k. The spacing between the zeroes of
the highest order function $u_{N,p}$ appearing in the regularized solution $\tilde{f}$
essentially determines the achievable resolution limits, for a given signal-
to-noise ratio. Since in first approximation the zeroes of $u_{N,p}$ are equidistant in the variable ln r, we can define a resolution ratio δ analogous to
(12). In the case where $P(r)$ is the characteristic function of the interval
$[1,\gamma]$, this ratio is given by $\delta_0 = \gamma^{1/P}$. We have computed δ for various sets
of discrete data and observed that often this resolution ratio is not much
larger than the ratio δ_0 given by (12) in the case of complete data, at least
for E/ϵ greater than 10^{+2}. The lack of information due to limited data is
then compensated by the a priori knowledge of the support of the unknown
size distribution.

CONCLUDING REMARKS

The singular system appears to be a useful tool for inverting experimental light scattering data. Formula (25), when truncated appropriately as
a function of the signal-to-noise ratio, provides a practical way of re-
trieving the particle size distribution from any finite set of measurements.
For a given experimental arrangement, the singular system can be computed
once for all and prestored in the computer. The data inversion becomes then
very fast, so that efficient software packages for automatic data processing
could be developed on the basis of formula (25).

A priori knowledge about the localization of the solution is to be used
in order to compensate somehow the lack of complete data. As explained, this
a priori knowledge can be most conveniently expressed by means of a smooth
profile function which suppresses undesirable edge effects in the computed
solutions. The method leads to good reconstructions from simulated data
provided that one does not pretend to go beyond the resolution limits inherent to the problem. These reslution limits can be easily assessed for a
given noise level, by means of the singular system. Examples of computed size
distributions can be found elsewhere[3] and numerical results will be discussed
during the oral presentation of the paper. It appears from these results
that the reconstructions are not very sensitive to the actual analytical
shape of the profile function. A good guess of its width is however essential
for correct recovery.

Let us finally mention that the present method can be extended to Mie scattering kernels and can be modified in view of a better description of the experiment, including e. g. the integration effect of the detectors, provided that the relationship between the measured data and the particle size distribution can still be considered as linear.

ACKNOWLEDGEMENTS

C. De Mol in "Chercheur qualifié" of the Belgian National Fund for Scientific Research. This work has been partly supported by the NATO Grant No. 463/84, by EEC contract No. ST 29 − 0089 − 3 and by the "Ministero della Pubblica Istruzione" (Italy)

REFERENCES

1. H. C. van de Hulst, "Light Scattering by Small Particles", Dover, New York (1981).
2. K. S. Shifrin and A. Ya. Perel'man, The determination of the spectrum of particles in a dispersed system from data on its transparency, <u>Optics and Spectroscopy</u> 15:285 (1963).
3. M. Bertero, C. De Mol and E. R. Pike, Particle size distributions from spectral turbidity: a singular−system analysis, <u>Inverse Problems</u> 2: 247 (1986).
4. M. Bertero, P. Boccacci, C. De Mol and E. R. Pike, Extraction of poly-dispersity information in photon correlation spectroscopy; this conference.
5. A. N. Tikhonov and V. Y. Arsenin, "Solutions of Ill−posed Problems", Wiley, New York (1977).

SMALL ANGLE LIGHT SCATTERING PATTERNS

FROM MICROMETER- SIZED SPHEROIDS

Jean- Claude Ravey

Laboratoire de Physico- Chimie des Colloides
LESOC - UA CNRS 406- Université Nancy I
BP 239- 54506 Vandoeuvre lès Nancy Cedex- France

INTRODUCTION

The scattering intensity functions $i_1(\vartheta)$ depending of one
scattering angle ϑ do reflect the peculiar optical, morphological and
orientational properties of the scatterer. This is particularly true for
larger scatterer sizes (e.g. in the micrometer size range), where these
angular intensity functions exhibit a rather complex succession of
extrema of various amplitude at larger and larger ϑ.

The two- dimension scattering patterns, i.e. as functions of two
angles (ϑ and γ), which contain much more information should be, a
fortiori, still more useful for the determination of the properties of
the scattering particle. More specifically, it could even be conceived
that for the scatterer evaluation, we have to investigate the 2D
patterns within a much more reduced range (but $0 \leqq \gamma \leqq 180$) than it
must be done with the classical $i_1(\vartheta)$.

The simplest example of 2D pattern is the Fraunhofer diffraction
pattern from apertures of various forms. Thus, from the shape of the
central portion of the diffraction pattern from an elliptical aperture
(i.e. the respective angular locations of the first intensity minima in
the two principal directions in the pattern) the two principal
dimensions of the ellipse may be calculated.

Now, considering the small angle scattering patterns by micrometer
sized spheroids, and extending the former suggestion made by Dandliker,
Maron et al. for spheres (1-6), we could wonder whether the size and
shape of the scatterer are obtainable from the mere delineation of the
$I(\vartheta,\gamma)$ curves corresponding to the first maxima and minima of the
scattered intensity (see fig.1).

Indeed such a size and shape determination is commonly attempted in
the field of bioparticles in aqueous suspensions. For example, the two
principal dimensions of red blood cells are often tentatively evaluated
by this scattering method when they are submitted to shear gradient
(7-9), for a study of their deformability and clinical applications. For
that purpose the small angle scattering patterns are interpreted in
terms of the Fraunhofer diffraction by apertures corresponding to the
projection area of the particles into the plane perpendicular to the

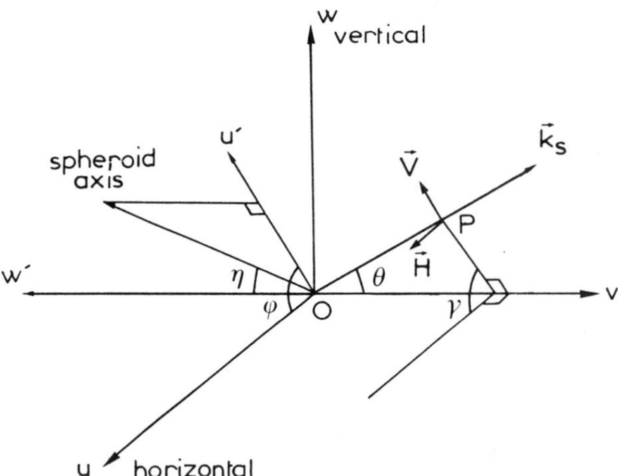

Fig. 1- The direction of the incident wave is along the
v axis. The scattering direction is defined by
ϑ and γ . η and φ determine the orientation of
the spheroidal scatterer.

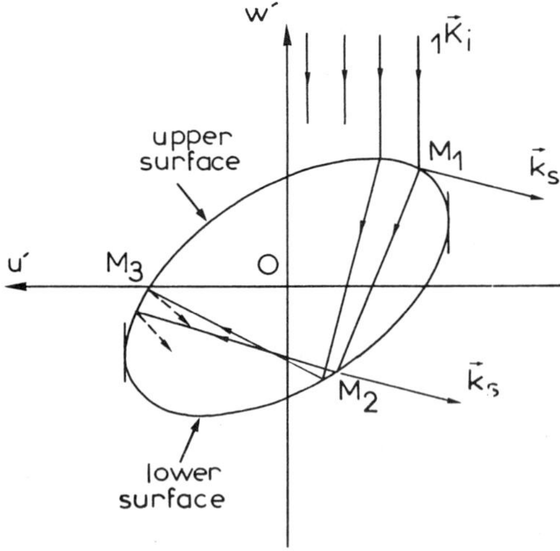

Fig. 2- Propagation of a particular elementary beam
incident upon a scatterer. K_s is the scattering
direction.

direction of the incident wave. However, a closer examination of these actual experimental small angle patterns shows that the rings corresponding to the first secondary maxima do not close around the central bright spot (10), as they should within the classical diffraction approximation (see figs.7,9). This is particularly true when He-Ne lasers are used, since then the true absorption of oxygenated haemoglobin for the wavelength λ =6300 Å is very low: clearly, such particles cannot be likened to small opaque screens. On the other hand, with a particle volume of about 100 μ^3 and such a wavelength, the particle size parameters are typically $\alpha = 2\pi D/\lambda$ = 50 (D being a characteristic dimension). And since in aqueous media the relative refractive index is about m=1.04, the classical Rayleigh- Gans- Debye approximation is clearly non valid.

Therefore, we have to use either the exact electromagnetic solution of the scattering problem by spheroids, or to derive another more suitable approximation.

Exact numerical solutions of the wave equation do exist, at least in principle (11-14). But for such large values of the size parameters, their use is rather difficult, at least for systematic calculations, given the large number of parameters of interest: there are two angular scattering parameters, different sizes and shapes of the particle with various orientations and various optical properties (the real and imaginary parts of the refractive index m , etc...). On the other hand, we believe the size parameters are not large enough to support a right use of the Ray Optics approximation, even if the separate purely diffraction term is added (15).

Hence, we have chosen to develop calculations based on a Physical Optics Approximation, which will be proved very useful for the case of the scattering by micrometer sized particles, and involving quite reasonable computation times, at least as far as non depolarized components are concerned and very large scattering angle ($\vartheta \sim 180°$) are not considered (16-18).

In the present communication, we shall first summarize the principles of this approximation, giving a few numerical data for the aim of comparison with results from exact theories. Then we shall discuss about the combined influences of the orientation, size, shape and refractive index on the angular location of the first extrema of the scattering intensity as a function of ϑ . And lastly, we shall present a few bidimensional patterns whose evolution will be discussed in line with these above influences.

THE PHYSICAL OPTICS APPROXIMATION

Principle of the method

The basis of the method consists of a direct integration of the Maxwell's equations: through a surface integral, the electromagnetic field $\vec{E}_p$ scattered at every point P of space may be exactly expressed in terms of the electromagnetic surface fields ($\vec{E}$ and $\vec{H}$) over a closed surface S. The Stratton- Chu method (19,20) is a solution of this type. For the radiated electric field we have the following expression:

$$(1) \quad -4\pi \, \vec{E}_p = \int_S [\, i\omega\mu \, (\vec{n} \times \vec{H}).\,\phi \; + \; (\vec{n} \times \vec{E}) \times \vec{\nabla}\phi \; + \; (\vec{n}.\vec{E})\vec{\nabla}\phi \,] \; dS \; + \; \frac{1}{i\omega\varepsilon} \int_C \vec{\nabla}\phi \; \vec{H}.\vec{dl}$$

where $i, \omega, \mu, \varepsilon$ have their familiar electromagnetic meaning, $\vec{n}$ the unit vector locally normal to the considered surface element dS. The curvilinear integral takes into account the posible occurence of discontinuities in $\vec{E}$ and $\vec{H}$. The scalar quantity ϕ is such that:

$$\phi = (1/r) \, \exp(2\,\pi i r/\lambda \;)$$

where r is the distance separating the observation point P from the integration element dS. In the Fraunhofer zone (P far removed from S) one obtains:

$$\vec{\nabla}\phi = i \, \vec{k}_s \; \phi \; 2\pi \, / \, \lambda$$

where $\vec{k}_s$ is the unit vector along the direction of the scattered wave.

Now, S being identified to the surface of a regular shaped scatterer, generally that surface field is not known and must be evaluated by some approximate methods. For that purpose, we used an approximation dealing only with the boundary conditions: the Physical Optics Approximation. We assume that, at any point of S, the unknown field is the same as it would be if the Fresnel's laws were applied on the plane tangent at that given point. Therefore the corresponding surface integral giving the scattered far field $\vec{E}_p$ can be evaluated for any scatterer if its surface has no edges, no sharp points, i.e. if the radius of curvature R at any point is not too small compared with the wavelength. As underlined by Beckmann (21), the error so incurred is of the order of $(\lambda^2/R)^2$, so that the method should be suitable for sufficiently large scatterers. However let us emphasize the importance of the orientation of the scatterer: as a matter of fact the contribution to $\vec{E}_p$ from a particular point M of the surface is proportional to the field in that point. Thus that contribution is generally very small for all the points where the incident rays are tangent to the surface. As a result, if the orientation of the particle is such that these points coincide with the points of smallest radius of curvature, the validity of our approximation should be greatly improved. And conversely, if these points of high curvature are strongly illuminated, our approximation coud be much less valid.

In order to perform the surface integral, the linearly polarized plane wave incident upon the upper surface (the illuminated one) of a scatterer is decomposed into small parallel elementary beams. In principle, the method only requires the use of Fresnel's laws for plane waves at each point of the scatterer surface which is illuminated by these elementary incident beams and/or by beams resulting from refraction and successive reflections inside the particle (figure 2).

As we are interested in scatterers whose refractive index may be complex, the plane waves to be considered may be inhomogeneous plane waves (22). Moreover in order to account for the finite curvature of the scatterer surface, a correction has been introduced concerning the non planar character of the waves inside the scatterer. As the inner elementary beams can no longer be cylindrical but are assumed to present the usual astigmatic structure of light pencils, amplitude and phase of the corresponding waves are modified by applying the principle of conservation of the flux of the electromagnetic energy along a tube of rays and by adding a jump in phase of $\pi/2$ after a possible passage through a focal line.

The diffraction phenomenon

Eq.1 may be regarded as an analytical formulation of the Huygens-Fresnel principle. It is then possible to investigate the Fraunhofer diffraction by an aperture from our formulation as follows: first we set the refractive index of the scatterer equal to one, and second we have to consider only the contribution to $\vec{E}_p$ due to a half surface of the scatterer, for example its illuminated part. Of course, in case of the scatterer is a spheroid (of semi axes a and b) whose revolution axis is along the direction of the incident beam, it has been verified that we get the intensity distribution characterized by the well known function $[J_1(x)/x]^2$, where $x =(2\pi a/\lambda).\sin\vartheta$,for any ellipticity. For a spheroid of arbitrary orientation, the calculated intensities from the upper surface correspond to the diffraction by an elliptical aperture which is the projected area of the spheroid on a plane perpendicular to the incident wave direction.

The anomalous diffraction

In that approximation, the wave incident upon the scatterer which induces the radiation propagates through it without noticeable deviation and undergoes a phase shift due to the optical path length (23). From our formulation, this situation may be reproduced by several means, not necessarily strictly equivalent. For example, for m close to one, we could perform the whole of the calculations, but keeping the directions of the transmitted and incident waves identical. However, such a method still remains too complex. We prefer a simpler alternative way. Consider again a scatterer whose refractive index is taken equal to one as far as the surface fields are concerned. Accordingly the rays are not deviated, no reflection occurs and the Fresnel's laws indicate that the field strengths (the phase being omitted) are constant everywhere. Without further modification, the net result of the double integration over the entire surface of the spheroid would be zero. Now if in addition we take into account the true value of the refractive index m in the bulk of the scatterer, a correct calculation of the phase shifts and of the attenuation is performed. It is of course restricted to small values of the imaginary part of the possibly complex refractive index.

Comparison to exact results

Many comparisons have already been performed concerning the scattering efficincy factors, the forwardscattering and the backscattering by some spherical and spheroidal particles (16,17). In the present paper we want to biefly re-examine a few data for the angular intensity functions $V_v(\vartheta)=i_1(\vartheta)$. Let us recall that for a spheroid whose ellipticity is p we have to define two size parameters α and β such that $\beta/\alpha =p$.

One example for spherical particles is given in fig.3, the size parameter being $\alpha=40$, and the relative refractive index m=1.2. It must be noted the very good agreement between P.O. Approximation and Mie theory. In particular, instead of the first <u>minimum</u> predicted by exact and P.O.A. calculations for $\vartheta =8°$, in both the diffraction and RGD theories the curve $V_v(\vartheta)$ exhibits a maximum.

Figure 4 displays the angular distribution $V_v(\vartheta)$ for a prolate sheroid not very elongated (p=2) with a size parameter α equal to 9. The direction of the incident wave is along the revolution axis of the sheroid ($\eta =0$) so that $V_v(\vartheta)$ is identical to the intensity usually called i_1. The dashed curve is from our Physical Optics

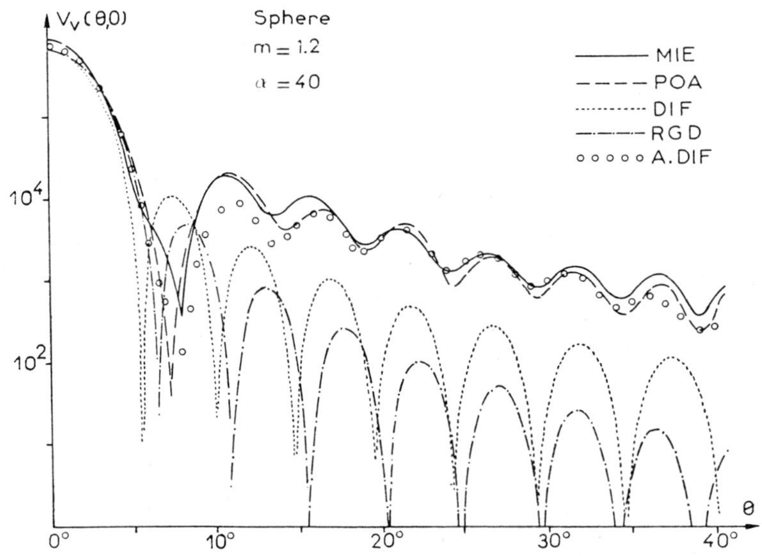

Fig. 3- The angular intensity function $V_v(\vartheta)$ for a
        dielectric sphere with α=40 in Mie theory,
        P.O. Approximation, RGD theory, Diffraction and
        Anomalous Diffraction approximations.

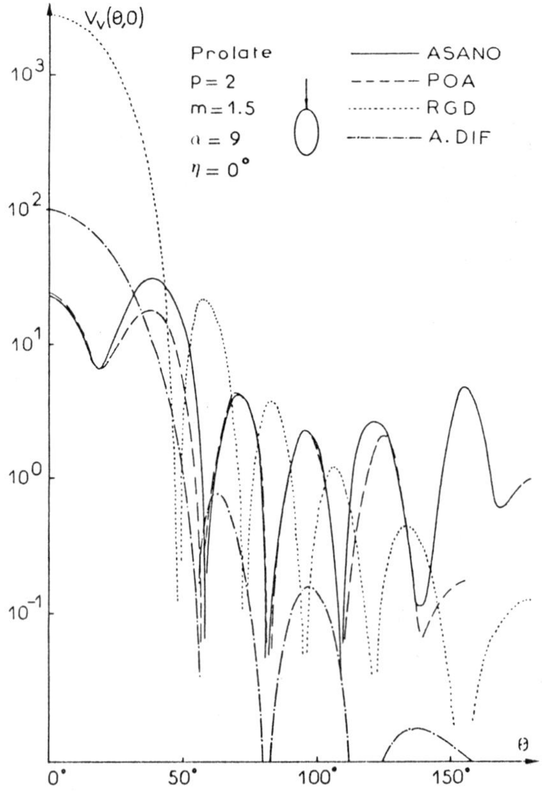

Fig. 4- The angular intensity function $V_v(\vartheta)$ for a
        prolate spheroid for nose-on incidence, accor-
        ding to several types of calculation.

Approximation. The dotted curve is the result of the diffraction theory.
The solid curve is from Asano's exact theory and at $\vartheta=37°$, it shows an
intensity peak whose value is larger than $I(\vartheta=0)$. Such a surprising
peak has been first observed by Barber and Yeh (12) and is due to
favorable interference between the diffracted and the transmitted light
in the forwardscattering direction.

It is worth-noting that our approximation accounts for this
intensity peak although its value is slightly smaller than the exact
one. In addition, it can be seen that the agreement between our theory
and the rigorous one is rather satisfactory for scattering angles up to
150° and that the diffraction is clearly inadequate for describing the
scattering of this dielecric spheroid. In other respects, it is
interesting to note that the field of applicability of this Physical
Optics Approximation based on the use of the refraction and transmission
coefficients for a plane surface (the local tangent plane) seems rather
wide since the smallest curvature radius of our spheroid is about 2. In
spite of the smallness of this value, the agreement between exact and
approximate theories is good, except for the backscattering direction.
It is limited neither to this particular spheroid nor to this particular
orientation and it indicates that our approximation may be useful for
scattering calculations.

To sum up, and from many other comparative calculations, it appears
that results from P.O.Approximation are generally in good agreement with
the exact calculations (when they exist), if the following requirement
is fulfilled: the radius of curvature R_c at any point of the
strongly illuminated surface of the scatterer (i.e. where a non
negligible surface field exists) must be greater than about λ . For a
sheroid, it is not possible to determine such an α minimum in that
case, since the relevant radii of curvature depend on the size, shape
and orientation. The above minimal condition may be fulfilled even for a
small spheroid if adequately oriented.

The spheroids may be dielectric as well as conducting ones. As a
general rule, the larger the spheroid or the radius of curvature, the
better the approximation is. Contrary to the exact theories (EBCM, Asano
and Mie) this is an advantage since our method does not need any series
expansion which could be more or less quickly convergent according to
the size of the scatterer, and thus do not present computational
instabilities even for values as large as 50 or 60. In fact our results
seem generally valid as far as the scattering in the forward direction
and the first peaks are concerned, and their application to small angle
light scattering problems should be particularly fruitful.

THE FIRST EXTREMA IN THE RADIATION PATTERN

Coming back to spherical particles, the computations have been
performed for particle size parameters α between 25 and 55. The range
of (relative) refractive index values m covered by our analysis extends
from $m \rightarrow 1$ to $m=1.25$, by steps of 0.005. The computations have been
performed with Mie equations and also with the physical optics
approximation (PO) and the anomalous diffraction approximation (AD)
(18). The results have also been compared with values obtained with the
Fraunhofer diffraction (FD) and Rayleigh-Gans (RG) approximations, which
are independent of m. Typical variations of $\log(i_1(\vartheta))$ for different
values of m are shown in Fig.5, for the particle size parameter $\alpha=28.7$.
It appears at once that a phenomenon occurs which parallels that obseved
by Kerker (5) and which could be guessed from the results of Patitsas
(24). As m increases the first minimun and maximum flatten out an
disappear. For $m=1.142$ (and $m=1.261$, etc.) the second minimum becomes
the first one.

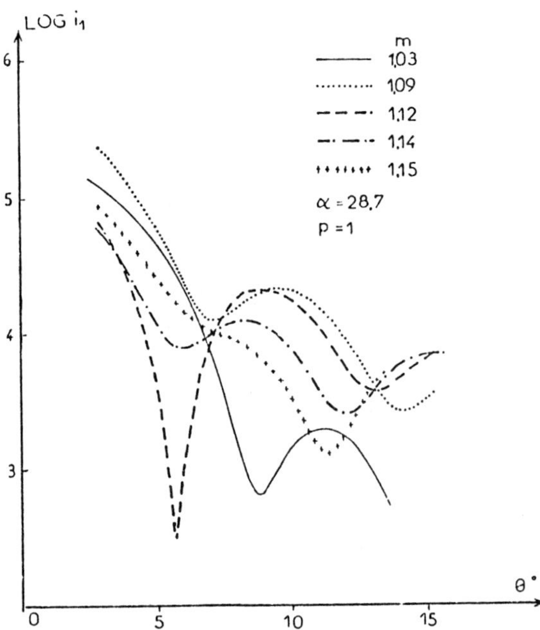

Fig. 5- Small angle radiation pattern $V_v(\vartheta)$ for a sphere
with $\alpha=28.7$ for various values of the refractive
index m.

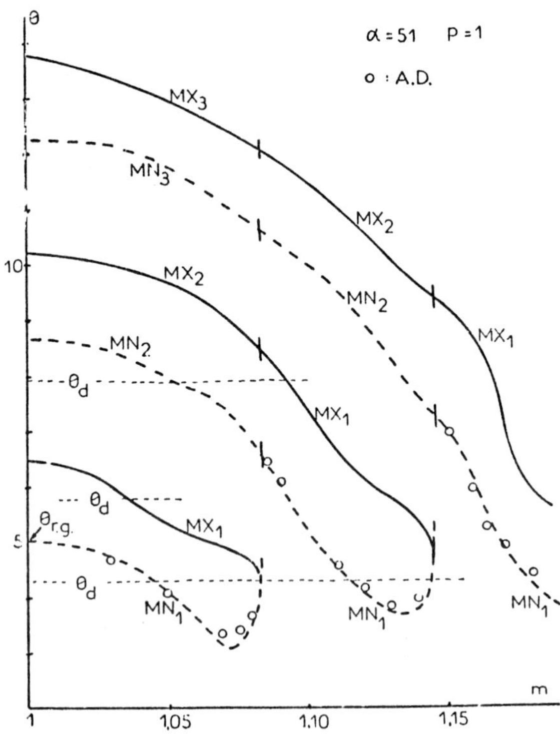

Fig. 6- The m variation of the angular position of the first
minima (MN) and maxima (MX) for the sphere with
$\alpha=51$. ϑ_{rg} and ϑ_d are relative to RGD and diffraction
theories. Circles are for Anomalous Diffraction.

The curves of Fig.6 correspond to a larger sphere of size parameter $\alpha=51$, where the variations of the angular positions of the first extrema (minima and maxima) are represented, and where the successive jumps are clearly brought in light. As proof of the validity of the PO approximation, we can check that the approximate values are practically identical to the exact values (Mie results). Although a very small difference can still be perceived for the smallest sphere, the Mie and PO results are indiscernible for larger spheres (Fig.6). But much more striking is the fact that the much less involved anomalous diffraction theory practically leads to the exact values as far as the angular position of the extrema is concerned (of course, this does not generally hold for the absolute level of the scattered intensity). When $m \rightarrow 1$, we get values which tend to be coincident with the results of the Rayleigh-Gans-Debye approximation, as expected: the positions of these extrema are those of the function:

$$(2) \qquad I_{RG} \sim J^2_{3/2}(x)/x^3$$

($J_\nu(x)$ is the Bessel function of order ν)
with $x=2\,\alpha\sin(\vartheta/2)$ (these positions are given by x=4.493, 7.725, 10.904... and x=5.763, 9.095,..., respectively for the minima and maxima). Then as m increases, the angular position of the first minimum decreases, crossing the position given by the Fraunhofer diffraction, and then increases again just before the jump occurs. Most interesting is the fact that the first maximum and minimum coalesce at almost exactly the position which would correspond to the first minimum in the Fraunhofer diffraction. Let us recall that the positions of the extrema of the diffraction pattern are those of the function:

$$(3) \qquad I_{FD} \sim (2J_1(z)/z)^2$$

with $z = \alpha\sin\vartheta$. (These positions are given by z=3.832, 7.016, 10.174... and z = 5.136, 8.417, respectively, for the minima and maxima.)

Then, for a given size parameter, the first minimum can be exactly calculated from the simple above equation only for a set of discrete values of the refractive index. It cannot be concluded, however, that for such particular values of m, the actual radiation pattern will be identical to that of a diffraction phenomenon. Indeed when the coincidence holds simultaneously for the first minimum, it does not hold simultaneously for the other minima and maxima.

Coming back to the problem of the size determination of a particle from the knowledge of the angular position of the extrema, now we can say that the rash use of the Eq.3 would generally lead to erroneous values, the error being as large as 100 % if the refractive index is not far from a particular jump value.

For the case of spheroidal particles, all the curves exhibit exactly the same features as those described for the case of spherical particles, which can be summarized as follows (18):

-The curves oscillate in a regular way about the values which correspond to the extrema in the Fraunhofer diffraction pattern of the elliptical aperture whose delineation is that of the projection of surface particles onto the plane perpendicular to the incident beam. If we characterize the spheroid by its two principal size parameters αandβ, such that p =β÷α is the ellipticity and β always indicates the dimension along the revolution axis, the Fraunhofer diffraction is still

described   by the function of Eq. (3); but
$z = \alpha \sin\vartheta$, or $z = \alpha \sin\vartheta \cdot [\, p^2 + (1-p^2)\cos^2\eta\,]^{1/2}$,
depending on whether the revolution axis is in the vertical or in the
horizontal plane. Here η is the angle between this axis and the
direction of the incident beam and the observation plane is the
horizontal one.
- When $m \rightarrow 1$, the values quite naturally become equal to Rayleigh-Gans
values: they still can be calculated from Eq. (2), but with x expressed
as:

$$x = 2\alpha \, \sin(\vartheta/2)\sqrt{1 + (p^2 - 1)\cos^2\phi}$$

where ϕ is the angle between the incident beam and the exterior
bissectrix of the scattering angle.
-Jumps regularly arise for definite sets of values of the refractive
index. Clearly, their frequency depends on the size, the shape, and the
orientation. But a close examination of our set of results shows that
the only parameter governing the presence of these jumps is in fact the
optical thickness of the particle for the ray which passes through its
center, i.e. the maximum (half) thickness τ along the direction of the
incident beam:

$$\tau = \beta\sqrt{p^2 + (1 - p^2)\cos^2\eta}$$

In other words, as far as the jumps are concerned, any spheroid (α, β) in
any orientation state (η) behaves exactly as a sphere whose diameter is
equal to the maximum thickness of the particle along the direction of
the incident wave. In more general terms, the m_j values of the
refractive index for which the j-jump occurs can be calculated according
to the following relation (18):

$$(4) \qquad \frac{\beta(m_j - 1)}{\sqrt{p^2 + (1 - p^2)\cos^2\eta}} = 4.17 + 3.38(j - 1)$$

a very simple result, and quite unexpected at first sight.

THE SMALL ANGLE 2-D SCATTERING PATTERNS

From the above discussion about jumps in the extrema of $V_v(\vartheta,\gamma$
$=0)$, we can infer that the small angle 2D patterns outlook will be very
sensitive to the combined size/ shape/ orientation/ refractive index
effects, particularly when the Eq. (4) holds.

Examples of patterns are given in Figs.7-11, where are drawn a few
iso- intensity contours as a function of ϑ,γ . Since here we are
primarily interested in the general outlook of the patterns, we have not
sytematically reported on these figures all the values of the intensity
$V_v(\vartheta,\gamma)$. As a matter of fact, we have expressed these values on
the decimal logarithmic scale. Typically, $\log_{10}(i_1(0))$ is about
5.2 for systems of figures 7,8, while the iso- level contours correspond
$\log_{10}(V_v(\vartheta,\gamma))$ in the range 4 to 2.5. For systems in figures
9-11, $\log_{10}(i_1(0))$ is about 5.5 to 6 according to refractive
index, and the contour levels range from 5 to 4.

First, Figs.7,8 are related to a spheroid whose parameter sizes are
$\alpha=16$ and $\beta=48$, (p=3) and m =1.20. But in Fig.7 the orientation of the
axis of the spheroid is perpendicular to the incident beam ($\eta=90,\varphi=0$),
while in fig.8, this axis is titled by the angle $\eta=45°$. Hence, the
maximum tickness (eq.4) has been changed in such a way the first jump

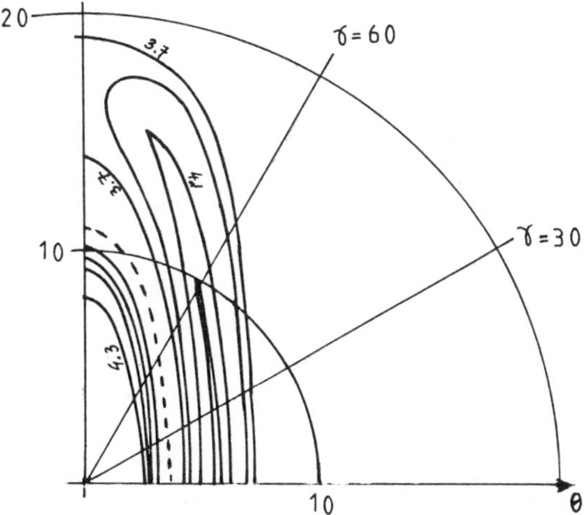

Fig. 7- Small angle 2D pattern $V_v(\vartheta,\gamma)$ for a spheroid with $\alpha=16$ and $\beta=48$, oriented perpendicularly to the incident beam ($\eta=90$, $\varphi=0$).

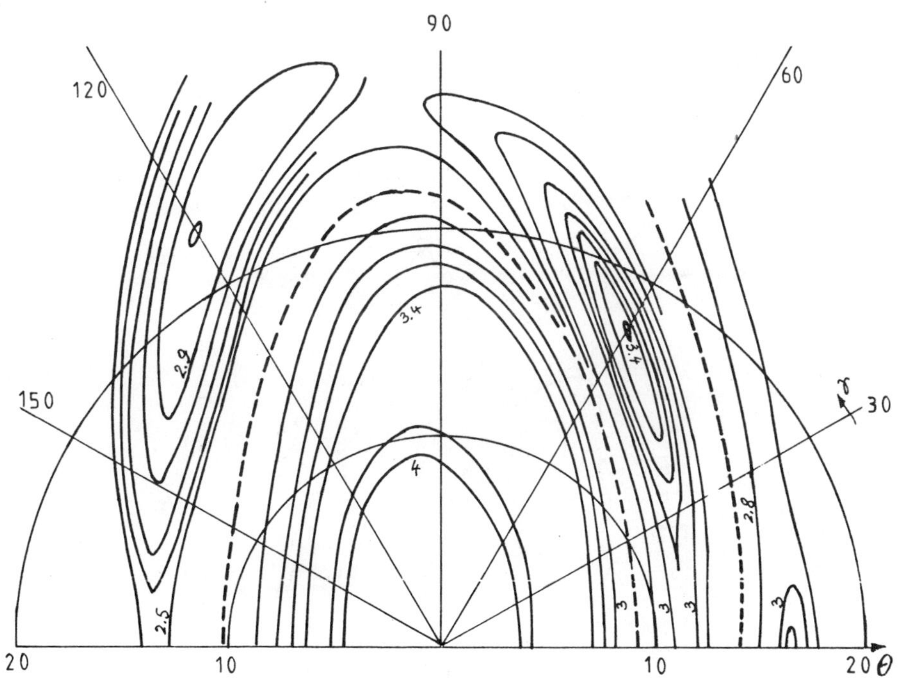

Fig. 8- Same as for fig.7, but the orientation is $\eta=45$, $\varphi=0$. The broken lines represent the minima of the scattered intensity.

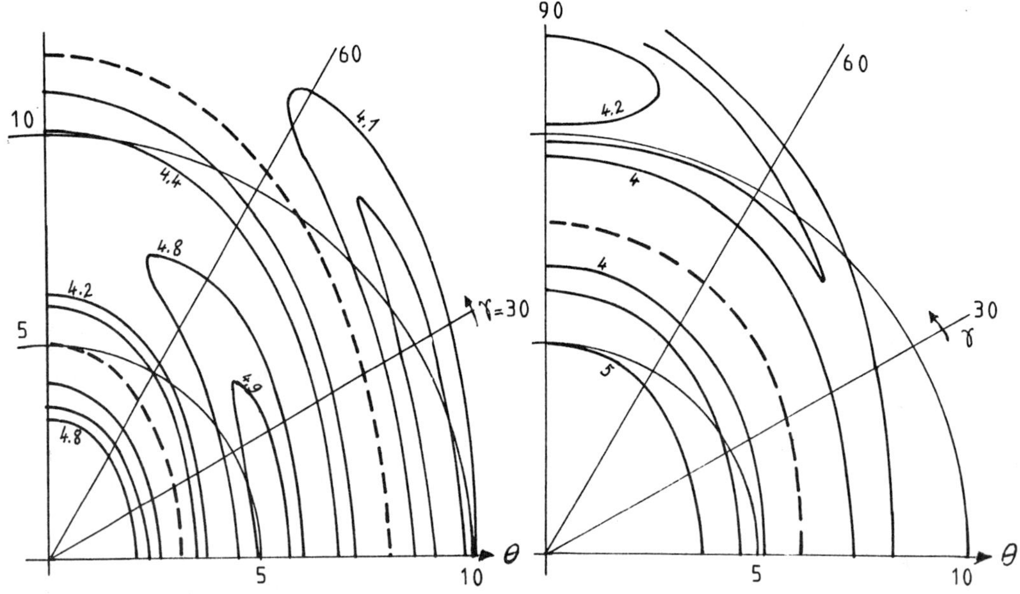

Fig. 9- Small angle pattern for
a spheroid with α=34 and β=51
at perpendicular orientation: m=1.10

Fig. 10-   m = 1.14

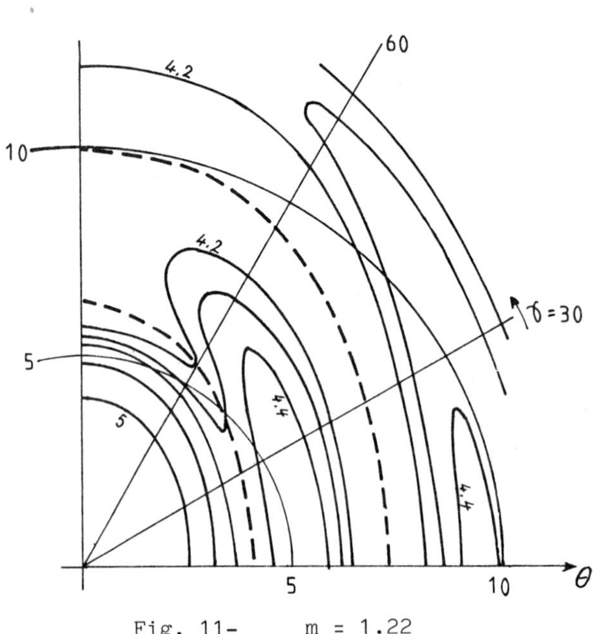

Fig. 11-   m = 1.22

has occured. More specifically, the locations of the first minima are
3.5° and 11.5° when η = 90°, and the first rings do not close around the
central spot but present maximum reinforcements in the plane γ = 0. On
the other hand, the corresponding locations for the tilted particles are
9.1° and 21°, and the maxima on the first secondary rings exist for γ =±
60°.

Therefore, a direct evaluation of the size of the particle from a
simplistic use of diffraction formulas would lead to quite erroneous
results, with an error greater than 100 %.

A parallel evolution can be obtained at constant particle size and
orientation, but in changing the relative refractive index. Figs.9-11
concern slightly elongated spheroids, with α= 34 and β= 51, oriented
perpendicularly to the incident been (η = 90, φ = 0).

For m = 1.10, the secondary "rings" are still not closed around the
central spot, and the maxima are located in the plane γ = 0.
Incidentally, let us emphasize that the experimental patterns from
erythrocytes oriented and deformed along flow lines exactly look like
the theoretical spectra of fig.9. The principal locations of the first
minima are ϑ= 3.15° and 5°. When m = 1.14, a jump has been encountered,
and the maxima are situated on the unclosed "rings" in the plane γ =
90°. Besides, the locations of the principal minima are about 6° and
8°(fig.10). Going on, with m = 1.22, they become 4° and 6.3°, and the
pattern tends to recover the outlook for the system with m = 1.10
(fig.11).

CONCLUSION

There is a perfect correlation between the occurence of the jumps
for $V_v(\vartheta$, γ = 0) and $V_V(\vartheta$, γ = 90°). Moreover, whathever the
<u>absolute</u> value of the angular position of the two principal minima (γ =
0, γ = 90), their ratio is rougly constant and equal to the reciprocal
of the ellipticity of the spheroid, at least when its axis is
perpendicular to the incident beam. A more complete discussion of this
result will be published elsewhere. Hence, at any rate, an approximate
determination the <u>shape of the scatterer</u> may be performed from a mere
inspection of the anisometry of the central spot, as for the case of the
Fraunhofer diffraction by opaque screens. But, generally, the <u>size</u>
cannot be obtained so easily: we have also to take into account the
actual value of the refractive index and to refer to abaques describing
the occurence of the jumps in the extrema.

To conclude, we have to emphasize that the small angle patterns of
micrometer- sized particles should be investigated for any γ between 0
and 180°, by considering <u>several</u> first "rings". Indeed, the location
of the extrema on these rings do supply quite valuable information
concerning the actual orientation of the scatterer relatively to the
direction of the incident beam in relation to its refractive index.

REFERENCES

1 - Dandliker,W.B., J.Amer.Chem.Soc. 72, 5110 (1950)
2 - Nakagaki,M., and Heller, W., J.Chem.Phys. 32, 835 (1960)
3 - Maron,S.H., Pierce,E., and Elder,M.E., J.Colloid Sci., 18, 391
      (1963)
4 - Maron,S.H., and Elder,M.E., J.Colloid Sci., 18, 107 (1963)

5 - Kerker,M., Farone,W.A., Smith,L.B., and Matijevic,E., J. Colloid Sci., 19, 193 (1964)

6 - Heller,W., and Nakagaki,N., J.Chem.Phys., 64, 4912 (1976)

7 - Bessis,M., and Mohandas,N., Blood Cells, 1, 307 (1975)

8 - Plasek,J., and Marik,T., Appl. Opt., 21, 4335 (1982)

9 - Guillot,M., Ravey,J.C., Guerlet,B., Mazeron,P., and Stoltz,J.F., J.Biophys. Med. Nucl., 4, 225, (1980)

10- Ravey,J.C., and Mazeron,P., Biorheology, Suppl. I, 287 (1984)

11- Waterman,P.C., Phys. Rev., D3, 825, (1971)

12- Barber,P., and Yeh,C., Appl.Opt., 14, 2864 (1975)

13- Asano,S., and Yamamoto,G., Appl.Opt., 14, 29, (1975)

14- Asano,S., and Sato,M., Appl.Opt., 19, 962 (1980)

15- Hodkinson,J.R., and Greenleaves,I., J.Opt.Soc.Amer., 53, 577 (1963)

16- Ravey,J.C., and Mazeron,P., J.Optics, 13, 273 (1982)

17- Ravey,J.C., and Mazeron,P., J.Optics, 14, 29 (1983)

18- Ravey,J.C., J.Colloid Interface Sci., 105, 435 (1985)

19- Stratton,J.A., Electromagnetic theory, McGraw Hill, New york, 1941

20- Silver,S., Microwave antenna theory and design, McGraw Hill, New York, 1947

21- Beckmann,P., The depolarization of electromagnetic waves, The Golem Press, Boulder (Colorado), 1968

22- Bell,E.E., Handbuch der Physik, (Bd XXV, 1-58), Springer Verlag, Berlin, 1967

23- Van der Hulst,H.C., Light scattering by small particles, Wiley, New york, 1957

24- Patitsas,A.J., J.Colloid Interface Sci., 46, 266 (1974)

LASER BEAM SCATTERING BY INDIVIDUAL SPHERICAL PARTICLES :

NUMERICAL RESULTS AND APPLICATION TO OPTICAL SIZING

Bruno Maheu, Gérard Gréhan, and Gérard Gouesbet

Laboratoire d'Energétique des Systèmes et Procédés
UA CNRS 230 - INSA de Rouen - BP 08
76130 Mont-Saint-Aignan - France

Optical sizing measurements and, more generally, optical diagnostics require some good knowledge of light-matter interaction. In our laboratory, we are working on both the theory and the applications of light scattering. Another paper presented in this symposium[1] is devoted to the recent developments of the theory of beam scattering by spherical, isotropic and homogeneous particles. This theory is called Generalized Lorenz-Mie Theory (GLMT).

In the present paper, we want to detail the numerical results computed from the theory and their application to optical sizing. The extension of the GLMT to the case of an arbitrary location of the scatterer being very recent, the numerical computations have been up to now limited to the previous state of the theory, i.e. the case where the scatterer is located at the beam waist center, with an extension to the special case of the near forward angle.

The paper is organized as follows : the first section focuses on theoretical results with a brief recall of the beam description and of the main formulae of the GLMT in the special case where the scatterer is located at the beam waist center. Then we give a few numerical values of the main coefficients appearing in the GLMT. In the second section, we discuss the localized interpretation of the new coefficients introduced in the GLMT and show that tedious computations can be avoided by this localized approximation to the GLMT. Finally, in the third section, our predictions are compared with experimental data and with Hamelin's work for the near forward angle.

I - GENERALIZED LORENZ-MIE THEORY

The description of laser beam scattering can be treated as a generalization of plane wave scattering. The Lorenz-Mie Theory (LMT)[2] thoroughly describes the scattering of a plane wave by a spherical, isotropic, homogeneous particle. Since 1980 we have been developing a theory called Generalized Lorenz-Mie Theory (GLMT) for the scattering of a laser beam (simple model for a real laser beam)[3-8]. This GLMT is now suited to account for the scattering of such a beam by any spherical, isotropic, homogeneous and non-magnetic scatterer having any arbitrary location with respect to the beam[1].

The reader will refer to ref 1 for detailed explanations and, in the following lines, we will limit ourselves to a brief recall of (i) the description of the beam (ii) the main formulae of the GLMT for the special case of a scatterer located at the beam waist center. We will then give a few numerical results for the main coefficients of the GLMT.

I.1 - <u>The Beam Description</u>

The description of the beam is given in a detailed form in refs 1, 5 and 6. It follows Lax et al [9] and Davis [10].

The laser beam is modelled as a Gaussian beam since laser users often work with the TEM_{00} Gaussian mode of the beam and the expressions of the field components are (for $x_0 = y_0 = z_0 = o$) :

$$E_r = E_0 \Psi_0 \cos\varphi \sin\theta \left(1- \frac{2Q}{l} r \cos\theta\right) \exp(-ikr \cos\theta) \qquad (1)$$

$$E_\theta = E_0 \Psi_0 \cos\varphi \left(\cos\theta + \frac{2Q}{l} r \sin^2\theta\right) \exp(-ikr \cos\theta) \qquad (2)$$

$$E_\varphi = - E_0 \Psi_0 \sin\varphi \exp(-ikr \cos\theta) \qquad (3)$$

$$H_r = H_0 \Psi_0 \sin\varphi \sin\theta \left(1- \frac{2Q}{l} r \cos\theta\right) \exp(-ikr \cos\theta) \qquad (4)$$

$$H_\theta = H_0 \Psi_0 \sin\varphi \left(\cos\theta+ \frac{2Q}{l} r \sin^2\theta\right) \exp(-ikr \cos\theta) \qquad (5)$$

$$H_\varphi = H_0 \Psi_0 \cos\varphi \exp(-ikr \cos\theta) \qquad (6)$$

where the notations are those of (1).

The above description of the Gaussian beam is called "order L" (L for "lowest"). Although the order L description of the beam does not rigorously comply with Maxwell's equations, it has been shown [6] that the involved relative errors are small and cannot be detected except in "exotic" situations. Actually these errors are of the order of s^2 for field components, s being a small dimensionless number equal to $1/kw_0$ (k wavenumber and w_0 beam waist radius). Due to the theoretical lower limit for w_0, i.e. the wavelength value, s is always smaller than 0.15 and usually much smaller. Note that $1/s = kw_0$ can be called "the beam waist parameter" since it is analogous to the "particle size parameter" (wavenumber x radius).

A cruder approximation to the actual beam can be obtained by neglecting also the s-terms in the field components : this approximation is called the L^- order and corresponds to beams where the axial components of the electromagnetic field are completely neglected. This case had been dealt with in our first papers [3-4]. In the L^- order of the beam description, the expressions of the field simply become :

$$E_r = E_0 \Psi_0 \cos\varphi \sin\theta \exp(-ikr \cos\theta) \qquad (7)$$

$$E_\theta = E_0 \Psi_0 \cos\varphi \cos\theta \exp(-ikr \cos\theta) \qquad (8)$$

$$E_\varphi = - E_0 \Psi_0 \sin\varphi \exp(-ikr \cos\theta) \qquad (9)$$

$$H_r = \frac{H_0}{E_0} \frac{\sin\psi}{\cos\psi} E_r \tag{10}$$

$$H_\theta = \frac{H_0}{E_0} \frac{\sin\psi}{\cos\psi} E_\theta \tag{11}$$

$$H_\varphi = - \frac{H_0}{E_0} \frac{\cos\psi}{\sin\psi} E_\varphi \tag{12}$$

Of course, our choice of a Gaussian description of the beam is not the unique possibility. In particular, other authors develop descriptions referring to other beam modes. These beam modes have been described by Kogelnik and Li [11] and paraxial approximations with their corrections are given by authors like Agrawal and Pattanayak [12], Couture and Belanger [13] or Takenaka et al [14]. The question of the choice for the beam model is not a crucial point since the GLMT is well suited to handle various beam descriptions. Most of the formulae of the GLMT do not depend on the beam description. It is then possible to adapt the theory to non-Gaussian beam descriptions.

I.2 - The Formulation of the Theory

Starting from a given beam model, the GLMT uses Bromwich Scalar Potentials to express the scattered electromagnetic field components in a spherical coordinate system. In the special case where the spherical scatterer is located in the middle of the beam waist, the general formulae of ref 1 can be simplified since we can put $x_0 = y_0 = z_0 = 0$. For example, we obtain the following relations for the r, θ, φ components of the far field :

$$E_r = H_r \to 0 \tag{13}$$

$$E_\theta = \frac{iE_0 \exp(-ikr)}{kr} \cos\varphi \sum_{n=1}^{\infty}{}' \frac{2n+1}{n(n+1)} g_n \left[a_n \tau_n(\cos\theta) + b_n \pi_n(\cos\theta) \right] \tag{14}$$

$$E_\varphi = \frac{- iE_0 \exp(-ikr)}{kr} \sin\varphi \sum_{n=1}^{\infty}{}' \frac{2n+1}{n(n+1)} g_n \left[a_n \pi_n(\cos\theta) + b_n \tau_n(\cos\theta) \right] \tag{15}$$

$$H_\varphi = \frac{H_0}{E_0} E_\theta \tag{16}$$

$$H_\theta = - \frac{H_0}{E_0} E_\varphi \tag{17}$$

In the above equations, the unique difference between LMT and GLMT lies in the presence of the correcting coefficients g_n. The other terms are exactly those of the LMT. The functions $\tau_n(\cos\theta)$ and $\pi_n(\cos\theta)$ are

Legendre functions defined in ref 1. The values of these functions depend on the direction θ of the point where the scattered field is measured.

The other coefficients appearing in eqs (13-17) can be divided into scattering and beam terms. On the one hand, a_n and b_n are the amplitude coefficients of the LMT (see ref 15 for instance). They can be called scattering terms as they only depend on the scattering particle properties (size parameter, complex refractive index) without any dependence on the incident beam. On the other hand, the g_n coefficients are related only to the incident beam. The expression of these coefficients are the following :

$$g_n = \frac{2n+1}{\pi n (n+1)} \frac{1}{(-1)^n i^n} \int_0^{\pi} \int_0^{\infty} ikr \sin^2 \theta \ f \ \exp(-ikr\cos\theta)$$

$$\Psi_n^1 (kr) \ P_n^1 (\cos\theta) \ d\theta \ d(kr) \qquad (18)$$

where the functions ψ_n^1 and P_n^1 are respectively spherical Bessel functions and associated Legendre polynomials and f is a basic radial function related to the amplitude profile of the illuminating beam.

Other expressions of the GLMT are also very close to those of the LMT with the presence of a g_n coefficient in each term of the involved series.

Numerical computations being of common use for the LMT, their extension to the GLMT only requires the computation of the new g_n coefficients.

I.3 - <u>Numerical Computations</u>

The spherical Bessel functions of eq (18) are highly oscillatory functions. Therefore, the numerical computation of the g_n coefficients is not an easy task and requires efficient multiple integration routines as well as long computing times. We have carried out the g_n computations using QB01AD routine (Harwell library) and IBM 3090 computers of the CIRCE (Centre InterRégional de Calcul Electronique - Orsay - France), computing times ranging from 30 seconds up to several hours CPU. These computations first have been made for the L^- order description of the beam with :

$$f = iQ \exp\left[- iQ \left(\frac{r \sin\theta}{w_0}\right)^2\right] = \Psi_0 \quad (19)$$

We recently extended our computations to the L order description of the beam when :

$$f = \Psi_0 \left(1 - 2Qs \frac{r \cos\theta}{w_0}\right) \qquad (20)$$

The results corresponding to a few values of n are summarized in the first and second columns of table 1. The wavelength is equal to 0.5 µm and the beam waist diameter ($2w_0$) to 10 µm. Results for other values of the subscript n have been published in refs 7,8,17 in the case of order L^- with the non explicitly stated approximation $iQ = 1$. Note that a few previously published results have been recomputed without this approximation for the present paper, their new value being given in Table 1. In Table 1, we can notice a nearly perfect agreement between the orders L and L^-. The relative difference remains less than 0.5 %. This value should be compared to the global accuracy of each g_n computation which is about 1 %. Although attainable, a better accuracy would have led to a prohibitive cost of the integration. Accordingly, the comparison between L and $\bar{L}$ orders shows that the $\bar{L}$ order is still a good approximation of the actual beam and that the L order description only brings few refinements to this approximation.

Once the g_n values are known, the computation of the scattering parameters (such as the scattered intensity at a given point, the scattering and extinction cross-sections or the radiation pressure cross-section) can be carried out for particles of arbitrary size and/or complex refractive index provided that they are located at the beam waist center. The adaptation of existing Mie codes to the computation of laser beam scattering is very easy. Our own computations have been carried out with a modified version of the Supermidi code[18].

But the critical point is in fact the cost of the g_n computations. This cost is unaffordable in most cases when systematic studies are required. This difficulty is a powerful motivation for the research of simplified g_n coefficients. This also explains the interest in the localized interpretation of the g_n which enables very easy computations as it will be shown in the following section.

Table 1 : Comparison between exact g_n values for order L and $\bar{L}$ and approximate exponentials. The exact values are computed from integral (18) with kr-integration range $[10^{-3}, 10^{5}]$. The relative differences are defined with respect to the order L value (waist diameter : 10 μm, wavelength : 0.5 μm).

Coefficient	Order L	Order $\bar{L}$		Approximate exponential	
	value	value	relative difference (%)	value	relative difference (%)
g_1	0.995	0.995	0.0	0.999	+ 0.4
g_2	0.996	0.995	− 0.1	0.998	+ 0.2
g_{30}	0.788	0.787	− 0.1	0.790	+ 0.3
g_{60}	0.394	0.394	0.0	0.396	+ 0.5
g_{80}	0.192	0.193	+ 0.5	0.194	+ 1.0
g_{100}	0.077	0.077	0.0	0.077	0.0

II - LOCALIZED INTERPRETATION OF THE GENERALIZED LORENZ-MIE THEORY

The difficult and expensive computation of integral (18) can be avoided by a very simple and efficient physical interpretation based on Van de Hulst's localization principle[16,19].

II.1 - <u>The Localization Principle</u>

The localization principle links a geometrical light ray with each term in Mie series of eqs (13-17) provided that each ray has a minimal width relative to its length and to the wavelength of the light. In the case of plane wave scattering, Van de Hulst [19] states that 'a term of order n corresponds to a ray passing the origin at a distance $(n+1/2)\lambda/2\pi$'. According to this principle we have interpreted each g_n coefficient as the amplitude of the ray hitting the particle at the above-mentioned distance from the beam axis [16].

At the waist, the beam profile is Gaussian. Thus, the approximate expression for g_n is :

$$g_n = \exp-(((n+1/2)\lambda/2\pi)^2/w_o^2) \qquad (21)$$

The computation of the above exponential is quite instantaneous and, injecting the g_n value in the GLMT, we obtain the localized approximation to the GLMT.

II.2 - <u>Comparisons Between Localized Approximation to the GLMT and Exact GLMT</u>

The validity of the localized approximation to the GLMT has been confirmed in several ways [16], the most direct one being the comparison between the exact values of the coefficients g_n computed from (18) and the approximate values from (21). The third column of table 1 shows the approximate exponential values of the g_n.

The agreement of the exponential values with the numerical integration at order L is very good and the relative differences are less than the accuracy of the integration.

This agreement provides a firm validation of the localized approximation.

II.3 - <u>Discussion of the Localized Approximation</u>

A detailed discussion of the localized approximation on the base of both mathematical arguments and partial integrations of (18) allows a firmer physical interpretation of the g_n coefficients in agreement with the physical meaning of the localization principle. The result of the discussion can be summarized by the following statement : the physical information contained in each g_n is concentrated in a narrow region at a distance from the beam waist center depending on the value of n.

Let us develop each argument of the discussion successively.

The mathematical argument relies on the asymptotic behaviour of Bessel functions for large values of the integer n. When (n+1/2) is greater than the size parameter of the scatterer, the contribution of the corresponding Bessel functions to the scattered intensities rapidly becomes negligible. Conversely, the main contribution comes from lower integer values corresponding, through the principle of localization, to rays hitting the scatter center or passing close to it.

The numerical argument can rely on computation of (18) for finite ranges of integration. As a matter of fact, numerical integrations never range from zero to infinity but are carried out for finite (large) ranges of integration. This remark holds for the argument (kr) in (18), the actual integration range being $[0, kr_{max}]$. If we plot the result of

the integration as a function of the upper limit kr_{max} of the
integration range, we obtain the curves of fig. 1 for the order L^-. Let
us consider one of the curves, i.e. the result of the integration of a
given g_n for increasing values of kr_{max}. It can be seen that the result
of the integration first remains negligible, then grows steeply and
finally reaches its asymptotic limit. In other words, the main
contribution to the numerical value of a given g_n coefficient is located
in a narrow range of kr values i.e. in a narrow region of the beam.
Hence, each g_n can be thought of as a localized quantity.

Furthermore, the kr-region where a given g_n is localized is
strongly linked to the value of the integer n. For instance, if we
define the limits of the localization kr-region by the 2 % and the 50 %
values of the g_n coefficient, we find the following regions : about
[0.6,2.4] for g_1/ [3.5,5] for g_3/ [6,9] for g_6/ [30,40] for g_{30}/ [45,65]
for g_{45}/ [80,105] for g_{80} and [100,150] for g_{100}. These values show that
the g_n coefficients can be thought of, not only as localized, but more
precisely as localized in a kr-region close to the n value. This point
is a striking quantitative assessment of the statements of the
localization principle.

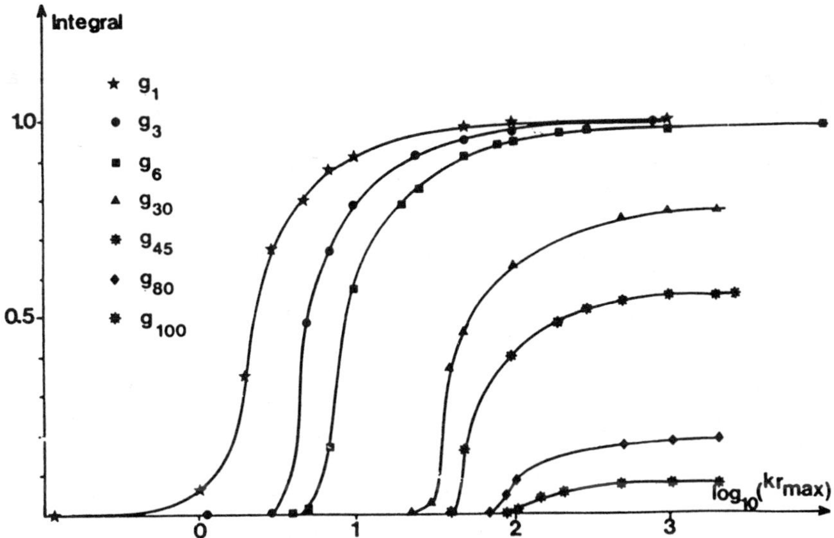

<u>Fig. 1</u> : Result of integration (18) as a function of kr_{max}, the upper kr
integration range (L^- order waist diameter : 10 μm, wavelength :
0.5 μm).

III - COMPARISONS WITH EXPERIMENTAL DATA

This section is devoted to a comparison between experimental data
and GLMT-computations for a particle located on the incident beam axis.
These computations are carried out by extending (at least provisionally)
the localized approximation discussed in sections I and II.

To this purpose, the most typical experiment is the study of light
scattered by one particle in optical levitation. The principle of an
optical levitation experiment is very simple. One spherical particle is
held in equilibrium between its weight and the vertical pressure
radiation acting on it, while the horizontal pressure radiation balance
is due to the gradients of energy density in one section of the laser
beam.

Details of the experimental set up have been given by different authors (20,21,22).

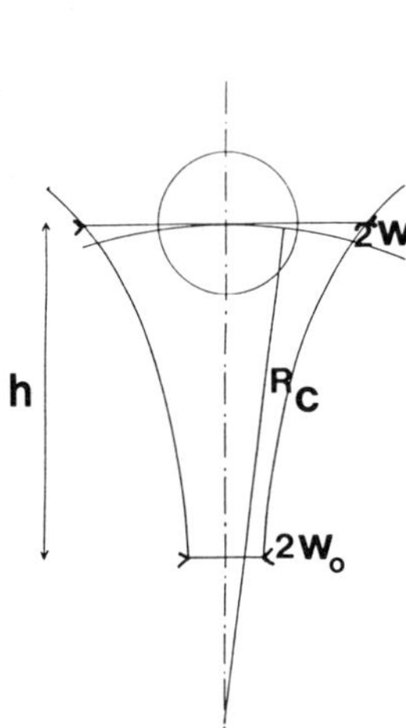

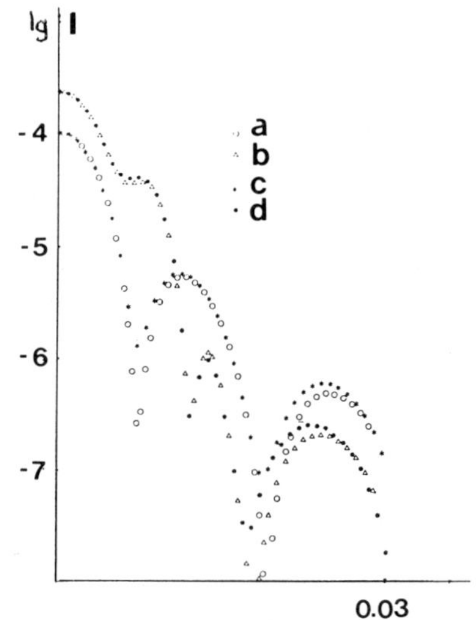

Fig. 2 : Precise view of a particle in the laser beam.

Fig. 4 : Comparison between GLMT and diffraction theory, a and b diffraction theory for h equal 1 and 10 mm respectively. c and d GLMT for h equal 1 and 10 mm respectively.

Figure 2 shows a precise view of the particle in the laser beam.

The particle is trapped in the laser beam, at a distance h above the waist.

The electric field of the laser beam, in the transversal plane containing the center of the particle has been taken from Kogelnik and Li[11] :

$$E = (w_o/w) \exp (-ik(h + \rho^2/2R_c - \psi/k)) \exp-(\rho^2/w^2) \qquad (22)$$

The local beam radius w, at $1/e^2$ intensity is :

$$w = w_o (1 + (2h/kw_o^2)^2)^{1/2} \qquad (23)$$

The curvature radius R_c of the wave front and the complementary phase angle ψ are respectively :

$$R_c = h (1 + (kw_o^2/2h)^2) \qquad (24)$$

$$\psi = \mathrm{arctg} (2h/kw_o^2) \qquad (25)$$

ρ is the distance to the axis.

The light being scattered in all directions by the particle, two regions must be considered : (i) one region where the scattered light

interfers with the incident laser beam, (ii) one region where only the
scattered light is collected.

In order to interpret measurements by means of the localized
principle, two methods have been used. In the first one presented below,
we tentatively generalize the relation (21) by writing :

$$g_n = \exp\!\left(-\left(\left((n+1/2)\lambda/2\pi\right)^2/w^2\right)\right) \tag{26}$$

This method is used to study the scattered light for large
scattering angles. Note that relation (26) could be discussed. If the
localized approximation for a waist location is well founded (sections
I-II), this approximation for an axis location as used in this section
III is actually in current development.

In the second method, all the scattering angles will be studied
including the near-forward direction.

III.1 - <u>First Interpretation of the Optical Levitation Experiment</u>

Figure 3 refers to a glass sphere with a diameter d ~ 21 μm and a
real refractive index m ~ 1,552 ± 0.015. The beam waist was measured to
be 2 w_o = 13.1 μm. The scattering angle θ ranges from 0° to 18°. The
interpretation of the measurements was obtained by curve fitting. The
GLMT provides a curve for each d,m,w and λ set. We considered that the
m- and λ-values were precisely known and we adjusted the d- and w-values

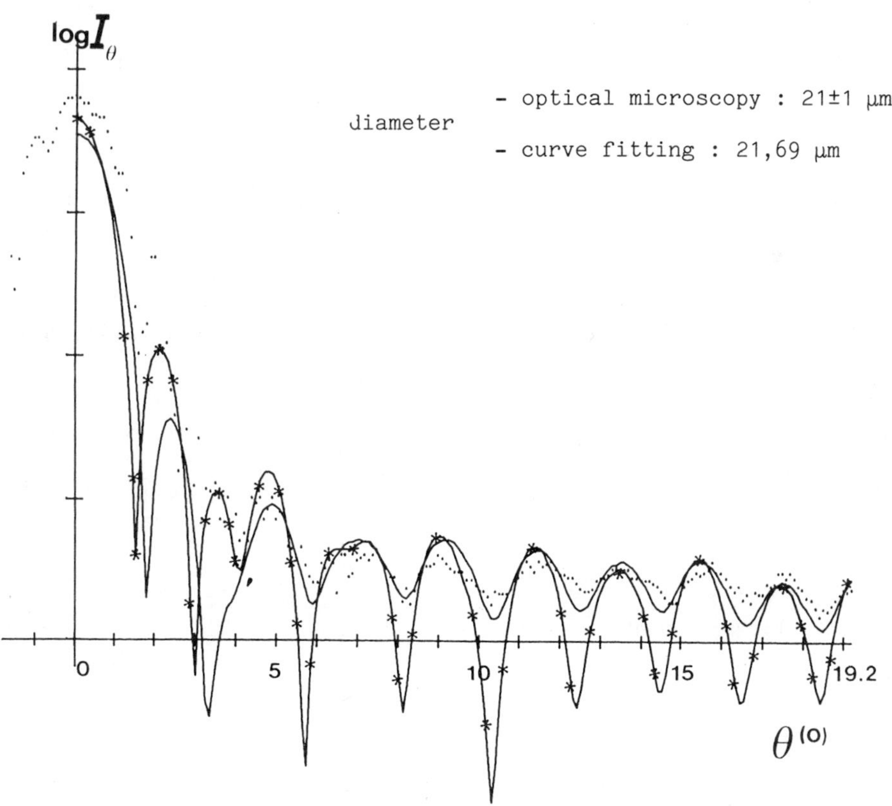

<u>Fig. 3</u> : Comparisons of LMT (*) and GLMT with experiments
(λ = 0.5145 μm).

to obtain the best fit. Fortunately, the influence of each one of these two parameters is very different. The particle diameter affects the θ-locations of the lobes and their adjustment is very sensitive. Conversely, the value of the beam radius modifies the contrast of the lobes.

The curve for the same particle lit by a plane wave is also given on figure 3. The agreement between experimental data and GLMT is very good, while pure LMT clearly cannot reproduce these data.

It should be pointed out that no valid comparison can be made between the present GLMT and experiments concerning the near forward scattering ($0° < \theta < 3°$) because of the interference between the scattered light and the laser beam. This problem will be explained in the next section.

III.2 - Second Interpretation of the Optical Levitation Experiment

For a thorough description of the interaction between a laser beam and one particle located anywhere on the beam axis, we must consider the interaction between the incident beam and the scattered wave.

In the past we carried out this work in the case of a particle lit by a plane wave [23] with the program HOLOMIDI.

We have adapted Holomidi [23] to take the laser beam, and the g_n coefficients into account.

According to the localized interpretation, and using (22) we now tentatively write the g_n's as following :

$$g_n = w_o/w \ \exp(-i \ k(h+((n+1/2)\lambda/2\pi)^2/2R_c-\psi/k))) \ .$$

$$\exp-(((n+1/2)\lambda/2\pi)^2/w^2) \tag{27}$$

Then, the scattered amplitude is added together with the amplitude of the direct laser beam at the detection location. Intensities are given by Poynting's vector modulus :

$$S = \frac{1}{2} \ \mathrm{Re} \ (E \wedge H^*)$$

To assess the validity of this interpretation, at least in the near forward direction, a comparison can be made with the computations from Hamelin [24] and Chevaillier et al [25] (diffraction theory).

Figure 4 compares the intensities computed by GLMT and by the diffraction theory for the near forward direction, the glass sphere having a diameter of 43,6 μm, the beam waist radius being 29,6 μm and the wavelength 0.6238 μm.

The evolution of the diffracted intensity, when the particle is moving in the laser beam, is very similar in both approaches. Note that our computations predict a slightly higher diffracted intensity (except for the first lobe) than does Hamelin's approach. This might come from the fact that Hamelin takes into account neither the reflected nor the refracted light.

CONCLUSION

In this paper, we have shown that the g_n coefficients which appear in the GLMT can be easily computed by using the Localization Principle.

This possibility has been established by comparison of the results of the Localization Principle with :
(i) exact computations of the g_n (for a particle waist location)
(ii) comparison with experimental data (for a particle axis location)
(iii) Hamelin's result, based on diffraction, for the near forward angle, and also a particle axis location.

Therefore, not only does the localized interpretation allow us to interpret various scattering experiments but it also opens the way towards more extensive studies of the influence of the spatial distribution of energy in a laser beam and towards potential applications to optical particle sizing.

AKNOWLEDGMENTS

We are grateful to Dr Souillard for correcting the english of the paper and to Ms Grébonval for typing it.

Computations have been partly made possible thanks to the Centre de Calcul Electronique of Orsay, France, with use of routines from the Harwell library.

REFERENCES

1. G. Gouesbet, B. Maheu and G. Gréhan, "Scattering of a Gaussian beam by a sphere using a Bromwich formulation : case of an arbitrary location", Symposium on Optical Particle Sizing : Theory and Practice, Rouen, France, 12-15 May 1987.
2. G. Mie, "Beiträge zur Optik Trüber Medien, speziell Kolloidaler Metallösungen", Ann. der Phys., 25, 377-452 (1908).
3. G. Gréhan, "Nouveaux progrès en théorie de Lorenz-Mie. Application à la mesure de diamètres de particules dans des écoulements", Thèse, Rouen (1980).
4. G. Gouesbet and G. Gréhan, "Sur la généralisation de la théorie de Lorenz-Mie", J. Opt., 13, 2, 97-103 (1982).
5. G. Gouesbet, G. Gréhan and B. Maheu, "Scattering of a Gaussian Beam by a Mie Scatter Center Using a Bromwich Formalism", J. Opt., 16, 2, 83-93 (1985).
6. G. Gouesbet, B. Maheu and G. Gréhan, "The Order of Approximation in a Theory of the Scattering of a Gaussian Beam by a Mie Scatter Center", J. Opt., 16, 5, 239-247 (1985).
7. B. Maheu, G. Gréhan and G. Gouesbet, "Diffusion de la lumière par une sphère dans le cas d'un faisceau d'extension finie : 1re partie. Théorie de Lorenz-Mie Généralisée : les coefficients g_n et leur calcul numérique", Congrès du GAMS, Paris, Nov. 1986.
8. G. Gréhan, B. Maheu and G. Gouesbet, "Localized Approximation to the Generalized Lorenz-Mie Theory and its Application to Optical Sizing", ICALEO'86, Arlington, VA, Nov. (1986).
9. M. Lax, W.H. Louisell and W.B. Mc Knight, "From Maxwell to Paraxial Wave Optics", Phys. Rev. A, 11, 4, 1365-1370 (1975).
10. L.W. Davis, "Theory of Electromagnetic Beams", Phys. Rev. A, 19, 3, 1177-1179 (1979).
11. H.W. Kogelnik and T. Li, "Laser Beams and Resonators", Appl. Opt., 5, 1550-1566 (1966).
12. G.P. Agrawal and D.N. Pattanayak, "Gaussian Beam Propagation beyond the Paraxial Approximation", J. Opt. Soc. Am., 69, 4, 575-578 (1979).
13. M. Couture and P.A. Belanger, "From Gaussian Beam to Complex-Source-Point Spherical Wave", Phys. Rev. A, 24, 1, 355-359 (1981).

14. T. Takenaka, M. Yokota and O. Fukumitsu , "Propagation of Light Beams beyond the Paraxial Approximation", J. Opt. Soc. Am. A, 2, 6, 826-829 (1985).

15. M. Kerker, "The Scattering of Light and Other Electromagnetic Radiations", Academic Press, New-York and London (1969).

16. G. Gréhan, B. Maheu and G. Gouesbet, "Scattering of Laser Beams by Mie Scatter Centers : Numerical Results Using a Localized Approximation", Appl. Opt., 25, 19, 3539-3548 (1986).

17. B. Maheu, G. Gréhan and G. Gouesbet, "Generalized Lorenz-Mie Theory : First Exact Values and Comparisons with the Localized Approximation", Appl. Opt., 26, 1, 23-25 (1987).

18. G. Gréhan, G. Gouesbet and C. Rabasse, "The Computer Program Supermidi for Lorenz-Mie Theory and the Research of One-to-one Relationships for Particle Sizing", Symposium on Long Range and Short Range Optical Velocity Measurements, Institut Franco-allemand de St Louis, 15-18 Sept., Proceedings, XVI 1-XVI 10 (1980).

19. H.C. Van de Hulst, "Light Scattering by Small Particles", J. Wiley and Sons, New-York (1957).

20. A. Ashkin, "Acceleration and Trapping of Particles by Radiation Pressure", Physical Review Letters, 24, 4, 156-159 (1970).

21. G. Roosen, "La Lévitation Optique de Sphère", Journal canadien de Physique, 57, 9, 1260-1279 (1979).

22. G. Gréhan and G. Gouesbet, "Optical Levitation of a Single Particle to Study the Theory of Quasi-Elastic Scattering of Light", Applied Optics, 19, 15, 2485-2487 (1980).

23. F. Slimani, G. Gréhan, G. Gouesbet and D. Allano, "Near-Field Lorenz-Mie Theory and its Applications to Microholography", Applied Optics, 23, 22, 4140-4148 (1984).

24. P. Hamelin, "Application de la Diffusion Lumineuse à la Métrologie des Particules en Ecoulement Diphasique Dispersé", Thèse de l'Institut National Polytechnique de Toulouse (1986).

25. J.P. Chevaillier, J. Fabre and P. Hamelin, "Forward Scattered Light Intensities by a Sphere Located anywhere in a Gaussian Beam", Applied Optics, 25, 7, 1222-1225 (1986).

EXTRACTION OF POLYDISPERSITY INFORMATION

IN PHOTON CORRELATION SPECTROSCOPY

M.Bertero[o], P.Boccacci[o], C.De Mol[+] and E.R.Pike[*]

(o) Dipartimento di Fisica dell'Università di Genova
and Istituto nazionale di Fisica Nucleare,
I- 16146 Genova, Italy
(+) Département de Mathématique, Université Libre de
Bruxelles ,B-1050 Bruxelles, Belgium
(*) Dept. of Physics, King's College, London WC2R 2LS
and RSRE, Great Malvern, WR14 3PS, England

INTRODUCTION

The polydispersity problem comes about when using laser light scattering to measure the Brownian motion of macromolecules and hence, essentially, to measure their size. The spectrum of intensity fluctuations of
light scattered by a suspension of identical particles can be shown to be
of Lorentzian shape with half-width proportional to the diffusion constant of the molecules in suspension. A pratical technique, now widely
used,measures, by special high-speed digital hardware, the autocorrelation function of the digital stream of detected single-photon events
(photon correlation spectroscopy, or PCS), which is related by the Wiener
Khintchine theorem to the spectrum and thus gives rise to an exponential
function of decay constant proportional to the inverse of the diffusion
constant. In a typical experiment, 10^7 or more scattered single photons
will by analysed using digital circuitry and a digital read-out of the
exponential correlation function will be available with accuracy of the
order of 0.1 % on each point. These points are chosen to cover a few decay times in either a linear or more preferably an integrated logarithmic
sampling scheme. The data reduction problem in this monodisperse case,
namely fitting a single exponential curve, poses no serious problem. When
particles of a number of different molecular sizes, however, are present
- the polydisperse case - each size fraction gives rise to its own exponential curve and hence the total scattering produces an autocorrelation
function which is a sum of exponentials and the data reduction problem becomes one of Laplace transform inversion.

There is a large literature on the numerical inversion of the Laplace
transform but a new approach was initiated by McWhirter and Pike[2] , who
introduced the eigenvalues and eigenfunctions of the transformation and
were thus able to construct an " information theoretic " inversion proce-

dure based on the well-known Nyquist sampling ideas of standard information theory. This approach led to consider a sampling scheme for a model
inversion in which the reconstruction was attempted with a resolution decreasing exponentially with increase of the radius parameter[3]. This is
just the equivalent of the Nyquist sampling of Fourier theory when the
problem is dilationally rather than translationally invariant. The corresponding method of inversion, based on simple linear least-squares procedure, is known as the " exponential sampling method ".

As experience with this method built up it became apparent that the
resolution achievable, although not high, was significantly better than
that predicted by the eigenvalue spectrum. The origin of this discrepancy was shown by Bertero et al.[4] to be due to an implicit resctriction of
the support of the reconstructed solution: in practice finite limits must
be set for numerical calculations and these, of necessity, add some " a
priori " knowledge of the position and extent of the solution. This knowledge can be used explicitly to calculate the extra resolution possible by
an extension of the method of eigenfuction expansions known as singular
function expansions, well known in pure mathematics since the last century but little used in physics until quite recently. This method is so
flexible that it can also be used for the problem with discrete data and
for adding several kinds of " a priori " information in terms of profile
functions. The main applications of this method to Laplace transform inversion will be described in this paper.

RESOLUTION LIMITS IN LAPLACE TRANSFORM INVERSION

The problem of determining the resolution limits in Laplace inversion
was solved by McWhirter and Pike[2] using the eigenvalues and eigenfunctions of the transformation. Here we follow an equivalent approach which
is based on the use of the Mellin transform. Consider the first kind integral equation

$$g = Af \tag{1}$$

where A is an integral operator having the following form

$$(Af)(p) = \int_0^{+\infty} K(pt)\, f(t)\, dt \ , \quad 0 < p < +\infty \ . \tag{2}$$

When $K(x) = exp(-x)$ the integral operator A is just the Laplace transformation. In general, if the function $K(x)$ is real and such that $x^{-1/2}K(x)$
is integrable, then the operator A is bounded and self-adjoint in the
space $L^2(0, +\infty)$.

The Mellin transform of a square integrable function is defined by [5]

$$\hat{f}(\omega) = \int_0^{+\infty} f(t) t^{-\frac{1}{2}+i\omega} dt \tag{3}$$

and, using well-known results [5,4], the solution of equations (1),(2) can
be written in the following form

$$f(t) = \frac{1}{2\pi} \int_{-\infty}^{+\infty} \frac{\hat{g}(-\omega)}{\hat{K}(-\omega)} t^{-(\frac{1}{2}+i\omega)} d\omega \tag{4}$$

where $\hat{g}(\omega)$ and $\hat{K}(\omega)$ are the Mellin transforms of $g(p)$ and $K(x)$ respectively . The inversion formula (4) however, is affected by numerical in-

stability since $\left|\hat{K}(\omega)\right| \to 0$ when $|\omega| \to \infty$. In particular, in the case of the Laplace transform we have

$$\left|\hat{K}(\omega)\right|^2 = \left|\Gamma(\tfrac{1}{2} + i\omega)\right|^2 = \frac{\pi}{\cosh(\pi\omega)} \quad . \tag{5}$$

In order to get stable approximate solutions one must use regularization techniques wich have been developed for the treatment of ill-posed problems [6].The most simple of these techniques, wich is also known as optimum filtering, is the following. If we assume that the data function $g(p)$ is corrupted by additive, zero-mean, white noise with power spectrum ϵ^2 and that the unknown solution $f(t)$ is also from a zero-mean white noise process with power spectrum E^2, then optimum filtering consists in restricting the integral in eq. (4) to the interval $[-\omega_0, \omega_0]$ where ω_0 is the solution of the equation

$$\left|\hat{K}(\omega_0)\right|^2 = \frac{\pi}{\cosh(\pi\omega_0)} = \left(\frac{\epsilon}{E}\right)^2 \quad . \tag{6}$$

Let us denote by $\tilde{f}_0(t)$ the approximate solution provided by the procedure above . Then, in the case of noise free data, it is easy to show that the following relation holds true between $\tilde{f}_0(t)$ and the true solution $f(t)$[7]

$$\tilde{f}_0(t) = \int_0^{+\infty} M_0(t,s)f(s)ds \tag{7}$$

where

$$M_0(t,s) = \frac{\sin\left[\omega_0\, ln(t/s)\right]}{\pi(ts)^{1/2}ln(t/s)} \quad . \tag{8}$$

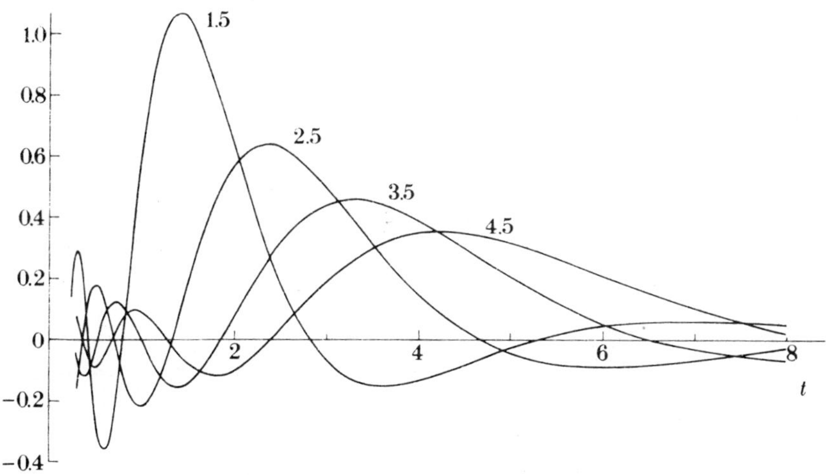

Fig. 1. Behaviour of the function $M_0(t,a)$ (eq. 8), for different values of a. The value of ω_0 taken in these computations is 4.98, corresponding to $E/\epsilon = 10^3$ in (6).

The kernel $M_0(t,s)$ is the "impulse response function" describing the total effect of both the instrument used for the measurement of the Laplace transform and the subsequent recovery procedure provided by the optimum filtering. When $f(t) = \delta(t-a)$ the recovered function is indeed $\tilde{f}(t) = M_0(t,a)$. The function (8) for several values of $s = a$ is plotted in Fig.1 in the case $E/\epsilon = 10^3$. The typical behaviour is that the principal peak becomes broader and lower as a increases. The width of the peak can be taken as a measure of the resolution achievable for the given value of signal-to-noise ratio E/ϵ. An extension of the Rayleigh criterion, which is well-known in optics, consists in taking the position of the first zero to the right of the main peak as a resolution distance. We find a resolution distance increasing with a. In fact, thanks to the dilation properties of the Laplace transform, we have a constant resolution ratio δ_0 which is defined as the ratio between the position of the zero and the value of a. The meaning of δ_0 is that two exponentials of values t_0, t_1 cannot be resolved unless $t_1 \geq \delta_0 t_0$. From eq. (8) we find that

$$\delta_0 = exp(\pi/\omega_0) \ . \tag{9}$$

The value of δ_0, as a function of E/ϵ, can be easily computed using eqs. (6) and (9). We have $\delta_0 = 2.44, 1.88, 1.63$ respectively for $E/\epsilon = 10^2, 10^3, 10^4$.

FINITE LAPLACE TRANSFORM INVERSION

As we remarked in the Introduction it is possible to obtain an improvement of the resolution ratio defined in the previuos section if we use " a priori" knowledge of the location and extent of the solution. This is equivalent to consider the problem of inversion of the finite Laplace transform which is defined as follows

$$(Lf)(p) = \int_a^b e^{-pt}f(t)dt, \quad 0 < p < \infty \tag{10}$$

where $a > 0$ and $b < \infty$ are given numbers. The operator L is compact from $L^2(a,b)$ into $L^2(0,+\infty)$ and therefore one can introduce the singular system of L, i.e. the set of the triples $\{\alpha_k; u_k, v_k\}$ which solve the following coupled equations

$$Lu_k = \alpha_k v_k \ , \quad L^* v_k = \alpha_k u_k \tag{11}$$

where L^* is the adjoint operator given by

$$(L^* g)(t) = \int_0^{+\infty} e^{-tp} g(p)dp \ . \tag{12}$$

It is easy to show, using the dilation invariance of the Laplace transform, that the singular values α_k depend only on the ratio $\gamma = b/a$. As a consequence we can restrict ourselves to the case

$$a = 1 \ , \quad b = \gamma \ .$$

Then the singular values α_k are the square roots of the eigenvalues of the self-adjoint integral operator $\tilde{L} = L^* L$ given by

$$(\tilde{L}f)(t) = \int_1^\gamma \frac{f(s)}{t+s}ds, \quad 1 \leq t \leq \gamma \tag{13}$$

and the singular functions u_k are the eigenfunctions of $\tilde{L}$ associated with

$\alpha_k{}^2$. Analogously the singular functions v_k are the eigenfunctions of the integral operator $\hat{L} = LL^\star$

$$(\hat{L}g)(p) = \int_0^{+\infty} \frac{e^{-(p+q)} - e^{-\gamma(p+q)}}{p+q} g(q) dq \tag{14}$$

associated again with the eigenvalues α_k^2.

An important property of these singular functions, which has been proved recently [8], is that the singular functions u_k are also the eigenfunctions of the following self-adjoint, second order differential operator

$$(\tilde{D}u)(t) = -\left[(t^2 - 1)(\gamma^2 - t^2)u'(t)\right]' + 2(t^2 - 1)u(t) \tag{15}$$

the boundary conditions being provided by the requirement that $u(t)$ is bounded at $t = 1$ and $t = \gamma$. Analogously the singular functions v_k are eigenfunctions of the following fourth-order differential operator

$$(\hat{D}v)(p) = [p^2 v'']'' - (\gamma^2 + 1)[p^2 v'(p)]' + (\gamma^2 p^2 - 2)v(p) \ . \tag{16}$$

This property is analogous to a well-known property of the prolate spheroidal wave functions of Slepian and coworkers and makes possible the accurate computation of a large number of singular functions of the finite Laplace transform [9]. Some results are plotted in Fig. 2 and Fig. 3.

In terms of the singular system of L, the inversion formula of the finite Laplace transform, i.e. the solution of the equation $g = Lf$, is given by

$$f(t) = \sum_{k=0}^{+\infty} \frac{1}{\alpha_k}(g, v_k)u_k(t) \tag{17}$$

where (g, v_k) is the usual scalar product of $L^2(0, +\infty)$. This formula is affected by numerical instability as eq. (4) since the singular values α_k tend to zero when $k \to \infty$. When γ is small they tend to zero very rapidly indeed. For instance we have the following results [4] in the case $\gamma = 5$

$$\alpha_0 = 0.875 \quad \alpha_1 = 0.193 \quad \alpha_2 = 0.038 \quad \alpha_3 = 0.007 \quad .$$

The method of optimum filtering can also be applied to eq. (17) and the result is that we must keep only the singular values satisfying the condition

$$\alpha_k \geq \epsilon/E \tag{18}$$

which is analogous to eq. (5). We denote by K_s the maximum value of the index of the singular values satisfying condition (18), so that $K_s + 1$ is the number of terms we keep in eq. (17) (number of degrees of freedom).

Now it is possible to prove [8] that $u_k(t)$ has exactly k zeros in the interval $[1, \gamma]$; furthermore, as follows from numerical computations, these zeros are approximately equidistant in the variable $ln(t)$. As a consequence we take as resolution elements the intervals between adjacent zeros of the singular function of highest order, i.e. $k = K_s$ and we define as a resolution ratio δ_s the quantity given by [4]

$$\delta_s = \gamma^{1/K_s} \ . \tag{19}$$

This quantity can be computed by means of numerical calculations of the singular values of L and compared with the resolution ratio δ_0 defined in the previuos section. For instance in the case $\gamma = 5$ we have found

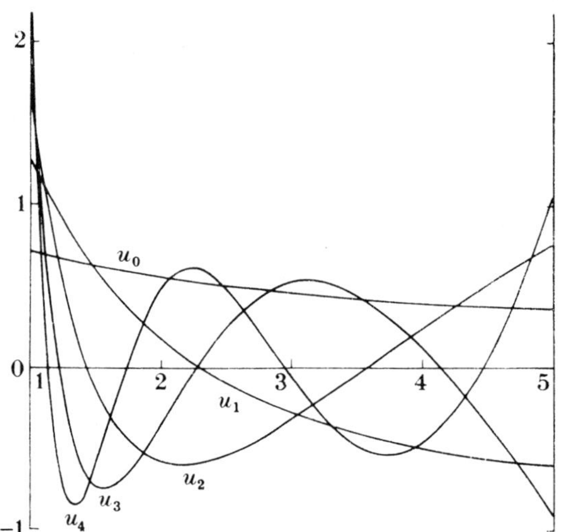

Fig. 2. Singular functions $u_k(t), 1 \leq t \leq \gamma, k = 0, 1, \ldots, 4$ in the case $\gamma = 5$.

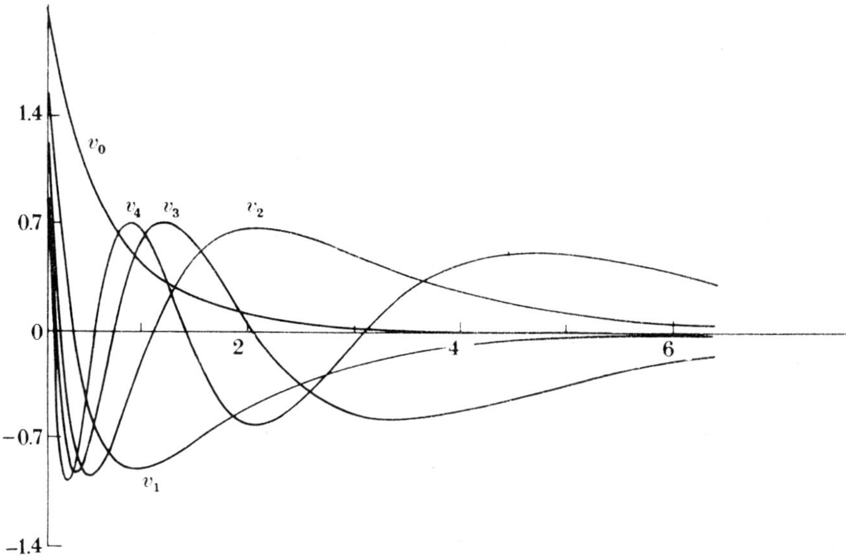

Fig. 3. Singular functions $v_k, k = 0, 1, \ldots, 4$ in the case $\gamma = 5$.

$\delta_s = 1.74, 1.44, 1.32$ for $E/\epsilon = 10^2, 10^3, 10^4$. These figures demonstrate the beneficial effect of the knowledge of the location of the solution on the resolution limits.

THE CASE OF DISCRETE DATA

In a pratical problem the Laplace transform is not known for all the values of p but only at a finite set of points $p_1, p_2, \ldots, p_N$ and therefore we have to solve the problem

$$g_n = (L_N f)(p_n) = \int_1^\gamma e^{-p_n t} f(t) dt \; ; \quad n = 1, \ldots, N \tag{20}$$

where the g_n are the measured values of the Laplace transform and L_N is the operator transforming the unknown function $f(t)$ into a vector of a euclidean space. Then the adjoint operator is given, by

$$(L_N^\star \mathbf{g})(t) = \sum_{n-1}^{N} w_n g_n e^{-p_n t} \tag{21}$$

where g is the vector with components g_n and the w_n are the weights defining the scalar product in the data space [7].

The advantage of the method of singular function expansions is that it can be applied also to the present case. We denote by $\{\alpha_{N,k}; u_{N,k}, \mathbf{v}_{N,k}\}$ the singular system of L_N, i.e. the set of the solutions of the coupled equations

$$L_N u_{N,k} = \alpha_{N,k} \mathbf{v}_{N,k} \; , \quad L_N \mathbf{v}_{N,k} = \alpha_{N,k} u_{N,k} \; . \tag{22}$$

The computation of this system is quite easy since the operator $\tilde{L}_N = L_N L_N^\star$ is just a matrix $N \times N$ (it is related to the Gram matrix of the exponentials $exp(-p_n t)$ [7]) and therefore one can reduce solution of the problem (22) to a standard eigenvalue problem for a symmetric matrix.

Now the solution of minimal norm of problem (20) is given by an expansion similar to (17)

$$f^\dagger(t) = \sum_{k=0}^{N-1} \frac{1}{\alpha_{N,k}} (\mathbf{g}, \mathbf{v}_{N,k}) u_{N,k}(t) \tag{23}$$

and also this solution is affected by numerical instability if N is too large. Since it is possible to show, by means of numerical computations, that the singular values $\alpha_{N,k}$ can provide good approximations of the singular values α_k defined in the previuos section, we conclude that the method of optimum filtering will select the same number of terms in eq. (17) and eq. (23).

We have considered two data-point distributions [7]: a set of N equidistant points : $p_n = c + d(n-1)$, characterized by two parameters, the position c of the first point and the distance d between adjacent points, and a set of N points forming a geometric progression : $p_n = c\Delta^{n-1}$, characterized again by two parameters, the position c of the first point and the dilation factor Δ, giving the ratio between adjacent data points.

The conclusion of several numerical computations, directed towards the optimum determination of the number and location of the data points, is that the points forming a geometric progression provide much better approximations of the singular values than the equidistant points. Furthemore the number of points required is much smaller in this case than in the case of equidistant points.

As an example of our results we give the first three singular values
for the case $\gamma = 5$, obtained using 5 geometrically spaced data points [7]

$$\alpha_{N,0} = 0.874 \ , \quad \alpha_{N,1} = 0.191 \ , \quad \alpha_{N,2} = 0.038 \ .$$

Comparing with the values of α_k, always for $\gamma = 5$, given in the previous
section we find an excellent agreement.

INVERSION IN WEIGHTED SPACES

The method described in the previous sections, namely the inversion
of the finite Laplace transform using optimum filtering, has some unsat-
isfactory features. In practical situations the support of the unknown
function is not known exactly although some estimations can be made from
the experimental data since the derivatives of the Laplace transform $g(p)$
at the point $p = 0$ are related to the moments of the unknown function $f(t)$.
The first and second moments provide information about the "localization"
of $f(t)$ but not a precise estimation of its support.

A second and more serious difficulty originates from the behaviour
of the singular functions $u_k(t)$ at the edges of the support, the interval
$[1,\gamma]$. As follows from Fig. 2 , these functions are quite large at $t = 1$ and
$t = \gamma$ and therefore they are large precisely in those regions where the
unknown function is presumably small. This fact gives rise to spurious
and troublesome edge effects.

A way to overcome these difficulties is to look for solutions of the
Laplace transform inversion in a weighted L^2 space. In such a way we are
not forced to a rigid choice of the support, we can use the information
contained in the knowledge of the lower moments of the unknown function
and we can force the solution to be small in those regions where it is presumed to be small.

We can formulate the problem as follows [10] : find a solution of eq.
(1), (2) satisfying the condition

$$\int_0^{+\infty} \frac{f^2(t)}{P^2(t)} dt < +\infty \tag{24}$$

where $P(t)$ is a given profile function (for instance a gamma distribution
or a log-normal distribution) whose first and second moments have been
adapted to the corresponding moments of $f(t)$.

If we put $f(t) = P(t)\Phi(t)$ then the inversion of the Laplace transformation in the weighted L^2-space (24) is equivalent to the inversion in
$L^2(0,+\infty)$ of the following operator

$$(K\Phi)(p) = \int_0^{+\infty} e^{-pt} P(t)\Phi(t) dt \ . \tag{25}$$

This operator is compact (more precisely of the Hilbert-Schmidt class) if
the profile function $P(t)$ satisfies the following condition

$$\int_0^{+\infty} \frac{P^2(t)}{t} dt < +\infty \ . \tag{26}$$

Then one can introduce again the singular system of K and use optimum
filtering for determining a stable approximate solution of the problem.

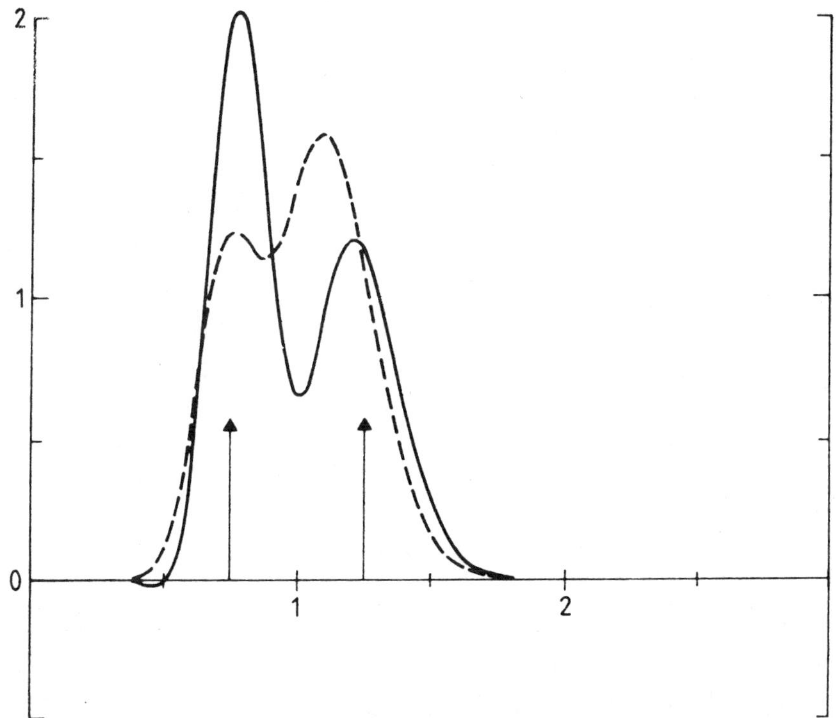

Fig. 4. Reconstruction of $f(t) = c_1\delta(t - t_1) + c_2\delta(t - t_2)$, with $c_1 = c_2 = 0.5$, $t_1 = 0.75, t_2 = 1.25$ using singular functions corresponding to $\beta = 16$ in eq. (27), in the case of signal-to-noise ratio of the order of 10^2 (broken curve) and the order of 10^3 (full curve). The arrows indicate the positions of the delta functions.

We have performed several computations , also in the case of discrete data with a uniform or geometric distribution of the data points, assuming that $P(t)$ is a gamma distribution (also called Schulz function)

$$P(t) = \frac{\beta^\beta}{\Gamma(\beta)} t^{\beta-1} e^{-\beta t} \tag{27}$$

whose first moment is unity and whose second central moment is given by $Q = 1/\beta$. Several numerical experiments have been performed in order to relate Q to the experimental polydispersity factor. The application of the method to experimental data on coagulating vesicle suspensions has provided satisfactory results [11]. Here we only give an example (see Fig. 4) of a reconstruction obtained with this method. Notice that the ratio between the values of the two exponentials is $t_2/t_1 = 1.66$ and that it is possible to resolve these exponentials already in the case of a signal-to-noise ratio $E/\epsilon = 10^2$.

ACKNOWLEDGMENTS

This work has been partly supported by NATO Grant No. 463/84, by EEC contract No. STE J- 0089 - 3 and by Ministero della Pubblica Istruzione, Italy. C. De Mol is " Chercheur qualifié " of the Belgian National Fund for Scientific Research.

REFERENCES

1. H.Z. Cummins and E.R. Pike (eds.), "Photon Correlation Spectroscopy and Velocimetry", Plenum, New York (1977)

2. J.G.McWhirter and E.R. Pike, On the numerical inversion of the Laplace transform and similar Fredholm integral equations of the first kind , *J. Phys. A*, 11:1729 (1978)

3. N.Ostrowsky, D. Sornette, P. Parker and E.R. Pike , Exponential sampling method for light scattering polydispersity analysis, *Opt. Acta*, 28:1059 (1981)

4. M.Bertero, P. Boccacci, and E.R. Pike, A singular value analysis of the Laplace transform inversion in the presence of noise , *Proc. R. Soc. Lond.*, A 383:15 (1982)

5. E.C. Titchmarsh, "Introduction to the theory of Fourier integrals", Clarendon, Oxford (1948)

6. A.N. Tikhonov, and V.Y. Arsenine, "Solutions of ill-posed problems", Wiston/Wiley, Washington (1977)

7. M. Bertero, P. Brianzi, and E.R. Pike, The effect of sampling and truncation of data on the Laplace transform inversion, *Proc.R. Soc. Lond.*, A 398:23 (1985)

8. M.Bertero, and F.A. Grünbaum, Commuting differential operators for the finite Laplace transform, *Inverse Problems*, 1:181 (1985)

9. M.Bertero, F.A. Grünbaum, and L. Rebolia, Spectral properties of a differential operator related to the inversion of the finite Laplace transform, *Inverse Problems*, 2:131 (1986)

10. M. Bertero, P. Brianzi, and E.R. Pike, Laplace transform inversion in weighted spaces,*Inverse Problems*, 1:1 (1985)

11. M. Bertero, P. Brianzi, E.R. Pike, G. de Villiers, K.H. Lan and N. Ostrowsky, Light scattering polydispersity analysis of molecular diffusion by Laplace transform inversion in weighted spaces, *J.Chem.Phy.*, 82:1551 (1985)

PARTICLE SIZE DISTRIBUTIONS FROM FRAUNHOFER DIFFRACTION

M.Bertero[o], P.Boccacci[o], C.De Mol[+] and E.R.Pike[*]

(o) Dipartimento di Fisica dell'Università di Genova
and Istituto nazionale di Fisica Nucleare
I- 16146 Genova, Italy
(+) Département de Mathématique, Université Libre de
Bruxelles ,B-1050 Bruxelles, Belgium
(*) Dept. of Physics, King's College, London WC2R 2LS
and RSRE, Great Malvern, WR14 3PS, England

INTRODUCTION

It is well known [1] that the mean intensity per steradian diffracted by
a random distribution of N opaque particles is given, in the Fraunhofer
region, by

$$I(s) = I_0 \sum_{i=1}^{N} \frac{a_i^2 J_1^2(ka_i s)}{s^2} \quad , \tag{1}$$

where a_i is the radius of the particle i, k is the wave-vector of the radiation and s is the scattering angle. In the absence of noise an annular
detector or segment of an annulus will respond with a current $g(s)$ proportional to $sI(s)$. If we proceed now to the limit of a continous distribution
of radii $p(t)$, $t = ka$, we find for $f(t) = t^3 p(t)$, the volume distribution of
the opaque particles, the following integral equation

$$g(s) = \int_0^{+\infty} K(st)f(t)dt \tag{2}$$

where

$$K(x) = \frac{J_1^2(x)}{x} \quad . \tag{3}$$

In the presence of noise which is white, additive and constant per
unit area of detector surface, it may be convenient to take as data function the total current per unit area which is proportional to $I(s)$. Therefore if the unknown function is $f(t) = t^4 p(t), t = ka$, we find again an integral equation of the type (2) where now

$$K(x) = \frac{J_1^2(x)}{x^2} \quad . \tag{4}$$

These integral equations can be investigated using a method developed
by McWhirter and Pike[2], if one assumes that the diffraction pattern $g(s)$
is given for any value of s. In this way one can estimate the resolution
limits achievable for a given noise level[3]. Equation (2), however, holds
true only in the case of small angles, normally $s \simeq \sin(s)$; secondly the
form used for the diffraction pattern of an opaque particle is valid only
when the radius a of the particle is much greater than the wavelength λ,
say $a > 10\lambda$. Therefore we must replace equation (2) by the following one

$$g(s) = \int_{t_0}^{t_1} K(st)f(t)dt, \quad s_0 \leq s \leq s_1 \; . \tag{5}$$

This integral equation can be investigated using the method of singular
system analysis [4]. The results obtained show that the knowledge of the finite support of the distribution can more than compensate the loss of experimental knowledge of diffraction data due to sampling and truncation.

Since the problem is mildly ill-conditioned, the number of singular
functions corresponding to singular values greater than the noise level,
i.e. the singular functions used for inversion, is rather large. As a
consequence, the reconstruction of a narrow distribution can take rather
large negative values while the physical solution must be positive. The
positivity constraint, however, implies the use of nonlinear optimiza-
tion methods which are time consuming from the computational point of
view. For this reason we introduce filtering methods which are still linear and have the interesting property of producing positive solutions in
the absence of noise. In the case of real data the restored solution can
have negative values but these can only be an effect of the noise. The
method can be easily implemented. Numerical simulations show that nearly
positive solutions can be obtained in the case of few percent error on the
data.

THE EIGENFUNCTION APPROACH

Equation (2),(3) or (2),(4) belongs to a class of first kind Fredholm
integral equations which also include the problem of Laplace transform
inversion. The latter, which is basic for the extraction of polydispersity information in photon correlation spectroscopy, has been discussed
by us in a paper published in these Proceedings [5], hereinafter referred to
as I.

In the absence of noise, the solution of equation (2) is given by

$$f(t) = \frac{1}{2\pi} \int_{-\infty}^{+\infty} \frac{\hat{g}(-\omega)}{\hat{K}(-\omega)} t^{-(1/2+i\omega)} d\omega \tag{6}$$

where $\hat{g}(\omega)$ and $\hat{K}(\omega)$ are the Mellin transforms of $g(s)$ and $K(x)$ respectively (see I for the definition of the Mellin transform we are using).

In the case of Fraunhofer diffraction $\hat{K}(\omega)$ can be evaluated by means
of the following expression of the Weber-Schafheitlin integral

$$\int_0^{+\infty} x^{y-3} J_1^2(x) dx = \frac{2^{y-3}\Gamma(3-y)\Gamma(y/2)}{\Gamma^2(2-y/2)\Gamma(3-y/2)} \; . \tag{7}$$

For instance, in the case of the kernel (4) we have [3]

$$\hat{K}(\omega) = \frac{2^{-5/2+i\omega}\Gamma\left(\frac{5}{2}-i\omega\right)\Gamma\left(\frac{1}{4}+i\frac{\omega}{2}\right)}{\Gamma^2\left(\frac{7}{4}-i\frac{\omega}{2}\right)\Gamma\left(\frac{11}{4}-i\frac{\omega}{2}\right)} \tag{8}$$

and $|\hat{K}(\omega)| \sim |\omega|^{-3}$ when $|\omega| \to \infty$. An analogous expression holds true in the case of the kernel (3). The modulus of $\hat{K}(\omega)$ can be computed and it gives the eigenvalue spectrum of the corresponding integral operator [2,3].

Since $\hat{K}(\omega)$ tends to zero at infinity, the integral (6) does not converge in the case of noisy data. Then an approximate stable solution can be obtained by means of the optimum filtering technique, discussed in I, which consists in restricting the integral of equation (6) to the interval $[-\omega_0,\omega_0]$ where ω_0 is the solution of the equation

$$|\hat{K}(\omega_0)| = \frac{\epsilon}{E} \tag{9}$$

and E/ϵ is the signal-to-noise ratio. As also discussed in I, this implies the existence of a resolution ratio δ_0 given by

$$\delta_0 = \exp\left(\frac{\pi}{\omega_0}\right) \ . \tag{10}$$

By combining (8),(9) and (10) one can compute the resolution ratio for various values of the signal-to-noise ratio. For instance we have found $\delta_0 = 1.72, 1.27, 1.12$ for $E/\epsilon = 10^2, 10^3, 10^4$ respectively [3]. As an example we might ask how many size fractions can be resolved in the range $2 - 100\ \mu m$ when the signal-to-noise ratio is 10^2. Since in this case the resolution ratio is 1.72, it follows that between 7 and 8 fractions, on a geometric - progression scale, may be resolved in this range. This number rises to 16 if the signal-to-noise ratio is improved by a factor of 10.

THE CASE OF LIMITED DATA

The approach described in the previous section, even if useful for estimating the resolution limits inherent to the inversion of Fraunhofer diffraction data, cannot be used in pratice since it requires everywhere defined data. Now the data are truncated and sampled and therefore the more general method of singular system analysis must be used [4]. This analysis runs parallel to the analyses of the inversion of the Laplace transform described in I.

The first step is that the integral equation (2) is replaced by the integral equation (5). The lower and upper bounds s_0, s_1 describe the effect of the truncation of the data, while t_0, t_1 describe the "a priori" knowledge of the "localization" of the size distribution. The integral operator involved in this equation

$$(Af)(s) = \int_{t_0}^{t_1} K(st)f(t)dt, \quad s_0 \leq s \leq s_1 \tag{11}$$

is a compact operator from $L^2(t_0, t_1)$ into $L^2(s_0, s_1)$ and we can introduce its singular system $\{\alpha_k; u_k, v_k\}$. As in the case of the Laplace transform, considered in I, the singular values depend only on $\gamma = t_1/t_0$ in the limiting case $s_0 = 0$, $s_1 = \infty$. This result is still approximately true if the range where the data are given is sufficiently large. This means that the quantity $t_0 s_0$ must be smaller than the first zero of $J_1(x)$ and the ratio s_1/s_0 must be much larger than γ.

Also the problem with discrete data can be treated along the same lines used for the Laplace transform problem. In such a case one can introduce the singular system $\{\alpha_{N,k}; u_{N,k}, v_{N,k}\}$ of the operator

$$(A_N f)(p_n) = \int_1^\gamma K(p_n t) f(t) dt \; ; n = 1, \ldots, N \tag{12}$$

which transforms a function of $L^2(1, \gamma)$ into a N-dimensional vector. It is possible to prove [4] that, when the number of data points tends to infinity and the distance between adjacent points tends to zero, the singular values of A_N approach the singular values of A. As in I we have considered both uniform and geometric sampling of the data. Details on the numerical results can be found in Bertero et al.[4]. Here we just give a few qualitative results. For a given value of γ the singular values of the Fraunhofer problem tend to zero more slowly than the singular values of the Laplace transform and therefore the corresponding number of degrees of freedom is greater in the case of Fraunhofer than in the case of Laplace. The behaviour of the singular functions is qualitatively the same in the two cases - see Fig. 2 of I for a plot of the singular functions u_k - and in particular the singular function u_k has exactly k zeros in the interval $[1, \gamma]$. This property can be used in order to define, as in I, a resolution ratio δ_s given by

$$\delta_s = \gamma^{1/K_s} \tag{13}$$

where $K_s + 1$ is the number of degrees of freedom. Using this equation and the computed values of the α_k we have found that, in the case $\gamma = 5$, $\delta_s = 1.6, 1.26$ for $E/\epsilon = 10^2, 10^3$ respectively. These results have been obtained using $s_0 t_0 = 0.16, t_1/t_0 = 24$. If we compare these figures with the figures given in the previous section, we find that the knowledge of the finite support of the distribution can more than compensate the loss due to the finite support of the data.

POSITIVITY AND WINDOWING

In order to explain our method for restoring positivity of the solution in the inversion of Fraunhofer data, we first come back to the equation (2). This equation can be transformed into a first kind convolution equation if we perform the following change of variables

$$t = e^y \; , \; s = e^{-x} \; , \; F(y) = e^y f(e^y)$$

$$G(x) = g(e^{-x}) \; , \; H(x) = K(e^{-x}) \; . \tag{14}$$

Then the equation (2) becomes

$$G(x) = \int_{-\infty}^{+\infty} H(x - y) F(y) dy \tag{15}$$

and in the case of noise free data its solution can be given in terms of
Fourier transform

$$F(y) = \frac{1}{2\pi} \int_{-\infty}^{+\infty} \frac{\hat{G}(\omega)}{\hat{H}(\omega)} e^{iy\omega} d\omega \qquad (16)$$

where $\hat{G}, \hat{H}$ are the Fourier transforms of G, H respectively. Notice that
equation (16) can be obtained from equation (6) by means of the change of
variables indicated above.

Now the optimum filtering consists in restricting the integral (16)
to some bounded interval $[-\omega_0, \omega_0]$. A more general approach for regular-
izing equation (16) can be obtained by defining appropriate solutions in
terms of filter or window functions $\hat{W}_\alpha(\omega)$ (depending on a parameter α,
which is called regularization parameter) as follows [6]

$$\tilde{F}_\alpha(y) = \frac{1}{2\pi} \int_{-\infty}^{+\infty} \hat{W}_\alpha(\omega) \frac{\hat{G}(\omega)}{\hat{H}(\omega)} e^{iy\omega} d\omega \quad . \qquad (17)$$

Notice that, in the absence of noise, the relation between the filtered
solution $\tilde{F}_\alpha(y)$ and the true solution $F(y)$ is given by

$$\tilde{F}_\alpha(y) = \int_{-\infty}^{+\infty} W_\alpha(y - y') f(y') dy' \qquad (18)$$

where the "impulse response function" $W_\alpha(y)$ is the inverse Fourier transform of $\hat{W}_\alpha(\omega)$.

In this framework, optimum filtering can be defined as follows: let
$\chi_\alpha(\omega)$ be the characteristic function of the interval $[-1/\alpha, 1/\alpha]$, i.e. the
function which is 1 on this interval and zero elsewere; then the optimum
filtering, as defined in I, corresponds to taking

$$\hat{W}_\alpha(\omega) = \chi_\alpha(\omega) \; , \; \alpha = \frac{1}{\omega_0} \qquad (19) \quad .$$

The corresponding "impulse response function", given by

$$W_\alpha(y) = \frac{\sin(\omega_0 y)}{\pi y} \qquad (20)$$

is not positive and has large negative side-lobes for large values of ω_0.
But if we consider the triangular window

$$\hat{W}_\alpha(\omega) = (1 - \alpha|\omega|)\chi_\alpha(\omega) \; , \; \alpha = \frac{1}{\omega_0} \qquad (21)$$

then we have

$$W_\alpha(y) = \frac{\omega_0}{2\pi} \frac{\sin^2(\omega_0 y/2)}{(\omega_0 y/2)^2} \qquad (22)$$

Since the "impulse response function" is positive, from equation (18) we
deduce that $\tilde{F}_\alpha(y) \geq 0$ whenever $F(y) \geq 0$ and this method provides an approximate positive solution at least in the absence of noise. Notice however that positivity has been obtained at the price of a loss of resolution. For a given value of ω_0, indeed, the width of the function (22) is
twice the width of the function (20). However, more terms may be used in
the triangular window case to reduce this factor of two without apparently
compromising the signal-to-noise ratio.

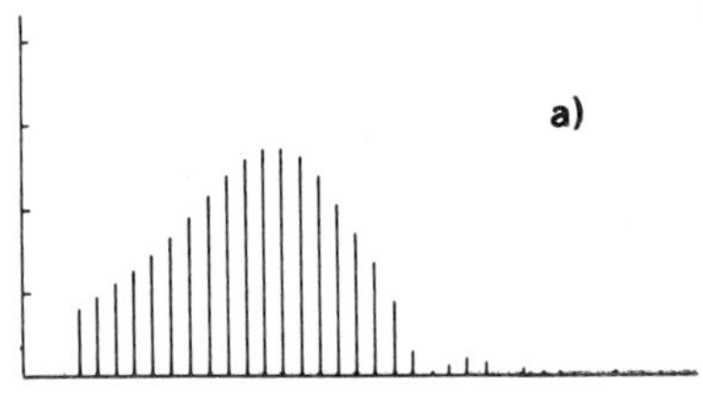

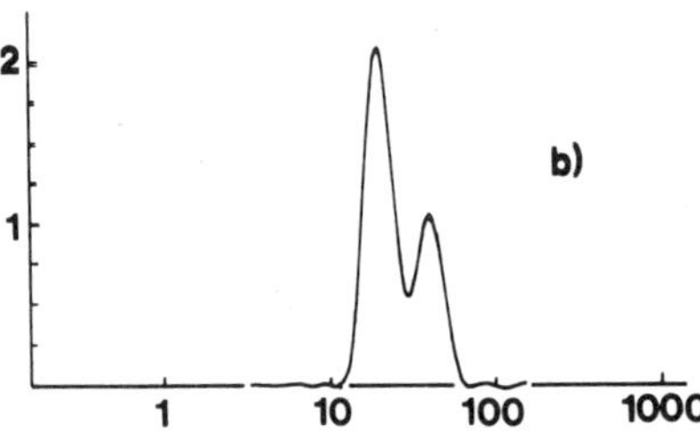

Fig. 1. A particle size distribution consisting of two delta functions concentrated at $t_1 = 20\lambda, t_2 = 40\lambda$ (λ = wavelength of the incident radiation) with weights respectively 1 and 0.5. a) the data vector ; b) the restored distribution. The computed area is 1.51.

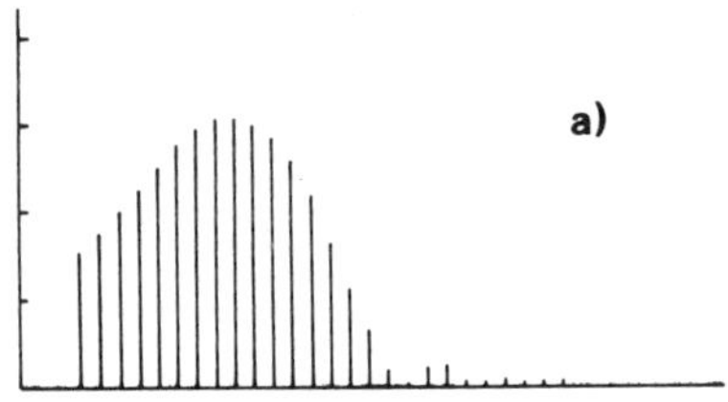

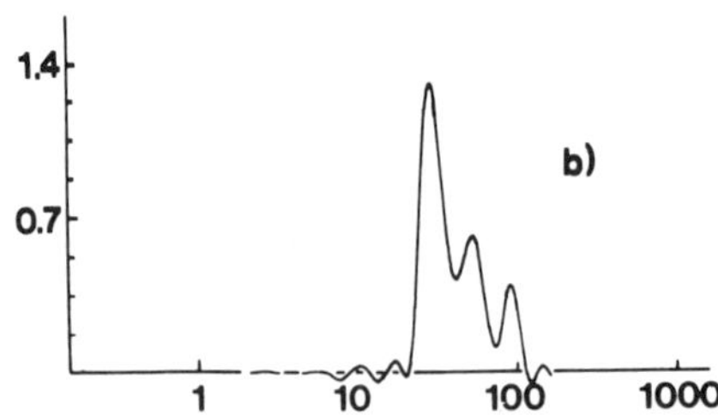

Fig. 2. A particle size distribution consisting of three delta functions concentrated at $t_1 = 30\lambda, t_2 = 52\lambda, t_3 = 87\lambda$ (λ = wavelength of the incident radiation) with weights respectively 1, 0.5 and 0.25. a) the data vector ; b) the restored distribution. The computed area is 1.76.

Another window function which provides a nearly positive "impulse response function" $W_\alpha(y)$ is the Hamming window, which is well-known in signal processing

$$\hat{W}_\alpha(\omega) = \frac{1}{2}[1 + cos(\pi\alpha\omega)]\chi_\alpha(\omega) \ , \ \alpha = \frac{1}{\omega_0} \ . \tag{23}$$

The corresponding "impulse response function" is not exactly positive but has approximately the same width of the function (22) and much smaller side lobes. For this reason this window can be more convenient than the triangular window.

The windowing method outlined above can be easily extended to the method of singular function expansions which must be used in the case of limited data.

In Fig. 1 and Fig. 2 we give two examples of reconstructions obtained using a Hamming window combined with the singular function expansions. These correspond to the case $\gamma = 130$ and we have used 31 data points forming a geometric progression in to the interval $[s_0, s_1]$ with $s_1/s_0 = 130$ (s_0 is approximatively 10^{-2}). In the figures we plot $F(y)$ as a function of $y = \ln t$ (equation (14)). The wavelength λ is taken as the unit of the size t of the particles. The reconstructions have been obtained using 23 singular functions and , as we have checked by means of several simulations, they are not significantly affected by a noise of 1% on the data. In the case of the reconstruction of three delta functions the ratio between two adjacent delta functions is 1.7, which is just the resolution ratio estimated in the case $E/\epsilon = 10^2$.

ACKNOWLEDGMENTS

This work has been partly supported by NATO Grant No. 463/84, by EEC contract No. STE J- 0089 - 3 and by Ministero della Pubblica Istruzione, Italy. C. De Mol is " Chercheur qualifié " of the Belgian National Fund for Scientific Research.

REFERENCES

1. L.P. Bayvel and A.R. Jones, "Electromagnetic Scattering and its Applications ", Applied Science Publishers, London (1981)
2. J.G.McWhirter and E.R. Pike, On the numerical inversion of the Laplace transform and similar Fredholm integral equations of the first kind , *J. Phys. A* , 11:1729 (1978)
3. M. Bertero, and E.R. Pike, Particle size distributions from Fraunhofer diffraction: I. An analytic eigenfunction approach, *Opt. Acta*, 30:1043 (1983)
4. M. Bertero, P. Boccacci, and E.R. Pike, Particle size distributions from Fraunhofer diffraction: II. The singular value spectrum, *Inverse Problems*, 1:111 (1985)
5. M. Bertero, P. Boccacci, C. De Mol, and E.R. Pike, Extraction of polydispersity information in photon correlation spectroscopy, published in these Proceedings.
6. M. Bertero, P. Brianzi, C. De Mol, and E.R. Pike, Positive regularized solutions in electromagnetic inverse scattering, in " Proc. Int. URSI Symp. on Electromagnetic Theory", Budapest (1986).

CALCULATION OF CALIBRATION CURVES FOR THE PHASE DOPPLER TECHNIQUE: COMPARISON BETWEEN MIE THEORY AND GEOMETRICAL OPTICS

S.A.M. Al-Chalabi[1], Y. Hardalupas[2], A.R. Jones[1], and A.M.K.P. Taylor[2]

Departments of Chemical[1] and Mechanical[2] Engineering
Imperial College of Science and Technology
London SW7 2BX, England

INTRODUCTION

The phase-Doppler technique (Durst & Zaré, 1975; Bachalo & Houser, 1984) measures the diameter and velocity of *spherical* particles simultaneously, with the spatial and temporal resolution of a laser-Doppler anemometer. The range of diameters and particle concentrations which can be measured range from a few to several hundred micrometers and up to about 10^{10} particles/m^{-3}. Instruments based on this principle have a number of advantages over others based on the laser-Doppler anemometer. One is that the technique is not based on the intensity of the scattered light and the measurements are therefore insensitive to random beam attenuation by either particles which lie outside the measurement volume or by obscuration of windows in a test-section (Hardalupas et al., 1986). There is therefore also no need to account for any effect on the size information of the Gaussian intensity of the incident light beams, in contrast to systems which measure visibility, for example (e.g. Yeoman et al. 1982).

In contrast to the velocity information, the calibration curve is dependent on the location and the separation of the detectors. The calibration curve can be calculated, rather than measured, although there are at least three fundamentally different ways in which the calculation can be performed. These are based on the Mie solution to scattering theory and approximations derived from geometrical optics.

Previous work

The simplest way to calculate the calibration curve is to assume that the scattering of light by the sphere is due to either refraction or reflection, depending on which mode of scattering is presumed to be dominant (e. g. Durst & Zaré, 1975; Saffman et al., 1984; Bauckhage & Floegel, 1984; Drain, 1985). This approach results in a one-line equation which describes a calibration curve that turns out to give good agreement with experiment for diameters larger than about 10 μm (e. g. Bachalo & Houser 1984; Hardalupas 1986; Lightfoot & Negus, 1986). This equation is not valid, however, for a wide range of detector locations and the predicted linear dependance of phase difference with diameter does not hold true for particles smaller than approximately 10 μm.

A more complicated calculation results if it is considered that the scattered light is due to the superposition of reflected, refracted and diffracted light (e.g. van de Hulst, 1981). No new principles are involved in extending the calculation to two incident beams, which is the situation for a phase-Doppler anemometer (Hardalupas 1986). The calculation must, however, be performed on a computer but Glantschnig & Chen (1981) have proposed an elegant simplification to the analysis, valid up to about 60° of scatter angle, which reduces the amount of computational effort required to obtain numerical results. Calculations based on the geometrical optics approximation predict the regions of space in which the formulae of, for example, Saffman et al. (1984) remain valid, at least for the forward scatter direction, and explain the origin of the oscillations in the calibration curve for small diameters (Hardalupas & Taylor, 1987). These calculations can show the influence of practical details on the calibration curve, such as the size of the collecting aperture and of the polarisation in the incident laser beams, and are an improvement on the one-line formulae.

Solutions obtained from geometrical optics are, however, an approximation to the exact solution to the scattering of electromagnetic waves from a spherical dielectric obtained by Mie. In the past, Mie's solutions have been used to calculate the visibility of a Doppler signal in the context of particle sizing (e.g. Jones, 1974) and to calculate the calibration curves of a phase-Doppler anemometer (Saffman et al., 1984).

There are at least two incentives to persisting with the approximate calculation of geometrical optics rather than the exact Mie solutions. One is that much less computational effort is required to obtain a calibration curve using the geometrical optics approximation than from Mie's solution. The other is that geometrical optics describes the physics of the scattering process with greater insight and this is useful in understanding the qualitative form of the calibration curve, as we demonstrate below. The work of Ungut et al. (1981) and Glantschnig & Chen (1981) has demonstrated by comparison with Mie's solution that, at least for the *intensity* distribution of the scattered light from *one* incident laser beam, the approximation is good for particles larger than about 2μm and for angles of scatter less than 60° from the direction of propagation of the beam.

The current Contribution

The purpose of this work is to evaluate the accuracy of predictions of the calibration curve of a phase-Doppler anemometer calculated from the geometrical optics approximation by comparison with Mie solutions, for water droplets in air. We shall consider rays which undergo up to one internal reflection in the geometrical optics approximation. Calibration curves are presented for scatter at 30°, 100°, 120°, 133°, 150°, 164° and 175° to the axis of the laser-Doppler anemometer. A secondary purpose is to examine the potential of the phase-Doppler technique in the side- and backscatter directions.

THEORETICAL ANALYSIS

We do not describe the details of the calculation in detail here, to save space, and refer to other publications for further information.

Mie Solutions

The theoretical treatment has been described by Hong & Jones (1976). It assumes that the particles are spherical, homogeneous and isotropic, and are illuminated by two plane monochromatic waves. In some of the results presented below, Gaussian integration is performed over rectangular apertures of the collection lenses, using 10 points in each direction.

Geometrical Optics Approximation

The scattered light is assumed to be obtainable as a superposition of reflected, refracted and diffracted light and for angles of scatter from each beam less than about 60° the details of the calculation have been described by Hardalupas (1986). For larger angles of scatter from water droplets, we allow for light scattered by three paths:

(i) external reflection scatters light in the interval $0 < \theta < 180^{\circ}$;

(ii) light which is twice refracted scatters over the interval $0 < \theta < 82.8^{\circ}$;

(iii) light which is internally reflected once scatters between $138.0^{\circ} < \theta < 180^{\circ}$. We note here that there are two potential paths for light to reach the interval $138.0^{\circ} < \theta < 165.6^{\circ}$ the region occupied by the main rainbow.

These three paths are conveniently summarised in figure 1, which is taken from van de Hulst (1981), and we have included an additional path to which we shall refer later. In the first two modes of scattering, there is a closed-form analytic expression for the angle of incidence on the sphere, given the angle θ at which light is scattered (see Glantschnig & Chen, 1981). In the third case, the equation relating the angle of scatter to the angle of incidence is a transcendental function which we solved iteratively. In common with Glantschnig & Chen, we have made extensive use of van de Hulst's formulae for the change of phase along each path. Where we present results which are averages over rectangular aperture of the collecting lens, we have integrated numerically over a grid of 41 by 7 points. For convenience we show in figure 2 the definitions of the angles which we shall use below.

APPLICATIONS

For the purposes of this work, we divide the interval of scattering angle $0 < \theta < 180^{\circ}$ into six separate regions, depicted in figure 1, depending on the combination of paths that light can take to reach a given scatter angle. We remind the reader that the scatter angle from each beam is different for a given observation direction and we define, for convenience, the mean scatter angle as the angle from the optical axis of the anemometer. In what follows, the crossing angle of the beams is 10° and the wavelength

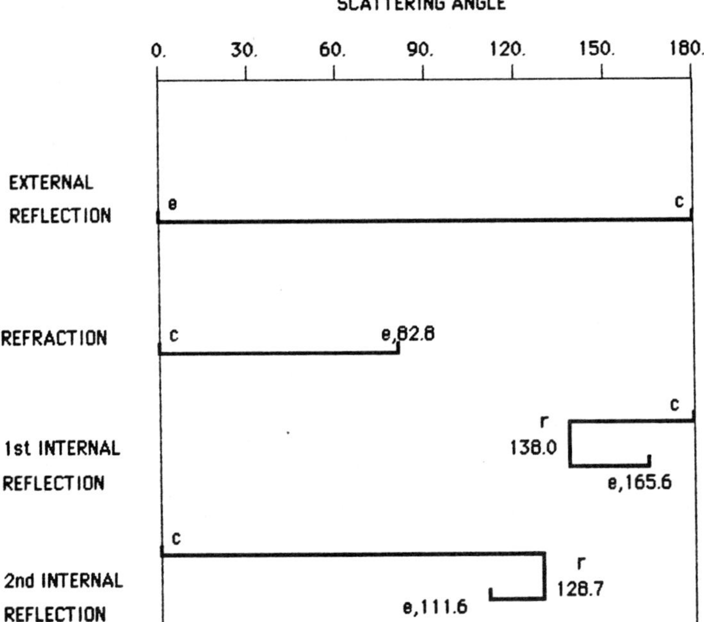

Figure 1   Classification of scatter angle into six intervals according to the combination of ray paths across a water droplet. Diagram is after van de Hulst (1981).

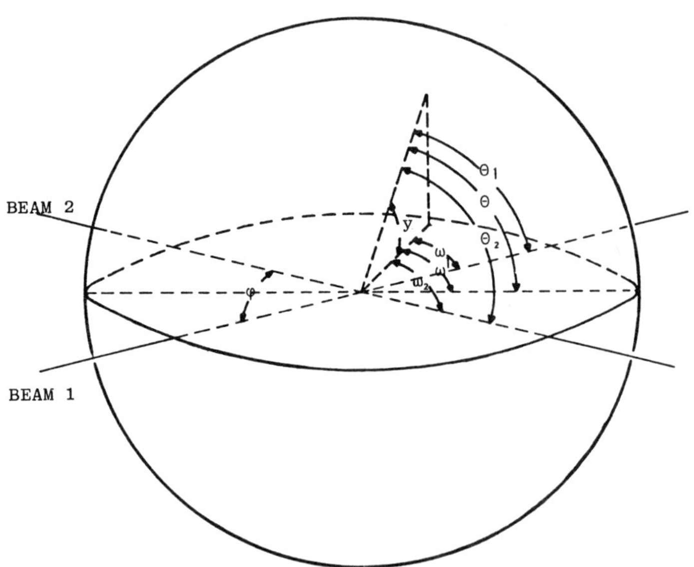

Figure 2   Definitions of angles of elevation, y, rotation, ω, and mean scatter angle, θ, from optical axis of anemometer.

of light is 632.8 nm. We now consider each of these regions in turn.

$\underline{0 < \theta < 82.8^\circ}$ Within this interval, light is scattered by surface reflection and by refraction through the droplet: we have included diffraction in our calculations but the influence is negligible beyond about 25° and we, in common with others, seek to avoid the effects of diffracted light. Figure 3 (a) shows the calibration curve for two observers who are located at an angle of elevation, relative to the plane which contains the two incident laser beams, of $y = 30^\circ$ and an angle of rotation, relative to the plane of symmetry of the anemometer, of $\omega = 5.3^\circ$ and 1.0° (see figure 2 for the definition of these angles). The curve is predicted by the geometrical optics approximation and high frequency oscillations are evident . The direction of polarisation of the incident beams is normal to the plane which contains the two beams: the amplitude of the oscillations is larger if the direction of polarisation lies in this plane. If only refracted light were present, then the calibration curve would be the straight line shown in the same figure. When the calculation is repeated for both the geometrical optics approximation and Mie's solution, but now taking into account the finite size of the apertures on the collecting lens, the results are as shown in figure 3 (b). The oscillations largely disappear in both graphs and the "integrated" curves coincide with the straight line in figure 3 (a) because of the shapes we have chosen for the apertures of the detectors and because the reflected light is weaker than the refracted light. We have chosen rectangular apertures with dimensions $\delta\omega = 0.95^\circ$ and $\delta y = 9.5^\circ$ which correspond to 50 mm x 5 mm slits placed at 300 mm from the measuring volume of the anemometer. These are representative of the dimensions that we currently use in the laboratory. The oscillations which remain in the Mie calculation at Mie size parameters $\pi d/\lambda \equiv 180$, where d is the diameter of the particle and λ is the wavelength of light, are probably due to insufficient Gaussian integration because it is certain that this effect is absent in practical calibration curves.

These figures show that straight line calibration curves can be obtained in directions of about 30° relative to forward scatter, at least for particles larger than 10 μm and for practical apertures, and that the Mie and geometrical optics solutions are in close agreement. This give us confidence in our calculation methods and in the neglect of the ray path which contributes light in this interval after two internal reflections. The implications for the phase-Doppler technique are, first, that oscillations in the calibration curve are due to interference between two or more paths and that, secondly, we can expect qualitatively similar results over most of this scattering interval.

$\underline{82.8 < \theta < 111.6^\circ}$ The geometrical optics approximation shows that the light which occurs in this interval is due to external reflection only, since we have decided not to include light scattered by two internal reflections. If this is true, then we expect to see small oscillations in the calibration curve because there is but one path to this interval. Figure 4 compares the calculated phase from the Mie and geometrical optics solutions for

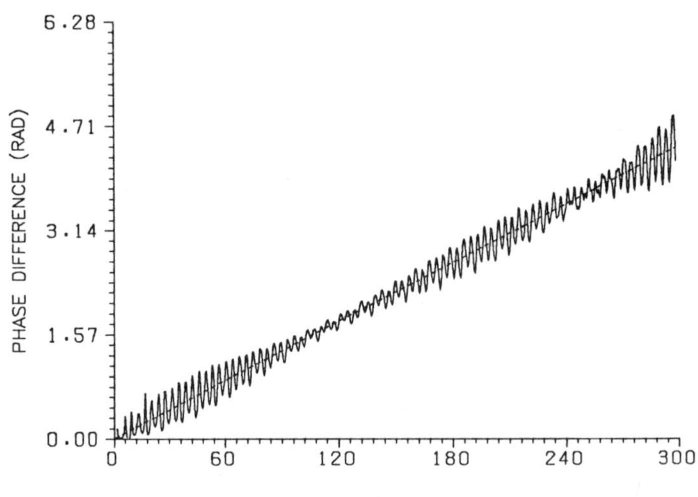

Figure 3(a) Oscillations in the calibration curve due to use of point apertures for a mean scatter angle of 30°. Solid line is full geometrical optics approximation; straight (dashed) line is the calibration curve due to refracted light only. Abscissa is Mie size parameter, $\pi d/\lambda$, and ordinate is in radians.

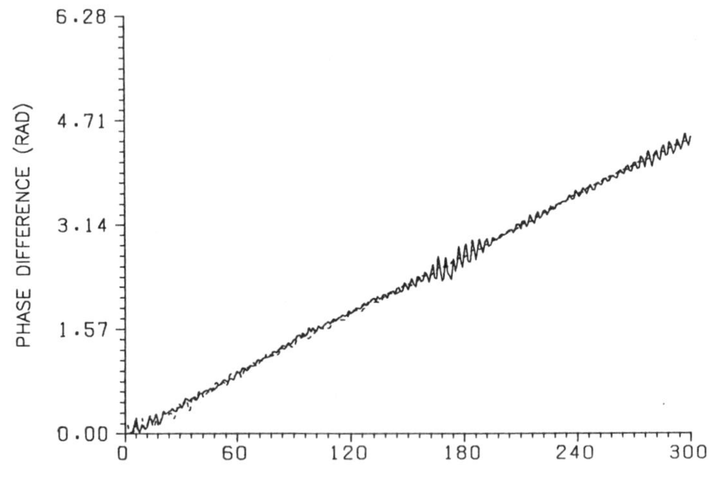

Figure 3(b) Reduction in the oscillations in figure 3(a) owing to inclusion of finite aperture size, 0.95° x 9.5°, in the calculations. Mie's solution is the solid curve; geometrical optics approximation is dashed curve.

one observer located at a mean scatter angle between the two beams of 100°, corresponding to $\omega = 178^\circ$ and $y = 80^\circ$. We have not calculated the phase difference for two observers because of the relatively large cost of evaluating the Mie solution and since we loose no vital information in doing so. For these reasons we have also not integrated over apertures since it is demonstrated above that the reduction in the amplitude of the oscillations can be anticipated by comparison with the prediction from a single path.

The discontinuous straight line is the calibration curve due to external reflection and the solid line is Mie's solution: the amplitude of the oscillations is indeed small and will be reduced when the integrating effect of the detector apertures is included. The direction of polarisation is unimportant for this location. The disadvantage associated with the collection of light scattered in this interval is that the change in phase with particle size is slow and it is therefore difficult to obtain good resolution. A commercially available system does use this interval but the necessary resolution is achieved by separating the photodetectors by relatively large angles which preclude mounting the detectors in a single, integrated unit.

$\underline{111.6 < \theta < 128.7^\circ}$ The paths here are, in principle, no different from those in the preceeding heading. However, figure 1 shows that the ray which undergoes two internal reflections gives rise to the second rainbow in this interval (see van de Hulst for a complete discussion) and, since rainbows contain appreciable energy and large interference effects, we expect that the calibration curve will be qualitatively different. Figure 5 confirms this expectation for a mean scatter angle of 120°: our geometrical optics solution remains a straight line with a small gradient but the Mie solution shows large amplitude oscillations around this line. Again, the effect of rotation of the direction of polarisation by 90° is qualitatively unimportant. It is unlikely that the integration effects of the apertures will damp the oscillations as effectively as in figure 3 (a + b) and, given a choice, the preceeding interval (82.8° to 111.6°) is more likely to result in a single-valued calibration curve.

$\underline{128.7 < \theta < 138.0^\circ}$ Geometrical rays arise in this region from external reflection or from those which undergo three, or more, internal reflections. Generally, the larger the number of internal reflections, the weaker is the emerging ray and it would be reasonable to expect that the calibration curve should be dominated by the effects of the externally reflected rays alone. Figure 6 shows the surprising result that, for a mean scatter angle of 133°, the amplitude of the oscillations is violent, although the "mean" trend is around the solution predicted from geometrical optics. Our explanation for this unexpected result is that this interval is narrow (9.3°) and, as a consequence, the scatter angle from each beam is close to either the main or second rainbow angles. As we have demonstrated already, the rainbow angles generate large oscillations in the calibration curves due to the associated interference patterns.

The rainbow angles are functions of the refractive index of the particle. The main and second rainbow angles for water are 138.0° and 128.7° respectively: for an index of 1.5 the

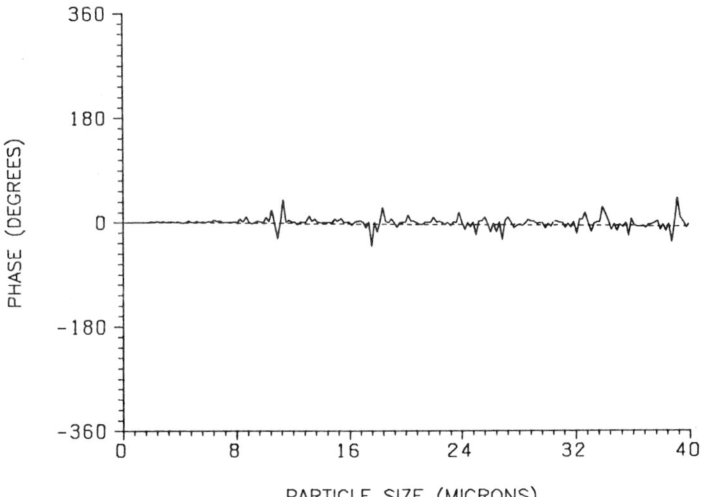

Figure 4   Variation of phase (degrees) with particle diameter (μm) at a mean scatter angle of 100° showing comparison between Mie's solution (solid line) and approximation based on externally reflected light only (dashed line).

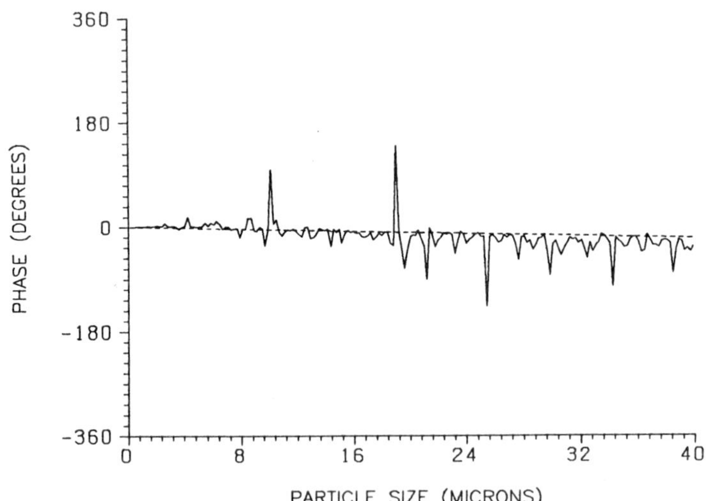

Figure 5   Variation of phase (degrees) with particle diameter (μm) in the second rainbow, mean scatter angle of 120°, showing comparison between Mie's solution (solid line) and approximation based on externally reflected light only (dashed line).

corresponding angles are 157.2° and 97.4°. The interval between the rainbows is much larger for the latter case and it is possible that the calibration curve is smoother than shown in figure 6. This aspect deserves more study than we have been able to devote to it here.

<u>138 < θ < 165.6°</u> This is the region which is occupied by the main rainbow and covers a range of angles which is potentially useful for measurements near backscatter. The paths which contribute to light scattered in this direction are external reflection and two separate branches from one internal reflection. Figure 7 (a) shows that both geometrical optics and the Mie solution predict both high frequency oscillations and large amplitude excursions over two separate ranges of size, between 4 and 8 μm and between 20 and 30 μm. It is not apparent why the sign of the phase for the first range is different for the two solutions but it is not of great practical significance because it is the amplitude of the oscillations which is important.

The effect of the external reflection on the calibration curve is shown by comparing figure 7 (b), which is due to internally reflected light only for the geometrical optics approximation, with figure 7 (a): the high frequency oscillations are now absent. The excursions between 4 and 8 μm and between 20 and 30 μm remain prominent, however, and must be due to interference between the two branches of light in the rainbow. The calibration curve is multi-valued and it is unlikely that integration over an aperture will remove this effect in the way in which the "high frequency" oscillations in figure 3 (a) were. The underlying structure of the calibration curve therefore precludes useful measurement for a mean scattering angle of 150°.

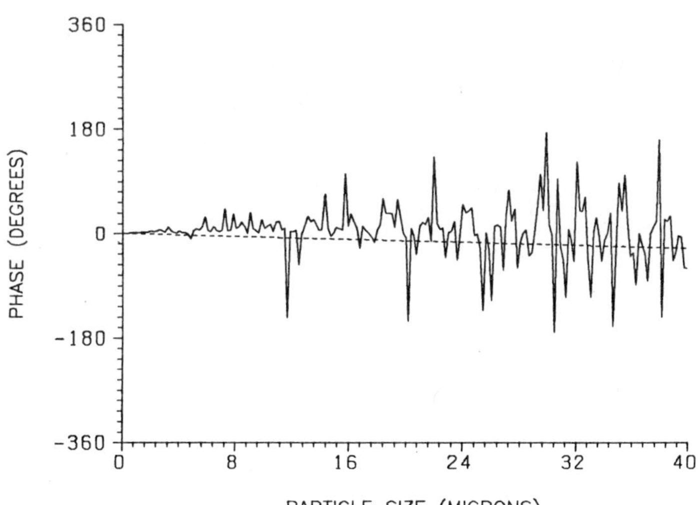

Figure 6   Variation of phase (degrees) with particle diameter (μm) between the second and main rainbow angles, mean scatter angle of 133°, showing comparison between Mie's solution (solid line) and approximation based on externally reflected light only (dashed line).

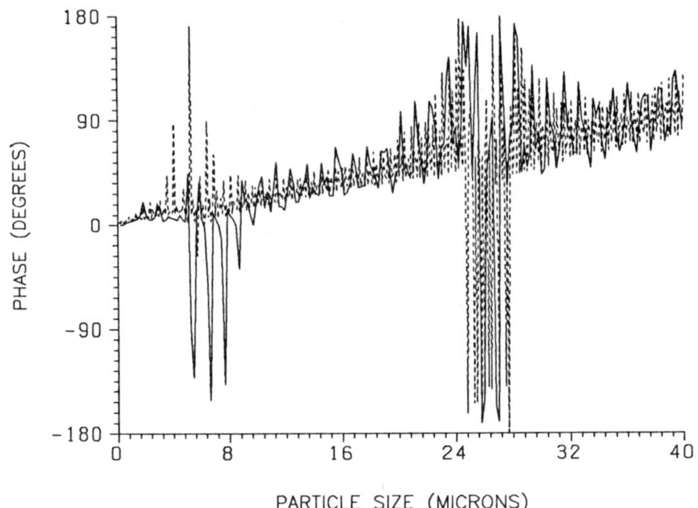

Figure 7(a) Variation of phase (degrees) with particle diameter (μm) within the main rainbow, mean scatter angle of 150°, showing comparison between Mie's solution (solid line) and geometrical optics approximation (dashed line).

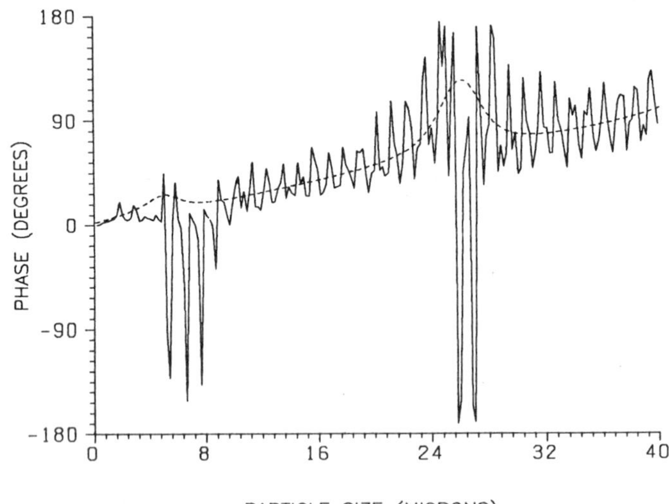

Figure 7(b) Comparison of Mie's solution (solid line) of figure 7(a) with approximation based on internally reflected light only (dashed line).

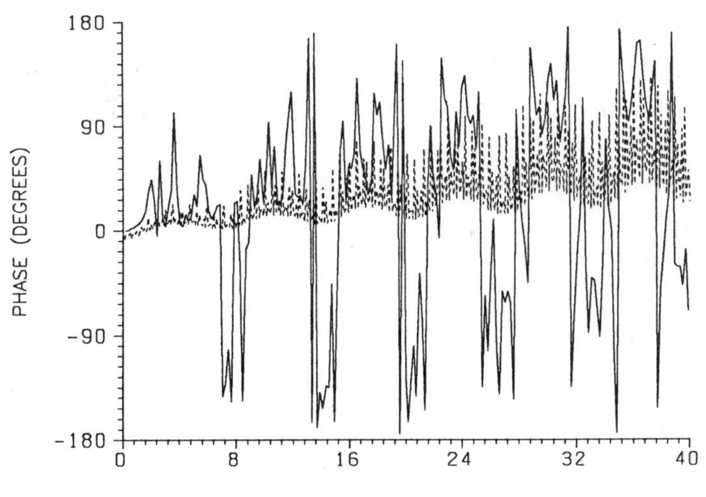

Figure 8(a) Variation of phase (degrees) with particle diameter (μm) at edge of main rainbow , mean scatter angle of 164°, showing comparison between Mie's solution (solid line) and geometrical optics approximation (dashed line).

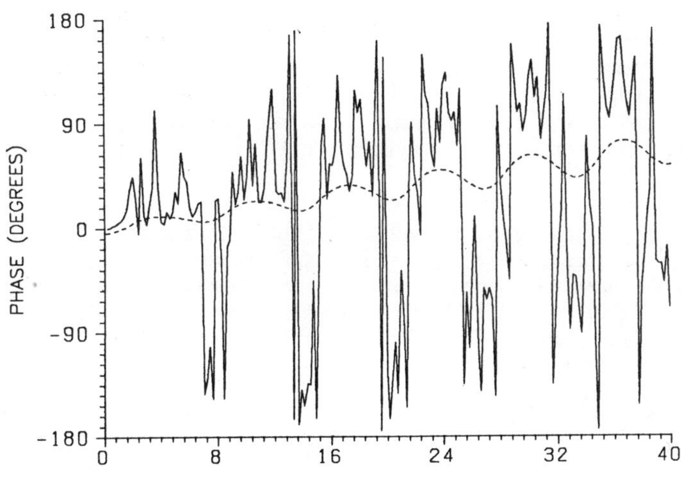

Figure 8(b) Comparison of Mie's solution (solid line) of figure 8(a) with approximation based on internally reflected light only (dashed line).

The strength of the externally relected light is weaker than the internally reflected light over a large part of the main rainbow, but for scatter angles greater than 158° this is no longer true for one branch of the internally reflected light (van de Hulst, 1981). We have, therefore, calculated the calibration curve for a scatter angle of 164° and the results are shown in figures 8 (a) and (b) for vertical polarisation; similar results are obtained for either polarisation direction. The Mie solution and the geometrical optics approximation both show large amplitude, high frequency oscillations but quantitatively the Mie solution gives the greater amplitude for these. When the effect of the externally reflected light is removed, figure 8 (b), it is evident that the internally reflected light gives rise to a quasi-periodic calibration curve with a period which matches the oscillations of the Mie theory well. The practical implication is that, even if the high frequency oscillations are removed through the intagration effect of the apertures of the detectors, the underlying rainbow pattern will remain and a single-valued calibration curve cannot be obtained. It is likely that the amplitude of the oscillations in figure 8 (b) will be smaller if the crossing angle of the incident beams is reduced. The penalty incurred is that the slope of the calibration curve is reduced and may ultimately lead to an unacceptably low resolution.

As a final note on this interval, we remark that if the refractive index of the particle were 1.5, than the main rainbow angle contains three branches, rather than the two shown in figure 1, and that thus the calibration curve corresponding to figure 8 (b) is likely to have higher frequency, larger amplitude oscillations.

$\underline{165.6 < \theta < 180°}$ In this interval it is known that the approximation used in geometrical optics is no longer valid because of the existence of so-called "edge rays" (van de Hulst) and we have not attempted to produce calibration curves based on the geometrical approximation. Figure 9 is the calibration derived from Mie's solution for a mean scatter angle of 175° and it is clear that no useful calibration curve is attainable in backscatter.

The work of Negus & Drain (1982) suggests that there are advantages to using circularly polarised incident light in establishing viable calibration curves for visibility in backscatter. It is likely that similar advantages can also be realised for the phase-Doppler technique and this is a promising direction for future work.

CONCLUSIONS

The following summarises our conclusions:

   1. This work compares the accuracy of calculations of the calibration curves for the phase-Doppler anemometer using the geometric optics approximation by reference to Mie's solution. The geometric optics approximation we have used considers

paths up to and including the first internal reflection for water droplets.

2. The geometric optics approximation agree quantitatively with Mie's solutions for all mean scatter angles except for locations at or between the main and second rainbow, where the agreement remains qualitatively correct and in the backscatter direction, where edge rays are important.

3. The finite aperture of the detector optics is important in damping high-frequency oscillations in the calibration curve and it is cheaper of computer time to do this with the geometric optics approximation than with Mie's solution.

4. We draw the tentative conclusion that it is not possible to establish a useful calibration curve for mean scatter angles larger than about 100° for water droplets, at least for crossing angles of 10° and linear polarisation of the incident beams.

5. We suggest that future work be directed towards establishing whether there are any improvement can be made through the use of circularly polarised light in backscatter directions.

ACKNOWLEDGEMENTS

Y. H. and A. M. K. P. T. wish to ackowledge the support of Professor J. H. Whitelaw, of the Department of Mechanical Engineering. This work has been made possible by financial support from SERC (S. A. M. A.), a SERC/CEGB co-funded grant (Y. H. & A. M. K. P. T.) and by the award of a Royal Society Research Fellowship (A. M. K. P. T.).

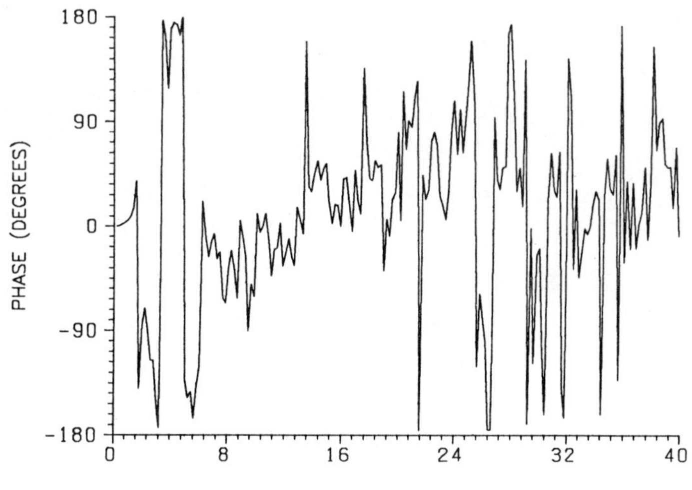

Figure 9   Variation of phase (degrees) with particle diameter (μm) from Mie's solution at a mean scatter angle of 175°.

REFERENCES

Bachalo, W. D. and Houser, M. J., 1984, Phase/Doppler spray analyzer for simultaneous measurements of drop size and velocity distributions, _Optical Engineering_, 23: 583-590.

Bauckhage, K. and Floegel, H. H., 1984, Simultaneous Measurement of droplet size and velocity in nozzle sprays. _2nd Intl. Symposium on Applications of Laser Anemometry to Fluid Mechanics_, Paper 18.1, Lisbon.

Drain, L. E., 1985, Laser Anemometry and particle sizing, _Intl. Conf. on Laser anemometry - advances and application_, Organised and sponsored by the Univ. of Manchester, Manchester, M13 9PL, Paper 12, pp 17-30.

Durst,F. and Zaré,M., 1975, Laser Doppler measurements in two phase flows, in: "_The accuracy of flow measurements by laser Doppler methods_", pp 403-429. Proceedings LDA Symposium, Copenhagen, P O Box 70, DK-2740 Skovlunde, Denmark.

Glanntschnig, W. J. & Chen, S.-H., 1981, Light scattering from water droplets in the geometrical optics approximation, _Applied Optics_, 20:2499-2509.

Hardalupas, Y., 1986, Phase-Doppler anemometry for simultaneous particle size and velocity measurements. Report FS/86/14, Fluids Section, Deptartment of Mechanical Engineering, Imperial College, London.

Hardalupas, Y. & Taylor, A. M. K. P., 1987, The identification of LDA seeding particles by the phase-Doppler technique, _Experiments in Fluids_ : to appear.

Hardalupas, Y., Taylor, A. M. K. P. & Whitelaw, J. H., 1986, Depth of field considerations in particle sizing using the phase-Doppler technique, _3rd Intl. Symposium on Applications of Laser Anemometry to Fluid Mechanics_, Paper 18.4, Lisbon.

Hong N. S. & Jones, A. R., 1978, Some aspects of light scattering in laser fringe anemometers, _J. Phys. D: Appl. Phys._, 11:1963-1967.

Jones, A. R., 1974, Light scattering by a sphere situated in an interference pattern, with reference to fringe anemometery and particle sizing, _J. Phys. D: Appl. Phys._, 7:1369-1376.

Lightfoot, N. S. and Negus, C. R., 1986, Comparison of the visibility and phase particle sizing techniques and their application to fuel sprays, _3rd Intl. Symposium on Applications of Laser Anemometry to Fluid Mechanics_, Paper 20.4, Lisbon.

Negus, C. R. & Drain, L. E., 1982, Mie calculations of the scattered light from a spherical particle traversing a fringe pattern produced by two intersecting laser beams, _J. Phys. D: Appl. Phys._ 16:375-402.

Saffman,M., Buchhave, P. and Tanger, H., 1984, Simultaneous measurement of size, concentration and velocity of spherical particles by a laser Doppler method, _2nd Intl. Symposium on Applications of Laser Anemometry to Fluid Mechanics_, Paper 8.1, Lisbon.

Ungut, A., Grehan, G. & Gouesbet, G., 1981, Comparisons between geometrical optics and Lorenz-Mie theory, _Applied Optics_, 20: 2911-2918.

van de Hulst, H. C.,1981,"Light scattering by small particles", Dover Publications, Inc., New York.

Yeoman, M. L., Azzopardi, B. J., White, H. J., Bates, C., J., Roberts, P. J., 1982, Optical development and application of a two colour LDA system for the simultaneous measurement of particle size and particle velocity, _in_ :_Engineering applications of laser velocimetry_, pp. 127-135, Winter annual meeting, ASME, Pheonix, Arizona.

PARTICLE SIZING OF POLYDISPERSE SAMPLES BY MIE-SCATTERING

Otto Glatter and Michael Hofer

Institut für Physikalische Chemie
Universität Graz, Heinrichstraße 28
A-8010   Graz, Austria

ABSTRACT

A procedure for the computation of size distributions of
polydisperse systems from elastic light scattering data is
presented. Precise solutions are obtained if the shapes of the
particles are known, if the size can be expressed by a single
parameter and if it is possible to calculate the shape factor
Φ (h,m,R) for the particles, as it is the case for spheres and
spheroids. The method is not restricted with respect to the
range of the refractive index m. However, wrong estimates for
this index lead to severe errors in the results. Approximate
solutions can be found for different globular particles by an
evaluation as a distribution of spheres. The range of applicability of the method depends on the experimental set-up, but is
in most cases in the size range from 100 nanometers to several
microns, i.e., it lies in the gap between X-ray or neutron
small-angle scattering and Fraunhofer diffraction. The inverse
scattering problem is solved with the modified Indirect Fourier
Transform method representing the solution as a series of
equidistant cubic B-splines. The regularization method incorporated in the procedure uses a stabilization parameter that
can be determined directly from the data. The solution is
nearly free from oscillations typical for ill-conditioned problems and it is largely independent of the actual number of
splines. A series of simulated experiments is used to show the
merits and limitations of the method.

INTRODUCTION

The possibilities of interpretation of elastic light scattering data in real space of monodisperse spherical systems[1]
and for monodisperse nonspherical and inhomogeneous systems[2]
have been shown recently. For a certain size range - depending
on the relative refractive index m and on the shape of the particles - it is possible to transform elastic light scattering
data into real space using the Indirect Fourier Transformation
method[3-6]. The resulting distance distribution function allows
a classification of the structure of the suspended monodisperse

particles[6-7]. The internal structure (radial polarization density profile) of inhomogeneous particles can be computed from the distance distribution function with a convolution square root technique, if the particles are close to spherical symmetry. All these methods can only be applied to monodisperse particles not smaller than 200 nm.

Many systems in the size range between 100 nm and several microns, like emulsions, show a distinct polydispersity in size. Assuming a certain shape - like spheres for emulsions - one is interested in a determination of the size distribution. Such an investigation is known as particle sizing. The most common optical methods for particle sizing are quasi-elastic light scattering[8-10] and Fraunhofer diffraction[11]. There exist several commercially available systems. The angular dependence of the scattered intensity is not taken into account with quasi-elastic light scattering - particle sizers with a fixed scattering angle of 90 degrees. Such instruments can be used with high accuracy only for particles smaller than a few hundred nanometers. Fraunhofer diffraction starts with a size of several microns if the refractive index is low. We will show in this paper that elastic light scattering experiments fill the gap between the two established methods. It works well in a size range between 100 nm and some microns even for low refractive index.

A similar method for particle sizing with X-rays and neutron small-angle scattering experiments has been developed a few years ago[12]. The angle dependent scattered intensity is the integral over the contributions from the different particles to the specified angle. The computation of the size distribution form the scattered intensity is called inverse scattering problem. A modified version of the Indirect Fourier Transformation method has been used to solves this problem for small-angle scattering data. With small-angle experiments it is possible to investigate particles in a size range from several nanometers up to about 100 nm. This size range is comparable to the one for quasi-elastic light scattering experiments. The main differences of the two methods are the facts that quasi-elastic light scattering experiments are much faster, but small-angle scattering experiments allow a higher resolution.

For elastic light scattering experiments we have to take into account the so-called Lorenz-Mie theory[13,14] in order to describe the dependence of the scattered intensity on the size and on the refractive index. With the Extended Boundary Condition Method it is possible to calculate these functions for spherical and nonspherical particles[15-19]. The solution of the inverse problem for light scattering data is impeded by severe truncation effects. These effects are determined by the experimental set-up (Θ_{min} and Θ_{max}) and by the wavelength of the light λ_o.

The possibilities and limitations for the determination of size distribution functions from elastic light scattering experiments have been investigated recently[20]. Data produced from known distributions by computer simulations are used to show the capabilities and limitations of the method.

THEORY

Size distributions can be calculated from polydisperse systems, if all particles have the same relative refractive index m (same material) and if the particles have the same shape but different size defined by one size parameter R. Approximative solutions are possible if the shapes differ or if the shape is not known precisely. Three different size distributions are common. The number distribution $D_n(R)$ gives the number of particles in the size interval R and R+dR, the distribution $D_m(R)$ is the mass or volume in the same interval and the intensity distribution $D_i(R)$ defines the scattering intensity at zero angle originating from the size interval. The scattered intensity of light from all particles I(h) and the size distributions are related by the following integrals:

$$I(h) = \int_0^\infty D_n(R)\, \Phi(h,m,R)\, dR \qquad [1]$$

$$I(h) = \int_0^\infty D_m(R)\, \Phi(h,m,R)\, \Phi(0,m,R)^{-1/2}\, dR \qquad [2]$$

and

$$I(h) = \int_0^\infty D_i(R)\, \Phi(h,m,R)\, \Phi(0,m,R)^{-1}\, dR$$

or

$$I(h) = \int_0^\infty D_i(R)\, \Phi_0(h,m,R)\, dR \qquad [3]$$

where h is the length of the scattering vector, $\Phi(h,m,R)$ is the scattering function of a particle with size R and refractive index m and

$$\Phi_0(h,m,R) = \Phi(h,m,R)\, \Phi(0,m,R)^{-1} \qquad [4]$$

is the same function but normalized to one at zero angle. The parameter R can be any characteristic dimension of the particle, in the case of spheres it is the radius. In a very rough approximation one can give an analytical relation between the different size distributions:

$$D_i(R) \cong R^3 D_m(R) \cong R^6 D_n(R) \qquad D_m(R) \cong R^3 D_n(R) \qquad [5]$$

This relation is valid only in the Rayleigh-Debye-Gans limit $2kR(m-1) << 1$, i.e., for small particles with very low refractive index. $\Phi(0,m,R)$ deviates from R^6 with increasing size and refractive index[21]. Eq. [5] can only be used for a qualitative understanding but not for quantitative computations. Elastic light scattering experiments should not be performed with a highly collimated beam in order to avoid a dependence of the scattering curve on the intensity distribution in the waist[22].

The integral equations [1-3] cannot be solved analytically because there exists no general analytical expression for $\Phi(h,m,R)$. Therefore it is impossible to find an inverse transformation as it is possible in the Rayleigh-Debye-Gans regime (for references see[12]). Numerical methods have to be applied to solve the inverse scattering problem. The Indirect Fourier Transformation method is a general approach that permits a large number of parameters to be estimated. As a priori information we use the assumption that the size distribution is a band-limited function, i.e.

$$D(R) \neq 0 \qquad \text{only for} \qquad R_{min} \leqq R \leqq R_{max} \qquad [6]$$

where R_{min} may be equal to zero. The effect of wrong band limits will be discussed in the next section.

The approach starts with a linear expansion for the size distribution D(R), omitting the index in D(R) means that there is no difference for the three types of distributions:

$$D(R) = \sum_{j=1}^{N} c_j \, \chi_j (R) \qquad\qquad [7]$$

where N is the number of basis functions $\chi_j(R)$. This number is usually not larger than 30. Cubic B-splines are used as basis functions[23-25]. They are defined as convolution products of a step function and represent smooth curves with a minimum second derivative. They overlap only with six neighbours as they differ from zero in the range of four knots. The distance of the knots defines the possible resolution of the distribution, the splines are interpolating between these knots. The expansion coefficients c_j are the unknowns.

Inserting the linear expansion [7] into the integral equation [1] to [3] leads to the approximation curve in the reciprocal space $I_A(h)$:

$$I_A(h) = \sum_{j=1}^{N} c_j \, \psi_j (h) \qquad\qquad [8]$$

where the $\psi_j(h)$ are the B-splines transformed into reciprocal space. Using Eq. [1] we get

$$\psi_j(h) = \int_0^\infty \chi_j (R) \, \Phi (h,m,R) \, dR \qquad\qquad [9]$$

Similar expressions can be found with Eqs. [2] and [3]. The coefficients c_j can be determined by the constrained least squares condition

$$L + N_{c'} = \sum_{i=1}^{M} [I_{exp}(h_i) - \sum_{j=1}^{N} c_j \, \psi_j (h_i)]^2 / \sigma (h_i)^2 +$$

$$+ \kappa \sum_{j=1}^{N-1} (c_{j+1} - c_j)^2 = Min \qquad\qquad [10]$$

where M is the number of data points and $\sigma(h_i)^2$ is the variance of the i-th point. The constraint $N_{c'}$ is necessary to avoid artificial oscillations. The optimum value of the stabilizing parameter κ can be determined directly from the point of inflexion method[3,4]. This is an important advantage of the procedure. In many algorithms for the solution of such illposed problems it is necessary to use additional a priori information, or the parameter must be selected by a weak quality condition for the solution.

Condition [10] leads to the so-called normal equations which can be solved using a conventional matrix inversion routine[26]. It is possible to calculate the error propagation, i.e., to state how the errors of the sampling points $I_{exp}(h_i)$ given by $\sigma(h_i)$ are transmitted to the coefficients c_j and to any linear combination of them[26,3].

The resolution of details in the size distribution function D(R) is limited by the distance of the knots of the B-splines DRB. The maximum number of coefficients, N, and the resolution DRB are limited by the termination effect (h_{min} and h_{max}) and by the statistical accuracy of the data $\sigma(h_i)$. There exists no analytical expression for the determination of the resolution. The sampling theorem of the Fourier transformation[27] can be used as a first approximation. The real limits can be estimated only with simulations as it will be shown in the next section.

The quality of the solution of the inverse problem can be improved in many cases by the additional constraint of non-negative solutions[28]. Such a constraint does not change essentially the resulting solution but avoids oscillations around the abscissa. Series of tests have to be performed before this constraint will be used in the standard evaluation routine. Other running activities concerning the use of singular value decomposition[28] did not yet lead to improved solutions for our inverse problem.

RESULTS

In this section we describe the most important results from a series of test calculations. We have performed all these simulations in the following way: starting from a given size distribution we calculate the corresponding scattering function in a certain h-range. Different limiting h-values can change the results a little, but the main features are not changed appreciably. Normally distributed noise of 1 to 3% has been added to the theoretical values in order to simulate the final data points. These simulated scattering functions have been transformed into real space using the Indirect Fourier Transformation method. The resulting size distributions have to be compared with the given theoretical distributions. All functions are plotted with arbitrary amplitude scaling factors.

Rayleigh-Debye-Gans Approximation

In a first test one can try to use the Rayleigh-Debye-Gans approximation for the scattering function of a sphere in Eqs. [1] - [3] in order to avoid the time consuming exact Mie-computations.

The corresponding tests show that the Rayleigh-Debye-Gans approximation cannot be used for polydisperse samples. Even in the case of m = 1.05 one gets artificial peaks at low R-values typical for such an approximation. The errors of the Rayleigh-Debye-Gans approximation increase with the m-value. This means that the correct expressions for $\Phi(h,m,R)$ have to be used in Eqs. [1] - [3].

Influence of the Refractive Index

It is evident from the above findings that the refractive index plays an important role for the evaluation of light scattering data from polydisperse samples. Our test example uses a bimodal distribution of spheres centered at R = 400 and 600 nm with a peak height ratio of 2:1 with a refractive index m = 1.2. The statistical accuracy of the data points is one percent. The evaluation in the size range $135 \leq R \leq 1050$ nm with the correct m-value of 1.2 leads to the full line in Fig. 1. The result is very close to the exact theoretical distribution. The results for slightly modified m-values (1.15 and 1.25) are shown as dashed lines in the same figure. The mean deviation

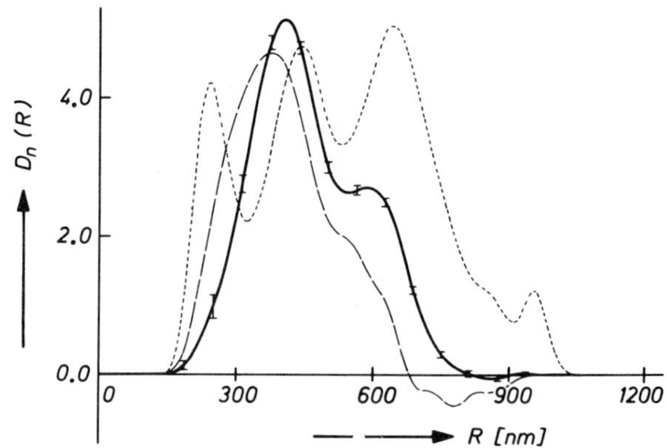

Fig.1   Bimodal Gaussian number distribution of spheres with
        m=1.2, centered at R = 400 and 600 nm with σ = 70.7
        and a peak height ratio 2:1. Full line: evaulation with
        the correct m-value. Dashed lines: evaluation with in-
        correct m-values (----- m=1.15, ——— m=1.25).

per point (MD) is equal to 1.09 for the right m-values, but we
get MD = 6.5 for m = 1.25 and MD = 1.36 for m = 1.15, i.e., the
deviations in the fit are larger if we try to evaluate with a
too low m-value. This corresponds to the results in Fig. 1,
where it can be seen clearly that the result for m = 1.25 is
better than the result for m = 1.15, but both results are far
from the correct solution. This example demonstrates that the
knowledge of the correct refractive index is very important
for the evaluation of elastic light scattering data from polydisperse samples.

Effect of Size Restriction

        The maximum value R_{max} for the size distributions is mainly
determined by the minimum scattering angle Θ_{min} or h_{min}. We
have restricted our calculations to $\Theta_{min} = 12°$ or $h_{min} =$
0.00321 nm^{-1}. We therefore used $R_{max} = 1050$ nm for all our
calculations. This value will be discussed later. The natural
limit for R_{min} is equal to zero. Let's look at a distribution
that has no significant contributions at low R-values like the
Gaussian in Fig.2 centered at R = 600 nm with σ = 100. This
function drops below 1% of the maximum near R = 300 and R =
900 nm. The solution from data with 1% error for $0 \leq R \leq 1050$ nm
is shown as a full line in Fig.2. This solution with MD = 1.0371
deviates from the Gaussian in the range $0 \leq R \leq 400$ nm and the
deviations are within the error bars. The solution for $180 \leq$
$R \leq 1050$ nm is shown as dashed-dotted line in the same figure.
It deviates from the full line only in the range $180 \leq R \leq$
350 nm. The quality of the fit is the same (MD = 1.0372), i.e.,
the size restriction does not eliminate any essential information
from input data. Such restrictions can help to reduce instabilities in the mathematical precedure. They are allowed only if
the distribution is equal or very close to zero in this range.
This is indicated by the fact that the MD-value is the same
for the restricted solution.

126

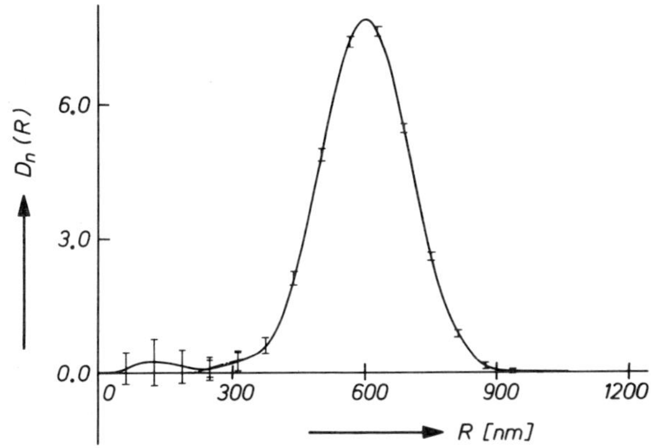

Fig.2   Effect of size restriction demonstrated with a Gaussian
         at R = 600 nm. Full line: result without restriction,
         dashed-dotted line: solution for 180 ≤ R ≤ 1050 nm.

Resolution

        The number of splines N and the distance between the knots
DRB defines the resolution of the distribution. A decreasing
distance DRB leads to increasing stability problems. How to
find the right number for N or DRB, how does the solution depend on these numbers? It has already been shown in the original
papers that the Indirect Fourier Transformation method can
produce a correct solution for a relatively wide range of N,
the number of basis functions in [7] . This number must be
large enough in order to guarantee a sufficient representation of the features of the distribution. If the number is too
small, the MD-value is high and can be lowered by an increase
of N. Once a good representation is reached, one can increase
N over a wide range without changing the result appreciably.
If we use a Gaussian like in Fig.2 we can use 10, 20 or 30
splines to fit the intensity distribution and we get results
which are identical within the graphical resolution of the
figure with MD-values varying between 1.05 and 1.004. If we
would use N >> 30 we would introduce crucial stability problems in most cases.

Small and Large Particles

        For many applications it is very important to know if it
is possible to find a very small number of large particles in
a suspension of many small particles. For this purpose we have
performed the following test. A Gaussian distribution of small
particles, centered at R = 150 nm with σ = 31.6 and height 10^4
and a second Gaussian at R = 600 nm with σ = 100 and height
1.0 are combined to a number distribution. The corresponding
results from data with 3% noise for 0 ≤ R ≤ 800 nm is shwon in
Fig.3. The Gaussian at 150 nm is well reproduced, the large
particles cannot be seen in this representation (amplitude
ratio 10^4:1!). No good (stable) solutions can be found for
R_{max} > 800 nm.
        Large particles produce a high scattered intensity. We
can therefore hope to find them in the intensity distribution.
The theoretical intensity distribution for our example is shown

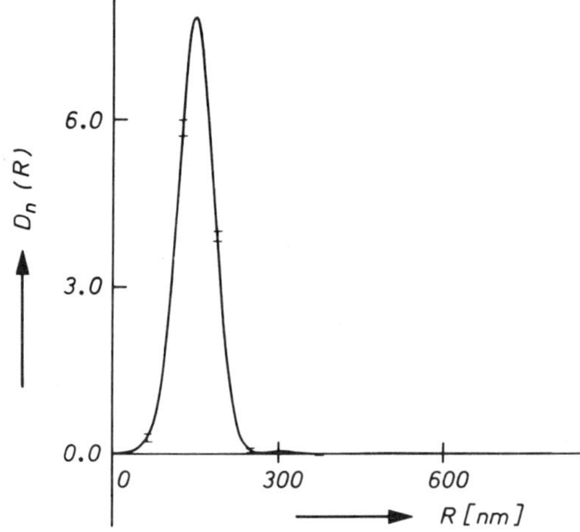

Fig.3 Bimodal Gaussian number distribution centered at R =
       150 nm with σ = 31.6 and height 10^4 and at R = 600 nm
       with σ = 100 and height 1. The full line shows the
       resulting distribution from data with 3% noise in the
       range 0 ≦ R ≦ 800 nm.

as a dashed line in Fig.4. The position of the maximum of the
large particles is at about 700 nm. This shift is caused by
the increasing weighting factor (see Eq.[5]). The solution
from the simulated scattering data as described in Fig.3 is
shown as full line in Fig.4. This result demonstrates that a
very small number of large particles (~100 ppm) can be found
in the intensity distribution. The error bars are large, but
the bimodal distribution is well resolved.

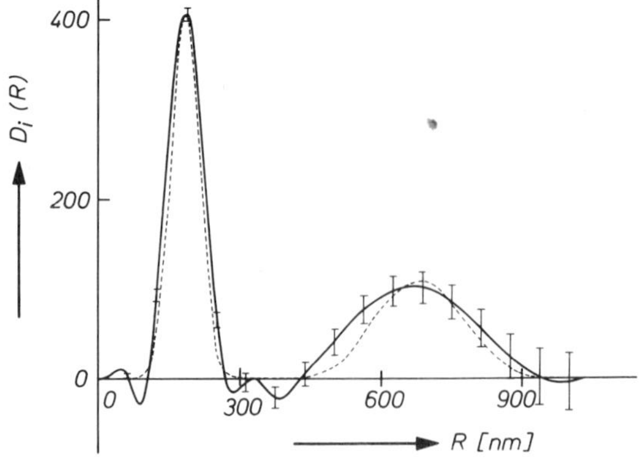

Fig.4 Intensity distribution calculated from the number dis-
       tribution in Fig.3. Dashed line: theoretical distribu-
       tion, full line: result from simulation with 3% noise.

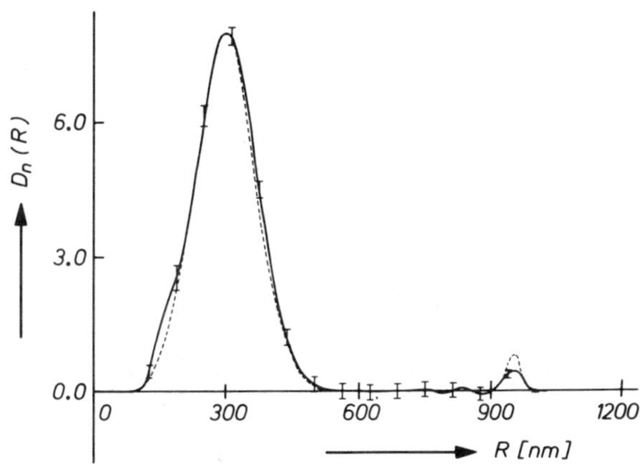

Fig.5   Theoretical bimodal Gaussian number distribution cen-
        tered at R = 300 nm with σ = 70.7, height 1.0 and at
        R = 950 nm with σ = 10, height 0.1 (dashed line) and
        the result from simulated data in the range 20° = Θ =
        150° with 1% noise for 90 $\leq$ R $\leq$ 1050 nm (full line).

Maximum Size

     There exists no clear sampling limit for the transforma-
tions [1] to [3] as it can be given in the case of the Fourier
transform[27]. An additional problem for practical applications
is the fact that the real limits for a value of R_{max} depend
highly on the accuracy of the data and not only on h_{min}. In
order to find the real limits one has to simulate the experimental situation (h_{min} and σ (h_i)) for a typical problem. In
Fig.5 we show the results of such an example out of a series
of test calculations. The simulated number distribution of
spheres consists of two Gaussians centered at R = 300 nm with
σ = 70.7 and height 1.0 and at R = 950 nm with σ = 10 and height
0.1. The scattering data have been simulated with an accuracy
of 1% in the range Θ_{min} = 20° (h_{min} = 0.0053) to Θ_{max} = 150°.
The theoretical distribution is shown as a dashed line in
Fig.5 together with the solution for 90 $\leq$ R $\leq$ 1050 nm as a full
line with error bars. The peaks are well resolved and at the
correct positions with the correct widths, but the height ratio
is not correct. Here we have the limiting situation that we can
see the existence of small and large particles but cannot
evaluate the correct number. From the Fourier sampling limit
we would get a size restriction to D_{max} = $2R_{max}$ = 600 nm, in
our example we have a D_{max}-value of about 2000 nm! The MD-value
of 0.95 indicates a good fit to the data. Even if we do not
know the precise factor relating the maximum dimension to the
inverse of h_{min}, the reciprocity between the two quantities
does exist. This means that we can increase the resolution
R_{max} if we decrease h_{min}.
     We have seen in Fig.3 that we can resolve a narrow distri-
bution with a relative width σ /R_{max} = 0.035. If the distributions are extremely narrow (like delta-functions) the solution
oscillates around the abscissa. These oscillations indicate
narrow distributions as they do not exist in the case of broad
peaks.

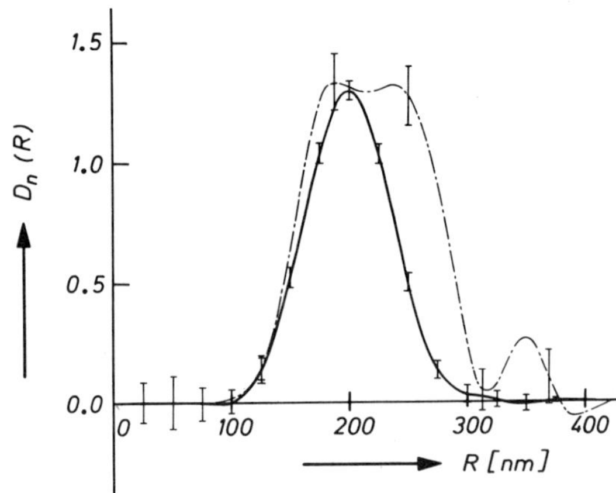

Fig.6   Gaussian number distribution of prolate ellipsoids
        (semi axes R,R and 2R) centered at R = 200 nm with
        σ = 35. The evaluation as a distribution of prolate
        ellipsoids for 0 ≤ R ≤ 400 nm gives the full line. An
        approximative solution as a distribution of spheres for
        70 ≤ R ≤ 800 nm is shown as a dashed-dotted line.

Nonspherical Particles

So far, tests were performed on spherical particles. As
already mentioned, the method is not restricted to spherical
particles. If particles of different size can be described by
a single size parameter R and if it is possible to calculate
the functions Φ (h,m,R), our procedure can be used to calculate the size distribution from the elastic light scattering
experiments.

The computation of the correct functions Φ (h,m,R) may be
sometimes impossible or at least very time consuming. In many
applications it is not of importance to have a very precise
information about the size, but it is only important to know
the average size of the particles, or it is important to know
if there exist different groups with different size (multi-
modal distributions).

Nonspherical globular particles can be evaluated as distributions of spheres. There exists a theoretical proof in the
Rayleigh-Debye-Gans limit that an ellipsoid can be represented
by a certain size distribution of spheres[29].

We take a distribution of prolate ellipsoids (m = 1.1,
minor semi-axis = R, major semi-axis = 2R). The distribution
is a Gaussian centered at R = 200 nm with σ = 35. The simulated
data have an accuracy of 1%. An evaluation in the range 0 ≤ R ≤
400 nm as prolate ellipsoids with the correct axial ratio of
1:2 gives the full line in Fig.6 which coincides well with the
theoretical profile. The evaluation of the same data set as
spheres in the range 70 ≤ R ≤ 800 nm results in the dashed-
dotted line. The peak is skewed to larger R-values and the
functions show decreasing oscillations around zero in the range
300 ≤ R ≤ 800 nm. The MD-value for both calculations is close
to one.

Fig.6 demonstrates that globular nonspherical particles
give reasonable distributions for spheres, i.e., it is not
necessary to know the correct shape of the particles in a poly-

disperse system. The more the shape deviates from the sphere,
the more the distribution is skewed and broadened, but peaks
do appear at the correct position.

CONCLUSION

Size distributions of polydisperse samples can be evaluated from elastic light scattering experiments if the distribution can be described by a single size parameter. The shape of
the particles have to be assumed or known. A priori information
from other sources is necessary for this assumption. A simultaneous determination of size and shape of polydisperse systems
is not possible with this method. On the other hand it is not
necessary to know the shape precisely. Different globular particles can be evaluated as a distribution of spheres. Experimental problems and first applications will be discussed in a
separate paper[30]. The most important facts are as follows.
The solution must be diluted to low concentration in order
to avoid interparticular interferences and multiple scattering.
The solvent (in most cases water) has to be produced free from
dust and micro air bubbles. The low refractive index of air
(m = 0.75) would give rise to strong contributions to the
scattering curve.
The size restriction of the method depends on the experimental set-up. The instrument should allow the measurement over
a wide angular range from a few degrees up to about 150°. The
use of a He-Ne laser as a light source (λ_0 = 632.8 nm) would
help to minimize h_{min}. The possible size range reaches in any
case from about 100 nm to some microns. The refractive index is
arbitrary, but it is important to use the correct value for
the evaluation.
Within the field of elastic scattering experiments the
method lies exactly in the gap between X-ray or neutron small-
angle scattering and Fraunhofer diffraction. Quasi-elastic
light scattering particles sizing methods have a partial overlap with our method, they are faster but the resolution is
lower. A comparative study of the elastic light scattering and
quasi-elastic light scattering particle sizing methods is in
preparation.

ACKNOWLEDGMENTS

This work was supported by the österreichischer Fonds zur
Förderung der wissenschaftlichen Forschung. The computation of
the shape factors for different Mie-scatterers were performed
with FORTRAN algorithms sent to us by Prof. P.W.Barber. We
have to thank him for this generous cooperation. We also thank
Prof. J.Schurz for the stimulation to do this work and to
Mrs. B.Mueller for the drawing of the figures.

REFERENCES

1. O. Glatter, M. Hofer, Ch. Jorde and W.-D. Eigner, Inter-
      pretation of Elastic Light Scattering Data in Real Space,
      J. Colloid Interface Sci., 105:577 (1985).
2. O. Glatter and M. Hofer, Interpretation of Elastic Light
      Scattering Data in Real Space II: Nonspherical and In-

homogeneous Monodisperse Systems, J. Colloid Interface Sci., In Press.

3. O. Glatter, Data Evaluation in Small Angle Scattering: Calculation of the Radial Electron Density Distribution by Means of Indirect Fourier Transformation, Acta Physica Austr., 47:83 (1977).

4. O. Glatter, A New Method for the Evaluation of Small-Angle Scattering Data, J. Appl. Cryst., 10:415 (1977).

5. O. Glatter, Evaluation of Small-Angle Scattering Data from Lamellar and Cylindrical Particles by the Indirect Transformation Method, J. Appl. Cryst., 13:577 (1980).

6. O. Glatter, Data Treatment and Interpretation, in: "Small Angle X-ray Scattering," O. Glatter and O. Kratky eds., Academic Press, London (1982).

7. O. Glatter, The Interpretation of Real-Space Information from Small-Angle Scattering Experiments, J. Appl. Cryst., 12:166 (1979).

8. B. Chu, "Laser Light Scattering," Academic Press, New York (1974).

9. B.E. Dahneke, "Measurements of Suspended Particles by Quasi-Elastic Light Scattering," Wiley & Sons, New York (1983).

10. R. Pecora, "Dynamic Light Scattering," Plenum, New York (1985).

11. L.P. Bayvel and A.R. Jones. "Electromagnetic Scattering and its Applications," Applied Science Publ., London (1981).

12. O. Glatter, Determination of Particle Size Distribution Functions from Small-Angle Scattering Data by Means of the Indirect Transformation Method, J. Appl. Crtyst., 13:7 (1980).

13. G. Mie, Beiträge zur Optik trüber Medien, speziell kolloidaler Metallösungen, Ann. Phys., 25:377 (1908).

14. M. Kerker, "The Scattering of Light and other Electromagnetic Radiation," Academic Press, New York (1969).

15. P.W. Barber and C. Yeh, Scattering of electromagnetic waves by arbitrarily shaped bodies, Appl. Opt., 14:2864 (1975).

16. P.W. Barber and D.-S. Wang. Rayleigh-Gans-Debye applicability to scattering by nonspherical particles, Appl. Opt., 17:797 (1978).

17. D.-S. Wang and P.W. Barber, Scattering by inhomogeneous nonspherical objects, Appl. Opt., 18:1190 (1979).

18. D.-S. Wang, C.H.H. Chen, P.W. Barber and P.J. Wyatt, Light scattering by polydisperse suspensions of inhomogeneous nonspherical particles, Appl. Opt., 18:2672 (1979).

19. P.W. Barber, D.-S. Wang and M.B. Long, Scattering calculations using a microcomputer, Appl. Opt., 20:1121 (1981).

20. O. Glatter and M. Hofer, Interpretation of Elastic Light Scattering Data in Real Space III: Determination of Size Distributions of Polydisperse Systems, J. Colloid Interface Sci., In Press.

21. M. Hofer "Analysis of Elastic Light Scattering from Preresonant Mie Particles in Real Space," Thesis, University Graz (1987).

22. C. Yeh, S. Colak and P.W. Barber, Scattering of sharply focused beams by arbitrarily shaped dielectric particles: an exact solution, Appl. Opt., 21:4426 (1982).

23. T.N.E. Greville, "Theory and Applications of Spline Functions", Academic Press, New York (1969).

24. J. Schelten and F. Hoßfeld, Applications of Spline Functions to the Correction of Resolution Errors in Small-Angle Scattering, J. Appl. Cryst., 4:210 (1971).

25. O. Glatter and H. Greschonig, Approximation of Titration

Curves and other Sigmoidal Functions by Proportionally spaced Cubic B-splines, <u>Microchim. Acta</u>, In Press.
26. S. Brandt, "Statistical and Computational Methods in Data Analysis," North-Holland Publ. Comp., Amsterdam (1970).
27. R. Bracewell, "Fourier Transform and its Applications," McGraw Hill, New York (1965).
28. C.L. Lawson and R.J. Hanson, "Solving Least Squares Problems," Prentice-Hall, Englewood Cliffs N.J. (1974).
29. P. Mittelbach and G. Porod, Zur Röntgenkleinwinkelstreuung verdünnter kolloidaler Systeme VII. Die Berechnung der Streukurven von dreiachsigen Ellipsoiden, <u>Acta Physica Austr.</u>, 15:122 (1962).
30. M. Hofer, J. Schurz and O. Glatter, Determination of Particle Size Distributions of Oil-Water Emulsions by Elastic Light Scattering, In Preparation.

OPTIMAL SCALING OF THE INVERSE FRAUNHOFER DIFFRACTION PARTICLE SIZING

PROBLEM: THE LINEAR SYSTEM PRODUCED BY QUADRATURE

E.D. Hirleman

Laser Diagnostics Laboratory
Mechanical and Aerospace Engineering Department
Arizona State University
Tempe, Arizona 85287

ABSTRACT

Solution of the linear system of equations obtained by discretization
and numerical quadrature of the Fredholm integral equation describing
Fraunhofer diffraction by a distribution of particles is considered. The
condition of the resulting system of equations depends on the
discretization strategy. However, the specific set of equations is shown
to depend on the discretization scheme used for the scattering angle domain
(the number, positions and apertures of the detectors) and for the size
domain (the number and extent of the discrete size classes). The term
scaling is used here to describe particular formulations or configurations
of the scattering angles and size classes, and a method for optimally
scaling the system is presented. Optimality is determined using several
measures of the condition (stability) of the resulting system of linear
equations. The results provide design rules for specifying an optimal
photodetector configuration of a Fraunhofer diffraction particle sizing
instrument.

INTRODUCTION

The near-forward scattering of light by particles large compared to
the wavelength λ is described by Fraunhofer diffraction theory. Under
those conditions the intensity $i(\theta)$ scattered at some angle θ in the far-
field is given by:

$$i(\theta) \ = \ i(0) \ \left(\frac{2J_1(\alpha\theta)}{\alpha\theta}\right)^2 \tag{1}$$

where α is the particle size parameter $\pi d/\lambda$, J_1 is the Bessel function of
first kind and first order, $i(0)$ is the on-axis ($\theta=0$) scattered intensity,
and the dimensions of the intensity i in Eq. (1) are power per unit solid
angle. Since the total diffracted energy must be equal to the energy
incident within the particle projected area, Eq. (1) can be integrated over
all scattering angles to determine:

$$i(0) \ = \ I_{inc} \ \frac{\lambda^2}{16\pi^2} \ \alpha^4 \tag{2}$$

where the on-axis scattered intensity i(0) has units w/sr and the incident intensity I_{inc} is in w/m^2. A distribution of particles, described by a probability density such that $n(\alpha)d\alpha$ is the number of particles between α and $\alpha+d\alpha$, would scatter as:

$$i(\theta) = \frac{I_{inc}\lambda^2}{16\pi^2} \int_0^\infty \alpha^4 \; \left(\frac{2J_1(\alpha\theta)}{\alpha\theta}\right)^2 \; n(\alpha)d\alpha \qquad (3)$$

The particle sizing problem requires that Eq. (3) be inverted, i.e. to provide information on the unknown particle size distribution $n(\alpha)$ using measurements of scattered light which can be related to $i(\theta)$. Now the number distribution $n(\alpha)$ is of not necessarily the function of interest; rather it may be some moment such as the area or volume distribution function that is desired. For example, in fuel spray combustion it is the distribution of volume (energy content) and area (surface area for evaporation) over droplet size which are most important. If the area distribution function $a(\alpha)$ was of interest, it could obviously be calculated after the fact from the solution $n(\alpha)$ obtained by the inversion process, but it is our objective here to consider the possibility that use of some moment of $n(\alpha)$ directly in Eq. (3) might improve the overall performance of the inversion. In fact, we propose to formulate Eq. (3) in the form:

$$i(\theta)\theta^a = C_1 \int_0^\infty \frac{J_1^2(\alpha\theta)}{\theta^{2-a}\alpha^{b-2}} \; \alpha^b n(\alpha)d\alpha \qquad (4a)$$

where a and b are variable scaling parameters of the inverse Fraunhofer problem. We also define some new variables related to the scaling of the problem:

$$i_a(\theta) = i(\theta)\theta^a \qquad (4b)$$

$$n_b(\alpha) = n(\alpha)\alpha^b \qquad (4c)$$

Further, we note that photodetectors do not in general measure $i(\theta)$ directly, but rather optical power which is intensity integrated over some finite detector aperture. A linear photodiode array with equal area detector elements (or a translating pinhole-detector assembly) would have an output equal to $i(\theta)$ to within a constant, or a=0 in Eq. (4a).

A kernel function k with scaling parameters is also defined:

$$k_{ab}(\alpha\theta) = \frac{J_1^2(\alpha\theta)}{\theta^{2-a}\alpha^{b-2}} \qquad (5)$$

so that the system Eq. (4a) then becomes

$$i_a(\theta) = \int_0^\infty k_{ab}(\alpha\theta)n_b(\alpha)d\alpha \qquad (6)$$

which is known as a nonhomogeneous Fredholm integral equation of first kind as introduced by the Swedish mathematician Fredholm in about 1900. Now our ability to accurately determine the particle size distribution will depend

on the performance of the numerical solution to Eq. (6), which, in turn,
depends on the exact form of the kernel function k and therefore the
scaling of the problem through the parameters a and b. Another artifact of
scaling the problem will be the particular discrete angles θ where
scattering data will be measured and the discrete sizes α where the size
distribution will be determined (here we consider interpolation between
discrete solution points as data analysis rather than as an inherent part
of the solution process).

The objective of this paper is to consider the optimal scaling of the
problem posed by Eq. (6) particularly with respect to methods which involve
solutions of a set of linear equations obtained by discretization and
numerical quadrature of Eq. (6). The dominant element of the quadrature
process is an instrument function matrix (typically n by n where n is the
number of optical sensors used) which requires inversion. We will study
the condition numbers and eigenvalues of the instrument matrices as a
function of scaling parameters over which the experimentalist has some
control. We assume that optimizing the properties of the instrument
function matrix will be useful regardless of the specific numerical
algorithm which might be used to obtain a final solution. Optimal
formulation of the problem will be considered from eigenfunction and
singular function points-of-view (i.e. where the integral equation (6) is
solved without invoking a discretization procedure) in subsequent
publications.

Inversion of Fredholm Integral Equations

There are a number of methods for inverting integral equations in
general, and Fredholm equations in particular. These can be divided into:
integral transforms, i.e. analytical inversion of Eq. (6) after Koo and
Hirleman (1986); reduction to a discrete linear system by numerical
quadrature after Twomey (1963) and Rust and Burris (1972) ; and methods
using functional analysis involving expansion of both the measured and
unknown functions in terms of eigenfunctions after Bertero and Pike (1983),
singular functions after Bertero at al (1985), or Fourier series after
Lanczos (1964). In this paper we consider the solution of the linear
system obtained by numerical quadrature, i.e. solving the linear system of
m equations in n unknowns (m $\geq$ n) obtained by discretizing $i_a(\theta)$ into m
discrete values and approximating the integral in Eq. (6) by quadrature.
First discretizing in θ obtains:

$$i_a(\theta_1) = \int_0^\infty k_{ab}(\alpha\theta_1)n_b(\alpha)d\alpha$$

$$i_a(\theta_2) = \int_0^\infty k_{ab}(\alpha\theta_2)n_b(\alpha)d\alpha \tag{7}$$

$$\cdots \cdots \cdots \cdots \cdots \cdots$$

$$i_a(\theta_m) = \int_0^\infty k_{ab}(\alpha\theta_m)n_b(\alpha)d\alpha$$

Now, there is no quadrature error in arriving at Eqs. (7), i.e. any
solution $n_b(\alpha)$ to Eq. (6) will satisfy Eqs. (7) exactly. There may be,
however, many additional functions $n_b(\alpha)$ which satisfy Eqs. (7) but are not
solutions to Eq. (6).

Since photosensors in general integrate over a finite aperture, actual measurements will be averages (possibly weighted over some region of θ in the neighborhood of some representative angle θ_i, such that:

$$i_a(\theta_i) = \int_0^\infty i_a(\theta)w_t(\theta,\theta_i)\,d\theta \tag{8}$$

where $w_t(\theta,\theta_i)$ is a weighting or windowing function which describes the photodetector aperture. Here, w_t is normalized such that the integral of $w_t\,d\theta$ is 1.0. Then for the ith detector,

$$i_a(\theta_i) = \int_0^\infty w_t(\theta,\theta_i)\,d\theta \int_0^\infty k_{ab}(\alpha\theta)n_b(\alpha)\,d\alpha \tag{9}$$

For example, a linear photodiode array with equal area pixels would give a response proportional to $i(\theta)$, or a = 0 in Eq. (6). If each pixel subtended an angle represented by $\Delta\theta$, then the weighting function for uniform responsivity elements with no edge effects would be rectangle functions:

$$w_t(\theta,\theta_i) = \begin{cases} 1, & |\theta-\theta_i| \le \Delta\theta/2 \\[2mm] 0, & |\theta-\theta_i| > \Delta\theta/2 \end{cases} \tag{10}$$

Annular ring detector elements with equal $\Delta\theta$ spaced elements would respond as $i(\theta)\theta = i_1(\theta)$, or a=1, with the same window functions as given in Eq. (10).

Now, a similar discretization can be performed on the continuous variable α. Defining α_j as a size representative of the jth discrete size class, then the weighting function $w_s(\alpha,\alpha_j)$ to be used in quadrature is such that

$$i_a(\theta) \simeq \sum_{j=1}^{n} \left[\int_0^\infty w_s(\alpha,\alpha_j)k_{ab}(\alpha\theta)\,d\alpha\right] N_b(\alpha_j) \tag{11}$$

where the integral of $w_s(\alpha,\alpha_j)\,d\alpha$ is normalized to 1.0 for all j and N_b is representative of the bth moment of the total number of particles in the jth size class such that $N_b(\alpha_j)$ is the integral of $n_b(\alpha)\,d\alpha$ over the neighborhood of α_j. Equation (11) is an approximation, and the error in the approximation (i.e. the quadrature error) increases as the actual distribution function $n_b(\alpha)$ within the jth size class range diverges from the assumed form for $w_s(\alpha,\alpha_j)$. The quadrature error will, in general, vary inversely with n. Integration of Eq. (11) over the detector aperture results in:

$$i_a(\theta) \simeq \int_0^\infty w_t(\theta,\theta_i)\,d\theta \left(\sum_{j=1}^{n}\left[\int_0^\infty w_s(\alpha,\alpha_j)k_{ab}(\alpha\theta)\,d\alpha\right]\right) N_b(\alpha_j) \tag{12}$$

which can be rearranged:

$$i_a(\theta_i) \simeq \sum_{j=1}^{n} \int_0^\infty \int_0^\infty w_t(\theta,\theta_i)[w_s(\alpha,\alpha_j)k_{ab}(\alpha\theta)d\alpha]d\theta \; N_b(\alpha_j) \tag{13}$$

From which we identify a representative kernel matrix element $k_{ab}(\theta_i,\alpha_j)$ as:

$$k_{ab}(\theta_i,\alpha_j) \equiv \int_0^\infty \int_0^\infty w_t(\theta,\theta_i)w_s(\alpha,\alpha_j)k_{ab}(\alpha\theta)d\alpha d\theta \tag{14}$$

The system of equations can then be written in matrix form as:

$$I = K\ N \tag{15}$$

where I is a vector of length m with elements $i_a(\theta_i)$, N is a vector of length n with elements $N_b(\alpha_j)$, K is an m×n matrix with elements defined by Eq. (14) and I, K and N depend on the scaling parameters a and b but the subscript notation has been dropped. Now, the performance of the particle sizing procedure reduces to solution of the linear system Eq. (15). Clearly the properties of the matrix K, in particular the condition number, determine the performance. The condition number of the instrument function matrix K is, in turn, determined by the scaling and formulation of the problem. The independent variables which affect K are: the scaling powers a and b from Eq. (2); the positions of the discrete values of θ_i and α_j from Eqs. (7) and (11); and the weighting functions w_s and w_t from Eq. (13). The problem of optimally scaling the inverse Fraunhofer diffraction is then to select these independent instrument variables to optimize the information obtained from the measurements.

For our purposes here, we consider values of the following parameters or functions to be known constraints:

1.   D,   the dynamic range or maximum bandwidth of the scattering angles covered by the sensors, i.e. $D=\theta_m/\theta_1$; and

2. e_i, the noise characteristics of the measurements of $i_a(\theta_i)$.

The design variables are:

1. w_t, the aperture windowing function of the detectors;

2. w_s, the weighting function for contributions over the finite bandwidth of the discrete size classes;

3.   m, the number of detectors;

4.   the scaling law which dictates the positions θ_i of the detectors;

5.   n, the number of independent size classes;

6.   the scaling law which dictates the spacing of the α_j;

7.   a, the moment of $i(\theta)$ measured;

8.   b, the moment of $n(\alpha)$ used in the solution.

Intuition might suggest that increasing the number of angles m at which scattering measurements are made would always increase the amount of information obtained. While it is true that the maximum number of independent sizes n at which $N(\alpha_j)$ could be determined has an upper limit of m(number of equations $\geq$ number of unknowns in Eq. (15)), measurement errors put another bound on the number of useful independent measurements and size distribution determinations. The presence of measurement errors can be included as:

$$I = KN + E \qquad (16)$$

where elements e_i of the m-vector E describe the measurement error in each value of I. The addition of E in Eq. (16) elucidates the inherent ill-posed nature of Eq. (1) as manifested by the fact that very different solution vectors N may satisfy Eq. (16) to within the unknown error components e in the measurements. On one hand one would like to increase the degrees of freedom in Eq. (16) to increase the information extracted about N, but the ill-posed natures of the problem will increase the instabilities and therefore the uncertainty in measured N as n increases. Optimal formulation involves balancing these competing demands.

The solution of Eq. (16) in the most straightforward form would involve obtaining the inverse of K, but unfortunately for reasonable numbers of degrees of freedom K becomes very ill-conditioned. The presence of the error vector E also complicates the problem over conventional solution of a system of linear equations. The error term generally means that many vectors N solve Eq. (16) to within the noise, and hence solution of Eq. (16) is generally posed as searching among these solutions N for one that is best in some sense. Methods for the solution are discussed in detail by Twomey (1963) and Rust and Burris (1972). In this paper we will not concern ourselves with the specific algorithm used to solve Eq. (16), but rather will consider optimizing the K matrix so that the performance of whichever method is used will be optimized. We consider the condition number of K to be the objective function which should be minimized.

THE CONDITION OF A MATRIX

Particle size analysis in the present context comprises solution of Eq. (16), i.e. determination of N from measured I and known K. if the experimental errors were zero, then the problem would simply reduce to finding the inverse of the K matrix. In the presence of noise, there will generally be a large number of vectors N which satisfy Eq. (16) to within the uncertainty E in I. The problem then becomes selection of one of the many "solutions" of Eq. (16) which is best in some sense, but it is still the properties of k which determine the magnitude of the differences between the ensemble of N vectors which are solutions. Instability, or the property of being ill-posed, is related to the desire that the solution be relatively insensitive to small changes (e.g. experimental error) in the measurement vector. In other words, we are interested in the proportionality constant c for:

$$\frac{\|\Delta N\|}{\|N\|} = c \frac{\|\Delta I\|}{\|I\|} = c \frac{\|E\|}{\|I\|} \qquad (17)$$

where $\|$ indicates some vector norm, for example the Holder norm defined as:

$$\|X\|_1 \equiv \sum_{i=1}^{n} |X_i| \qquad (18)$$

Now, if a small relative change ΔI in the measured vector I gives a large ΔN, then the solution would be unstable. The primary manifestation of the instability would be that a large array of very different N solution

vectors would satisfy the system equations to within the experimental
error. Now, we can also write:

$$I + \Delta I = K(N + \Delta N) \tag{19}$$

Further, consider the extrema of the quantity $\|KN\|$ which would give
the largest error, i.e. a worst case relative error would occur, since
$\Delta I = K\Delta N$ where

$$\frac{\|K\Delta N\|}{\|\Delta N\|} = \text{a minimum} \tag{20}$$

and

$$\frac{\|KN\|}{\|N\|} = \text{a maximum} \tag{21}$$

which can be combined and expressed as

$$\text{cond}(K) \geq \frac{\|\Delta N\|}{\|N\|} \cdot \frac{\|I\|}{\|\Delta I\|} \tag{22}$$

where:

$$\text{cond}(K) = \frac{\max \|KN\|}{\min \|KN\|} \quad , \quad \|N\| = 1 \tag{23}$$

Note that the value of the condition number of the inverse diffraction
matrix K depends on the specific norm used, but for particle sizing the
Holder 1—norm in Eq. (18) is useful, such that

$$\|X\|_1 = \sum_{i=1}^{n} |X_i| = 1 \tag{24}$$

where the absolute value can be dropped in Eq. (24) since particle
populations must be positive. The condition number can be calculated when K
is nonsingular from:

$$\text{cond}(K) = \|K^{-1}\| \, \|K\| = C_1 \tag{25}$$

and Forsythe et al (1977) give an algorithm for estimating the r.h.s. of
Eq. (25) which does not require explicit determination of the inverse of
the K matrix. Other arguments could be made which would indicate that the
ratio of the largest to the smallest eigenvalues or singular values of a
matrix are good indicators of the condition:

$$\text{cond}(K) \geq \frac{\max \left| \lambda_i(K) \right|}{\min \left| \lambda_i(K) \right|} \tag{26}$$

where λ_i are the eigenvalues of the matrix K.

RESULTS AND DISCUSSION

For our further purposes here we assume a square linear system, such
that the number of sensors m equals the number of independent size classes
n. For a given set of design parameters, the cond(K) will increase with n
as shown in Fig. 1. The maximum allowable condition number will depend on
the uncertainty in N which can be tolerated, and hence there will be some
maximum number of degrees of freedom n which can be supported. For
example, if the noise to signal ratio given by $\|E\|/\|I\|$ was 1%, and a laser

diffraction system matrix **K** had a condition number of 10, then the relative
uncertainty in the elements of the solution **N** would be approximately 10
times 1% or 10%. Clearly, for any desired level of uncertainty in **N**,
minimizing the condition number of **K** will optimize the instrument system.
For that reason we are interested in the behavior of the cond(**K**) for given
n as the scaling or design variables a, b, w_t, w_s, sensor positions, and
size class positions are varied.

A standard approach for any sensor geometry would be to specify and
associate each size class with one sensor such that:

$$\theta_i \alpha_{n+1-i} = \beta_{opt} \qquad (27)$$

where θ_i and α_i are representative of the ith detector and size class
respectively, such that $k(\theta_i \alpha_{n+1-i})$ from Eq. (5) is a maximum. With this
assumption, only the sensor positions are variable, subject to the assumed
detector bandwidth D. We consider nonoverlapping detectors (as would be
the case for a monolithic photodiode array), and two designs where the
detectors are evenly distributed in either θ or $\log\theta$ space. For a linear
spacing,

$$\theta_{i+1} = \theta_i + \Delta\theta \qquad (28)$$

and for log spacing:

$$\theta_{i+1} = \delta\theta_i \qquad (29)$$

where δ is a constant.

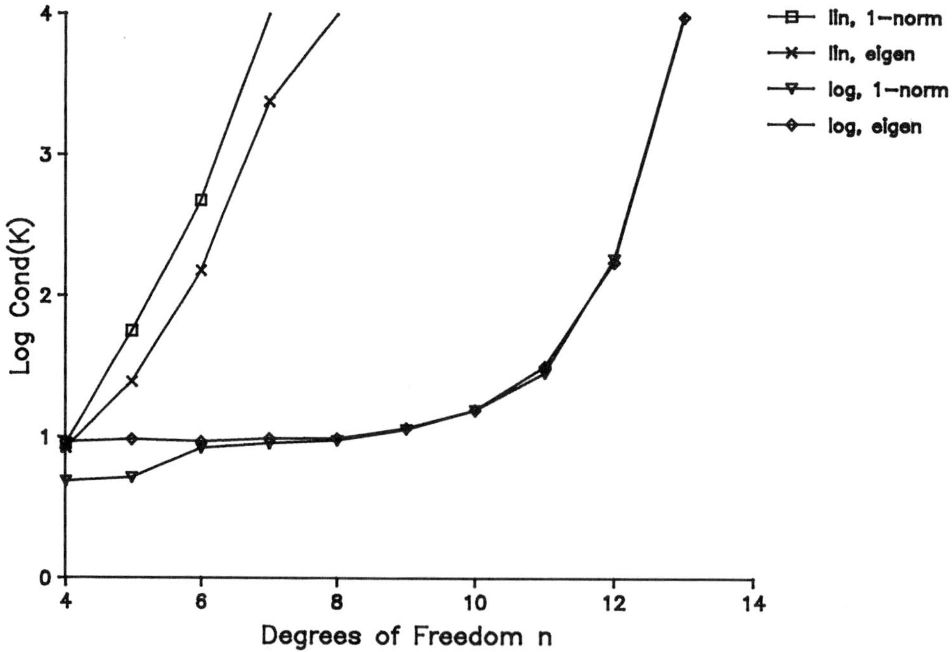

Figure 1. Plot of the condition number of the Fraunhofer diffraction
matrix **K** for linear- and log-scaled detectors with bandwidth D = 100,
scaling parameters a = 2 and b = 2, and rectangular weighting functions w_t
and w_s. The condition numbers were calculated based on the 1-norm using
Eq. (25) and the eigenvalue extrema from Eq. (26).

As demonstrated in Fig. 1, log spacing gives smaller condition
numbers than linear spacing, and this behavior was observed for all values
of the remaining design parameters we studied. This detector configuration
was also studied by Bertero et al (1985), and an analogous sampling scheme
was found to advantageous for quasi-elastic light scattering methods by
Ostrowsky et al (1981). For that reason we consider only log-spaced
detectors, and now focus on the implications of some special scaling laws.
First, if the parameters a and b are chosen subject to the condition:

$$a + b = 4 \tag{30}$$

we note that the kernel function k in Eq. (5) is symmetric where
$k(\alpha,\theta) = k(\theta,\alpha)$. Further, if the detector aperture function is specified
such that:

$$w_t(\theta,\theta_i) = w_t(\theta/\theta_i) \tag{31}$$

and the quadrature weighting function such that:

$$w_s(\alpha,\alpha_j) = w_s(\alpha/\alpha_j) \tag{32}$$

then matrix K will be symmetric such that elements are related by
$k_{ij} = k_{ij}$. Another very important result of the scaling laws comprised of
Eqs. (27,29,30,31, and 32) is that K has the very unique Toeplitz or Hankel
form (see Appendix B). This means that there are only 2n-1 independent
elements in K, and elements on all the cross-codiagonals are recurrent.

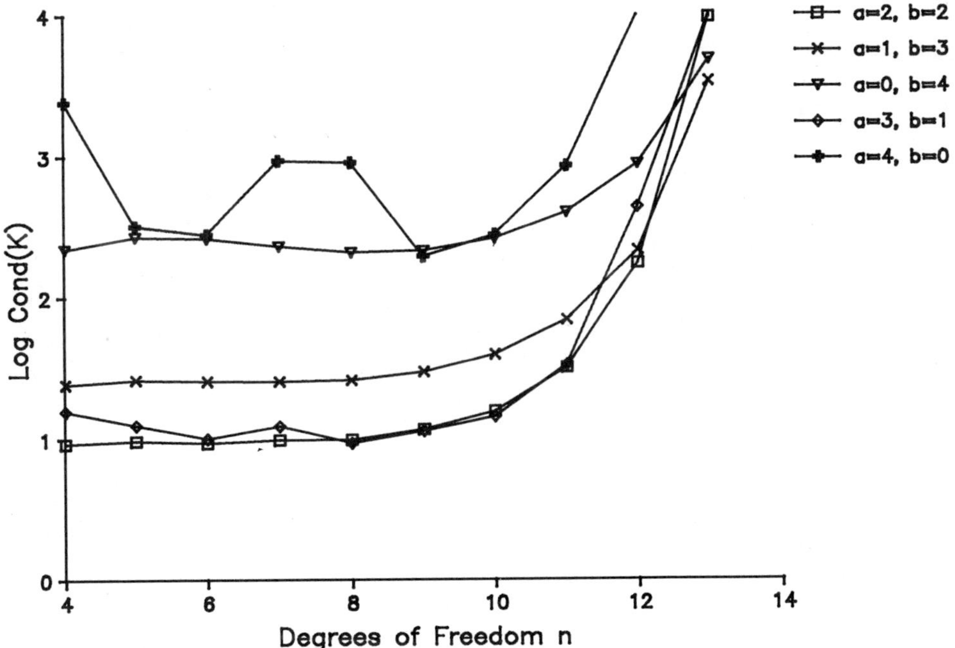

Figure 2. Plot of the condition number of the Fraunhofer diffraction
matrix K for log-scaled detectors with bandwidth D = 100 and rectangular
weighting functions w_t and w_s for the indicated scaling parameter values.
The condition numbers were calculated based on the eigenvalue extrema from
Eq. (26).

Scaling a laser diffraction instrument to obtain Toeplitz form of the instrument function matrix has some obvious advantages. For example, a Toeplitz matrix requires $O(n)$ storage elements as opposed to $O(n^2)$, and as discussed by Kohar (1969) and Bunch (1985) a Toeplitz linear system can be solved with $O(n^2)$ operations as compared with $O(n^3)$ for a general system. However, of interest here is the stability of the inversion as indicated by the condition number. Fig. 2 shows the condition numbers for several combinations of a and b, and two things are clear. First, combinations of a and b satisfying Eq. (30) give systems of linear equations with the best condition numbers. Secondly, the specific configuration where:

$$a = b = 2 \tag{33}$$

is optimal.

For the case $b = 2$, the solution vector is the second moment of the number distribution, or the area distribution function. Similarly, for $a = 2$, the second moment of the intensity is the measurement vector to be used in the inversion, and it is seen from Appendix A that a log-scaled annular ring detector will give signals which approximate $i_2(\theta) = i(\theta)\theta^2$ averaged over the finite width detector. In that case, as discussed above, elements on the cross diagonal are equal and are the maximum value of all elements. Further, the sum of any column of K in that case represents the fraction of the total energy diffracted by a particle which is collected by the detector array, and hence are of order 1.0. Therefore, when the solution vector is on an area basis and log-scaled ring detectors are used, the K matrix is inherently equilibrated. This is not surprising since particles diffract light on a projected area basis.

CONCLUSIONS

Design rules for achieving optimal scaling of the inverse Fraunhofer diffraction particle sizing problem obtained by discretization and numerical quadrature have been derived. The optimal formulation, in the sense that the stability of the linear equation solution is maximized, is best obtained by log-scaled annular ring detectors with the particle area distribution function as the solution vector. Additional advantages of utilization of the scaling laws presented here include the fact that an n×n instrument function matrix has only $O(n)$ unique elements and can be inverted with $O(n^2)$ operations.

REFERENCES

Bertero, M. and Pike, E.R., 1983, 'Particle size distributions from Fraunhofer diffraction: an analytic eigenfunction approach", *Optica Acta*, 30:1043-1049.

Bertero, M., Boccacci, P. and Pike, E.R., 1985, "Particle size distributions from Fraunhofer diffraction: the singular-value spectrum", *Inverse Problems*, 1:111-126.

Bunch, J.R. and Parlett, B.N., 1971, "Direct methods for solving symmetric indefinite systems of linear equations," *SIAM J. Numer. Anal.*, 8:639-55.

Bunch, J.R., 1985, "Stability of methods for solving Toeplitz systems of equations,", *J. Assoc. Comp. Mach.*, 6:349-64.

Forsythe, G.E.. Malcolm, M.A. and Moler, C.B., 1977, *Computer Methods for Mathematical Computations*, Prentice-Hall, Inglewood Cliffs, NJ.

Koo, J.K. and Hirleman, E.D., 1986, "Comparative study of laser
     diffraction analysis using integral transform techniques: factors
     affecting the reconstruction of droplet size distributions," Paper
     86-18, Spring Meeting, Western States Section, Combustion Institute,
     April 28, 1986, Banff, Canada.

Lanczos, C., 1964, *Applied Analysis*, Prentice Hall, Englewood Cliffs, NJ.

Ostrowsky, N., Sornette, D., Parker, P., and Pike, E.R., 1981,
     "Exponential sampling method for light scattering polydispersity
     analysis," *Optica Acta*, 28:1059-70.

Rust, B.W. and Burris, W.R., 1972, *Mathematical Programming and the the
     Solution of Linear Equations*, Elsevier, New York.

Twomey, S., 1963, "On the numerical solution of Fredholm integral
     equations of the first kind by the inversion of the linear system
     produced by quadrature", *J. Assoc. Comput. Mach.*, 10:97-101.

Zohar, S., 1969, "Toeplitz matrix inversion: the algorithm of W. F.
     French,", *Int. J. Systems Sci.*, 9:921-34.

ACKNOWLEDGMENTS

This research was sponsored by the Air Force Office of Scientific
Research, Air Force Systems Command, USAF, under Grant Number
AFOSR-84-0187, Dr. Julian Tishkoff, program manager. The U.S. Government
is authorized to reproduce and distribute reprints for Governmental
purposes notwithstanding any copyright notation thereon.

APPENDIX A: WEIGHTING FUNCTION MODELS FOR SOME DETECTOR GEOMETRIES

Since scattering measurements for Fraunhofer diffraction particle
sizing are generally made in the back focal plane of a transform lens, we
need the equation $r=f\theta$ which relates scattering angle θ to the radial
position r in the focal plane through the focal length f. Thus an aperture
of radius r in the transform plane subtends a solid angle $\pi\theta^2=\pi r^2/f^2$.

Now for spherical particles the diffraction pattern is axisymmetric
and the solid angle subtended by an annular detector at θ of width $d\theta$ where
$d\theta$ is small and second-order terms in $d\theta$ are neglected is:

$$\Omega = \pi(\theta+d\theta)^2 - \pi\theta^2$$

$$= \pi\theta^2 + 2\pi\theta d\theta + \pi(d\theta)^2 - \pi\theta^2$$

$$= 2\pi\theta d\theta \tag{A.1}$$

So we see that the transduced signal S_{lin} from the differential (linearly
spaced) annular rings with finite $\Delta\theta$ spacing is:

$$S_{lin}(\theta) \simeq i(\theta)\Omega_{lin}(\theta) \simeq i(\theta)2\pi\theta\Delta\theta \tag{A.2}$$

In other words, uniformly spaced, equal width ring detectors respond as a=1 in Eq. (4a).

For log-scaled annular ring detectors, such that the ratio of the limiting scattering angles is a constant δ over all detectors:

$$\Omega \simeq \pi(\delta\theta)^2 - \pi\theta^2$$

$$\simeq \pi\theta^2(\delta^2-1) \tag{A.3}$$

Thus the energy collected by log-spaced annular detectors is given by:

$$S_{log}(\theta) \simeq i(\theta)\Omega_{log}(\theta) \simeq i(\theta)\theta^2\pi(\delta^2-1) \tag{A.4}$$

which indicates that log-spaced annular detectors respond as a=2 in Eq. (4a).

APPENDIX B: TOEPLITZ AND RELATED MATRICES

Mathematical operations with matrices which have some form of degeneracy can be considerably simplified compared to those with general matrix forms. For example, symmetric n×n matrices which have the property that the elements $k_{ij}=k_{ji}$ require less computer storage than a general nonsymmetric matrix and can be inverted more efficiently using the algorithm suggested by Bunch and Parlett (1971). A persymmetric matrix is one which is symmetric about the secondary or cross diagonal (running from upper right to lower left) such that $k_{ij}=k_{n+1-j,n+1-i}$. A Toeplitz matrix is one in which elements on the main diagonal and all codiagonals are recurrent, i.e. $k_{i+1,j+1}=k_{ij}$. Toeplitz matrices have only 2n-1 independent elements as do Hankel matrices which have $k_{i-1,j+1}=k_{ij}$ such that elements on all co-secondary diagonals are equal. A Hankel matrix can be transformed into Toeplitz form through either pre- or post-multiplication by matrix J which has ones along the main secondary diagonal and zeros elsewhere (i.e by reversing the order of either rows or columns). As expected, there are very efficient algorithms for solving Toeplitz systems of linear equations which utilize the special properties as discussed by Zohar (1969) and Bunch (1985). The Toeplitz structure allows solution of linear systems with $O(n^2)$ operations as opposed to $O(n^3)$ with $O(n)$ storage as opposed to $O(n^2)$.

OPTIMAL FILTERING APPLIED TO THE

INVERSION OF THE LAPLACE TRANSFORM

Douglas A. Ross

University of Colorado at Denver
Department of Electrical Engineering
1100 14th Street
Denver, Colorado, U.S.A. 80202

1. INTRODUCTION

The laser scattering measurement of the Brownian motion of particles
suspended in a colloid may be modeled by the integral equation

$$g(t) = \int_{o}^{\infty} G(\gamma) \exp(-\gamma t) \, d\gamma$$

which is the Laplace transform. In this equation $g(t)$ is the
autocorrelation of the electric field of scattered light, and $G(\gamma)$ the
linewidth distribution describing the particle size distribution of the
colloid, with the property $G(\gamma) \geq 0$ for all γ .

The Laplace transform is an ill-conditioned integral equation. This
means that an estimate of $g(t)$ which contains noise will cause noise in $G(\gamma)$
which has infinite variance. There is an extensive literature dealing with
inversion of the Laplace transform in the context of laser measurement of
Brownian motion[1,2,3,4,5,6]. It has been observed that highly accurate
estimates of $g(t)$ yield very limited information about the linewidth
distribution[7].

The purpose of this paper is to introduce the method of optimal filtering[8] to reduce the effect of noise in the inversion. Using the eigenfunction spectrum of the Laplace transform, the noise spectrum which results
from statistical noise in $g(t)$ can be filtered to minimize its effect. An
optimal filter can be derived which minimizes the mean square difference between a filtered estimate and the true linewidth distribution. The resulting
variance is finite, in contrast to an infinite variance using an unfiltered
estimate of the linewidth distribution.

The optimal filter is the solution to a Wiener-Hopf equation, requiring
the form of $G(\gamma)$ to be known. Since in practice a set of data representing
an estimate of $g(t)$ contains no aprior information about the linewidth distribution, it may be impossible to define the optimal filter. This problem
was taken up by Kalman[9,10,] whose work led to the technique of adaptive filtering. The application of adaptive filtering to the inversion of the
Laplace transform will be the subject of a seperate paper. However, since
the mean square error in determining the linewidth distribution from a noisy

estimate of g(t) is the same whether an optimal or adaptive filter is used, the results of this paper represent the smallest error variance that can be obtained and thus provide a basis of comparision of all methods of inversion of the Laplace transform.

2. INVERSION BY DIRECT INTEGRATION

The Laplace transform may be inverted by direct integration using

$$G(\gamma) = \int_o^\infty g(t)\, K(\gamma t)\, dt$$

where

$$K(x) = \frac{1}{2\pi} \int_{-\infty}^\infty \frac{x^{-j\mu}}{\Gamma(1-j\mu)}\, d\mu \qquad\qquad (x>0)$$

The Kernal function has the following properties

$$\int_o^\infty \exp(-\gamma t)\, K(\gamma' t)\, dt = \delta(\gamma - \gamma)$$

$$K(x) \xrightarrow[x \to \infty]{} \frac{1}{\pi}\, \exp(\tfrac{\pi}{2} x)\, \sin(x)$$

Note that the kernal function is not integrable, indicating that the Laplace transform represents an ill-conditioned integral equation. The convergence of the inversion integral depends on g(t) approaching zero rapidly enough, which cannot occur if g(t) contains noise.

The effect of noise may be seen as follows. Suppose $\tilde{g}(t) = g(t) + \tilde{n}(t)$ is an estimate of g(t) with zero mean noise $\tilde{n}(t)$. Then

$$\tilde{G}(\gamma) = G(\gamma) + \int_o^\infty \tilde{n}(t)\, K(\gamma t)\, dt$$

The variance of the estimate of $G(\gamma)$ is

$$\sigma_G^2(\gamma) = E\left\{ \left[\tilde{G}(\gamma) - G(\gamma) \right]^2 \right\} = \int_o^\infty \int_o^\infty C_n(t,t') K(\gamma t) K(\gamma t')\, dt dt'$$

where $C_n(t,t') = E\{\tilde{n}(t)\tilde{n}(t')\}$ is the autocovariance of $\tilde{g}(t)$. As a simple example suppose $\tilde{n}(t)$ is "white" uncorrelated noise so that

$$C_n(t,t') = \frac{\eta_o}{2}\, \delta(t-t')$$

Then

$$\sigma_G^2(\gamma) = \frac{\eta_o}{2} \int_o^\infty \left[K(\gamma t) \right]^2 dt = \infty$$

since $K(\gamma t)$ is not integrable. There is zero probability, or maximum uncertainty, that $\tilde{G}(\gamma) = G(\gamma)$ in the presence of noise.

3. SPECTRAL DECOMPOSITION OF THE LAPLACE TRANSFORM

The spectral decomposition of the Laplace transform, using the appropriate eigenfunctions, is analagous to the Fourier decomposition of the traditional time invariant linear system. The filter, as used here, must operate on the eigenfunction spectrum. The eigenfunctaions are solutions to

$$\lambda(\mu)\Phi^*(t,\mu) \; = \; \int_0^\infty \Phi(\gamma,\mu) \; \exp(-\gamma t)d\gamma$$

It may be verified by direct integration that

$$\Phi(\gamma,\mu) \; = \; \frac{\overline{\lambda(\mu)}}{\sqrt{2\pi\gamma\Gamma\left(\frac{1}{2}-\mu\right)}} \; \gamma^{-j\mu} \; = \; \frac{1}{\sqrt{2\pi\gamma}} \; \exp\left[\; j\,\frac{1}{2}\,\theta(\mu) \; \right] \gamma^{-j\mu} \quad (\gamma>0)$$

with eigenvalue spectrum

$$\lambda(\mu) \; = \; \left| \; \Gamma\left(\frac{1}{2}-j\mu\right) \; \right| \; = \; \sqrt{\frac{\pi}{\cosh(\pi\mu)}}$$

and

$$\theta(\mu) \; = \; \arg\left\{ \Gamma\left[\frac{1}{2}+j\mu\right] \right\} \; = \; \mu\Psi\left(\frac{1}{2}\right) + \sum_{n=0}^{\infty} \left[\frac{\mu}{n+\frac{1}{2}} - \tan^{-1}\left(\frac{\mu}{n+\frac{1}{2}}\right) \right]$$

where[11] $\Psi(\frac{1}{2}) \; = \; -1.963510026021423\ldots\ldots$

The Laplace transform has a continuous spectrum of eigenvalues over $-\infty < \mu < \infty$. The eigenvalues are extremely small for large μ since

$$\lambda(\mu) \; \xrightarrow[|\mu| \longrightarrow \infty]{} \; \sqrt{2\pi} \; \exp\left(-\frac{\pi}{2} |\mu| \right)$$

and this property is the source of ill-conditioning. Orthogonality of the eigenfuncations is

$$\int_0^\infty \Phi(t,\mu) \; \Phi^*(t,\mu')dt \; = \; \delta(\mu-\mu')$$

$$\int_0^\infty \Phi(t,\mu) \; \Phi^*(t',\mu)d\mu \; = \; \delta(t-t')$$

The Laplace transform may be inverted using these eigenfunctions as follows. Let

$$G(\gamma) \; = \; \int_{-\infty}^\infty H(\mu) \; \Phi(\gamma,\mu) \; d\mu$$

where $H(\mu)$ is the spectrum of $G(\gamma)$. Then

$$H(\mu) \; = \; \frac{1}{\lambda(\mu)} \int_0^\infty g(t) \; \Phi(t,\mu) \; dt$$

If g(t) is known then G(γ) may be obtained from its spectrum. Also note that if g(t) is real then H($-\mu$) = H*(μ).

Other properties of the eigenfunctions may be easily verified. The kernal function is represented by

$$K(\gamma t) = \int_o^\infty \frac{\Phi(\gamma,\mu)\ \Phi(t,\mu)}{\lambda(\mu)}\ d\mu$$

The spectrum of the kernal function is $\lambda^{-1}(\mu)$ which is unbounded as $|\ \mu\ | \longrightarrow \infty$. Similarly

$$\exp(-\gamma t) = \int_o^\infty \lambda(\mu)\Phi(\gamma,\mu)\Phi(t,\mu)d\mu$$

Parseval's theorem is

$$\int_o^\infty G^2(\gamma)d\gamma = \int_{-\infty}^\infty |\ H(\mu)\ |^2\ d\mu$$

or

$$\int_o^\infty g^2(t)dt = \int_{-\infty}^\infty \lambda^2(\mu)|H(\mu)|^2\ d\mu$$

4. DERIVATION OF THE WIENER-HOPF EQUATION

If the estimate $\tilde{g}(t)$ = g(t) + $\tilde{n}$(t) contains zero mean noise then the spectrum of $\tilde{G}(\gamma)$ is

$$\tilde{H}(\mu) = H(\mu) + \frac{\tilde{N}(\mu)}{\lambda(\mu)}$$

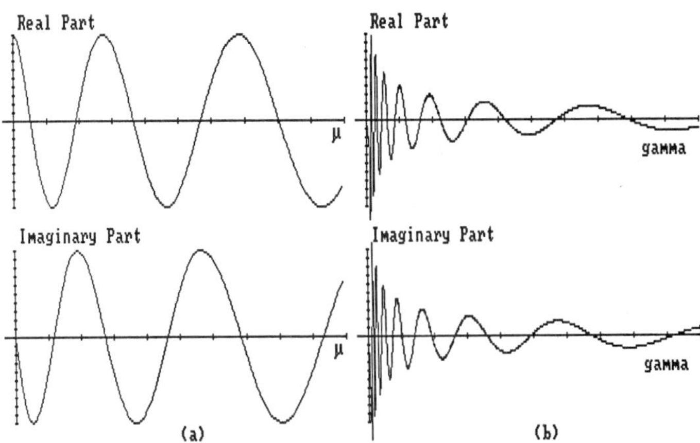

Fig. 1 The real and imaginary parts of the eigenfunctions of the Laplace transform versus (a) μ; (b) γ.

where

$$\tilde{N}(\mu) = \int_0^\infty \tilde{n}(t) \, \Phi(t,\mu) \, dt$$

Thus

$$\tilde{G}(\gamma) = G(\gamma) + \int_0^\infty \frac{\tilde{N}(\mu)}{\lambda(\mu)} \, \Phi(\gamma,\mu) \, d\mu$$

and the variance of $\tilde{G}(\gamma)$ is

$$\sigma_G^2(\gamma) = \int_{-\infty}^\infty \int_{-\infty}^\infty \frac{C_N(\mu,\mu')}{\lambda(\mu)\lambda(\mu')} \, \Phi(\gamma,\mu) \, \Phi^*(\gamma,\mu') \, d\mu \, d\mu'$$

where

$$C_N(\mu,\mu') = E\{\tilde{N}(\mu) \, \tilde{N}^*(\mu')\} = \int_0^\infty \int_0^\infty C_n(t,t') \, \Phi(t,\mu) \, \Phi^*(t,\mu') \, dt \, dt'$$

In the case of white uncorrelated noise

$$C_n(t,t') = \frac{\eta_o}{2} \, \delta(t-t')$$

gives

$$C_N(\mu,\mu') = \frac{\eta_o}{2} \, \delta(\mu-\mu')$$

and

$$\sigma_G^2(\gamma) = \frac{\eta_o}{2} \int_{-\infty}^\infty \frac{|\Phi(\gamma,\mu)|^2}{\lambda^2(\mu)} \, d\mu = \infty$$

In the presence of white noise in $\tilde{g}(t)$ the spectrum of the resulting noise in $\tilde{G}(\gamma)$ varies as

$$\lambda^{-2}(\mu) = \frac{\cosh(\pi\mu)}{\pi}$$

causing an infinite variance.

In order for the variance of $\tilde{G}(\gamma)$ to be finite suppose $F(\gamma,\mu)$ is some filter and let

$$\tilde{G}_F(\gamma) = \int_{-\infty}^\infty \tilde{H}(\mu) \, F(\gamma,\mu) \, \Phi(\gamma,\mu) \, d\mu$$

be a filtered estimate of $G(\gamma)$. The mean square difference between $\tilde{G}_F(\gamma)$ and $G(\gamma)$ is

$$D^2(\gamma) = E\{[\tilde{G}_F(\gamma)-G(\gamma)]^2\} = \underbrace{[G_F(\gamma)-G(\gamma)]^2}_{\text{bias or distortion}} + \underbrace{\sigma_{G_F}^2(\gamma)}_{\text{noise variance}}$$

The optimal filter $F_0(\gamma,\mu)$ minimizes the mean square difference.

It may be shown by the variational method[12] that the optimal filter satisfies the Wiener-Hopf equation

$$\int_{-\infty}^{\infty} \frac{C_N(\mu,\mu')}{\lambda(\mu)\lambda(\mu')}\, F_0(\gamma,\mu)\Phi(\gamma,\mu)\,d\mu = H^*(\mu')\left[G(\gamma) - \int_{-\infty}^{\infty} H(\mu)F_0(\gamma,\mu)\Phi(\gamma,\mu)\,d\mu\right]$$

The optimal filter depends on the autocovariance of the noise spectrum, which is developed in the next section.

5.   STATISTICAL NOISE IN ESTIMATING g(t)

Usually $\tilde{g}(t)$ represents a time average estimate of $g(t)$. Suppose $\tilde{x}(t)$ is a zero mean stationary process with autocovariance $g(t)$ and let

$$\tilde{g}(t) = \frac{1}{T}\int_{0}^{T} \tilde{x}(t')\tilde{x}(t'-t)\; dt'$$

In the case of photon correlation this integral is replaced by a sum of products and $\tilde{x}(t)$ is the number of photons detected in a sample increment centered at t. The noise in $\tilde{g}(t)$ is

$$\tilde{n}(t) = \tilde{g}(t) - g(t) = \frac{1}{T}\int_{0}^{T} [\tilde{x}(t')\tilde{x}(t'-t)-g(t)]\; dt'$$

with zero mean and autocovariance

$$C_n(t,t') = \int_{0}^{T}\int_{0}^{T} [E\{\tilde{x}(t_1)\tilde{x}(t_1-t)\tilde{x}(t_2)\tilde{x}(t_2-t')\} - g(t)g(t')]\; dt\; dt'$$

To evaluate the autocovariance of the noise the fourth order moment of $\tilde{x}(t)$ must be known. Since in practice fourth order statistics are rarely measured, the properties of $C_n(t,t')$ may not be known.

If $\tilde{x}(t)$ is a stationary normal process then

$$E\{\tilde{x}(t_1)\tilde{x}(t_1-t)\tilde{x}(t_2)\tilde{x}(t_2-t')\} = g(t)g(t') + g(t_1-t_2)g(t_1-t_2-t+t')$$

$$+ g(t_1-t_2+t')g(t_1-t_2-t)$$

giving

$$C_n(t,t') = \frac{1}{T}\int_{-T}^{T} \left(1 - \frac{|x|}{T}\right) [g(x)g(x-t+t') + g(x-t)g(x+t')]\; dx$$

Note that the noise is non-stationary since $C_n(t,t')$ is not just a function of t-t'.

In order to simplify the above further, the Fourier transform of the autocovariance combined with the spectral density of $\tilde{x}(t)$, $S(f)$, gives

$$F\{C_n(t,t')\} = 2 \cos(2\pi ft') \, S(-f) \int_{-\infty}^{\infty} S(f') \, \text{sinc}^2[\pi(f'-f)T] \, df'$$

where $\text{sinc}(x) = \sin(x)/x$. The convolution of the spectral density of $\tilde{x}(t)$ with a sinc squared spectrum represents the effect of the truncation error of the interval T used in the time average estimat of $g(t)$. The effective bandwidth of the sinc squared spectrum is $\Delta f = 1/T$. Typically $M = 10^6$ samples and $\Delta T = 1$ ms gives $T = 1000$ seconds and $\Delta f = 10^{-3}$ Hz, and the bandwidth is infintesimal compared with the frequency variation of the spectral density. Thus

$$\int_{-\infty}^{\infty} S(f') \, \text{sinc}^2[\pi(f'-f)T] \, df' \doteq S(f) \int_{-\infty}^{\infty} \text{sinc}^2[\pi(f'-f)T] \, df' = \frac{1}{T} S(f)$$

Since $S(-f) = S(f)$, a property of any spectral density,

$$F\{C_n(t,t')\} \doteq \frac{2}{T} \cos(2\pi ft') \, S^2(f)$$

or

$$C_n(t,t') \doteq \frac{4}{T} \int_{0}^{\infty} S^2(f) \, \cos(2\pi ft) \, \cos(2\pi ft') \, df$$

This may also be written as

$$C_n(t,t') = \frac{1}{T} g_2(t-t') + \frac{1}{T} g_2(t+t')$$

where $g_2(t) = g(t)*g(t)$ and $*$ denotes convolution. The autocovariance of statistical noise may be accurately approximated as consisting of two parts. The first, $g_2(t-t')$, is stationary while the second, $g_2(t+t')$, is non-stationary.

The autocovariance of the noise spectrum may be determined using the simplified form for $C_n(t,t')$ and evaluating the various integrals. The result is

$$C_N(\mu,\mu') = \frac{\pi}{j4T} \frac{\exp\left[-j\frac{1}{2}\theta(\mu) + j\frac{1}{2}\theta(\mu')\right]}{\lambda(\mu)\lambda(\mu')\cos\left(\frac{\pi}{4} + j\frac{\mu\pi}{2}\right) \cos\left(\frac{\pi}{4} - j\frac{\mu'\pi}{2}\right)} \frac{\int_{0}^{\infty} g_2(t)t^{-j\mu+j\mu'} \, dt}{\Gamma(1-j\mu+j\mu')\sinh\left[\frac{\pi}{2}(\mu-\mu')\right]}$$

Considering the various factors which make up $C_N(\mu,\mu')$ the term $\sinh\left[\frac{\pi}{2}(\mu-\mu')\right]$ gives a simple pole at $\mu = \mu'$. The residue of this pole is

$$\lim_{\mu' \longrightarrow \mu} j2\pi(\mu-\mu') \, C_N(\mu,\mu') = \frac{2}{T} \int_{0}^{\infty} g_2(t)dt = \frac{1}{T} S^2(0)$$

Table 1. Variation of the bias and rms error in estimating $G(\gamma)$
using the optimum filter.

T/t_c	$\left[1 - \dfrac{G_F(\gamma)}{G(\gamma)}\right] * 100$	$\dfrac{D(\gamma)}{G(\gamma)} * 100$
1	88.89 %	94.28 %
10	44.44	66.67
100	7.407	27.22
1000	0.7937	8.909
10000	0.07994	2.827
100000	0.007999	0.8944
1000000	0.0008	0.2828

6. DERIVATION OF THE OPTIMAL FILTER

The Wiener-Hopf equation satisfied by the optimal filter may be
simplified using the pole at $\mu = \mu'$. The result is

$$\frac{1}{T} S^2(0) \frac{F_0(\gamma,\mu')\Phi(\gamma,\mu')}{\lambda^2(\mu')} = H^*(\mu') \left[G(\gamma) - \int_{-\infty}^{\infty} H(\mu)F_0(\gamma,\mu)\Phi(\gamma,\mu)\,d\mu \right]$$

The solution to this equation is

$$F_0(\gamma,\mu)\ \Phi(\gamma,\mu) = \frac{G(\gamma)}{\frac{1}{T}S^2(0) + g_2(0)}\ \lambda^2(\mu)H^*(\mu)$$

It may also be shown that

$$G_F(\gamma) = \frac{g_2(0)}{\frac{1}{T}S^2(0) + g_2(0)}\ G(\gamma)$$

$$\sigma_G^2(\gamma) = \frac{\frac{1}{T}S^2(0)\ g_2(0)}{\left[\frac{1}{T}S^2(0) + g_2(0)\right]^2}\ G^2(\gamma)$$

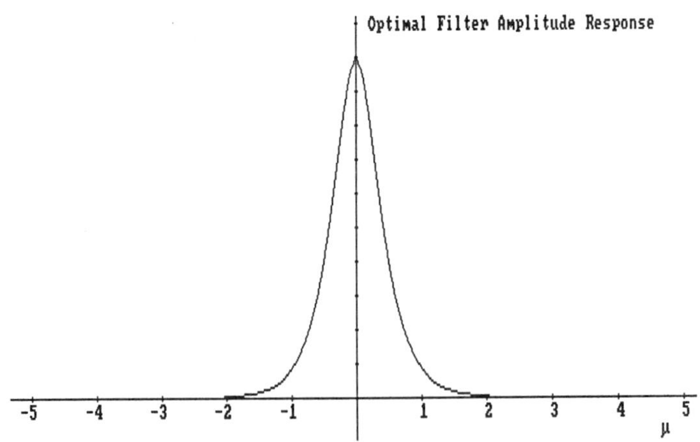

Fig. 2. The optimal filter in the example $g(t) = \exp(-\alpha t)$.

and

$$D^2(\gamma) = \frac{\frac{1}{T} S^2(0)}{\frac{1}{T} S^2(0) + g_2(0)} G^2(\gamma)$$

The optimal filter estimate of $G(\gamma)$ has the property that

$$\lim_{T \to \infty} \tilde{G}_F(\gamma) = G(\gamma)$$

with absolute certainty (unit probability) since

$$\lim_{T \to \infty} D^2(\gamma) = 0$$

7. SAMPLE CALCULATION

Consider the example of Brownian motion by particles all of the same size. Then $g(t) = \exp(-\alpha t)$ $(t>0)$ giving $g_2(0) = 1/2\ \alpha$ and $S(0) = 2/\alpha$. In this case $G(\gamma) = \delta(\gamma-\alpha)$ and the optimal filter is

$$F_0(\gamma,\mu) = \frac{2\alpha}{1 + \dfrac{8}{\alpha T} \cosh(\pi\mu)}\ \pi\ G(\gamma)$$

The amplitude response of the filter is shown in Fig. 2. The bias and mean square difference are

$$1 - \frac{G_F(\gamma)}{G(\gamma)} = \frac{D^2(\gamma)}{G^2(\gamma)} = \frac{1}{1 + \dfrac{\alpha T}{8}}$$

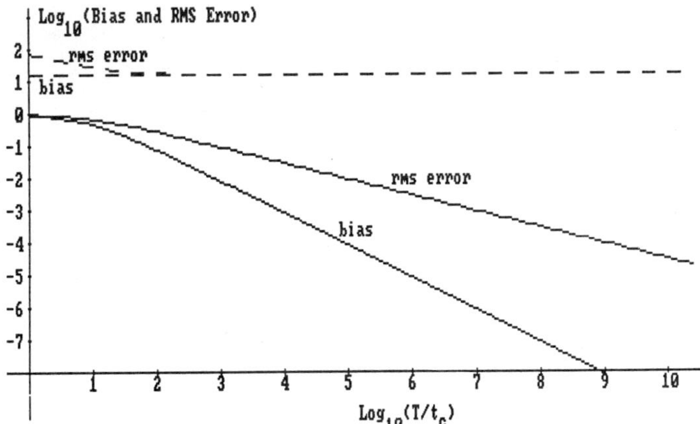

Fig. 3. The variation with T/t_c of the bias and rms error in estimating $G(\gamma)$ using the optimum filter (solid line) and the low pass filter (dashed line).

The variation of the bias and rms error with αT are plotted in Fig. 3 and values calculated in Table 1.

These results show that the bias and mean square error using the optimal filter depend on the ratio of the correlation time, $t_c = 1/\alpha$, to the duration of the time average estimate of $g(t)$, T. A good estimate is obtained if $T \gg t_c$. The mean square statistical noise is

$$\sigma_n{}^2(t) = \frac{1}{T} g_2(0) + \frac{1}{T} g_2(2t)$$

The largest value occurs at $t=0$ giving

$$\sigma_n{}^2(0) = \frac{2}{T} g_2(0)$$

Thus

$$G_F(\gamma) = \frac{1}{1 + 8\sigma_n{}^2(0)} G(\gamma)$$

and

$$D^2(\gamma) = \frac{8\sigma_n{}^2(0)}{1 + 8\sigma_n{}^2(0)} G^2(\gamma)$$

The variation of the bias and mean square error with $\sigma_n{}^2(0)$ is shown in Fig. 4.

8. NON-OPTIMAL FILTERING

The idea of filtering data representing the Brownian motion of particles in a colloid has been used by other authors, but not explicitly stated. For example, in representing the problem by a matrix equation which is inverted by singular value decomposition, the matrix eigenvalue spectrum is truncated when the eigenvalues are too small[7]. This is equivalent to using an ideal low pass filter, which of course is non-optimal.

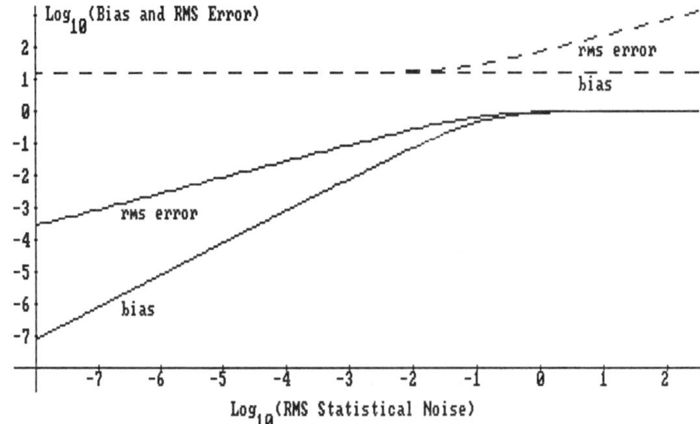

Fig. 4.   The variation with the rms statistical noise in $g(t)$ of the bias and rms error in estimating $G(\gamma)$ using the optimal filter (solid line) and the low pass filter (dashed line).

The effect of such a non-optimal filter may be seen as follows. Suppose

$$F(\gamma,\mu) = G(\gamma) \begin{cases} 1 & , \ |\mu| \le \mu_o \\ 0 & , \ |\mu| > \mu_o \end{cases}$$

where μ_o is the filter's bandwidth. Then the filtered estimate of $G(\gamma)$ is

$$\tilde{G}_F(\gamma) = G(\gamma) \int_{-\mu_o}^{\mu_o} \tilde{H}(\mu) \ \Phi(\gamma,\mu) \ d\mu$$

and

$$\sigma^2_{G_F}(\gamma) = G^2(\gamma) \int_{-\mu_o}^{\mu_o} \int_{-\mu_o}^{\mu_o} \frac{C_N(\mu,\mu')}{\lambda(\mu)\lambda(\mu')} \ \Phi(\gamma,\mu) \ \Phi^*(\gamma,\mu') \ d\mu d\mu'$$

Evaluation of this utilizing the pole of $C_N(\mu,\mu')$ as in section 6 gives

$$\sigma^2_{G_F}(\gamma) = \frac{1}{T} \ S^2(0) \ \frac{\sinh(\pi\mu_o)}{\pi^3 \gamma} \ G^2(\gamma)$$

In the example of section 7 $g(t) = \exp(-\alpha t)$ gives

$$\frac{D^2(\gamma)}{G^2(\gamma)} = 4\mu_o^2 + \frac{1}{T} \ S^2(0) \ \frac{\sinh(\pi\mu_o)}{\pi^3 \alpha}$$

This expression has no minimum. Thus although the mean square error in estimating $G(\gamma)$ using an ideal low pass filter is finite, there is no optimium choice of bandwidth. The mean square error in this case, together with that obtained by an optimal filter is plotted in Fig. 3 and Fig. 4. It may be seen that this non-optimal filter provides a relatively poor estaimate of the linewidth distribution. These results agree qualitatively with results pre-sented by other authors[7], who have not addressed the problem of optimal filtering.

9. DISCUSSION AND CONCLUSIONS

The optimal filter and other results of this paper are derived assuming a time average estimate based on a zero mean, stationary, normal process $\tilde{x}(t)$ with autocovariance $g(t)$. Filtering refers to the spectral decompostion of $g(t)$ using the eigenfunctions of the Laplace transform, rather than the Fourier spectra of these quantities. The results represent an optimal or minimum variance filter which is the solution to a Wiener-Hopf equation. The particular case discussed in section 6 and in Fig. 3 and Table 1 shows that the error in an optimal filter estimate of the linewidth distribution depends only on T/t_c. Provided the duration of the time average is long compared with the correlation time of $g(t)$ the error is small. For example, with $T/t_c = 10^4$ the error is less than 3%. In the singular value decomposition inversion of the Laplace transform some authors have used a form of non- optimal filter, an ideal low pass. Although such a non-optimal filter gives a finite error variance, there is no optimum choice of filter bandwidth, and the results with such a filter are far less accurate than those obtainable with an optimum filter.

These results suggest that in contrast to the assertions contained in
previous papers there is no limit to the information about the linewidth
distribution that can be obtained, except the obvious limit set by the
amplitude of the statistical noise in g(t). The reduction of noise by
optimal filtering provide a basis of comparison of all methods of inversion
of the Laplace transform.

REFERENCES

1. Johnson, R.P.C. and Ross, D.A., _Analysis of Organic and Biological
    Surfaces_, Edited by P. Echlin, Chapter 20: Laser Doppler Microscopy
    and Fiber Optic Doppler Anemometry, 507-527, John Wiley and Sons
    (1984).

2. Ross, D.A. "Laser Particle Sizing by Orthogonal Polynomial,"
    Proceedings of the Fourth International Conference on Photon
    Correlation Techniques in Fluid Mechanics, 15 (1980).

3. Dhadwal, H.S. and Ross, D.A. "Size and concentration of Particles in
    Syton using the Fibre Optic Doppler Anemometer, FODA," Journal of
    Colloid and Interface Science, 76,2,478-489 (1980).

4. B. Chu, Esin Gulari and Erdogan Gulari, "Photon Correlation
    Measurements of Colloidal Size Distributions. II. Details of
    Histogram Approach and Comparison of Methods of Data Analysis,"
    Physica Scripta,19,476-485 (1979).

5. McWhirter, J.G. and Pike, E.R., "On the Numerical Inversion of the
    Laplace Transform and Similar Fredholm Integral Equations of the
    First Kind," Pure A Math. Gen., 11, 9, 1729-1745 (1978).

6. Bertero, M., Boccacci, P and Pike E.R., "On the recovery and resolution
    of exponential relaxation rates from exponential data: a singular-
    value analysis of the Laplace transform inversion in the presence of
    noise." Proc. R. Soc. London Ser. A (GB), 383, 1784, 15-29 (1982).

7. Chu, B., Ford, J.R. and Dhadwal H.S., "Correlation function profile
    analysis of polydisperse macromolecular solutions and colloidal
    suspensions," Methods Enzymol., 117, _Enzyme Structure, Part J_,
    256-297, Academic Press (1985).

8. Norbert Wiener, _Extrapolation, interpolation, and smoothing of
    stationary time series, with engineering applications_, Technology
    Press of MIT (1949).

9. R.E. Kalman, "A new Approach to Linear Filtering and Prediction
    Problems," ASME Transactions, 82D (1960).

10. R.E. Kalman and R.C. Bucy, "New Results in Linear Filtering and
    Prediction Theory," ASME Transactions, 83D (1961).

11. _Handbook of Mathematical Functions_, Edited by M. Abramowitz and I.E.
    Stegun, National Bureau of Standards, Tenth Printing (1972).

12. Papoulus, A., _Probability, Random Variable, and Stochastic Processes_,
    First Eddition, McGraw-Hill, Inc. (1965).

MODELING OF MULTIPLE SCATTERING EFFECTS IN

FRAUNHOFER DIFFRACTION PARTICLE SIZE ANALYSIS

E.D. Hirleman

Laser Diagnostics Laboratory
Mechanical and Aerospace Engineering Department
Arizona State University
Tempe, Arizona 85287

ABSTRACT

A model for the direct problem of calculating the forward scattering
signature of a multiple scattering medium is presented. The new formulation is optimized for integration into schemes for reconstructing the particle size distribution from laser diffraction (forward scattering) signatures obtained from optically thick media. The analysis is valid for media
where the particle sizes and interparticle spacings are large (relative to
the wavelength and the particle size, respectively) such that Fraunhofer
diffraction theory adequately describes the properties of the forward scattered light from individual scattering events. The simulated performance
of laser diffraction particle sizing instruments was then studied using
predictions of the scattered light signatures which would be measured by
laser diffraction instrument under multiple scattering conditions. The results were compared with experimental data and theoretical calculations
based on other models.

NOMENCLATURE

a – albedo, ratio of the scattering cross–section to the total
extinction cross–section of a particle, i.e. the fraction of the
incident energy intercepted by a particle which is scattered rather
than absorbed

a_f – forward scattering albedo, ratio of forward scattering cross–
section to total extinction cross–section for a particle, $a_f = 0.5$
in the geometric optics case, independent of particle composition

f_n – probability that a photon will be scattered (in the forward direction) exactly n times while passing through a medium

h scattering redistribution function

C_{abs} – optical absorption cross–section of a particle (m^2/particle)

C_{ext} – optical extinction cross–section of a particle (m^2/particle)

b – optical depth (dimensionless)

C_{sct} – optical scattering cross–section of a particle (m^2/particle)

L – scattering phase function which is the discrete angular
distribution function for scattered light normalized to 1.0

n – the number of particles in a finite volume

$\langle n \rangle$ – the expected number of particles in a finite volume

P_n – the probability that exactly n particles are in a finite volume

T –   transmittance of a medium, the probability that a photon will tra-
           verse a medium without being scattered or absorbed

Subscripts

det,i   refers to the i_{th} detector
fwd   forward scattering
inc   incident, for radiation incident on a particle
sct   scattered
x   refers to x component in cartesian coordinate system
y   refers to y component in cartesian coordinate system
z   refers to z component in cartesian coordinate system

Superscripts

/   the prime superscript indicates quantity is in local light
         scattering coordinate system rather than inertial system

Greek

γ direction cosines of scattered rays
ℓ the length of the medium (m)
ϕ azimuthal scattering angle in local coordinate system
Φ azimuthal scattering angle in inertial coordinate system
ρ particle number density (particles/m^3)
θ scattering angle in local coordinate system
Θ scattering angle in inertial coordinate system

INTRODUCTION

Particle and droplet size distributions, being parameters of fundamental importance, should be priority measurement objectives for intelli-
gent sensors in next-generation propulsion systems. Unfortunately there
are a number of problematic scientific issues limiting the development of
laser light scattering particle sizing instruments capable of on-line,
autonomous, and self-diagnosing operation in hostile environments. The
objective of this research is to contribute to the scientific knowledge
base necessary to characterize and extend the applicability of laser
diffraction instruments under these conditions. One major concern is the
effect of multiple scattering on the performance of laser diffraction
instruments, both from an error detection and hopefully an error correction
point of view. In this paper we present an efficient method for predicting
the angular distribution and other relevant properties of near-forward
scattered light from a dispersion of spherical particles under conditions
where multiple scattering is significant. The formulation is optimized for
integration into schemes for the inverse problem of determining the size
distribution from light scattered by a multiple scattering medium.

The scattering of incident light by particles significantly larger
than the wavelength can be described by geometrical optics where the solution of Maxwell's equations can be reduced to the modeling of refraction,
reflection, and diffraction. In the near-forward direction, the contributions of reflective and refractive scatter are small and Fraunhofer
diffraction theory is adequate to describe the aggregate scattered light
properties when the photons generally undergo one or less scattering events
before leaving the medium. There are, however, two practical situations
where this analysis is inadequate. These are 1) when the interparticle
spacing is so small that the scattering characteristics for a particle are
dependent on the position and sizes of adjacent particles, and 2) when the
optical depth of the medium is large enough that a significant number of

160

photons are scattered more than once before reaching the detector(s). Both
of these phenomena have been termed multiple scattering in the literature,
but in this paper we restrict ourselves only to consideration of the latter
case where individual scattering events are described by the scattering
from an isolated spherical particle, but that the physical extent of the
aerosol is large enough that most photons encounter more than one particle.

A common optical configuration for a Fraunhofer diffraction particle
sizing instrument utilizes a monolithic photodiode array detector with
annular ring detector elements at the back focal plane of a scattered light
collection lens as shown in Fig. 1. The most common ring detector geometry
is that originally designed by Recognition Systems Inc., where the rings
increase in thickness as distance from the center of the detector increases
thereby compressing the dynamic range of the scattering signals as
discussed by Swithenbank et al. (1977) and Hirleman (1984). The
theoretical development presented here is an efficient method for
predicting the scattering signature on such an axisymmetric detector from
an arbitrary size distribution of isolated, spherical particles for cases
where at least a few tenths of a percent of the incident light is
transmitted without undergoing a scattering event.

Previous work on the general problem of multiple light scattering is
presented by van de Hulst (1980). Felton et al. (1984, 1985) have
developed a theory for the diffraction regime of interest here which used
the adding method discussed by van de Hulst (1980) to determine the
scattering characteristics of an optically thick medium by summing the
effects of a series of thin slabs (i.e. multiple scattering not important
within a thin slab). Felton et al. (1985) discretized the problem in the
azimuthal direction and assumed that photons scattered out of the cone of
scattering angles subtended by the detector outer radius could not be
rescattered back into the detectors. Gomi (1986) used a similar discrete
ordinates approach to determine correction factors for some cases of
Fraunhofer diffraction particle sizing. The unique analysis reported here
uses the method of successive orders combined with some aspects of the
discrete ordinates approach applied to the special conditions relevant to
Fraunhofer diffraction particle size analysis using ring detectors.

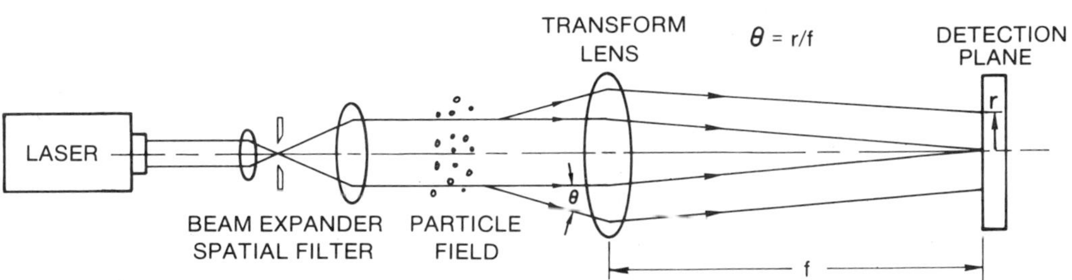

Figure 1. Schematic of Laser Diffraction Particle Sizing System.

Theoretical predictions for angular scattering distribution, transmittance,
and the size distribution obtained from a scattering inversion assuming
single scattering are presented. The results are compared for some
specific size distribution functions and detector configurations discussed
by previous authors.

THEORY OF MULTIPLE SCATTERING

The scattering of light by isolated spherical particles can be
described by Lorenz-Mie theory as presented by van de Hulst (1957) and
Kerker (1966). This solution to Maxwell's equations is rather involved but
well understood and commonly used. For situations where the particles are
significantly larger then the wavelength, the scattering process can be
accurately modeled using geometric optics (reflection and refraction)
coupled with Fraunhofer diffraction. In general, predictions of the
aggregate scattering signature from an ensemble of particles can be
described by linear superposition of the single scattering patterns. There
are, however, two situations where single scattering theory is not
adequate:

1.   where the interparticle spacing is so small that the scattering
      characteristics of a particle depend on the positions and sizes of
      adjacent particles (dependent scattering); and
2.   where the optical path (extent in the direction of the incident
      radiation) is so large that a significant number of photons are
      scattered more than once before exiting the medium and reaching the
      detector(s).

Both of these phenomena have been termed "multiple scattering" in the
literature, but for this paper we define "multiple scattering" to include
only the second phenomenon and restrict our attention to situations of
independent scattering where each scattering event can be described by the
characteristics of scattering from an isolated particle. We then consider
the fate of individual photons which may undergo a series of independent,
single scattering events before detection. The forward problem, that of
predicting the angular profiles of the multiple-scattered radiation, is
quite difficult. Even more difficult, then, is the inverse problem of
determining the particle size distribution from light scattering
measurements which are perturbed by multiple scattering.

The complexity of the multiple scattering model required for a
particular system depends on several factors, the most important of which
involves the scattering characteristics of the particles. The radiative
transfer equation discussed by van de Hulst (1980) is applicable to the
general problem where the single scattering phase function L depends on
both θ and ϕ and the particle properties (number density, size
distribution) are spatially nonuniform. In the present context of laser
diffraction measurements of relatively large spherical particles we are
able to make the following general assumptions:

1.   The interparticle spacing is large such that individual scattering
      events can be described as scattering by isolated particles
      (independent scattering).
2.   The ensemble averaged scattering characteristics produce a
      diffraction pattern which is axisymmetric about the optical axis
      (i.e. independent of azimuth). This condition occurs for either
      spherical particles or a large number of randomly oriented
      nonspherical particles.
3.   The particles are randomly and uniformly distributed in space and
      the macroscopic medium properties are independent of time.

Further, we make the following assumptions which are somewhat unique to
this problem:

4.   Particles are large compared to the wavelength, which in turn
       implies that geometric optics (i.e. the superposition of reflection,
       refraction, and Fraunhofer diffraction) can be used to model the
       scattering process. Further, we assume that reflected and refracted
       light is not scattered back into the detectors which implies that
       anomalous diffraction is also neglected.
5.   The only scattered light of significance here is in the near-forward
       direction and is described by Fraunhofer diffraction theory.
       Further, this implies that the optical pathlength of multiple
       scattered photons does not deviate significantly from that of a
       linear path parallel to the optical axis.

Now considering that the objective here is to model the performance of
laser diffraction instruments we will assume that the detector geometry is
comprised of axisymmetric ring elements. This, coupled with the assumption
of axisymmetric scattering eliminates the azimuthal scattering angle ϕ as a
significant parameter, though it will appear as an intermediate variable.

Photon transport phenomena in multiple scattering media can be
analyzed using several methods after van de Hulst (1980). First, the
radiative transport equation can be solved using, for example, expansion of
the integrand terms. Secondly there are methods which discretize the
scattering angles with varying degrees of freedom as in the four flux model
of Maheu and Gouesbet (1986). The properties of optically thick layers can
be obtained by analyzing the angular reflectance and transmittance of thin
(single scattering) layers and then combining layers using doubling or
adding. Monte Carlo methods provide a rather simple but computationally
inefficient and successive order approximations are recommended by
van de Hulst (1980), Hartel (1940), Tully (1980), and Poole et al (1981).

The unique problem of multiple scattering from particles large
compared to the wavelength (i.e. the laser diffraction problem) has been
considered by Tully (1980) using successive orders, and Felton et al.
(1984, 1985) and Gomi (1985,1986) using a combination of discrete ordinates
and the adding method. In this paper we purpose a unique formulation which
combines the method of successive orders with a discrete ordinates approach
adopted specifically to the axisymmetric ring detector configuration used
in most laser diffraction systems.

Scattering Orders

Consider the propagation of a photon into a medium populated with
randomly positioned monodisperse spherical particles in Fig. 2 and define
the extinction cross-section C_{ext} such that a photon striking a particle
within the area C_{ext} will be either scattered or absorbed. Now the
probability that the photon can traverse the medium without being scattered
or absorbed (i.e. the photon is transmitted) is equivalent to the
probability that there are no particles within a right-circular cylinder of
area C_{ext} and length ℓ with axis coincident with the incident photon path.
For randomly distributed particles we recognize that the occupancy of the
cylinder is governed by a Poisson distribution:

$$P_n = \frac{\langle n \rangle^n \; e^{-\langle n \rangle}}{n!} \tag{1}$$

where P_n is the probability that exactly n particles are in the volume
where $\langle n \rangle$ is the expected number of particles. We note that

$$\langle n \rangle = \rho \, C_{ext} \ell \tag{2}$$

where ρ is the particle number density. Now the transmittance T is equivalent to the probability that there are no particles in the cylinder such that:

$$T = P_o = e^{-\rho \, C_{ext} \ell} \tag{3}$$

Note that the exponent in Eq. (3) is generally termed the optical depth b

$$b \equiv \rho \, C_{ext} \ell \tag{4}$$

We now consider the probability that a photon is scattered in the *forward* direction exactly once which, unfortunately, is not identical to the probability that exactly one particle is in the cylinder of length ℓ for two reasons:

1. The photon will undergo a discontinuous journey due to the redirection after each scattering event. Therefore, the probability of exactly one forward scattering event requires the first particle uncounted to be at position ℓ_1, the event at ℓ_1 to be a forward scattering event, and exactly zero particles be in the cylinder of length ℓ_2 where

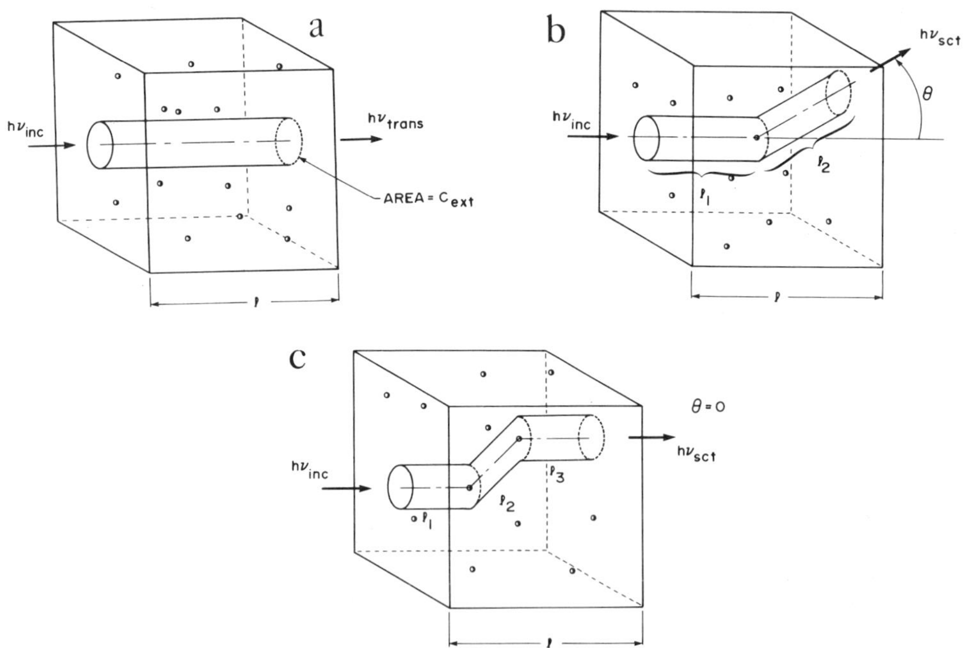

Figure 2. Photon paths through a particulate medium. A scattering or absorption event will occur if a particle is within a cylinder of cross-sectional area C_{ext} centered on the photon path. a) Photon is transmitted unscattered. b) Photon undergoes scattering event at distance ℓ into the medium and is redirected into angle θ from the incident photon direction. c) Photon undergoes two scattering events and in this case emerges parallel to the incident direction. This photon would be measured as part of the transmitted (unscattered beam) by a laser diffraction system, and would cause an erroneous indication of the transmission.

164

$$\ell = \ell_1 + \ell_2 \cos\theta \tag{5}$$

Note that only for small angles θ will:

$$\ell = \ell_1 + \ell_2 \tag{6}$$

2.   The photon may be absorbed (or scattered out of the forward direction).

The first problem means that the photon path length depends on the scattering angle which is a stochastic variable. However, under the assumption of large particles the scattering is predominantly in the near-forward direction and $\cos\theta \approx 1$ such that the photon path is always approximately given by the medium thickness ℓ. (Monte Carlo results to be discussed later indicate that this assumption is in error by typically less than 1% for size distributions typical of sprays).

To consider the second problem we define the single scattering albedo a such that the scattering and absorption cross-sections are given by:

$$C_{sct} = a \cdot C_{ext} \tag{7}$$

$$C_{abs} = (1-a) \cdot C_{ext} \tag{8}$$

We now consider that in geometric optics regime the extinction cross-section is exactly twice the particle projected area:

$$C_{ext} = 2 \cdot (\pi D^2/4) = \pi D^2/2 \tag{9}$$

Phenomenologically the two particle areas in Eq. (13) can be considered first, photons incident on the particle cross-section are refracted or reflected and secondly, the diffracted light which is also equal to that incident on one particle projected area following Babinet's principle. Thus for totally absorbing particles:

$$C_{sct} = C_{abs} = a \cdot C_{ext} = 0.5 \cdot C_{ext} \tag{10}$$

For nonabsorbing particles where $C_{abs} = 0$, from a forward scattering perspective we might still consider $a = 0.5$ writing that:

$$C_{sct,fwd} = 0.5 \cdot C_{sct} = 0.5 \cdot C_{ext} \tag{11}$$

If we assume that any light scattered out of the forward direction (i.e. any light not diffracted but rather reflected or refracted) is diffusively scattered and will, therefore, have a negligible probability of being scattered back into the forward direction, then we can consider a forward scattering albedo $a_f = 0.5$ in the geometric optics limit.

The probability that a photon undergoes exactly one scattering event requires that there be exactly one particle in the cylinder of length ℓ and that the photon be forward scattered (rather than absorbed, reflected, or refracted) by that one particle. Defining now the variable f_n to represent the probability that a particle undergoes exactly n *forward* scattering events and then exits the medium (i.e. is not absorbed in the last event) we have:

$$f_1 = a_f b e^{-b} \tag{12}$$

and

$$f_2 = \frac{a_f^2 b^2 \, e^{-b}}{2} \tag{13}$$

from which we generalize:

$$f_n = \frac{(a_f b)^n \, e^{-b}}{n!} \tag{14}$$

Plots of the probability f_n for various values of the optical depth b are given in Fig. 3.

Scattering Signature (Angular Distribution)
____

The next task is to predict the angular scattering distribution, as would be measured by a laser diffraction instrument, for a general multiple scattering medium. We have from Eq. (11) the probability distribution for scattering of any order n, and defining $S_n(\Theta)$ as the probability that a photon scattered exactly n times will finally propagate within a finite scattering angle range represented by Θ we can write:

$$S(\Theta) = \sum_{i=0}^{\infty} f_n \cdot S_n(\Theta) \tag{16}$$

where S without a subscript represents the composite scattering signature which is the superposition of contributions from all scattering orders. The coordinate systems are shown in Fig. 4. The final task is then to determine the $S_n(\Theta)$ to utilize in Eq. (16) to calculate the laser diffraction instrument response. We discretize the scattering angles into conical shells centered on the optical axis subtending a range of scattering angle Θ for the jth discrete angle range $\Theta_{i,j}$ to $\Theta_{o,j}$ where the subscripts i and o represent inner and outer Θ limits respectively. This configuration is tailored for the assumed case of axisymmetric scattering and a detector configuration as shown in Fig. 1.

Consider Fig. 4 where we show an isolated particle situated at the origin of an inertial coordinate system such that the Z axis coincides with the optical axis of the incident laser beam. A photon is incident on the particle at an angle Θ_{inc} from the optical axis assuming, without loss of generality, that the incident photon is in the X-Z plane such that the azimuthal incident angle Φ_{inc} is zero. The incident photon is, in general, traveling in the Θ_{inc} direction following redirection during a previous scattering event. Now under the conditions of independent scattering we note that the angular probability distribution for the photon following the scattering event with the particle at the origin in Fig. 4 is given by the discrete single scattering phase function L (θ', ϕ') where the scattering angles θ' and ϕ' must be in a relative or local light scattering coordinate system with the optical (Z') axis colinear with the incident photon direction.

Now in order to model the radiative transfer particularly with respect to a laser diffraction detection system which is centered on the Z axis we must determine Θ_{sct} and Φ_{sct} of the photon in the inertial or instrument coordinate system after the scattering event. We can write the following expressions for the direction cosines γ_x, γ_y, and γ_z in the inertial system:

$$\gamma_x = \cos\theta' \, \sin\Theta_{inc} + \sin\theta' \, \cos\phi' \, \cos\Theta_{inc} \tag{17}$$

$$\gamma_y = \sin\theta' \, \sin\phi' \tag{18}$$

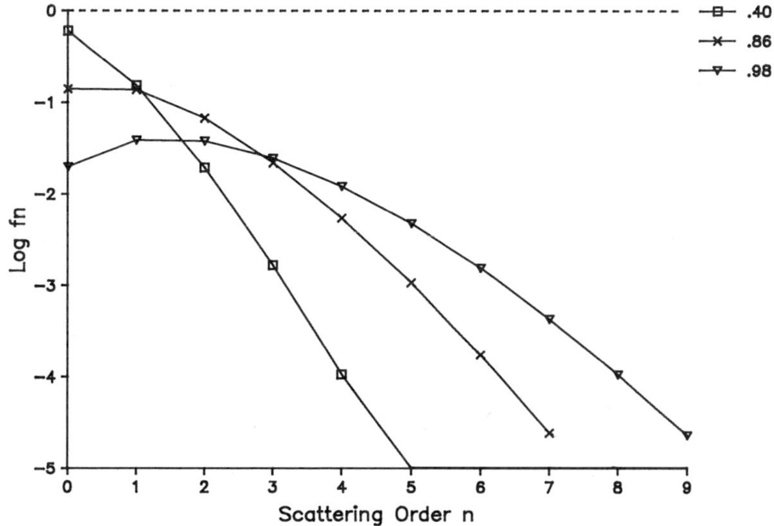

Figure 3. The probability f_n that a photon will be forward scattered exactly n times vs. the scattering order n for forward scattering albedo a_f = 0.5 for various values of extinction.

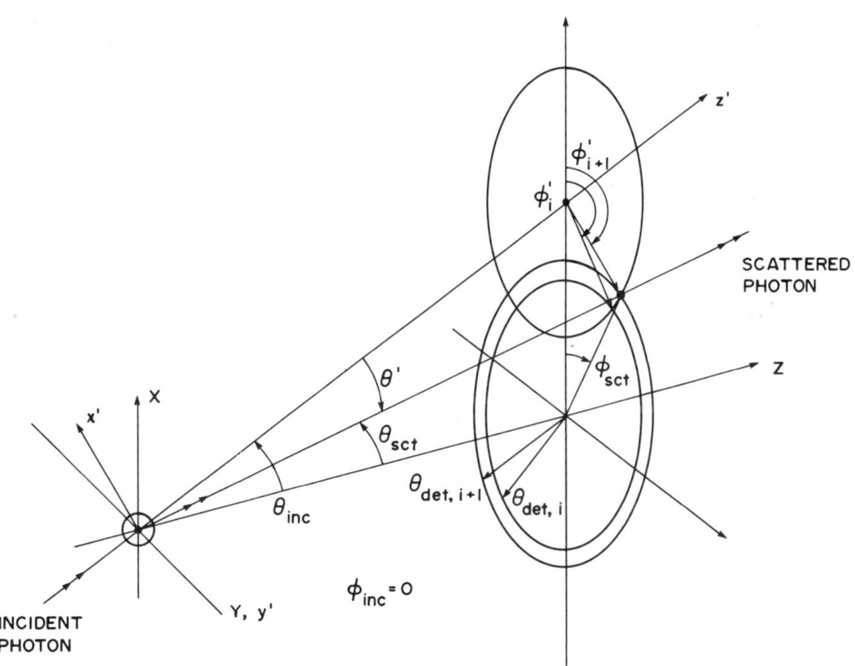

Figure 4. Scattering coordinate system for multiple scattering. A photon is incident on a particle situated at the origin of the inertial coordinate system. The scattering process and the accompanying phase function is defined in the light scattering coordinate system defined by the incident photon. The photon is scattered at angles θ' and ϕ' which, after coordination transformation, is transposed into angles O_{sct} and Φ_{sct} in the inertial XYZ system, and all energy scattered between ϕ' and $-\phi'$ would be collected by detector elements out to the ith element.

167

$$\gamma_z = \cos\theta' \cos\Theta_{inc} - \sin\theta' \cos\phi' \sin\Theta_{inc} \tag{19}$$

from which we obtain the final scattering angles:

$$\Theta_{sct} = \tan^{-1} \left((\gamma_x{}^2 + \gamma_y{}^2)^{1/2}/\gamma_z \right) \tag{20}$$

$$\Phi_{sct} = \tan^{-1} (\gamma_x/\gamma_y) \tag{21}$$

Now if the scattering event of interest is the nth for this particular photon, then the scattering order signature $S_n(\Theta_{sct})$ for photons forward scattered exactly n times can be found from:

$$S_n(\Theta_{sct}) = \sum_{\text{all }\Theta_{inc}} h(\Theta_{sct}, \Theta_{inc}) S_{n-1}(\Theta_{inc}) \tag{22}$$

where h is the scattering redistribution function or the probability that a photon incident in direction Θ_{inc} (for any Φ_{inc}) will leave the next scattering event traveling in the direction Θ_{sct}. Inherent in Eq. (22) is the assumption that a photon traveling at any Θ will, after the next scattering event, have the same angular distribution function regardless of the number of previous scattering events it had undergone. Now the scattering redistribution function h is given by:

$$h(\Theta_{sct}, \Theta_{inc}) = \sum_{\text{all }\theta'} h'(\Theta_{sct}, \theta', \Theta_{inc}) L(\theta') \tag{23}$$

where the local redistribution function h' is the probability that a photon incident at Θ_{inc} and scattered at an angle θ' in noninertial light scattering coordinate system will, after transformation, have a final direction represented by Θ_{sct}. Note that:

$$h'(\Theta_{sct}, \theta', \Theta_{inc}) = (\phi_i - \phi_{i+1})/\pi \tag{24}$$

where ϕ_i and ϕ_{i+1} are the values of ϕ which satisfy Eqs. (17–21) for $\Theta_{sct} = \Theta_{det,i}$ and $\Theta_{det,i+1}$ respectively. Note that ϕ_i and ϕ_{i+1}, and h' in turn depend only on the detector geometry and not on the light scattering properties of the particles. Thus for a given detector the m^3 values of h' (where m is the number of discrete detector elements required for sufficient resolution) can be calculated once and stored. The redistribution function h is determined by the detector geometry (through h') and the single scattering phase function $L(\theta')$ as given in Eq. (25):

$$L(\theta') = C_2 \int_0^\infty D^2 \left[J_0^2(\beta_i) + J_1^2(\beta_i) - J_0^2(\beta_{i+1}) - J_1^2(\beta_{i+1}) \right] \rho(D)dD \tag{25}$$

where $L(\theta')$ is the fraction of energy diffracted between angles $\theta'_{det,i}$ and $\theta'_{det,i+1}$ which are the limiting angles for the finite detector element represented by θ', and

$$\beta_i = \frac{\pi D}{\lambda} \theta'_{det,i} \tag{26}$$

$$\beta_{i+1} = \frac{\pi D}{\lambda} \theta'_{det,i+1} \tag{27}$$

Typical plots of $L(\theta')$ are shown by Swithenbank et al (1976).

We now consider again Eq. (22) and note that $S_0(\Theta)$ represents the angular distribution of the incident radiation. Further, for collimated incident radiation at $\Theta=0$ (as for the typical laser diffraction application) we observe that

$$S_1(\Theta) = h(\Theta, 0)\, S_0(\theta) = L(\theta) \tag{28}$$

where the inertial Θ and the light scattering θ' coordinate systems are coincident in this special case. Equations (22) and (28) comprise a recursion relation which can be used to calculate S_n for successive orders, once S_{n-1} is known. Some example calculations of S_n are plotted in Fig. 5. Note that as the scattering order increases, the most probable scattering angle for a photon also increases and therefore a smaller fraction of the total scattered energy is actually collected by the detector which has a limited maximum collection angle.

Returning to Eq. (22) and considering the previously discussed discretization of scattering space into m discrete incident and scattering angles, we define the scattering vectors S and S_n such that the jth elements correspond to $S(\theta_j)$ and $S_n(\theta_j)$ respectively. Further, define the redistribution matrix H such that element i,j is $h(\theta_i, \theta_j)$ where the sct and inc notations are implicit. Eq. (22) then becomes simply:

$$S_n = HS_{n-1} \tag{29}$$

$$S_{n+1} = HS_n = H{-}HS_{n-1} \tag{30}$$

so we observe that:

$$S_n = H^n S_0 = H^n S_{inc} \tag{31}$$

Returning to Eq. (16) we write

$$S(\theta) = \sum_{i=0}^{\infty} \frac{(a_f b)^n e^{-b}}{n!} S_n(\theta) \tag{32}$$

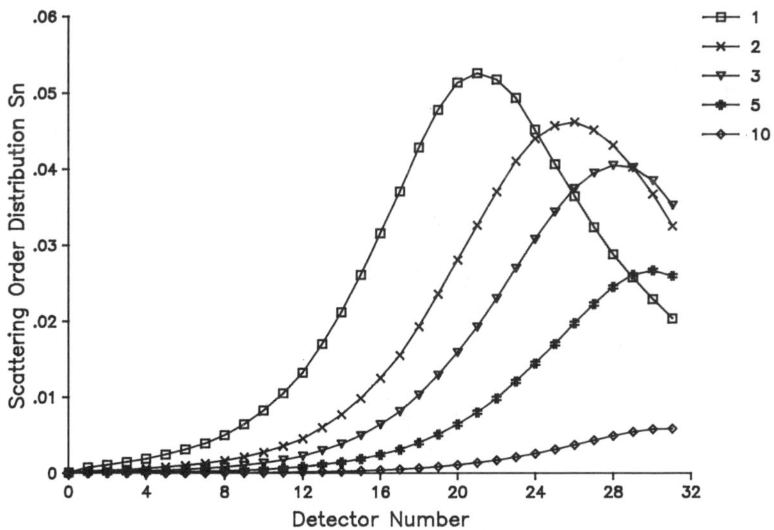

Figure 5. Scattering distribution function S_n vs. detector number for various scattering orders. The plotted value for any detector number represents the fraction of those photons scattered exactly n times and which would strike the indicated detector. Calculations are for the RSI/Malvern detector, f.l. = 300mm, and a Rosin Rammler particle size distribution with X = 26, N = 2.9.

which can be written

$$S = \exp(-b)\left[S_{inc} + a_f bHS_{inc} + \frac{(a_f b)^2}{2} H^2 S_{inc} + \ldots\right] \qquad (33)$$

$$S = \exp(-b)[1 + a_f bH + (a_f b)^2 H^2 + \ldots] \, S_{inc} \qquad (34)$$

which for $a_f = 0.5$ in the diffraction regime gives:

$$S = \exp(-b) \, \exp(0.5 \, b \cdot H) \, S_{inc} \qquad (35)$$

This equation greatly simplifies the direct calculation of multiple scattering signatures, and also provides a very efficient means for obtaining particle size distributions by inverting multiple scattering diffraction data.

RESULTS AND DISCUSSION

Calculations were made for a Rosin-Rammler particle size distribution with X=26 and N=2.9 in order to compare with other work. The RSI/Malvern detector with dimensions reported by Hirleman et al. (1984) was subdivided into 140 total detectors with a maximum delta radius of 0.143 mm. A buffer zone of 20 additional rings was added to the outside of the detector. Calculations based on the present formulation are shown in Fig. 6 for various extinctions up to 0.975. Plots of S(Θ) for the discrete scattering angles corresponding to the RSI/Malvern ring detector with a 300mm lens are given in Fig. 7.

Composite scattering patterns for particle media of varying optical depths are shown in Fig. 7. Again for larger optical depths, the average photon is scattered more than once (see Fig. 3) and the energy is shifted to larger scattering angles which produces an artifact of smaller

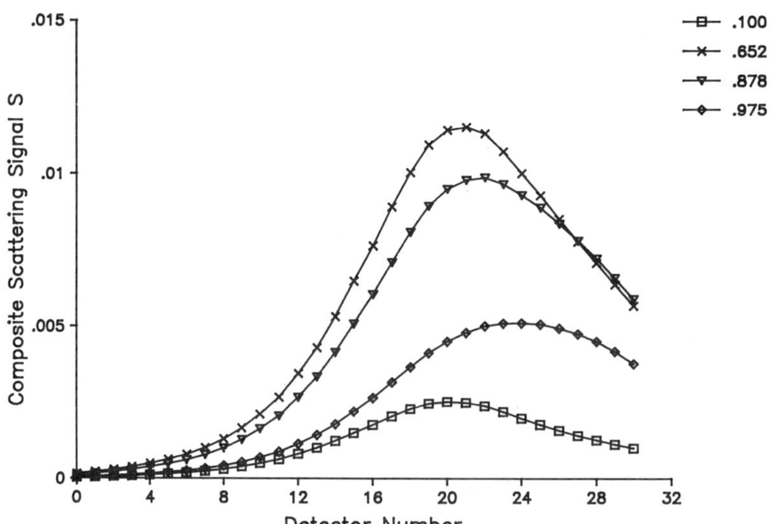

Figure 6. Composite Scattering distribution function S vs. detector number calculated for various extinctions. The plotted value for any detector number represents the fraction of incident photons which would strike the indicated detector. Calculations are for the RSI/Malvern detector, f.l. = 300mm, and a Rosin Rammler particle size distribution with X = 26, N = 2.9.

particles. Also plotted in Fig. 7 are data from Felton et al. (1984) and
the results of a Monte Carlo simulation also developed as a part of this
work to provide an independent check. Note that the data from the
techniques show some significant variations.

The source of the discrepancy is also shown in Fig. 7. The model of
Felton et al. (1984,1985) assumed that photons which were scattered at
angles larger than the outermost detector could not be rescattered back
into the detector θ range. This assumption significantly affects the
results under some conditions. We added 20 additional buffer rings outside
of the actual active detector elements (which increased the effective size
of the detector by a factor of about 1.25) to account for some of those
photons which do get scattered back into the detector range. The Monte
Carlo results (the highest curve of Fig. 7) are equivalent to that obtained
with a very large buffer zone (greater than the outer detector radius).
Thus the data from the lowest curve in Fig. 7 do not consider photons
rescattered into the detector field of view. The middle curve of Fig. 7
accounts for photons which are scattered out less than 25% of the detector
radius and then are rescattered back into the field of view. These
additional photons preferentially strike the outer detectors (because they
were already out at large angles) and cause an increase in the predicted
scattering at the outer rings. Finally, when all photons are considered
(i.e. larger buffer zones) the curve will increase further on the outer
rings until the results asymptotically approach the upper curve in Fig. 7.
Note that neglecting photons scattered out of and then back into the
detector gives a low estimate of the effects of multiple scattering.

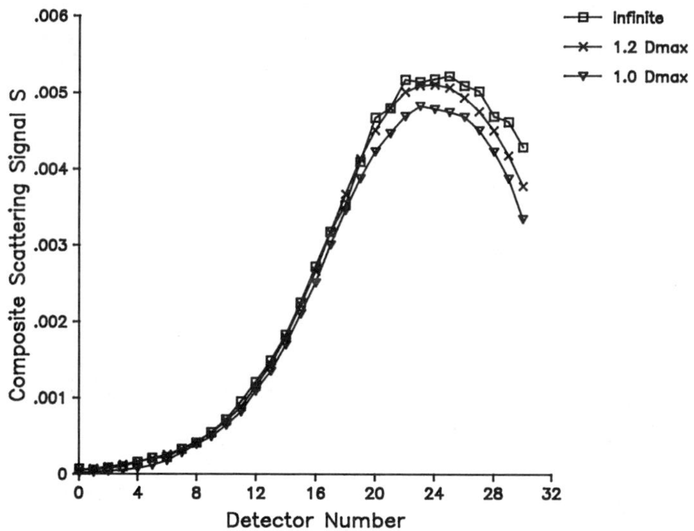

Figure 7. Composite scattering distribution function S for extinction
= 0.975 with X = 26 um and N = 2.9 for the RSI/Malvern detector with
f.l. = 300 mm. The lower curve is obtained if only those photons which are
never scattered into an angle which exceeds 1.0 times the maximum angle
subtended by the detector are considered. This assumption matches that
made by Felton et al. (1984,1985), and Hamidi and Swithenbank (1986). The
middle curve is obtained if photons are neglected only when scattered into
angles greater than 1.2 times the maximum. The upper curve is obtained
with the present model when all photons are considered, and independent
Monte Carlo calculations agree with the discrete ordinates/successive order
approach used here.

A generalization is possible, as we should stop adding additional buffer rings when $h(\Theta_{\text{det outer ring}}, \Theta_{\text{last buffer ring}})$ is acceptably small. This condition, in turn, depends on the single scattering phase function since smaller particle sizes will require larger buffer zones because the photons are scattered into large angles. As a further study we have plotted predicted composite scattering signatures with previously

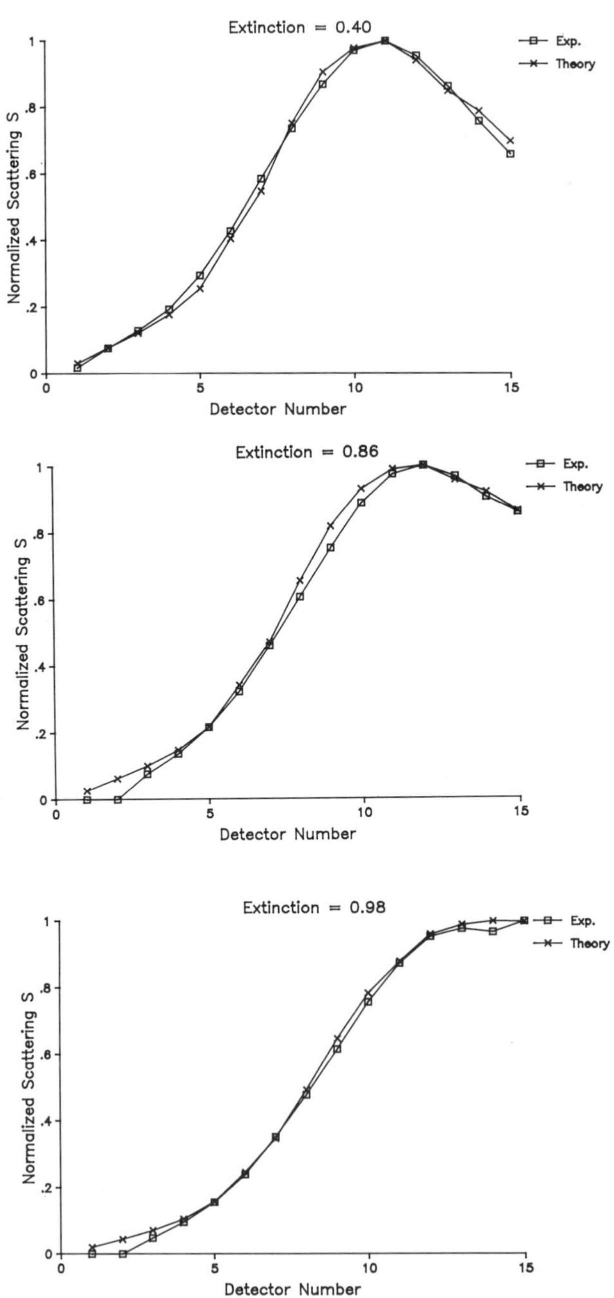

Figure 8. Plot of measured scattering signatures for various levels of extinction from the 689 kPa data of Dodge (1984) compared with theoretical calculations from this study.

unpublished experimental data obtained by Dodge (1984) and used in the
preparation of the referenced paper. Here a series of effectively
identical nozzles were lined up in a laser diffraction instrument sample
volume to study the effect of increasing extinction. The nominal (low
extinction) values used here were $X = 37.6\mu m$ and $N = 1.94$ which
corresponded to $SMD_0 = 20.6$ um for the 689 kPa run condition with 45° solid
cone nozzles. The measured scattering signatures in Fig. 8 (obtained by
adding adjacent signals and normalizing) show reasonable agreement with our
calculations giving further substantiation to this work. The effect of
multiple scattering on laser diffraction measurements of the volume medium
diameter $(D_{v,05})$ for Rosin-Rammler software is shown in Fig. 9. As
reported in previous work by Dodge (1984) and Felton et al. (1984),
multiple scattering can cause significant changes in apparent particle
sizes as measured by laser diffraction instruments.

The Inverse Problem

The laser diffraction particle sizing problem is to measure the
composite forward scattering signature from a particle field and perform a
mathematical inversion to obtain an estimate of the size distribution. Now
in general the media are optically thin so that the higher orders of f_n of
Eq. (14) are negligible. In that case the measured composite scattering
distribution is equal to $S_1(\Theta)$, and the inversion of S_1 has been studied by
many authors. We propose to formulate the multiple scattering inversion
problem as that of obtaining an estimate of $S_1(\Theta)$ from a measured signature
scattering composite $S(\Theta)$ followed by the determination of $\rho(D)$ from $S_1(\Theta)$.

All terms on the right hand side of Eq. (16) are then unknowns, but
we have information on the f_n from a measurement of the transmission.
There is, however, a complication in that the Beer-Lambert Law of Eq. (3)
must be corrected for forward scattered energy which enters the
transmission detector giving an underestimate of the true optical depth.

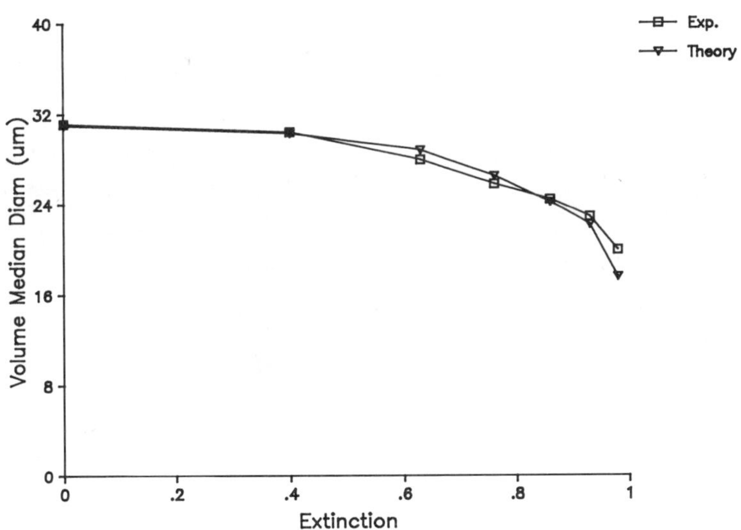

Figure 9. Dependence of the volume medium diameter $D_{v,0.5}$ as measured by a
laser diffraction particle sizing instrument on optical extinction from the
689 kPa data of Dodge (1984). Rosin-Rammler inversion software was used in
both the actual measurements and the inversions of scattering signatures
calculated used the multiple scattering model developed here.

<u>Scaling Improvements</u>

In another paper in this volume Hirleman (1987) proposed a method for optimal scaling of the Fraunhofer diffraction problem. In particular, it was found that the detector geometry should be such that the annular ring elements should be distributed uniformly in log Θ space, i.e. such that the ratios of inner, outer and mean angles are constant within and between all adjacent elements. That scaling improves the performance of inversion schemes as discussed by Hirleman (1987). Of interest here, however, is the fact that log-scaling of the detector produces useful simplifications of the multiple scattering formulation as well. Battistelli et al. (1985) considered the amount of multiple-scattered energy received by an on-axis circular aperture. It was found that the probability that a photon of any scattering order would fall within an aperture radius r at some distance D form a source is equal to the probability for detector radius kR at a distance kD where k is some positive real constant. It follows, then, that the same similarity law applies to the energy within an annular ring and we can write:

$$S_n(k\Theta)\Big|_{\ell=kD} = S_n(\Theta)\Big|_{\ell=D} \cdot k^{n-2} \exp(-(k-1)b) \tag{36}$$

where the small angle relation r=fΘ has been implemented.   Similarly, for variations in optical depth:

$$S_n(\Theta)_{b=kb_0} = S_n(\Theta)_{b=b_0} \cdot k^n \exp\ (-(k-1)b) \tag{37}$$

It is then clear that geometric (logarithmic) rather than arithmetic scaling of the detector elements which optimizes the inverse single scattering problem as discussed by Hirleman (1987) are needed here as well. Since the optical depth of a medium is generally unknown, the significant data compression possibilities of the scaling laws in Eqs. (36) and (37) for the inverse Fraunhofer diffraction problem under multiple scattering conditions are clear.

CONCLUSIONS

A new formulation for the calculation of scattering signatures for distributions of spherical particles in optically thick media has been presented. The model uses successive order multiple scattering for discrete annular detectors and has been validated using independent methods. This technique was developed for future incorporation into inversion algorithms for determining particle size distributions in multiple scattering media.

ACKNOWLEDGEMENTS

This research was sponsored by the Air Force Office of Scientific Research, Air Force Systems Command, USAF, under Grant Number AFOSR-84-0187, Dr. Julian Tishkoff, program manager. The U.S. Government is authorized to reproduce and distribute reprints for Governmental purposes notwithstanding any copyright notation thereon.

The author is also grateful to Lee G. Dodge of Southwest Research Institute for providing previously unpublished data on multiple scattering.

REFERENCES

Battistelli, E., Bruscaglioni, P, Ismaelli, A. and Zaccanti, G., 1985, "Use of two scaling relations in the study of multiple-scattering effects on the transmittance of light beams through a turbid atmosphere," *J. Opt. Soc. Am. A.*, 2:903-11.

Dodge, L.G., 1984, "Change of calibration of diffraction-based particle sizers in dense sprays," *Optical Engineering*, 23:626-630.

Felton, P.G., Hamidi, A.A., and Aigal, A.K., 1984, "Multiple scattering effects on particle sizing by laser diffraction," Report No. 431 HIC," Dept. of Chemical Engineering, Univ. of Sheffield, England.

Felton, P.G., Hamidi, A.A., and Aigal, A.K., 1985, "Measurement of drop size distribution in dense sprays by laser diffraction," *Proceedings of the 3rd International Conference on Liquid Atomization and Spray Systems*, Institute of Energy, London.

Gomi, H., and Hasegawa, K., 1984, *Int. J. Multi. Flow*, 10:653.

Gomi, H., 1986, "Multiple scattering correction in the measurement of particle size and number density by the diffraction method," *Applied Optics*, 25:3552-3558.

Hamidi, A.A and Swithenbank, J., 1986, "Treatment of multiple scattering of light in laser diffraction measurement techniques in dense sprays and particle fields," *J. Inst. Energy*, 59:101.

Hartel, W., 1940, *Licht*, 10:141.

Hirleman, E.D., 1984, "Particle sizing by optical, nonimaging techniques," *Liquid Particle Size Measurement Techniques*, ASTM Publication STP 848, ed. by J.M. Tishkoff, R.D. Ingebo and J.B. Kennedy, pp. 35-60, American Society of Testing Materials, Philadelphia.

Hirleman, E.D., 1987, "Optimal scaling of the inverse Fraunhofer diffraction particle sizing problem: the linear system produced by quadrature", this volume.

Kerker, M., 1969, *The Scattering of Light and Other Electromagnetic Radiation*, Academic Press, New York.

Maheu, B. and Gouesbet, G., 1986, "Four-flux models to transfer equations: special cases," *Applied Optics*, 25:1122-1128.

Poole, L.R., Venable, D.D. and Campbell, J.W., 1981, "Semianalytic Monte Carlo radiative transfer model for oceanographic radar systems," *Applied Optics*, 20:3653-3656.

Swithenbank, J., Beer, J.M., Taylor, D.S., Abbot, D. and McCreath, C.G., 1977, "A laser diagnostic technique for the measurement of droplet and particle size distribution," *Experimental Diagnostics In Gas Phase Combustion Systems*, Progress in Astronautics and Aeronautics Vol. 53, ed. B.T. Zinn, AIAA, New York.

Tully, D.B., 1980, "Multiple scattering theory for small angle light scattering," in *Practical Electro-Optical Instruments and Techniques*, SPIE 255:114.

van de Hulst, H.C., 1957, *Light Scattering by Small Particles*, John Wiley and Sons, New York.

van de Hulst, H.C., 1980, *Multiple Scattering*, Vols. I and II, Academic Press, New York.

LDV-SIGNAL-ANALYSIS FOR PARTICLE VELOCITY

AND SIZE DETECTION USING GEOMETRICAL OPTICS

Thoma Borner*, and LIdong Zhan

Lehrstuhl für Strömungsmechanik
Universität Erlangen-Nürnberg
Egerlandstr. 13, 8520 Erlangen, FRG

ABSTRACT

The paper presents a three dimensional direct geometrical simulation of laser light beams split into
differential light rays being treated as waves, interacting
with a spherical particle to study light scattering
characteristics depending on all relevant measuring and
particle parameters of a LDV measurement system.

INTRODUCTION

Laser light scattering from particles is porgressively
being used for particle velocity and size detection. Lorenz-
Mie theory /1/ gets widely employed to describe or analyse
particle light scattering characteristics. Extensions that
treat particles in a typical dual focus laser-Doppler velocimeter (LDV) environment have been reported in the literature
including two beams /2,3/, Gaussian intensity particle
illumination /4,5/.

However, for large particles with non-symetrical
illumination and time resolved trajectories through an LDV-
probe volume light scattering properties cannot be obtained
easily. To still explain some of the light scattering phenomena, geometrical optics in different approaches have been
employed in comparison to Lorenz-Mie theory /6,7,8/ to
demonstrate its potential accuracy.

The understanding and interpretation of light scattering phenomena in two-phase flows was the major objective of
many researchers studying signal visibility and amplitude
(eg.) /9/, phase-Doppler for particle size detection eg.
/10,11,12,13,14,17/ or the tripple peak technique using
particle residence time eg. /15,16/.

---

* Now at WALTHER & CIE AG, 5000 Köln 80, P.O.Box 850380

In this paper a three-dimensional direct geometrical simulation of laser light beams split into differential rays interacting with a spherical particle is employed to study light scattering characteristics depending on all the relevant device and particle parameters of an LDV-measurement. To this end each of the two laser beams is subdivided into differential rays according to the Gaussian intensity distribution that are traced through the particle to a specified detector plane. Simultaneously an equation for the electrical field amplitude vector $\vec{E}$ is solved for each ray so that the Doppler-shift of the laser light and the phase of the ray depending on its path length is computed. Particles of given size, refractive index and trajectory are passed through the measuring volume formed by two crossed laser beams storing the light intensity distribution in space for each time step. All differential rays are computed tracing them to the particle, in the particle and away from the particles. Reflected and refracted contribution of $\vec{E}$ are processed according to Fresnel's formulas. Internal reflections are followed untill $I_R = I_{max}/e^2$.

The procedure proposed predicts light scattering phenomena from particles devided into three stages:

1. Set-up of laser beams and differential ray definition.

2. Generation of ray and wave equations to describe the light particle interaction in 3D-space and time, selecting desired space angles to be stored.

3. Integration of rays on specified detector surfaces to analyse intensity in time (LDV-signal, phase shift, frequency) or intensity in space (fringe pattern).

The following section focusses on results obtained with the procedure and compares phenomenologically the signal properties reported in the literature. Namely for large particles, e.g. paritcles of a similar size as the LDV measuring volume, this method proves to have a high potential. As reported in /10,18/ the standard LDV formula does not hold, still all the effects can hardly be included into the proposed analytical solution. /5/ and /11/ show that the fringe distance in space is strongly nonlinear. If the Doppler-shift is properly included the fringe pattern will also be non-symmetrical. This has a significant impact on required positions of detectors for the phase detection method.

It should be notified here that Lorenz-Mie theory gives very meaningful results for laser light scatter on a spherical or elipsoidal particle, however, the computational effort for large particles is tremendous. Some, for small particles hardly effective parameters, as non-symmetric particle trajectory through the measuring volume or non-homogenious illumination cannot be neglected for large particles. The relative motion within a fluid particle may have to be taken into account /18/.

BASIC EQUATIONS AND SOLUTION PROCEDURE

The analysis of light scattering from a spherical particle into differential rays by means of a numerical simulation of a laser-Doppler velocimetry (LDV) measurement is demonstrated in the present paper.

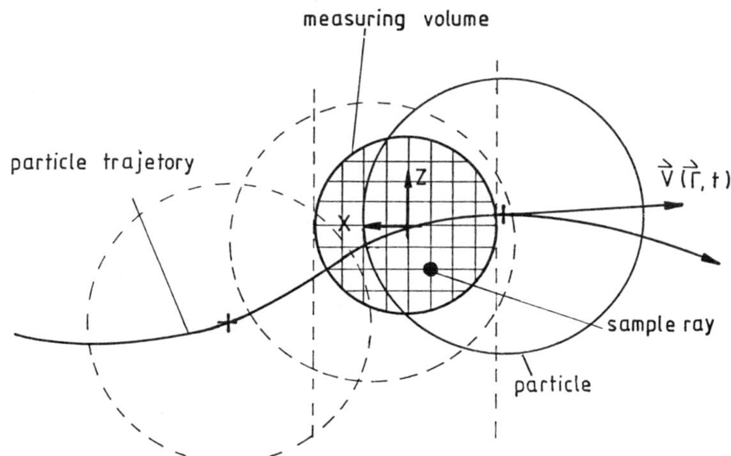

Fig. 1: Front view of measuring volume with particle passing non-symmetrically.

The procedure requires a geometrical description of the optical path of each light ray

$$(\vec{r}_k - \vec{r}_{ok}) \times \vec{s}_k = 0 \tag{1}$$

and an equation for the electrical field amplitude vector of the wave equation:

$$E_k(\vec{r},t) = \mathrm{Re}(E_{ok} \cdot \exp\{2\pi i f_k (t - \vec{r} \cdot \vec{s}_k/c) + \phi_k(\vec{r})\}) \tag{2}$$

Where $\vec{r}_{ok}$ is the starting point of the differential ray k, $\vec{r}_k$ is a location vector anywhere in space and $\vec{s}_k$ is the propagation or unit vector of the ray. $\vec{E}_k$ stands for the local instantaneous electrical field amplitude vector that is evaluated for the real part of the wave equation containing the frequency f_k, time t, speed of light c and the phase ϕ_k. In the following the index k will be omitted.

Figure 1 gives an impression on the complexity of a large particle interacting with a sample ray being diverted through space. The effect of defraction is neglected in all geometrical optices approaches and would be most pronounced in the above particle position, however, the particle size is always kept large compared to the laser light wave length.

The employed laser beams of diameter d, total power P each are supposed to have a symmetrical, concentric Gaussian intensity distribution. A set-up coordinate system is being used indexed $\vec{m}_{x'}$, $\vec{m}_{y'}$, $\vec{m}_{z'}$, providing the laser beam intensity distribution in the z'/y' plane as indicated in Figure 2.

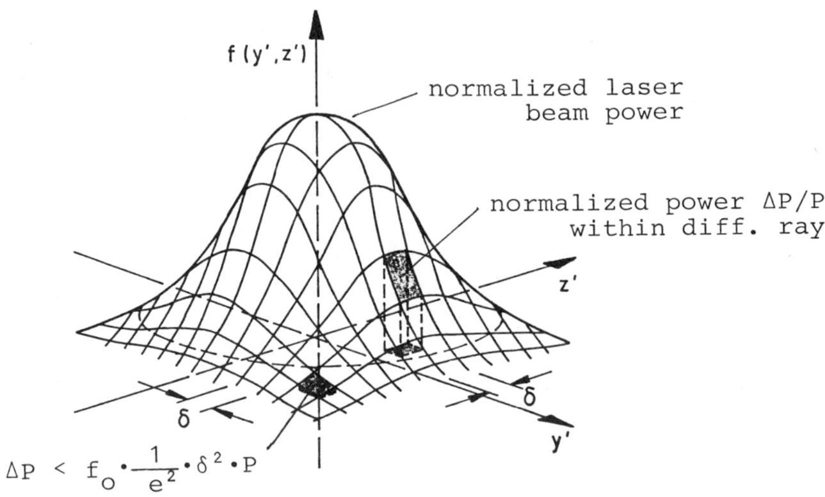

Fig. 2: Light intensity distribution in laser beam

It is assumed that x' is the propagation direction defined by the unit vector $\vec{s}' = (1,0,0)$.

$$f(y',z') = 8/\pi d^2 \cdot \exp\{-8(y'^2 + z'^2)/d^2\} \qquad (3)$$

The Gaussian intensity beam is split into individual rays incremented by δ with the coordinates

$$\vec{r}' = (x',y',z') = (0,m\delta,n\delta); \qquad -N,M \leq n,m \leq +N,M \qquad (4)$$
$$\sqrt{(M^2 + N^2)} \cdot \delta \leq d/2$$

Each ray has a given power ΔP

$$\Delta P = f(\vec{r}') \cdot \delta^2 \cdot P \qquad (5)$$

that permits to deduce the power density of each ray and obtain the time mean amplitude of the electrical field vector $\vec{E}'$ /20/.

$$\overline{W} = f(\vec{r}') \cdot P = \overline{|\vec{E}^2|} \cdot \varepsilon/4\pi \qquad (6)$$

It can be shown that mean electrical field amplitude and maximum amplitude $\vec{E}_o'$ relate.

$$\overline{|\vec{E}|^2} = \overline{|\vec{E}_o|^2}/2 = (E^2_{ox'} + E^2_{oy'} + E^2_{oz'})/2 \qquad (7)$$

Considering that for electromagnetic waves where only the x-component of $\vec{E}$ and $\vec{H}$ is equal to zero, we obtain the unit vector $\vec{u}'$ for a known polarisation ratio $U_{yz} = (E_{oy'} / \sqrt{E^2_{oz'} + E^2_{oy'}})$:

$$\vec{u}' = (u'_x, u'_y, u'_z) = (0, U_{yz}, \sqrt{(1-U^2_{yz})}) \qquad (8)$$

$$\vec{E}' = |\vec{E}'_o| \cdot \vec{u}' \qquad (9)$$

For further computation the electrical field vector $\vec{E}'$, the propagation unit vector $\vec{s}'$ and the point of origin $\vec{r}_o'$ of a ray is transformed to the main coordinate system by a matrix M that accounts for the laser beam crossing angle ψ for a chosen system.

Main coordinate system

The main coordinate system as shown in Figure 3 is arranged so that the laser beams lie in the x = 0 plane while each beam is inclined $\psi/2$ respectively $-\psi/2$ versus the y-axis.

$$M_1 = \begin{pmatrix} 1 & 0 & 0 \\ 0 & \cos(\psi/2) & \sin(\psi/2 + \pi/2) \\ 0 & \sin(\psi/2) & \cos(\psi/2 + \pi/2) \end{pmatrix} \qquad (10)$$

$$M_2 = \begin{pmatrix} 1 & 0 & 0 \\ 0 & \cos(-\psi/2) & \sin(-\psi/2 + \pi/2) \\ 0 & \sin(-\psi/2) & \cos(-\psi/2 + \pi/2) \end{pmatrix} \qquad (11)$$

Multiplying with the above matrices the transformed quantities in the main coordinate system write: $\vec{E}_{o1}$, $\vec{E}_{o2}$, $\vec{s}_{o1}$, $\vec{s}_{o2}$, $\vec{r}_{o1}$, $\vec{r}_{o2}$. The index one and two refers to the laser beam one or beam two that the ray belongs to.

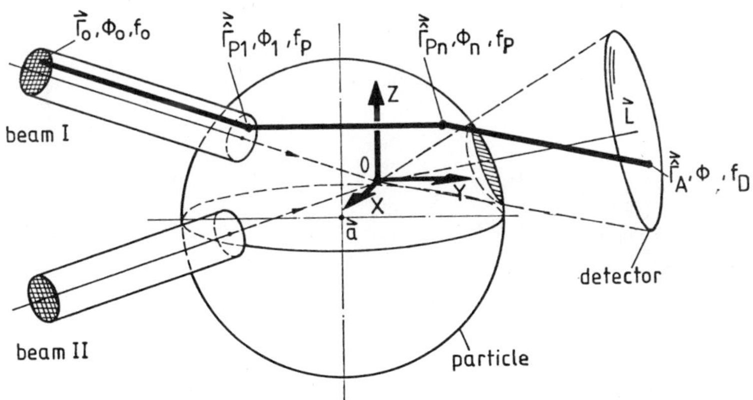

Fig. 3: Main coordinate system defining rays of two laser beams, a spherical particle and an aperture as a detector plane

For the further processing variables in the main coordinate system are non-indexed. As soon as a ray intersects with the particle/fluid interface at $\vec{r}_p$ the following conditions have to be fullfilled.

$$(\hat{r}_{px} - a_x)^2 + (\hat{r}_{py} - a_y)^2 + (\hat{r}_{pz} - a_z)^2 = (d_p/2)^2 \tag{12}$$

$$s_y \cdot \hat{r}_{px} - s_x \cdot \hat{r}_{py} = s_y \cdot r_{ox} - s_x \cdot r_{oy} \tag{13}$$

$$s_z \cdot \hat{r}_{px} - s_x \cdot \hat{r}_{pz} = s_z \cdot r_{ox} - s_x \cdot r_{oz} \tag{14}$$

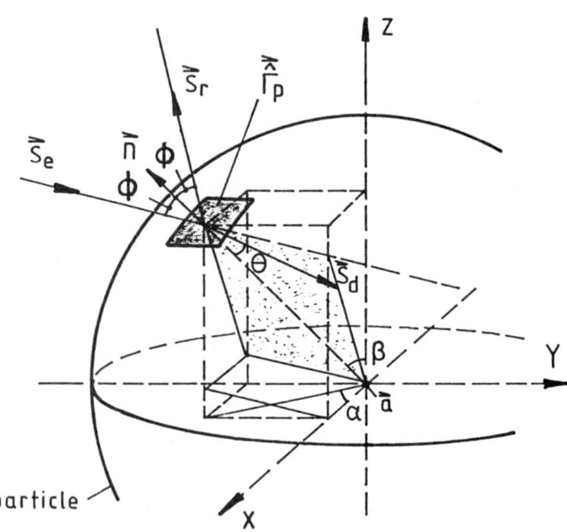

Fig. 4: Ray/particle interaction and definition of normal vector $\vec{n}$ on surface element

The following relationships can be derived:

$$\vec{n} = (\vec{\hat{r}}_p - \vec{a})/|\vec{\hat{r}}_p - \vec{a}| = (\sin\beta \cdot \cos\alpha,\ \sin\beta \cdot \sin\alpha,\ \cos\beta) \tag{15}$$

$$\vec{n} \cdot \vec{s}_e = |\vec{n}| \cdot |\vec{s}_e| \cdot \cos(\pi - \phi) = \cos(\pi - \phi) \tag{16}$$

$$\phi = \pi - \arccos(\vec{n} \cdot \vec{s}_e) \tag{17}$$

$$\theta = \arcsin(n_{12} \cdot \sin\phi) \tag{18}$$

where $\vec{a}$ is the particle unit point, $\vec{s}_e$ is the unit vector of the present ray and $\vec{r}_p$ is the intersection point of the ray on the particle surface. For the further tracing of the incident light, namely splitting intensities into reflected $\vec{s}_r$ and refracted components $\vec{s}_d$ it is switched to a local coordinate system.

Local coordinate system

As often as a ray with the unit vector $\vec{s}_e$ interacts with a surface element on the particle, a new local coordinate system with the base vectors $\vec{m}_1$, $\vec{m}_2$, $\vec{m}_3$ its origin at $\vec{r}_p$ is generated to compute reflected $\vec{s}_r$ and refracted $\vec{s}_d$ contributions $\vec{E}_r$, $\vec{E}_d$ of the light intensity.

Where

$$\vec{m}_1 = (\vec{s}_e \times \vec{n})/(|\vec{s}_e \times \vec{n}|) \tag{19}$$

$$\vec{m}_2 = (\vec{m}_1 \times \vec{m}_3)/(|\vec{m}_1 \times \vec{m}_3|) \tag{20}$$

$$\vec{m}_3 = \vec{n} \tag{21}$$

Now the propagation vectors of an individual ray can be obtained for the <u>reflected</u> contribution:

$$
\begin{aligned}
\vec{s}_r \cdot \vec{s}_e &= \cos(\pi - 2\phi) \\
\vec{s}_r \cdot \vec{n} &= \cos\phi \\
\vec{s}_r \cdot \vec{m}_2 &= 0
\end{aligned}
\tag{22}
$$

for the <u>refracted</u> contribution

$$
\begin{aligned}
\vec{s}_d \cdot \vec{s}_e &= \cos(\phi - \theta) \\
\vec{s}_d \cdot \vec{n} &= \cos(\pi - \theta) \\
\vec{s}_d \cdot \vec{m}_2 &= 0
\end{aligned}
\tag{23}
$$

Figure 4 shows that the refracted and the reflected beam will proceed in a plane defined by the particle center position $\vec{a}$ and the intersection point $\vec{r}_p$ and the unit vector of the incident ray $\vec{s}_e$. The beam will stay in this plane also after multiple internal reflections.

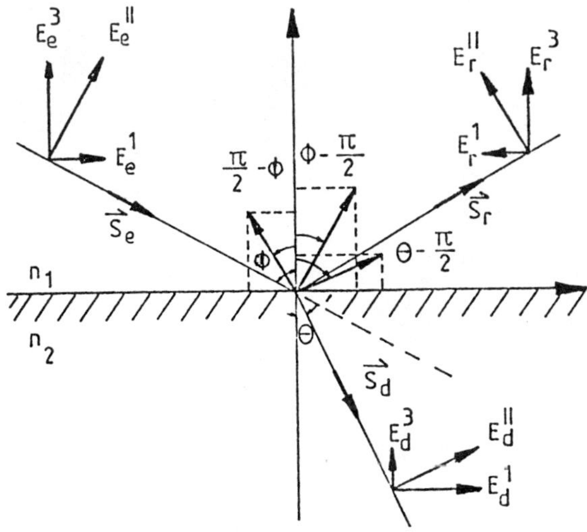

Fig. 5: Local coordinate system

In the local coordinate system see Figure 5, the amplitude vector $\vec{E}_e$ of the incident ray can be divided into its components $E_e^{\perp} = E_{ey}$ perpendicular to the surface and $E_e^{\parallel} = +E_{ex}/\cos\phi = E_{ez}/\sin\phi$ parallel to the surface giving the equivalent components for reflected part $E_r^{\perp}$, $E_r^{\parallel}$ and the refracted part $E_d^{\perp}$, $E_d^{\parallel}$:

$$E_d^{\parallel} = \frac{2\sin\theta \cdot \cos\phi}{\sin(\phi+\theta) \cdot \cos(\phi-\theta)} \cdot E_e^{\parallel} \tag{24}$$

$$E_d^{\perp} = \frac{2\sin\theta \cdot \cos\phi}{\sin(\phi+\theta)} \cdot E_e^{\perp} \tag{25}$$

$$E_r^{\parallel} = \frac{\tan(\phi-\theta)}{\tan(\phi+\theta)} \cdot E_e^{\parallel} \tag{26}$$

$$E_r^{\perp} = \frac{\sin(\phi-\theta)}{\sin(\phi+\theta)} \cdot E_e^{\perp} \tag{27}$$

These equations can be converted to describe the new electrical field vector:

$$\vec{E}_e = (E_e^{\parallel} \cdot \sin(-\pi/2+\phi), \; E_e^{\perp}, \; E_e^{\parallel} \cdot \cos(-\pi/2+\phi)) \tag{28}$$

$$\vec{E}_r = (E_r^{\parallel} \cdot \sin(\pi/2-\phi), \; E_r^{\perp}, \; E_r^{\parallel} \cdot \cos(\pi/2-\phi)) \tag{29}$$

$$\vec{E}_d = (E_d^{\parallel} \sin(-\pi/2+\theta), \; E_d^{\perp}, \; E_d^{\parallel} \cdot \cos(-\pi/2+\theta)) \tag{30}$$

Multiplying with an appropriate transformation matrix the above vectors can be retransformed to the main coordinate system to be employed for further computation e.g. new particle interaction, path to detector plane.

Frequency

Figure 6 shall explain the Doppler light frequency shift of a ray originating from the system Σ_0 at rest, received by a scatter Σ_p that moves at a finite speed $\vec{V}_p$ and emits the light that is received by a detector system Σ_D also at rest.

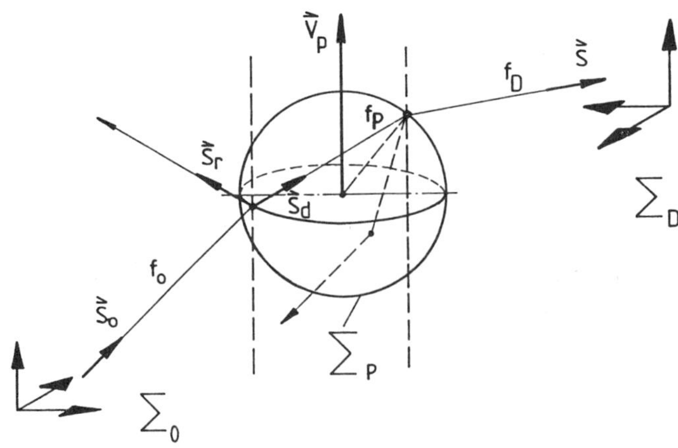

Fig. 6: Doppler light frequency shift including a light source Σ_0, a scatterer Σ_p and a detector Σ_D.

The light gets shifted between system Σ_o and Σ_p

$$f_p = f_o \cdot (1 - \vec{V}_p \cdot \vec{s}_o / c) \tag{31}$$

and between system Σ_p and Σ_D

$$f_D = f_p \cdot (1 + \vec{V}_p \cdot \vec{s} / c) \tag{32}$$

The total light frequency shift f_D then reads

$$f_D = f_o \cdot (1 - \vec{V}_p \cdot \vec{s}_o / c) \cdot (1 + \vec{V}_p \cdot \vec{s} / c)$$
$$\simeq f_o \cdot \{ 1 - \vec{V}_p \cdot (\vec{s}_o - \vec{s}) \}, \quad |\vec{V}_p| << c \tag{33}$$

where f_o is the original light frequency, $\vec{V}_p$ the speed of the particle, $\vec{s}_o$ the unit vector of the incident ray, $\vec{s}$ the scattered ray and c the speed of light. It should be emphasized here that f_D is not the frequency usually analysed for velocitiy detection, while f_{LDV} is the so-called Doppler-frequency.

$$f_{LDV} = f_{o_1} \cdot \{ 1 - (\vec{s}_{o_1} - \vec{s}_1) \cdot \vec{V}_p / c \} -$$
$$- f_{o_2} \cdot \{ 1 - (\vec{s}_{o_2} - \vec{s}_2) \cdot \vec{V}_p / c \} \tag{34}$$

The beating frequency f_{LDV} is several orders of magnitude smaller than f_D.

<u>Integration</u>

Defining a round plane in space with a centerpoint $\vec{L}$ and a diameter D acting like the light collecting apperture of an LDV-system, rays can be selectively collected in space and time. If its normal vector is directed towards the origin of the main coordinate system every ray/surface intersection point $\vec{r}_A$ has to fullfill

$$(\vec{r}_A - \vec{L}) \cdot \vec{L} = 0 \tag{35}$$

Whether a ray hits the surface and contributes to the light integrated is determined by

$$\vec{r}_A \cdot \vec{s} > 0 \tag{36}$$

$$|\vec{r}_A - \vec{L}| \leq D/2 \tag{37}$$

If the normal vector of the detector plane would not point towards the origin of the optical set-up an experiental set-up would not be aligned properly.

For the vector summation of all $\vec{E}_k$ contributions from all rays hitting a detector surface per time increment the phase at the intersection point $\vec{r}_A$ has to be known. Figure 3 shows a ray passing through a transparent particle including the frequency f_o being the unshifted freqnency of the laser, f_p being the frequency of the light shifted once inside the particle and the freqnency f_D of the light leaving the particle being shifted twice. Equation (39) is used to trace the local phase Φ through space knowing the local frequency and the initial phase Φ_o of the electrical field vector is defined according to the set-up conditions $\Phi = 0$, $\vec{r} = \vec{r}_o$, and $\vec{s} = \vec{s}_o$

$$\Phi_o = f_o \cdot (\vec{r}_o \cdot \vec{s}_o)/c \qquad (38)$$

The total phase difference between the origin and the detector phase can be split into three parts including the path length and the freqnency within each section of the ray.

$$\Phi = - f_o \cdot (\vec{r}_p - \vec{r}_o) \cdot \vec{s}_o/c - f_i \cdot \sum_{j=1}^{n} (\vec{r}_p^{\,j-1} - \vec{r}_p^{\,j}) \cdot \vec{s}_j/c -$$
$$- f_D \cdot (\vec{r}_A - \vec{r}_p^{\,n}) \cdot \vec{s}/c \qquad (39)$$

The three terms in the equation take care of the path length and the frequency f_o from the origin Φ_o to the particle surface, the path length (including multiple reflection) and frequency f_p within the particle and the path length and frequency f_D between the particle surface and the detector plane.

A resultant electrical field summation vector $\vec{E}_t$ can be obtained according to equation (40)

$$\vec{E}_t(\vec{r}_A, \hat{t}) = \sum_{k=1}^{n} \vec{E}_{ok} \cdot \cos\{2\pi f_o (1 + (\vec{s}_k - \vec{s}_{ok}) \cdot \vec{V}_p/c) \cdot$$
$$\cdot (\hat{t} - \vec{r}_A \cdot \vec{s}_k/c) + \Phi_k\} \qquad (40)$$

Here n denotes the number of rays k crossing the detector plane at time $\hat{t}$, while $\vec{V}_p$ is the particle velocity and $\vec{s}_k$ the unit vector of the ray. The light intensity $I_t(\hat{t})$ can then be obtained by evaluating eq. (41).

$$I_t(\hat{t}) = \frac{c}{4\pi} \sqrt{\frac{\varepsilon}{\mu}} \, |\vec{E}_t(\hat{t})|^2 \qquad (41)$$

Still some averaging and smoothing has to be done similarly to what a photo-detector will do to recognize a Doppler-frequency rather than respond to instantaneous light intensity changes of very high frequency. To this end each of the wave equations is averaged over approximately 50 cycles of the laser light frequency. If this is repeated for sufficiently many time steps within the passage time of the particle T_p using sufficiently many rays to reach a "quasi" continuous light distribution in spaces, the intensity distribution in time will give you a Doppler-signal and the distribution in spaces a typical fringe pattern.

VERIFICATION AND RESULTS

The numerical simulation package described above was run on an HP 1000, A-700 Micro-Computer approximately requiring 0.4 MB memory, 0.5 MB direct access disc and 80 MB magnetic tape storage space. The limited storage capacity of the system required to set limits to (i) the time resolution (ii) the spacial resolution and (iii) the discredisation of the beams since all rays and wave equations are stored in space and time to allow repetitive processing of the same scattered light. Depending on the actual problem the priorities can be set accordingly. Figure 7 gives an overview on different test runs of a bubble $n_{12} = 0.75$ rising symmetrically through a measuring volume.

The particle diameter $d_p = 1$ mm, the laser beam crossing angle $\psi = 5°$, the center point of the forward scatter receiving optics at $\dot{L}$ with a diameter $D = 10$ mm, beam diameter $d = 0.2$ mm and its power $P = 5$ watt, wave length $\lambda = 628$ nm were prescribed, changing the time resolution ΔT of the particle passage through the measuring volume. Typical signal shapes reported in the literature e.g. /15,16/ could be reproduced, fully modulated also in the side peaks that are dominated be reflected light, see Figure 7a.

Figure 7b provides a better resolution of the so-called Doppler-frequency f_{LDV} that typically diviated 5 % from these standard LDV formular $f = 2 V_{pz} \cdot \sin \psi / \lambda$ for the above geometry, see also /18/. Figure 7c shows the effect of the discrete rays that sum up and depending on the present number of rays produce steps in the intensity time plot. The effects of particle size, receiving apperture size and detector position could be reproducted as reported in the literature.

In /17/ Ohba and Matsuyama report on the displacement of the LDV DC signal contribution in time received by two photodetectors in space to size particles. Figure 8 gives an overview on the DC-displacement observed in time by three differently located photodetectors (8a-8c) for a bubble $d_p = 0.5$ mm.

Comparison of results for differently sized particles shows that the linear relationship for particle size and delay of DC-signal contributions holds. It could also be shown that if the detectors are brought closer together this method is equivalent to the phase-Doppler technqiue /10-14/, however, in this case the determination of phase difference is much more precice than finding a DC-maximum off set.

Figure 9 gives an example of LDV-signals from three detectors closely placed together in forward scatter direction to evaluate the phase shift $\Delta \Phi$ depending on their spacial orientation and the size of the particle. In this instance overlapping lenses of $\Delta L = 3$ to 5 mm were chosen. Linear dependence of phase shift and sizes could be shown within 6 % accuracy.

CONCLUSION

The authors provide a general purpose package for the prediction of scattered light intensity in space and time from spherical particles. A new approach of numerically simulating an LDV optical set-up was tested and proved reliable for a number of problems in scattered light

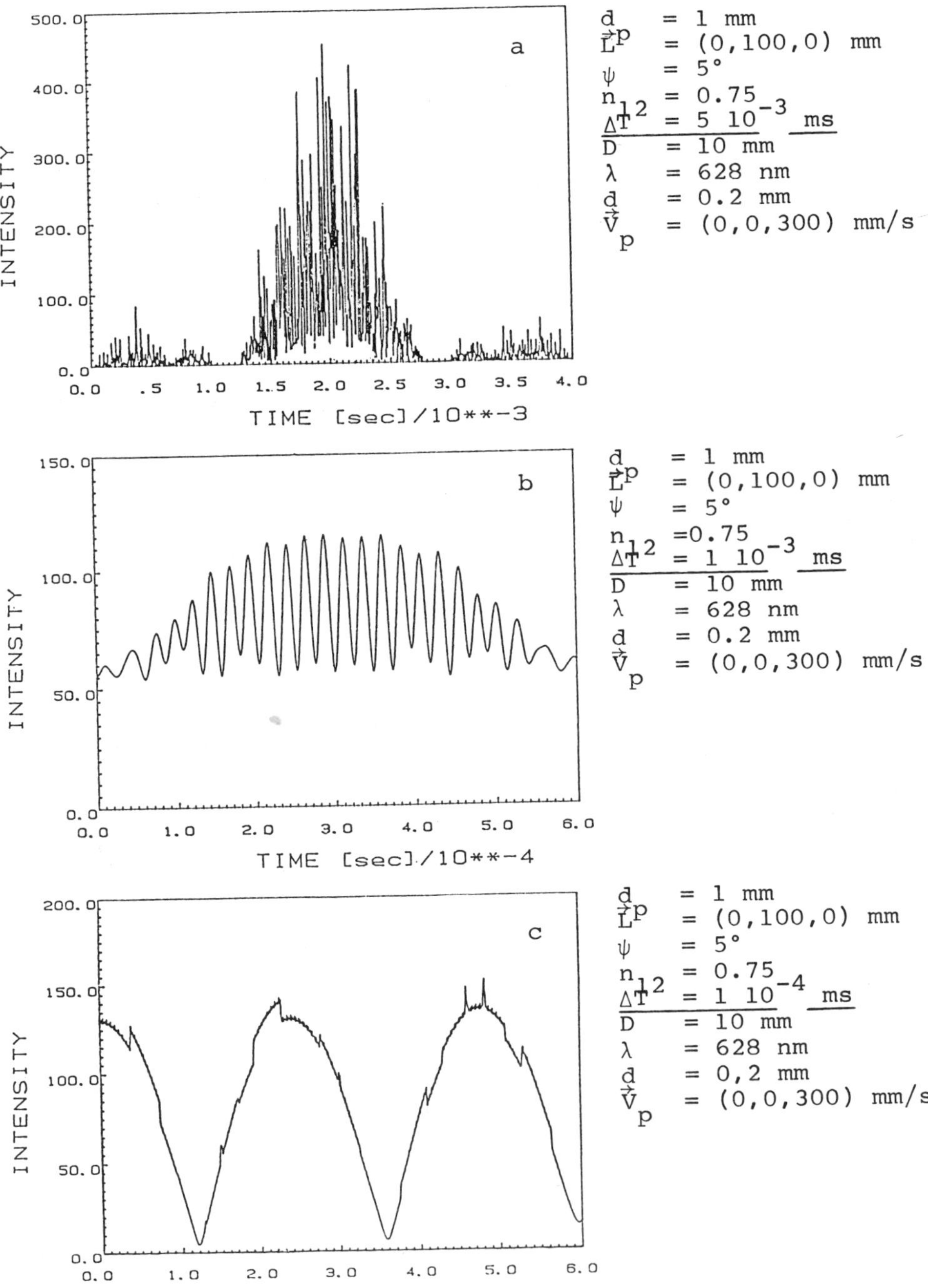

Fig. 7: Time resolution analysis for verification.

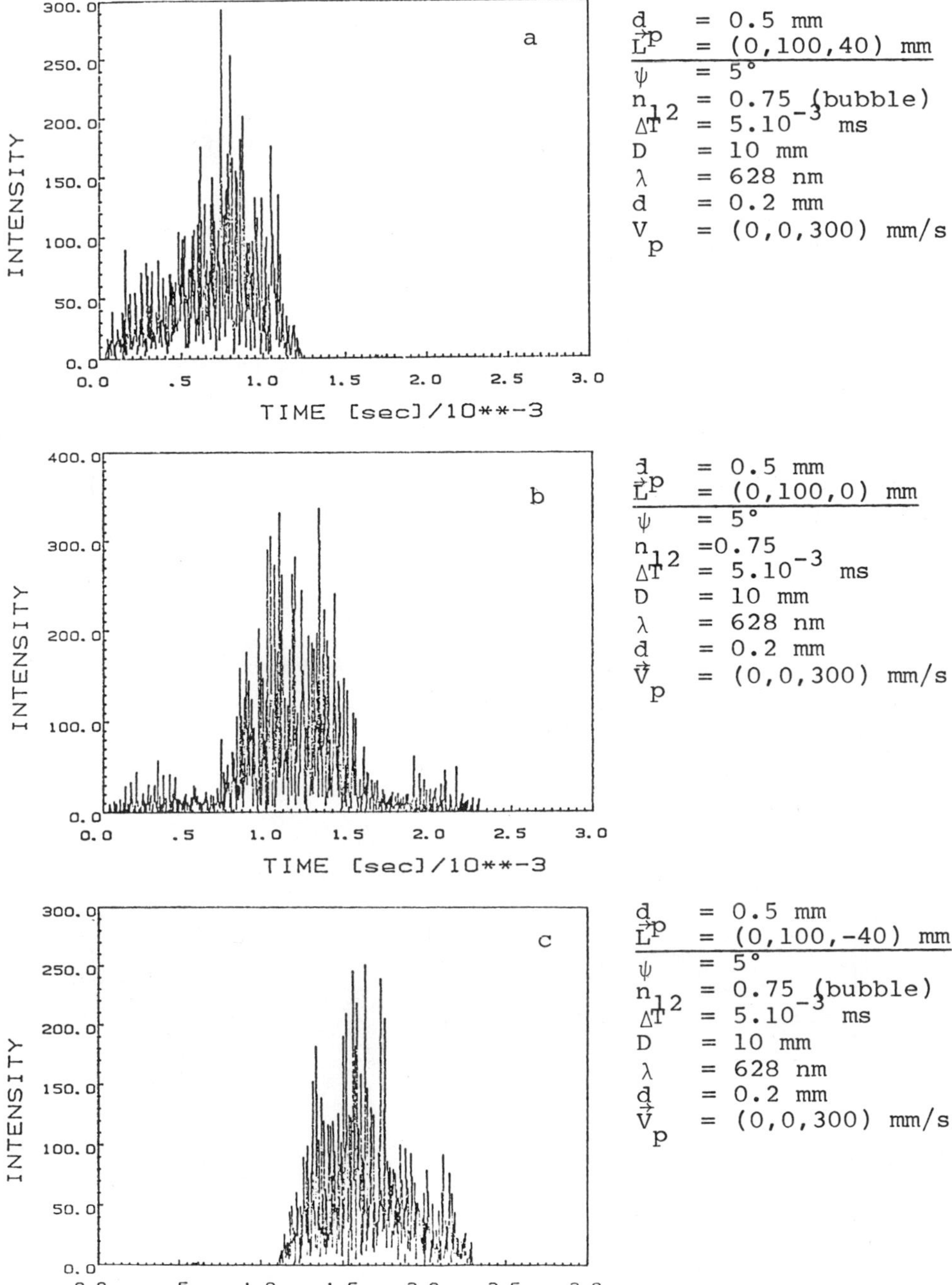

Fig. 8: Displacement of LDV-DC signal contribution in time observed by three photodetectors.

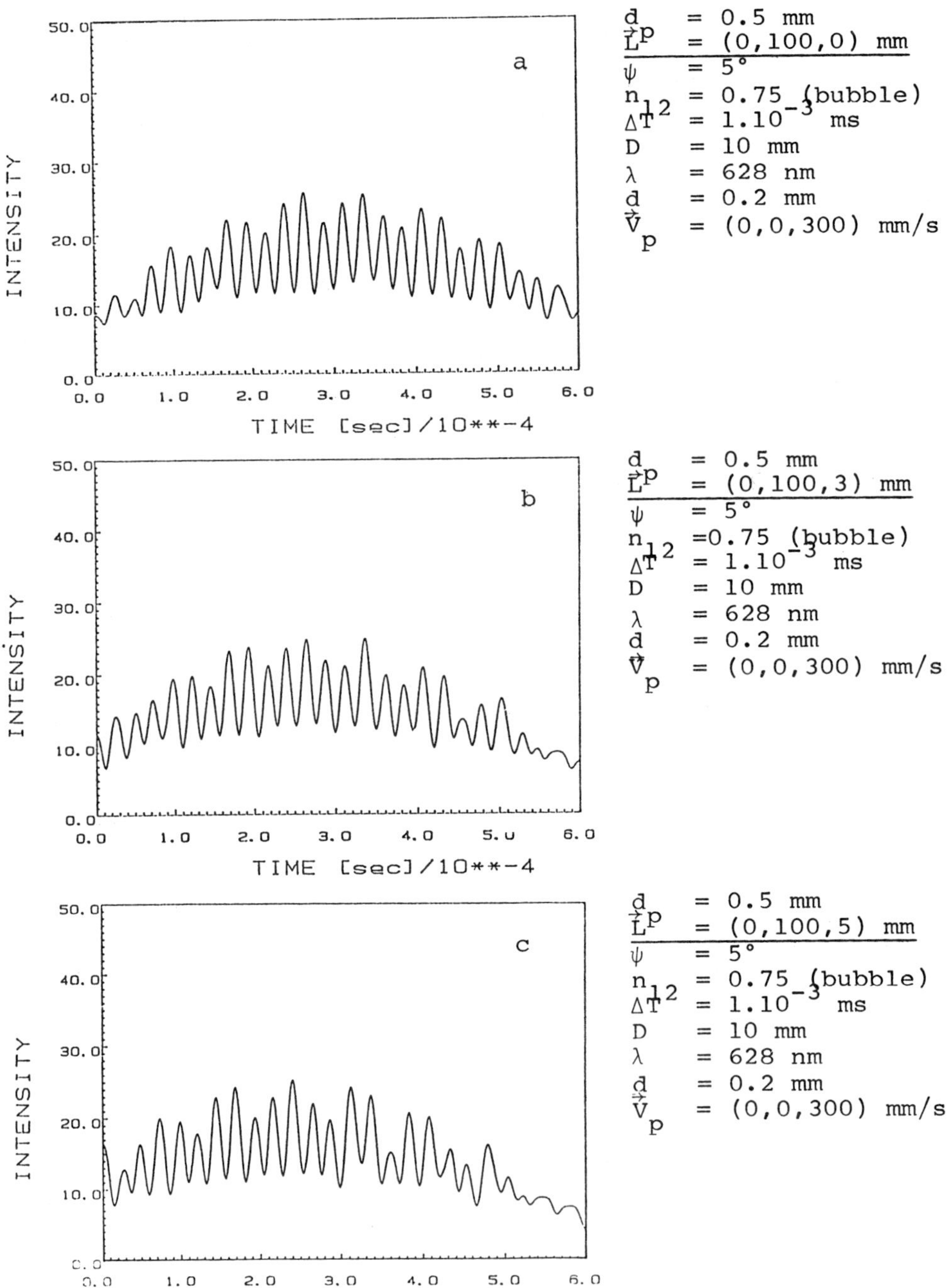

Fig. 9: Phase-Doppler simulation using three detectors.

analysis for particles. To this end differential rays are
traced towards and through a particle and are subsequently
integrated over a given detector plane and/or for a given
time. The procedure allwos to take all relevant features of
large particle light scattering into account: relative refractive index, particle velocity, parameters of LDV-
systems, internal circulation of fluid particles, any
non-linear and/or non-steady passage through the elysoidal
measuring volume.

The potential of the technique is its possible application to (i) the simulation of various LDV-systems e.g. dual
beam, reference beam, phase doppler etc. (ii) their opti-
mization and (iii) detailed biasing analysis.

ACKNOWLEDGEMENTS

The authors would like to acknowledge the financial
support of the Deutsche Forschungsgemeinschaft (DFG) to perform this work at the Lehrstuhl für Strömungsmechanik,
University of Erlangen-Nürnberg. Support for Lidong Zhan was
granted by the Chinese government.

REFERENCES

1. G. Mie, Beiträge zur Optik trüber Medien, speziell
   Kolloidaler Metallösungen, Ann. Phys., Vol. 25, No.
   3, p. 377 (1908).
2. W.P. Chu, D.M. Robinson, Scattering from a moving
   spherical particle by two crossed coherent plane
   waves, Appl. Optics, Vol. 16, No. 3, p. 619 (1977).
3. D.M. Robinson, W.P. Chu, Diffraction analysis of
   Doppler signal characteristics for cross beam laser
   Doppler velocimeter, Appl. Optics, Vol. 14, No. 9,
   p. 2177 (1975).
4. G. Gouesbet, G. Graham, B. Mahen, Scattering of a
   Gaussian beam by a Mie scatter center using a Brown-
   wich formalism, J. Optics, Vol. 16, No. 2, p. 83
   (1985).
5. D.G. Fergusson, I.G. Currie, Theoretical evaluation
   of LDA-techniques for two-phase flow measurements,
   in Proc. of the 3rd Int. Symp. on Appl. of LDA to
   Fluid Mechanics, Lisbon, Portugal (1986).
6. H.C. van de Hulst, in "Light scattering by small
   particles", John Wiley & Sons, New York (1957).
7. W.J. Glantschnig, S. Chen, Light scattering from
   water droplets in the geometrical optics approximation, Appl. Optics, Vol. 20, p. 2499 (1981).
8. J.R. Hodkins, I. Greenleaves, Computations of light-
   scattering and extinction by spheres according to
   diffraction and geometrical optics and some comparison with the theory, J. of the Opt. Soc. of
   America, Vol. 53, No. 5, p. 577 (1963).
9. W.M. Farmer, Observation of large particles with a
   laser interferometer, Appl. Optics, Vol. 3, p. 610
   (1974).
10. F. Durst, M. Zaré, Laser-Doppler measurements in
    two-phase flows, SFB-report No. 80/TM/63, Univer-
    sity of Karlsruhe (1975).

11. W.D. Bachalo, Method for Measuring the size and Velocity of spheres by a dual beam light scatter interferometer, _Appl. Optics,_ Vol. 19, p. 403 (1980).
12. W.D. Bachalo, M.J. Houser, Analysis and testing of a new method for drop size measurement using light scatter interferometry, Report no. NAS3-23684, NASA Lewis Research Center (1983).
13. M. Saffman, P. Buchhave, H. Tanger, Simultaneous measurement of size, concentration and velocity of spherical particles by a laser-Doppler method, Proc. 2nd. Int. Symp. on Apl. of LDA to Fluid Mech. (1984).
14. Y. Hardalupas, Phase-Doppler anemometry for simultaneous particle size and velocity measurement, Imp. College of Science and Technology, Report no. FS/86/14 (1986).
15. W.W. Martin, A.H. Abdelmessik, J.J. Liska, F. Durst, Characteristics of laser-Doppler signals from bubbles, _Int. J. of Multiphase Flow,_ No. 7, p. 439 (1980).
16. A. Brankovic, I.G. Currie, W.W. Martin, Laser-Doppler mesurements of bubble dynamics, _Phys. Fluids,_ Vol. 27, No. 2, p. 348 (1984).
17. K. Ohba, H. Matsuyama, Simultaneous measurement of size and velocity of large particles, Proc. of the 3rd. Int. Symp. on Appl. of LDA to Fluid Mechanics, Lisbon, Portugal (1986).
18. A. Chartellier, J.L. Achard, Limitations of the classical LDA formula for velocity measurements for large particles in two-phase suspension flows, _Phys. Chem. Hydrodynamics,_ Vol. 6, No. 4, pp. 463 (1985).
19. S.S. Sadal, R.E. Johnson, Stokes flow past bubbles and drops partially coated with thin films, _J. Fluid Mech.,_ Vol. 129, p. 237 (1983).
20. F. Durst, A. Melling, J.H. Whitelaw, "Principles and Practice of laser-Doppler Anemometry", Academic Press, London (1981).

EMPLOYMENT OF LIGHT SCATTERING INFORMATION
TO LAY OUT OPTICAL MEASURING SYSTEMS FOR
MEASUREMENTS OF PARTICLE PROPERTIES

Franz Durst
Lehrstuhl für Strömungsmechanik
Universität Erlangen-Nürnberg
Egerlandstr. 13, 8520 Erlangen, FRG

SUMMARY OF CONTENTS

The present paper provides a review of the employment of theoretical
information on light scattering to lay out optical measuring systems for
measurements of particle properties. Optical particle velocity, particle
size and particle concentration measurements are considered and special
instrumentation requirements are pointed out. It is shown that the most
stringthend requirements exist for the measurements of the particle
velocity. For this, laser-Doppler anemometry is employed resulting in
high frequency signals yielding special requirements for photodetectors.
These are explained in detail, since they define the final signal-to-
noise ratio obtainable from measuring systems. It is shown that the
employment of different photodetectors for particle velocity and particle
size measurements yields an optimum for the lay out of the detection
system. For laser-Doppler measurements, the special properties of lasers,
e.g. the existence of discrete axial modes, do also need to be considered
for high particle velocity measurements. They usually do not effect measurements at low particle velocities.

Using the above information, in conjunction with computations of
light scattering intensity and phase permits an entire concept for the
lay out of optical systems to be introduced for particle velocity, size
and concentration measurements. The major considerations of this lay out
concept are summarized and references that contain more details are provided. Various systems that were designed according to this concept are
shown and examples of measurements presented. Indications of further improvements are given and suggestions for more detailed measurements are
made.

INTRODUCTION

The lecture will provide conclusions that came out of the long time
the author and his collaborators have spent in the field of optical
instrument developments, relating mainly to laser-Doppler anemometry but
also to particle sizing and particle concentration measurements. Relevant
information from this work is provided in form of slide panels with short
explanations. This will support the lecture being presented at the
Symposium on Optical Sizing: Theory and Practice in Rouen, France.

The present lecture:

o looks at combined optical systems to measure particle properties (velocity, size and concentration)
o treats light scattering from a measuring volume containing
  two light beams
o theoretically treats photodetectors and their influence on
  SNR of measuring signals
o considers lasers and the influence of their axial modes on
  LDA-measurements
o provides instructions on the lay out of optical systems
o introduces optical systems and shows results

SUMMARY OF LECTURE

The above slide provides a summary of the lecture and states the
various points that will be treated in detail.

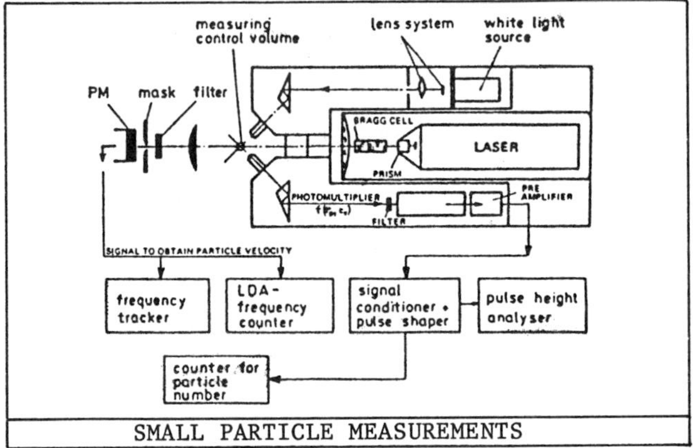

SMALL PARTICLE MEASUREMENTS

For measurements of small particles, the above system is recommended. It consists of a combined laser-Doppler, white-light source
optical system.

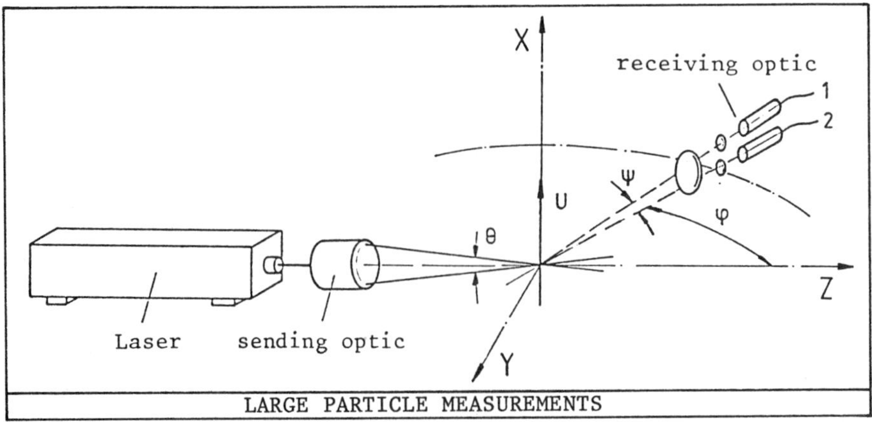

LARGE PARTICLE MEASUREMENTS

For measurements of large particles, the recommended system embraces
a laser-Doppler anemometer and a Doppler phase particle sizer.

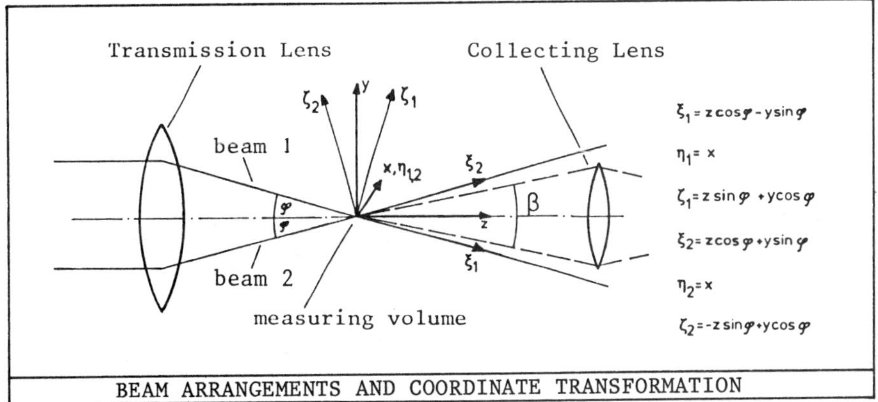

BEAM ARRANGEMENTS AND COORDINATE TRANSFORMATION

It is assumed that the measuring control volume is penetrated by two laser beams of equal light intensities possessing distributions as given below:

$$I(\zeta^2 + \eta^2) = \frac{2}{\pi s^2} \frac{P_{\mathrm{L}}}{2} \exp\left[-2\frac{\zeta^2+\eta^2}{s^2}\right]. \tag{1}$$

Employing the above given coordinate transformation yields

$$I_1 = \frac{2}{\pi s^2} \frac{P_{\mathrm{L}}}{2} \exp\left[\frac{-2}{s^2}(x^2 + y^2\cos^2\phi + z^2\sin^2\phi + yz\sin(2\phi))\right],$$

$$I_2 = \frac{2}{\pi s^2} \frac{P_{\mathrm{L}}}{2} \exp\left[\frac{-2}{s^2}(x^2 + y^2\cos^2\phi + z^2\sin^2\phi - yz\sin(2\phi))\right]. \tag{2}$$

For the present considerations it is sufficient to consider the particle motion along the y-axis, along which $I_1 = I_2$ read

$$\begin{aligned}I_1(x=0,\ z=0)\\ I_2(x=0,\ z=0)\end{aligned} = \frac{2}{\pi s^2} \frac{P_{\mathrm{L}}}{2} \exp\left[-2\frac{y^2\cos^2\phi}{s^2}\right]. \tag{3}$$

The higher intensity that is scattered by a small particle in a direction (Θ, Φ) is given below. Different Θ- and Φ-values are given in the relationship below, due to the different angles of the two incident beams

$$(I_s)_1 = \frac{I_1}{k^2 r^2} G_1(\theta_1, \phi_1),$$

$$(I_s)_2 = \frac{I_2}{k^2 r^2} G_2(\theta_2, \phi_2), \tag{4}$$

The quantities $G_1(\Theta_1, \Phi_1)$ and $G_2(\Theta_2, \Phi_2)$ are computed using Mie's theory of light scattering.

The two scattered light waves may be computed as follows:

$$(E_s)_1 = \sqrt{(2(I_s)_1)}\cos[\omega_1 t + \Psi_1],$$

$$(E_s)_2 = \sqrt{(2(I_s)_2)}\cos[\omega_2 t + \Psi_2]. \tag{5}$$

The superposition of both light waves yields:

$$E_s = (E_s)_1 + (E_s)_2 = \sqrt{(2(I_s)_1)}\cos[\omega_1 t + \Psi_1] + \sqrt{(2(I_s)_2)}\cos[\omega_2 t + \Psi_2] \tag{6}$$

More details are provided in ref. /1/.

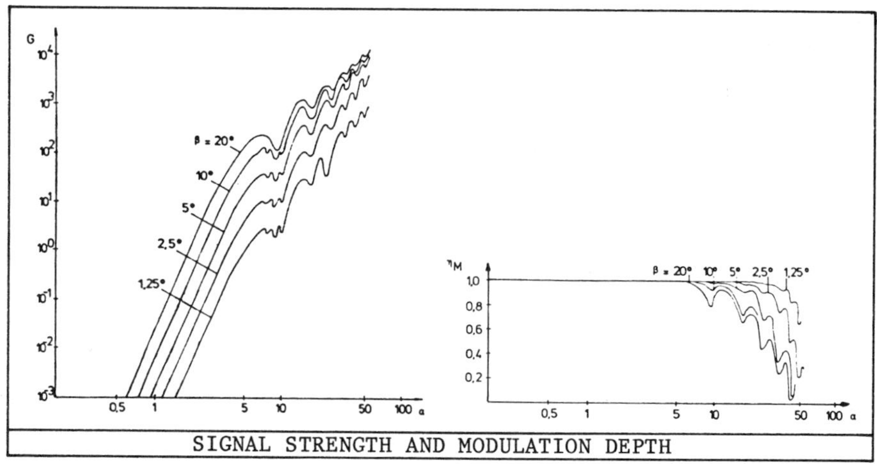

For the two superimposed light beams in equation (6), the light intensity in a direction of observation can be computed to be:

$$I_\text{s} = E_\text{s}^2 = \{\sqrt{(2(I_\text{s})_1)}\cos[\omega_1 t + \Psi_1] + \sqrt{(2(I_\text{s})_2)}\cos[\omega_2 t + \Psi_2]\}^2$$

$$= 2(I_\text{s})_1 \cos^2[\omega_1 t + (\Psi_\text{s})_1] + 2(I_\text{s})_2 \cos^2[\omega_2 t + \Psi_2] + 4\sqrt{[(I_\text{s})_1 (I_\text{s})_2]} \tag{7}$$

$$\cos[\omega_1 t + (\Psi_\text{s})_1] \cos[\omega_2 t + \Psi_2].$$

Since the two light frequencies ω_1 and $\omega_2 \gg (\omega_1 - \omega_2)$, the above relationship can be simplified to yield the signal that will be detectable by these photodetectors employed in optical measuring systems

$$I_\text{s} = (I_\text{s})_1 + (I_\text{s})_2 + 2\sqrt{((I_\text{s})_1 (I_\text{s})_2)}\cos[(\omega_1 - \omega_2)t + \Psi_1 - \Psi_2]. \tag{8}$$

Integration over the entire detection angle of the receiving optical system yields the instantaneous light power received by the photodetector

$$P_\text{s} = (P_\text{s})_1 + (P_\text{s})_2 + 2 \int_\Omega \sqrt{((I_\text{s})_1 (I_\text{s})_2)}\cos[(\omega_1 - \omega_2)t + \Psi_1 - \Psi_2]\, d\Omega$$

or rewritten with: $(\omega_1 - \omega_2) = 2\pi f_\text{D}$ \hfill (9)

$$P_\text{s} = [(P_\text{s})_1 + (P_\text{s})_2]\{1 + \eta_\text{M}\cos(2\pi f_\text{D} t + \Psi)\}$$

The total light power scattered at each position in the measuring volume can be computed using generally applicable computer programs for light scattering of particles yielding the following properties of the resultant optical signal:

Modulation Depth:

$$\eta_\text{M} = \frac{(P_\text{s})_\text{max} - (P_\text{s})_\text{min}}{(P_\text{s})_\text{max} + (P_\text{s})_\text{min}}. \tag{10}$$

Signal Strength:

$$P_\text{s} = \tilde{G} \cdot \frac{I}{k^2}; \tag{11}$$

The quantity $\tilde{G}$ represents a normalized light power predicted from Mie-theory after integration over the detection angle of the scattered light beam. The above diagrams show typical values of G and η_M for given optical arrangements. Further information can be found in ref. /1/ or can be computed for other optical systems using the computer program described in ref. /2/.

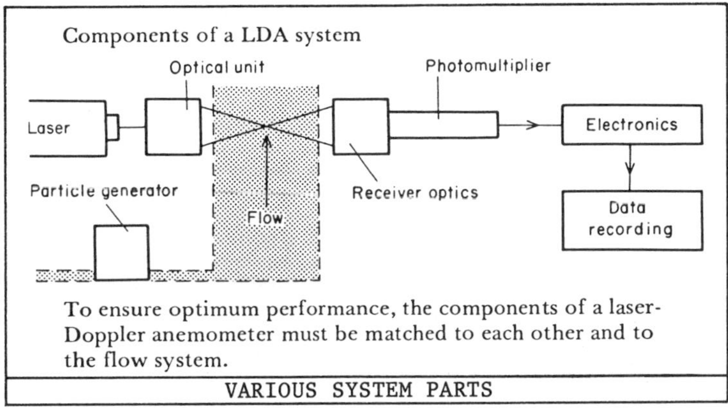

To ensure optimum performance, the components of a laser-Doppler anemometer must be matched to each other and to the flow system.

VARIOUS SYSTEM PARTS

Any optical system to measure particle properties consists of transmission optics, receiving optics, signal processing electronics and data acquisition systems. For optimum performance, it is essential that the various parts are matched to operate in such a way that the best possible signal-to-noise ratio results and measurements can be carried out with the required speed. Matching usually also ensures ease of operation of the resultant system.

It is the task of the transmission optics to provide the required laser power and to assure the light intensity distribution needed in the measuring control volume.

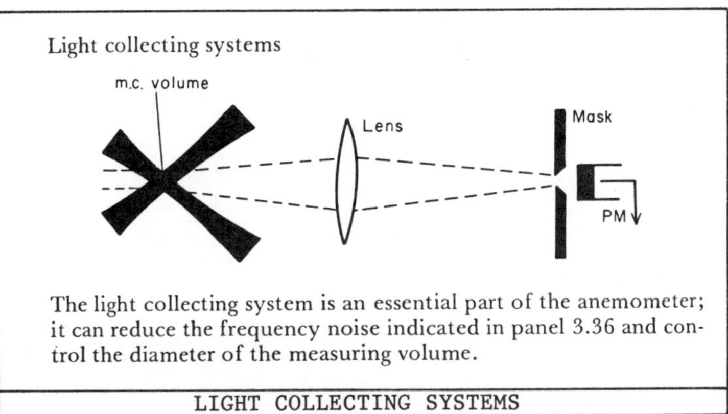

The light collecting system is an essential part of the anemometer; it can reduce the frequency noise indicated in panel 3.36 and control the diameter of the measuring volume.

LIGHT COLLECTING SYSTEMS

Considering the receiving optics, it is essential that only those parts of the measuring control volume enter into the photodetector that provide the relevant information. Appropriate masking is required in order to assure that information from spurious particles results that passes the light beams of the transmission optics in regions outside of the measuring control volume. In most optical systems, the extensions of the measuring control volume is defined by a combination of transmission and receiving optics.

In the following consideration, it is assumed that all the light that comes into the receiving part of the optical system and enters the mask in front of the photomultiplier, will be available for signal detection. This detection usually means a transformation of an optical signal into an electrical signal. This transformation needs further considerations.

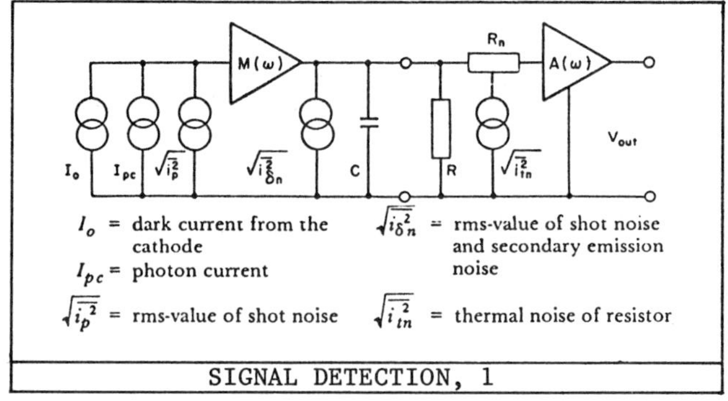

I_o = dark current from the cathode

I_{pc} = photon current

$\sqrt{\overline{i_p^2}}$ = rms-value of shot noise

$\sqrt{\overline{i_{\delta n}^2}}$ = rms-value of shot noise and secondary emission noise

$\sqrt{\overline{i_{tn}^2}}$ = thermal noise of resistor

SIGNAL DETECTION, 1

Photodiodes, Avalanche-photodiodes and photomultipliers are photo-detectors that are extensively employed in optical measuring systems for particle properties. All these photodetectors can be analytically treated using the idealized operational diagram shown above. This diagram con-tains all the essential components of the detector and introduces their influence on the signal detection, i.e. on the resultant signal-to-noise ratio. The cathode current i_{Kath} results from the detected photon current also yielding a shot noise contribution that can be given as

$$i_{\mathrm{sn}}^2 = 2ei_{\mathrm{Kath}}\Delta f. \tag{12}$$

Most photodetectors possess an internal amplification which also acts on the shot noise present after the cathode:

$$M^2 i_{\mathrm{sn}}^2 = 2ei_{\mathrm{Kath}}\Delta f \cdot M^2. \tag{13}$$

In general, the internal amplification introduces an additional noise contribution which is usually expressed through an excess noise factor

$$i_{\mathrm{en}}^2 = (F-1)M^2 i_{\mathrm{sn}}^2. \tag{14}$$

Further noise contributions are to be expected from the inherently pre-sent dark noise current:

$$i_{\mathrm{d}}^2 = 2ei_{\mathrm{D}}\Delta f. \tag{15}$$

The noise due to the load resistance R_A is given by the Nyquist formula:

$$i_{R_A}^2 = 4kT\frac{1}{R_A}\Delta f. \tag{16}$$

Similarly the noise contribution due to the input resistance to the pre-amplifier is:

$$u_{R_{\mathrm{pa}}}^2 = 4kTR_{\mathrm{pa}}\Delta f, \tag{17}$$

This voltage acts on the impedance formed by the anode capacitance C_A and the load resistance R_A:

$$Z = \frac{1}{\dfrac{1}{R_A}+i\omega C_A} \tag{18}$$

yielding the following noise spectrum:

$$i_{R_{\mathrm{pa}}}^2(\omega) = 4kTR_{\mathrm{pa}}\Delta f\left(\frac{1}{R_A^2}+\omega^2 C_A^2\right). \tag{19}$$

Integration over Δf yields:

$$i_{R_{\mathrm{pa}}}^2 = 4kTR_{\mathrm{pa}}\Delta f\left(\frac{1}{R_A^2}+\frac{4}{3}\pi^2\Delta f^2 C_A^2\right). \tag{20}$$

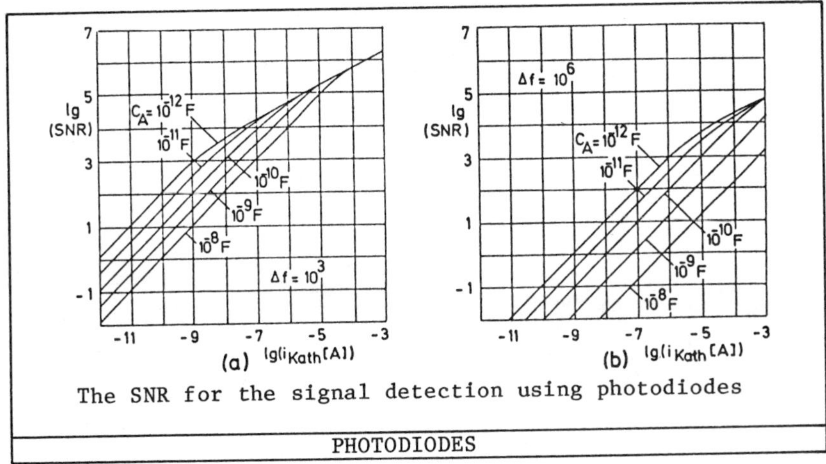

The SNR for the signal detection using photodiodes

PHOTODIODES

The various noise contributions can be combined to yield the final expression that will be present on the signal across the load resistance:

$$i_{n_{ges}}{}^2 = 2ei_{Kath}\Delta f M^2 F + 2ei_D\Delta f + 4kT\frac{1}{R_A}\left(1 + \frac{R_{pa}}{R_A} + \frac{4}{3}\pi^2 R_{pa}R_A\Delta f^2 C_A{}^2\right)\Delta f. \qquad (21)$$

Using this expression allows the signal-to-noise ratio of the detected optical signals to be defined:

$$SNR = M\,\frac{i_{Kath}}{i_{n_{ges}}} \qquad (22)$$

Photodiodes

Applying the above formulae to photodiodes requires to introduce M = 1 (no internal amplification) and F = 1 (no excess noise). This yields the following final formula for the SNR:

$$SNR = i_{Kath}\bigg/\left[2e(i_{Kath} + i_D)\Delta f + 4kT\frac{1}{R_A}\left(1 + \frac{R_{pa}}{R_A} + \frac{4}{3}\pi^2 R_{pa}R_A\Delta f^2 C_A{}^2\right)\Delta f\right]^{1/2}. \qquad (23)$$

The evaluation of this for $i_D = 10^{-10}$ A and for $R_{pa} = 100\,\Omega$ yields signals as those given in the slide panel. To compute these these resistance R_A was chosen to yield a signal passage characteristic such that the highest Doppler-frequency only shows a signal amplitude alternation of 1%.

A comparison of the two figures in the above slide panel shows that the signal-to-noise ratio variation with the cathode current shows two characteristic ranges. In one range, the signal-to-noise ratio varies proportional to the cathode current and in the other one it varies proportional to the square route of the cathode current. The first range is limited by the noise produced by the various resistances indicated in the circuit diagram and, hence, the noise goes proportional to the cathode current. The square route variation is the region where the shot noise dominates. This region is reached at high cathode currents.

A comparison of both diagrams also shows that the required bandwidth of the detection system reduces the signal-to-noise ratio. This is apparent from equation (23).

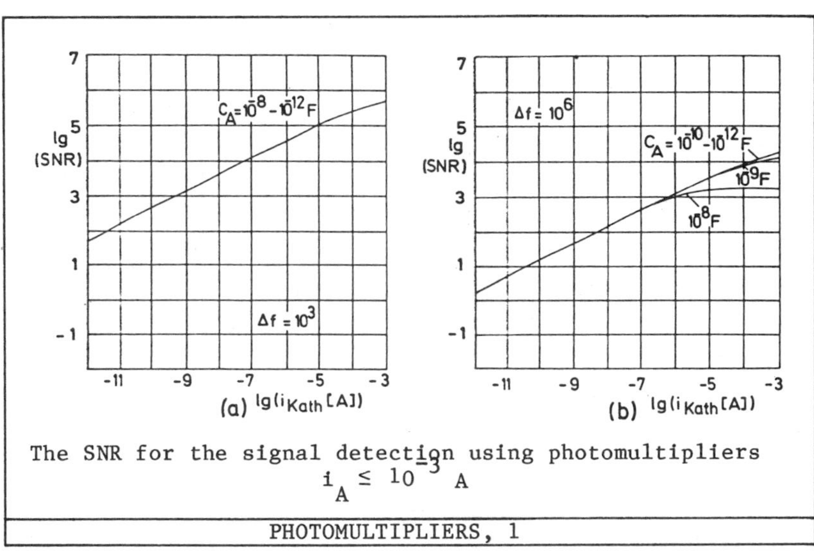

The SNR for the signal detection using photomultipliers
$i_A \leq 10^{-3}$ A

PHOTOMULTIPLIERS, 1

Photomultipliers

For photomultipliers the final relationship for the SNR reads as follows, since $M = \delta^K$ (δ = amplification pro diode and K = number of diodes) and $M^2 F = \delta^{2K+1} - \delta^K/(\delta-1)$:

$$\text{SNR} = i_{\text{Kath}}\delta^K \Big/ \Big[2ei_{\text{Kath}}\frac{\delta^{K+1}-\delta^K}{\delta-1}\Delta f + 2ei_D\Delta f + 4kT\frac{1}{R_A}\Delta f$$
$$\left(1 + \frac{R_{\text{pa}}}{R_A} + \frac{4}{3}\pi^2 R_{\text{pa}}R_A\Delta f^2 C_A{}^2\right)\Delta f \Big]^{1/2} \quad (24)$$

The above diagrams are for $R_A = R_{\text{pa}} = 100\ \Omega$ and $i_D = 10^{-10}$ A. The maximum internal amplification was taken to be 10 t and $K = 10$.

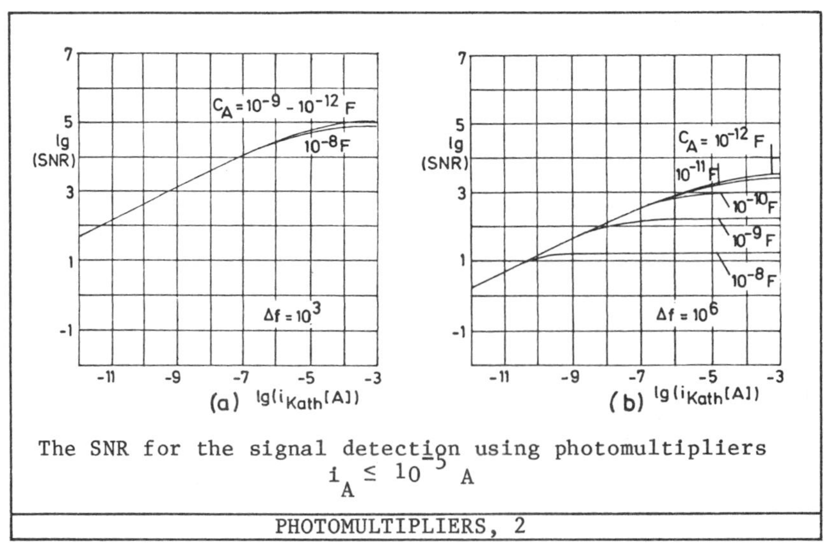

The SNR for the signal detection using photomultipliers
$i_A \leq 10^{-5}$ A

PHOTOMULTIPLIERS, 2

Due to limitations imposed onto the maximum possible anode current, photomultipliers show a limiting value on the maximum detectable SNR. The lower the anode capacitance the higher this limit will be.

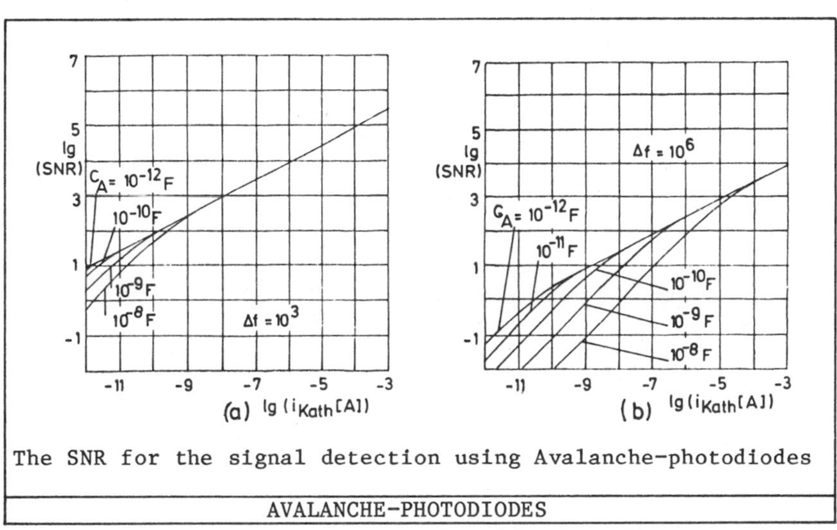

Avalanche-Photodiodes

For Avalanche-photodiodes the excess noise factor can be given as $F = M^x$ with $x \approx 0.3$ to 1.0. This yields the following formula for the SNR of these photodetectors:

$$\text{SNR} = Mi_{\text{Kath}} \Big/ \left[2ei_{\text{Kath}}M^2M^x\Delta f + 2ei_{\text{D}}\Delta f + 4kT\frac{1}{R_{\text{A}}}\left(1 + \frac{R_{\text{pa}}}{R_{\text{A}}} + \frac{4}{3}\pi^2 R_{\text{pa}}R_{\text{A}}\Delta f^2 C_{\text{A}}{}^2\right)\Delta f\right]^{1/2}. \quad (25)$$

For $x = 1$, $M = 100$ and $i_{\text{D}} = 10^{-9}$ A and for R_{pa} and R_{A} having values similar to those used for the other detectors, the above diagrams result.

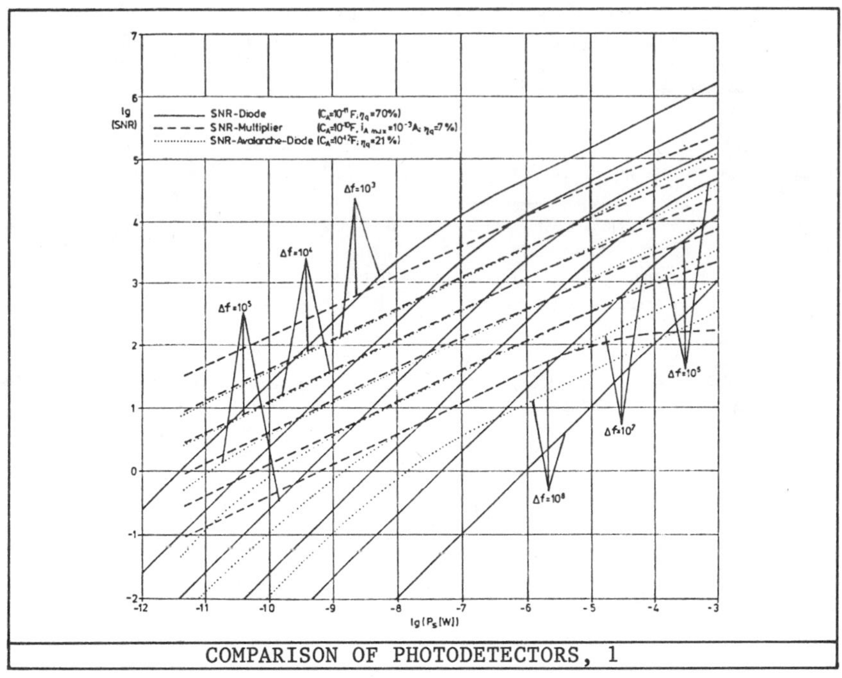

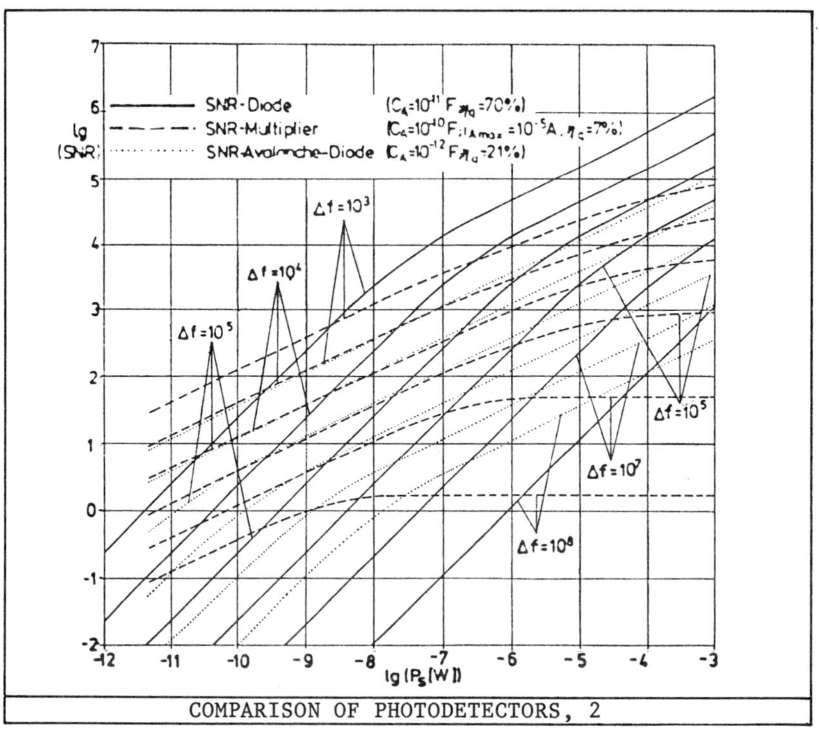

COMPARISON OF PHOTODETECTORS, 2

A comparison of the various photodetectors is given in the last two diagrams employing typical properties for each detector type. The latter diagram is for a photomultplier similar to those that are nowadays employed in commercial LDA-systems. From this diagram, we can conclude for small bandwidth optical units, that photodiodes become superior to photomultipliers and Avalanche-photodiodes when large scattered light powers are available. For wide bandwidth systems, this is also the case. For such systems, the diagram indicates that Avalanche-photodiodes obtain photomultipliers that are superior to those of photomultipliers. The superiority starts to come in already at low light scattering powers. Hence, Avalanche-photodiodes should be given more trials in LDA-systems.

If combined instruments are used to measure the particle velocity via laser-Doppler anemometry and the particle size via amplitude detection systems, the bandwidth of the velocity measuring and the particle sizing parts of the optical unit are different. A laser-Doppler requires usually the higher bandwidth. A tenth has a reduced signal-to-noise ratio. The employment of two photodetectors is therefore optimum providing the higher signal-to-noise ratio for the size measuring system by employing a reduced bandwidth of the detection electronics.

It should be stressed that in the above diagrams the Avalanche-photodiode was introduced with a very high excess noise factor and also with a quantum efficiency which is lower than the quantum efficiency of cathode material that are nowadays available. This stresses even more the fact that Avalanche-photodiodes should be given closer considerations as photodetectors for wide bandwidth optical detection systems.

Since for optical arrangements, the light intensity I is known from equation (3) and the light power P_s can be computed from equation (11), the signal-to-noise ratio of optical signals can be obtained from the above diagrams.

ELECTRONICS

Regarding the electronic systems employed in laser-Doppler anemometry, the counter system and photon-correlator based on electronic units
are considered here. These units, like other electronic signal processing
systems in laser-Doppler anemometry, possess an upper and lower signal
strength limit between they work satisfactorily. These limits are given
in the slide panel below and are explained in ref. /3/ together with the
signal strength of the scattered light of particles for a 1 Watt laser
power and for forward scatter. The data for the scattering light computations were:

          Laser power: 1 Watt
          Distance of detection lens: 2 m
          Radius of detection aperture: 0.05 m
          Intersection angle of beam: 2°
          Diameter of measuring volume: 0.6 mm

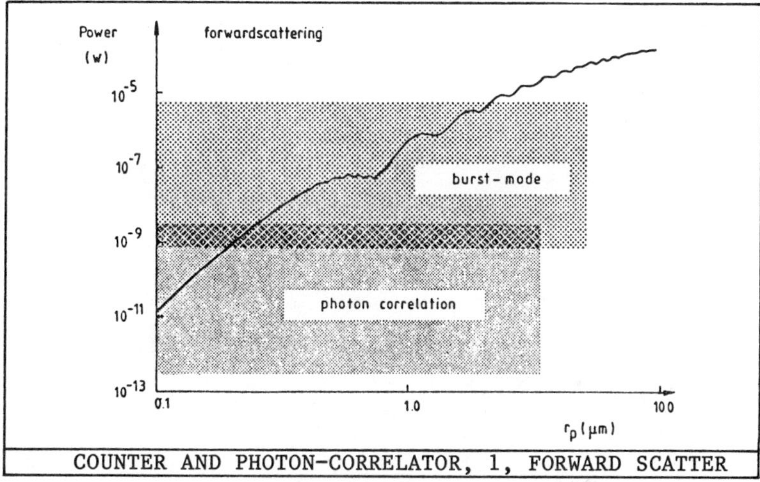

COUNTER AND PHOTON-CORRELATOR, 1, FORWARD SCATTER

In the backward scatter the same information looks as that provided
in the next slide panel. Comparing both slide panels permits the correct
electronic processing system for forward and backward scattering measurements to be chosen.

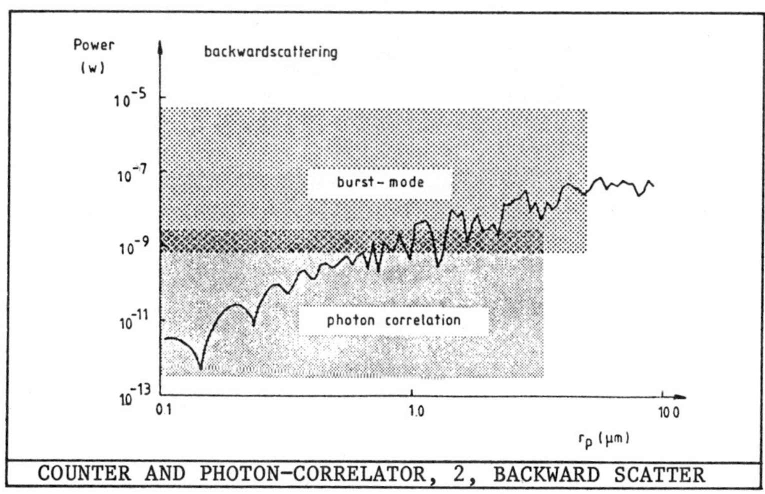

COUNTER AND PHOTON-CORRELATOR, 2, BACKWARD SCATTER

The above slide panels show that the correct choice of electronic signal processing systems depends on the availability of particles. In the atmosphere the size distribution of naturally available particles is:

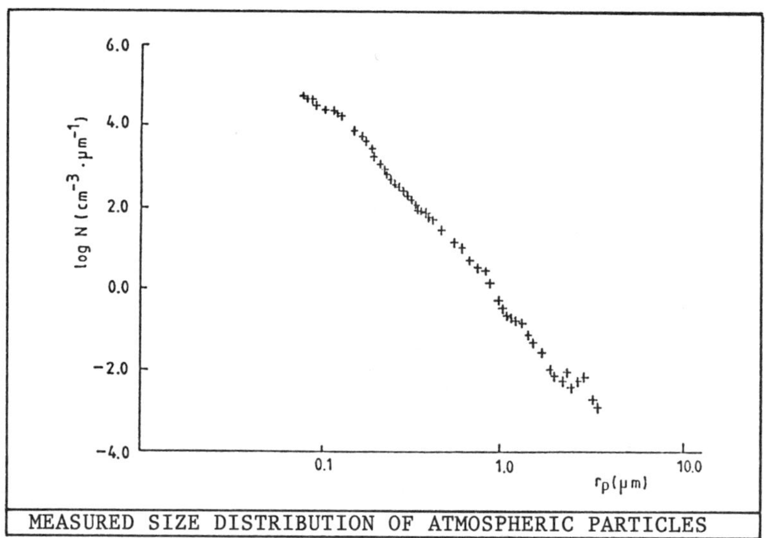

MEASURED SIZE DISTRIBUTION OF ATMOSPHERIC PARTICLES

Light Sources

If one carries out predictions of SNR for optical signals from light scattering particles one usually finds for low Doppler frequencies good agreement between predicted and experimental results. For higher Doppler frequencies, usually no agreement is obtained and the question arises what causes the differencies with higher Doppler frequencies. An answer to this question has been given in ref /4/, where it is shown that the axial modes of lasers can cause additional signal contributions that possess discrete frequencies. The description below summarizes the results given in ref. /4/.

It is well known that conventional gas lasers, as they are employed in optical light scattering experiments, exhibit various axial laser modes. Hence, the output of such lasers may be described as:

$$(\varepsilon_s)_1 = \sum_{n=1}^{M} \frac{a_{1n}(t)}{R_p} \exp\left[-i\left(\frac{\pi 2R_p}{\lambda_n} - \omega_n t - \phi_n\right)\right], \tag{26}$$

$$(\varepsilon_s)_2 = \sum_{m=1}^{M} \frac{a_{2m}(t)}{R_p} \exp\left[-i\left(\frac{\pi 2R_p}{\lambda_m} - \omega_m t - \phi_m\right)\right]. \tag{27}$$

or the scattered light waves from a moving particle read:

$$(\varepsilon_s)_1 = \sum_{n=1}^{M} \frac{a_{1n}}{R_p}\left(\exp - i\left\{\frac{2\pi}{\lambda_n}(R_p)_0 - (\phi_n)_0 - \omega_n t - \frac{2\pi}{\lambda_n}\{U_p\}_i \cdot [\{l_1\}_i - \{k_p\}_i] \cdot t\right\}\right), \tag{28}$$

$$(\varepsilon_s)_2 = \sum_{m=1}^{M} \frac{a_{2m}}{R_p}\left(\exp - i\left\{\frac{2\pi}{\lambda_m}(R_p)_0 - (\phi_m)_0 - \omega_m t - \frac{2\pi}{\lambda_m}\{U_p\}_i \cdot [\{l_2\}_i - \{k_p\}_i] \cdot t\right\}\right). \tag{29}$$

The intensity at the surface element of the photodetector may be expressed as:

$$I_Q = \frac{1}{T}\int_0^T |R[(\varepsilon_s)_1] + R[(\varepsilon_s)_2]|^2 dt, \tag{30}$$

where $[R(\varepsilon_s)_k]$ is the real part of the complex scattered light field.

Taking into account the fact that the integration time of the photodetector is much larger than the inverse of the light frequency, permits the following expression to be derived:

$$I_Q = \sum_{j=1}^{M} \frac{a_{1j}^2 + a_{2j}^2}{R_p^2} + \sum_{n=1}^{M} \sum_{m=1}^{M} \frac{a_{1n}a_{1m} + a_{2n}a_{2m}}{R_p^2}$$
$$\times \cos\left\{2\pi(R_p)_0\left(\frac{1}{\lambda_m} - \frac{1}{\lambda_n}\right) + [(\phi_m)_0 - (\phi_m)_0] + 2\pi(\nu_m - \nu_n)t\right\}$$
$$+ \sum_{n=1}^{M}\sum_{m=1}^{M} \frac{a_{1n}a_{2m}}{R_p^2} \cos\left[2\pi(R_p)_0\left(\frac{1}{\lambda_m} - \frac{1}{\lambda_n}\right) + [(\phi_m)_0 - (\phi_n)_0]\right.$$
$$\left. + 2\pi t\left((\nu_m - \nu_n) - \{U_p\}_i\left\{\frac{1}{\lambda_m}[\{k_p\}_i - \{l_2\}_i] - \frac{1}{\lambda_n}[\{k_p\}_i - \{l_1\}_i]\right\}\right)\right]. \tag{31}$$

This equation can be simplified if the frequency difference between the two beams is small enough that the following relationship holds:

$$\lambda = \lambda_m = \lambda_n, \text{ i.e., } (\nu_m - \nu_n) \ll \tfrac{1}{2}(\nu_m + \nu_n). \tag{32}$$

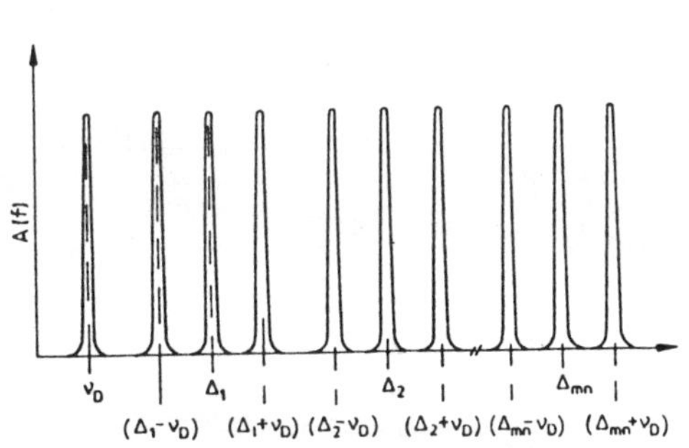

Δ_1 is the frequency difference between adjacent mode frequencies

$$\Delta_1 = \nu_{m+1} - \nu_m \qquad \Delta_k = \nu_{m+k} - \nu_m$$

Sketch of spectral distribution of energy of LDA signal components obtained from multiaxial-mode laser beams. (This sketch does not contain the experimentally observed amplitude differences of the various LDA frequency components.)

AXIAL LASER MODES

The final expression therefor is

$$I_Q = \sum_{j=1}^{M} \frac{a_{1j}^2 + a_{2j}^2}{R_p^2} + \sum_{n=1}^{M}\sum_{m=1}^{M} \frac{a_{1n}a_{1m} + a_{2n}a_{2m}}{R_p^2} \cos\{[(\phi_m)_0 - (\phi_n)_0]$$
$$+ 2\pi t(\nu_m - \nu_n)\} + \sum_{n=1}^{M}\sum_{m=1}^{M} \frac{a_{1n}a_{2m}}{R_p^2} \cos\left([(\phi_m)_0 - (\phi_n)_0]\right.$$
$$\left. + 2\pi t\left\{(\nu_m - \nu_n) + \frac{1}{\lambda}\{U_p\}_i[\{l_2\}_i - \{l_1\}_i]\right\}\right). \tag{33}$$

This equation shows several terms: the first would be present even if there is no coherence between the various axial modes, the second shows the beating between the signals from the various modes in each individual beam, and the third term gives the beating of signals obtained from the various modes in the two beams. Introducing the sensitivity vector:

$$\{\mathbf{n}\}_i = \{\mathbf{l}_2\}_i - \{\mathbf{l}_1\}_i \tag{34}$$

permits the photomultiplier current for a single axialmode laser to be written as:

$$i_a = \text{const}\,\frac{(a_{1j}^2 + a_{2j}^2)}{R_p^2}\left\{1 + \frac{a_{1j}a_{2j}}{(a_{1j}^2 + a_{2j}^2)} \times \cos\left[2\pi t\,\frac{1}{\lambda}\{U_p\}_i\{n\}_i\right]\right\}. \tag{35}$$

This equation shows the relationship that is usually used in laser-Doppler anemometry; therefore, the signals from the various modes are neglected. This derivation suggests that this neglect requires further consideration.

For multimode laser output the total signal may be written as

$$i_a = \text{const}\left(\sum_{j=1}^{M}\frac{a_{1j}^2 + a_{2j}^2}{R_p^2} + \sum_{n=1}^{M}\sum_{m=1}^{M}\frac{a_{1n}a_{1m} + a_{2n}a_{2m}}{R_p^2}\cos[\Delta\phi_{m,n}\right.$$

$$\left. + 2\pi(\nu_m - \nu_n)t] + \sum_{n=1}^{M}\sum_{m=1}^{M}\frac{a_{1n}a_{2m}}{R_p^2}\cos\left\{\Delta\phi_{m,n}\right.\right. \tag{36}$$

$$\left.\left. + 2\pi t\left[(\nu_m - \nu_n) + \frac{1}{\lambda}\{U_p\}_i\{n\}_i\right]\right\}\right).$$

This equation shows that multimode lasers yield not only the signals for $m = n$ usually considered in laser-Doppler anemometry but also the terms that contain the differences between the various mode frequencies and sums and differences of these frequencies and the Doppler frequency. Considering the mode frequencies, we may write the frequency of the second term in equation (36) as

$$(\nu_m - \nu_n) = (m - n)\frac{c}{2L} = \Delta_{mn}\frac{c}{2L}. \tag{37}$$

In addition, the frequencies of the last term in equation (36) may be written as

$$\nu_{s,\Delta} = \left|\pm |\Delta_{mn}|\frac{c}{2L} + \frac{1}{\lambda}\{U_p\}_i\{n\}_i\right|. \tag{38}$$

Measurements of high frequency LDA signals require the use of high frequency response photodetectors and preamplifiers; therefore, the detection electronics might not only respond to the Doppler frequency,

$$\nu_D = \nu_{s,0} = \left|\frac{1}{\lambda}\{U_p\}_i\{n\}_i\right|, \tag{39}$$

but also to higher frequencies such as

$$\nu_{s,1} = \left|\pm\frac{c}{2L} + \frac{1}{\lambda}\{U_p\}_i\{n\}_i\right|\nu_{s,2} = \left|\pm\frac{c}{2L} + \frac{1}{\lambda}\{U_p\}_i\{n\}_i\right|. \tag{40}$$

This equation shows clearly that detection of high frequency LDA signals will become difficult when the signal to be measured has the same magnitude as the signal resulting from the difference of the Doppler frequency and the first mode difference frequency. Equation (40) shows that the total output of a high frequency response photomultiplier has a spectrum as shown in the above slide panel. when a photomultiplier with a high frequency response is used to detect LDA signals from a multiaxial-mode laser light source.

In ref. /4/, the following electronic system was employed to experimentally verify the occurence of higher frequency signals caused by axial modes.

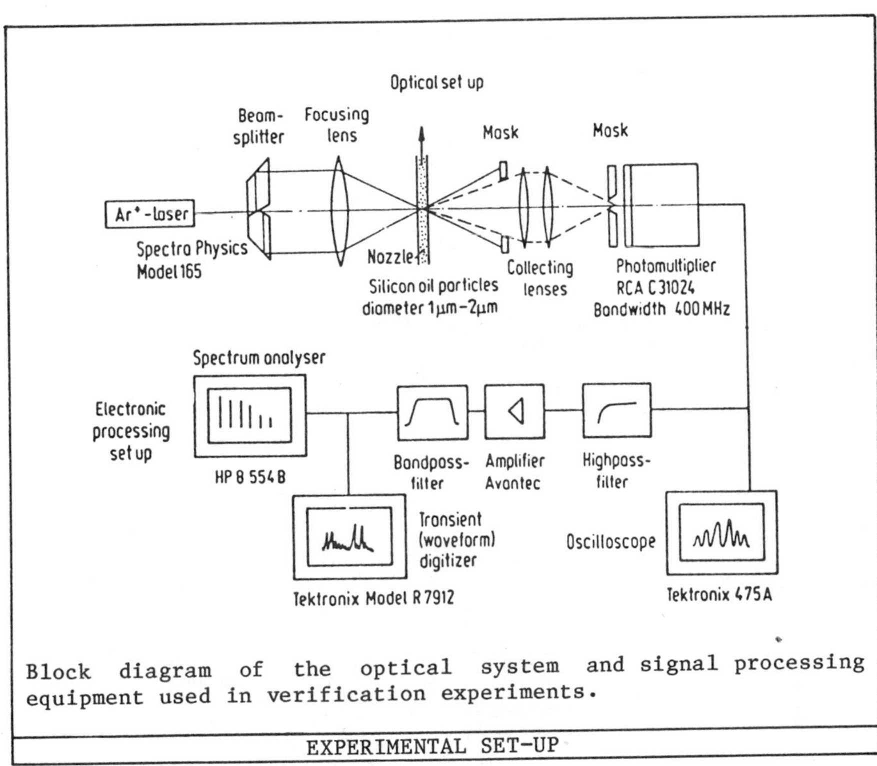

Block diagram of the optical system and signal processing equipment used in verification experiments.

EXPERIMENTAL SET-UP

Frequencies as given below resulted:

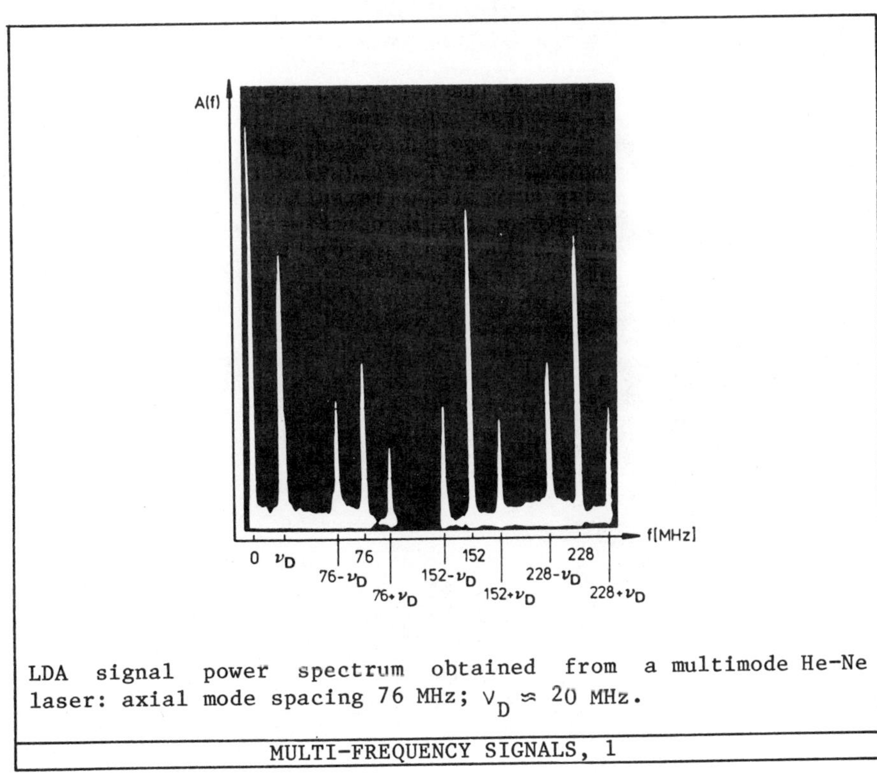

LDA signal power spectrum obtained from a multimode He–Ne laser: axial mode spacing 76 MHz; $\nu_D \approx 20$ MHz.

MULTI-FREQUENCY SIGNALS, 1

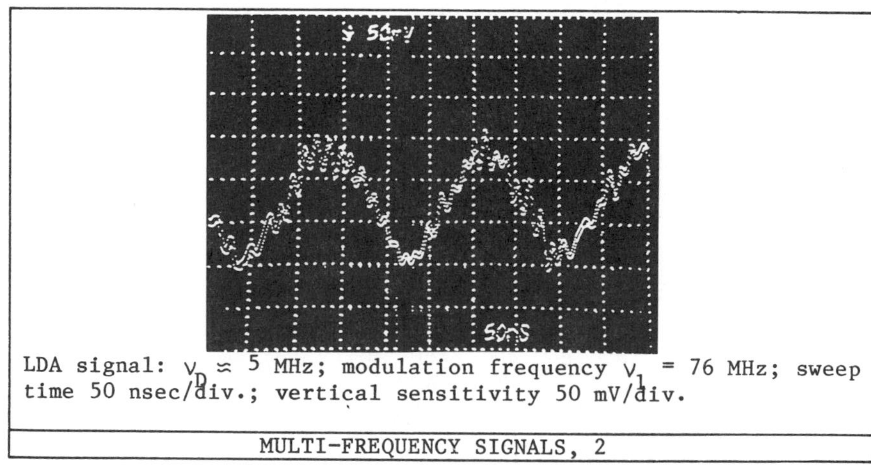

LDA signal: $\nu_D \approx 5$ MHz; modulation frequency ν_1 = 76 MHz; sweep
time 50 nsec/div.; vertical sensitivity 50 mV/div.

MULTI-FREQUENCY SIGNALS, 2

APPLICATION

The application of laser-Doppler anemometry in wind tunnels has
special requirements that are briefly summarized below for a system that
was designed and built for the 37.5 m² wind tunnel of Volkswagen AG in
Wolfsburg.

LDA-measurements for automotive investigations require the application of backscatter optical systems to allow measurements near all parts
of the automobile to be investigated.

To obtain analog signals at the photomultiplier output of a laser-
Doppler anemometer requires the application of particles in the range of
1 μm and larger. To ensure that these are available in sufficient number
involves the employment of proper seeding devices. Such seeding devices
can be successfully used in small wind tunnels with cross sectional
dimensions typically 1 x 1 m². In large wind tunnels, like the 37.5 m²
Volkswagen wind tunnel I, it is not practicable to seed the whole wind
tunnel flow. Screen clogging, contamination of the balance pads or even
in the balance were considered to be major problems if the complete wind
tunnel flow is artificially seeded. Local seeding devices upstream of the
measuring position would disturb the flow and would require special probe
and traversing mechanism. Based on these recognitions particle seeding on
our wind tunnels should be avoided and the LDA-system should be designed
to work on naturally available particles.

LDA-SYSTEM IN 37.5 m² WIND TUNNEL OF VOLKSWAGEN AG

A 1-D system should be considered first. An extension to a 2-D or even 3-D system should be investigated later. Developing a new technique as done here should not start with the most complicated 2-D or 3-D systems, but rather establish the principal technique in considering a 1-D system.The 1-D system should be positioned outside the wind tunnel jet in such a way that X- and Z-components could be measured (one at a time) with minimum changes to the set up.

The system should allow for the measurements of the complete flow field around full size cars in the 37.5 m² Volkswagen wind tunnel I. This requires a suitable mechanism for traversing the measuring point. In addition the system should be transportable between wind tunnels, for example to perform measurements in the 6 m² Volkswagen wind tunnel II.

Based on the above requirements, the system has the following features: backscatter system, 1-D system, no seeding required and a traverse range which allows the measurement of the flow field around cars in the 37.5 m² Volkswagen wind tunnel I.

The above slide panel shows the complete set up for testing in the 37.5 m² Volkswagen wind tunnel I and the following slide panel shows the optical system in detail. By shifting the frequencies of both laserbeams (with Bragg cells) the direction of the wake flow can be determined.

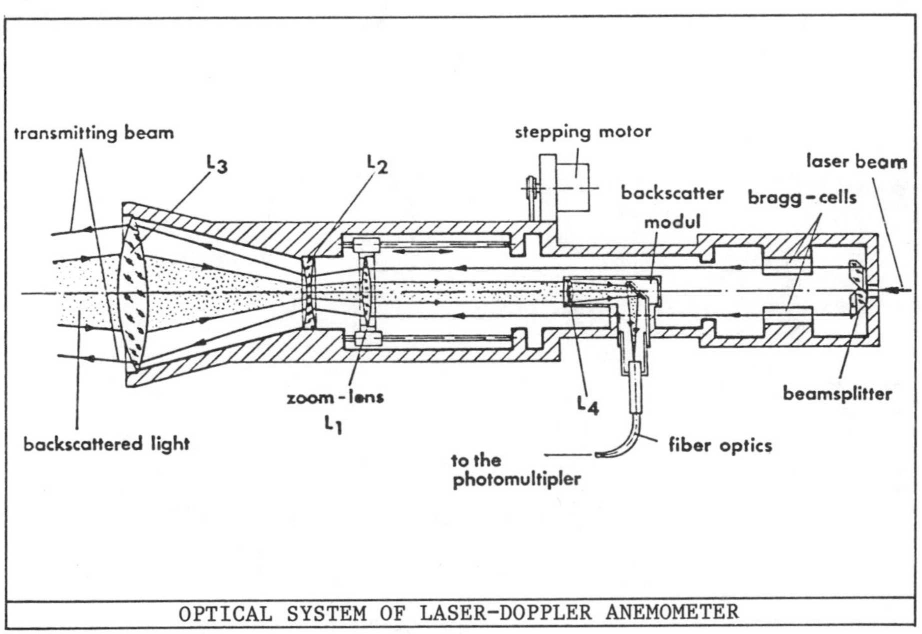

OPTICAL SYSTEM OF LASER-DOPPLER ANEMOMETER

LDA backscatter systems usually result in low scattered and detected light powes due to the long distance involved and also due to the fact that the backscattered light is less intense by a factor of 500 to 1000 compared to forward scattered light. Power is particularly low if naturally available particles are used as flow tracers. This results in laser-Doppler signals consisting of low photon rates. Signals of this kind show an "average photon rate" $\dot{N}_p$ which varies with time according to the following equation:

$$\dot{N}_p = \frac{P_S}{h\nu} = \dot{N}_o \exp\left[-\left(\frac{t-t_i}{\Delta\tau_i}\right)\right]\{1 + \eta_{M,i} \cdot \cos\left[2\pi\left(f_D + f_{SH}\right)\left(t-t_i\right)\right]\}$$

In this equation, P_S is the scattered light power, h the Planck constant, ν the light frequency, N the maximum photon rate in the signal, η_M the modulation depth of the signal, f_D the acquired Doppler frequency and f_{SH} the shift frequency. The time t_i indicates the arrival time of i-th particle and $\Delta\tau_i$ indicates the signal duration.

The time-varying photon rate expressed by this equation can be used to detect the Doppler frequency contained in the above equation. For this purpose, the individual photons are detected by a special photomultiplier to produce an electron pulse train which can also be described by the above equation.

Feeding this pulse train into a digital autocorrelator yields an auto-correlogram from which the Doppler frequency can be deduced. A simple Fourier transform can be applied if the turbulence level is above several percent (e.g. in the wake of the car) or by matching the resultant autocorrelation function by the following expression that can be deduced from the above equation:

$$G(\tau) = A \exp\left[-\left(\frac{\tau}{\Delta\tau}\right)^2\right] \int_{-\infty}^{+\infty} p(u)\left\{1 + \frac{\eta^2}{2} \cos\frac{2\pi u\tau}{\Delta X}\right\} du + B$$

A is a normalizing constant, p(u) is the probability density distribution of the local velocity field, η is the modulation depth of the autocorrelation function, ΔX is the fringe spacing and B the contribution from background noise.

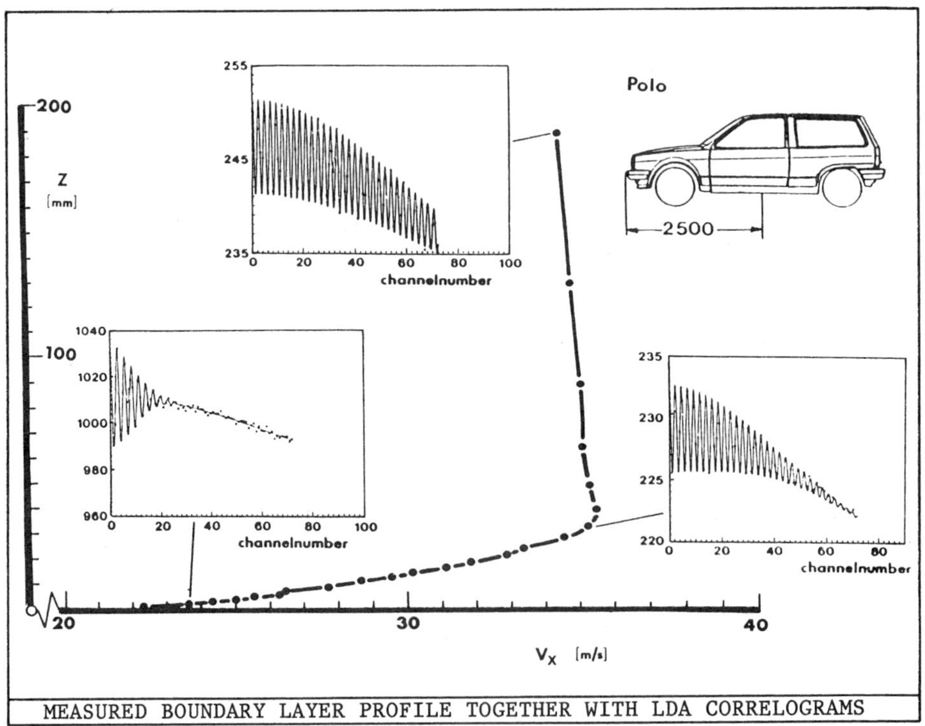

MEASURED BOUNDARY LAYER PROFILE TOGETHER WITH LDA CORRELOGRAMS

The above slide panel shows typical correlograms obtained during boundary layer measurements on a car body. The points are the measured photon signals, the lines are the model functions as described above. The frequencies in the correlograms are a measure of the local mean velocity, the turbulence intensity is determined from the damping of the curves.

210

To obtain entire velocity profiles, the measuring volume is traversed up to 4.5 m distance from the optics under computer control in the Y-direction using the zoom system of the optics. The traverse in the vertical Z-direction is carried out by step motors up to heights of 3.0 m. For different X-positions the whole system is moved along the flow direction. Turning the optical system by 90 degrees allows X-component measurements. The Y-component could be measured by setting up the system above the wind tunnel jet with the optical axis in Z-direction.

The above slide panels shows a typical boundary layer profile measured at the centerline of the roof of a car. This boundary layer profile exhibits features of a strongly turbulent flow. From the next slide panel one can see that the boundary layer thickness in the roof region is 30 mm, and that the region of velocity overshoot above the car is only slowly decreasing with the downstream distance. The increase in turbulence level towards the wall agrees with observations in other turbulent boundary layers in low subsonic flow.

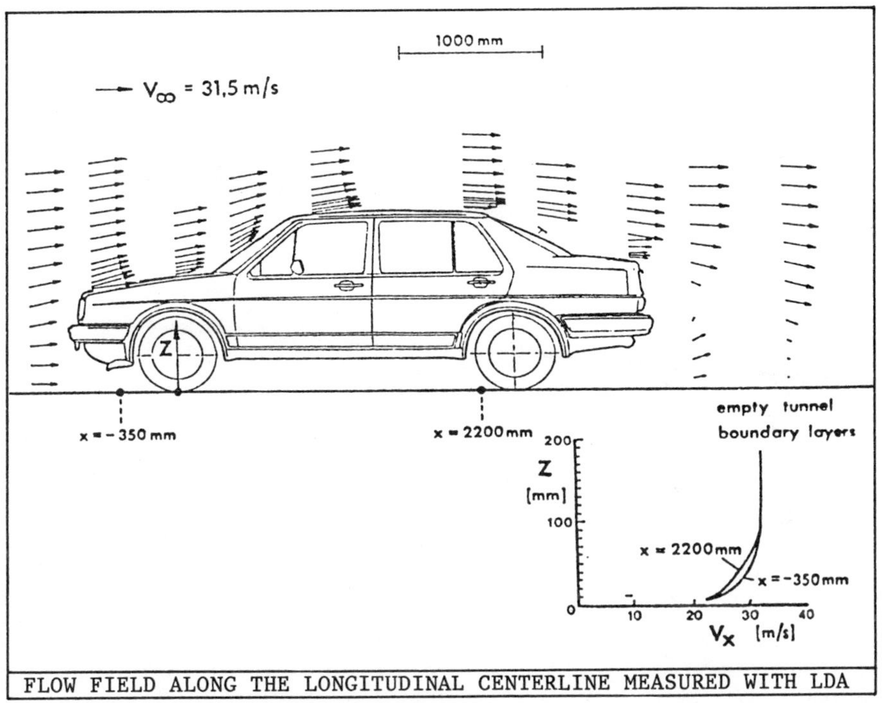

FLOW FIELD ALONG THE LONGITUDINAL CENTERLINE MEASURED WITH LDA

The above slide panel shows the LDA-measured flow field around the VW-Jetta car. Some interesting observations are:

o In front of the car the upward directed flow extends more than 2 m above the ground. There seems to be nowhere in front of the car a downward flow component.

o The flow profiles in front of the windshield, especially in the boundary layer, indicate separation tendencies.

o The separation region above the trunk extends nearly to the rear end edge of the trunk.

o The recirculation zone behind the car seems to extend down to the tunnel floor.

o The boundary layer on the tunnel floor underneath the car is much thinner than it is at the same place in the tunnel without the car. This could be an indication that floor boundary layer effects are misinterpreted if the empty tunnel boundary layer is only considered.

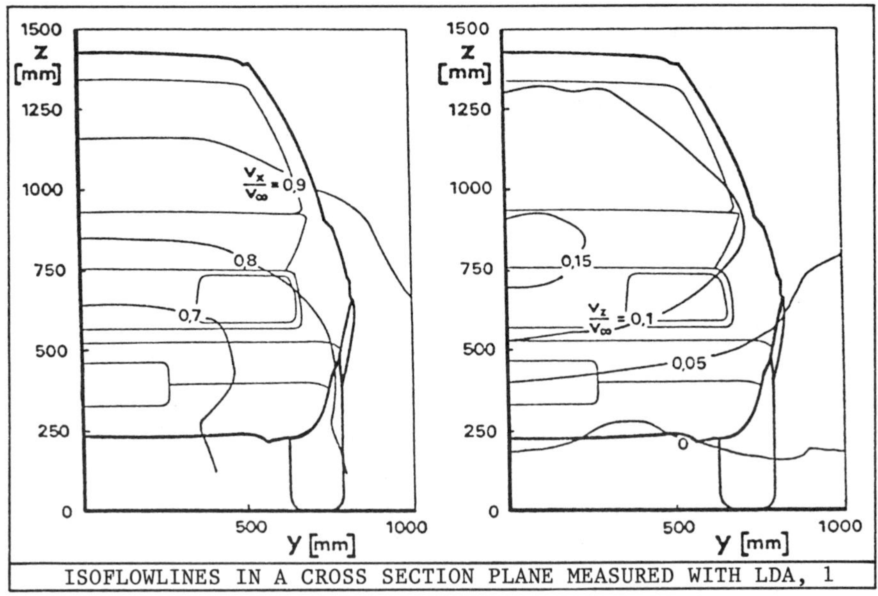

In the above slide panel flow isolines are shown in a Y-Z-plane in front of the car. Here it can be seen that the upward flow extends over the whole width of the car. Also no region on the lower part with a downward component was found.

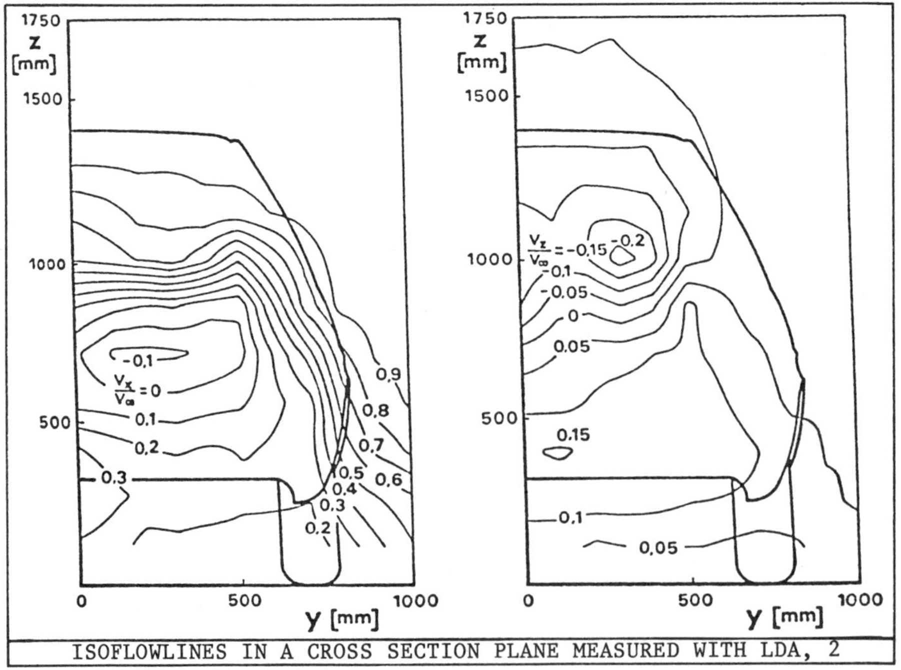

212

In the above slide, flow isolines in a Y-Z-plane immediately behind the car are shown. The region behind the car shows very low flow velocities. The region with real return flow is very small. The V-Z-component shows that the separation line between upward and downward flow is near the height of the trunk rear edge. The flow coming from under the car already has an upward direction 100 mm above the tunnel floor.

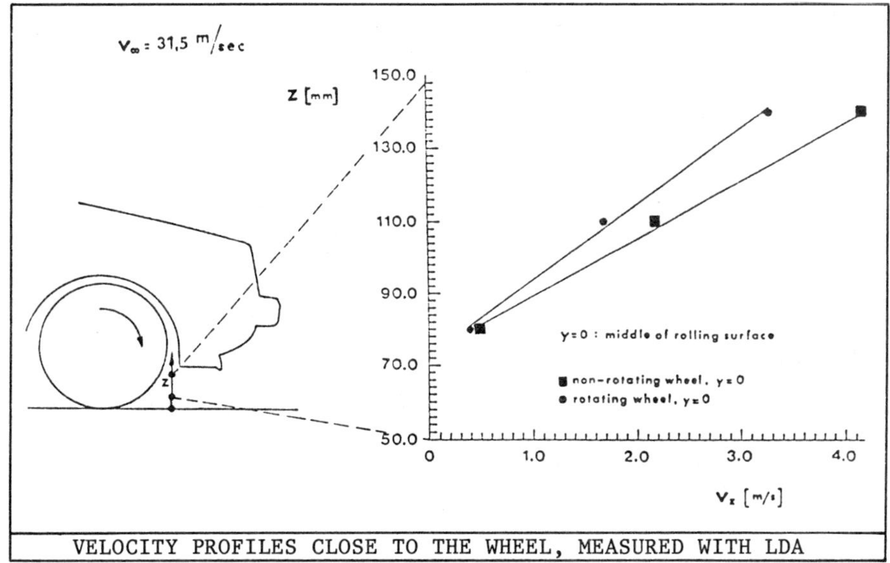

VELOCITY PROFILES CLOSE TO THE WHEEL, MEASURED WITH LDA

In the above slide panel the influence of rotating wheels on the flow in the vicinity of the wheel is shown. On the VX-component, not shown heree, the influence is negligible. The influence on the upward component is only significant very close to the wheel. The overall qualitative flow field in the wheel area is not changed by the rotating wheel.

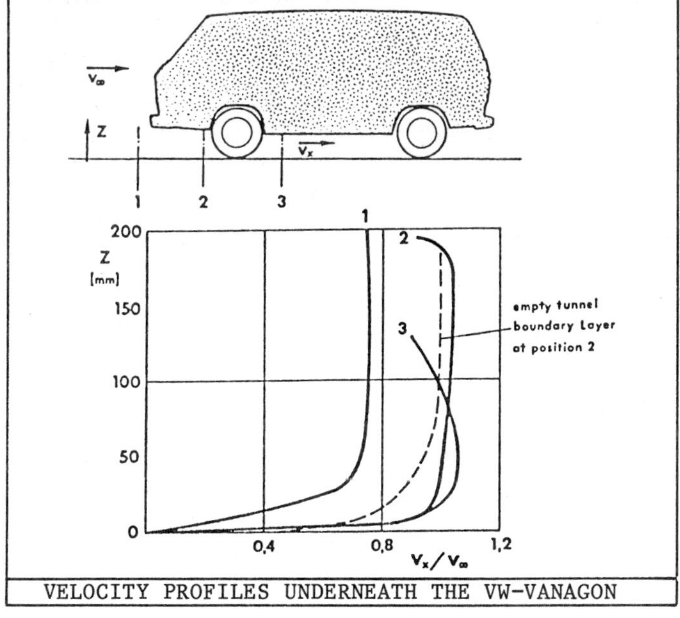

VELOCITY PROFILES UNDERNEATH THE VW-VANAGON

In the above slide panel detailed velocity profiles measured underneath the VW-Vanagon are shown. Similar as a former slide panel showed, an acceleration associated with a decrease in the wind tunnel floor boundary layer is observed in the frontal part. Behind the frontal axes a small wake extending up to position 3 is observed.

Further typical applications for the LDA-system are wind tunnel calibrations. The complete calibration of the new 6 m² VW wind tunnel II was done with the help of the described LDA.

The LDA-system, developed jointly by the University of Erlangen and Volkswagen, is suitable for use in large automotive wind tunnels. It can be used without particle seeding.

With the LDA-system, detailed and precise information of the flow fields around different cars was obtained. This was possible even with the 1-D system used up to now. By suitable positioning of thee optical system, 2-D or even 3-D information can be gained provided there is no time dependence. The system is used in both of the Volkswagen wind tunnels.

Besides complicated handling, the measurements with the LDA are time consuming. Up to 45 sec are used for one measuring point. The main application, therefore, has been confined to research projects, such as developing computational codes.

Further improvements of the system, such as a 2-D or even 3-D system, more comfortable X-traversing and reduction of measurement time are presently under investigation.

For example the entire optic was designed to easily permit its extension to simultaneous two-channel measurements by inserting a beam splitter arrangement that yields two pairs of perpendicularly located blue and green laser beams that can be focussed into the measuring control volume by the same zoom optics shown in one of the above slide panels. The extension of the optical system to three dimensions requires a second light emitting system.

A reduction of measuring time can be achieved by extending the electronic system and taking into account new developments in the photon-correlator field that has made photon-correlators available with effective correlation sampling times of 10 ns and less. This yields an increase in the effective maximum Doppler-frequency that can be fully utilized to extend laser-Doppler anemometry, operating with natural particles to higher particle velocities.

CONCLUSIONS

The present paper summarized information relevant to the usage of theoretical results of light scattering from particles to lay out optical systems for measurements of particle properties. Such results enter into the lay out of the electronics and define the signal processing system. The lay out considerations were applied to a laser-Doppler anemometer system for wind tunnel measurements and the functioning of the system was demonstrated.

LITERATURE

/1/   F. Durst und K.F. Heiber, "Signal-Rausch-Verhältnisse von Laser-Doppler-Signalen", OPTICA ACTA, Vol. 24, No. 1, pp. 43-67, 1977.

/2/   F. Durst, M. Macagno und G. Richter, "Light Scattering by Small Particles: Refined Numerical Computations", Report SFB 80/TM/195, Universität Karlsruhe, 1981.

/3/   F. Durst, F. Ernst und J. Völklein, "Laser-Doppler-Anemometer-System für lokale Geschwindigkeitsmessungen in Windkanälen: Systemauslegung und Verifikation", accepted for publication in ZFW.

/4/   D. Dopheide and F. Durst, "High frequency laser-Doppler measurements using multiaxial-mode lasers", Applied Optics, Vol. 20, No. 9, pp. 1557-1570, 1981.

MEASUREMENT OF PARTICLE ASYMMETRY USING CROSS-CORRELATION TECHNIQUES

John G Rarity

Royal Signals and Radar Establishment
St Andrews Road
Malvern, Worcs, WR14 3PS, England

INTRODUCTION

Most static scattering[1] techniques measure orientation averaged scattering factors of particles in suspension. Information on particle shape is difficult to obtain from such results. Similarly, the technique of dynamic light scattering will conventionally measure an 'equivalent sphere' hydrodynamic radius of the particles in suspension although for highly asymmetric particles (ie long rods) an estimate of the rotational diffusion constant can be obtained from multiangle linewidth measurements[2,3]. For small particles with only slight asymmetry the change in dynamic scattering linewidth with angle becomes comparable with that seen due to sample polydispersity alone and information on rotational diffusion and shape cannot be obtained from the data.

The light scattered at any instant from an individual asymmetric particle is distributed asymmetrically about the illumination axis. Optical single particle counters that measure this asymmetry are being developed[4]. These techniques are limited to large particle studies where enough scattered light can be collected in the finite residence time. Furthermore, single particle techniques tend to be invasive, involving the removal of a sample for measurement.

When a small number of asymmetric particles is illuminated the light scattered will still be distributed asymmetrically. The random orientations of the separate particles will, however, dilute the effect. The intensity of light scattered will fluctuate due to the changing numbers of particles in the scattering volume, due to their changing orientations (under Brownian rotation) and due to the changing phase relationships between fields scattered by different particles. This last source of fluctuation, interparticle interference fluctuations, is what is normally studied by dynamic light scattering techniques[5]. The timescale of these large fluctuations is related to the average time taken for a particle to diffuse a distance comparable to the wavelength coupled with a term due to rotational diffusion for asymmetric particles. It is this coupling of the similar rotational and translational diffusion timescales that makes direct study of rotational diffusion and hence particle asymmetry difficult using conventional dynamic light scattering.

The interparticle interference fluctuations can be eliminated by using a detection system that resolves distances much smaller than the typical interparticle distances but not smaller than the particles themselves. Such a system ensures incoherent addition of intensities from different particles but full coherent addition of fields within the particles themselves. The number and orientation driven intensity fluctuation effects now dominate and can be studied directly using correlation techniques (see Theory section).

Number fluctuations occur on timescales similar to the time taken for a particle to diffuse across the scattering volume (typically seconds) while orientation fluctuations occur on much shorter timescales (typically milliseconds). Separation of these two sources of fluctuation is thus relatively easy. The rotational signal can in general be analysed to yield two parameters: a visibility and a timescale. From this one can extract, in principle, a two parameter description of particle size/shape. We demonstrate this principle here in the measurement of the radii of a sub-micron "ellipsoid-like" cross-linked polystyrene particle suspension.

When there are no asymmetric particles in the sample the rotational signal disappears. We discuss the potential of these number fluctuation techniques for detection of small amounts of asymmetry in a system focussing on the detection of low order aggregates in a suspension of nominally spherical particles.

THEORY

The random intensity fluctuations are quantified using correlation techniques (typically digital correlators coupled with photon counting detectors), the intensity autocorrelation function being defined as

$$G^{(2)}(\tau) \;=\; \langle I(o)I(\tau)\rangle \tag{1}$$

where τ is a delay time between intensity measurements and the angular brackets indicate a time and ensemble average for stationary systems. In practice, for these number fluctuation studies, full incoherent detection is obtained by using two detectors viewing the sample at different angles (θ_1, θ_2) with angular separation $(\Delta\theta)$ larger than the coherence angle θ_c which is given by the ratio of the wavelength λ to the sample volume dimension L

$$\Delta\theta > \theta_c \;\approx\; \lambda/L \tag{2}$$

Using the output from these two detectors the normalised intensity cross correlation function $g^{(2)}(\theta_1, \theta_2, \tau)$ can be measured where

$$g^{(2)}(\theta_1, \theta_2, \tau) \;=\; \frac{\langle I(0_1, \theta)I(\theta_2, \tau)\rangle}{\langle I(\theta_1)\rangle\langle I(\theta_2)\rangle} \tag{3}$$

For incoherent detection $(\Delta\theta > \theta_c)$ $g^{(2)}$ can be shown to have the form[6]

$$g^{(2)}(\theta_1, \theta_2, \tau) = 1 + \frac{1}{\langle M\rangle} g_N(\tau)g_R(\theta_1, \theta_2, \tau) \tag{4}$$

Here $\langle M\rangle$ is the mean number of particles in the scattering volume and its inverse determines the visibility of the number fluctuation correlation $g_N(\tau)$ and the rotational correlation $g_R(\theta_1, \theta_2, \tau)$. $g_N(\tau)$

decays from unity to zero on a timescale comparable with the time taken
for a typical particle to cross the volume. For Brownian particles of
diameter 0.2μm and scattering volume dimension L ~ 5μm this time is of
the order of a few seconds. $g_R(\tau)$ contains all information on internal
dynamics and orientation fluctuations. Such fluctuations occur
typically on millisecond timescales hence the decays of g_R and g_N are
clearly distinguishable and easily separated. This is one of the main
advantages of the cross-correlation technique.

For Rayleigh-Gans-Deby (RGD) particles with spherical top symmetry
$g_R(\tau)$ has the form[7].

$$g_R(\theta_1,\theta_2,\tau) = 1 + \sum_{\substack{L=2 \\ \text{even}}}^{\infty} P_L(\cos\Delta\theta/2)\, A_L\, e^{-L(L+1)D_R(\tau)} \tag{5}$$

where P_L is the Legendre polynomial of order L and for small $\Delta\theta$
$P_L(\cos\Delta\theta/2) \approx 1$. The timescale of the decay is dependent on the
rotational diffusion constant D_R and the relative contributions of the
Lth order amplitude factor A_L. The A_L are given more explicitly by

$$A_L = \frac{1}{2L+1}\; \frac{a_L(\theta_1)a_L(\theta_2)}{a_o(\theta_1)a_o(\theta_2)} \tag{6}$$

where $a_o(\theta)$ is the orientation averaged intensity scattered by the
particle (ie the intensity/particle measured in static light scattering
experiments). All a_L are coefficients of an expansion, in terms of the
Legendre polynomials, of the instantaneously scattered intensity at time
t, $i(\theta,t)$ from a single Brownian rotating particle[8].

$$i(\theta,t) = \sum_L a_L(\theta)P_L(\cos\beta(t)) \tag{7}$$

where β is the angle between the particle symmetry axis and the familiar
scattering vector $\underline{K}$, the difference between the scattered wavevector $\underline{K}_s$
and the incident wavevector $\underline{K}_i$. This scattering geometry is illustrated
in Figure 1.

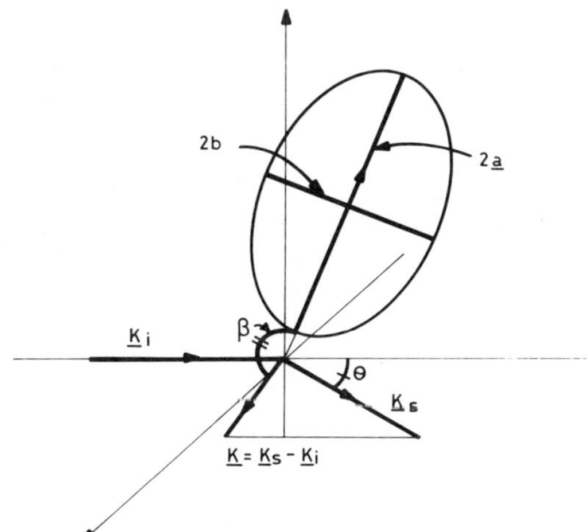

Figure 1. Scattering geometry
for a randomly oriented prolate
ellipsoid. θ, the scattering
angle, is the angle between
incident and scattered
wavevectors $\underline{K}_i$ and $\underline{K}_s$. β is
defined as the angle between the
particle major axis $\underline{a}$ and
scattering vector $\underline{K} = \underline{K}_s - \underline{K}_i$.

For the case of the RGD doublet a simple analytic form for a_L occurs[9].

$$a_o = 1 + j_o (Kr_{12})$$

$$a_L = (2L+1) j_L (Kr_{12})$$

(8)

where K is the modulus of the scattering vector and r_{12} is the centre to centre separation of the spheres.

More generally the orthogonality of the Legendre polynomials allow us to calculate a_L from

$$a_L(\theta) = \frac{2L+1}{2} \int_{-1}^{1} i(\theta,Z) P_L(Z) dZ$$

(9)

with $Z = \cos\beta$. Such projections are tabulated for powers of Z up to Z^{12} [10]. Hence one can calculate a_L by expanding $i(\theta,Z)$ in powers of Z up to order L (P_L is a polynomial of order L) and summing the products of the tabulated projections and the coefficients of the expansion.

For an ellipsoid of rotation with radius b and length 2a (see Figure 1) $i(\theta,t)$ is given by [1]

$$i(\theta,t) = \frac{j_1^2(u(t))}{u^2(t)} cV^2$$

(10)

where c is a constant and V is the particle volume. $j_1(u)$ is the first order spherical Bessel function and

$$u(t) = Kb [1+(q^2-1)\cos^2 (t)]^{\frac{1}{2}}$$

where we define the axial ratio $q = a/b$. Expansion of this function in powers of $\cos^2 \beta$ (Z^2) have been evaluated allowing calculation of a_L. For brevity we refer the reader to the original publication[8] for the details of the rather complicated analytical form of a_L.

We specialise here to small K and/or b and show A_2 and A_4 as a function of $U_o = Kb < 3$ for various q in Figure 2. Clearly A_4 (and thus higher terms) is negligible for the range of U_o shown hence g_R shows a single exponential decay.

$$g_R(\tau) \approx 1 + A_2 e^{-6D_R\tau} . P_2(\cos\Delta\theta/2)$$

(11)

The requirement that Kb < 3 and $A_2 > .02$ (to allow measurement) limits the range of particle sizes one can study where this approximation is valid. With a typical experimental apparatus scattering angle θ can range between $30°$ and $135°$, laser sources such as the Krypton ion system emit in the wavelength range $\lambda = 400-700$nm and K is given by

$$K = |\underline{K}| = 4\pi n/\lambda \sin\theta/2$$

(12)

Refractive index n is of order 1.33 hence K can range between $6\mu m^{-1}$ and $30\mu m^{-1}$. Division of the ordinate of Figure 2 by these values of K leads to a scale reading in microns. From this we can see that this approach should be applicable for particles as small as 30nm radius (q > 1.3) and as large as 500nm radius (q < 1.3). That is particles with length between 80nm and 1.3µm. For optimum choice of scattering angle a prior estimate of particle hydrodynamic 'equivalent sphere' radius can be made using the conventional dynamic light scattering technique.

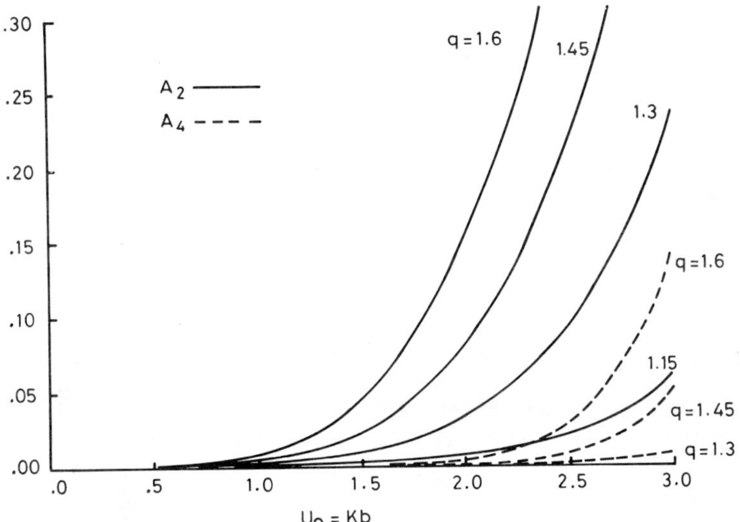

Figure 2. A_2 and A_4 contributions to g_R (equation 5) for a prolate ellipsoid radius b and various axial ratios q. A typical value of K for 90° scattering angle, λ = 633nm and n = 1.33 is 18.6 μm^{-1}. Division of the ordinate U_0 by this value yields a scale reding in microns.

From a linear fit to

$$\ln[g_R(\tau)-1] - \ln P_2(\cos\Delta\theta/2) = A_2 - 6D_R\tau \qquad (13)$$

one can obtain estimates of A_2 and D_R in a single measurement. From measurements of D_R one can calculate an equivalent sphere rotational radius r'.

$$r' = \left(\frac{K_B T}{8\pi\eta D_R}\right)^{1/3} = b\left(\frac{q}{P_R(q)}\right)^{1/3} \qquad (14)$$

where K_BT is the Boltzmann energy and η is solution viscosity. $P_R(q)$ is
the Perrin correction which for prolate ellipsoids is given by

$$P_R(q) = \frac{3q^2}{2(q^4-1)} \left\{\frac{(2q^2-1)}{q(q^2-1)^{\frac{1}{2}}} \; \ell n \; q+(q^2-1)^{\frac{1}{2}}\right\} \tag{15}$$

A_2 can be calculated along the b, q contour defined by equation 14
with the measured value of r' to obtain estimates of b and q where A_2
matches the measured value.

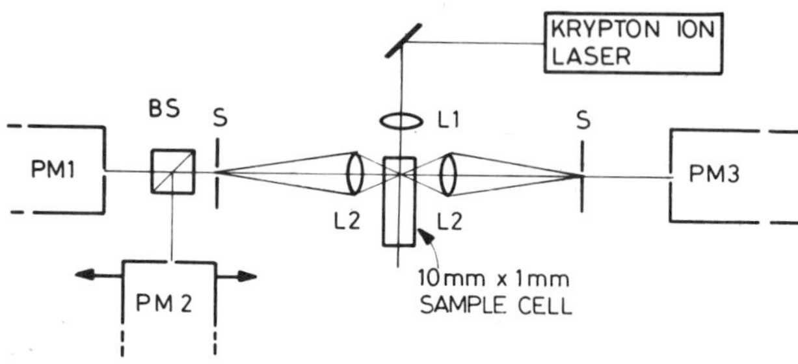

Figure 3. Schematic of the apparatus. L1 and L2 are microscope lenses,
S are slits, BS is a beam splitter and PM1, 2 and 3 are photon counting
photomultipliers. See text for dimensional details.

EXPERIMENT

Measurement of ellipsoid radii

To ensure visibility of the number fluctuation term a small
scattering volume must be used with relatively dilute samples to ensure
<M> small (see equation 4). In this experiment a volume of about
3×10^{-9} ml was used although with high magnification optics much smaller
scattering volumes can be created ($\sim 10^{-11}$ ml[6]). The scattering volume
was defined by focussing the beam from a Krypton ion laser into a sample
cell using a 32mm focal length lens. Images of the scattering volume
were transferred to 200μm slits via X6 (0.15NA) microscope objectives.
Two imaging systems were used on opposite sides of the cell at 90°
scattering angle. A schematic of this set up is shown in Figure 3. In
one arm the resulting speckle pattern behind the slit was viewed by two
photon counting photomultipliers via a beamsplitter. These detectors
were adjusted to ensure they viewed different speckles by minimising the
interference fluctuation signal obtained in cross-correlation from a
relatively concentrated suspension of small spherical particles. A
single photomultiplier viewed the opposed slit. Two values of $\Delta\theta$ can
thus be accessed by such an apparatus; $\Delta\theta \approx 0$ ($P_2\cos(\Delta\theta/2)\approx1$) when same
side detectors (PM1 and PM2 in Figure 3) are used and $\Delta\theta = 90°$

$(P_2(\cos\Delta\theta/2) = -\tfrac{1}{2})$ when opposed detectors (PM1 and PM3) are used to measure the cross-correlation function. As $P_2(\cos 90)$ is negative the latter method leads to anticorrelation effect. A routine apparatus would need only two detectors, most probably in the first (same side) configuration, as no extra information is gained by opposed detectors when the approximation in equaiton 11 is applicable (alignment of the two slits on exactly the same scattering volume can be difficult).

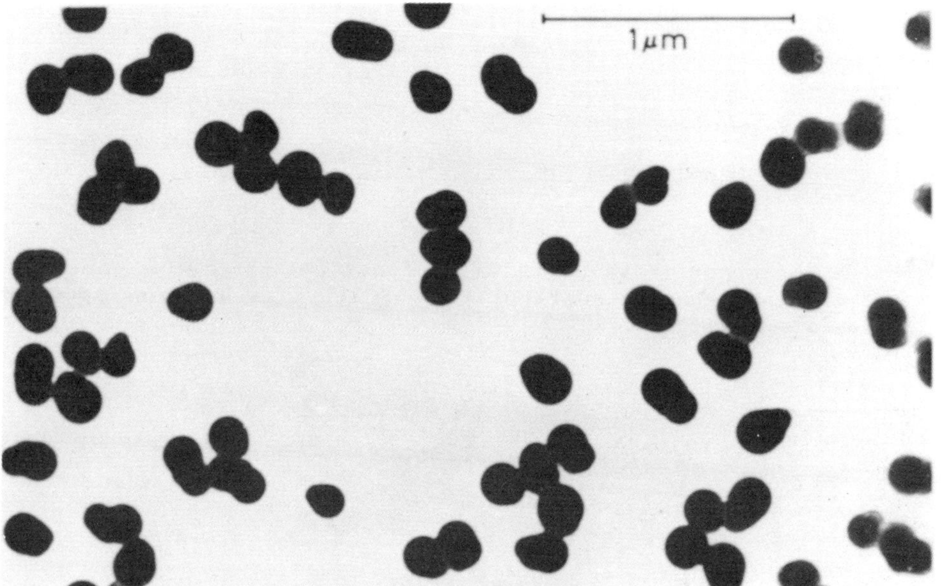

Figure 4. Electron Micrograph of the "ellipsoidal" latex studied.

An ellipsoid like particle (see electron micrograph reproduced in Figure 4) was used to demonstrate the method. Cross-correlation functions were collected using a Malvern Instruments K7023 single bit (scaled) correlator. Careful sample preparation was essential to ensure dust free samples as the presence of dust can seriously distort the number fluctuation signal. Typical measurements of $g^{(2)}(\tau)$ are shown in Figure 5. The short time rotational signal is clearly seen superimposed on the slow number fluctuation decay. An estimate of $g_R(\tau)$ was obtained by dividing $(g^{(2)}(\tau) - 1)$ by a straight line fit to the long time decay. A semi logarithmic plot of $(g_R(\tau) - 1)$ versus τ is shown in Figure 6. A linear fit to this data provides an estimate of A_2 and D_R (see equation 13). Using the simultaneous equation technique described in the previous section we obtained values for b and q. By averaging several measurements

$$q = 1.52 \pm .05$$
$$b = 81 \pm 4\text{nm}$$

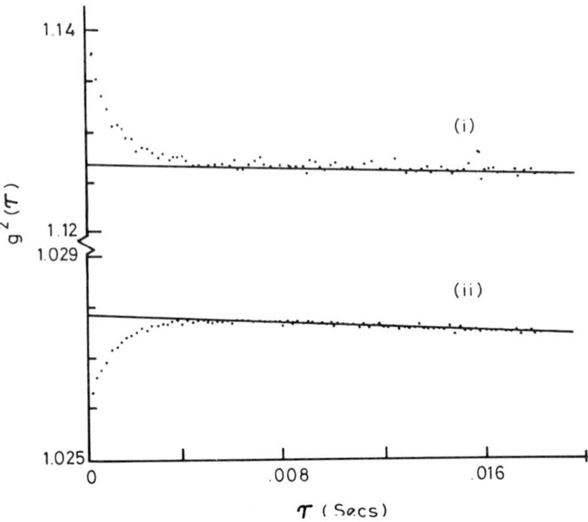

Figure 5. Measured cross-correlation functions $g^{(2)}(\tau)$ for same side (i) and opposed detector configurations. The slow decay of $g_N(\tau)$ is fitted by a straight line.

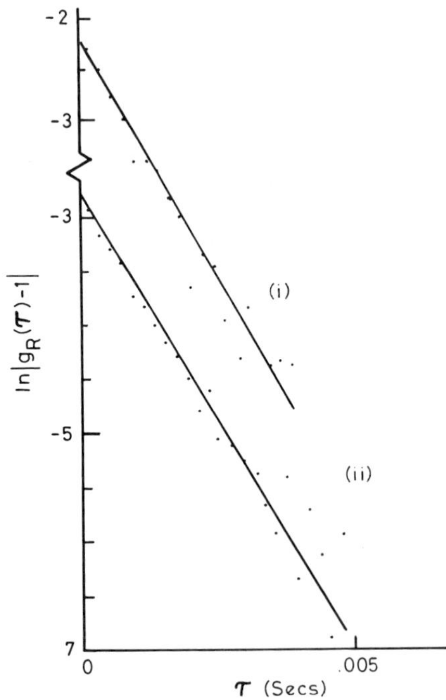

Figure 6. $[g_R(\tau)-1]$ obtained from the data in Figure 5 plotted semilogarithmically versus τ.

From the electron micrographs (EM's) we estimated

$$q = 1.47$$
$$b = 85nm$$

Due account was taken of the fact that the EM is a two dimensional projection of a three dimensional randomly oriented particle. Clearly, there is good agreement between the two techniques.

Detection of sample aggregation

Low levels of aggregation can often be present in colloid samples. This cannot usually be detected by the conventional photon correlation technique apart from a slight increase in the measured second cumulant or polydispersity factor[5]. The sensitivity of the cross-correlation technique to low levels of asymmetry can often show up small numbers of doublets in apparently unaggregated samples. The formula for the cross-correlation function for doublets is given by equaitons 5, 6 and 8. In general more than one component will contribute to the summation shown in equation 5. However, the intercept ($\Delta\theta = 0$) can be calculated using the formula[10].

$$\sum_{L=0,\ \text{even}} (2L+1) j_L^2(Z) = \tfrac{1}{2}(1+j_o(2Z)) \tag{16}$$

and coupling the time behaviour into a general function $f(\tau)$ we can write

$$g_R(\tau) = 1 + A_T f(\tau) \tag{17}$$

with

$$A_T = \frac{x\left[\tfrac{1}{2} + \tfrac{1}{2}j_o(2Kr_{12}) - j_o^2(Kr_{12})\right]}{1 + xj_o(Kr_{12})(2 + (j_o Kr_{12}))} \tag{18}$$

where x is the number fraction of doublets in the system.

For larger values of Kr_{12} all j_o terms become small and

$$A_T \simeq \tfrac{1}{2}x \tag{19}$$

Given the detection criterion $A_T \geqslant .02$ quoted earlier we expect to be able to estimate doublet contamination down to 4% levels. As a demonstration of the technique Figure 7 shows several measurements of this intercept made using the same side detector arrangement described previously. The sample studied was a spherical silica colloid of diameter 272nm, as measured by conventional photon correlation spectroscopy, suspended in hexane at various temperatures. The samples were prepared by filtering the diluted stock suspension through a $0.5\mu m$ filter hence residual aggregates should be dominated by doublets. With this assumption and the estimated value of Kr_{12} in equation 18 ($Kr_{12} = 5.85$ here) the fraction x can be calculated and is shown on the right hand axis in Figure 7. The experiment was carried out to test the stability of this colloid in hexane at high temperatures. Such samples have shown a 'gas-like' to 'liquid-like' phase transition with rising temperature at high concentrations[11] and one might have expected some increase in aggregate levels in these dilute systems. This is clearly not seen, probably because the phase transition occurs at much higher temperatures for these concentrations.

Three separate sample preparations are shown indicating a reasonable degree of reproducibility.

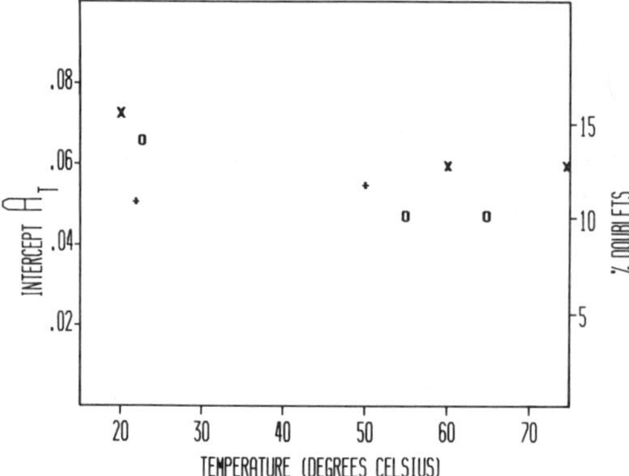

Figure 7. Total intercept of $(g_R(\tau)-1)$, A_T as a function of temperature. Sample is a spherical silica colloid diameter 272nm suspended in hexane. The data indicate the presence of low order aggregates in the sample and the equivalent doublet fraction is shown on the right hand axis.

DISCUSSION

Throughout this work we have implicitly assumed that the Rayleigh-Gans-Debye scattering approximation[1] holds. For particles of size greater than 100nm this is not strictly valid. For doublets an approximation where interparticle interference effects are calcualted in the RGD approximation has been shown to be valid for much larger particles as long as individual particle scattering factors are described by the full Mie theory[12]. We assume this work on angle averaged scattering to apply equally well in our experiments and expect intercepts and decay time to approximate well the RGD results quoted for Brownian rotating doublets.

In the case of the ellipsoids the agreement between experiment and theory is rather more surprising in that angular average scattering of spheres of similar size (125m radius) differs by up to 20% from the RGD result. However the intercept measured arises in a ratio of the

226

intensity factors a_2 and a_0 (see equation 6) and we expect deviations from the RGD approximation to partially cancel. Similarly, distortion of the temporal behaviour of g_R should not be large as first order Mie corrections could be well approximated by sums of even order Legendre polynomials (as in equation 7). This theory has not been tested for ellipsoids of larger dimension than those studied here.

A further extension of the work with general doublets involves calculation of the initial decay of the function $f(\tau)$ (equation 17) using the explicit values of the a_L given in equaiton 8. Such measurements should allow one to determine whether the signal arises from doublets, higher order aggregates or optically asymmetric singlet particles. In a preliminary analysis of the data taken on the silica samples studied here we mesured decay rates similar to those expected for the L = 4 term which does in fact dominate for doublets of this size ($Kr_{12} \approx 5.85$). One must note however that the rotational diffusion constants for compact trim ers and tetragonal aggregates will not differ greatly from doublets.

The cross-correlation technique has now been applied to a wide variety of samples including

(i) Study of the asymmetry of high order "fractal" aggregates.
(ii) Measurement of rotational diffusion of rod shaped virus particles[6].
(iii) Measurement of polymer flexing motions[14,15].
and (iv) Study of many particle ordering effects in concentrated liquid like suspensions[16,17].

We have further demonstrated here that the technique can be applied both in the study of contamination by low order aggregates and in the measurement of the radii of low axial ratio ellipsoids of revolution. Extensions of the theory to cover the case of the general ellipsoid[8] have been published. Two principle decay times should appear in the low K approximation. However, resolution of closely spaced exponentials is difficult and one expects limited extra information to be obtainable for this case.

ACKNOWLEDGEMENTS

The author would like to thank Dr B Vincent and Dr J Edwards of Bristol University for provision of the silica latex and Professor R H Ottewill of Bristol and Dr S M Underwood of Dulux Australia for the preparation and Electron micrograph sizing of the ellipsoid sample used in this work.

REFERENCES

1. M. Kerker, "The Scattering of Light and other Electromagnetic Radiation", Academic, New York (1969).
2. H. Z. Cummins, F. D. Carlson, T. J. Herbert, and G. Woods, Translational and Rotational Diffusion Constants of Tobacco Moasic Virus from Rayleigh Linewidths, Biophys. J., 9:519 (1969).
3. T. A. King, A. Knox, and J. D. G. McAdam, Translational and Rotational Diffusion of Tobacco Mosaic Virus from Polarised and Depolarised Light Scattering, Biopolymers 12:1917 (1973).
4. See relevant papers in this book.

5.   B. B. Weiner, Particle Sizing Using Photon Correlation Spectroscopy in "Modern Methods of Particle Size Analysis", H. G. Barth ed., Wiley Interscience, New York (1984) pp93-116.

6.   W. G. Griffin and P. N. Pusey, Anticorrelations in Light Scattered by Nonspherical Particles, Phys Rev Letts, 43:1100 (1979).

7.   C. V. Heer, Intensity Cross-Correlations in Light Scattered by the Rotational Motion of Macromolecules, Optics Comms, 38:110 (1981).

8.   J. G. Rarity, Measurement of the Radii of Low Axial Ratio Ellipsoids using Cross-Correlation Spectroscopy, J. Chem. Phys. 85:733 (1986).

9.   J. G. Rarity and K. J. Randle, Measurement of the Intensity Cross-Correlation Function of Light Scattered by Brownian doublets, Optics Comms, 50:101 (1984).

10. M. Abramowitz and I. A. Stegun, Handbook of Mathematical Functions, U.S. GPO, Washington, DC (1965).

11. C. Cowell, R. Li-In-On and B. Vincent, Reversible Flocculation of Stearically Stabilised Dispersions, J. Chem. Soc. Faraday Trans. I, 74, 337 (1978).

12. A. Lips and S. Levine, Light Scattering by Two Spherical Rayleigh Particles over all Orientations, J. Coll and Interf Sci 33:455 (1970).

13. J. G. Rarity and P. N. Pusey, Light Scattering from Aggregating Systems: Static, dynamic (QELS) and number fluctuations, in "On Growth and Form", H. E. Stanley and N. Ostrowski eds, Martinus Nijhoff, Boston (1986).

14. Z. Kam and R. Rigler, Cross-Correlation Laser Scattering, Biophys. J. 39:7 (1982).

15. P. N. Pusey, Calculation of the Fourth-Order Correlation Function of a Polymer Coil, J. Chem. Phys., 18:1950, (1985).

16. B. J. Ackerson and N. A. Clark, Cross-Correlation Intensity Fluctuation Spectroscopy Applied to Colloidal Suspensions, Faraday Discuss Chem Soc 76:00 (1983).

17. P. N. Pusey and J. G. Rarity, Measurement of Higher-Order Correlation Functions by Intensity Cross-Correlation Light Scattering, J. De Physique C9:43 (1985).

LAMBDA DEPRESSION/OVERLAY HISTOGRAM ANALYSIS OF POLY(ACRYLATE) AS A
FUNCTION OF ADDED SALT

Kenneth S. Schmitz and Jae-Woong Yu

Department of Chemistry
University of Missouri-Kansas City
Kansas City, Missouri   64110

INTRODUCTION

Dynamic light scattering (DLS) techniques monitor the decay of spontaneous fluctuations in the polarizability of the medium. Any process that results in a change in polarizability can therefore be examined by DLS. The time dependence of these processes can be extracted from the autocorrelation function [C(t)] or spectral density [S(ω)] of the scattered light. If the light scattered by the supporting medium (solvent and salt) can be neglected, then the molecular (solute) correlation function for the self-beat (homodyne) correlation function can be expressed as,

$$G(K,t) = [g_1(K,t)]^2 + B + \varepsilon(t) \tag{1}$$

where

$$g_1(K,t) = \sum_i a_i(K)\exp(-\gamma_i t) \tag{2}$$

is the molecular correlation function, B is the baseline, $\varepsilon(t)$ represents random fluctuating noise, $a_i(K)$ is the amplitude for the ith decay process which has a decay rate of γ_i, and K is the scattering vector with magnitude

$$K = (4\pi n/\lambda)\sin(\theta/2) \tag{3}$$

where n is the index of refraction, λ is the wavelength of the incident light, and θ is the scattering angle. The amplitude associated with each decay process is of the general form,

$$a_i(K) = N'(\partial n/\partial C_i)^2_{T,\mu'}(M_i C_i/N_A)S(Kd_i,Kr_i) \tag{4}$$

where

$$N' = [\sum_j(\partial n/\partial C_j)^2_{T,\mu'}(M_j C_j/N_A)S(Kd_j,Kr_j)]^{-1} \tag{5}$$

is the normalization constant, $(\partial n/\partial C_i)_{T,\mu'}$ is the index of refraction increment of species i at constant chemical potential of all other components, M_i is the molecular weight of component i at a concentration C_i, N_A is Avogadro's number, and $S(Kd_i,Kr_i)$ is the scattering form factor for the ith component, where d_i represents the difference in internal coordinates of scattering centers located on the ith particle and r_i is the distance of separation between a scattering center on the ith particle and a scattering center on another particle in the solution. To the first order correction term in the concentration, the structure factor $S(Kd_i, Kr_i)$ can be expressed as the sum of "self" and "pair" terms,

$$S(Kd_i, Kr_i) = P(Kd_i) + \sum_j C_j [(\partial n/\partial C_j)/(\partial n/\partial C_i)]_{T,\mu'} S'(Kr_i) \qquad (6)$$

where the self term is the particle form factor,

$$P(Kd_i) = (1/n_i)^2 < \sum_{i'} \sum_{i''} \exp[iK(d_{i'} - d_{i''})] > \qquad (7)$$

for the n_i scattering units in the ith polymer, and the pair term,

$$S'(Kr_i) = < \sum_i \sum_j (1/N_i N_j) \sum_{i'} \sum_{j'} (1/n_i n_j) \exp[iK(r_i^{cm} - r_j^{cm} + r_{i'} - r_{j'})] > \qquad (8)$$

for the N_m polymers of type "m" with the center-of-mass located at r_m^{cm}. The general form for the scattering amplitudes $a_i(K)$ is thus written as,

$$a_i(K) = (\partial n/\partial C_i)^2_{T,\mu'} (M_i C_i/N_A) P(Kd_i)$$

$$\{1 + \sum_j C_j [(\partial n/\partial C_j)/(\partial n/\partial C_i)]_{T,\mu'} [S'(Kr_i)/P(Kd_i)]\} \qquad (9)$$

where the normalization parameter N' defined by Eq. (5) is omitted for convenience. Clearly, the interpretation of the relative amplitudes as defined by Eq. (9) is somewhat complicated if the relative number density C_i is desired. For a system containing only one type of particle, the relative amplitude expression takes on the form,

$$a_i(K) = M_i C_i P(Kd_i) \{1 + \sum_j C_j [S'(Kr_i)/P(Kd_i)]\} \qquad (10)$$

where the indices i and j reflect the polydispersity in molecular weight of the preparation. There are two systems that permit further simplification of Eq. (10): the spherical particle; and the flexible particle with one contact point with other particles.

The Spherical Particle

In the case of the spherical particle the center-of-mass and the internal coordinates can be averaged independently of each other. This is because a rotation about the center-of-mass does not alter the distribution of the internal scattering centers. One can therefore write,

$$(1/n_i n_j) < \sum_{i'} \sum_{j'} \exp[iK(r_{i'} - r_{j'})] >$$

$$= [(1/n_i) < \sum_{i'} \exp(iKr_{i'}) >][(1/n_j) < \sum_{j'} \exp(-iKr_{j'}) >]$$

$$= [P(Kd_i)]^{1/2} [P(Kd_j)]^{1/2} \qquad (11)$$

and

$$S''(Kr_i^{cm}) = (1/N_i N_j) < \sum_i \sum_j \exp[iK(r_i^{cm} - r_j^{cm})] > \qquad (12)$$

The relative amplitudes now have the general mathematical form,

$$a_i(K) = M_i C_i P(Kd_i)\{1 + \sum_j C_j S''(Kr_i^{cm})[P(Kd_j)/P(Kd_i)]^{1/2}\} \qquad (13)$$

We compare Eq. (13) for the polydisperse system with the corresponding expression for a monodisperse solution of spheres,

$$a(K) = MCP(Kd)[1 + CS''(Kr^{cm})] \qquad (14)$$

The Flexible Polymer/ One Contact Point

Zimm[1] approached the scattering problem for a flexible polymer by assuming that there is only one point of contact between the polymer and one of its neighbors. If the point of contact is with the scattering unit at r_i^{cu} and r_j^{cu} for polymers i and j, then one can write

$$r_{i'} = r_i^{cu} + \Delta r_{i'} \qquad (15)$$

where $\Delta r_{i'} = r_{i'} - r_i^{cu}$ is the distance from the center-of-mass to the contact unit in the ith polymer. An important point to be made is that the single internal coordinate $r_{i'}$ is now expressed in terms of the difference between <u>two</u> internal coordinates of the same polymer. The interparticle structure factor $S'(Kr_i)$ is now expressed as,

$$S'(Kr_i) =$$

$$< \sum_i \sum_j (1/N_i N_j)\exp[iK(r_i^{cm} + r_{i'}^{cu} - r_j^{cm} - r_{j'}^{cu})]> P(K\Delta r_i)P(K\Delta r_j) \qquad (16)$$

where the average in the brackets < > is over all contact points as well as center-of-mass locations. The particle form factors are defined by Eq. (11) where $d_i \rightarrow \Delta r_i$. The relative amplitudes now become,

$$a_i(K) = M_i C_i P(Kd_i)[1 + \sum_j C_j S'''(Kr_i)P(Kd_j)] \qquad (17)$$

where

$$S'''(Kr_i^{cm}) = < \sum_i \sum_j (1/N_i N_j)\exp[iK(r_i^{cm} + r_{i'}^{cu} - r_j^{cm} - r_{j'}^{cu})] > \qquad (18)$$

The amplitude for the monodisperse system is,

$$a(K) = MCP(Kd)[1 + CS'''(Kr_i^{cm})P(Kd)] \qquad (19)$$

Flexible Polyelectrolytes

One can infer from the form of Eqs. (13), (14), (17), and (19) that the amplitudes of flexible polymers at finite concentrations and multiple contact points depends upon the molecular form factor in a complex way, where the concentration correction term depends upon the form factor that reflects the number of direct spatial contacts. The situation for polyelectrolytes is further clouded in that the interactions between the segments of different chains is long range and therefore multiple intersegment interactions prevail. It is for this reason that the form of $S'(Kr_i)$ given by Eq. (8) is retained for the center-of-mass and intersegmental terms cannot be separated.

THEORY

The primary objective in the analysis of multiple decay correlation
functions is to obtain the amplitude profile of the decay processes as a
function of the decay rates. Once this is obtained, then further analysis
of these amplitudes in terms of the number distribution, for example, is
achieved through formulae discussed in the previous section. We now review
methods currently used to analyze polydispersity in the correlation function
and introduce a new method, lambda depression analysis, which may be of
value in the analysis of decay functions with a narrow spread in decay
rates.

Cumulant Analysis

Koppel[2] introduced a series expansion method for the analysis of paucity disperse correlation functions. The normalized correlation function
$g_1(K,t)$ [cf. Eq. (2)] is represented as,

$$\ln[g_1(K,t)] \simeq 1 - K_1 t + (1/2)K_2 t^2 - (1/6)K_3 t^3 + \dots \tag{20}$$

where the cumulants K_n are given by

$$K_n = \{\partial^n \ln[g_1(K,t)]/\partial t^n\}_{t=0} \tag{21}$$

Asymptotic Analysis

Schmitz and Pecora[3] proposed a method of analysis that is similar to
the cumulant method except that correlation functions obtained at various
data collection interval (Δt) are utilized. The correlation functions are
analyzed as a single exponential decay function with a characteristic decay
rate $1/\tau_c$. In the limit $\Delta t \to 0$, one has,

$$1/\tau_c \simeq \langle 1/\tau \rangle - [(\langle 1/\tau^2 \rangle - \langle 1/\tau \rangle^2)/2]N_c \Delta t + \dots \tag{22}$$

for the $N_c + 1$ point correlation function, where

$$\langle 1/\tau^n \rangle = \sum_i a_i(K)(1/\tau_i)^n \tag{23}$$

for the homodyne function $g_1(K,t)$. It is emphasized that the asymptotic and
cumulant methods of analysis are mathematically identical, hence these two
methods differ only in their applications. The cumulant method is applied
to only one decay function, hence the "true" $t \to 0$ limit is achieved only
for the short data collection intervals. Indeed, it can be shown through
the analysis of highly polydisperse correlation functions that the values
of the cumulants are strongly dependent upon the data collection interval.

Laplace Transform Methods/CONTIN and Histograms

The series expansion methods briefly reviewed above are strictly valid
for paucity disperse correlation functions. Indeed, they can provide only
the average values of the decay rates and higher moments without any detailed information about the distribution of the decay rates. To obtain
more detailed information it is desirable to invert Eq. (1) to obtain the
$a_i(K)$ as a function of γ_i. We therefore express $G(K,t)$ as a continuous
function of γ,

$$G(K,t) = [\int_0^\infty G(\gamma)\exp(-\gamma t)d\gamma]^2 + B + \varepsilon(t) \tag{24}$$

The presence of the random noise term $\varepsilon(t)$ renders the inversion of Eq. (24)
an ill-posed problem. In other words, the time constants for the decay

232

function cannot be greater than the rate at which the noise fluctuates.
McWhirter and Pike[4] and Ostrowsky et al.[5] showed that the maximum amount
of information that can be extracted from $G(\gamma)$ is when exponentially spaced
values of γ are employed, viz.,

$$\gamma_n = \gamma_{n-1} \exp(\pi/\omega_{max}) \tag{25}$$

where ω_{max} is the truncated value of the frequency as determined by the
noise level of $\varepsilon(t)$.

CONTIN is a packaged program which is available upon request[6]. CONTIN
assumes a continuous distribution of amplitudes, and is composed of a fixed
core of 53 subprograms and 13 USER programs. The memory requirement is
200 Kbytes. This program is non-interactive, where the "best fit" solution
is provided after a series of calculations. The calculations involve two
cycles, the first is an unweighted analysis of the data and then a weighted
analysis of the data.

Fletcher and Ramsay[7] developed an overlay histogram method with exponential sampling procedure for Laplace inversion of Eq. (24). In this procedure the central frequency is estimated from the cumulant analysis method.
The program is user interactive, where the central frequency can be changed
and the number of steps and width of each step are the input parameters.
Several histograms are generated, where they differ in the central frequency
but have the same number of steps and step width. The family of histograms
is then normalized and a composite histogram is generated by an overlay
procedure.

Exponential Depression Analysis

Isenberg and Small[8] developed a sensitive method for analyzing closely
spaced decay rates of fluorescence decay data. We define an exponentially
depressed function $F(\lambda,t)$ by

$$F(\lambda,t) = G(K,t)\exp(-\lambda t) \tag{26}$$

where λ is an arbirary constant. There are two important properties of
Eq. (26) when compared to Eqs. (1) and (2). First, the relative magnitudes
of the baseline and noise are depressed. [In photon correlation spectroscopy the baseline can be subtracted, hence one can set $B = 0$ in Eq. (1).]
Second, the relative amplitudes of the decay processes are not altered
whereas the decay rates are shifted by a known amount, λ. The nth moment
of the exponentially depressed function is defined as,

$$M_n(\lambda, T) = \int_0^T t^n F(\lambda,t)\, dt \qquad (n = 0, 1, 2, \ldots) \tag{27}$$

where $T = N_c \Delta t$ is the total time window for the correlation function. By
changing variables $(\gamma_i + \lambda)t = x_i$ with the definition $\gamma_0 = 0$, one obtains

$$M_n(\lambda,T) = \left[\Sigma\, a_i(K) Y_i(n)/(\gamma_i + \lambda)^{n+1} \right] + [B+\varepsilon(t)] Y_0(n)/\lambda^{n+1} \tag{28}$$

where

$$Y_i(n) = \int_0^{x_i(max)} x_i^n \exp(-x_i)\, dx_i \tag{29}$$

It is recognized that $Y_i(n) \to n!$ when $x_i(max) \to \infty$. Operationally, however,
the value of $x_i(max)$ is determined from the time window of the correlation
function, viz.,

$$x_i(max) = -\ln[F(\lambda,N_c \Delta t] \tag{30}$$

One must therefore perform the numerical integration described by Eq. (29) to use in Eq. (28).

The primary advantage of using the function $F(\lambda,t)$ instead of $G(K,t)$ is that the (residual) baseline and the random noise can be suppressed. The initial effect of this suppression is to decrease the breadth of the distributions of decay modes obtained by the Laplace transform techniques. Another advantage is in the analysis of systems with two or more decay modes that have comparable decay rates, such as might occur for monomer-dimer-trimer solutions of bovine serum albumin. This capability of the exponential depression technique, i.e., to distiguish between a small number of discrete and closely spaced relaxation processes, is limited only by the number of moments that can be calculated from the prescription given by Eq. (27). One can, in principle, discern n+1 decay rates from the n moments and the normalization condition $[\Sigma\, a_i(K) = 1]$.

We define an apparent decay rate on the basis that $F(\lambda,t)$ is a single exponential decay function. According to Eq. (28), the apparent decay rate for the nth moment is,

$$k_a(n) = [Y(n)/M_n(\lambda,T)]^{1/(n+1)} - \lambda \qquad (31)$$

where it is assumed that the baseline has been suppressed. Clearly, if the decay function is truly a single exponential decay, then the same value for $k_a(n)$ will be obtained regardless of the value of n. A plot of $k_a(n)$ versus λ should therefore be a straight line. [The λ-plots are also a way to obtain unique decay rates for closely spaced exponential decay functions, where the "local flatness" is a criterion for the correct value of $k_i(n)$, hence γ_i.][8]

METHODS

Simulated correlation functions comprised of 70 exponential decay functions were generated for the purpose of comparison of the various analysis methods and also to assess the effect of noise in the interpretation of the numerical results. The amplitudes of the decay functions were generated by

$$a_i = A\exp\{-B[(y - \langle y\rangle)/\langle y\rangle]^2\} \qquad (32)$$

where A is an artibrary constant that denotes the relative contribution of a family of function that constitutes a particular mode, B is an adjustable parameter the determines the width of the distribution, and $\langle y\rangle$ is the average value of y, being either the relaxation rate or the relaxation time for the particular mode. In the present series of functions, the correlation function was generated as a bimodal function with the relative contributions $A_f = 0.67$ (fast mode) and $A_s = 0.33$ (slow mode). The relaxation times for the fast mode where linearly spaced over the range 100 μsec $\leq \tau_f \leq$ 200 μsec and likewise the slow mode over the range 200 μsec $\leq \tau_s \leq$ 800 μsec. The other characteristic parameters were: $B_f = 26.96$, $\langle\tau_f\rangle = 150$ μsec; $B_s = 8.32$, $\langle\tau_s\rangle = 500$ μsec. Each mode contained 35 exponential decay functions. The correlation functions were then generated by Eq. (1), where the amplitudes a_i were normalized to unity, the baseline was B = 1, and the random noise was generated with a random noise generator with the constraint that the maximum value for any one point was 0.2. After the noise was generated for each point, the set of values was adjusted so that $\langle\varepsilon(t)\rangle = 0$ prior to the final incorporation in $G(K,t)$. To assess the effect of the added random noise these analyses were also performed with $\varepsilon(t) = 0$. [The root-mean-square deviation of the noise level used in the simulated correlation functions was 0.056.]

The poly(acrylate) used in these studies was kindly provided by Dr. Kim. The molecular weight as determined by viscosity measurements was 10^7

Daltons.[9] The concentration of the final solutions was $\sim$ 20 ppm. The different KCl solutions were made by direct dilution using 9 parts of the stock poly(acrylate) solution to 1 part of the KCl solution.

The quasielastic light scattering facility was previously described.[10] The samples were gravity-flow filtered into previously rinsed cylindrical cells. The 488 nm line of a Spectra Physics argon ion laser was the incident radiation, and the scattering angles were over the range 30° - 90°. The correlation functions were obtained with a Langley-Ford 1096 Correlator.

Representative correlation functions used in the simulated study and for poly(acrylate) are shown in Figure 1. It is important to note that the poly(acrlyate) correlation functions more closely resemble noiseless simulated correlation functions.

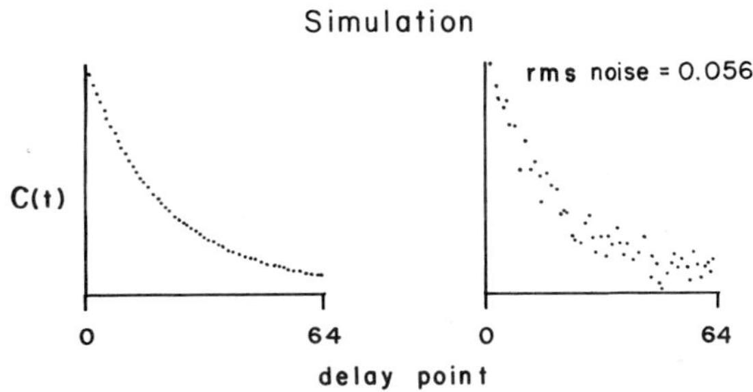

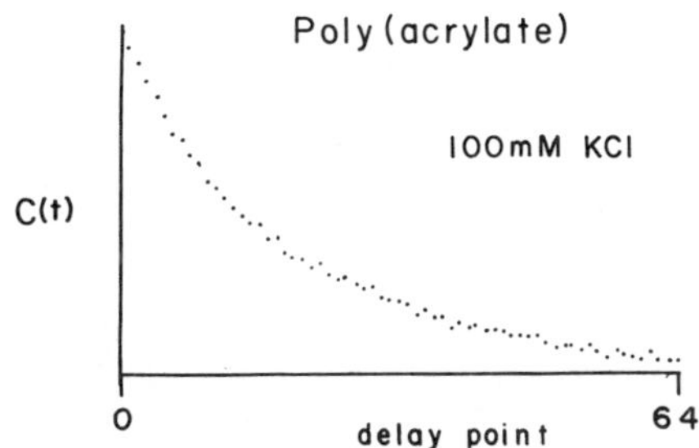

Figure 1. Representative Correlation Functions

Top: The simulated correlation functions were composed of two modes each having 35 exponential decay functions. The function on the left does not have added noise, whereas the function on the right has added noise with a root-mean-square value of 0.056 [cf. Eq. (1)].

Bottom: Correlation function for the sodium salt of poly(acrylate), where T = 25°C, θ = 40°, and Δt = 300 μsec.

RESULTS

Overlay histograms were determined for 3-step histograms by varying the central value (initially estimated by the first cumulant) and step width, or multiplier for the exponential sampling width. As shown in Figure 2, the overlay histogram for the simulated correlation function without added noise reproduces the input distribution of amplitudes (shaded area in Figure 2) quite well. The value of $\dot{\gamma}$ at the maximum of $G(\gamma)$ was estimated by extrapolation of the "edges" of each mode, where γ_{max} is defined as the point of intersection of the two lines (cf. Figure 2). It is concluded that the overlay histogram procedure of Fletcher and Ramsay[7] can accurately descern the peak locations of a bimodal distribution for correlation functions <u>without high levels of noise</u>. It is also concluded that the extrapolation procedure provides a reasonable estimate (within 10–20%) of the location of the maximum in $G(\gamma)$ for each mode.

The effect of noise on the overlay histogram procedure was examined using correlation functions with a relatively high noise level (root-mean-square deviation of 0.056). The λ-depressed functions $F(\lambda,t)$ are shown in Figure 3 along with λ-plots for the first five moments [cf. Eq. (31)]. The fact that the λ-plots are not superimposed is a clear indication of a high degree of polydispersity in the relaxation times. It is noted that $F(\lambda,t)$ is defined by the homodyne correlation function [cf. Eq. (1)], hence the decay rates shown in the λ-plots have been adjusted by the factor of 2.

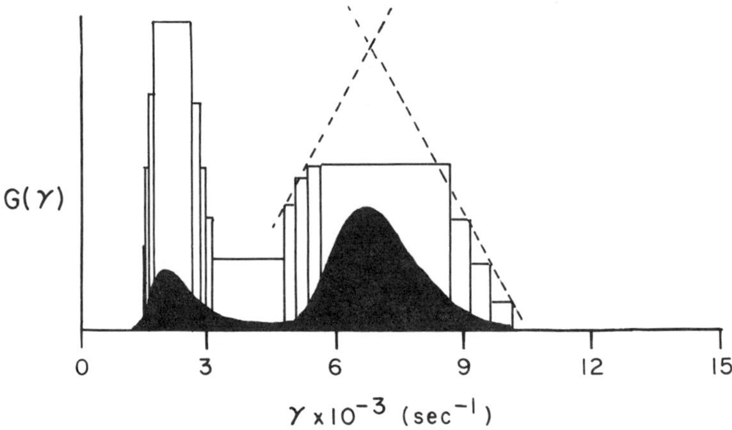

Figure 2. Overlay Histogram Analysis of Simulated Correlation Functions

The overlay histogram with exponential sampling procedure of Fletcher and Ramsay[7] was used on a noiseless correlation function [$\epsilon(t) = 0$ in Eq.(1)] as shown above. The solid area represents the actual distribution of amplitudes generated by Eq. (32) using the parameters: 100 μsec $\leq \tau \leq$ 200 μsec, B = 26.96, and A = 2; and 200 μsec < τ < 800 μsec, B = 8.32, and A = 1. Because the distribution in decay rates is linear in the relaxation time, one has for the respective distributions $<\tau>$ = 150 μsec and $<\tau>$ = 500 μsec. The number of exponential functions in each mode was 35. Because of the exponential sampling method, the width of each histogram increases with an increase in the decay rate. An estimate of the location of the maximum in $G(\gamma)$, i.e., γ_{max}, was determined by extrapolation of the edges, as shown in the figure.

236

Overlay histogram/λ-depression analysis of a correlation function with noise
is shown in Figure 4. The effect of the noise is to introduce a "repulsive"
interaction between the two peaks in the bimodal distribution. That is, the
fast decay mode lies at a higher frequency and the slow decay mode lies at
a lower frequency range than the corresponding input distributions. The histogram of the λ-depressed function tends to correct for this apparent repulsion,
but clearly values of λ greater than 4000 s^{-1} are needed (for this function)
to obtain a reliable estimate of the average decay rates. [Note that $\gamma + \lambda/2$
is the correct variable since $G(K,t)$ is a homodyne function.]

   Representative overlay histograms for the poly(acrylate) system are
shown in Figure 5. In all cases examined in which the correlation function
decayed to 15-25% of its initial value, these histograms exhibited two relaxation modes.

   The results of the different methods of analysis of the simulated correlation functions are summarized in Table 1. The results of the histogram
analysis of the poly(acrylate) system are summarized in Table 2, along with
the theoretical calculations of persistence lengths [total persistence length
L_p and electrostatic persistence length L_{el}] employed in the interpretation
of the relaxation modes.

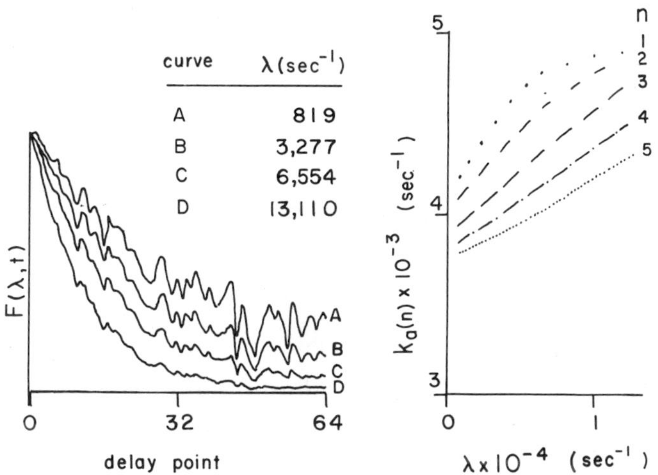

Figure 3. λ-Depression Analysis of Simulated Correlation Functions with Noise

DISCUSSION

   In order to interpret the two relaxation modes in the histograms of
the poly(acrylate) system, we first examine the expected value of diffusion
coefficient for a flexible coil of 10^7 Daltons molecular weight. The contour
length L is 2.5×10^5 Å on the basis of a length density 40 gm/Å.[11] For coils,
   The correlation function shown at top-right in Figure 1 was analyzed by
the λ-depression method of Isenberg and Small[8]. The λ-plot (right) for the
first 5 moments of the first-order decay rate indicates polydispersity.

Correlation functions with noise were analyzed by the overlay histogram method, where the root-mean-square displacement of noise was 0.056. In the histograms above, the dotted curves are the actual distributions of the amplitudes for the two modes.   As indicated in the top curve, the effect of

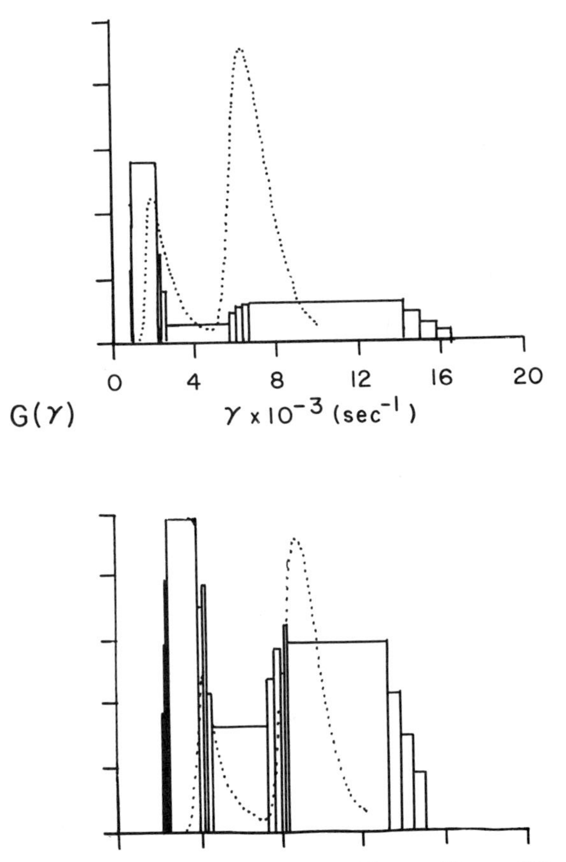

Figure 4.   Overlay Histogram/Lambda Depression Analysis

noise is to broaden the histogram, where the two peaks can be described as having a "repulsive" interaction. The bottom curve is a histogram for a λ-depressed function (λ = 4000 s^{-1}, where the factor of 2 in the above plot corrects for the homodyne form of G(γ). The dotted curves are shifted by 2000.

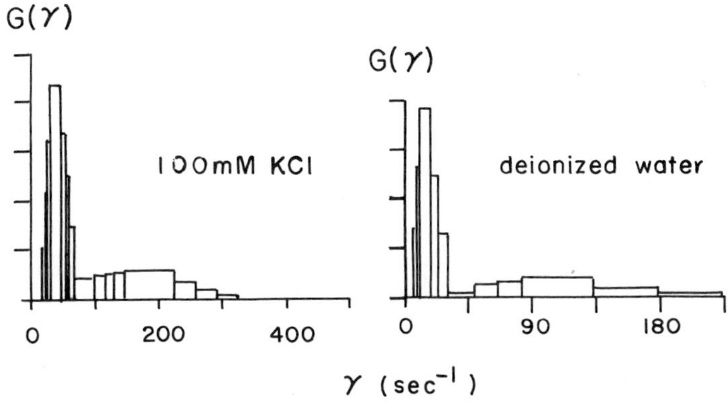

Figure 5. Overlay Histograms of Poly(sodium acrylate)

The above histograms were calculated from correlation functions for poly(sodium acrylate) at 25°C and a scattering angle θ = 40°.

Table 1. Comparison of the Decay Rates[*] Obtained from the Analysis Methods

r.m.s. noise	Theoretical			Cumulant order		λ-Plot (λ = 0)	Overlay Histogram			
		mode					λ = 0		λ = 4000	
	overall	fast	slow	1st	3rd		fast	slow	fast	slow
0.0	5234	6785	2130	5235	5235	4600	6780	1985	----	----
0.056	5234	6785	2130	6460	5835	4100	9560	1324	8220[a]	1235[a]

[*] in units of 1/sec
[a] corrected for the λ/2 shift

Table 2. Flexible Coil Parameters for Poly(sodium acrylate)

C_s (M/L)	Apparent Hydrodynamic Radius[*]			Persistence Length[*]			
	3rd order cumulant	Histogram		L_p^a	L_{el}^b	L_{el}^c	L_{el}^d
		fast	slow				
$7 \times 10^{-5 e}$	6120	3153	30310	270	260	14600	9635
5×10^{-2}	----	2448	39550	162	152	204	13
1×10^{-1}	3590	1955	14340	102	92	102	7

[*] units are in angstroms
[a] estimated from Eq. (33).
[b] estimated from Eq. (33) and $L_p = L_o + L_{el}$ with L_o = 10 Å[11].
[c] estimated from Eq. (35)
[d] estimated from Eq. (34)
[e] calcualted on the basis of a polyion concentration of 20 ppm, an average molecular weight of 10^7 Daltons, an average charge spacing of b = 1.8 Å at a level of 75% neutralization due to counterion condensation, and a mass density of 40 g/Å.

$$R_H = 0.27[2LL_p]^{1/2} \qquad \text{(hydrodynamic radius)} \qquad (33)$$

where L_p is the persistence length, which for polyelectrolytes is composed of an electrostatic part (L_{el}) and an intrinsic part (L_o), $L_p = L_o + L_{el}$. Assuming $L_p = L_o = 10$ Å[11], one estimates for a lower limit to the hydrodynamic radius, $R_H = 2236$ Å. This indicates that the _fast_ mode in the histogram plot is to be associated with the translational diffusion of the poly(acrylate).

According to the theories of Odijk[12] and Skolnick and Fixman[13],

$$L_{el} = \begin{cases} 1000/16\pi b^2 C_s N_A & (b/\lambda_B > 1) \qquad (34) \\ 1000/16\pi\lambda_B^2 C_s N_A & (b/\lambda_B < 1) \qquad (35) \end{cases}$$

where b is the average spacing between charges and $\lambda_B = e^2/\varepsilon kT$ is the Bjerrum length, where e is the electron charge and ε is the bulk dielectric constant. Eq. (35) attempts to take into consideration counterion condensation[14] of the Manning type.[15] In calculating L_{el} in the deionized water solution, the counterions from the poly(acrylate) are the only contirbutors to C_s, which is estimated to be 7×10^{-5} M/L on the basis of total ionization and the concentration of the monomeric units. It is clear from the numbers given in Table 2 that the estimated value of L_{el} from the data lies between L_{el} computed from Eqs. (34) and (35) for the relatively high salt conditions. It is also clear that _neither_ of the above equations describes the experimentally estimated value of L_{el} under conditions of zero added salt. We believe this discrepancy is due to a failure in the theory, where it is applied outside the limits of the approximations inherent in the theory. The approxiamtions leading to Eqs. (34) and (35) are: (1) small deriavtions from rodlike behavior; and (2) that the electrostatic contribution to the bending modulus act _along_ the contour length of the polyion, i.e., proportional to the curvature of the polyion. Theoretical calculations of the free energy difference between the rod-bent structures[16,17] indicate L_{el} is proportional to $1/C_s$ only when the intrinsic persistence length is longer than the Debye-Huckle screening length, $\lambda_{DH} = (1000/8\pi N_A\lambda_B I_s)^{1/2}$, where N_A is Avogadro's number and I_s is the ionic strength. Physically, the condition $L_o/\lambda_{DH} > 1$ assures that the electrostatic range of interaction does view the flexible polyelectrolyte as a rodlike structure. This is probably why the Odijk/Skolnick-Fixman theory works well for DNA ($L_o \sim 450$ Å) but not for very flexible polyions. [Calculations of the relative probability indicates a 90° bend is half as probable as a rod[16,17].]

REFERENCES

1. B. H. Zimm, J. Chem. Phys. 16:1093 (1948).
2. D. E. Koppel, J. Chem. Phys. 57:4814 (1972).
3. K. S. Schmitz and R. Pecora, Biopolymers 14:521 (1975).
4. J. G. McWhirter and E. R. Pike, J. Phys. A: Math. Nucl. Gen. 11:1729 (1978)
5. N. Ostrowsky, D. Sornette, P. Parker, and E. R. Pike, Optica Acta, 28:1059 (1981).
6. S. W. Provencher, Computer Phys. Comm. 27:229 (1982).
7. G. C. Fletcher and D. J. Ramsay, Optica Acta 30:1183 (1983).
8. I. Isenberg and E. W. Small, J. Chem. Phys. 77:2799 (1982).
9. O.-K. Kim, T. Long, and F. Brown, Poly. Comm. 27:71 (1986).
10. K. S. Schmitz, M. Lu, N. Singh, and D. J. Ramsay, Biopolymers 23:1637 (1984).
11. T. Kitano, A. Taguchi, I. Noda, and M. Nagasawa, Macromolecules 13:57(1980).
12. T. Odijk, J. Polm. Sci. Polm. Phys. Ed. 15:477 (1977).
13. J. Skolnick and M. Fixman, Macromolecules 10:944.
14. T. Odijk and A. C. Houwaart, J. Polm. Sci. Polm. Phys. Ed. 16:627 (1978).
15. G. S. Manning, Quart. Rev. Biophys. 11:179 (1978).
16. K. S. Schmitz, J.-W. Yu, D. Browning, and J. D. Tate, Polm. Preprint (1987) (in press).
17. K. S. Schmitz, "An Introduction to Dynamic Light Scattering", Academic Press, Orlando (to appear).

MEASUREMENT OF SMALL POLYDISPERSITIES BY PHOTON CORRELATION SPECTROSCOPY

P. N. Pusey and W. van Megen*

Royal Signals and Radar Establishment
St Andrews Road
Malvern, Worcestershire, WR14 3PS, England

INTRODUCTION

Photon correlation spectroscopy (PCS) of scattered laser light is possibly the most widely used technique for sizing submicron particles suspended in liquids.[1] Its advantages are that it is reasonably rapid and, being based on measurement of the particles' natural Brownian motion, non-invasive. A disadvantage of the method, as normally practiced, is its relative insensitivity to polydispersity which precludes the study of narrow particle size distributions (PSD). In this brief report we describe a relatively new method of analysing PCS data to provide significant information about the distribution of slightly polydisperse spherical particles; a more detailed account can be found in Reference 2.

CONVENTIONAL ANALYSIS OF PCS DATA

For a dilute suspension of spherical particles PCS measures $g(Q,\tau)$, the time correlation function of the amplitude of the scattered light field:

$$g(Q,\tau) = \frac{\int I(QR)\ P(R)\ \exp[-D(R)Q^2\tau]\ dR}{\int I(QR)\ P(R)\ dR} . \tag{1}$$

Here:

τ is the correlation delay time;

R is the particle radius;

*Department of Applied Physics, Royal Melbourne Institute of Technology, Melbourne, Victoria, Australia.

$Q = (4\pi/\lambda) \sin(\theta/2)$, the scattering vector, where λ is the wavelength of light in the suspension and θ the scattering angle;

$P(R)$ is the particle size distribution, the distribution of particle radii $[\int P(R)\, dR = 1]$;

$I(QR)$ is the intensity scattered at angle θ by a particle of radius R;

$D(R)$ is the translation diffusion constant of a particle of radius R.

The diffusion constant is given by the Stokes-Einstein relation

$$D(R) = \frac{kT}{6\pi\eta R} \qquad (2)$$

where k is Boltzmann's constant, T the temperature and η the viscosity of the liquid in which the particles are suspended.

The data analysis procedure commonly used is to write Equation (1) in terms of an inverse radius variable R^{-1} and its related distribution $P'(R^{-1})$ and to recognise that the measured quantity $g(Q,\tau)$ is the Laplace transform, with respect to R^{-1}, of $I(QR)\, P'(R^{-1})$. Thus a numerical Laplace inversion is performed to recover $I(QR)\, P'(R^{-1})$ which, with prior knowledge (or assumptions) concerning the form of $I(QR)$, can be converted into the particle size distribution $P(R)$. As is well known, the difficulty with this approach is that the Laplace transformation is ill-conditioned: small unavoidable statistical errors in $g(Q,\tau)$ are converted into large errors in $P(R)$. For this reason the technique is insensitive to small polydispersities. In simple terms, the sum of several exponentials with similar decay constants (Equation (1)) looks much like an "average" single exponential.

The latter point can be quantified by a moment expansion.[3] The exponential in Equation (1) is expanded about its mean value to give

$$g(Q,\tau) = \exp[-\overline{D}(Q)\, Q^2\tau][1 + V\, \overline{D}(Q)^2\, Q^4\, \tau^2/2 + \ldots] \quad , \qquad (3)$$

where the average diffusion coefficient $\overline{D}(Q)$ is given by

$$\overline{D}(Q) = \int I(QR)\, P(R)\, D(R)\, dR \Big/ \int I(QR)\, P(R)\, dR \quad , \qquad (4)$$

and the variance factor V by

$$V = \frac{\overline{D^2(Q)}}{\overline{D}(Q)^2} - 1 \qquad (5)$$

with

$$\overline{D^2(Q)} = \int I(QR)\, P(R)\, D^2(R)\, dR \Big/ \int I(QR)\, P(R)\, dR \quad . \qquad (6)$$

For a monodisperse system $[P(R) = \delta(R-R_o)]$ $V = 0$; thus the quantity V is, at the same time, a measure of the departure of $g(Q,\tau)$ from a single exponential and a measure of the polydispersity.

We define the polydispersity σ by

$$\sigma = \left(\overline{R^2} - \overline{R}^2\right)^{\frac{1}{2}} / \overline{R} \tag{6}$$

where

$$\overline{R^n} = \int R^n \, P(R) \, dR \quad . \tag{7}$$

For narrow PSDs, $\sigma \leq 0.2$ say, it can be shown[2] that

$$V \approx \sigma^2 \quad . \tag{8}$$

Now the typical error in an experimental measurement of V is rarely smaller than ± 0.01. Thus this method of analysing PCS data cannot distinguish a particle suspension with distribution of size having a width $\sigma = 0.1$ ($V = 0.01$) from a monodisperse suspension: the light scattering correlation function $g(Q,\tau)$ cannot be distinguished from a single exponential. Even in the case $\sigma = 0.2$ ($V = 0.04$) the polydispersity would only just be detected. For example, the latter parameters correspond to a mixture of two species with size ratio 1.5.

DETECTION OF SMALL POLYDISPERSITIES

For the case of particles which show at least one minimum in the accessible angular profile of scattered light, polydispersities much smaller than 0.1 can be detected from the variation with angle (or scattering vector) of the average diffusion coefficient $\overline{D}(Q)$ (Equation (4)).

As an example we consider the case of homogeneous spheres for which the Rayleigh-Gans-Debye (RGD) scattering approximation holds so that

$$I(QR) = Q^{-6}(\sin QR - QR\cos QR)^2 \tag{9}$$

and which have a Schulz PSD:

$$P(R) = \frac{R^Z}{\Gamma(Z+1)} \left(\frac{Z+1}{\overline{R}}\right)^{Z+1} \exp\left[-\frac{R}{\overline{R}}(Z+1)\right] \quad , \tag{10}$$

for which

$$\sigma^2 = (Z+1)^{-1} \quad . \tag{11}$$

For these functions we can calculate analytically (see Reference 2 for details) both $I_T(Q)$, the total intensity scattered by the polydisperse suspension,

$$I_T(Q) = \int I(QR) \, P(R) \, dR \quad , \tag{12}$$

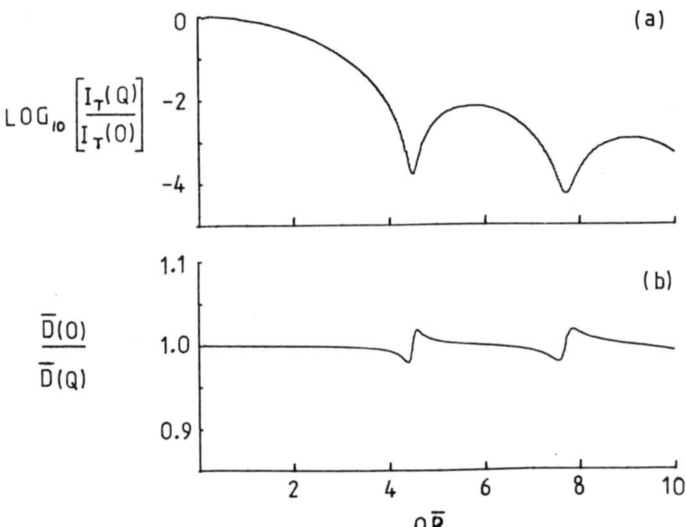

Figure 1.   (a) Normalised total intensity (Equation (12)) and
(b) Normalised inverse average diffusion coefficient
(Equation (4)) as function of scattering vector Q. A
Schulz particle size distribution with polydispersity
$\sigma = 0.02$ has been used.

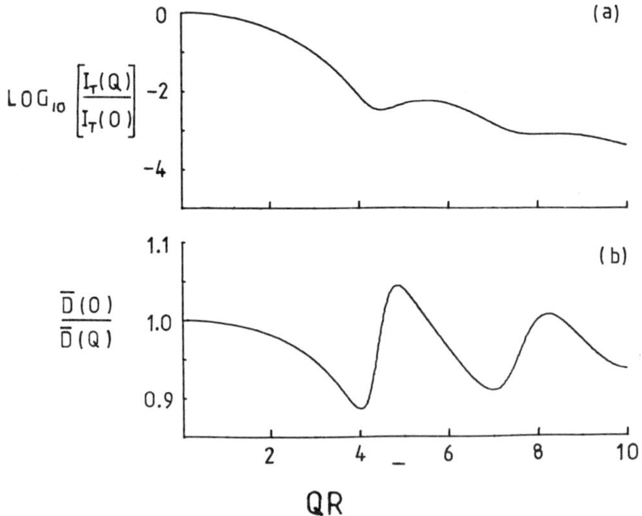

Figure 2.   Same as Figure 1 except $\sigma = 0.10$.

and $\overline{D}(Q)$ (Equation (4)). In Figures 1 and 2, $I_T(Q)/I_T(0)$ and $\overline{D}(0)/\overline{D}(Q)$ are plotted for $\sigma = 0.02$ and $\sigma = 0.1$.

In Figure 1a the intensity differs from the well-known RGD form for monodisperse spheres mainly in that, because of the small polydispersity ($\sigma = 0.02$), the intensity does not extend fully to zero at the minima. In Figure 2a ($\sigma = 0.1$) the intensity minima are much more severely attenuated. In Figures 1b and 2b we see that, in the region of the intensity minima, $\overline{D}(0)/\overline{D}(Q)$ undergoes a characteristic "swing" whose magnitude increases with increasing polydispersity. It is this phenomenon which can be exploited to measure small polydispersities.

The form of the curves in Figures 1b and 2b can be understood as follows. At small values of Q, $Q\overline{R} \rightarrow 0$, a particle of radius R scatters light in proportion to R^6 (Equation (9)). Thus, from Equations (2) and (4),

$$\overline{D}(0) = kT/6\pi\eta(\overline{R^6}/\overline{R^5}) \quad . \tag{13}$$

As scattering angle θ, or Q, is increased the scattering from all particles decreases. However the scattering by the larger particles falls relatively more rapidly. As the first intensity minimum is approached, the larger particles stop scattering first. Thus the measured average radius, $kT/6\pi\eta\overline{D}(Q)$, is weighted more heavily by the smaller particles and it therefore decreases. However, on further increase of Q, the larger particles start to scatter again whereas the scattering by the smaller particles decreases to zero. Then $1/\overline{D}(Q)$ increases. For relatively small σ, ≤ 0.1, this process is repeated at subsequent intensity minima.

It can be seen in Figures 1b and 2b that, for small σ, the relative extent of the swing compared to the plateau value is roughly $\pm \sigma$. It can further be shown[2] that this remarkably simple result applies, not just for the Schulz distribution, but for all narrow, relatively symmetrical distributions. It is this dependence of the phenomenon on the first power of the polydispersity, rather than the σ^2 dependence found in other light scattering measures of polydispersity (eg Equation (8)), which gives the technique its sensitivity. Effectively the deep minimum in I(QR) acts as a narrow "inverse window" which is scanned through the PSD by changing scattering angle (or Q).

Figure 3 shows some experimental results. The particles were sterically-stabilised spheres of colloidal polymethymethacrylate,[4,5] suspended in dodecane. Standard PCS equipment was used with an argon ion laser operating at a wavelength in vacuo of 488 nm. The PCS data were analysed by the method of moments[3] in which $\ln g(Q,\tau)$ was fitted to a quadratic function of τ (see Equation (3)). The mean diffusion coefficient $\overline{D}(Q)$ was obtained from the first moment, the coefficient of the term linear in τ. In Figure 3 we plot the effective radius $\overline{R}(Q)$, defined by

$$\overline{R}(Q) = kT/6\pi\eta \ \overline{D}(Q) \quad . \tag{14}$$

Each data point is the average of two or three measurements of duration a few minutes. The solid line in Figure 3 is a theoretical curve for $\overline{R} = 185$ nm and $\sigma = 0.055$. It is seen that, while the experimental data have the same general form as the theory, there are differences larger than the typical experimental uncertainty ($\pm \sim 1\%$), particularly in the region of the maximum. It is possible that this is due to asymmetry of the particle size distribution (see (iii) below and the appendix of Reference 2). However the principle of the method is clearly demonstrated.

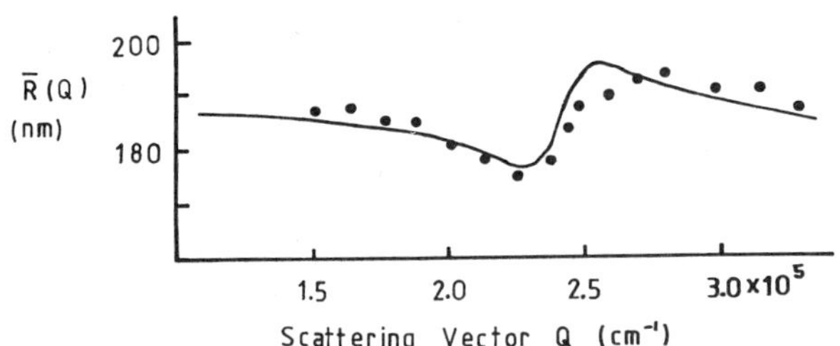

Figure 3.   Experimental results (points) for colloidal spheres of
polymethylmethacrylate. The effective radius $\bar{R}(Q)$
(Equation (14)) is plotted against scattering vector Q.
Solid line is theoretical prediction for a Schulz PSD with
$\bar{R}$ = 185 nm and σ = 0.055.

Several final points can be noted:

(i) Clearly the method outlined above will work only if at least one
minimum in the angular intensity profile is experimentally accessible.
In the RGD approximation the first minimum occurs at $Q\bar{R}$ = 4.49; with
typical values of laser wavelength etc this leads to the requirement $\bar{R} \geq$
150 nm.

(ii) Since $\bar{D}(Q)$ can, under favourable conditions, be measured with a
precision of at least 1% it should, by this method, be possible to
measure polydispersities as small as σ = 0.01. It should be realised,
however, that such narrow distributions would lead to very weak
scattering near the intensity minima so that very clean samples would be
required.

(iii) The theoretical results discussed above apply to relatively
symmetrical PSDs ie those, such as the Schulz distribution, having small
skewness. Remarkably it can be shown[2] that it should be possible to
measure the skewness of asymmetric distributions even in the limit $\sigma \to 0$.
Asymmetry of the PSD will appear as an asymmetry of the swing in $\bar{D}(Q)$
about its plateau value.

(iv) While, for computational simplicity, we have limited the discussion
here to the case of homogeneous spheres, in the RGD scattering
approximation, the principle of the method should apply to any particles
showing a significant minimum in their angular intensity profile. In
unpublished work we have calculated the effect numerically for Mie
scattering spheres and analytically for core-shell particles in the RGD
approximation.

(v) The ability to measure, non-perturbatively, polydispersities less
than 0.1 has already proved very valuable in our research on the
properties of model colloids.[4,5] It may also have industrial
applications, for example, in the manufacture of high-strength ceramics
by a process where concentrated suspensions of nearly monodisperse
spheres of the ceramic material are dried and sintered.[5,6]

REFERENCES

1.   H. G. Barth, Ed., "Modern Methods of Particle Size Analysis", Wiley-
Interscience, New York (1984).

2.   P. N. Pusey and W. van Megen, Detection of small polydispersities by photon correlation spectroscopy, J. Chem. Phys. 80: 3513 (1984).

3.   P. N. Pusey, D. E. Koppel, D. W. Schaefer, R. D. Camerini-Otero and S. H. Koenig, Intensity fluctuation spectroscopy of laser light scattered by solutions of spherical viruses: R17, QB, BSV, PM2 and T7. I. Light-scattering technique, Biochemistry 13: 952 (1974); D. E. Koppel, Analysis of macromolecular polydispersity in intensity correlation spectroscopy: The method of cumulants, J. Chem. Phys. 57: 4814 (1972).

4.   W. van Megen, R. H. Ottewill, S. M. Owens and P. N. Pusey, Measurement of the wavevector dependent diffusion coefficient in concentrated particle dispersions, J. Chem. Phys. 82: 508 (1985).

5.   P. N. Pusey and W. van Megen, Phase behaviour of concentrated suspensions of nearly hard colloidal spheres, Nature 320: 340 (1986).

6.   P. Calvert, A spongy way to new ceramics, Nature 317: 201 (1985).

OPTICAL EXTINCTION SPECTRA OF SYSTEMS OF SMALL METAL

PARTICLES WITH AGGREGATES

Michael Quinten and Uwe Kreibig

Universität des Saarlandes
FB 11 — Physik
D 66oo Saarbrücken, Germany

I. INTRODUCTION

Optical properties of well separated small spherical particles as
contained in diluted colloidal systems are well understood by Mie's[1]
scattering theory for electromagnetic waves. Additional aspects which are
not included in Mie's theory arise for colloidal systems where particles
form aggregates or clusters by lumping.
In the following we shall restrict ourselves on pure coagulation aggregates. This means that coalescence due to formation of grain boundaries between neighbouring particles will be excluded.
The optical properties of coagulation aggregates differ from those of
single particles due to electromagnetic interactions between all particles within an aggregate. The changes in the optical extinction spectra
(i.e. absorption and scattering losses) of such systems containing ensembles of N-particle aggregates with different particle number N can then
be described theoretically by treating one N-particle aggregate representative for each ensemble.
In this paper we shall present an investigation of the extinction properties of Au-particle systems containing coagulation aggregates.
For one selected sample we try a full quantitative comparison between experimental and theoretical results. Quantitative computations were performed by applying an electromagnetic multipole scattering theory for the
aggregates as a whole [2,3].
Special care was taken on the one hand, of the characterization and analysis of the sample and, on the other hand, of the reduction of the
approximations underlying the calculations.
As a result of the quantitative comparison, limitations of validity of
the used electromagnetic multipole interaction theory could be estimated.
In special, the importance of high order multipolar interactions [4] is
discussed.

II. SAMPLE PREPARATION AND TOPOGRAPHY

We shall present two different experimental methods to prepare coagulation aggregates in colloidal noble metal systems and give results of a
detailed examination of the sample topography based on the classification of the aggregate types observed by TEM analysis.
Colloidal gold particles in aqueous electrolytic solutions were prepared
by chemical reduction applying Zsigmondy's[5] preparation technique. These

primary systems have very low filling factors ($f \sim 10^{-6}$). They merely
contain single, well separated spherical particles which are known to be
surrounded by clouds of ions forming the electric double layer.
This electric double layer causes a repulsive electrostatic potential
between the particles, approximatively described by a screened Coulomb-
potential $V(r) = \exp(-\kappa r)/r$ which counteracts the van-der-Waals attraction. Particles coagulate if the repulsive potential becomes smaller
than the van-der-Waals attraction.
This can be achieved by addition of ions to the suspension which cause κ
to increase. Then the single particles tend to approach their neighbours
almost until contacting and eventually form aggregates with shapes varying from pseudofractal chainlike structures to massive lumps. Addition
of a stabilizing agent like gelatin stops the aggregation process since
gelatin forms coverings around the aggregates which prevent the system
from further coagulation.
Varying, both, the concentration c of added ions and the time t until
stabilization we were able to control the aggregation process and to produce systems where the aggregates varied from small to very large average sizes and from chainlike to compact shapes.
At low concentrations c the aggregation velocity is sufficiently small
to prepare samples with strongly different mean aggregate sizes by varying the aggregation time t from a few seconds to many minutes. These
samples were advantageous for our examinations since their optical properties could be attributed to only few different types of aggregates.
At high concentrations however, the TEM-micrographs of the suspensions
showed even for very short times t a broad aggregate size distributions.
In order to separate different kinds of aggregates such samples were cen-
trifugated in a ultracentrifuge after stabilization.
Both methods allow to reduce the width of aggregate size distributions.
The TEM-micrographs however showed that it was impossible to prepare
monodisperse aggregates, i.e. uniform aggregate sizes. They further showed that narrower size distributions were obtained with the first method.
One of these latter samples with extremely narrow distribution and small
mean aggregate size was selected and was analyzed in detail in order to
compare its optical extinction quantitatively with the computations
described in section IV.
This sample was prepared by addition of a low concentration of $CuSO_4$
($c = 1.5 \cdot 10^{-5}$-M) to a single particle hydrosol with mean diameter
$2R = 38$ nm and stabilizing with gelatin after $t \simeq 10$ s. Solid samples
as necessary for TEM were obtained by drying the solutions on thin carbon foils (thickness ≈ 20 nm). There is no additional aggregation during
this drying process as was controlled by varying the amount of added
stabilization agent.
Keeping the thickness of the dried films low, true two-dimensional arrangements could be prepared which allowed the unequivocal identification
of each aggregate. So, it is believed that the TEM results are representative for the aqueous colloidal system the optical spectra were obtained from.
On eight TEM micrographs we identified a total of 7000 particles forming
various kinds of aggregates.

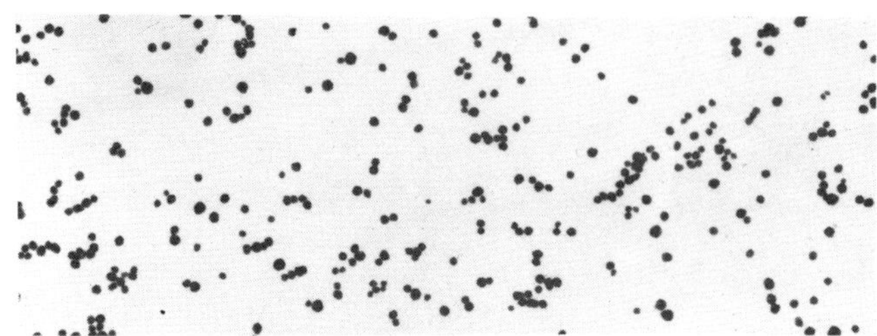

Fig.1. Au-particle aggregates : part of a TEM -
photograph ($2R = 38$ nm)

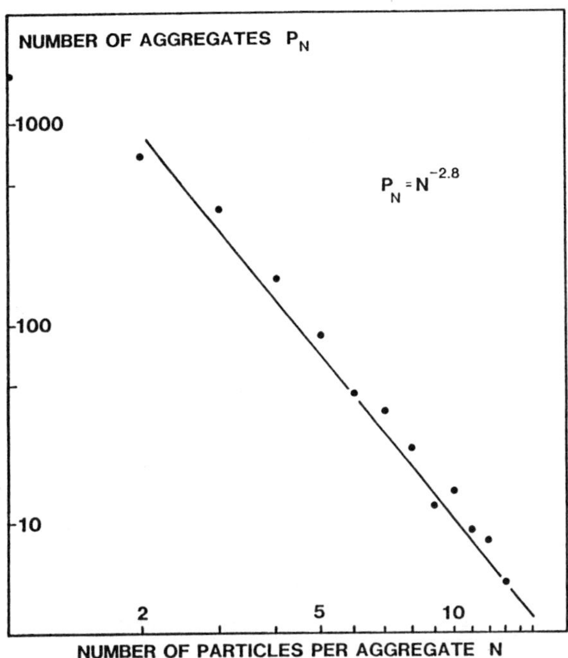

Fig.2: Probability of N-particle aggregates in the analyzed sample

We counted the amount P_N of aggregates with same particle number N. P_N follows a power law[6] $P_N = N^{-A}$ with A = 2.8 as illustrated in figure 2. The next step was to classify the shapes of these aggregates. For this purpose we introduced several kinds of highly symmetric aggregate forms for each N. The shapes identified on the TEM-micrographs were then allo-cated to the most similar of them. The result for N = 1 to N = 7 partic-les per aggregate (p.p.a.) is given in figure 3.

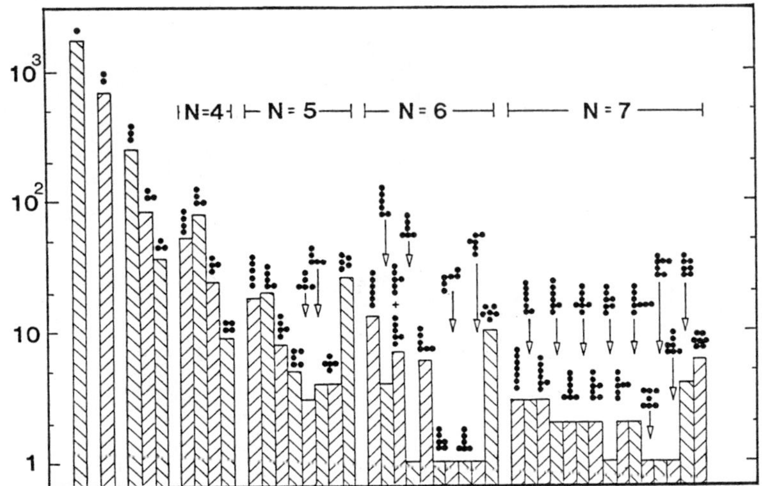

Fig.3: Size and shape histogram of the examined sample:
the sample aggregates were allocated to classes
of highly symmetric aggregate shown at the top
of the histogram columns.

III. OPTICAL PROPERTIES

In this section we present the optical extinction spectrum measured
from the diluted aqueous colloidal Au particle sample.
The spectrum was recorded for 4 eV $\geq$ hω $\geq$ o.5 eV. It is illustrated in
figure 4 from 1 eV to 3.5 eV. For comparison also the corresponding single particle spectrum is shown, which was measured before initiating the
aggregation process. This latter spectrum shows the plasmon polariton
band, typical for well isolated spherical Au particles. The peak position is 2.36 eV, the halfwidth of the low energy flank amounts to o.16 eV.
It corresponds quantitatively to the spectrum predicted by Mie's[1]
theory.
The aggregation process gives rise to strong changes of this spectrum as
is obvious from the comparison of both spectra in figure 4. In detail
the aggregated sample spectrum shows two clearly separated maxima with an
energy difference of ΔE = o.41 eV instead of the single Mie maximum. The
intensity ratio of both peaks is about 2:1. The long-wavelength flank
decreases very rapidly indicating that there exist only a few larger
aggregates in this sample[3]. This is also proved by our TEM analysis. The
following section will be devoted of the quantitative interpretation of
the measured peak spectrum.

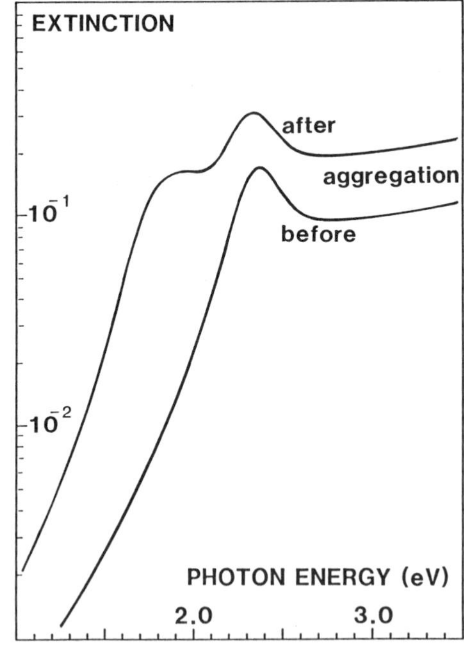

Fig.4. Measured optical extinction spectra
of the aqueous Au-particle sample :
a.) before b.) after the aggregation
process. The values of a.) are multiplied by a factor o.5 to separate
the spectra.

IV. CALCULATION OF AGGREGATE SPECTRA

In this section we describe the calculations of scattering of electromagnetic waves by N-particle aggregates consisting of isolated spherical particles. We apply the derived formulae on all the high-symmetric
aggregates introduced for the classification of the sample aggregates in
section II. Then we calculate the total optical extinction spectrum of
the analyzed sample by adding up the aggregate spectra wheighted with
the probabilities of the according sample aggregates.
The calculations base on phenomenological electrodynamics and are an extension of Mie's theory by taking into account the direct interaction
among the particles of an aggregate via electromagnetic scattering fields.
The advantage of this ansatz is that numerical results can be obtained
for special materials, as e.g. gold, by inserting their optical dielectric function, which then may be quantitatively compared with our experimental results.
The single particles were assumed to be monodisperse (diameter $2R$ = 38 nm)
and spherical and to touch almost their next neighbours in the considered
aggregates. For additional control computations this distance could be
enlarged. For the refractive index of the embedding medium we chose
n_0= 1.4 due to water plus stabilizing agent. Refractive index measurements showed that this value changes only slightly when stabilizing
agents and salts are added. For the dielectric function of Au we used
the data of Otter[7] and Krumm[8].
Analogous to the Mie-theory one has to determine the electric and magnetic fields of the incident wave and the waves inside and scattered by
the particles from Maxwell's equations. The interaction field at particle I, I = 1,...., N, is given by superimposition of the waves scattered
by all the particles $J \neq I$ of the aggregate.
The extension of Mie's theory now is to take into account this additional incident field in Maxwell's boundary conditions on the surface of
each particle I of the aggregate. The fields and their corresponding
elctromagnetic potentials Π (r, θ, ϕ) are developped into spherical har-
monics Y_{1m} (θ, ϕ) and Riccati-Bessel-functions ψ_1 (kr) and ζ_1 (kr):

$$\Pi^{inc} = 1/(k^2 r) \cdot \sum_{1=1}^{\infty} \sum_{m=-1}^{1} c_{1m} \cdot \exp(-i\vec{k}\vec{r}(I))\psi_1(kr)Y_{1m}(\theta,\phi) \qquad (1.1)$$

with $k^2 = (\omega^2/c^2)n_0^2$, n_0 = refractive index of the embedding
medium

$$\Pi^{sca} = 1/(k^2 r) \cdot \sum_{1=1}^{\infty} \sum_{m=-1}^{1} b_{1m}(I)\zeta_1(kr)Y_{1m}(\theta,\phi) \qquad (1.2)$$

$$\Pi^{ins} = 1/(\hat{k}^2 r) \cdot \sum_{1=1}^{\infty} \sum_{m=-1}^{1} a_{1m}(I)\psi_1(\hat{k}r)Y_{1m}(\theta,\phi) \qquad (1.3)$$

with $\hat{k}^2 = (\omega^2/c^2)n^2(\omega)$, $n(\omega)$ = refractive index of gold with
$n^2(\omega) = \varepsilon(\omega)$

$$\Pi^{int} = 1/(k^2 r) \cdot \sum_{1=1}^{\infty} \sum_{m=-1}^{1} (\sum_{J \neq I}^{N} \sum_{q=1}^{\infty} \sum_{p=-q}^{q} b_{qp}(J)A_{1m}^{qp})\psi_1(kr)Y_{1m}(\theta,\phi) \qquad (1.4)$$

Π^{int} of eq. (1.4) is the potential of the additional coupling field incident at particle I. The terms A_{1m}^{qp} transform the scattered wave of particle J into the coordinate frome of particle I[2,3,9,10]. They decrease as
$(R_{IJ})^{-(1+q+1)}$ with the particle distance R_{IJ} between particle I and J.

Therefore for large R_{IJ} the interaction potential vanishes and one obtains again the Mie-result for the scattering by single spheres. Inserting the four potentials in Maxwell's boundary conditions on the surface of particle I one has a set of linear equations which allows to determine the complex coefficients b_{lm} (I) of the scattering wave emitted by particle I[3]:

$$b_{lm}(I) \cdot \eta_1 + \sum_{J \neq I}^{N} \sum_{q=1}^{\infty} \sum_{p=-q}^{q} A_{lm}^{qp} \cdot b_{qp}(J) = -c_{lm} \cdot \exp(-i\vec{k}\vec{r}(I)) \cdot$$

$$\cdot \left(1 + i \frac{\vec{k}\vec{r}(I)}{kR} \cdot \frac{n(\omega)\psi_1(kR)\psi_1(\hat{k}R)}{n_0\psi_1(kR)\psi_1'(\hat{k}R) - n(\omega)\psi_1'(kR)\psi_1(\hat{k}R)} \right) \qquad (2)$$

where

$$1/\eta_1 = \frac{n(\omega)\psi_1(\hat{k}R)\psi_1'(kR) - n_0\psi_1'(\hat{k}R)\psi_1(kR)}{n(\omega)\psi_1(\hat{k}R)\zeta_1'(kR) - n_0\psi_1'(\hat{k}R)\zeta_1(kR)} \qquad (3)$$

is the polarizability of the 1-th multipole of one single particle as given by Mie, and the primes denote first derivatives of the functions

The optical extinction of the N-particle aggregate which is proportional to the imaginary part of the b_{lm} finally follows from the summation over all N particles.

$$K(\omega) = 3c^2/(2n_0\omega^2R^3) \cdot Im\left(\sum_{I=1}^{N} \sum_{l=1}^{\infty} \sum_{m=-l}^{l} (-1)^l b_{lm}(I) \right) \qquad (4)$$

Numerical computation of the extinction spectrum $K(\omega)$ of the N-particle aggregate is only possible if the infinite sums over l and q in eq.(2) are truncated at a maximum order L, i.e. if one restricts on a finite number of multipole orders of the coupling fields. Fuchs[4] and Claro[4] argued that the order of multipoles to be taken into account increases with decreasing particle-particle distance R_{IJ} and eventually runs to infinity for touching particles.
Our truncated calculations and their quantitative comparison with experiment will be a check of their arguments. The particles of our sample and so much the more the aggregates were too large to be treated in the quasistatic (long-wavelength) approximation and so, retardation in the particles and between the particles were fully considered.
First we restricted our computations on L = 1, i.e. we considered only dipole-dipole coupling. Comprehensive discussion of this case was given recently[3].
The next step was the extension to L = 2, i.e. we considered dipole-dipole, dipole-quadrupole and quadrupole-quadrupole coupling, and we will present in the following first results of this program.
For both approximations we computed the extinction spectra of the high-symmetric aggregate types of section II up to N= 8 p.p.a. In doing so the orientation of the aggregates respectively to the Poynting vector was taken into account by averaging over statistically distributed orientations. This corresponds to the case of the aqueous colloidal system. For a selected number of aggregates the resulting extinction spectra are given in figure 5.

Obviously in both cases the electromagnetic interaction between the particles of the aggregate causes a splitting of the single particle Mie-resonance into different extinction bands and minimum two extinction maxima with different resonance energies can be observed in all spectra. Both, positions and heights vary with aggregate size and shape and it is the low energy maximum which reacts most sensitively. It is shifted towards the IR when N increases. A simple interpretation of the structure in terms of excitation modes is only possible for special aggregate geometries[3].

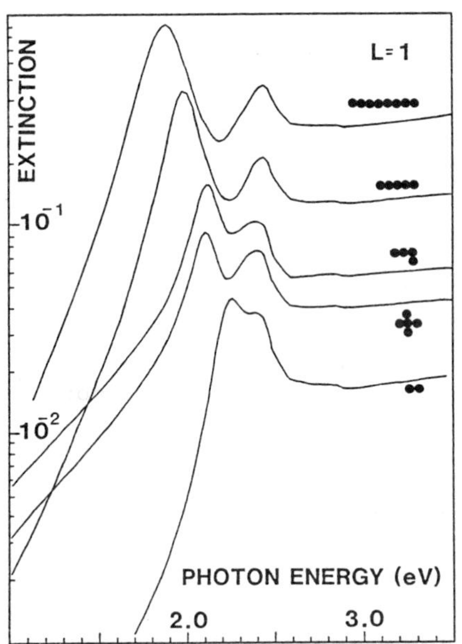

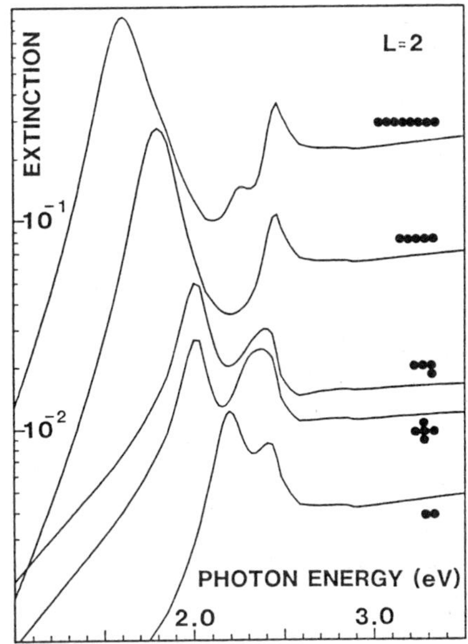

Fig.5. Computed extinction spectra for five selected high-symmetric aggregates :
left : dipolar interactions (L=1), right : dipolar and quadrupolar interactions. The spectra are shifted vertically for arbitrary amounts.

Comparing the splitting of the same aggregate in the L=1 spectrum and the L = 2 spectrum, one recognizes that the splitting increases when quadrupolar effects are included.
The total extinction spectrum to be compared with the measured one was determined from all such partial spectra by weighting with the corresponding numbers of aggregates (figure 3) and adding them all up. The spectra are illustrated in figure 6, again for L = 1 and L = 2, together with the corresponding single particle spectrum. The inclusion of quadrupolar effects obviously influences the high frequency peak only slightly, whereas the low frequency structures are shifted for about o.2 eV towards the IR.

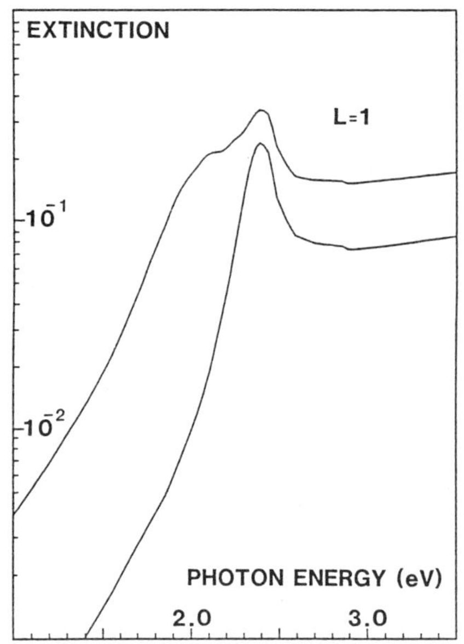

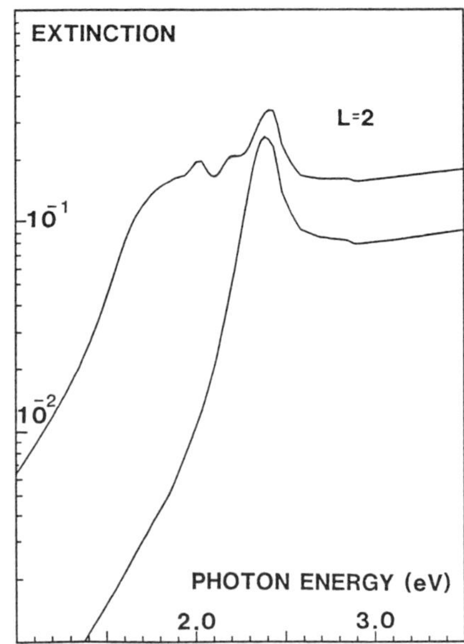

Fig.6. Computed total extinction spectra of the investigated sample:
left: dipolar interactions (L=1), right: dipolar and quadrupolar
interactions. For comparison the corresponding spectra of the
single particles are shown. They are shifted vertically by the
factor o.5.

V. COMPARISON AND DISCUSSION

In the last section we shall present the comparison of the computed
with the measured optical extinction spectrum for our analyzed sample.
In figure 7 a and b the experimental spectrum is plotted together with
the sample spectra computed for maximal multipole order L = 1 and L = 2,
respectively. The computed spectra are close to the measured spectrum,
though there is a somewhat better correspondence especially at the low
frequency side when dipole-quadrupole and quadrupole-quadrupole interactions are included.
Our conclusions are twofold:

The direct electromagnetic coupling theory including retardation
effects describes almost quantitatively the measured aggregate extinction spectrum. This statement refers to the absolute magnitude of the
extinction as well as to spectral features like the number of prominent peaks, their positions, their relative heights and finally the
steepness of the low frequency flank. It should be pointed out that
analogous calculations within the quasistatic approximation do not
yield comparingly good results.

The interaction in our sample is mainly due to the dipolar coupling.
Additional contributions in special at the low frequency part of the
spectrum come from quadrupolar coupling effects. Fuchs and Claro[4] predict that higher order multipoles also affect this spectral region.
The good correspondence of the measured and computed low energy flank
of our spectrum indicates that multipoles of higher order than quadrupolar do not have importance for the interpretation of aggregate sample
spectra when the single particle diameter remains lower than about
2 R = 6o nm.

Remaining differences between the measured and computed spectra can be
traced back to the following points:
The calculations were performed with data for the dielectric function
$\varepsilon(\omega)$ of bulk, monocrystalline material [7,8]. However, quantum size
effects and grain boundary scattering in the particles modify ε, especially its imaginary part. We plan an improvement by using data for ε derived by Kramers-Kronig-analysis of the measured extinction spectrum of
the isolated, single Au particles.

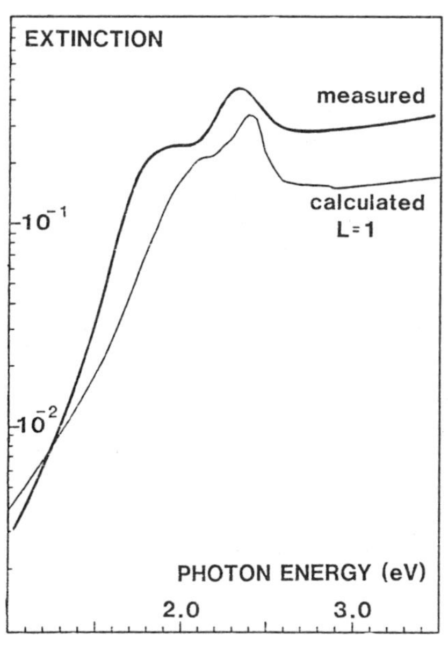

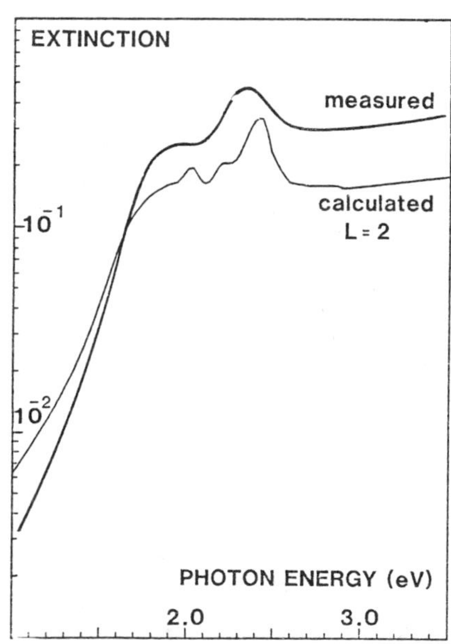

Fig.7. Comparison between the measured extinction spectrum and the computed total extinction spectra :
left : computed with dipolar interactions (L=1), right : computed
with dipolar and quadrupolar interactions (L=2). The computed curves are shifted vertically by the factor o.5.

Further, we supposed in our calculations touching neighbouring particles,
i.e. the distance D between their centers is D = 2R. The electron micrographs confirmed that the surface-surface separation is small in our
sample. On the other hand, grain boundaries between neighbouring particles, probably arising if the particles are in direct metallic contact
were observed only very rarely.
To control their influence we performed additional computations with
varied distance D. They show that distances up to D = 2.o5 R do not
effect essentially our results.
We therefore assume that there are thin interlayers of ions isolating
the particles from each other. Possibly, they are due to a secondary
potential minimum of the Hamaker-Derjaguin-Verwey-Overbeek-type[12].

REFERENCES

1. G. Mie, Ann.Phys.25 (1908) 377
2. J.M. Gérardy and M. Ausloos, Phys.Rev. B27 (1983) 6446
3. M. Quinten and U. Kreibig, Surface Sci. 172 (1986) 557
4. R. Fuchs, to be published
   F. Claro, Phys.Rev. B30 (1984) 4989
   F. Claro and R. Fuchs, Phys.Rev. B33 (1986) 7956
   R. Rojas and F. Claro, Phys.Rev. B34 (1986) 3730
   R. Fuchs and F. Claro, to be published in Phys.Rev. B
5. R. Zsigmondy, Das Kolloide Gold (Deuticke, Leipzig, 1925)
6. D. Schoenauer and U. Kreibig, Surface Sci. 156 (1985) 100
   U. Kreibig, M. Quinten, D. Schoenauer, physica scripta 34 (1986)
7. M. Otter, Z.Physik 161 (1961) 539
8. H. Krumm, Diploma work, Saarbrücken 1965
9. D. Langbein, Springer Tracts in Modern Physics, Vol.72 (Springer, Berlin, 1974)
10. B. Jeffreys, Geophys. J. Roy. Soc. 10 (1965) 141
11. U. Kreibig, Z.Physik 234 (1970) 307
12. J. Stauff, Kolloidchemie (Springer, Berlin, 1960)

PARTICLE SIZING OF SOOT IN FLAT PREMIXED HYDROCARBON

OXYGEN FLAMES BY LIGHT SCATTERING

H.Bockhorn[1],F.Fetting, A.Heddrich,
U.Meyer, and G.Wannemacher

Institut für Chemische Technologie
Technische Hochschule Darmstadt
Petersenstraße 20, 6100 Darmstadt, West-Germany

INTRODUCTION

Combustion of hydrocarbons in premixed flames is not complete under
fuel rich conditions. In this case fuel molecules are converted to the stable main products as well as to high molecular weight hydrocarbons and soot
particles with about 10^6 molecular mass units within a few milliseconds.
The soot particles form an aerosol within the burnt gases which can be defined by the volume fraction f_V, particle number density N and by particle
properties, e.g. the moments of the resulting particle size distributions.
Soot particles undergo coagulation while simultaneously new particles are
formed and the volume fraction of soot also increases due to reactions of
gaseous components with the surface of soot particles.

To investigate these processes, e.g. particle inception, surface
growth of particles and particle coagulation, usually the time dependent
developement of f_V , N and particle properties are measured. One method
for measuring these variables is the combination of light scattering and
extinction. Applying this method to sooting hydrocarbon flames one has to
take care for necessary corrections of the measured signals due to particular soot particle properties and characteristics of sooting flames. In
this work particle sizing by light scattering and extinction in sooting hydrocarbon oxygen flames has been investigated with respect to the particular situation in these flames. The experimental conditions were chosen to
provide particles in both the Rayleigh regime and Mie regime of scattering.

EXPERIMENTAL

The experimental arrangement for this investigation has been outlined
elsewhere[1,2]. Therefore, it is being discussed here only briefly.

The burner for the flat flames consisted of a 80 mm diameter brass
plate with about 1000 uniformely distributed holes of 1 mm diameter. The
burner was water cooled at the edges, the temperatures of the burner plate

---

Table 1. Experimental conditions and properties of the investigated flames.

Flame No.		1	2	3	4	5
Fuel		C_3H_8	C_2H_2	C_6H_6	C_3H_8	C_3H_8
$C/O-ratio$		0.80	1.10	0.85	0.70	0.70
Mole fraction diluent		--	0.55 Ar	0.45 Ar	--	0.056 H_2
Pressure	$P/mbar$	150	120	100	1013	1013
Cold gas velocity	v_0/cms^{-1}	8.1	20.4	24.4	2.2	2.2
Flame temperature	T_{max}/K	1840	1990	2145	≈ 1880	≈ 1900
Soot vol. fraction	$f_V^\infty \cdot 10^9$	25.8	23.7	36.4	1560	690

attaining values between 50° C and 80° C depending on the expertimental
conditions. The back flow of enthalpy to the burner amounted to 12% to 17%
of the total enthalpy release in the flames.

Data for the investigated flames are given in Table 1. The atmospheric
pressure flames were enclosed in short quartz cylinders placed on the
burner plate and stabilized with wire grids mounted 40 mm to 60 mm above the
burner. The low pressure flames were enclosed in a burning chamber which
had been equipped with all facilities for the optical measurements. The
burner could be moved relative to the optical axis of the light scattering
and extinction equipment which is similar to that used conventionally in
light scattering measurements, see e.g.[3] Two lasers with a wavelength of
488 nm and 633 nm respectively were being used. The laser beam was chopped
to eliminate background radiation from the flames. The scattered light was
analysed with respect to its polarization and detected with photomultipliers.

THEORETICAL BACKGROUND FOR EVALUATION OF MEASUREMENTS

The theory of light scattering and extinction is well established (see
e.g.[4,5,6]) and is reported here only to an extent necessary for elucidating
the evaluation procedure of the measurements. The measured quantity using
the equipment sketched roughly above is the monochromatic energy flux of
scattered light of wavelength λ, $F(\theta)$ which is given by

$$F(\theta) = Q \quad I_0 \quad \Delta V \quad \Delta\Omega \tag{1}$$

where ΔV is the scattering volume and $\Delta\Omega$ the solid angle aperture of the
detecting system. The monochromatic scattering coefficient Q which is the
monochromatic energy flux scattered at an angle θ per unit solid angle from
a unit volume for unit incident energy flux is obtained from the experimental data by means of a calibration procedure.

The approach from the Lorenz-Mie theory to the scattering coefficient
yields

$$Q = \frac{\lambda^2}{4\pi^2} \left[(i_{11} + i_{12})sin^2\phi \quad + \quad (i_{21} + i_{22})cos^2\phi \right] N \tag{2}$$

where ϕ means the polarization angle and i_{ij} are scattering functions which
are derived from the sqares of the amplitude functions of the scattering
matrix :

$$i_{ij} \quad = \quad |S_{ij}|^2 \tag{3}$$

The amplitude functions are expressed by infinite convergent series, e.g.

for isotropic homogeneous spheres:

$$S_{11} = \sum_{n=1}^{\infty} \frac{2n+1}{n(n+1)} \left[a_n \pi_n(cos\theta) \;+\; b_n \tau_n(cos\theta) \right] (-1)^{n+1} \tag{4}$$

$$S_{22} = \sum_{n=1}^{\infty} \frac{2n+1}{n(n+1)} \left[a_n \tau_n(cos\theta) \;+\; b_n \pi_n(cos\theta) \right] (-1)^{n+1} \tag{5}$$

In Eqs (4) and (5) a_n and b_n are the Mie coefficients which are solely dependent on the size parameter $\alpha = \pi d/\lambda$ and the refractive index $m = n - ik$ of the particles. These coefficients and the scattering angle dependent functions π_n and τ_n are generally computed by means of recursive formulas (see e.g.[4,5]). Another quantity measured with the equipment sketched above is the extinction coefficient k_{ext}:

$$k_{ext} \;=\; C_{ext} \; N \tag{6}$$

The extinction cross section C_{ext} for only elastic effects follows from the fundamental extinction relation[4] and the Lorenz-Mie theory:

$$C_{ext} = \frac{\lambda^2}{2\pi} \sum_{n=1}^{\infty} (2n+1) Re(a_n + b_n) \tag{7}$$

In sooting flames scattering and extinction is not only caused by elastic interactions with soot particles but also by scattering, absorption and fluorescence of gaseous components. Therefore, the measured quantities $F(\theta)$ and k_{ext} have to be corrected due to these effects. These interferences are particularly effective at incipient soot formation where the contributions of the soot particles to extinction and absorption are relatively small. It is supposed that the gaseous species in question are mainly polyaromatic hydrocarbons[3,7,8,9]. Hence, the profiles of extinction coefficients and scattering coefficients in the sooting flames investigated were corrected by those obtained in non sooting flames close to the sooting limit. Increase of absorption and scattering of gaseous components with increasing fuel concentration was allowed for by the increase of the concentrations of these higher hydrocarbons which were measured separately by probe measurements[2]. The corrections were substantial in those parts of the flames where the soot volume fraction was less than 5% to 10% of its final value.

RESULTS AND DISCUSSION

Rayleigh Regime of Scattering

The experimental conditions stipulated for flames 1 to 3 in Table 1 provide soot particle sizes in the Rayleigh regime of scattering ($\alpha \le 0.2$). Typical electron micrographs of soot particles from flame 1 are given in Fig. 1 for illustration. This figure also shows the particle size distribution for this experimental point. The histogram is based on evaluation of about 1000 particles on the electron micrograph (the experimental method for sampling soot particles from flames is discussed in [10]). From Fig. 1 follows that under these experimental conditions soot consists mainly of spheres which are size distributed. The size distributions approximating the measured histograms best are log-normal distributions. In Fig. 1 that log-normal size distribution which is the best fit is also given. Fitting was achieved by non-linear regression analysis according to the Marquardt procedure[11] with respect to the median value and standard deviation of the distribution.

The results from Fig. 1 suggest an isotropic homogeneous sphere scattering model for these experimental conditions. For that scattering problem the non diagonal elements of the scattering matrix vanish. From Eqs (2) to (7) follows considering only the first members of the series:

$$Q_{vv} = \frac{\lambda^2}{4\pi^2} \left| \frac{m^2-1}{m^2+2} \right|^2 \mu_6 N \tag{8}$$

and

$$k_{ext} = -\frac{\lambda^2}{\pi} \; Im \left\{ \frac{m^2-1}{m^2+2} \right\} \mu_3 N \tag{9}$$

where the μ_i are the i-th moments of the size distributions $P(\alpha)$ and the index vv denotes the polarization direction of scattered and incident light respectively. Equations (8) and (9) form a system of equations for particle number density N and first moment of the size distribution, e.g. mean value of α. The system is implicit with respect to μ_1. As in the Rayleigh regime of scattering Q_{vv} is independent on scattering angle, no further information is obtainable from Q_{vv} and k_{ext}. Hence, the other quantities , viz. the refractive index and the higher moments of the size distribution, have to be guessed. In this work a value $m = 1.57 - 0.56i$ was assumed according to[12]. The size distributions were supposed to be log-normal with a standard deviation $ln\sigma_g = 0.34$ corresponding to the results given in Fig. 1. This value is close to the value which follows from coagulation theory for a free molecular aerosol[10,13,14]. The effect of the width of the size distribution is illustrated in Fig. 2. As can be seen from Fig. 2 for a given Q_{vv} and k_{ext} the assumption of a monodisperse particle system brings about an error of 45% for N and 35% for the mean particle radius compared to a size distribution with $ln\sigma_g = 0.34$.

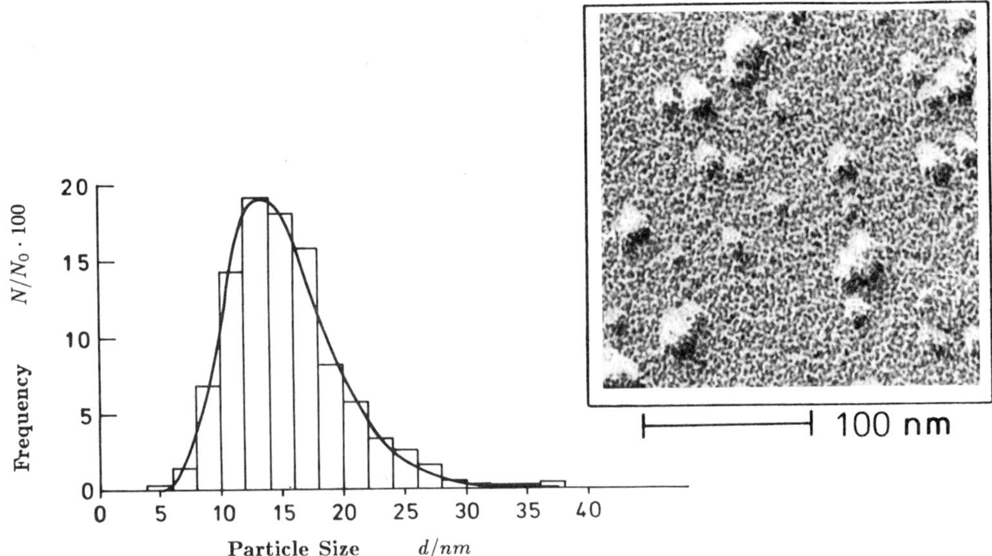

Fig.1. Typical electron micrographs and corresponding size distribution for soot particles from flame 1, 30 mm height above burner.

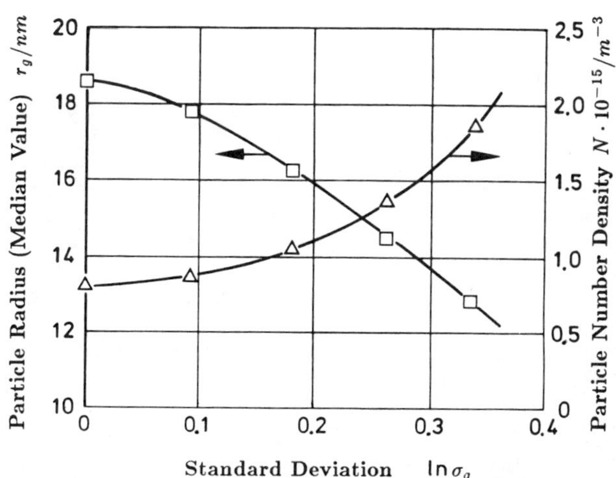

Fig.2. Variation of median particle radius and particle number density in dependence on standard deviation for given Q_{vv} and k_{ext}.

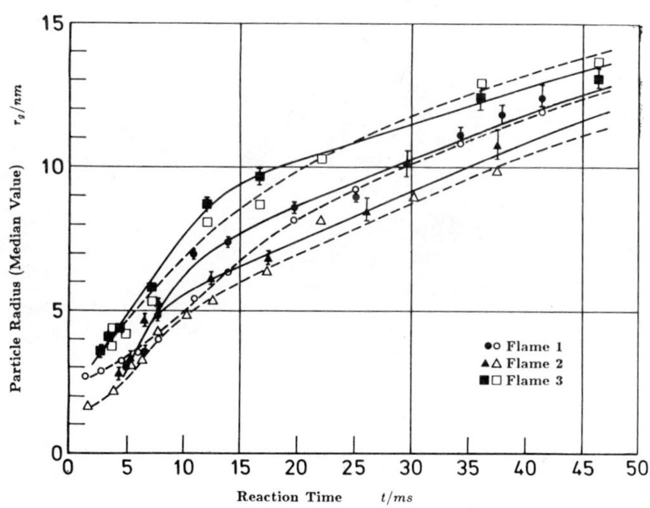

Fig.3. Median particle radii in dependence on reaction time. Open symbols: light scattering; full symbols: sampling technique.

The median values of particle radii from light scattering correspond very well with those from the sampling technique[10], if particle size distributions are considered. This is demonstrated in Fig. 3 where the median values of the particle radii are given for flames 1 to 3 in dependence on reaction time. The slight discrepancies between the results from both methods for small reaction times may be due to disturbances of the flame by the probe, discussed in [10].

Mie Regime of Scattering

For the normal pressure flames 4 and 5 the enhanced soot volume fractions result in soot particle sizes in the Mie regime of scattering. Typical electron micrographs of soot particles from flame 4 and the corresponding size distribution are shown in Fig. 4. The electron micrographs exhibit a more complex structure of the soot particles compared to those from flames 1 to 3 with $\alpha \approx 1$. Applying the isotropic homogeneous sphere scattering model implies the summation over some members of the series Eqs (4) and (5). This brings about a dependence of Q_{vv} on scattering angle and the evaluation of the angular distribution of Q_{vv} can be used to calculate the moments of the size distribution of α and the refractive index m of the soot particles. In this work the evaluation of the angular distributions of Q_{vv} was achieved by non-linear regression analysis[11] with respect to α_g and $ln\sigma_g$ of a log-normal size distribution as well as n and k. Again the resulting model equations derived from Eqs (2) to (5) are impicit with respect to α_g, $ln\sigma_g$, n and k.

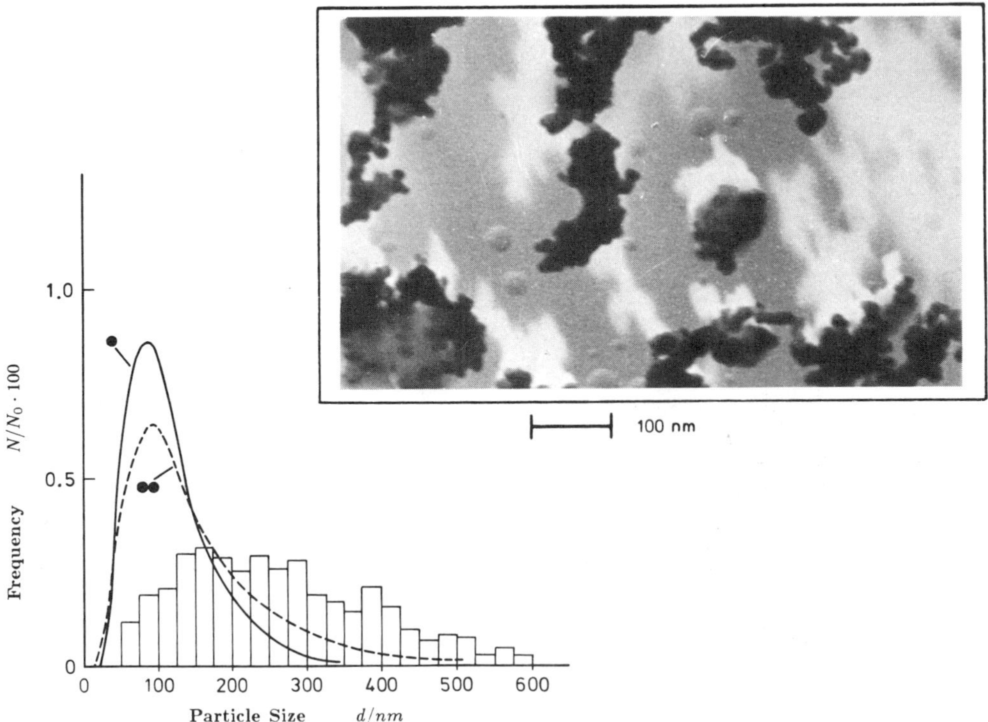

Fig.4. Typical electron micrographs and corresponding size distribution for soot particles from flame 4, 25 mm height above burner.
•: light scattering spheres; ••: light scattering two centre scatterer.

It is obvious from these equations that a number of combinations of α_g and $ln\sigma_g$ results in similar Q_{vv}. Hence, regression analysis yields a strong correlation between α_g and $ln\sigma_g$ which can be demonstrated by means of a chart of the sum of sqares for that regression problem in dependence on this two parameters, see Fig. 5. This chart represents a surface with a narrow valley with only slight gradients at the bottom. The correlations are of the form

$$\alpha_g = c_1 \sigma_g^{c_2} \tag{10}$$

where the constants have to be determined from a number of measurements for each flame (for example $c_1 = -0.22 + 1.59\alpha_{g,\sigma_g=1}$; $c_2 = -2.46$ for the soot formation zone of flame 4). Using Eq. (10), α_g can be determined with some additional information about σ_g which is drawn in this work from sampling sooting flames (compare Fig. 4) or coagulation theory.

The aggregate structure of soot particles sampled from flame 4 and 5 suggest other scattering models than the isotropic homogeneous sphere, e.g. spheroids or multiple centre scatterers. For a prolate spheroid all elements of the scattering matrix are different from zero and the solution of the scattering problem for this case[15] gives amplitude functions depending on particle properties as well as orientation of the spheroid. The exact electromagnetic solution for an absorbing spheroid with $\alpha_{sd} \approx 1$ and arbitrary orientation presents great mathematical difficulties and as far mostly approximations are proposed (compare[16,17]). Applying these approximations the scattering cross sections or the asymmetry ratio of the angular distributions of the scattering functions for spheroids compared to spheres can be calculated. Values for these quantities for soot particles are given for example in[3]. Pronounced effects however result only for spheroids with very large ratio of semiaxes. Supposing the soot aggregates given in Fig. 4 to be prolate spheroids a semiaxes ratio not larger than

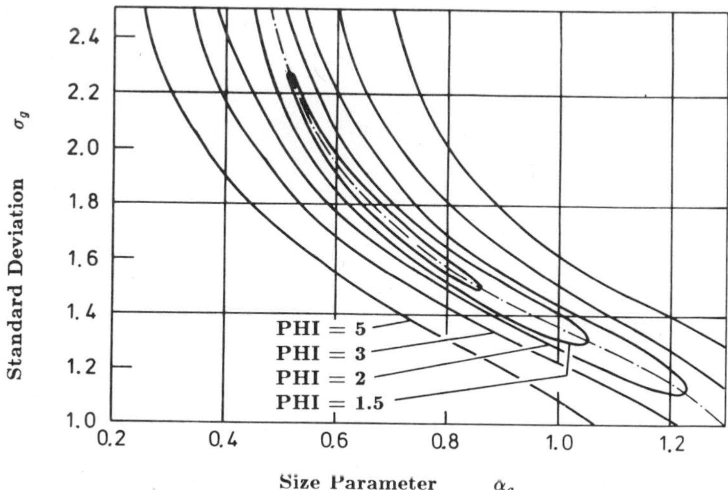

Fig.5. Sum of squares PHI from regression analysis of angular distribution of Q_{vv} in dependence on median size parameter and standard deviation (data correspond to flame 4, 40 mm height above burner).

2 could be expected. For the ratio of semiaxes in this order of magnitude
also the depolarization ratio $\rho_v = Q_{hv}/Q_{vv}$ for $\theta = 90^o$ is in the order of
1% maximum (see e.g.[3]) and is not a suitable measure for spheroid dimensions because of other depolarizing effects in the flames.

Another scattering model which can be more strictly handled is the
two centre scatterer. A solution for two randomly positioned spheres with
$\alpha_{2c} \leq 1$ is given in[18,19]. As in the case of spheroids all elements of the
scattering matrix are different from zero and the scattering functions
averaged over all possible orientations of the scatterers depend on particle properties, polarization angle, distance of the two scatterers and
wave length. The result is a depolarization of the incident light in the
same order of magnitude like the one for spheroids with a semiaxes ratio
of about 2 and - more pronounced - a less asymmetric angular distribution
of Q_{vv}. In this work the angular distributions of Q_{vv} from the soot aggregates from flame 4 and 5 have been evaluated with the Mie theory for
spheres. The resulting size parameters were corrected subsequently by a
correction factor beeing derived from the asymmetry ratio of a two centre
scatterer in dependence on $\alpha_{2c} = \alpha 2^{1/3}$ compared to that of spheres. The
correction factors applying the solutions of [18,19] are given in Fig. 6 for
soot particles with log-normal size distributions.

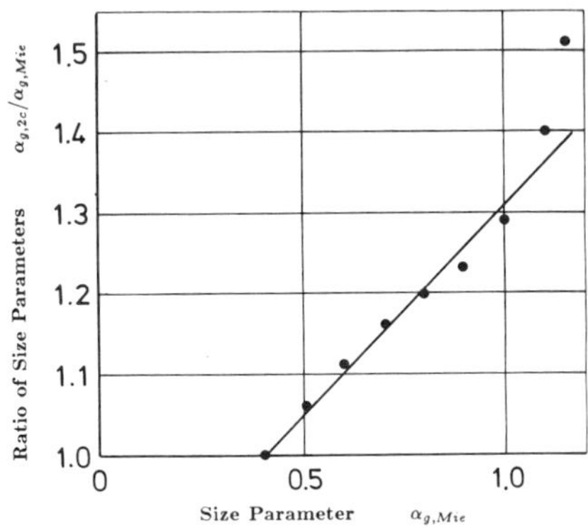

Fig.6. Correction factor for two centre scatterer model.

The median values of soot particles for flames 4 and 5 are given in
Fig. 7 in dependence on height above the burner. This figure also contains
the values obtained from the sampling technique. As can be seen from Fig.
7 the particle sizes - expressed as the diameter of particles with equivalent volume - from the sampling technique are larger than those from light

266

scattering employing the sphere scattering model. The two centre scatterer model yields better results. However, the discrepancies depend on the structure of the soot particles: Flame 5 with a lower final soot volume fraction caused by dilution with hydrogen produces more compact soot aggregates compared with flame 4. For this flame the results from the sampling technique and the light scattering measurements applying both models of scatterers are in better correspondence than those for flame 4.

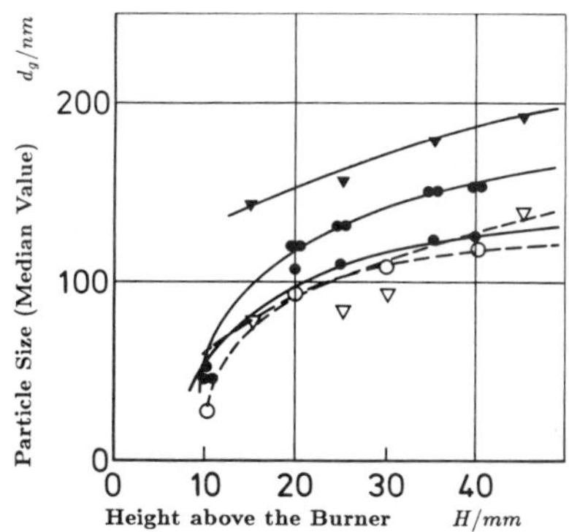

Fig.7. Median particle sizes in dependence on height above the burner. Full symbols: flame 4; open symbols: flame 5; ▽, ▼: sampling technique; ○,●: light scattering spheres: ●●: light scattering two centre scatterer.

Further results interpreting the angular distributions of Q_{vv} for the flames producing particles within the Mie regime of scattering are the real and imaginary part of the refractive index. These quantities are given in Fig. 8 in dependence on the height above the burner for flame 4. As can be seen from this figure the real part of the refractive index as well as the imaginary part show a dependence on reaction time which is very strong in the soot forming region of the flame and is levelled out towards the burnt gas region. This dependence indicates structural changes of the soot particles which can be explained in terms of the dispersion model for the refractive index[20,21]. The dispersion model for the optical constants of soot describes the combined contributions of both the bound and free electrons by the following dispersion equations:

$$n^2 - k^2 = 1 - \frac{e^2}{m_e \epsilon_0} \left[\sum_j \frac{Z_j(\omega_j^2 - \omega^2)}{(\omega_j^2 - \omega^2) + \omega^2 \gamma_j^2} - \frac{Z_{fr}}{(\gamma_{fr} + \omega^2)} \right] \qquad (11)$$

and

$$2nk = \frac{e^2}{m_e \epsilon_0} \left[\sum_j \frac{Z_j \omega \gamma_j}{(\omega_j^2 - \omega^2) + \omega^2 \gamma_j^2} + \frac{Z_{fr}}{(\gamma_{fr} + \omega^2)} \right] \qquad (12)$$

where e is the electron charge, m_e the electron mass, ϵ_0 the permissivity
constant, Z_j and Z_{fr} the bound and free electron number densities, γ_j and
γ_{fr} the damping constants of bound and free electrons, ω_j the resonance
frequency of bound electrons and ω the frequency of radiation. Equations
(11) and (12) describe the real and imaginary part of the refractive index in dependence on frequency of radiation and number density of bound
and free electrons. The wavelength dependent terms give a maximum for k
at $\omega = \omega_j$ and a maximum for n for $\omega \approx \omega_j$. Towards larger and shorter wave-
lengthes of radiation both quantities exhibit a steep decrease. The bound
and free electron number density terms bring about a slight increase in
k and a slight decrease in n with increasing number density of free electrons.

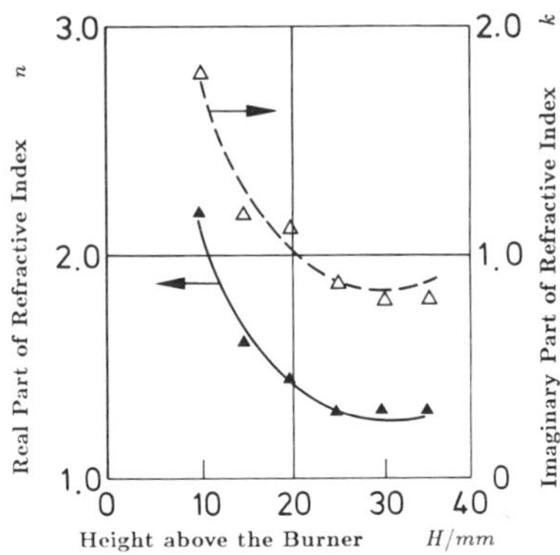

Fig.8. Real and imaginary part of refractive index in dependence on height
above the burner for flame 4.

Thus, the steep decrease of n and k in the soot forming region of the
flame may be interpreted as structural changes in the soot particles.
Thereby hydrocarbon molecules or radicals may be 'integrated' into primary soot particles extending the regions with graphite-similar structure and shifting the resonance frequency of the bound electrons. After
cease of the increase of the soot volume fraction the soot particles undergo a kind of tempering process in the burnt gas region of the flame which
causes graphitizing of the soot particles. This process increases the free
electron number density and brings about a slight increase in k and a further slight decrease in n as can be observed in the burnt gas region of the
flame.

SUMMARY AND CONCLUSIONS

Particle sizes of soot have been measured in flat premixed sooting
flames of various hydrocarbon oxygen flames by means of light scattering
and extinction. The experimental conditions of the flames investigated
provided particle sizes in the Rayleigh regime of scattering as well as in
the Mie regime of scattering. Different scattering models, e.g. spheres,
spheroids as well as two centre scatterers, have been applied depending on
the prevailing particle size and structure. The optical measurements were
compared with the results from a sampling technique involving the evaluation of electron micrographs of soot particles.

The particles in the sooting flames investigated are size distributed. The distribution function may be well approximated by log-normal
distributions. In the Rayleigh regime of scattering the standard deviations of the log-normal size distributions cannot be measured by light
scattering but are derived from particle sizing by means of the sampling
technique. Neglecting size distributions causes errors up to 35% for the
first moments of the size distribution and about 45% for the particle number density. If size distributions are considered, very good agreement
between light scattering and extinction measurements and measurements by
means of the sampling technique results for flames tne soot particles of
which are in the Rayleigh regime of scattering.

In the Mie regime of particle sizes the scattering coefficients are
angular dependent. Hence, the evaluation of the scattering measurements is
achieved by non-linear regression analysis of the angular dependent scattering signals. Non-linear regression analysis yields the optical properties, e.g. complex refractive index of the soot particles, and a correlation for the parameters of the size distribution. These correlations for
the parameters of the size distribution are used to calculate mean particle
sizes employing additional information from the sampling technique.

The flames producing soot particles in the Mie regime of scattering
give rise to particles with complex aggregate structure. For these particles a two centre scatterer model improves the consistency between the
results of light scattering and particle sampling. Additionally, in these
flames a dependency of the complex refractive index on particle age can be
worked out from the non-linear regression analysis of the scattering signals. The variation of the refractive index is discussed in terms of structural changes of the soot particles.

References

1. H.Bockhorn, F.Fetting, U.Meyer, R.Reck and G.Wannemacher, Measurement of the Soot Concentration and Soot Particle Sizes in Propane
   Oxygen Flames, in: 'Eighteenth Symposium (International) on Combustion', p. 1137, The Combustion Institute, Pittsburgh (1981)
2. H.Bockhorn, F.Fetting, G.Wannemacher and H.W.Wenz, Optical Studies
   of Soot Particle Growth in Hydrocarbon Oxygen Flames, in: 'Nineteenth
   Symposium (International) on Combustion', p. 1413, The Combustion
   Institute, Pittsburgh (1982)
3. A.D'Alessio, Laser Light Scattering and Fluorescence Diagnostics
   of Rich Flames Produced by Gaseous and Liquid Fuels, in: 'Particulate Carbon Formation During Combustion', p. 207, D.C.Siegla and
   G.W.Smith, ed., Plenum Press, New York (1981)
4. H.C. Van de Hulst, 'Light Scattering by Small Particles', John Wiley
   and Sons, New York (1957)

5. M.Kerker, 'The Scattering of Light and Other Electromagnetic Radiation', Academic Press, New York (1957)
6. A.R.Jones,Progr. Energy Combust. Sci., 5:73 (1979)
7. A.D'Alessio, A.DiLorenzo, A.Borghese, F.Beretta and S.Masi, Study
   of the Soot Nucleation Zone of Rich Methane-Oxygen Flames, in: 'Six-
   teenth Symposium (International) on Combustion', p. 695, The Combustion Institute, Pittsburgh (1977)
8. B.S.Haynes, H.Jander and H.Gg.Wagner Ber. Bunsenges. Phys. Chem.,
   84:585 (1980)
9. A.DiLorenzo, A.D'Alessio, V.Cincotti, S.Masi, P.Menna and
   G.Venitozzi, UV-Absorption, Laser Excited Fluorescence and Direct
   Sampling in the Study of the Formation of Polycyclic Aromatic Hydrocarbons in Rich CH_4/O_2 Flames, in: 'Eighteenth Symposium (International) on Combustion', p. 485, The Combustion Institute, Pittsburgh
   (1981)
10. H.Bockhorn, F.Fetting and A.Heddrich, in: 'Twenty-First Symposium
    (International) on Combustion', The Combustion Institute, Pittsburgh
    (1987)
11. D.M. Himmelblau, 'Process Analysis by Statistical Methods', John Wi-
    ley and Sons, New York (1970)
12. W.H.Dalzell and A.F.Sarofim, Trans. ASME, J. Heat Transfer, 91:100
    (1969)
13. F.S.Lai, S.K.Friedländer,J.Pich and G.M.Hidy, J. Coll. Interf. Sci.,
    39:395 (1972)
14. R.A.Dobbins and G.W.Mulholland, Combust. Sci.Technology, 40:175
    (1984)
15. S.Asano and G.Yamamato, Appl. Opt., 14:29 (1975)
16. A.F.Stevenson, J. Appl. Phys., 24:1134 and 1143 (1953)
17. W.Heller and A.Nagagaki, J. Chem. Phys., 60:3297 (1974) and 61:3619
    (1974)
19. S.Levine and G.O. Olaofe, J. Coll. Interf. Sci., 27:442 (1968)
19. G.Lips and S.Levine, J. Coll. Interf. Sci., 33:445 (1970)
20. S.Chippet and W.A.Gray, Combust. Flame, 31:149 (1978)
21. S.C.Lee and C.L.Tien, Optical Constants of Soot in Hydrocarbon
    Flames, in: 'Eighteenth Symposium (International) on Combustion',
    p. 1159, The Combustion Institute, Pittsburgh (1981)

The authors would like to thank the DEUTSCHE FORSCHUNGSGEMEINSCHAFT
for substantial financial support.

AN INSTRUMENT TO MEASURE THE SIZE, VELOCITY

AND CONCENTRATION OF PARTICLES IN A FLOW

Cecil F. Hess and Funming Li

Spectron Development Laboratories
3535 Hyland, Suite 102
Costa Mesa, CA 92626

ABSTRACT

A technique to measure the size, velocity and concentration of particles in a flow is discussed. An instrument to measure particles as small 0.5 µm moving at 1000 m/s was developed based on this technique. Two small beams of one color cross in the middle of two crossing larger beams of different color. The small beams, thus, define the middle of the larger beams, a region in which the intensity is almost constant. The particle size is obtained from the absolute intensity of the light scattered by particles crossing this uniform intensity region. Two velocity components are measured from the two independent fringe patterns.

The concentration is obtained from the probe volume size, and the size and velocity distributions of the particles. Results are presented for sprays of predictable characteristics and for polystyrene particles between 1.1 µm and 3.3 µm. It is shown that the method has an excellent size resolution and its accuracy is better than 10% of the particle size studied.

INTRODUCTION

A nonintrusive single-particle counter to measure the size, velocity and concentration of particles in a particle laden flow is described here. The method bases the size information on the absolute scattered light of individual particles crossing the middle of a Gaussian laser beam, as described in Reference (1). The velocity is obtained from the Doppler frequency arising when the particles cross an interference pattern of fringes. This method is referred to as the IMAX. Obtaining the particle size from the absolute scattered light has always been an attractive straightforward method due to its high dynamic range, and because it can be used with both liquid and solid particles, even in those cases where the particles are nonspherical and of an unknown index of refraction. Its use with lasers has been limited until recently due to their nonuniform intensity profile, typically Gaussian. The limitation stems from the fact that particles crossing through the center of the Gaussian beam scatter more light than those crossing through the edge, thus, appearing as larger particles. This ambiguity

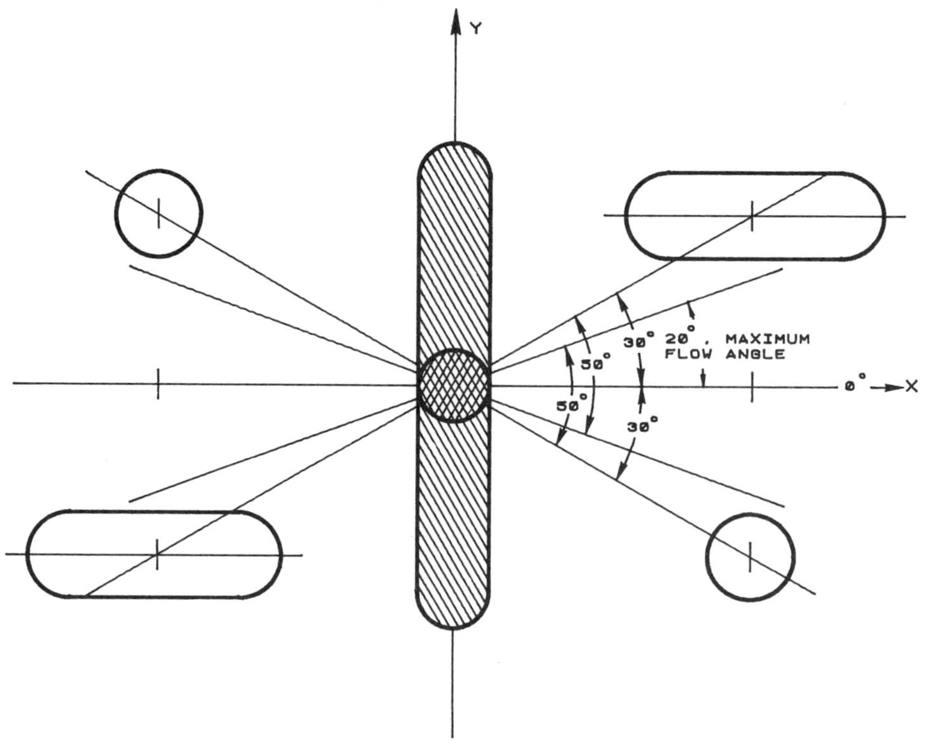

FIGURE 1. ELONGATED IMAX PROBE VOLUME.

has been overcome by identifying the middle of laser beams with the crossing of two smaller beams of different wavelength or polarization. Only particles traveling through the small beams and therefore, the middle of the large beams, are processed. The typical configuration of this method uses two blue (0.488 μm) beams crossing in the middle of two green (0.5145 μm) beams to produce two independent fringe patterns. Thus, the particle size and two components of the velocity are simultaneously obtained for each and every particle crossing the probe volume.

Very high speed electronics have been developed to measure particles as small as 0.5 μm traveling at 1000 m/s. The technique finds application in a variety of two-phase flows in which the spatial resolution of the size and velocity are required. Examples are sprays and aerosols, dust clouds and particle laden flows in which the velocity of the particles and gas are independently needed, and high speed wind tunnels in which the particle size is needed to verify if the particle followed the flow. Similar methods have independently been proposed and demonstrated by other researchers[2,3] as well as methods generically referred to as "top hat".[4]

In this work we describe the theoretical basis of the technique, an instrument developed based on it, and present results which validate the instrument performance.

DESCRIPTION OF OPTICAL TECHNIQUE

Figure 1 illustrates one of the possible probe volumes of this method. It is obtained by crossing two laser beams of wavelength λ_1 in the middle of two laser sheets of wavelength λ_2. Two independent fringe patterns are formed and only particles exhibiting an ac modulation from both fringe patterns, and meeting a coincidence criterium, are accepted as valid. The fringe patterns are chosen such that the particles cross $\pm$ 20° from the X axis and therefore, through the middle of the laser sheet. For totally random particle directions, the large probe volume must be circular as described in Reference (1). The receiver collecting the scattered light will normally have a spatial aperture (pinhole) which is schematically shown by the broken vertical lines. Thus, the effective blue and green probe volumes are almost identical. This is not a requirement, but it helps in the presence of high particle concentrations.

If we refer to the small beam as 1 and the large beam as 2, the intensity profiles in the probe volume can be spectrally separated and given by

$$I_1 = 2I_{o_1} \exp\left[\left(\frac{-2}{b_{o_1}^2}\right)\right](x^2 + y^2 + z^2\gamma^2/4)]\left[1 + \cos\frac{4\pi x \sin(\gamma_1/2)}{\lambda_1}\right], \quad (1)$$

and

$$I_2 = 2I_{o_2} \exp\left(\frac{-2 x^2}{b_{o_2}^2}\right)\left[1 + \cos\frac{4\pi x \sin(\gamma_2/2)}{\lambda_2}\right]. \quad (2)$$

Where it has been assumed that $z\gamma/2 \approx o$ (which is an excellent assumption since a pinhole in the receiver will limit z), and that the intensity of the large beams is only measured over the region defined by the small beams. I_o is the center intensity, γ the intersection angle, b_o the waist radius, λ the laser wavelength, and x, y, z are the coordinates. The intensity scattered by a particle is given by:

$$I_{s_1} = 2I_{o_1} K_1(d,n,\theta,\Omega,\lambda_1)\, G_1 \exp\left[\left(-\frac{2}{b_{o_1}^2}\right)(x^2 + y^2)\right]$$

$$\cdot \left[1 + \cos 2\frac{\pi\gamma x}{\lambda_1} \cdot V\right], \quad (3)$$

and

$$I_{s_2} = 2I_{o_2} K_2(d,n,\theta,\Omega,\lambda_2)\, G_2 \exp\left(\frac{-2x^2}{b_{o_2}^2}\right)$$

$$\cdot \left[1 + \cos 2\frac{\pi\gamma x}{\lambda_2} \cdot V\right], \quad (4)$$

where K is the scattering cross section which can be obtained by the Lorentz-Mie[5] theory, or for very large spherical particles using the geometric approximations described in Reference (6). K is, in general, a complex function of size d, the index of refraction n, the collection angle θ, the solid angle of collection Ω, the wavelength λ, and the polarization. G is the gain function of the instrument, and V is the visibility of the measured signal.

It should be pointed out that although the visibility is, in general, not an adequate sizing parameter[7], it represents an important

aspect of the Doppler signal. That is, it establishes the modulation of
the ac component. The size of the particle is obtained after low-pass
filtering the signal given by Equation (4). Thus, the cosine term is
cancelled out. The scattered intensity is then measured as a function
of x, and its peak value is registered (x = o). It is then obtained:

$$I_{s_2} = 2 I_{o_2} K_2 G_2 \ .$$

(5)

where I_{o_2} and G_2 are calibrated parameters of the instrument, and K_2 is
solved for the diameter d. The two velocity components are obtained
from the cosine terms of Equations (3) and (4) in the classical Doppler
way.

Notice that for particles larger than the wavelength, the scat-
tered light collected near on-axis ($\theta \approx o$) can be described using Fraunhofer diffraction theory. There, the scattering coefficient is given
by:

$$K = \frac{d^2}{4r^2} \int_{A_{lens}} \frac{J_1^2 (\alpha \sin \theta)}{\sin^2 \theta} \ dA ,$$

(6)

where $\alpha = \pi d / \lambda$ is the size parameter, J_1 is the Bessel function of the
first kind, and r the distance from the probe volume to the lens. Equation (6) shows that the scattered light is independent of index of
refraction and, furthermore, since diffraction is a function of the particle cross section, some irregularities in the particle shape can be
tolerated.

For particles with diameters near the wavelength, the scattering
cross section must be obtained solving the Mie scattering equations
numerically. The functional relationship between the diameter and K_2
can be quite complex. It is necessary to find the conditions under
which the ambiguities, if any, are within tolerable error margins. The
computations were made on an IBM AT computer. Parametric studies were
conducted to establish optimum experimental conditions. These parameters include the angle of collection (θ), the solid angle of collection (Ω) and index of refraction ($n_1 - in_2$).

Since the index of refraction of the particles in many applications may be quite different, it is important to establish conditions
which are less sensitive to these variations. Both real (n_1) and imaginary (n_2) parts of the refractive indices were varied to check the
sensitivity of these parameters. Figures 2 and 3 show the scattered
intensity as a function of the particle size parameter α for different
values of n_1 and n_2 at different scattered angles.

It was concluded from the above calculations that shallow angles
of collection ($\theta \leqslant 7°$) offer the most favorable conditions.

THE PROBE VOLUME

The probe volume is the product of the cross sectional area of
sensitivity A(d) times the sampling length. Its theoretical foundation
is described in Reference (1). A(d), as shown, is a function of diameter since particles that scatter light with large modulation (large
amplitude and visibility) are detectable over a larger region than those

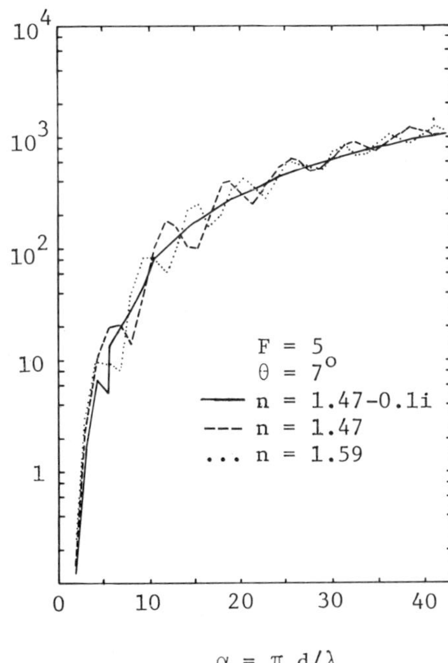

FIGURE 2. SCATTERED INTENSITY INTEGRATED OVER AN F #5 LENS CENTERED AT 7°.

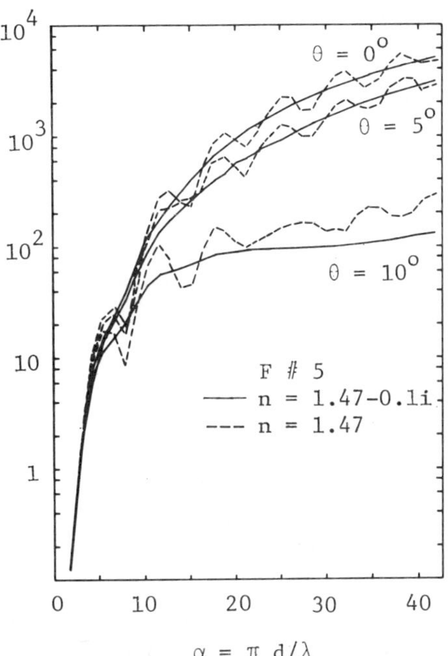

FIGURE 3. SCATTERED INTENSITY INTEGRATED OVER AN F #5 LENS CENTERED AT 0, 5° AND 10°.

with less modulation. As a result, the probability of detecting a particle is a function of its size. This probe volume is required to correct the counts of the size histograms and to obtain the concentration or number density of particles. In the IMAX technique, it is the probe volume of the small beams which is of interest since the intensity of the large probe volume remains constant throughout the measurement region. The cross sectional area of sensitivity, A(d), was verified experimentally using monodisperse droplets of known size. Since A(d) varies only with the coordinate y, it was only necessary to measure y experimentally. The z dimension was, however, checked for completeness. A Berglund-Liu droplet generator was mounted on a precision x-y micrometric traverse so that the droplets could be positioned anywhere in the probe volume. The orifice of the Berglund-Liu was kept very close to the probe volume to minimize errors due to the wander of the string of droplets. Data were also collected several times to further reduce the errors.

A very difficult part of this experiment was determining the edge of the probe volume for any particular size droplet. Theoretically, the edge of the probe volume is calculated by the expression given for y. Experimentally, there is a region near the edge of the PV where the rate of acquisition drops off. That is, some of the signals (droplets) are processed by the electronics and some are not. The electronics used here imposed an upper limit on the data rate of about 6 kHz, while the drops can be generated at a frequency of up to 60 kHz.

In defining the edge of the probe volume the data rate of signal acceptance was observed. One criterion corresponded to the location where the data rate dropped to about 90% of the maximum rate. Another criterion corresponded to a data rate very close to zero. Notice that, in principle, all the measured droplets are identical and are traveling exactly through the same trajectory. Therefore, the data rate should be either zero or have a fixed constant value. In actuality, the droplet trajectory can change by a few microns, therefore, causing some droplets to cross inside the probe volume while others cross outside. In addition, the droplets could vary in size, although we had no evidence of this.

The droplet generator was traversed from the position of peak intensity to the position of 90% data rate, and then zero data rate. The respective relative movements were recorded in each case. The 90% data rate corresponds to a conservatively small probe volume referred to in Figure 4 as y_{small}, while the zero data rate corresponds to a conservatively large probe volume and is indicated as y_{large}. Also, shown in Figure 4 are the average of y_{small} and y_{large}, and the theoretically predicted value.

Since the size range of droplets produced by the monodisperse droplet generator is rather limited, we extended the measurements by simulating the amplitude and visibility corresponding to different size particles. This was easily accomplished by masking the receiver. To further validate the probe volume algorithm some spray measurements were performed.

The following results correspond to a spray produced by a pressure nozzle (Spray Systems TG0.3 at 50 psi and 50 mm from the tip). These results are adequate to show trends and gross changes in the distributions. However, considerable transient variations were observed in both the size and velocity distributions. These variations were the result of changes in the spray pattern produced by the above-mentioned nozzle. Simple visual observations of the spray pattern indicated changes from conical to flattened sprays. Nevertheless, recognizing that a standard invariable spray is not available, we proceeded to make measurements that would allow us to test the probe volume algorithm. These measurements are shown in Figures 5a and 5b.

Figure 5a shows the raw data, while Figure 5b shows the probe volume corrected data. Comparing the raw and the probe volume corrected

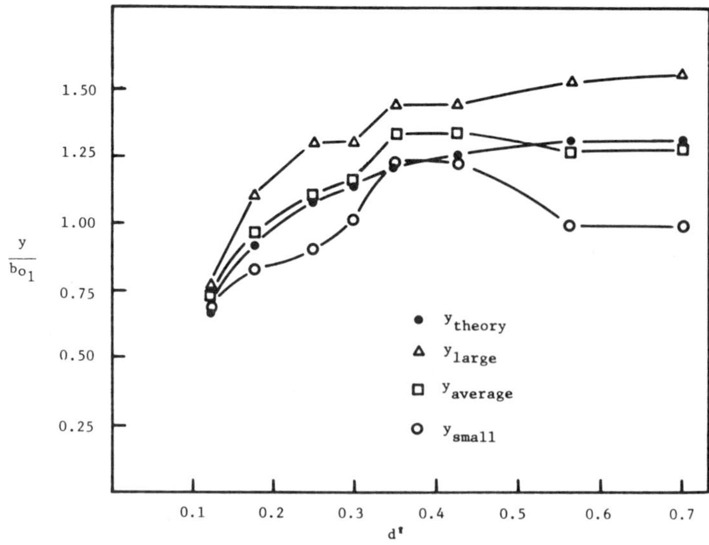

FIGURE 4. PROBE VOLUME AS A FUNCTION OF PARTICLE SIZE.

276

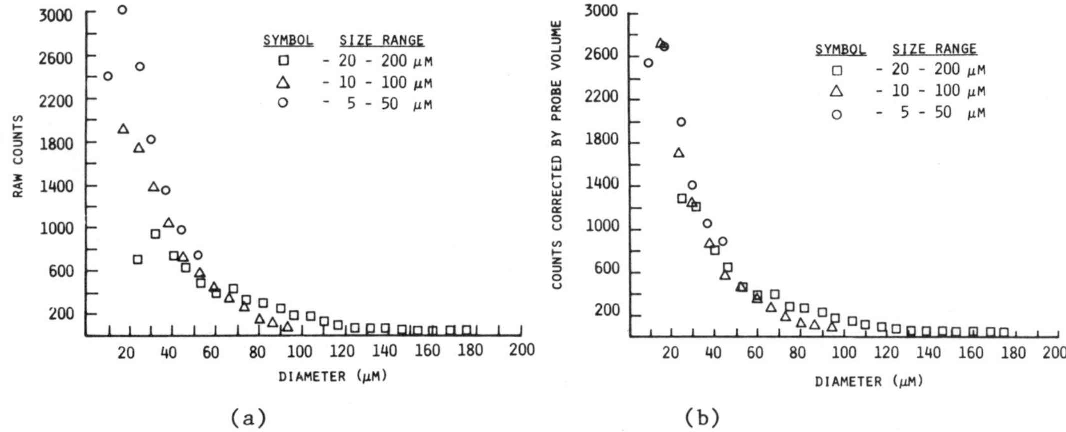

FIGURE 5.   IMAX MEASUREMENTS OF A SPRAY AT A RADIAL POSITION OF 0 mm.
a) RAW DATA;   b) PROBE VOLUME CORRECTED SIZE DISTRIBUTION.

data, it is quite apparent how signal detectability influences the
number of counts in the histograms. For instance, it is more difficult
to detect a 25 μm droplet in the size range of 20 to 200 μm than in the
size range of 5 to 50 μm. This change in sensitivity is what the probe
volume is all about, and it is characteristic of any optical technique
(for instance, in photography the data rate must be corrected by the
depth of field).

In order to test the resolution of the system, data were obtained
using three different size ranges: 5 to 50 μm, 10 to 100 μm and
20 to 200 μm. This is one of the most difficult self-consistency tests
imposed on any technique and most available techniques will show a shift
in the predicted data. IMAX shows excellent matching of the data in the
overlapping region, as illustrated in Figure 5b.

DESCRIPTION OF APPARATUS

An instrument has been developed to measure two velocity components and the size of particles as small as 0.5 μm traveling at 1000 m/s.
The peak amplitude dynamic range of this instrument is 1000 to 1. The
corresponding size range depends on the functional relationship between
the particle diameter and the scattered intensity. For particles larger
than 5 μm, the size dynamic range is about 30:1.

The transmitter is shown on Figure 6. Four beams from an argon-
ion laser are crossed to form a probe volume consisting of two crossing
blue beams in the middle of two crossing green sheets. The transmitter
is made of individual disks containing prealigned optical elements. All
the disks are compressed together by three rods to form a very rugged
unit which can be rotated as a whole to change the orientation of the
fringes when required. Fine lockable adjustments are provided to ensure
that the four beams cross at their waists and to correct for beam
steering resulting when the beams travel through thick glass windows.

The receiver is essentially a telescope with two photomultipliers
to collect the light scattered by particles crossing the sample volume.
This beam is split in two via a beam splitter and a mirror system; each
beam is focused on the pinhole of the corresponding PMT housing. The

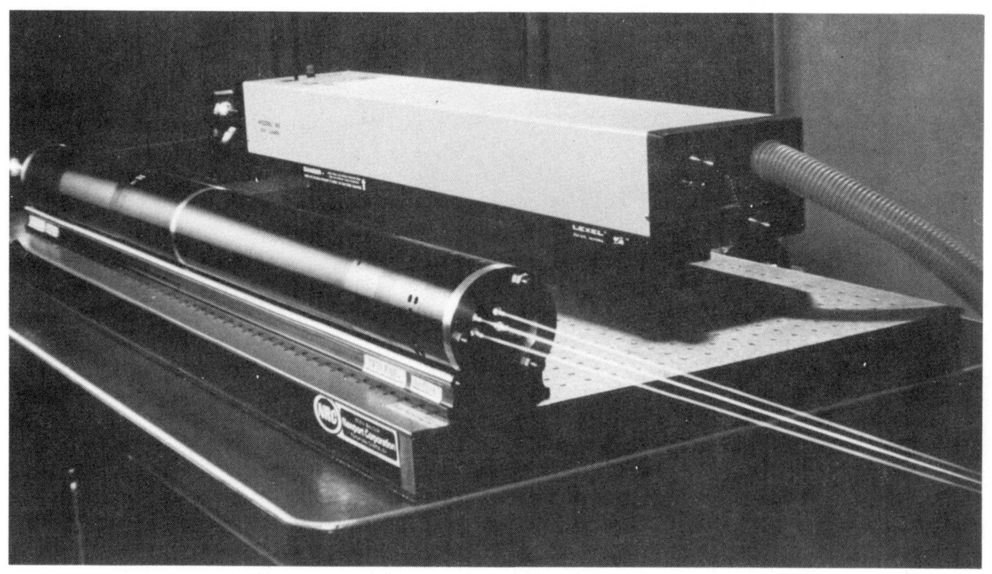

FIGURE 6.   TRANSMITTER WITH FOUR LASER BEAMS TO MEASURE
THE SIZE AND TWO VELOCITY COMPONENTS.

colors are separated by means of narrow-band filters internal to the PMT
housings. The outputs of the two PMTs are then electronically processed, and information about the size and velocity of individual
spheres crossing the probe volume is thus obtained. An electronic processor was developed for this purpose, and a dedicated microprocessor
was interfaced to store, display, and analyze the acquired data.

The sizing processor is built in a computer card and resides in
an IBM PC or AT. It can be interfaced to a variety of Doppler processors. The settings are addressed from the keyboard of the computer.
Thus, parameters such as PMT high voltage, velocity range, and threshold, are selected from the keyboard. The data rate depends on whether
the system is set to collect two velocity components with the proper
coincidence, or only one. In the first case, the data rate is about
20 kHz. Figure 7 shows the monitor output with the U and V velocity
components and the particle diameter. These three parameters, as well
as the time of arrival of each particle, are stored correlated in the
computer memory. Thus, relationships between size and velocity, as well
as turbulent characteristics, are readily obtainable.

RESULTS

The experiments reported here were conducted to validate the
performance characteristics of the instrument. This is a very difficult
task given the absence of a standard spray. It has been shown[7] that the
ability of an instrument to respond to a vertical string of monodisperse
drops is no guarantee of its performance. Furthermore, proving repeat-
ability in the measurement of an unknown spray, at best, may show that
the instrument remained aligned. Using the SMD as a sizing parameter
can be deceiving since it is highly influenced by the large drops and
does not show the ability of the instrument to respond to the full size
distribution.

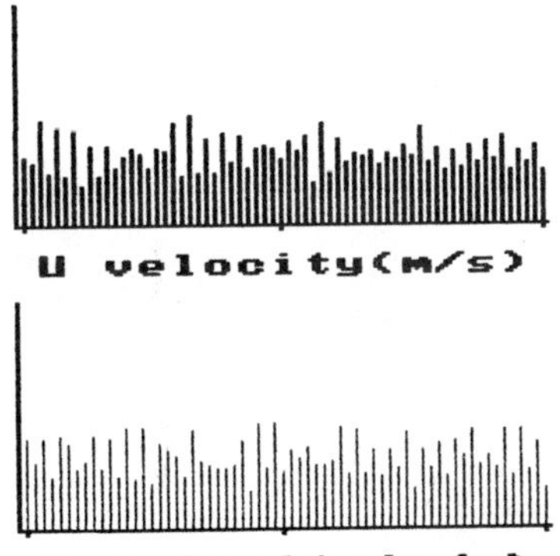

FIGURE 7. MONITOR DISPLAY OF THE SIZE AND
TWO VELOCITY COMPONENT HISTOGRAMS.

We conducted two kinds of validation experiments. In the first we sprayed dry polystyrene particles of known size and combined different size particles to produce sprays of prescribed characteristics. In the second, we used the Berglund-Liu droplet generator and introduced dispersion air to form a spray of monodisperse, bimodal and trimodal droplets. We basically repeated the experiments reported in Reference (8).

Polystyrene Particle Generator

We used polystyrene particles of 1.1 μm, 1.74 μm, 2.7 μm and 3.3 μm in diameter. These are latex particles made by Dow Chemicals of good size uniformity and of spherical shape. The particles come suspended in water with a concentration of 10% by weight. A few droplets of the particle suspension were introduced and diluted in a nebulizer. An air compressor provided the air flow to produce a mist carrying the polystyrene particles out of the nebulizer and into a heated chamber. There the water was evaporated and the particles were sprayed over the probe volume.

Different size particles could be introduced into the nebulizer, thus producing monodisperse, bimodal, trimodal and quadrumodal distributions. The compressor was also used to trap the particles after they passed through the probe volume to avoid contaminating the surrounding environment.

Results of the Polystyrene Sprays

The measurements reported here were obtained with a small breadboard using a 5 mW HeNe laser. The probe volume was formed using two orthogonal polarizations. [9]

The system was calibrated and tested with polystyrene latex particles of uniform and known size. The size and uniformity of the particles were checked with a microscope and agreed with the manufacturer's specifications.

Size and velocity distributions of the polystyrene particles flowing out of the heating chamber were obtained. First, we obtained the size corresponding to monodisperse particles, and we used it as calibration. We then proceeded to obtain bimodal size distributions. Figure 8a shows the distribution corresponding to 1.74 μm and 3.3 μm. The calibation was based on the 1.74 μm particles. The arrow with the 3.3 μm mark indicates the value predicted by the Mie calculations. In

Figure 8b an intermediate size of 2.7 μm was added and the calibration
high voltage was increased to decrease the size range. As before, the
calibration was based on the 1.74 μm, and the arrows point at the numerically predicted values. The 3.3 μm was measured very accurately, but
the measurement of the 2.7 μm particles was off by two bins. This error
was consistent and very repeatable. We attribute it to the oscillations
found in the scattering function. Figure 8c shows the distributions
corresponding to four different latex particles. As before, the calibration was obtained with the 1.74 μm particles. The results clearly
indicate that the instrument is very accurate and sensitive to changes
of a fraction of a micron.

<u>Results of the Water Sprays</u>

These measurements were obtained with the apparatus described in
Reference 1.

A vibrating orifice droplet generator was used to produce strings
and sprays of known size droplets. This generator produces a string of
droplets of equal size and spacing. These droplets can be dispersed
with external air to produce a spray of monodisperse droplets, or under
some dispersion conditions the primary droplets will collide and form
doublets and triplets.[8] The procedure used in these experiments was to
produce a string of large monodisperse droplets to calibrate the instrument. Smaller droplets were then produced by increasing the frequency
of vibration of the orifice, and with the dispersion air a spray of
these droplets was formed. The spray angle was ~10°, and the number
density was typically 500/cm^3. The calibration point was provided by a
string of 110 μm droplets produced with a flow rate of 0.21 cm^3/min and
a frequency of 5 kHz. Note that the droplet size can be precisely estimated given the flow rate and number of equal droplets produced. A
spray of primary droplets of 49 μm (produced with a flow rate of
0.21 cm^3/min and frequency of 56.9 kHz) was then produced with the
dispersion air. Figure 9(a) shows the measurements of the spray of
primary droplets. Figure 9(b) shows the measurements of a spray formed
of primary droplets and doublets. The sizes predicted, given the flow
rate and frequency of droplet generation, are 49 and 62 μm. The sizes
measured with the instrument were 46 and 57 μm, respectively.
Figure 9(c) extends the measurements of Figure 9(b) to the presence of
triplets. The predicted sizes in this case are 49, 62, and 70 μm. Note
that the measured diameters of the doublets and triplets are related to
the primary droplets by $2^{1/3}$ and $3^{1/3}$, respectively.

CONCLUSIONS

A method has been presented to measure the size and velocity of
particles in a flow. The particle size can range from about 0.5 μm to
5000 μm with a possible 30:1 dynamic range for any one configuration.
The method extends the well-established laser Doppler velocimetry to
simultaneously measure the size and velocity of individual particles
crossing the probe volume.

Results obtained with monodisperse, bimodal and trimodal sprays,
and with different size polystyrene particles demonstrate the high resolution and accuracy of the technique. The big advantages offered by
this system are: 1) It uncouples the size and velocity measurements
which permits choosing a fringe spacing to match the velocity without
interfering with the size measurement; 2) It uses absolute intensity to
measure the size, which leads to the measurement of irregularly shaped
particles by collecting and analyzing the diffracted component of the
scatterd light; 3) It measures simultaneously the size and velocity of
particles with high resolution over a broad dynamic range.

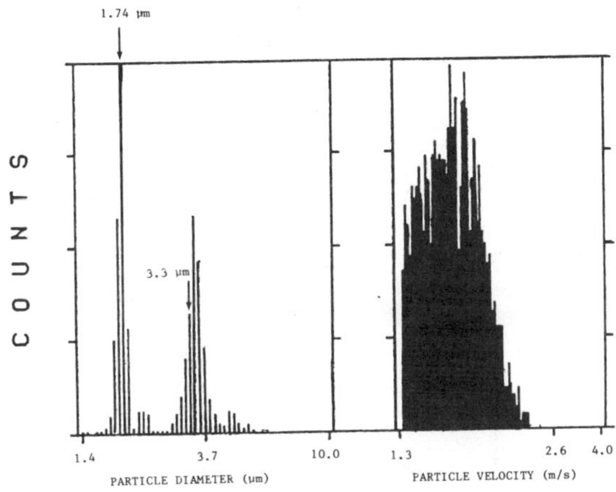

FIGURE 8a.   SIZE AND VELOCITY HISTOGRAMS OF 1.74 μm AND 3.3 μm POLYSTY-RENE SPHERES IN AIR.   PHOTOMULTIPLIER TUBE HIGH VOLTAGE = 500 VOLTS.

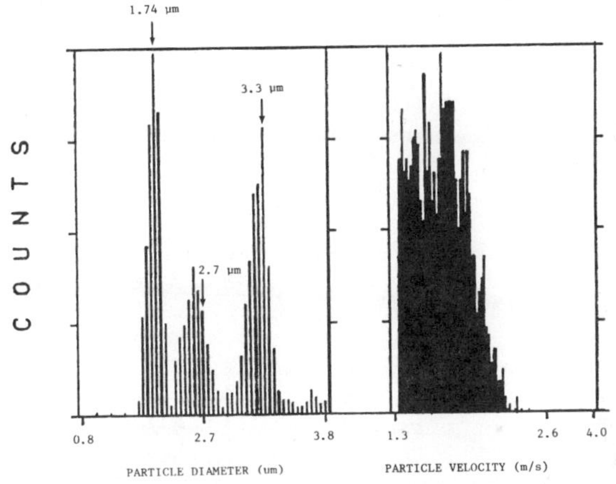

FIGURE 8b.   SIZE AND VELOCITY HISTOGRAMS OF 1.74 μm, 2.7 μm, AND 3.3 μm POLYSTYRENE SPHERES IN AIR.   PHOTOMULTIPLIER HIGH VOLTAGE = 550 VOLTS.

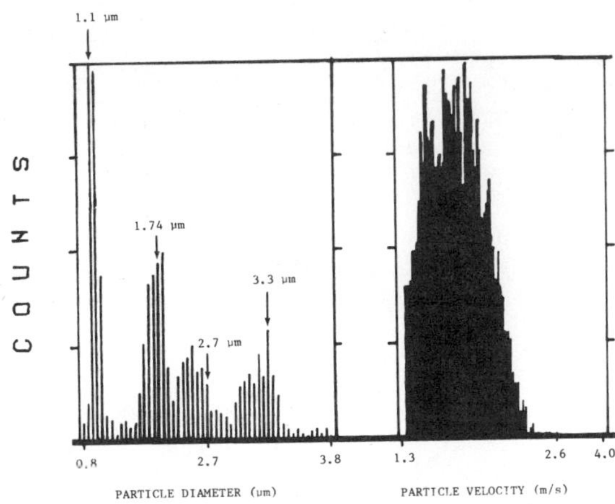

FIGURE 8c.   SIZE AND VELOCITY HISTOGRAMS OF 1.1 μm, 1.74 μm, 2.7 μm, and 3.3 μm POLYSTYRENE SPHERES IN AIR.   PHOTOMULTIPLIER HIGH VOLTAGE = 550 VOLTS.

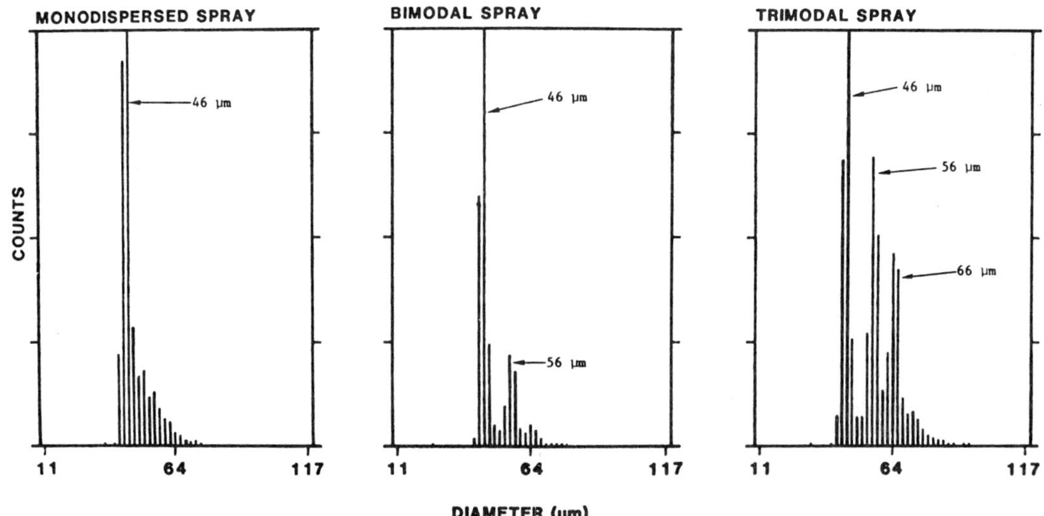

FIGURE 9.   IMAX DROPLET SIZE MEASUREMENTS.

REFERENCES

1. C. F. Hess, "Nonintrusive Optical Single-Particle Counter for
    Measuring the Size and Velocity of Droplets in a Spray," Applied
    Optics, Vol. 23, No. 23 (1984).
2. M. L. Yeoman, B. J. Azzopardi, H. J. White, C. J. Bates, and
    P. J. Roberts, "Optical Development and Application of a Two
    Color LDA System for the Simultaneous Measurement of Particle
    Size and Particle Velocity," in Engineering Applications of Laser
    Velocimetry, Winter Annual Meeting ASME, Phoenix, Arizona, 14-19
    Nov. 1982.
3. G. Gousbet, P. Gougeon, G. Gréhan, J. N. Le Toulouzan, N. Lhuissier,
    "Laser Optical Sizing, from 100 Å to 1 mm - Diameter and from
    0 to 1 Kg/m^3 Concentration," in AIAA 20th Thermophysics Confer-
    ence, 85-1083, Williamsburg, Virginia, 19-21 June 1985.
4. P. R. Ereaut, A. Ungut, A. J. Yule, and N. Chigier, "Measurement of
    Drop Size and Velocity in Vaporizing Sprays," in Proceedings
    Second International Conference of Liquid Atomization and Spray
    Systems, Madison, Wisconsin. (20-24 June 1982), p. 261.
5. H. C. van de Hulst, "Light Scattering by Small Particles" Wiley, New
    York, 1962, Chapter 9.
6. H. C. van de Hulst, "Light Scattering by Small Particles" Wiley, New
    York 1962, Chapter 12.
7. C. F. Hess, "A Technique Combining the Visibility of a Doppler Signal
    with the Peak Intensity of the Pedestal to Measure the Size and
    Velocity of Droplets in a Spray," in AIAA Twenty-Second Aerospace
    Sciences Meeting, 84-0203, Reno, Nevada, 9-12 January 1984.
8. B. Y. H. Liu, R. N. Berglund, and T. K. Argawal, "Experimental
    Studies of Optical Particle Counters," Atmos. Environ. 18, 717
    (1974).
9. C. F. Hess, "A Technique to Measure the Size of Particles in Laser
    Doppler Velocimetry Applications", Proceedings of the Interna-
    tional Symposium on Laser Anemometry, FED - Vol. 33, pp. 119-125.
    American Society of Mechanical Engineers, Winter Annual Meeting
    (1985).

THE PHASE DOPPLER METHOD: ANALYSIS AND APPLICATION

W. D. Bachalo

Aerometrics, Inc.
P.O. Box 308
Mountain View, California 94042

ABSTRACT

The measurement characteristics of the phase Doppler method are
presented. Features of the light scattering were analyzed using the
geometrical optics theory. In particular, the combined light scattering
by reflection and refraction were considered along with the parameters
affecting this problem. Bounds on the drop-to-beam diameter were
suggested to mitigate the problem. Several experiments including direct
comparisons to measurements by other instruments were described to
demonstrate the relative measurement accuracies. Mass flux measurements
which depend heavily upon the accurate determination of the diameters of
the largest drops were compared to sampling probe results and nozzle
flow rate. Number density measurements were compared to data obtained
using extinction and the Lambert-Beer Law. These data were in agreement
to within 5%. Detailed measurements in a complex turbulent two-phase
flow with strong recirculation were summarized. Velocity measurements
for discrete particle size classes were used to show the nature of the
particle response. In such flows, it was found that particles as small
as 10 um could behave quite differently from the gas phase. As an
example, at some locations the gas phase mean velocity was opposite in
direction to the dispersed phase. In general, the mean velocity and
angle of trajectory response of the drops was consistent with their
relative mass. Because of the need to obtain these data in difficult
environments, off-axis backscatter light detection was considered.
Comparative measurements between forward and backscatter were obtained
using monodispersed drops and sprays.

INTRODUCTION

The details of the interaction of particles with turbulent flows
are of interest in a wide variety of practical devices. For example.
the spray interaction with the turbulent airflow in gas turbines, cooling towers, and spray driers is of interest in terms of the efficiency
of the processes and the formation of products. Applications requiring
the predictable dispersion of various chemicals as sprays and aerosols
also necessitate the measurement of the particle size distributions and

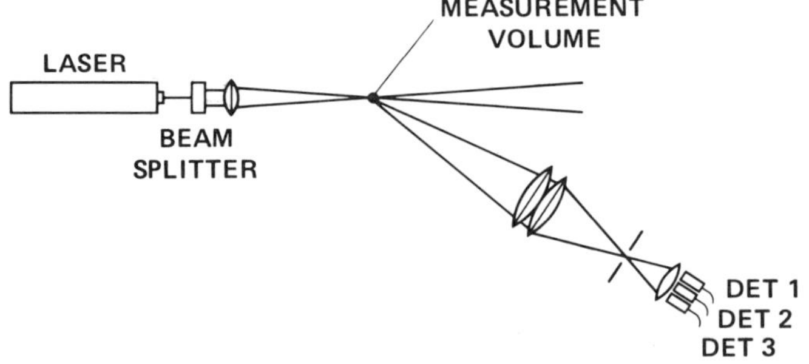

Fig. 1. Phase Doppler Particle Analyzer Optical Schematic.

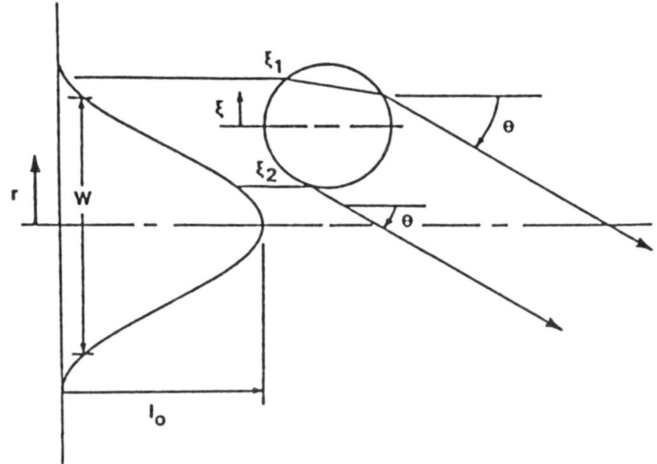

Fig. 2. Schematic Showing the Gaussian Intensity Beam
Incident Upon a Sphere.

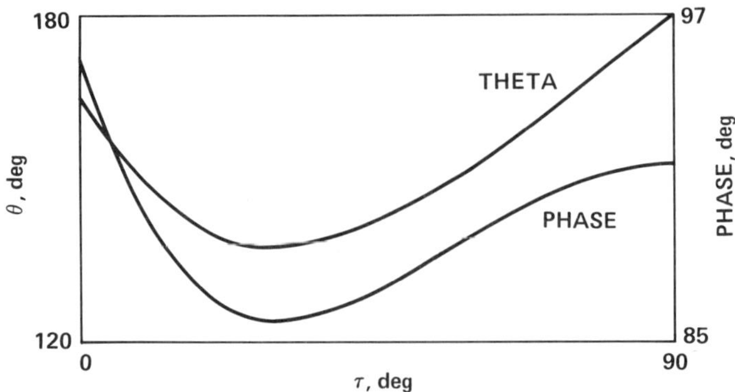

Fig. 3. Calculation of the Scattering Angle and Phase
Versus the Ray Incident Angle on the Sphere.

dynamics. Characterization of these resultant two-phase flows requires
the accurate measurement of the particle size distribution, simultaneous
size and velocity measurements of individual particles, angle of trajectory for each size class, particle number density, and volume flux.
In addition, the gas phase mean velocity and turbulence parameters need
to be obtained from the signals produced by particles on the order of a
micron. These measurements must be made in the presence of the
dispersed phase.

Over the past decade, a number of concepts have been proposed as
candidates for providing the aforementioned measurements. However, the
need to obtain reliable measurements in a range of environmental conditions including high number densities, complex turbulent flows, reacting
flows, and contaminated windows has rendered most of the concepts
impractical. In addition, the developed instrument must be relatively
easy to set up and operate. The need for frequent calibration and
exacting alignment has made some methods undesirable.

Nonetheless, the need to obtain information on the particle size
and dynamics has led to the development of instruments that could at
least provide some of the information outlined above. The small angle
light scatter detection methods considered by Dobbins et al.[1] and
Swithenbank et al.[2] have been favorably received because of their relative ease of operation, ability to measure irregular-shaped, as well as
spherical particles, and speed at which instruments based on the concept
can produce measurements. Disadvantages of the method are that it
obtains line-of-sight averaged measurements, loses sensitivity for
larger particles, and cannot provide information on the particle
dynamics.

Single particle counter methods based on light scattering cross
section with a deconvolution technique to overcome the problem of random
particle trajectories through a Gaussian beam, have been developed by
Yule et al.[3] and Holve and Self.[4] The method has the advantage of being
relatively insensitive to particle shape (although this has not been
explicitly demonstrated), measures the size distribution directly, and
can measure the average particle speed. Claims have been made that
particle size-velocity correlations can be obtained but, to our knowledge, such data have not been published. The method has the disadvantages of needing extremely careful alignment and beam quality control,
needs relatively frequent calibration, and is subject to errors produced
by beam and scattered light attenuation.

The more recently developed phase Doppler method offers the
potential for fulfilling most of the measurement needs for two-phase
flow research. Recognition of the method followed from the theoretical
description by Bachalo[6] of the dual beam light scattering at large off-
axis angles. This analysis revealed that the relative phase shift of
the light scattered by the mechanisms of reflection or refraction was
proportional to the drop size. With dual beam light scattering interferometry, the phase shift manifests as a fringe pattern formed by the
scattered light that has a spatial frequency inversely proportional to
drop size. Obtaining the particle size is reduced to measuring the
spatial frequency of the fringe pattern. As with the laser Doppler
velocimeter, the temporal frequency of the fringe pattern is proportional to the particle velocity.

In the following sections, the evaluation of the method in a number
of practical applications will be described. Measurements of particle
number density and volume flux will be demonstrated. The theoretical
considerations and actual spray measurements will be described.

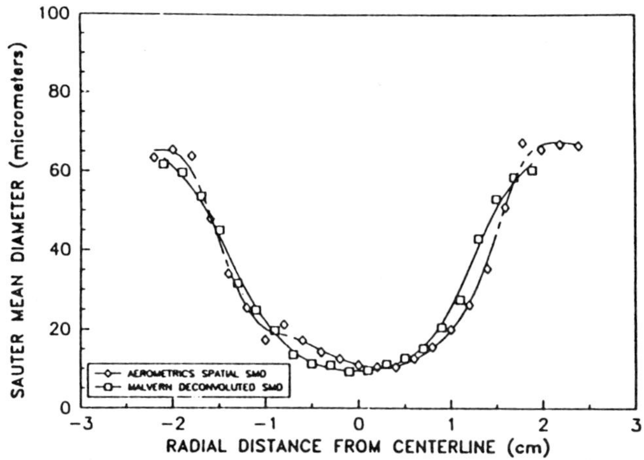

Fig. 4. Comparisons of SMD Measured by Aerometrics Phase Doppler Instrument and Malvern Laser Diffraction After Deconvolution (Ref. 16).

HAGO 50.0 60°P

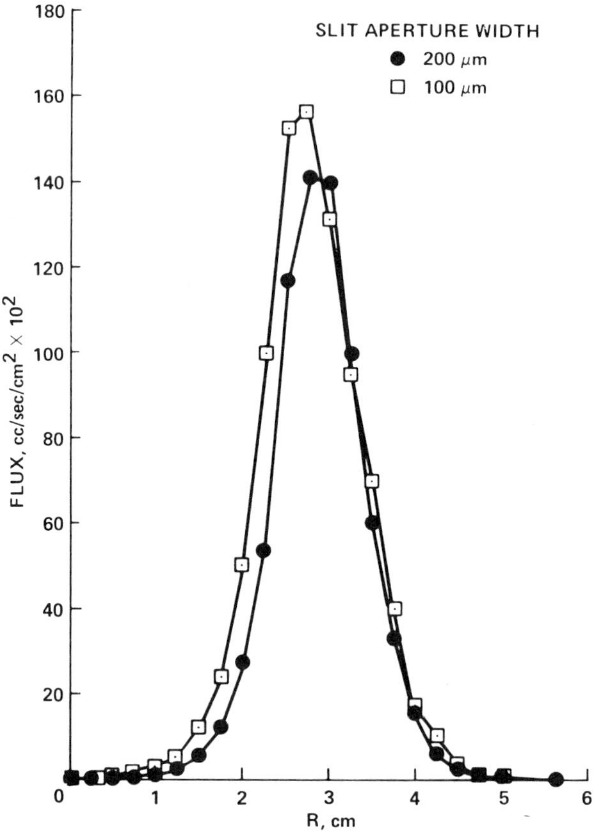

Fig. 5. Radial Variation of the Volume Flux.

THEORETICAL DISCUSSION

The general light scattering theory covering the phase Doppler method has been provided by Bachalo.[6] Since only spherical particles larger than the wavelength were considered, the geometrical optics theory was used. This approach had the significant advantage of providing direct insight to the light scattering mechanisms. For example, the relative effect of the reflected component on the off-axis forward scattered interference fringe pattern could be assessed. The need to use highly focused beams with diameters only two to three times the largest particle diameter in the distribution exacerbates this problem. The exact Lorenz-Mie theory was used by Pendleton[7] to verify the accuracy of the geometrical optics approach and to show that the method could be used to measure particles as small as 0.5 um. It is important to recognize that the Lorenz-Mie theory was derived for uniformly illuminated particles. When Gaussian beam intensity distributions are used to illuminate the particle and the beam is not at least an order of magnitude larger than the largest particle measured, this boundary condition is violated.

The phase Doppler method[5] utilizes an optical system as illustrated in Figure 1 which is approximately the same as an LDV except that three detectors are located at selected spacings behind a single receiver aperture. Particles passing through the intersection of the two laser beams scatter light which forms an interference fringe pattern in the surrounding space. The spatial frequency of the interference fringe pattern is inversely proportional to the particle diameter but also depends upon the laser wavelength, beam intersection angle, particle refractive index (unless reflected light is detected) and the location of the receiver. Dynamic measurements of the spatial frequency of the interference fringe pattern are achieved using pairs of detectors placed at appropriate locations and separations. Accurate measurements presuppose that the dominant light scattering mode is known.

The problem of detecting light scattered by reflection and refraction simultaneously was a concern from the beginning of the concept development. When the light scattering amplitudes by reflection and refraction are of a similar order of magnitude, the interference fringe pattern formed by the scattered light can become very complex. This condition occurs when beam diameters of only two to three times the drop diameter are used. Drop trajectories as shown in Figure 2 wherein the peak intensity is incident at a point on the particle that reflects to the receiver will produce scattered light intensities by refraction and reflection that are of similar amplitudes. Interference then occurs between the pairs of refracted rays, the pairs of reflected rays, the reflected from beam 1 and the refracted from beam 2 and vice versa, and the reflected and refracted light from each beam to form six superimposed fringe patterns. The refracted fringe pattern has a temporal phase variation that causes the fringes to appear to sweep in the direction of the drop. Fringes produced by reflection appear to move in the opposite direction. The interference between the reflected and refracted rays from each beam are stationary and are often referred to as the resonances on the scattering intensity distribution.

Extensive experimentation was conducted in the early stages of the development of the phase Doppler method to evaluate the extent of the problems associated with nonuniform particle light scattering. These experiments led to the development and implementation of proprietary techniques and logic to eliminate the occurrence of the problem.

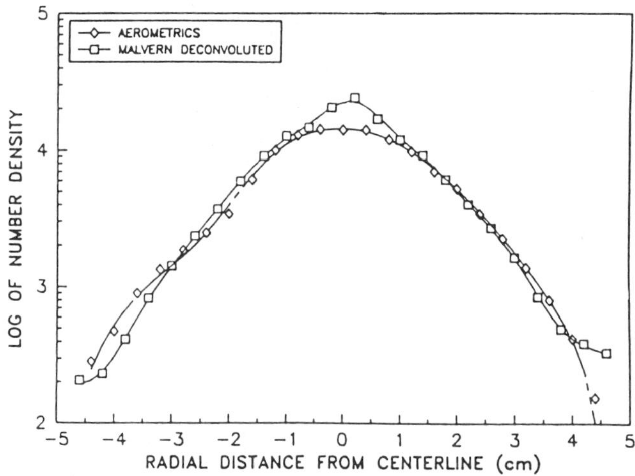

Fig. 6.   Comparison of the Number Density Measured by the Phase Doppler Instrument and Line-of-Sight Extinction After Deconvolution (Ref. 16).

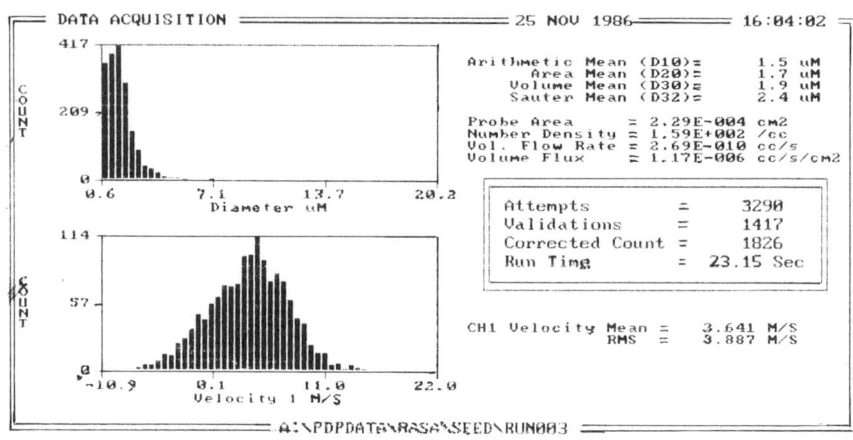

Fig. 7a.   Size Distribution Measured for Seed Particles.

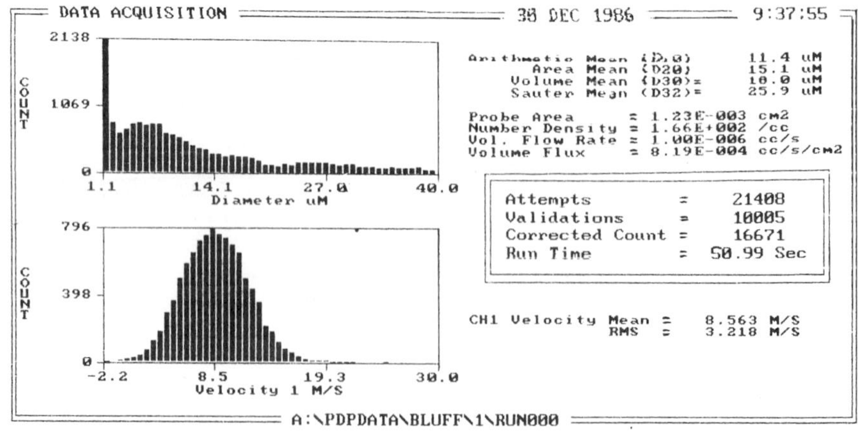

Fig. 7b.   Combined Spray and Seed Particles.

In many applications, it is desirable or even necessary to obtain the measurements using off-axis backscatter. According to the geometrical optics theory, the scattered light in the backscatter region will consist of the first surface reflection (p = 0) and internal reflection (p = 2). The angle between the incident ray and the pth emergent ray is given by

$$\Theta = 2(p\tau' - \tau) \qquad (1)$$

where τ and τ' are the angles between the surface tangent and the incident and emergent rays. In order to compute the relative phase shift, each emergent ray is referenced to a hypothetical ray (van de Hulst[8]) without phase change at the center of the sphere. Neglecting phase shifts of π at reflection and $\pi/2$ at focal lines, a simple geometric analysis results in the expression

$$\eta = 2\,\frac{\pi d}{\lambda}\,(\sin\tau - pm\,\sin\tau') \qquad (2)$$

for the phase shift with respect to the reference ray where m is the refractive index of the particle. For opaque or particles with significant absorption, the first surface reflection is dominant. With transparent materials and perpendicular polarization, the internal reflection is dominant. Which scattering mechanism is dominant is easily determined since the apparent direction of the fringes formed by the two mechanisms is in opposite directions. This may be seen with the use of a simple ray diagram for a sphere.

For internal reflection, p = 2, the relationship between Θ and τ results in a quadratic so that pairs of incident angles τ correspond to a single exit angle, Θ, Figure 3. This means that there can be two values of phase shift at individual points in the receiver plane. With dual beam light scattering, the interference becomes very complex since there will be interference between the rays from the individual beams and between the pairs of rays from the separate beams leading to six superimposed interference patterns. If the first surface reflection is significant, the number of superimposed interference fringe patterns increases accordingly.,

Our current work is involved with analyzing the interference in the off-axis backscatter direction. Measurements are also being conducted to demonstrate the reliability of this light scatter detection mode for obtaining particle size measurements. Originally, experiments were conducted to evaluate the analysis for pure first surface reflection and by internal reflection.[9] In the first case, monodispersed drops using dye to absorb the internal rays were measured. The resultant signal detected at 150° to the projected beam produced results in excellent agreement with the predictions. Transparent drops produced signals detected at 150° in reasonable agreement with the analysis for internal reflection.

The direction in which the scattered interference fringe pattern moves is critical to the measurements since the detection sequence over the three detectors determines the phase angle between signals. If a particle passes the probe in a reverse direction, it will produce a measurement of 360° - ϕ rather than ϕ, the signal phase angle. Such occurrences can produce large errors in the size measurement. To avoid this, frequency shifting must be used. As with conventional laser Doppler velocimetry, the frequency shifting is also required to allow a nearly equal probability of measurement for all particle trajectories, to reduce the frequency bandwidth, and resolve forward and reversed

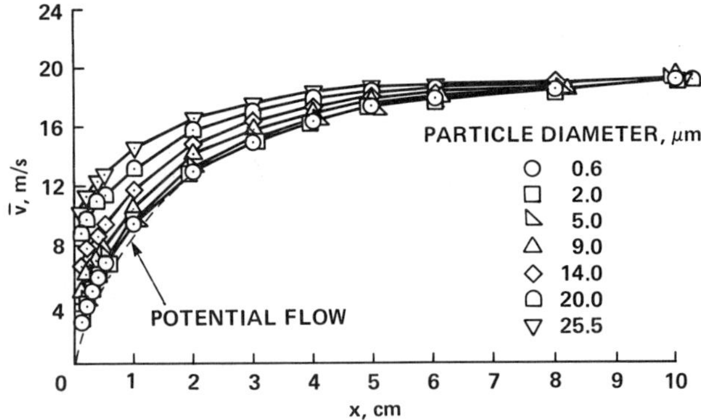

Fig. 8. Particle Response to a Strongly Decelerating Flow.

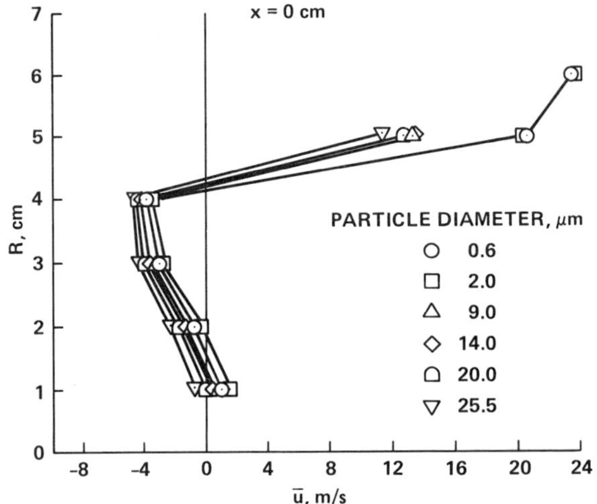

Fig. 9a. Radial Variation of Axial Velocity, x = 0.

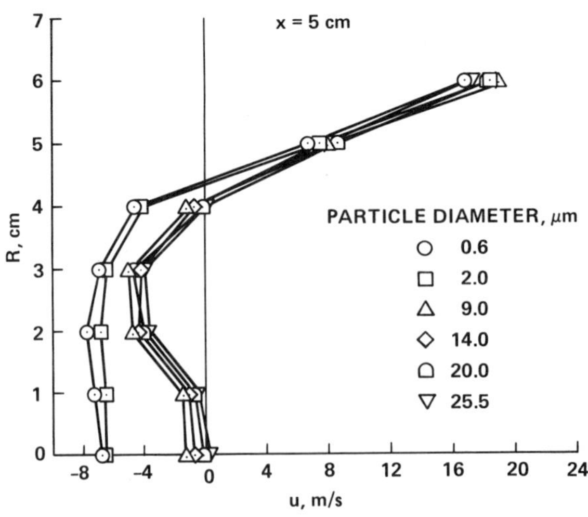

Fig. 9b. x = 5 cm.

velocities. In sprays, the smaller particles in the distribution may
have a significantly different velocity magnitude and direction than the
larger particles. Thus, the need for frequency shifting is of increased
importance.

In the Aerometrics' phase Doppler instrument, a rotating
diffraction grating is used for producing the frequency shift. This
device has the advantage of decoupling the beam deflection from the
shift frequency and can be set over a wide range of frequencies. An
encoder and computer interface was developed to set and continuously
control the rotational speed.

Any method associated with the LDV cannot escape the problems of
statistical or velocity bias. In highly turbulent two-phase flows, the
problem exists for each particle size class. A number of schemes have
been proposed for mitigating this bias. McLaughlin and Tiederman[10]
identified the problem and proposed a weighting function for reducing
the bias error. In uniformly seeded flows, Hoesel and Rodi[11] proposed a
probe volume residence time correction. For nonuniform particle number
densities, they suggested an inter-particle time-of-arrival correction
when the average particle time-of-arrival is small in comparison to the
time scale of the turbulence. Simpson and Chew[12] suggested a uniform
time-of-arrival sampling procedure to eliminate velocity bias and the
effects of nonuniform particle number density. Gould et al.[13] have
investigated the technique and applied the method to the correction of
highly turbulent flow measurements. Craig et al.[14] conducted experiments to quantify the error due to velocity bias and to evaluate the
correction methods. They concluded that the time-of-arrival correction
scheme could eliminate velocity bias from the data. However, they
concluded that constant time interval sampling was the only effective
means for removing bias due to nonuniformities in the particle number
density. This is unfortunate since the method requires approximately an
order of magnitude greater particle arrival rate to achieve a suitable
uniform time interval sampling.

The phase Doppler instrument measures the time-of-arrival of each
particle measured. Software has been developed to use this information
to quantify the correlation between particle time-of-arrival and velocity. The data can then be corrected for the velocity and nonuniform
time interval sampling and inter-particle time-of-arrival schemes.
However, these methods were not implemented in the data to be presented
in the subsequent section.

<u>Representative Measurements</u>

Because there is no definitive means for proving the particle size
measurement accuracy while operating in realistic environments, a series
of tests were conducted to obtain confidence in the technique. Basic
studies conducted with monodispersed drop streams do not provide
adequate information on the effects of beam distortion and coincident
particle occurrences in the sample volume. The only recourse available
was to make comparisons with measurements obtained from diverse
methods. These comparisons do not prove accuracy, but if properly
conducted, demonstrate that certain parameters of the spray are well
characterized.

In Figure 4, radial distribution of the SMD for a spray was
measured by Dodge[15] using an Aerometrics' Phase Doppler Particle
Analyzer and a Malvern small angle detection instrument. Because the
Malvern instrument produced a line-of-sight average measurement of the

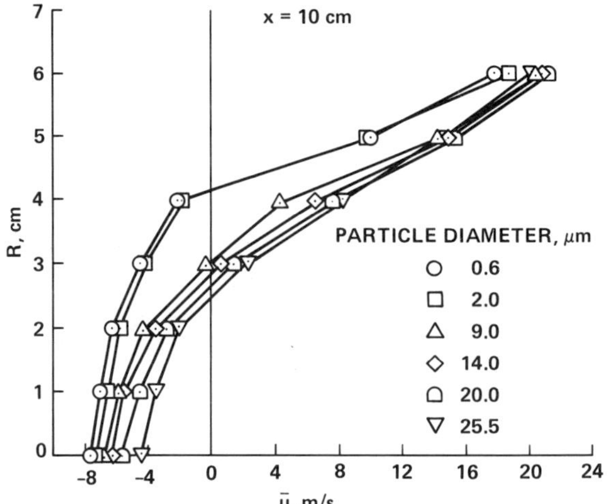

Fig. 9c.   x = 10 cm.

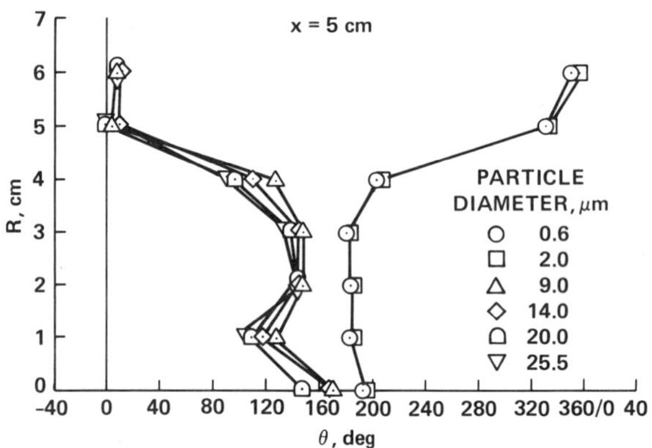

Fig. 10a.   Particle Angle of Trajectory for Representative
Size Classes, x = 5 cm.

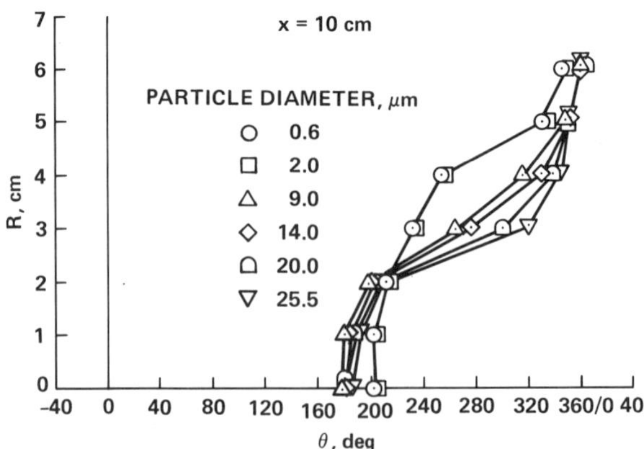

Fig. 10b.   x = 10 cm.

spatial size distribution, a deconvolution technique which assumed axial symmetry and followed a procedure similar to the well-known Abel inversion. The Aerometrics' instrument produced a temporal measurement of the size distribution. That is, the instrument counts particles passing the sampling cross section to produce a flux-dependent sample. As such, the number of particles counted in each size class depends upon their relative velocities. However, the measured velocity for each size class can be used to normalize the number of counts, thus removing the effect of the relative velocities. This forms a spatial distribution which depends upon the relative population of each size class in a volume. The conversion was made for the comparison in Figure 4. Considering that the scattering mechanisms and means for signal analysis are quite different, the agreement was reassuring.

Mass flux measurements, along with the particle size distribution, are of considerable importance in a number of practical applications. Accurate mass flux measurements requires the accurate measurement of at least the larger particles in the distribution and the cross section of the sample volume. A method has been developed to directly measure the sampling cross section for each particle size class each time a size distribution is obtained. This information is first used to remove the sample volume bias due to the Gaussian intensity distribution by correcting the number of samples in each class to be equivalent to that for a uniform sampling cross section. The measured cross section is then used with the number of samples and elapsed time for the accumulation of the sample to compute the flux.

Figure 5 shows the radial distribution of the flux for a pressure atomizer. A sampling probe was used at representative radial stations for comparisons to the phase Doppler data. These results were in agreement to within 5%. The spray was then assumed to be axisymmetric and the flux at each radial station integrated over each annular ring to obtain the flow rate. Comparison with the liquid flow into the atomizer also showed agreement to within 5%. This integration procedure has been incorporated in the system software.

Local particle number density is an additional parameter that is important to the description of the spray behavior. The phase Doppler instrument automatically reports the number density at each measurement point. Dodge[16] performed an independent determination of the number density using extinction measurements with the Lambert-Beer law and the deconvolution procedure to generate a radial distribution of the number density in a spray. Figure 6 shows the comparison between the two methods. Because the deconvolution procedure begins with a direct measurement at the outer edge and marches inward with the errors accumulating, the difference at the center was not surprising.

A concern with optical single particle counters is whether or not the small particles are being counted accurately. The good agreement in the number density strongly suggests that the counting is accurate. Whether or not the smallest particles are detected can also be assessed by introducing small seed particles on the order of one micrometer in diameter into a spray. This has the added benefit of determining whether seed particles necessary for measuring the gas phase velocity can be measured in the presence of relatively dense sprays. Figure 7a shows the measured size distribution of seed particles consisting of mineral oil droplets produced by a condensation generator. The seed particles were then introduced in the surrounding air and were entrained by a spray. The combined size distribution of the seed particles and spray is shown in Figure 7b. Even in relatively dense sprays, the seed particles could be measured.

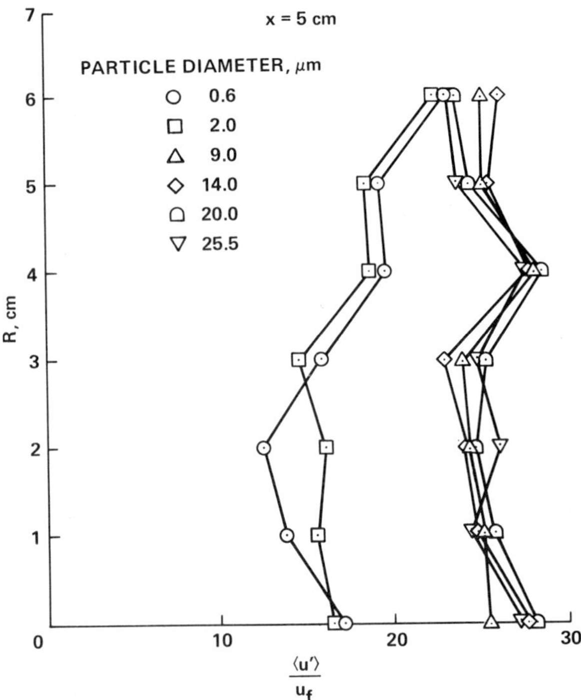

Fig. 11. Fluctuating (RMS) Axial Velocities.

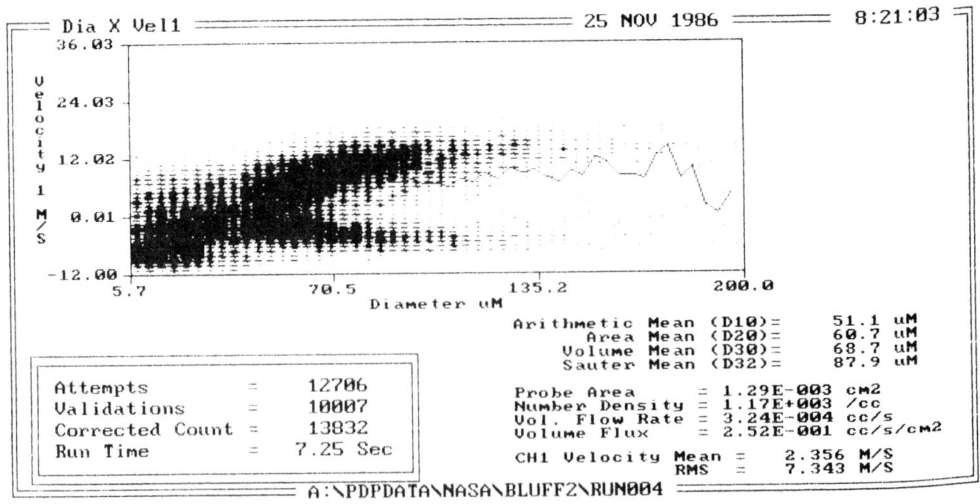

Fig. 12. Size-Velocity Correlation Showing the
Bifurcation of the Velocity Distribution.

Because the particle velocity measurements can be made with a high
confidence level, the velocity and size of particles obtained in a
decelerating flow could be used to infer accuracy in the size measurement. Experiments were conducted in the Aerometrics' two-phase flow
wind tunnel using a 63.5 mm diameter cylinder mounted transverse to the
flow and sprays generated upstream. The airflow speed was 20 m/s.
Measurements of drop size and velocity were obtained at axial stations
up to the cylinder. As seen in Figure 8, the particle lag was
consistent with the measured particle size. Although these results
cannot prove a high degree of accuracy because of the uncertainties in
the accuracy of the drag coefficient, it does show sufficient accuracy
for most practical applications. Furthermore, the method can be used to
validate the method in very dense sprays.

The experiments were of value in the converse sense in that
accurate particle size and velocity measurements can be used to
establish the particle drag coefficient, C_D. There is a need for a
better definition of the drag coefficient for liquid drops in the
presence of high density sprays and turbulent airstreams. These experiments are currently in progress at Aerometrics.

Extensive research is also being conducted on particle response and
two-phase turbulent flows. The importance of quantifying particle
response is amplified in two-phase flows since the seed particles used
to obtain the gas phase turbulence parameters must compete with signals
from the dispersed phase. Use of the largest seed particles possible
will serve to mitigate problems with the signal-to-noise ratio.

Detailed measurements of a flow field generated by a bluff body
mounted in the wind tunnel with spray injection on the leeward side were
performed.[17] The bluff body consisted of a circular disk 76 mm in
diameter. The airflow velocity was 21 m/sec. Data are presented for
discrete particle sizes of 0.6, 2, 5, 9, 14, 20, and 25 micrometers in
diameter which had corresponding mass ratios relative to a 1 um particle
of 0.2, 7, 125, 729, 2744, 8000, and 16581. The 0.6 and 2 um particles
were formed by the mineral oil, whereas, the larger particles were
water. Although water drops in the size range of 1 to 9 micrometers
were formed, these drops evaporated fairly rapidly and, hence, did not
have a sufficient population.

Radial distributions of the axial component of velocity are
presented in Figure 9. Figure 9a presents the radial profile at x = 0
(the nozzle exit plane). No data were available on the axis since this
was the exit of the atomizer. Outside of the spray cone, air entrained
by the spray carried the seed and water drops as large as 15 um in a
positive direction. Further out at R = 3 cm, the air had a relatively
strong reversed flow velocity. In that region, the largest drops had
the greatest reverse velocities. This was due to the deceleration of
the airflow by the bluff body. The larger drops had greater inertia and
thus, overshot the airflow. Many of these larger drops impacted the
leeward face of the disk. A local increase in the circulation at the
salient edge of the disk produced an increase in the reversed airflow
velocity. The velocity measurements based on the seed particles beyond
the edge of the disk were representative of the air velocity. Consequently, the drops lagged the airflow velocity by as much as 50%.

Downstream at 5 cm from the exit plane of the atomizer, Figure 9b,
the reversed airflow decelerated the drops emitted by the atomizer.
Measurements of these drops combined with drops following the reversed
airflow resulted in a mean velocity of approximately zero from the axis

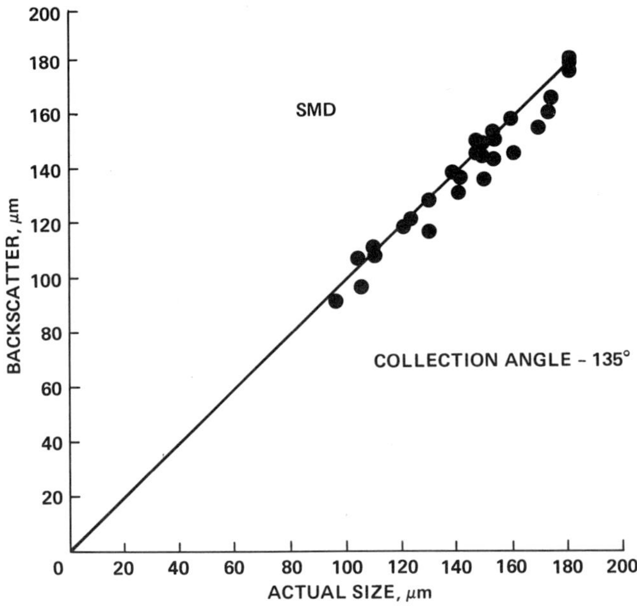

Fig. 13.   Measurements of Monodispersed Drops Using Off-axis Backscatter.

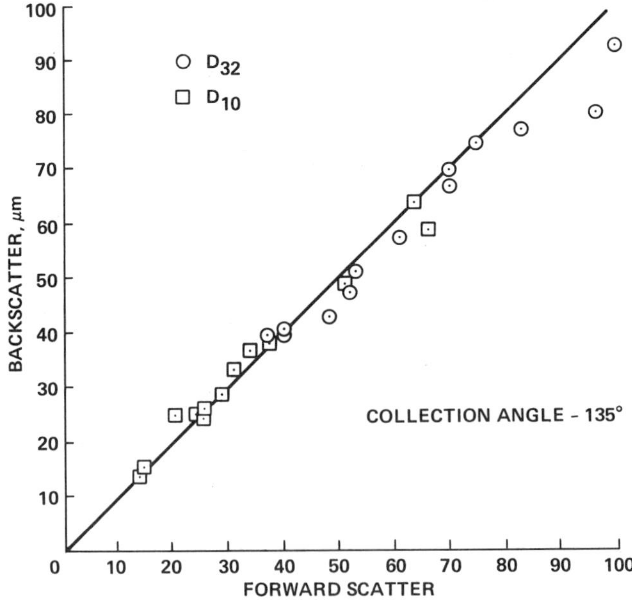

Fig. 14.   Comparisons of Spray Measurements Using Forward and Backscatter.

out to a radius of approximately 1 cm. The air velocity represented by
the seed particles had a much larger magnitude in the reversed direction
than the drops. This was due to the fact that the drops were injected
with a relatively high velocity and were subsequently decelerated and
returned in the negative direction by the drag forces. At 10 cm downstream of the atomizer, Figure 9c, the drops had a much greater negative
velocity than at the scan station. The majority of drops measured at
this location originated from further downstream. Radial velocity and
angle of trajectory data confirmed that the drops appearing at this
station were recirculated by the airstream.

Angle of trajectory results were computed from the axial and radial
velocity components for each size class, Figure 10. At the station 5cm
downstream of the atomizer, Figure 10a, the spray drops turned upstream
($\Theta = 180^{\circ}$ for reversed flow) at radii from 2 to 4 cm, but maintained an
outward angle of approximately 20°. In the same region, the seed
particles were turned inward by 20°. Both the seed particles and drops
had outward angles of 10° and 20°, respectively, at the outer radius.
The radial variations of the angle of trajectory at the 10 cm station,
Figure 10b, were similar for all particle sizes. However, the smaller
particles turned toward the centerline more readily than the larger
drops because of their greater momentum.

Distributions of the streamwise root-mean-square (rms) velocities
made dimensionless with the freestream velocity, $u_f = 21$ m/s, are
illustrated in Figure 11. In general, the spray drops had larger rms
values than the air (with some exceptions) throughout the flow.
Although the drops do not respond well to gas phase turbulence fluctuations, the injection characteristics and the mixing with recirculated
drops created the higher rms values. Evidence of this effect can be
seen in Figure 12 which is a plot of the size-velocity correlations.
The plot clearly shows a bifurcation in the drop velocities about zero
velocity. Two effects have occurred at this station: the convection of
particles with a near-zero streamwise velocity through the sample volume
is low; and the injected spray is passing recirculated drops moving in
the opposite direction.

Finally, measurements of monodispersed drops using an off-axis
backscatter angle of 135° are shown in Figure 13. This angle is not
far-removed from the angle of the rainbow, wherein the internally
reflected rays can have the same exit angle for two discrete points of
entry. Nonetheless, the measured results were in reasonable agreement
with the straight line passing through the origin. The particle
measurements using backscatter did appear to be more sensitive to drop
oscillations. When the monodispersed generator was adjusted to change
the phase of the oscillation at the measurement volume, a 10-15% change
in the measured size was reported. Measurements of sprays using off-
axis forward scatter were compared to the results obtained using off-
axis backscatter. At most locations, the results were in excellent
agreement, Figure 14; but at some radial locations, the Sauter mean
diameter differed by as much as 15%. The reason for this variance is
currently under investigation.

<u>Summary and Conclusions</u>

Characteristics of the phase Doppler technique for measuring the
size and velocity of spherical particles was reviewed. Because of the
intrinsic nature of the the light scatter detection, proper optical
design was found to be necessary if measurement errors resulting from
multi-component light scattering is to be avoided. Also, frequency

shifting was shown to have an added importance in the phase Doppler
instrument.

Because the method is relatively new, a great deal of effort has
been devoted toward verifying its performance in realistic environments. Monodispersed drop stream measurements, comparisons to data
obtained with other methods, and inferences from data obtained with
established techniques have been used. In general, the results have
demonstrated that the method, if properly implemented, can lead to a
reliable and robust instrument.

Backscatter measurements were compared with forward scattering
results. This work showed reasonable agreement but unexpected variations in some of the measurements will require more detailed analysis
before fully appreciating the reliability of this mode of detection.
Work is currently in progress to acquire the experimental information
needed to confirm the analysis.

REFERENCES

1.   R. A. Dobbins, L. Crocco, and I. Glassman, "Measurement of Mean
      Particle Sizes of Sprays from Diffractively Scattered Light,"
      Am. Inst. Aeronaut. Astronaut. J., Vol. 37, p. 1882, 1963.

2.   J. Swithenbank, J. M. Beer, D. S. Taylor, D. Abbot, and G. C.
      McCreath, "A Laser Diagnostic for the Measurement of Droplet
      and Particle Size Distribution," AIAA 14th Aerospace Sciences
      Meeting, Washington, D.C., Paper No. 76-79 (1977).

3.   A. J. Yule, N. A. Chigier, S. Atakan, and A. Ungut, "Particle Size
      and Velocity Measurement by Laser Anemometry," AIAA 15th
      Aerospace Sciences Meeting, Los Angeles, Paper No. 77-214
      (1977).

4.   D. Holve, and S. A. Self, "An Optical Particle Sizing Counter for
      In-Situ Measurements," Project SQUID Tech Report SU-2-PU
      (1978).

5.   W. D. Bachalo and M. J. Houser, "Phase/Doppler Spray Analyzer for
      Simultaneous Measurements of Drop Size and Velocity
      Distributions," Optical Engineering, Vol. 23, No. 5, p. 583-590
      (September/October 1984).

6.   W. D. Bachalo, "Method for Measuring the Size and Velocity of
      Spheres by Dual-Beam Light-Scatter Interferometry," Appl. Opt.,
      Vol. 19, No. 3 (1980).

7.   J. D. Pendleton, "Mie and Refraction Theory Comparison for Particle
      Sizing With the Laser Velocimeter," Appl. Opt., Vol. 21, No. 4
      (1982).

8.   H. C. van Hulst, "Light Scattering by Small Particles," Wiley, New
      York (1957).

9.   W. D. Bachalo and M. J. Houser, "Analysis and Testing of a New
      Method for Drop Size Measurement Using Laser Light Scatter
      Interferometry," NASA Contractor Report 174636 (1984).

10. D. McLaughlin and W. Tiederman, "Biasing Correction for Individual
     Realization of Laser Anemometer Measurements in Turbulent
     Flows," The Physics of Fluids, Vol. 16, No. 12, p. 2082 (1973).

11. W. Hoesel and W. Rodi, "New Biasing Elimination Method for Laser
     Doppler Velocimeter Counter Processing," Review of Scientific
     Instruments, Vol. 48, No. 7, p. 910 (1977).

12. R. L. Simpson and Y. T. Chew, "Measurements in Steady and Unsteady
     Separated Turbulent Boundary Layers," Laser Velocimetry and
     Particle Sizing, H. D. Thompson and W. H. Stevenson, Ed.,
     Hemisphere Publishing (1979).

13. R. D. Gould, W. H. Stevenson, and H. D. Thompson, "A Parametric
     Study of Statistical Velocity Bias," Proceedings, Laser
     Institute of America, Vol. 58, ICALEO (Nov. 1986).

14. R. R. Craig, A. S. Nejad, and E. Y. Hahn, "A General Approach for
     Obtaining Unbiased LDV Data in Highly Turbulent Non-reacting
     and Reacting Flows," AIAA Paper No. 86-0366 (1984).

15. L. G. Dodge, "Comparison of Drop-Size Measurements for Similar
     Atomizers," Special Report No. SwRI-8858/2, Southwest Research
     Institute, San Antonio (1986).

16. L. G. Dodge, D. J. Rhodes, and R. D. Reitz, "Comparison of Drop-
     Size Measurements Techniques in Fuel Sprays: Malvern Laser-
     Diffraction and Aerometrics Phase/Doppler," presented Central
     States Section, The Combustion Institute, NASA Lewis Research
     Center (1986).

17. W. D. Bachalo, R. C. Rudoff, and M. J. Houser, "Laser Velocimetry
     in Turbulent Flow Fields: Particle Response," presented AIAA
     25th Aerospace Sciences Meeting, Reno (1987).

Acknowledgements

The author gratefully acknowledges the support of this work by the
NASA Lewis Research Center, Contract No. NAS3-24844, Ms. Valerie Lyons,
Contract Monitor, and by the Air Force Office of Scientific Research,
Contract No. F49620-86-C-0078, Dr. Julian Tishkoff, Contract Monitor.

FRAUNHOFER DIFFRACTION BY RANDOM IRREGULAR PARTICLES

A.R. Jones

Chemical Engineering, Imperial College
London SW7 2BY, England

ABSTRACT

A statistical model is developed for Fraunhofer diffraction by random irregular particles. This is described in terms of a distribution function for radii and a correlation function in the surface. Results are presented for a variety of particle types, and the consequences to particle sizing discussed. Comparison with a "best" sphere suggests a radius based on the average $\langle R^n \rangle$ where n>3.

INTRODUCTION

Fraunhofer diffraction for sizing particles has become the basis for a variety of commercial instruments, where the analysis invariably assumes the particles to be spherical. This will be valid for liquid drops at low Reynolds number, but may not be when there is a significant relative velocity. It will rarely be true for solid particles.

Previous investigations of scattering by non-spherical particles have either considered an expansion of the surface in polynominals, or polyhedra (Schuerman, 1980; Mugnai and Wiscombe, 1986). These non-spherical (but regular) shapes may then be treated by a variety of numerical techniques, since the surface at any point can be predicted. Such models may be suitable for drops oscillating in a shear flow, for example.

However, it is a feature of true irregular particles that the surface cannot be predicted for any single position on any individual particle. The implication is that a statistical model is required.

In order to examine some of the principles involved, and to investigate the influence of irregularity on particle sizing, this paper describes a statistical model for scalar wave Fraunhofer diffraction. This is described in terms of a distribution function for radii and a correlation function in the surface.

THEORY

The diffraction integral in circular co-ordinates is (e.g. Born and Wolf, 1975)

$$\psi_p = \frac{1}{\lambda} \int_0^R \int_0^{2\pi} e^{-ik\rho \sin\theta \cos(\alpha-\phi)} \rho \, d\rho \, d\alpha \qquad (1)$$

where ψ_p is the amplitude at the point (r,θ,ϕ). A point within the circular aperture of radius R is given by (ρ,α). λ is the wavelength and $k=2\pi/\lambda$. We may replace equation (1) by an integral to infinity by the use of a transmission function such that

$$\psi_p = \frac{1}{\lambda} \int_0^\infty \int_0^{2\pi} T(\rho,\alpha) e^{-ik\rho \sin\theta \cos(\alpha-\phi)} \rho \, d\rho \, d\alpha$$

where

$$T(\rho,\alpha) = \begin{cases} 1 & \rho < R \\ 0 & \rho > R \end{cases}$$

For a non-circular aperture the same form may be used, but R becomes a function of angle.

We assume that there are many randomly positioned particles so that incoherent superposition applies. If $\psi_{p,n}$ is the amplitude due to the nth particle, the average intensity due to N particles is

$$\langle I \rangle = \frac{1}{N} \sum_{n=1}^{N} \left| \psi_{p,n} \right|^2$$

This results in

$$\langle I \rangle = \frac{1}{\lambda^2} \int_0^\infty \int_0^\infty \int_0^{2\pi} \int_0^{2\pi} \langle T(\rho_i,\alpha_i) T(\rho_j,\alpha_j) \rangle$$
$$\times e^{-iks\sin\theta \left(\rho_i\cos(\alpha_i-\phi)-\rho_j\cos(\alpha_j-\phi) \right)} \qquad (2)$$
$$\times \rho_i\rho_j \, d\rho_i d\rho_j \, d\alpha_i d\alpha_j$$

$T(\rho,\alpha)$ is unity to some value $\rho=R$ and is zero thereafter.

Hence

$$T(\rho,\alpha) = 1-U(\rho-R)$$

where R is a function of α and $U(\rho-R)$ is the unit step function

$$u(\rho-R) = \begin{cases} 0 & \rho < R \\ 1 & \rho > R \end{cases}$$

Now

$$\langle T(\rho,\alpha) \rangle = \int_0^\infty p(R)\left[1-U(\rho-R) \right] \, dR = \int_\rho^\infty p(R) dR \qquad (3)$$

where p(R) is the probability function describing the position of the edge. For a large number of randomly oriented particles p(R) will be independent of angle, as will $\langle T(\rho,\alpha) \rangle$. The same argument can be applied to $\langle T(\rho_i,\alpha_i) T(\rho_j,\alpha_j) \rangle$ and we may write.

$$\langle T(\rho_i,\alpha_i)T(\rho_j,\alpha_j)\rangle = \int p(R_i)\left[1-U(\rho_i-R_i)\right]\int \pi(R_i,R_j)\left[1-U(\rho_j-R_j)\right]$$
$$\times\, dR_i\,dR_j \qquad\qquad (4)$$

where R_i is the value of R at $\alpha = \alpha_i$, and $\pi(R_i,R_j)$ is the probability of R_j for any given R_i.

In the special case $\alpha_i = \alpha_j$ then

$$\langle T(\rho_i,\alpha_i)T(\rho_j,\alpha_j)\rangle = \int_o^\infty p(R_i)\left[1-U(\rho_i-R_i)\right]\left[1-U(\rho_j-R_i)\right]dR_i$$

If $\rho_i>\rho_j$ we can show that

$$\left[1-U(\rho_i-R_i)\right]\left[1-U(\rho_j-R_j)\right] = \left[1-U(\rho_i-R_i)\right]$$

and vice versa. It follows that

$$\langle T(\rho_i,\alpha_i)T(\rho_j,\alpha_j)\rangle = \begin{cases} \langle T(\rho_i)\rangle & \rho_i>\rho_j \\ \langle T(\rho_j)\rangle & \rho_i<\rho_j \end{cases}$$

When $\alpha_i = \alpha_j$ the two transmission functions are perfectly correlated. At the opposite extreme when they are uncorrelated they are independent and

$$\langle T(\rho_i,\alpha_i)T(\rho_j,\alpha_j)\rangle = \langle T(\rho_i)\rangle\langle T(\rho_j)\rangle$$

We may thus define a correlation function

$$C_{ij} = \frac{\langle T_iT_j\rangle - \langle T_i\rangle\langle T_j\rangle}{\langle T_s\rangle - \langle T_i\rangle\langle T_j\rangle} \qquad\qquad (5)$$

where $\langle T_s\rangle = \langle T_i\rangle$ for $\rho_i>\rho_j$ and $\langle T_s\rangle = \langle T_j\rangle$ for $\rho_i<\rho_j$ equation (2) now becomes

$$\langle I\rangle = A + B$$

where

$$A = \frac{1}{\lambda^2}\left|\int_o^\infty\int_o^{2\pi}\langle T_i\rangle\, e^{-ik\rho_i\sin\theta\cos(\alpha_i-\phi)}\rho_i\, d\rho_i\, d\alpha_i\right|^2$$

$$B = \frac{1}{\lambda^2}\int_o^\infty\int_o^\infty\int_o^{2\pi}\int_o^{2\pi} C_{ij}\, G_{ij}\, e^{-ik\sin\theta[\rho_i\cos(\alpha_i-\phi)-\rho_j\cos(\alpha_j-\phi)]}$$
$$\times\rho_i\rho_j\, d\rho_i\, d\rho_j\, d\alpha_i\, d\alpha_j$$

and

$$G_{ij} = \begin{cases} \langle T_i\rangle\left[1-\langle T_j\rangle\right] & \rho_i>\rho_j \\ \langle T_j\rangle\left[1-\langle T_i\rangle\right] & \rho_i<\rho_j \end{cases}$$

The correlation function will only be a function of the angular separation $\gamma = \alpha_j - \alpha_i$. It will also be independent of the direction in which γ increases, and symmetrical about the line $\gamma = 0$ to π. Then, C_{ij} may be expressed in terms of a Fourier series

$$C_{ij} = \sum_{\ell=1}^{\infty} C_\ell \cos \ell\gamma \qquad (6)$$

integration over α yields

$$A = \frac{4\pi^2}{\lambda^2} \left| \int_0^\infty \langle T_i \rangle \, J_0(k\rho_i \sin\theta)\rho_i \, d\rho_i \right|^2 \qquad (7)$$

$$B = \frac{\pi^2}{\lambda^2} \sum_{\ell=1}^{\infty} C_\ell \int_0^\infty \int_0^\infty G_{ij} \, J_\ell(k\rho_i \sin\theta) \, J_\ell(k\rho_j \sin\theta) \, \rho_i \rho_j \, d\rho_i \, d\rho_j$$

where $J_n(x)$ is a Bessel function.

EVALUATION OF THE CORRELATION AND EDGE PROBABILITY FUNCTIONS

The average radius is given by

$$\langle R \rangle = \int p(R) R \, dR$$

and

$$\langle R_i R_j \rangle = \int p(R_i) \, R_i \int \pi(R_i, R_j) R_j \, dR_i \, dR_j$$

$p(R_i)$ and $\pi(R_i, R_j)$ are the same functions as in equation (4) so the correlation function for R will be the same as that for T.

Hence

$$C_{ij} = \frac{\langle R_i R_j \rangle - \langle R \rangle^2}{\langle R^2 \rangle - \langle R \rangle^2}$$

Expanding the surface of any single particle as a Fourier series we have

$$R_n = a_{n,o} + \sum_{\ell=1}^{\infty} a_{n\ell} \cos \ell\alpha + \sum_{\ell=1}^{\infty} b_{n\ell} \sin \ell\alpha$$

writing $\gamma = \alpha_j - \alpha_i$ and averaging over all orientations

$$\langle R_i R_j \rangle = \sum_{\ell=0}^{\infty} \left[\langle a_\ell \rangle^2 + \langle b_\ell \rangle^2 \right] \cos \ell\gamma$$

Also $\langle R \rangle = \langle a_o \rangle$

and so

$$C_{ij} = \frac{\sum\limits_{\ell=0}^{\infty} \left[\langle a_\ell \rangle^2 + \langle b_\ell \rangle^2 \right] \cos \ell\gamma - \langle a_o \rangle^2}{\sum\limits_{\ell=0}^{\infty} \left[\langle a_\ell \rangle^2 + \langle b_\ell \rangle^2 \right] - \langle a_o \rangle^2}$$

Since $\langle b_o^2 \rangle = 0$ this may be written

$$C_{ij} = \sum_{\ell=1}^{\infty} C_\ell \cos \ell\gamma$$

where, since $C_{ij} = 1$ for $\gamma = 0$,

$$\sum_{\ell=1}^{\infty} C_\ell = 1$$

It also follows that $\int_o^{2\pi} C_{ij} \, d\gamma = 0$, which is confirmed by measurement.

Electron micrographs of quartz, sand and cement particles, as in figure 1, were traced on a computer screen. The centre of mass was calculated and the distance from this point to the edge obtained every ten degrees beginning along an arbitarily chosen direction. Each particle was independently analysed and scaled such that $\langle R \rangle = 1$. Thirty particles of each type were then analysed to yield $p(R)$, $\langle R \rangle$ and $\langle R^2 \rangle$. We note that

$$R_i R_j = R(\alpha_i) R(\alpha_i + \gamma)$$

and $\langle R_i R_j \rangle$ was obtained from this product using all thirty six values of α_i for each of the thirty particles.

To compare the model to the well-known result for spheres, the Airy function, circles were also traced onto the computer screen. However, due to inaccuracy in tracing the result was a reasonable representation of a rough sphere. The functions described above were obtained in the same way.

Results for $\langle T(\rho) \rangle$ and C_{ij} are shown in figures 2 and 3. As seen in figure 1, the cement particles are much smoother than the quartz with the sand lying between them, and $\langle T(\rho) \rangle$ develops a sharper edge in passing from quartz to sand to cement. As would be expected the rough circles produce a very sharp edge.

This sequence is not apparent in C_{ij}. If anything, the cement appears to lie between the quartz and the sand. The rough circles, stand apart where C_{ij} seems to be approximating a delta-function. The results of these measurements imply that C_{ij} and $\langle T(\rho) \rangle$ are independent.

Figure 1: Tracings of electron-micrographs of a) quartz
b) sand and c) cement dust.

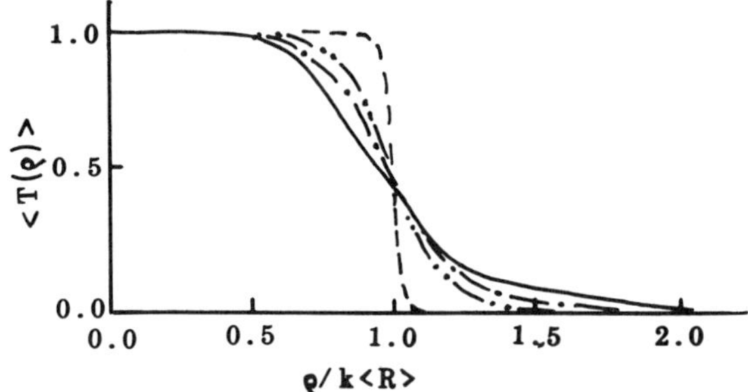

Figure 2: Transmission against radius: ——— quartz;
-.-. sand;   -..- cement; - - - circles.

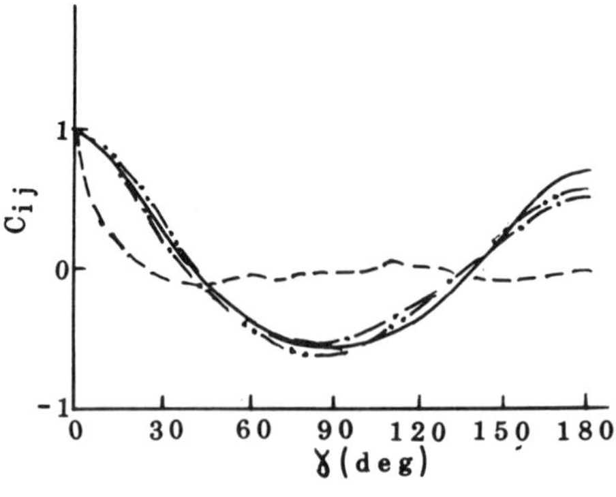

Figure 3: Correlation function against angular separation;
——— quartz; -.-. sand; -..- cement; ---- circles.

Finally the expansion coefficients C_ℓ were obtained by Fourier analysis of C_{ij}. The first twenty were calculated. This was sufficient to provide an excellent fit to the measured C_{ij}. The difference would not be discernible in figure 3.

RESULTS

Equation (7) was evaluated using gaussian quadrature. Comparisons were made of results based on 10,20 and 40-point integration. Ten point was inadequate. There were only small differences between 20 and 40-point, and in view of the computer time involved 20-point integration was used throughout.

Results for A, B and $\langle I \rangle$ are shown in figure 4. The first term (A) may be thought of as "incoherent" in the sense that it is a simple square and represents diffraction by the bulk of the particle without consideration of interaction between its various components. Similarly, B is "coherent" and takes into account this interaction via the correlation function. As the particles becomes smoother the term B becomes smaller while A increases and approaches the Airy function.

The influence of the coherent term on the diffraction pattern is clearly seen in figure 4(C). For quartz, the most irregular of the particles, the forward peak is slightly reduced but the second maximum is almost washed out. As the particles become smoother the forward peak and second maximum both increase, until for the rough spheres the curve is indistinguishable from the Airy pattern.

The question arises whether the diffraction pattern may be calculated by using p(R) as a size distribution. The result of integrating the Airey function over p(R) is included in figure 5, which is for sand It can be seen that this bears little similarity to the actual diffraion pattern.

The final question addressed was the influence of irregularity on particle sizing by Fraunhofer diffraction. To examine this, the function

$$I_D = \left[\frac{2J_1(\beta\kappa\sin\theta)}{\beta\kappa\,\sin\theta} \right]^2$$

was evaluated for various β and compared with the diffraction pattern by minimising

$$\int_0^{\theta_1} \left(\frac{I - I_D}{I}\right)^2 d\theta$$

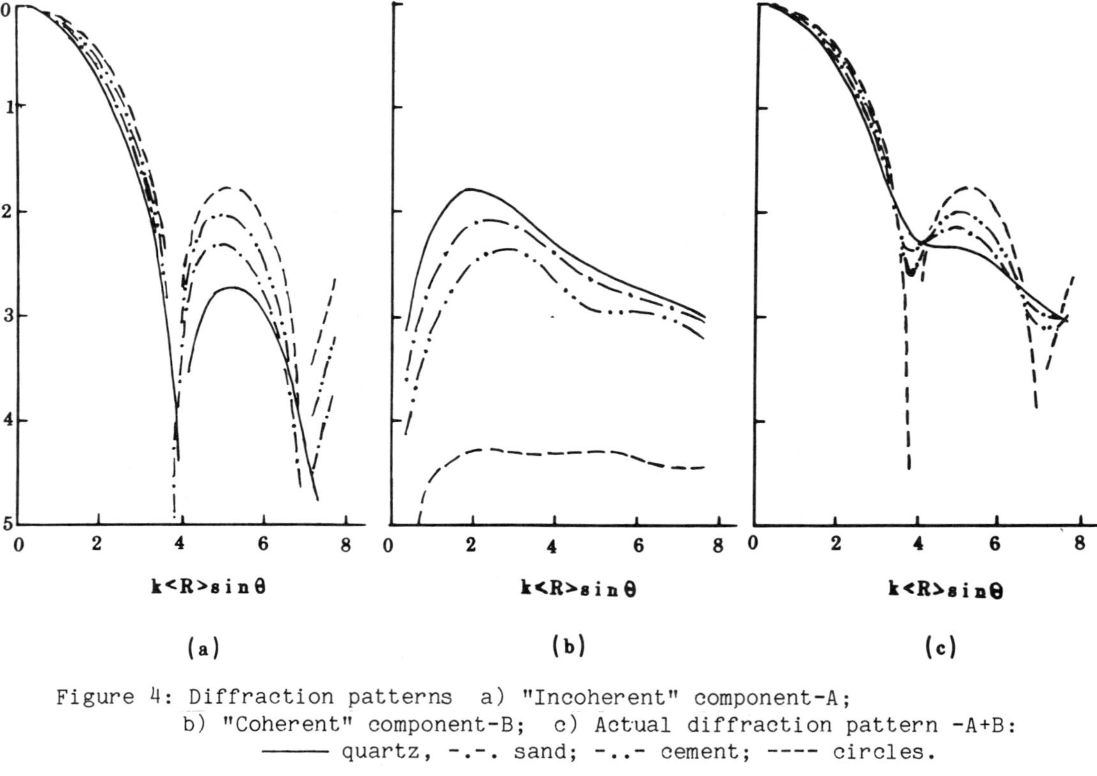

Figure 4: Diffraction patterns a) "Incoherent" component-A;
b) "Coherent" component-B; c) Actual diffraction pattern -A+B:
——— quartz, -.-. sand; -..- cement; ---- circles.

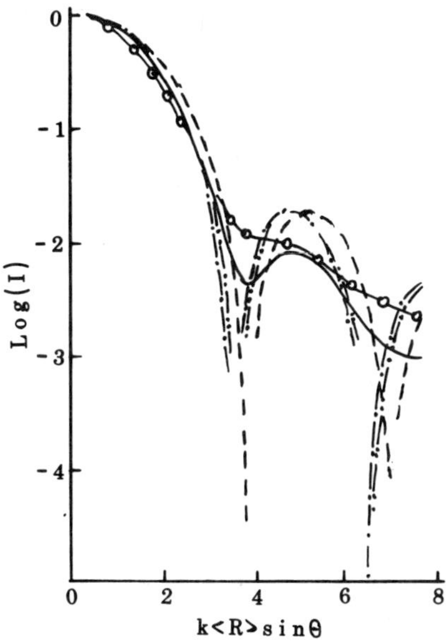

Figure 5: Comparison of diffraction pattern of sand with Airy
function: ———— sand; -.-. "Best fit" to first
minimum (β=1.077); -..- "Best fit" using all points
(β=1.068); --- Airy function (β-1);
-O- Integral of Airy function over edge distribution function.

It can be seen that performing the fit up to the first minimum of the
Airy pattern, or including all the points makes little difference. Also
over the bulk of the forward peak the fit is excellent, whereas the
second maximum is not fitted well. Thus, in all cases, the "best fit"
was obtained using angles only between zero and the first minimum. The
results are summarized in Table 1.

TABLE 1

Particle Type	β	n
Quartz	1.136	3.60
Sand	1.077	3.69
Cement	1.038	3.94
Circles	1.001	–

The particle size predicted by any individual instrument would
depend on the method of analysis. However, using this approach the
quartz would be oversized by 13.6%, the sand by 7.7% and the cement by
3.8%.

Finally, there has been discussion about whether for irregular particles a "best fit" is obtained using an equivalent area or equivalent volume sphere. This was estimated by using

$$\beta \langle R \rangle = \langle R^n \rangle^{1/n}$$

Equivalent area would yield n=2 and equivalent sphere n=3. In practice, n was found to be closer to 4. In view of the fact that close to θ=o $I\alpha R^4$ for a sphere, this is, perhaps, not surprising.

CONCLUSIONS

A scalar diffraction model has been developed for random irregular particles which relies on the probability distribution of radii and a correlation function in the surface. This has been used to estimate the influence on particle sizing instruments based on Fraunhofer diffraction. As expected, the forward peak of the diffraction pattern is not sensitive to shape, though it is estimated that particles may be oversized by up to of the order 10%.

The second maximum, however, is strongly sensitive to shape, as is the rest of the scattering polar diagram (Schuerman, 1980). Irregularity will have a serious influence on other methods of sizing (e.g. Seville et al, 1984). Further it is significant to radiative transfer and to the hydrodynamics of particles.

More general models are therefore required. Work up to 1980 is reviewed in the book by Schuerman (1980), and efforts continue on numerical calculation of well-defined shapes (e.g. Mugnai and Wiscombe, 1986). Recently there have been models for large rough spheres which combine Mie theory with geometrical optics of the surface (e.g. Bahar et al, 1986). However, these are restricted to particles which are essentially spherical in nature. At the opposite end of the scale, fractal theory and the average field approach have been applied to large open clusters of small particles such as occur in smokes (Berry and Percival, 1986). There is still considerable room for development of theoretical models applicable to irregular particles.

REFERENCES

Bahar, E., Chakrabarti, S. and Fitzwater, M.A., App. Optics, 25, 2530-2536 (1986)
Berry, M.V. and Percival, I.C., Optica Acta, 33, 577-591 (1986)
Born, M and Wolf, E., Principles of Optics, 5th Edn. (Pergamon Press, Oxford, 1975)
Mugnai, A and Wiscombe, W.J., App. Optics, 25, 1235-1244 (1986)
Schuerman, D.W. (Ed) Light Scattering by Irregularly Shaped Particles (Plenum Press, New York, 1980)
Seville, J., Coury, J., Ghadiri, M and Clift, R., 3rd Europ. Symp Particle Characterisation 305-322 (1984)

APPLICATION OF THE SHIFRIN INVERSION TO

THE MALVERN PARTICLE SIZER

L.P. Bayvel, J.C. Knight and G.N. Robertson

Department of Physics
University of Cape Town
Rondebosch 7700, South Africa

INTRODUCTION

The Malvern Particle Sizer was developed by J. Swithenbank et al.[1] in
1977, and has been extensively used since then for measuring the size
distribution of solid particles and droplets.   The reproducibility of
results obtained with the instrument is very good[2].   For many industrial
users, deviations of particle size measurements from nominal values is of
more importance than absolute results.

The instrument can be operated in several modes.   The Rosin-Rammler
and log-normal modes are well known[3].   These have the disadvantage,
however, that the functional form of the particle size distribution is
assumed a priori.

A further mode of operation is provided by the manufacturer, viz. the
"model-independent" mode, in which no prior assumption is made about the
nature of the size distribution.   Clearly this mode is to be preferred for
most purposes.   Unfortunately, neither the developers nor the manufacturers
of the instrument have revealed the algorithm which is used to obtain the
particle size distribution from the light scattering data.   A number of
people who have attempted to verify the performance of the instrument using
the model-independent programme have encountered discrepancies that cannot
be explained (or even properly investigated) without knowing the details of
the Malvern inversion algorithm[4-8].   Furthermore, the dependence of the
"model-independent" results on beam obscuration has not yet been
investigated systematically.   Published results[2,9] on the effect of varying
beam obscuration have all been obtained with the Rosin-Rammler programme;
and the manufacturer's recommendations for the use of the instrument, that
the beam obscuration should not exceed 50% in order to avoid multiple
scattering, are based on these results.

SHIFRIN'S INVERSION ALGORITHM

The aim of this paper is to present an alternative to the Malvern
"model-independent programme".   Our alternative programme assumes the
validity of the Fraunhofer diffraction approximation for the scattering of
light by small spherical particles, and neglects multiple scattering, but
makes no other assumptions at all.   The particle size distribution is
deduced from the light scattering data using an inversion technique due to
K.S. Shifrin.

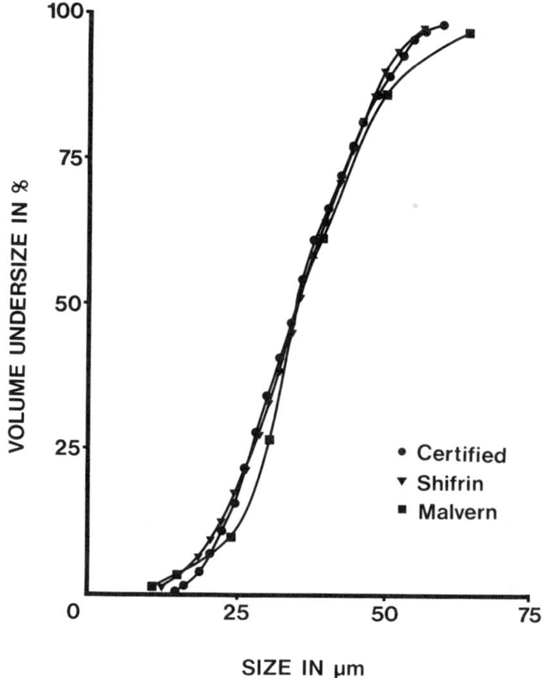

Fig. 1: Particle size distribution of the reference material 1003a (obscuration 20%).

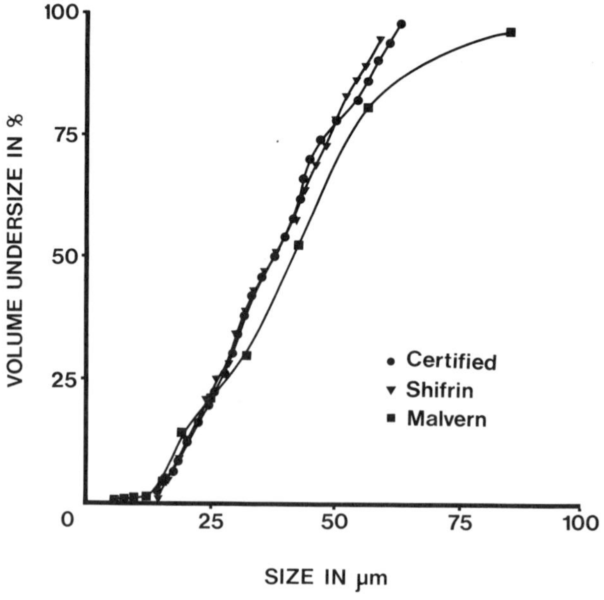

Fig. 2: Particle size distribution of the reference material 1004 (obscuration 20%).

Shifrin[10] shows that the differential scattering cross-section for a
spherical particle is given approximately by:

$$\left(\frac{d\sigma}{d\Omega}\right) \;=\; \left(\frac{a^2}{\theta^2}\right) J_1{}^2(2\pi a\theta/\lambda) \tag{1}$$

at small scattering angles θ, where a is the radius of the particle, λ is
the wavelength of the scattered light, and J_1 is a Bessel function of the
first kind.   It follows that the scattered intensity per unit solid angle
$I(\theta)$ from a particulate system with a particle size distribution $f(a)$ is:

$$I(\theta) \;=\; \frac{I_0}{\theta^2} \int_0^\infty J_1{}^2(x\theta)\, a^2\, f(a)\, da \tag{2}$$

where I_0 is the incident beam intensity and $x = 2\pi a/\lambda$.   Shifrin's inversion
technique[11] provides an exact and unique analytical solution of equation
(2), expressing $f(a)$ as a generalised Fourier-Bessel transform of $I(\theta)$:

$$f(a) \;=\; -\frac{\pi^2}{\lambda a^2} \int_0^\infty A(x\theta)\, \psi(\theta)\, d\theta \;, \tag{3}$$

where

$$A(x\theta) \;=\; x\theta\, J_1(x\theta)\, Y_1(x\theta) \;,$$

$$\psi(\theta) \;=\; \frac{d}{d\theta}\left[\theta^3 I(\theta)/I_0\right] \;,$$

and Y_1 is a Bessel function of the second kind.   From this it is easy to
derive the volume distribution $v(a)$, cumulative distribution $G(a)$, and
various means of the distribution functions (volume mean diameter, Sauter
mean diameter, etc.) without making any a priori assumptions about the form
of the distribution.   Furthermore, as has been shown by many authors[11-17],
the solution is very stable, since an integral transform is used, and
iterative linear algebra is avoided.   This is an important advantage in
adverse signal-to-noise conditions.

EXPERIMENTAL

The Malvern 2600 was checked prior to use by means of a reticule
developed by Laser Electro Optics Ltd.[18].   The Rosin-Rammler parameters $\overline{X}$
and N obtained with our instrument were within 2% of the certified values,
indicating that the factory calibration of the annular photodiode array
(to compensate for sensitivity variations) was entirely correct.

Experiments were performed using the following reference materials:

(1)   Calibrated glass beads (SRM 1003a and SRM 1004) supplied by the
      National Bureau of Standards (Washington, D.C., U.S.A.).

(2)   Particles of randomly shaped natural quartz (BCR 67 and 69) supplied
      by the EEC Bureau of Reference (Brussels, Belgium).

The glass beads are spherical although some (6% of 1004 and 1% of 1003a)
ranged from nearly spherical to ellipsoidal and fused.   The samples were
certified by microscopy.   The quartz particles are irregular in shape and
were certified by sedimentation analysis.

REFERENCE MATERIAL

NBS 1003a was supplied with number and volume cumulative undersize
percentages in the range 4-64 microns in 2 μm steps while NBS 1004 had a
volume percent undersize in the range 28-126 μm.   BCR 67 had a volume
percent undersize in nine size bands in the range of Stokes diameter

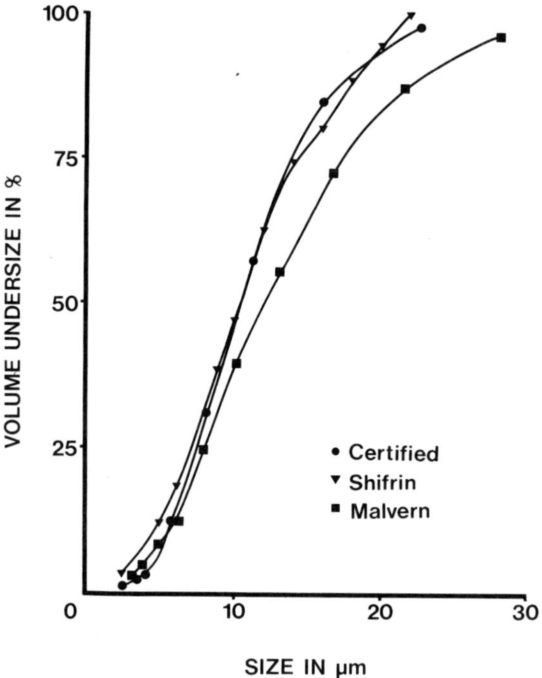

Fig. 3:   Particle size distribution of the reference material BCR 67 (obscuration 20%).

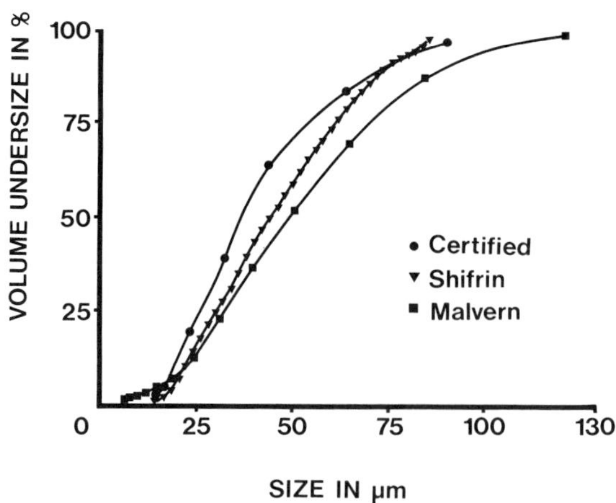

Fig. 4:   Particle size distribution of the reference material BCR 69 (obscuration 20%).

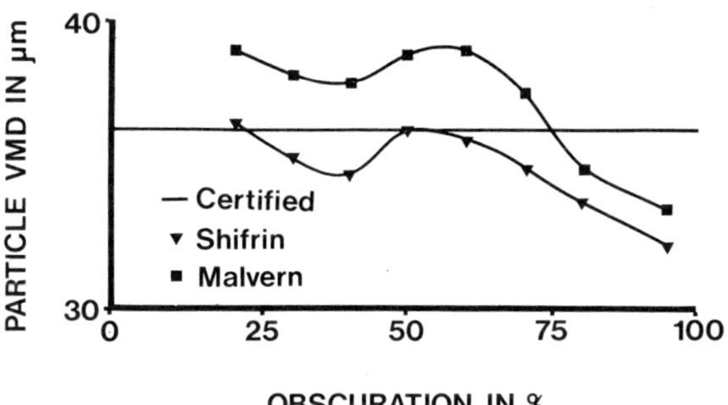

Fig. 5: Effect of obscuration on measured volume mean diameter of the reference material 1003a.

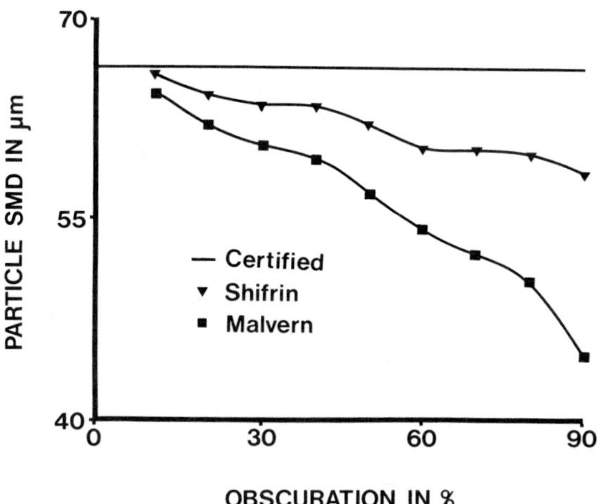

Fig. 6: Effect of obscuration on measured Sauter mean diameter of the reference material NBS 1004.

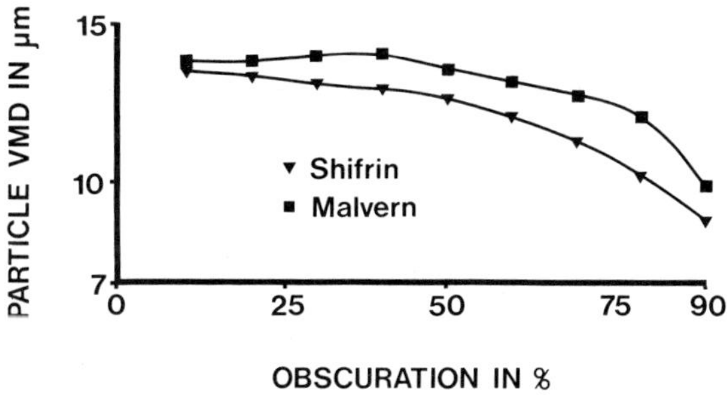

Fig. 7: Effect of obscuration on measured volume mean diameter of the reference material BCR 67.

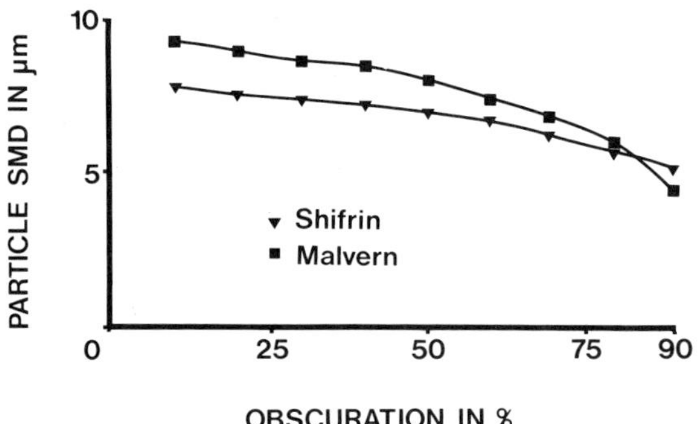

Fig. 8:   Effect of obscuration on measured Sauter mean diameter of the reference material BCR 67.

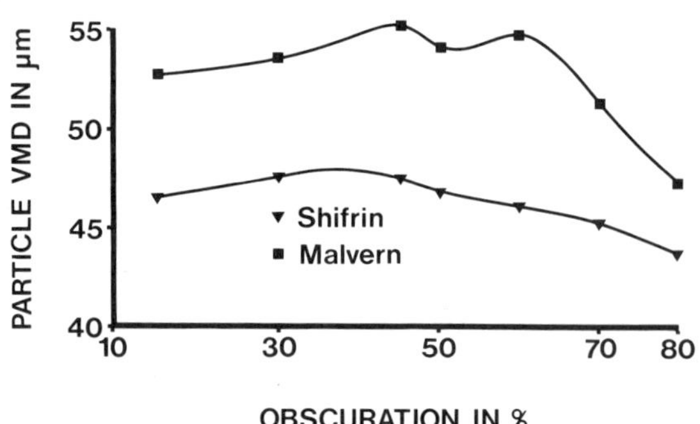

Fig. 9:   Effect of obscuration on measured volume mean diameter of the reference material BCR 69.

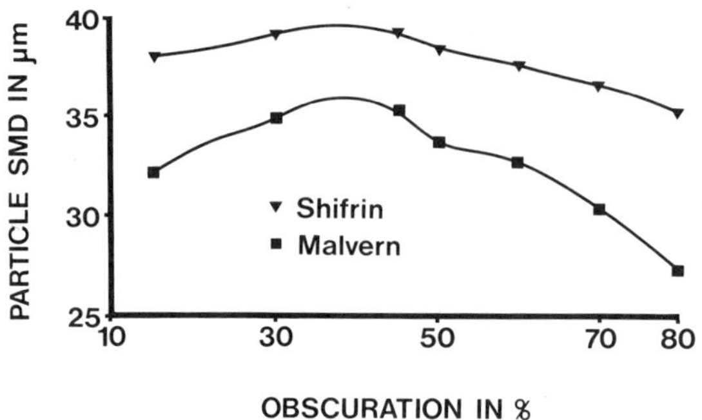

Fig. 10: Effect of obscuration on measured Sauter mean diameter of the reference material BCR 69.

2.4-32 μm, and BCR the same in seven size bands from 14 to 90 microns.

Following the recommendations of D. Watson[2], the reference material was thoroughly mixed, and sampled with a spatula. The dry material was added to clean deionised water so as to make an opaque solution. This solution was then agitated for a minute in an ultrasonic bath and a sample taken using a syringe. The sample cell, filled with clean deionised water, was placed in the sample volume region and a background reading taken, using 400 sweeps across the detectors. The sample solution was then slowly added to the sample cell, to the desired concentration. In the cell the sample was stirred with a magnetic stirrer.

The light intensity measurements were again made using 400 sweeps across the 30 detectors, and the previously recorded background scattering was subtracted. The corrected data were stored on floppy disc, and analysed both using the Malvern model-independent programme and using Shifrin's inversion method.

In order to apply Shifrin's method the data captured by the Malvern instrument were processed as follows. The areas of the annular detector elements in the Malvern particle sizer increase by about three orders of magnitude from the centre to the periphery, in order to compensate for the very rapid decrease in the intensity of the Fraunhofer diffraction pattern with angle, and thus to obviate any problems arising from the limited dynamic range of the photodiodes. The signal from each annular photodiode was therefore divided by its area in order to obtain the scattered light intensity at the angle subtended by the diode. The areas of the detector elements are given in the paper by Dodge[19]. The scattering angle θ_i is equal to r_i/f, where r_i is the average of the inner and outer radii of detector element number i and f is the focal length of the transform lens used. The resulting normalised intensity function $I(\theta_i)$, discretised at 30 points (determined by the geometry of the detector and the focal length of the transform lens used), was processed using Shifrin's algorithm and the particle size distribution functions f(a) and v(a) were obtained.

Measurements were made over a range of concentrations corresponding to obscuration values of 10% to 90%.

RESULTS

The graphs of the volume undersize distribution measured for all four reference materials at a concentration corresponding to 20% beam obscuration (which is recommended as ideal by the manufacturers[20]) are presented in figures 1 to 4. Three graphs are plotted on each figure: (i) the volume undersize (%) measured by the Malvern instrument using the model-independent programme; (ii) the volume undersize (%) calculated from the same light scattering data using Shifrin's inversion; (iii) the volume undersize distribution as certified by the Standards Bureaux.

It can be seen that the size distributions obtained using Shifrin's inversion are closer to the distributions certified by the NBS and the EEC BR than those obtained using the Malvern model-independent programme.

The effect on the derived particle size parameters of varying the obscuration is presented in figures 5 to 10. We choose to study the volume mean diameter and the Sauter mean diameter, since these are provided in the printout of the Malvern model-independent programme. Furthermore, the VMD of the NBS 1003a reference material has been determined by a large number of independent methods, including optical microscopy, air sedimentation, sedigraphy, the Coulter Counter, the Hiac-Royco counter, and the Microtrac. All these methods give values of 36 ± 1.3 μm for the VMD.

As can be seen from the curves, the values of VMD and SMD obtained
using Shifrin's inversion correspond more closely to the reference data
than do those obtained using the Malvern model-independent programme.

The particle size distribution curves and the various means derived
from them vary with beam obscuration for both inversion methods.
However, in most cases the effect of varying the obscuration is smaller
for Shifrin's inversion than for the Malvern model-independent programme.
In some cases the means of the distribution functions derived using
Shifrin's method change by less than 10% as one increases the obscuration
from low values up to 80% or even 90%. This implies that the generally
accepted criterion for the onset of significant multiple scattering, viz.
an obscuration of 50% or more, may need revision.

Similar trends are observed with non-spherical particles.

Finally, the Shifrin inversion allows one to calculate the particle
size distribution function for any desired increment in the argument.
The user is not restricted to a predetermined set of size bands (or
obliged to interpolate between them).

CONCLUSIONS

We conclude that the use of the Shifrin inversion algorithm, which
is described fully in the open literature, has a number of significant
advantages over Malvern's proprietary software in the context of optical
particle sizing research. It provides the research worker with a simple
and mathematically elegant alternative method of processing the light
scattering data, which is particularly suitable for investigating (and
extending) the limits of applicability of the Fraunhofer diffraction
particle-sizing technique.

REFERENCES

1.   J. Swithenbank et al., Prog. Astron. Aeron., 53: 421 (1977).
2.   D.J. Watson, Report M.A.R. 101, Malvern Instruments, 1982.
3.   P.G. Felton, In-stream measurement of particle size distribution.
       Presented at the Int. Symp. on in-stream measurements of
       particle solid properties. Bergen, Norway, 1978.
4.   G. Butters and A.L. Wheatley. In N. Stanley-Wood and T. Allen, Eds.,
       Particle Size Analysis, Wiley-Heyden, 425-436, Chichester, U.K.
       (1982).
5.   J. Seville et al., Particle Characterisation, 1: 45 (1984).
6.   A. Bürkholz and R. Polke, Particle Characterisation, 1: 153 (1984).
7.   D. Allano, P. Lisiecki and M. Ledoux, Report No. 488201 SAT 2/CT,
       ONERA/CORIA, UA CNRS No. 230, Rouen, France (1984).
8.   U. Tüzün and A. Farhappour, Particle Characterisation, 2: 104 (1985).
9.   L.G. Dodge, Optical Engineering, 23: 626 (1984).
10. K.S. Shifrin, Izvestiya USSR Academy of Sciences, Ser. Geography,
       14: 62 (1950).
11. L.P. Bayvel and A.R. Jones, "Electromagnetic scattering and its
       applications", Elsevier Applied Science Publishers, London and
       New York (1981).
12. J.B. Abiss, Theoretical aspects of the determination of particle size
       distribution from measurements of scattered light intensity.
       Technical Report 70151, Royal Aircraft Establishment,
       Farnborough, England (1970).
13. A.L. Fymat, Applied Optics, 17: 1677 (1978).
14. A.L. Fymat and K.D. Mease, Reconstructing the size distribution of
       spherical particles from angular forward scattering data. In:
       "Remote sensing of the Atmosphere: Inversion methods and
       applications", A.L. Fymat and V.E. Zuev, Eds., Elsevier,
       Amsterdam (1978).

15. A.L. Fymat and K.D. Mease, Applied Optics, 20: 194 (1981).
16. L.V. Ruscello and E.D. Hirleman, Determining droplet size
     distributions of sprays with a photodiode array.   Paper No.
     WWS/CI-81-49, Meeting of the Western States Section of the
     Combustion Institute, Tempe, Arizona, USA (1981).
17. L.P. Bayvel, A.R. Jones and P. Eisenklam, A light scattering
     instrument for measuring drop sizes.   Proceedings of the 2nd
     International Conference on Liquid Atomisation and Spray Systems,
     Madison, Wisconsin, USA, 329-334 (1982).
18. Laser Electro-optics Ltd., Tempe, Arizona, USA.
19. L.G. Dodge, Applied Optics, 23: 2415 (1984).
20. 2600 Particle Sizer User Manual, Malvern Instruments, Malvern,
     England.

MEASUREMENT OF SIZE DISTRIBUTION IN DENSE PARTICLE FIELDS

A.A. Hamidi and J. Swithenbank

Department of Chemical Engineering and Fuel Technology
University of Sheffield
Sheffield, S1 3JD, England

The problem of multiple scattering of light by dense particle
fields in a light scattering measurement technique is resolved. A
mathematical model for this phenomenon is developed which predicts the
light energy distribution produced by any given size distribution at any
obscuration. The mathematical analysis is carried out for Fraunhofer
and the anomalous diffraction theories. A systematic procedure is
established to correct for the effect of multiple scattering.
Excellent agreement with experimental results is obtained. The study
therefore enables the application of the measurement technique to very
dense particle fields.

1. - Introduction

A large number of chemical and physical processes of industrial
importance involve two-phase flows. The need for a better understanding of
these processes has led to the study of the interactions of the two phases
with each other. The two phases may be gas/liquid as in atomization, gas/
solid as in pulverized fuel furnaces or solid/liquid as in crystallization.
The particle or drop size distribution can have an important effect on
physical and/or chemical interactions taking place in the process and consequently many intrusive and non-intrusive techniques have been developed to
measure the size distribution of the dispersed phase [1,2]. Optical techniques employing lasers are amongst the most reliable techniques available
and one such technique is based on the measurement of the distribution of
light energy scattered at low angles in the forward direction [3-5].

These techniques invariably employ the mathematical analysis for
single scattering of the light by spherical particles which limits their
applicability to particle fields where the particle concentration is
sufficiently low so that single scattering phenomenon is by far the dominant
optical process. However, most practical two-phase flow processes involve
high particle densities and considerable multiple scattering takes place.
The presence of this phenomenon and the inherent limitations have been
pointed out by various research workers [6,7]. It has been concluded that
reliable results with this technique can be obtained provided that no more
than 50% of the incident laser light is scattered by the particles. Below
50% the effect of multiple scattering can be ignored, while for obscuration
greater the 50% the measured size distribution will appear broader and have

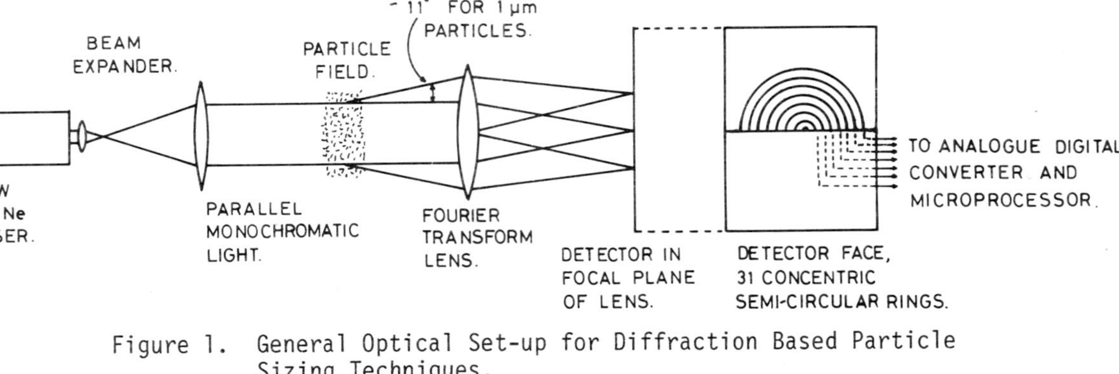

Figure 1. General Optical Set-up for Diffraction Based Particle Sizing Techniques.

a smaller mean diameter than the actual size distribution of the particle field.

A theoretical model is presented for the effect of multiple scattering which permits the application of the technique to particle fields where up to 98% of the incident light may be scattered. This corresponds to approximately a six-fold increase in the number concentration of the particle field in which the technique can be applied.

An experimental investigation has been carried out using spherical glass beads of known size distribution and the results have been compared with the findings of the theoretical model. For distributions which can be described by simple two-parameter size distribution models such as the Rosin-Rammler function, correction equations have been derived to compensate for the effect of multiple scattering [8]. A general correction procedure is described in this paper, which can be applied to any type of size distribution including the so-called 'model independent' distribution function used to represent multi-modal size distributions.

The analysis has been carried out for both Fraunhofer and the anomalous diffraction theories. The second theory is of greater relevance to processes such as sugar crystallization [9] and liquid/liquid emulsions, where the two phases have refractive indices which are close to each other.

2. - Theory

When a spherical particle is illuminated by a parallel beam of monochromatic, coherent light, a diffraction pattern is formed superimposed on the geometrical image. If the transmitted light falls on a Fourier transform lens, then the undiffracted light is focussed at a point on the axis in the focal plane and the diffracted light forms a far field pattern of rings around the central spot. This is the underlying principle of size measurement techniques which are based on the measurement of the small angle scattered light energy distribution. The optical components of the technique are shown in Figure (1).

2.1 Fraunhofer diffraction theory

The theory of light scattering by a single spherical particle has been discussed by Van de Hulst [10]. The application of Fraunhofer diffraction theory to particle sizing has been outlined by Swithenbank et al [5] who showed that for a collection of particles of different sizes which may be either stationary or moving, the light energy focussed on a ring bounded by radii s_1 and s_2 in the focal plane of the lens is the sum of the contributions from the individual particles:

$$E_{s_1,s_2} = C \sum_{i=1}^{m} \frac{W_i}{a_i} \left[J_0^2 \left(\frac{Ka_i s_1}{f} \right) + J_1^2 \left(\frac{Ka_i s_1}{f} \right) \right.$$

$$\left. - J_0^2 \left(\frac{Ka_i s_2}{f} \right) - J_1^2 \left(\frac{Ka_i s_2}{f} \right) \right] \qquad \ldots\ldots (1)$$

Here W_i is the weight fraction of the particles of radius a_i, C is a constant which depends on the laser power and $K=2\pi/\lambda$ is the wave number. If a detector array is used which consists of a set of semi-circular rings, then each of these rings will be associated with a characteristic particle size range, depending on the focal length of the collection lens.

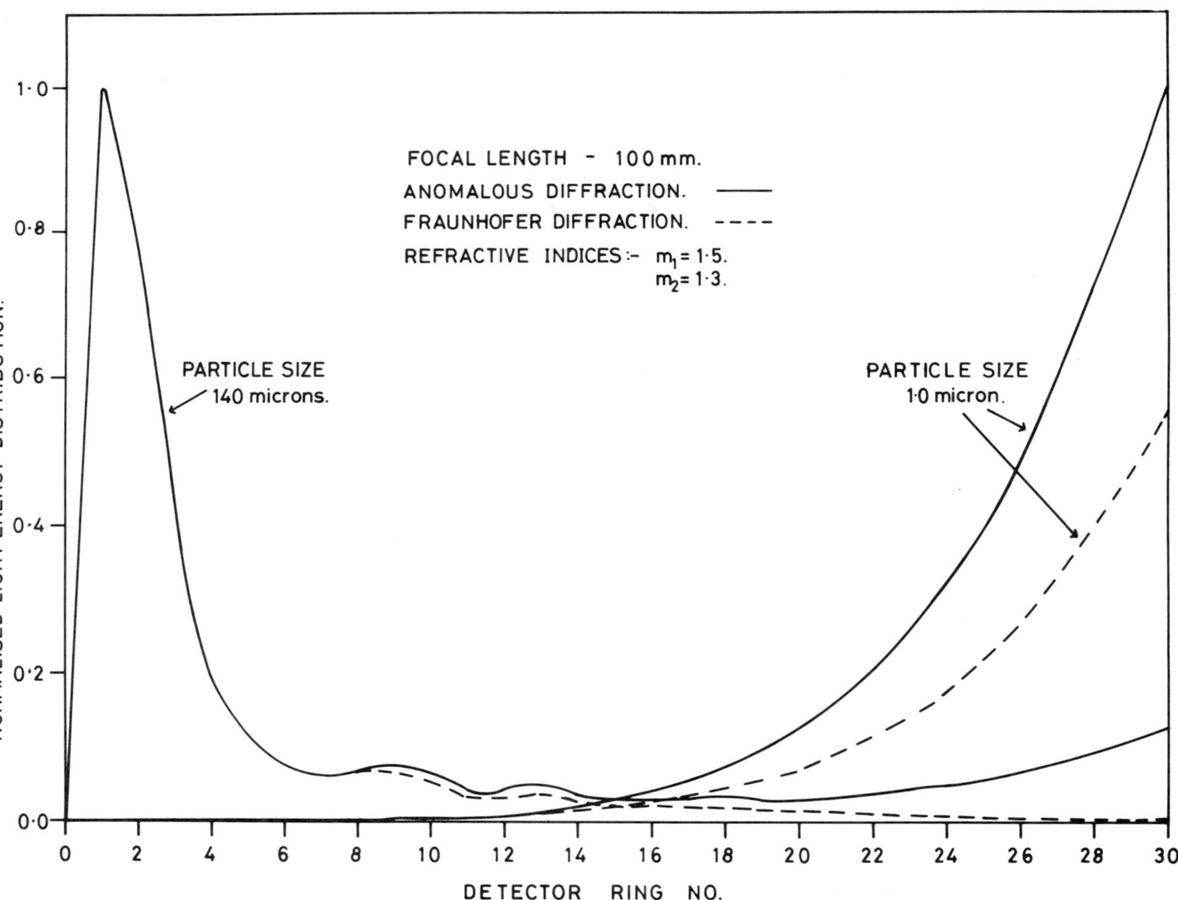

Figure 2.   Fraunhofer and Anomalous Diffraction Theories.

The total light energy distribution is the sum of the products of the energy distributions for each size range and the weight fraction in that size range. This can be expressed as:

$$E(I) = W(J).T(I,J) \qquad \dots \dots (2)$$

Here $T(I,J)$ is a two dimensional matrix, the elements of which define the scattered light energy distribution for each size class. These coefficients are calculated using equation (1) for different combinations of a_i, s_1 and s_2.

Obscuration of the light is the fraction of the incident light scattered by the particles. This can be shown to be:

$$\text{Obscuration} = 1 - \frac{I}{I_o} = 1 - \text{Exp}(-\tau\ell) \qquad \dots \dots (3)$$

where ℓ is the optical path through the particle field and τ is given by:

$$\tau = Q_s \sum_{i=1}^{m} N_i A_i \qquad \dots \dots (4)$$

For Fraunhofer diffraction it can be shown that the scattering efficiency Q_s is equal to two.

In deducing the size distribution from the measured light energy distribution, several size distribution models have been used to describe $W(J)$. For many size distributions with a single peak, two-parameter distribution functions such as the Rosin-Rammler or the Log-Normal distributions are sufficiently accurate to define the size distribution. For others, however, where the size distribution is either multi-model or near mono-size, the so-called 'model independent' option has been employed. The size distribution $W(J)$, is calculated iteratively such that:

$$\sum_{I=1}^{m} (E_{\text{measured}}(I) - W(J).T(I,J))^2 = \text{Minimum} \qquad \dots \dots (5)$$

2.2 Anomalous diffraction theory

Fraunhofer diffraction theory is applicable only if for the smallest size of the particle measurable using the technique, Q_s is equal to two. Q_s is dependent on the refractive index difference between the two phases. The smaller this difference, the larger is the size of the particle for which $Q_s = 2$. For a non-absorbing particle, Q_s can be computed from the following relationship:

$$Q_s = 2 - \frac{4}{\rho} \sin\rho + \frac{4}{\rho^2} (1 - \text{Cos } \rho) \qquad \dots \dots (6)$$

where ρ is the phase shift which is given by:

$$\rho = 2Ka|m_1 - m_2| \qquad \dots \dots (7)$$

Clearly, then, although the measurement technique is capable of measuring the scattered light energy distribution regardless of the refractive index difference, Fraunhofer diffraction theory cannot be applied to the mathematical analysis of the measured light energy distribution data if the sample contains a large fraction of particles for which Q_s is not equal to two.

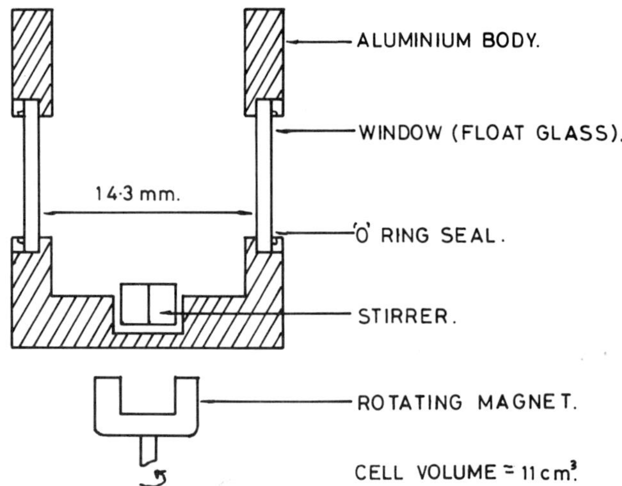

Figure 3.   Magnetically Stirred Cell used for Solid Particle Size Determination.

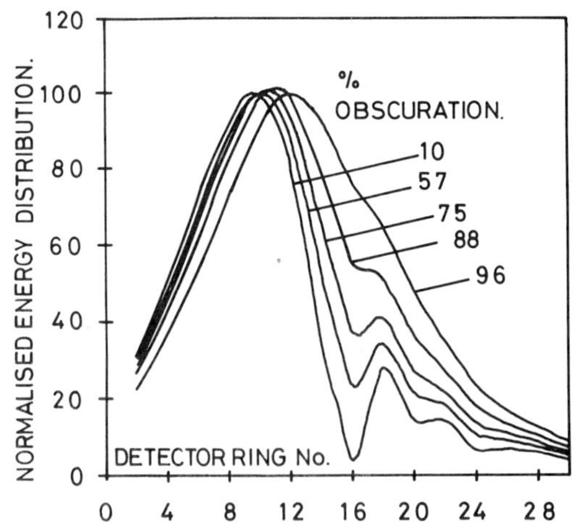

Figure 4.   Light Energy Distribution for a Monosize, (D=100 micron), Particle Field.

The theory of anomalous diffraction is discussed in great detail by
Van de Hulst. As a result of the mathematical analysis for anomalous
diffraction, a number of equations are derived which are analogous to
equation (1) for Fraunhofer diffraction. These equations can be used to
compute the values of the elements of the matrix T(I,J). Apart from this
operation, the rest of the mathematical analysis is identical to that
described for Fraunhofer diffraction.

Figure (2) illustrates the light energy distribution for two different
sizes of particles. Clearly for a particle of diameter equal to 140 microns
both theories produce the same data since Q_s is equal to two for both cases.
However, for the smaller particle size, 1 micron, there is a large difference present between Fraunhofer and anomalous diffraction.

2.3 The multiple scattering model

The multiple scattering model which has been developed is basically a
ray-tracing exercise. The particle field is divided into slices such that
each slice scatters the same fraction of the light incident to it; this
fraction is here taken to be 0.1. However, it can be shown that the fraction
can be as high as 0.5. The division of the particle field into slices each
of which scatters a small fraction of the incident light implies that within
each slice single scattering is the dominant phenomenon. It is further
assumed that the particles are non-absorbing and also that the unscattered
light and the scattered light leaving one slice become the incident light
for the next slice. For the first slice, therefore, the only incident light
is that of the parallel laser beam and the obscuration after the first slice
is 10%. Incident to the second slice, however, is the remainder of the
original beam ($0.9I_o$) plus the full angular spectrum of the scattered light
leaving the first slice. The obscuration after the second slice is 19%. It
is evident that by using a number of these slices in series, high obscuration can be achieved and the effect of multiple scattering can be simulated.

Clearly, the scattered light energy distribution leaving each slice
(N>1) is equal to the sum of the primary scattered light formed within this
slice as a result of the interaction of the original parallel beam with the
particles, plus the secondary scattered light component formed as a result of
the interaction of the scattered light leaving the previous slice with the
particles in this slice. An integration procedure has been developed which
enables the calculation of the secondary scattered light energy component.

The integration procedure involves consideration of each size band (K).
The primary scattered light energy distribution for each size band is
calculated from equation (1) for Fraunhofer diffraction theory or the analogous set of equations for the anomalous diffraction theory. The angular
distribution of this primary light energy distribution is divided into small
increments such that they correspond to the rings on the detector array used,
once the light energy contained within this angular increment has been
Fourier-transformed by the lens. The light contained within this increment
may then be considered as an incident beam of light offset from the optical
axis.

Therefore, part of the secondary scattered light energy distribution
which is due to this increment of the primary scattered light can be
computed using equation (1). If the integration is carried out throughout
the incident scattered light spectrum, the contribution to multiple scattering from size band K can be determined. By repeating this procedure for all
the size bands, a three dimensional matrix C(I,J,K) can be constructed. An
element of this matrix defines the fraction of the scattered light leaving

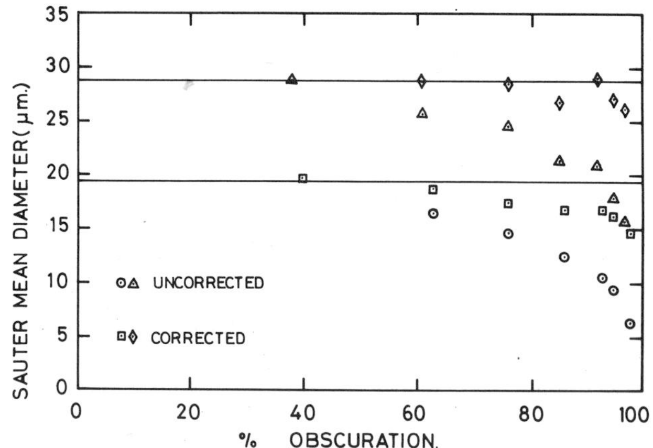

Figure 5.   Dodges Results (12). With and Without Correction for Multiple Scattering.

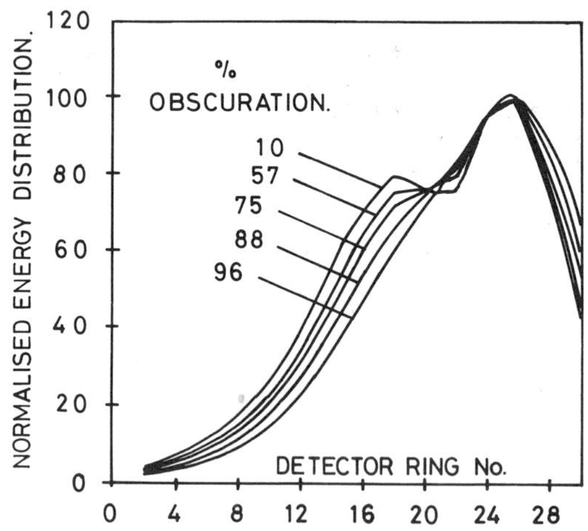

Figure 6.   Theoretical Light Energy Distribution for a Bimodal Distribution. (See Fig.7).

slice N which was destined to be focussed on the detector ring J but will be
focussed on ring I owing to the light being scattered by the particles in
size band K. If F is fraction of light scattered at each slice, in this
case F = 0.1, then for the case where I=J, C(I,J,K) = (1-F) irrespective of
the value of K. This simply follows from the condition that only the
fraction F of the light travelling towards ring I is multiply-scattered and
distributed among other rings J, (J≠I). Therefore the remainder of the
light, (1-F) goes on to be focussed on ring I.

The light energy distribution leaving the first slice, where only
single scattering of the original beam takes place, is therefore given by:

$$E(1,I) = \sum_{J=1}^{m} W(J).T(I,J).F \qquad \ldots \ldots (8)$$

The light energy distribution leaving the second slice, where multiple
scattering of the scattered light leaving the first slice takes place in
addition to the further primary scattering of the remainder of the original
beam leaving the first slice, is given by:

$$E(2,I)=I_o(1).E(1,I)+F\sum_{J=1}^{m} E(1,J) \sum_{K=1}^{m} W(K).C(I,J,K) \qquad \ldots \ldots (9)$$

In general, the light energy distribution leaving the Nth slice is given by:

$$E(N,I)=I_o(N-1).E(1,I)+F\sum_{J=1}^{m} E(N-1,J) \sum_{K=1}^{m} W(K).C(I,J,K) \ldots \ldots (10)$$

The first term on the left-hand side of equations (9) and (10) describes the
primary scattering of the incident light by the particles, while the second
term describes the multiple-scattering component. $I_o(N)$, which is the
fraction of the intensity of the original incident beam leaving slice N, is
given by:

$$I_o(N) = (1 - F)^N \qquad \ldots \ldots (11)$$

Therefore, in order to calculate the size distribution for dense
particle fields, where obscuration is greater than 50%, instead of solving
equation (5), the following equation should be solved:

$$\sum_{I=1}^{m} (E_{measured}(I) - E_{calculated}(N,I))^2 = \text{Minimum} \qquad \ldots \ldots (12)$$

The number of slices N is determined by the measured obscuration and the
value of F chosen in the analysis.

3. – Experimental study

The experimental work was carried out using suspensions of glass beads
(refractive index = 1.533) of known size distribution in suitable liquids.
Water (refractive index = 1.333) was used when the Fraunhofer diffraction
theory was employed in the data analysis. Mixtures of propan-2-ol
(refractive index = 1.378) and 1-methlynaphthalene (refractive index =
1.617) were used instead of water so that the refractive index of the continuous phase could be altered to study the effect of multiple scattering

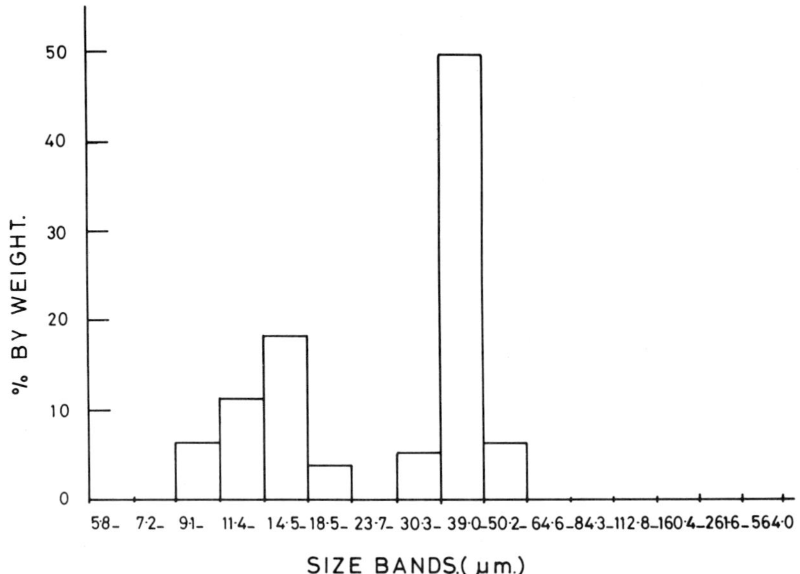

Figure 7. Measured Bimodal Size Distribution at 25% Obscuration.

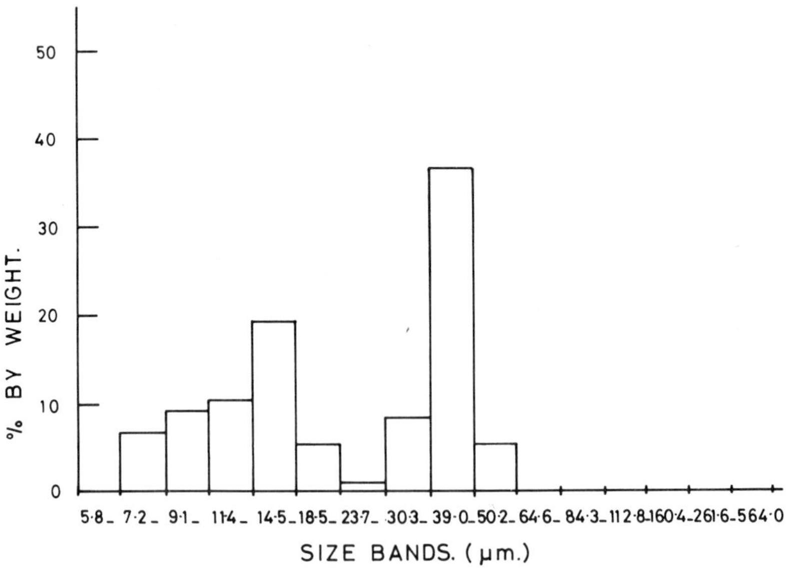

Figure 8. Apparent Size Distribution at 87% Obscuration, the Effect of Multiple Scattering is Ignored.

when anomalous diffraction was present also [9]. Figure (3) shows the magnetically-stirred cell used in these experiments.

The glass beads used were NBS standard reference materials 1003 and 1004 with size ranges from 5 to 30 micron and 37 to 105 micron, respectively. These two samples each gave size distributions with a single peak and could be defined by both the Rosin-Rammler and the Log-Normal distributions with a reasonable degree of accuracy. Latex spheres E19 and G7 were used as near monosize samples (13.8 and 40.8 micron respectively). By using a mixture of these two samples, a bimodal distribution could be produced and investigated.

The experimental procedure was the same regardless of which sample was under investigation. A very dense suspension of the sample was made. Readings of the light energy distribution were taken and the sample was diluted by removing 1 ml and replacing it by 1 ml of fresh liquid before another reading was taken. This procedure was repeated until the sample obscuration fell below 20%. The measured light energy distribution at any obscuration was used to calculate the apparent (Eq. 5) and the actual (Eq. 12) size distributions.

4. - Results

In order to illustrate the effect of multiple scattering on the light energy distribution for scattering from a monosized particle field, a mathematical simulation was carried out for particles of diameter 100 micron. Figure (4) shows the scattered light energy distribution at several sample obscurations. It is clear that for obscuration less than 60% the curves are similar in shape, which indicates that the effect of multiple scattering is negligible. However, for an obscuration greater than 60%, the scattered light energy distribution changes drastically. The secondary maxima and minima of the distribution disappear, resulting in broadening of the energy distribution and hence the apparent size distribution is no longer monosized. The peak of the energy distribution moves towards the larger scattering angles indicating an apparent reduction in the mean size. Similar light energy distribution curves can be calculated for any particle field of a given size distribution and refractive index difference.

In the earlier publications [8, 11], data was presented for a Rosin-Rammler size distribution and as a result of a parametric study it was clearly illustrated that the effect of multiple scattering is not only a function of the sample concentration but also of the actual size distribution of the particle field. Based on this study a number of correction equations for the measured mean size and the distribution width of the Rosin-Rammler and the Log-Normal distribution were derived.

The findings of the current investigation have been used to correct a set of data obtained by Dodge [12] who carried out an experimental study using an array of seven identical nozzles to simulate the effect of multiple scattering. This is shown in figure (5). Clearly the corrected values are much closer to the nominal mean diameter of the individual nozzles. Two recent studies, one employing a Monte Carlo simulation method [13] and the second a ray tracing procedure [14] similar to that developed by the present study, have shown very good agreement with the findings of the current investigation. This indicates that the general correction procedure developed in this investigation can be applied to dense particle fields.

In order to illustrate the application of the general correction procedure, a case study for a bimodal size distribution is presented here. Figure (6) shows the theoretical light energy distribution at several obscurations for a bimodal size distribution prepared by making a mixture of two near monosize Latex sphere samples. The actual size distribution

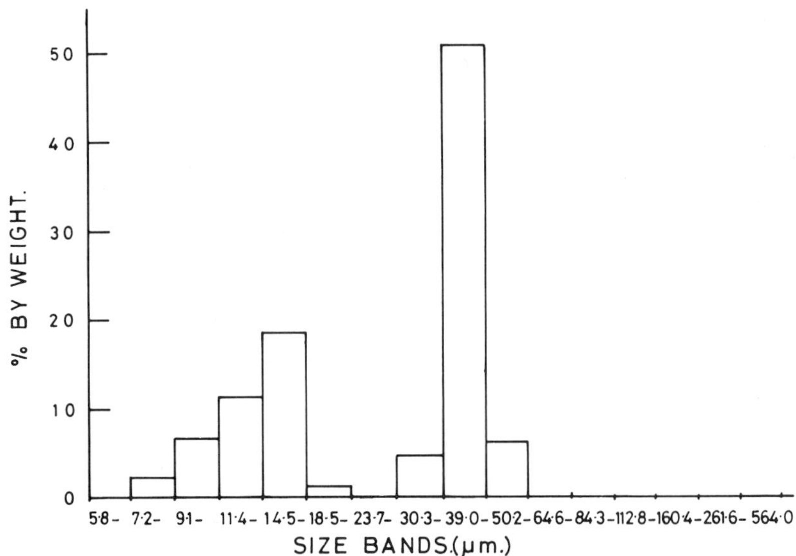

Figure 9.   Actual Size Distribution at 87% Obscuration,
Correction for Multiple Scattering is made.

measured at 25% obscuration is shown in Figure (7). From the light energy
distribution shown in Figure (6), the effect of multiple scattering is
clearly indicated. There is an increase in the width of the energy distribution and also the disappearance of the first peak of the distribution
which corresponds to the particles of larger size. Figures (8) and (9) show
the size distribution of the sample at 87% obscuration. Figure (8) is the
apparent size distribution produced when the measured light energy distribution is processed using equation (5) which ignores the effect of multiple
scattering. Comparison of Figure (8) and (7) illustrates the apparent
reduction in the mean size of the sample and the increase in the width of
the size distribution. The size distribution presented in Figure (9) on the
other hand, shows the calculated size distribution calculated when equation
(12) is used for the data analysis. Clearly this distribution now compares
favourably with the actual size distribution of the bimodal sample under
consideration. This analysis has been carried out for a number of different
size distributions and a number of particle fields with variation in the
refractive index difference. In general good agreement has been obtained.

Finally, three points are worth noting with regard to this study and
model. Firstly, the authors have often been asked if the division of the
particle field into slices of equal obscuration, each having a size distribution the same as the overall particle field, means that the particle field
itself must be homogeneous. This would clearly rule out the application of
the model to such problems as the analysis of sprays. A feature of
diffraction-based particle sizing is that the position of the particles has
no effect on the diffraction pattern provided that all the particles are
situated within the focal length of the collection lens. Therefore, the
question of the particle field homogeneity does not arise since this argument is also applicable to the multiple scattering model. Secondly, the
accuracy of the model and the general correction technique has been estimated
as a result of studies using the magnetically stirred cell. It must be noted

that the accuracy of the model also depends on the solution procedure used
for equation (12). As yet, no definite estimate of accuracy can be made
although experience has shown that the error is usually no more than 6%.
The final point is that a distinction must be made between a dense particle
field producing high obscuration and a dilute one with a long beam path
length through it giving a high obscuration. In the second case, although
high obscuration is obtained, multiple scattering may be negligible and
therefore the model developed in this study is not applicable.

5. - Conclusions

The mathematical model and the correction procedure developed for the
effect of multiple scattering in dense particle fields gives good agreement
with the experimental results. The technique was shown to be applicable to
both Fraunhofer and anomalous diffraction theories.

The study therefore enables the application of laser light diffraction
techniques to be used with great reliability for measurements in dense
particle fields where the obscuration of the incident light beam is greater
than 50% and approaching 100%.

Acknowledgment

Financial support towards this investigation by the Ministry of Defence,
United Kingdom, is gratefully acknowledged.

References

1. A. R. Jones, "A Review of Drop Sizing Measurement - The Application of
    Techniques to Dense Fuel Sprays," _Progress in Energy and Combustion
    Science_, Vol. 3 (1977) pp 225-234.
2. E. D. Hirleman, "Non-Intrusive Laser Diagnostics," _AIAA_, 18th Thermo-
    physics Conference, Montreal (June 1983).
3. J. Cornillault, "Particle Size Analyser," _Applied Optics_, Vol. 11, No. 2,
    (1972), pp 265-268.
4. R. A. Dobbins, L. Crocco and I. Glassman, "Measurement of Mean Particle
    Sizes of Sprays from Diffractively Scattered Light," _AIAA Journal_,
    Vol. 1, No. 8 (1963), pp 1882-1886.
5. J. Swithenbank, D. S. Beer, D. S. Taylor, D. Abbot and G. C. McCreath,
    "A Laser Diagnostic Technique for the Measurement of Droplet and
    Particle Size Distribution," _AIAA_ 14th Aerospace Science Meeting,
    Washington D.C. (1976).
6. C. Negus and B. J. Azzopardi, "The Malvern Particle Size Distribution
    Analyser: Its Accuracy and Limitations," _Harwell Internal Report_
    AERE-R9075 (Dec. 1978).
7. T. Yumanchi and Y. Ohyama, "A Study on the Measurement of Particle Size
    Distribution with Laser Diffraction Systems," _Bull.Jap.Soc.Mech.Eng._,
    Vol. 25, No. 210,(Dec. 1982), pp 1931-37.
8. P. G. Felton, A. A. Hamidi and A. K. Aigal, "Measurement of Drop Size
    Distribution in Dense Sprays by Laser Diffraction," _Proceedings of
    the Int. Conf. on Liquid Atomization and Spray Systems (ICLASS)_,
    Institute of Energy, London (1985).
9. D. J. Brown and E. J. Weatherby, Private Communication; to be presented
    at the inter. symp. on "Optical Particle Sizing: Theory and Practice,"
    Rouen, France (May 1987).
10. H. C. Van de Hulst, "Light Scattering by Small Particles," Dover
    Publications, Inc., New York (1981).
11. A. A. Hamidi and J. Swithenbank, "Treatment of Multiple Scattering of
    Light in Laser Diffraction Measurement Techniques in Dense Sprays and
    Particle Fields," _Journal of the Institute of Energy_ (June 1986),
    pp 101-105.

12. L. G. Dodge, "Change in Calibration of Diffraction Based Particle Sizers in Dense Sprays," <u>Optical Engineering</u>, Vol. 23, No. 5, (1984) pp 626-630.
13. E. D. Hirleman, Private Communication; to be presented at the Inter. Symp. on "Optical Particle Sizing: Theory and Practice," Rouen, France (May 1987).
14. H. Gomi, "Multiple Scattering Correction in the Measurement of Particle Size and Number Density by the Diffraction Method," <u>Applied Optics</u>, Vol. 25, No. 19, (Oct. 1986), pp 3552-3558.

APPENDIX I

Nomenclature

A_i	Area of particle with radius a_i
a	Radius of particle
$C(I,J,K)$	The multiple scattering matrix
$E, E(I), E(N,I)$	Light energy distribution
F	Fraction of light scattered in each slice
f	Collection lens focal length
I_o, I	Intensity of the parallel beam of light entering/leaving the particle field
$I_o(N)$	Intensity of the parallel beam of light leaving slice N
J_o, J_1	Bessel functions
K	The wave number
m_1, m_2	Refractive index of the two phases
N_i	Number of particles per unit volume
Q_s	Scattering efficiency
s_1, s_2	Radii of the photodetector ring at the focal plane of the lens
$T(I,J)$	Single scattering matrix
$W_i, W(J)$	Volume (weight) fraction in size class
λ	Wavelength of the light
ρ	Phase shift
τ	Particle field turbidity
ℓ	Optical path length

PARTICLE SIZING BY LASER LIGHT DIFFRACTION :

IMPROVEMENTS IN OPTICS AND ALGORITHMS

D. Kouzelis, S.M. Candel[*], E. Esposito, and S. Zikikout

Laboratoire E.M2.C du CNRS et de l'ECP
Ecole Centrale des Arts et Manufactures
Grande Voie des Vignes
92295 Châtenay-Malabry Cedex, France

INTRODUCTION

Many industrial processes like combustion of kerosene in gas turbines, production of cement, medical aerosols, insecticide sprays etc, involve the measurement and control of particle size distribution.

Among the many intrusive or non-intrusive techniques proposed in recent years, the most widely used is based on measurements of laser light diffraction by particles at small angles. This method is described for example by Felton et al[3], Swithenbank et al[4] and a commercial particle sizer based on this principle is available [5].

The main difficulty encountered in this technique arises from the inverse operation which provides the droplet size distribution from the measured diffraction pattern. The matrix used there is ill-conditioned and as a consequence small measurement errors lead to unacceptable solutions. Up to now this difficulty was overcome either by imposing the shape of the size distribution at the outset (for example a Rosin-Rammler distribution) and calculating two or three parameters to describe this distribution, or by performing the matrix inversion on a low resolution basis.

Tardieu et al[1,2] were among the first to develop a particle sizer based on the same optical principle but using a modern CCD array to measure diffracted light intensity. The use of a 1024 photodiodes array allows a high spatial sampling and a direct matrix inversion is then possible, without imposing a model of the size distribution. The inversion problem was solved by using a non negative Least Square Algorithm (NNLS) given by Twomey[11]. However, the performance of this system was limited by the low dynamic range obtained from the detector. To extend the dynamic range of the measurement it was found necessary to decrease the detector

____
* also ONERA 92320 Chatillon, France

thermal noise and to correct in some way for the non uniform exposure of
the array to the diffracted light.

The purpose of this paper is to describe the improvements brought on
the optical arrangement of Tardieu et al. with an emphasis on the
influence of noise sources. New approaches of the inversion problem are
also described. In those new procedures the diffracted light profile is
first smoothed then inverted with a conjugate gradient method or an
analytical integral expression.

NOMENCLATURE

a	: radius of particle	$[L]$
D	: particle diameter	$[L]$
$\Phi(a)$	: number size distribution function	$[L^{-1}]$
A'	: $\lambda f / 2\pi$	
f	: focal distance of Fourier lens	$[L]$
λ	: wavelength	$[L]$
θ	: scattering angle	(rad)
s	: distance from focal point	$[L]$ $\qquad \theta = s / f$

EXPERIMENTAL APPARATUS

When a spherical particle is illuminated by a parallel beam of
monochromatic, coherent light, a diffraction pattern is formed. This
pattern is large in comparison with the geometrical image. If a lens is
placed on the light path beyond the particle and if the image is observed
in the focal plane of the lens, then the direct light is focused to a
single point and the diffracted light forms a ring pattern around the
central spot, called the diffraction pattern (Fig. 1)[6,7,8,9].

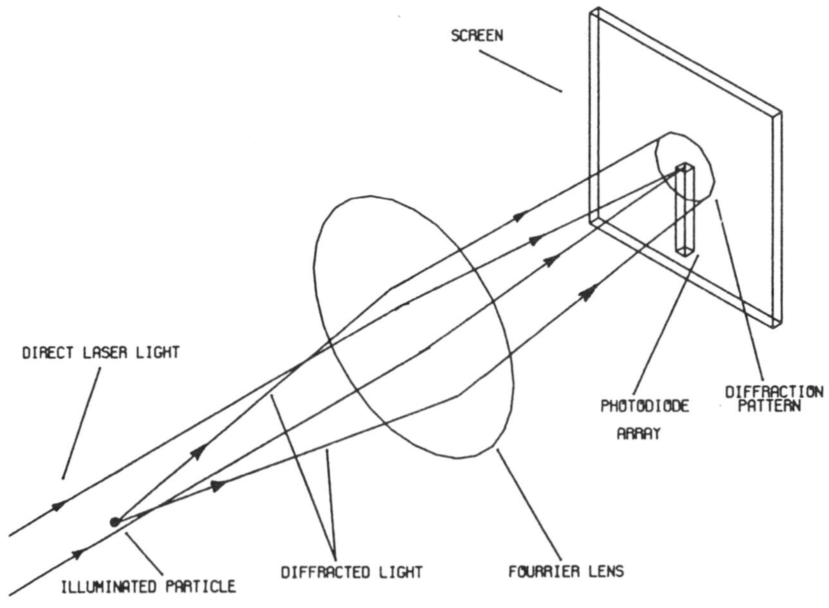

Figure 1. Light diffraction pattern formed by a small particle.

The diffracted light obtained in this way contains information on the particle size and this principle forms the basis of the system shown in Figure 2.

The light source (A) is an He-Ne 5 mW laser operating in the TEM_{00} mode at a wavelengh λ = 0.6328 µm. A filter of variable film thickness B is used to trim the laser power. This element is followed by a beam expanding telescope fitted with a spatial filter. The exit beam diameter measures about 8 mm. The parallel laser beam crosses the test section in two directions at right angles to the laser beam by two step by step motors (H, I).

The light scattered from the particles, and the direct beam cross the receiver lens (D), which operates as a Fourier transform lens, forming the far field diffraction pattern of the scattered light at its focal plane (Figure 3). The lens used here is 300 mm in focal length and is placed at 500 mm from the particle cloud.

The detector is a CCD-1024 photodiode array placed at right angle to the optical axis in the focal plane of the lens. This standard detector (THOMSON-CSF TH 7802) has 1024 active square diodes, each with a side width of 13 µm. The corresponding resolution is of 38 pl/mm and the nominal dynamic range is 4500/1.

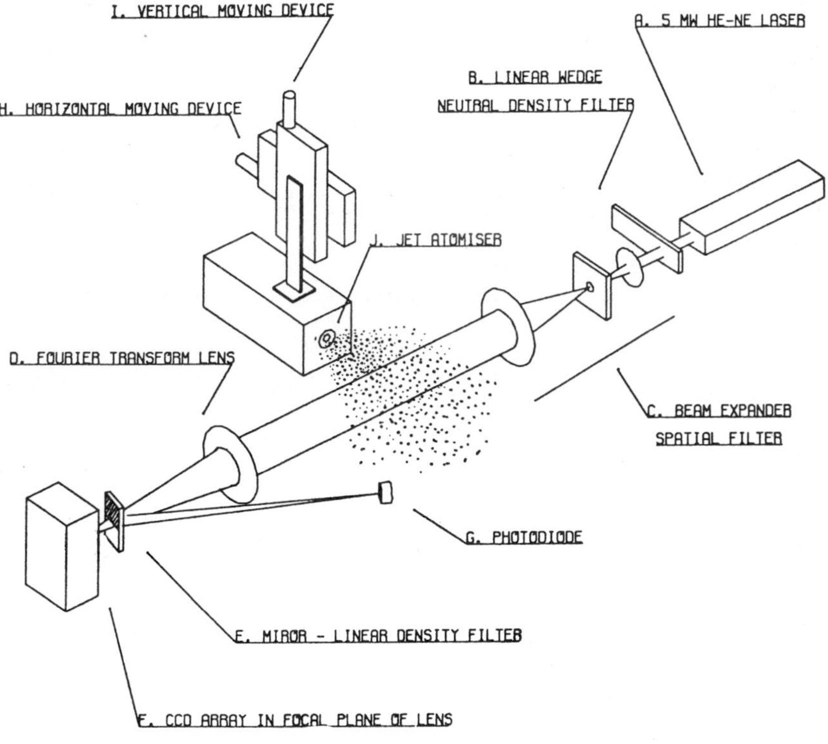

Figure 2. Optical set - up.

    A linear density filter (element E) is placed in front of the CCD
array. The upper half part of this filter has a high-reflection coating
which reflects the transmitted laser light beam on a single photodiode.
The direct laser power received from that photodiode, allows the
determination of the particle concentration. The measurement of the direct
light intensity yields the transmission map of the droplet cloud. The
linear density filter has a high attenuation factor near the optical axis
where the light intensity is high and a low attenuation at the bottom
where the level of diffracted light is low. This has the effect of
compressing the dynamic range of the system.

    The Fourier lens has the useful property that wherever the particle
is in the measuring volume (laser beam), its diffraction pattern is formed
at a finite distance, is stationary and centered on the optical axis.
Movements of the particle do not cause movement of the diffraction pattern
since the light diffracted at an angle θ, is focused at the same point on
the screen.

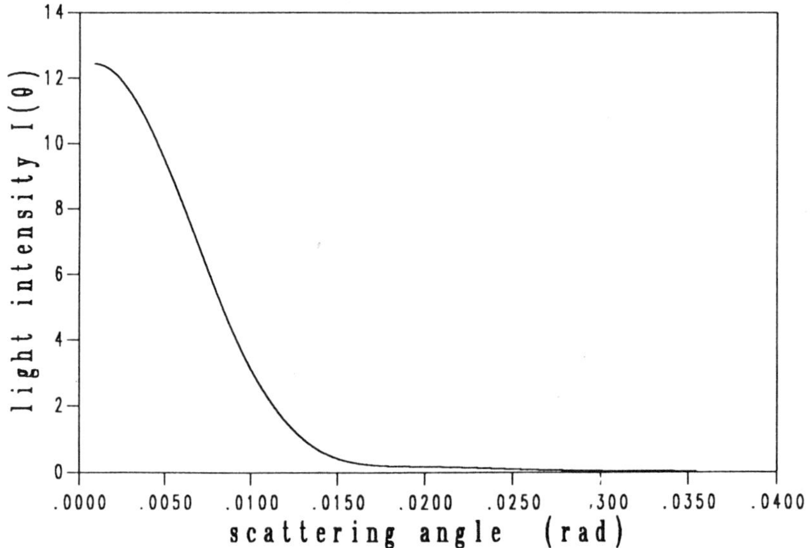

Figure 3. Typical diffracted intensity profile from a simulated
          Rosin-Rammler distribution.

    The scattered light measured by the photodiode array is the sum of
individual patterns formed from all particles simultaneously present in
the laser beam. Thus the system measures the integrated scattered light
corresponding to the particle distribution present in the beam during the
detector exposure.

DYNAMIC RANGE OF MEASUREMENTS

    If a matrix inversion is used to calculate the size distribution, a large dynamic range is required from the measurements[1,2]. To improve the dynamic range of the photodiode array the main noise sources are minimized[10] :

a) A thermoelectric cooling element provides a constant operating temperature of $-10°C$ which suppresses a large part of the thermal noise.

b) Random noise is reduced by acquiring and accumulating 100 data points (for each diode) in real time and thus dividing random noise by 10.

c) The data are transmitted to the A/D converter through a differential amplifier to reduce transmission noise.

d) Optical noise is diminished by placing the test table in a dark room and treating all optical surfaces with antireflection coating. The CCD-array is placed in an airtight box with a small nitrogen circulation. The droplet cloud is confined in a box comprising aerodynamic (jet blown) apertures.

In order to find the real diffracted pattern the array is first exposed to uniform light and then to the dark. If $R_u(i)$ and $R_o(i)$ respectively designate the response of the i-th diode under uniform light and in the dark and if $R(i)$ is the measured response then the corrected response is given by

$$r(i) = \frac{R(i) - R_o(i)}{R_u(i) - R_o(i)} \qquad (1)$$

THEORETICAL ASPECTS

The theoretical light intensity diffracted from a spherical particle of diameter $d = 2a$, at an angle θ is given by

$$I(\theta) = I_o \left[\frac{2 J_1(2\pi a \sin\theta / \lambda)}{2 \pi a \sin\theta / \lambda} \right]^2 \qquad (2)$$

where J_1 is the first order Bessel function and λ the wavelength. If the optical arrangement of Figure 3 is used, then on the focal plane

$$\sin\theta = s / f$$

where s is the distance from the optical axis and f is the focal distance.

The total light intensity distribution, diffracted from a cloud of particles is the sum of the weighted intensity distribution corresponding to each size range. This may be expressed as

$$I(s) = A \int_0^\infty a^4 \left[\frac{J_1(2\pi a s / \lambda f)}{2 \pi a s / \lambda f} \right]^2 \Phi(a) \, da \qquad (3)$$

where $I(s)$ designates the diffracted light intensity and $\Phi(a)$ the particle size distribution. In discrete form this relation may be written as a matrix equation:

$$[I(s)] = [M(s,a)] \; [\phi(a)] \qquad (4)$$

where

    $[I(s)]$ is a $[1024 \times 1]$ intensity matrix

    $[M(s,a)]$ is a $[1024 \times n]$ matrix which contains the coefficients defined from equation (3)

    $[\phi(N)]$ is a size distribution vector (n   values)

The inversion method should (1) impose minimal constraints and in particular should not use an a-priori particle distribution model (2) be sufficiently accurate, numerically stable and insensitive to measurements errors[11].

SIMULATIONS

The choice of a method fulfilling these criteria is based on numerical simulations of the problem. The droplet size distributions used are of the Rosin-Rammler type, defined as follows :

$$\phi(a) = a^t \cdot e^{-a^s} \qquad (5)$$

The corresponding diffraction pattern is calculated from Eq. 3 by assuming that the size distribution is continuous and linear in each size range :

$$I(s) \simeq \sum_{j=1}^{n-1} \left[\int_{\alpha_j}^{\alpha_{j+1}} \kappa \, a^2 \left[\frac{J(2\pi as/\lambda f)}{s} \right]^2 \phi(a) \, da \right] \qquad (6)$$

where

$$a \in [a_j , a_{j+1}] , \qquad \phi(a) = \alpha_j a + \beta_j$$

and

$$\alpha_j = \frac{\phi(a_{j+1}) - \phi(a_j)}{a_{j+1} - a_j} \qquad \beta_j = \frac{a_{j+1}\phi(a_j) - a_j \phi(a_{j+1})}{a_{j+1} - a_j} \qquad (7)$$

Equation (6) becomes

$$I(s_i) = \sum_{j=1}^{n-1} \left[\int_{\alpha_j}^{\alpha_{j+1}} \left[\frac{\phi(a_{j+1}) - \phi(a_j)}{a_{j+1} - a_j} + \frac{a_{j+1}\phi(a_j) - a_j \phi(a_{j+1})}{a_{j+1} - a_j} \right] a^2 \frac{J_1^2(2\pi as/\lambda f)}{s_i^2} \, da \right] \qquad (8)$$

Figure 3 gives an example of such a diffraction pattern. Noise is added to the simulated pattern and the dynamic range is assumed to be finite, with an upper limit equal to 8000.

INVERSIONS

A. <u>Conjugate gradient inversion</u>

The matrix system (4) is usually overdetermined and the number of equations m, is greater than the number of radii n. The size distribution $\Phi(a)$ is then obtained by minimizing

$$\sum_{j=1}^{m} \left(\sum_{i=1}^{n} M_{ji} \, \phi_i - I_j \right)^2 \qquad\qquad (9)$$

The least-squares solution which minimizes this expression is given by

$$[\,\phi\,] = \left[\; {}^t[M] \; [M] \right]^{-1} \; {}^t[M] \; [I] \qquad\qquad (10)$$

where $^t[M]$ is the transposed matrix of $[M]$

Now let

$$^t [M] \quad [M] = [A] \qquad \text{and} \qquad ^t [M] \quad [I] = [B]$$

Then Equation (10) may be written in the form :

$$[B] = [A] \; [\phi]$$

with $[A]$ a positive-definite n x n matrix.

Let us consider the quadratic form :

$$J_A ([\phi]) = \frac{1}{2} \, {}^t[\phi] \; [A][\phi] - {}^t[\phi][B]$$

If $[\phi*]$ minimizes J_A then :

$$d \, J_A \left([\phi*] \right) = 0 \qquad\qquad \text{or} \qquad\qquad [A] \; [\phi] - [B] = 0$$

In order to find $[\phi*]$, we use the Stiefel-Hesdenes algorithm[12]. If $\underset{\sim}{u}_0$ is a starting vector $(\underset{\sim}{u}_0 = 0)$, $\underset{\sim}{r}_0 = [B] - [A] \; \underset{\sim}{u}_0$, $\underset{\sim}{V}_0 = \underset{\sim}{r}_0$, then for i = 0, 1, 2, ... n-1 one computes

$$\underset{\sim}{u}_i = \underset{\sim}{u}_{i-1} + \frac{^t\underset{\sim}{r}_{i-1} \; \underset{\sim}{V}_{i-1}}{^t\underset{\sim}{V}_{i-1} \; [A] \; \underset{\sim}{V}_{i-1}} \; \underset{\sim}{V}_{i-1}$$

$$\underset{\sim}{V}_i = \underset{\sim}{r}_i - \frac{^t\underset{\sim}{r}_i \; [A] \; \underset{\sim}{V}_{i-1}}{^t\underset{\sim}{V}_{i-1} \; [A] \; \underset{\sim}{V}_{i-1}} \; \underset{\sim}{V}_{i-1}$$

$$\underset{\sim}{r}_i = \underset{\sim}{r}_{i-1} + \frac{^t\underset{\sim}{r}_{i-1} \; \underset{\sim}{V}_{i-1}}{^t\underset{\sim}{V}_{i-1} \; [A] \; \underset{\sim}{V}_{i-1}} \; \underset{\sim}{V}_{i-1}$$

$\underset{\sim}{u}_n$ will be the solution of the system. In practice this iteration should be carried from i=0 to n-1. However a good approximate solution is offten obtained for i < n, avoiding rounding errors.

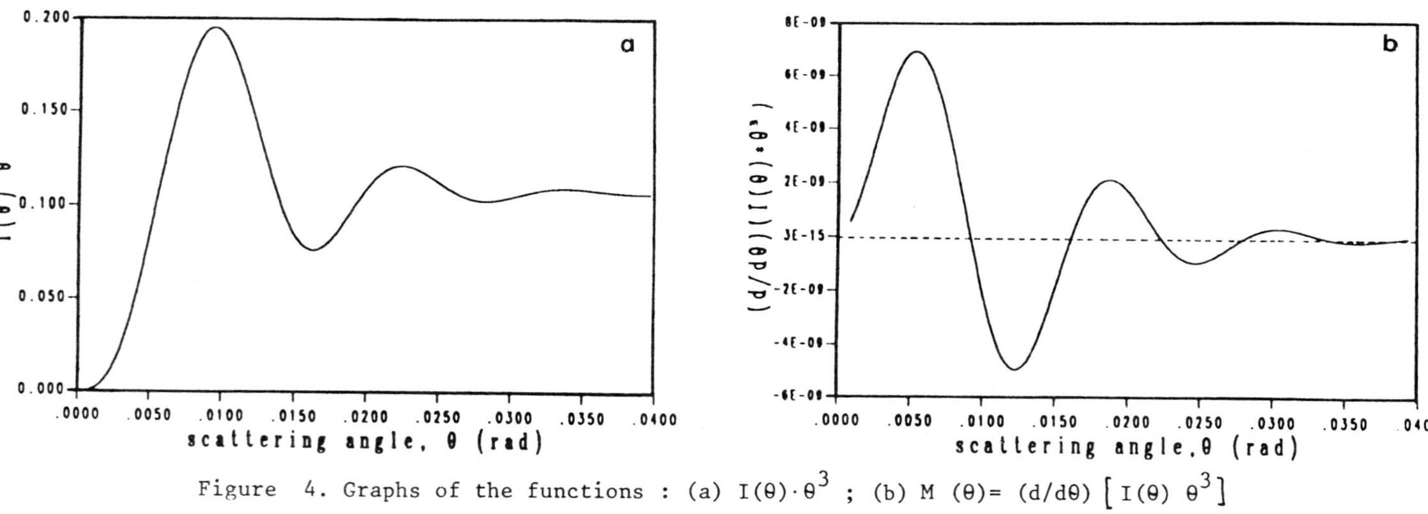

Figure 4. Graphs of the functions : (a) $I(\theta) \cdot \theta^3$; (b) $M(\theta) = (d/d\theta)\left[I(\theta)\,\theta^3\right]$

B. <u>Analytic inversion method</u>

The inversion method is based on a result of Chin[13] and Shifrin[14].
It is indicated in these references that equation (3) can be inverted
analytically to yield ϕ (a) independently of any prior assumption or
modelling of the distribution

$$\phi\ (\ a\ )\ =\ -2\pi\ /\ a^2 A' \int_0^\infty M(s)\ A(a\ s)\ ds \qquad (\ 12\ )$$

where $\quad A\ (a\ s)\ =\ a\ s\quad J_1\ (as)\ Y_1\ (as)$

$$M\ (\ s\ )\ =\ d/ds\ (\ s^3\ I(s)) \qquad\qquad (\ 13\ )$$

and $\quad A' \qquad =\ \lambda\ f\ /\ 2\ \pi$

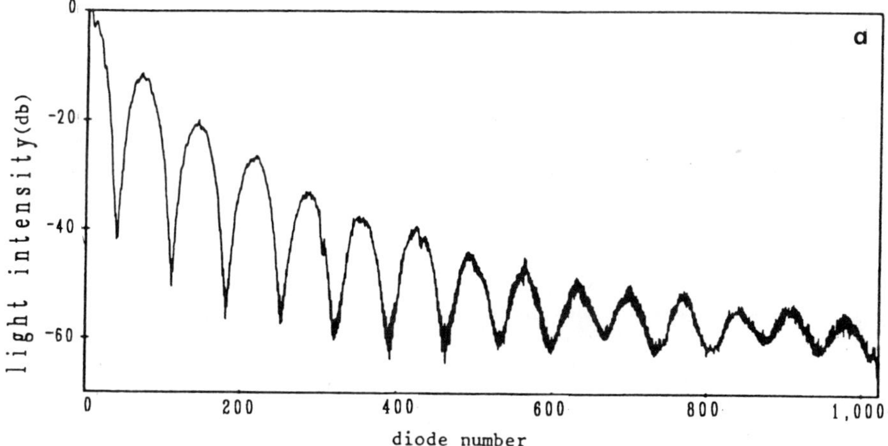

Figure 5. (a). Example of the diffracted intensity profile from a 100 um
pinhole.

By assuming a lower and an upper limit for the particle sizes (typically
from 4 μm to 300 μm in diameter), equation (12) may be cast in the
following discrete form

$$\phi\ (a_j)=\ -2\pi\ /\ a^2 A' \sum_{i=1}^m M(s_i)\ A(a_j s_i)\ ds \qquad (\ 14\ )$$

This expression is only useful if the derivative appearing in equation (3)
is not contaminated by noise. Differentiation of s^3 I(s) may be carried
out after some numerical filtering and smoothing (Fig. 5).

EXPERIMENTAL RESULTS

Several tests were carried out both on the inversion methods and on
the experimental apparatus to determine the accuracy and reliability of
the system.

343

First, the theoretical diffracted intensity was calculated for
different diameter pinholes, then the inversion methods were applied to
the obtained profiles. Figure 5a gives an example of the diffracted
intensity profile from a 100 µm pinhole. The result obtained with the
analytical inversion method is displayed in Fig. 5b. Despite some widening
of the resulting peak, the average value is well centered and the sidelobs
have a relatively low amplitude.

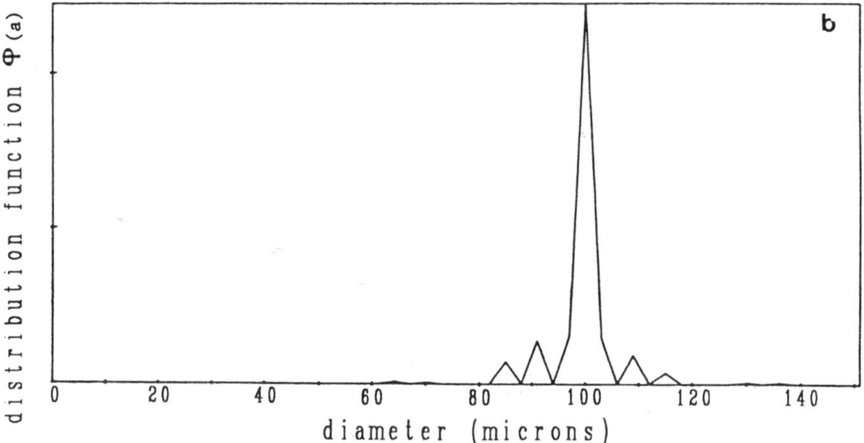

Figure 5. (b). Result of analytical inversion method, applied on this profile.

Figure 6 is a comparison between a theoretical Rosin-Rammler
distribution and the calculated distribution obtained from the
corresponding simulated diffracted intensity profile (similar to that
shown in fig. 4) using the conjugate gradient algorithm. The agreement
between both distributions is good. Some erroneous peaks appear in the
large diameter range. Although their importance is very small in a number
size distribution, the corresponding error might not be quite negligible
in a mass averaged interpretation of the results, and efforts should be
made to suppress them.

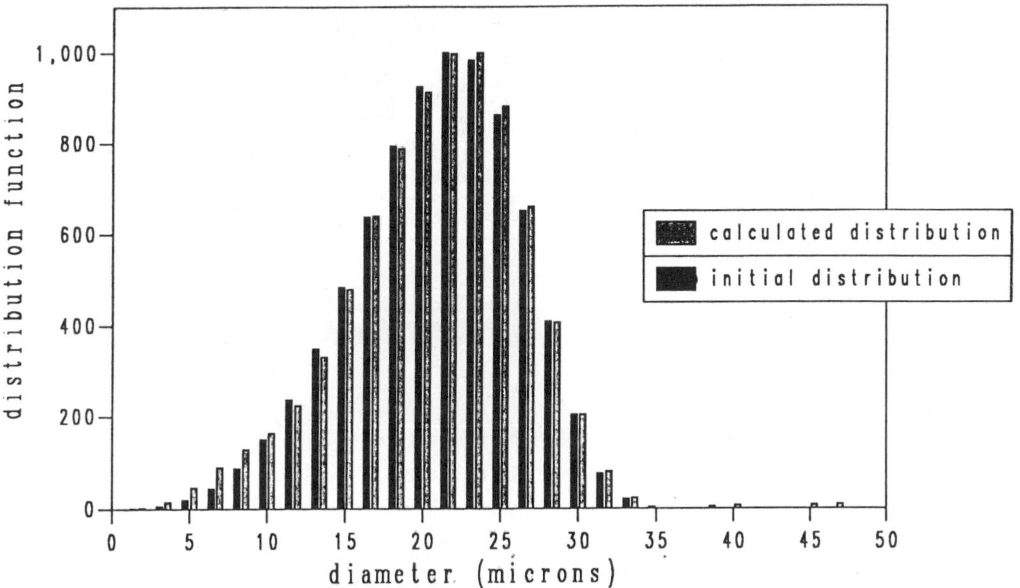

Figure 6. Theoretical and calculated distributions using the conjugate
gradient algorithm.

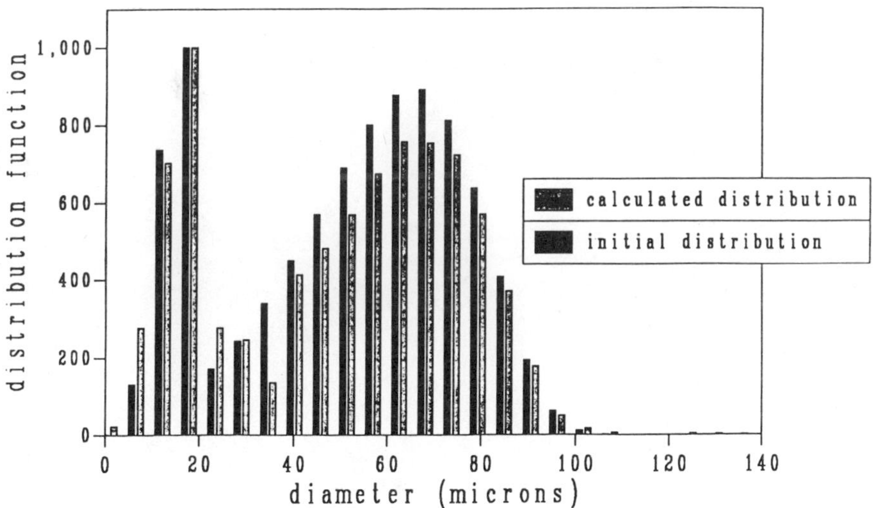

Figure 7. Result of the conjugate gradient algorithm, applied on a twin
Rosin-Rammler distribution.

One of the difficulties encountered with constrained or low resolution inversion methods is their inability to deal with bimodal distributions (such as may be encountered in the starting phase of gas turbine combustion chambers, when an auxiliary starter injector is used to help ignition). Figure 7 shows the result of a conjugate gradient inversion in the case of a twin Rosin-Ramler simulated distribution. The agreement is very good and the two main peaks of the original distribution are well situated and well defined in the inversion results.

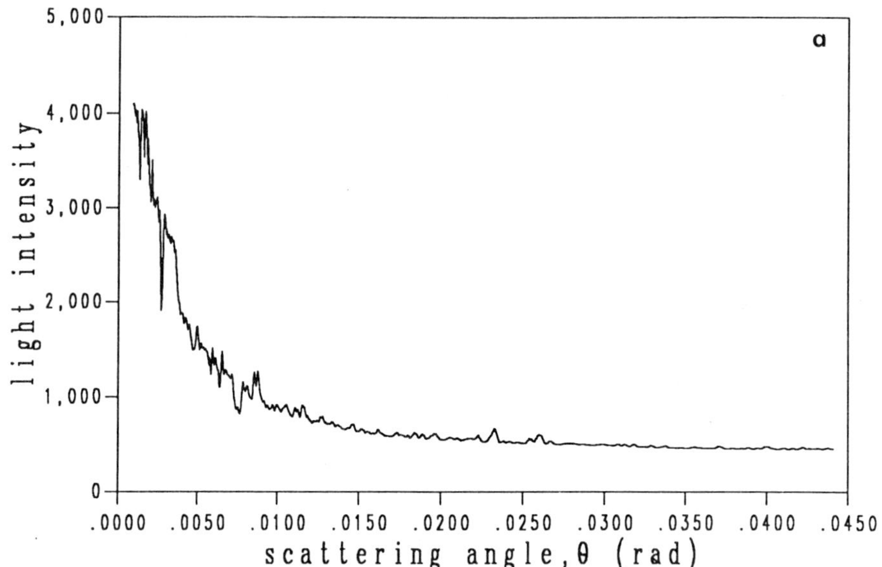

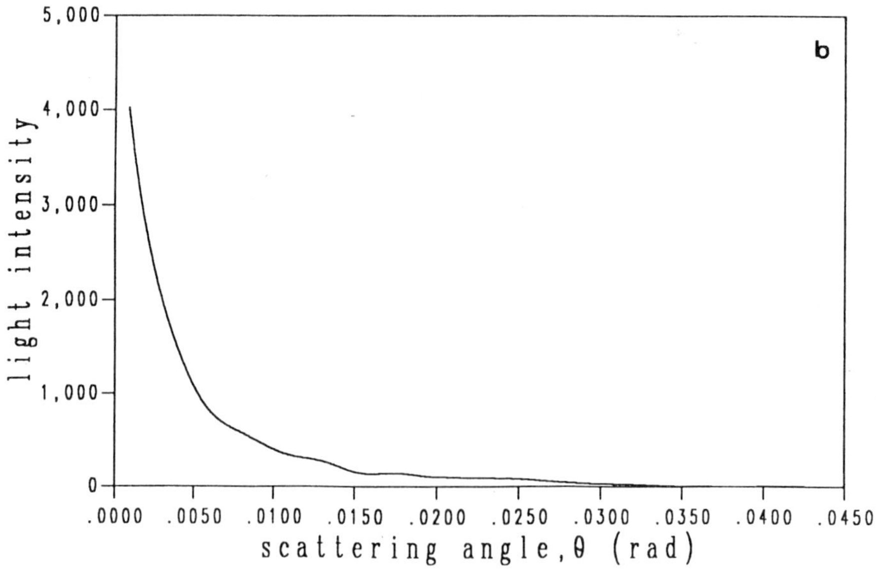

Figure 8. (a) Experimental intensity profile from an airblast atomizer water spray.
(b) Filtered and smoothed intensity profile.

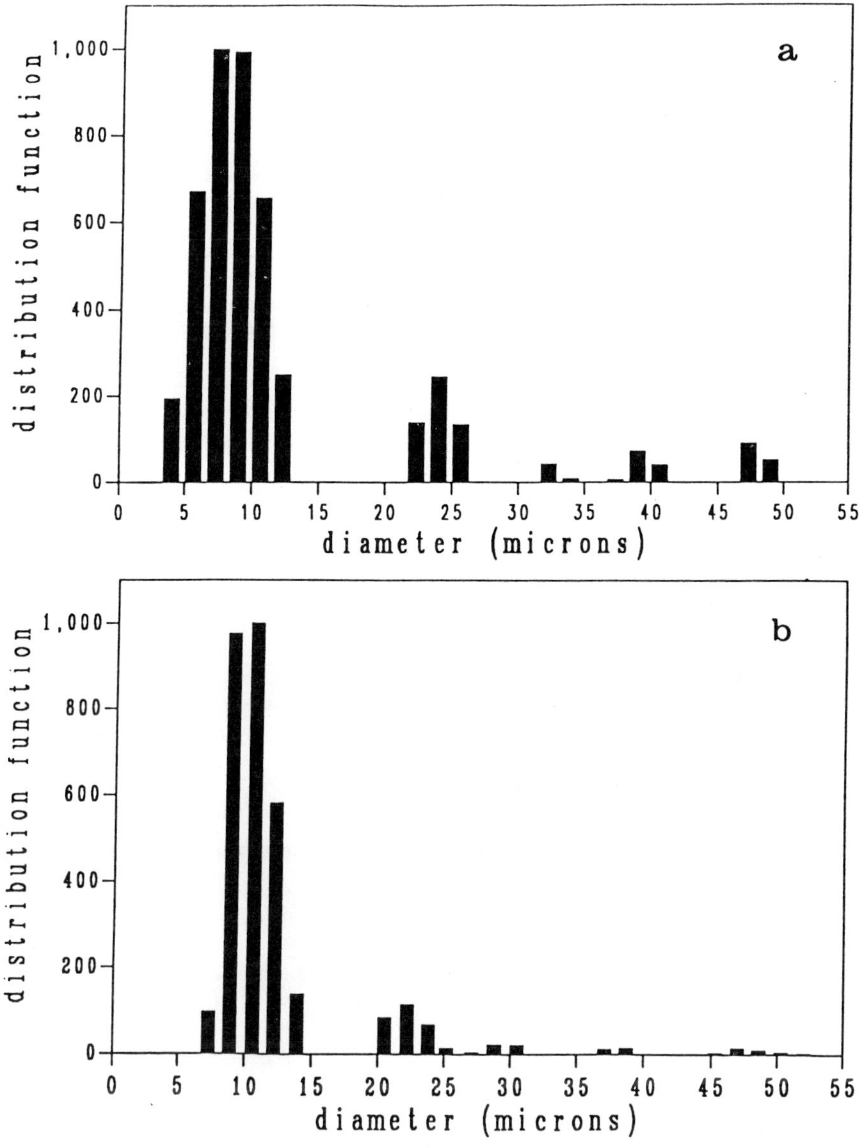

Figure 9. Calculatted distributions from the experimental intensity
profile of Fig. 8 ; (a)Using the conjugate gradient algorithm ; (b) using analytical expression method.

The optical set up of Figure 2 was used to obtain an experimental
diffracted intensity profile from water sprayed by an   airblast
atomizer. The measurements are performed at 75 mm from the injector face
and on the optical axis. The profile was filtered and smoothed (Fig. 8b)
then both inversion methods were successively applied.

Figure 9a shows the result obtained with the conjugate gradient
algorithm. The main peak is well defined and shows a maximum around 8 um
in diameter. Smaller peaks appear in the large diameter range. The method
itself does not allow a definite answer but their quasi periodic
appearance suggests that they are non physical.

Figure 9b gives the calculated distribution obtained from the same experimental curve with the analytical inversion method. The main peak is very similar to that obtained with the conjugate gradient algorithm, with an average value around 10 μm in diameter. Small periodic peaks also appear in the large diameter range, due to the numerical instabilities of the system.

The average distribution obtained from the comparison of the results of both inversion methods can be considered as the best droplet distribution available from the measured diffracted intensity profile (Fig. 11).

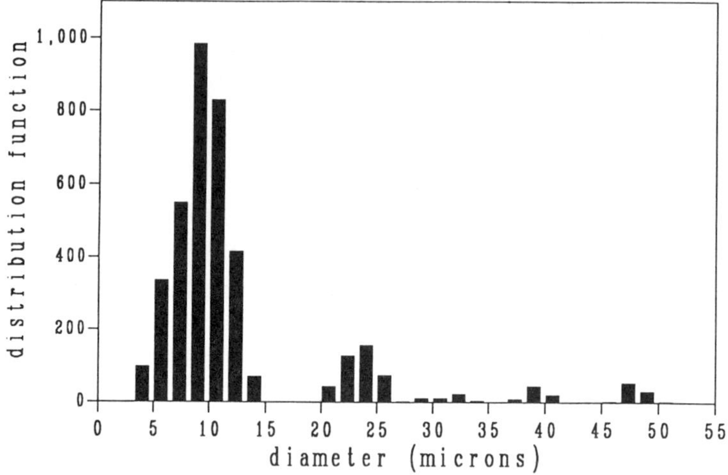

Figure 10. Average distribution obtained from the comparison of both methods.

CONCLUSION

The cylindrical symmetry of the diffraction pattern formed by a Fourier lens (Fig. 1), allows the use of a <u>linear</u> solid-state detector which provides a high resolution sampling of this pattern. Also, the absence of moving parts (mirrors, detectors, etc) during data acquisition and the signal integrating capability of the array are useful properties for particle sizing by Laser Light Diffraction. However, the photodiode array has not proved to be a very sensitive detector, so special efforts have been necessary to increase its performance.

Two different inversion algorithms have been developped and tested to obtain the droplet distributions from the measured diffracted light intensity profiles. The results obtained are quite similar but the comparison of the distributions calculated by both methods does not yet yield useful informations to suppress non physical peaks due to experimental or rounding errors. Efforts are currently being made to improve the experimental procedure and the effective dynamic range of the measurements.

Data acquisition by the analog to digital converter of a PDP 11/23 computer is fast (typically 1 second for the acquisition of a diffraction pattern), but numerical calculations of the particle size distribution are much longer (typically 3 minutes for 30 particle classes for each method). Recent improvements in microcomputer performance, when applied to the problem, will drastically reduce computing time.

REFERENCES

1. A. Tardieu, Développement d'une méthode optique de diffraction pour déterminer les tailles des gouttelettes dans un jet pulvérisé, Thèse de Docteur-Ingénieur, Ecole Centrale des Arts et Manufactures, Paris, 1983.
2. A. Tardieu, S.M. Candel, Droplet size distribution from diffracted light intensities, Proc. of 9th Intern. AIAA Progr. in Astronaut. and Aeronaut.,Vol. 95, P. 736-749, 1984.
3. P. G. Felton, Measurements of particle/droplet size distribution by a laser diffraction technique, Proc. of 2th European Symp. on Particle Caracterisation, PARTEC, Nuremberg, 1979.
4. J. Swithenbank, J.M. BEER, D.S. Taylor, D. Abbot, and G.C. Mc Greath, A laser diagnostic technique for measurement of droplet and particle size distribution, AIAA 16 : 79 (1976).
5. "Malvern particle sizer", Series 1800 and 2600, Malvern Instruments, England.
6. L.P. Bayvel, A.R.Jones,"Electromagnetic scattering and its applications" Applied Science Publishers, London (1981).
7. M. Kerker, "The scattering of light", Academic Press, New York (1969).
8. G. Goodman, "Introduction to Fourier optics", Mc Graw Hill (1968).
9. E. Hecht and A. Zajac, "Optics", Addison-Wesley Publishing Co (1974).
10. G. Horlick, Caracteristics of photodiode arrays for spectrochemical measurements, Appl. Spectroscopy 30, no 2 (1976).
11. S. Twomey, "Introduction to the mathematics of inversion in remote sensing and indirect measurements", Elsevier Publ. Co, New York (1977).
12. E. Stiefel, "Einfuhrung in die numerische mathematik", B.G. Teubner Verlagsgesellschaft, Stutgard.
13. J.H. Chin, C.M. Sliepcevich, M. Tribus, Particle size distribution from angular variation of intensity of forward scattering light at very small angles, J. Chem. Phys. 59 : 841 (1955).
14. K.S. Shifrin, A. Ya. Pelerman, Opt. Spectr. (URSS) 15 : 285-9 (english transl. 1963).
15. Proc. of the 2th Intern. Conf. on Liquid Atomization and Spray Systems,   June 20-24, Madison, Wisconsin, USA (1982).
16. J. Cornillault, Particle size analyser, Appl. Optics 11 : 265 (1972).

SHAPE, CONCENTRATION AND ANOMALOUS DIFFRACTION EFFECTS

IN SIZING SOLIDS IN LIQUIDS

D. J. Brown, E. J. Weatherby and K. Alexander

Department of Chemical Engineering and Fuel Technology
Sheffield University
Mappin Street, Sheffield, S1 3JD, UK

Small-angle laser light diffraction has been used to measure the size
distribution and relative concentration of solid particles in liquids. The
technique has been extended from spherical to non-spherical convex
particles, to the determination of absolute concentration and, finally, to
systems where the values of refractive index of solid and liquid are close
together.

INTRODUCTION

The technique of particle sizing by small-angle laser light diffraction has been applied to measure the size distribution and concentration of
solid particles suspended in liquids.

This involved firstly, a correction for shapes which are other than
the sphere assumed in the computer software program used with the laser
sizer. The correction procedure was validated by a study of the sizing of
solid spheres, cubes and octahedra. Secondly, the absolute concentration
of particles was obtained by combining information from the undiffracted
laser light beam with data from the diffraction rings which gave relative
concentration of particles of various sizes. Finally, a correction was
made for the anomalous diffraction of the light beam which occurred when
the refractive index of the solid was near that of the liquid. An understanding of anomalous diffraction was obtained by performing experiments in
which the liquid refractive index was below, equal to or greater than that
of the solid.

There follows an account of the theory of Fraunhofer and anomalous
diffraction. This theory is combined with the Beer-Lambert law and a
theorem of Cauchy to permit the calculation of the absolute concentration
of particles of different shape. The experimental testing, verification
and application of the theory are then described.

THEORY OF DIFFRACTION

A Fraunhofer diffraction pattern may be formed when a spherical
particle is illuminated by a parallel beam of monochromatic coherent light.
If a lens is placed in the light path after the particle and a screen is

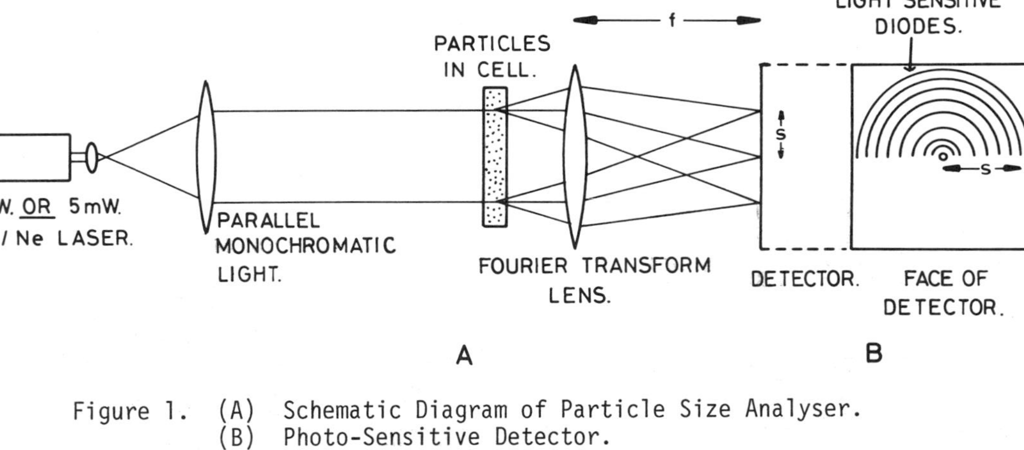

Figure 1. (A) Schematic Diagram of Particle Size Analyser.
         (B) Photo-Sensitive Detector.

placed at the focal plane of the lens, then the undiffracted light is focussed to a point on the axis and the diffracted light forms a 'far field' Fraunhofer pattern of rings around the central point (Figure 1).

For spherical particles with diameter greater than the wavelength of the illuminating radiation, the diameter of the diffraction pattern is inversely proportional to the particle diameter. (For particles smaller than the light wavelength, Mie theory must be used). The technique may be used to measure particle diameters as small as 1 µm using a He/Ne laser (wavelength $\lambda = 0.6328$ µm). As it is difficult to infer particle size distribution from measurements of light intensity, the light energy distribution is used.

<u>Fraunhofer diffraction</u>

The light energy within any ring on the focal plane bounded by radii s_1, s_2, is related to the energy E falling on a particle of radius a by the equation

$$L_{s_1,s_2} = E \{ J_0^2(kas_1/f) + J_1^2(kas_1/f)$$

$$- J_0^2(kas_2/f) - J_1^2(kas_2/f) \} \tag{1}$$

where J_0, J_1 are Bessel functions, f is the focal length of the lens and $k = 2\pi/\lambda$.

Now E is the energy falling on the particle and is proportional to the cross-sectional area of the particle, so that

$$E = C' N\pi a^2 \tag{2}$$

But the weight of spherical particles, W, is related to the number of particles, N, by

$$N = 3W/4\rho\pi a^3 \tag{3}$$

Thus

$$E = C' \frac{3W}{4\rho a} = C'' \frac{W}{a} \tag{4}$$

For a collection of particles of different sizes the light energy falling on any ring in the focal plane is the sum of the contributions from every particle. Therefore,

$$L_{s_1,s_2} = C'' \sum_{i=1}^{M} \frac{W_i}{a_i} \{ J_0^2 \left(\frac{ka_i s_1}{f} \right) + J_1^2 \left(\frac{ka_i s_1}{f} \right)$$

$$- J_0^2 \left(\frac{ka_i s_2}{f} \right) - J_1^2 \left(\frac{ka_i s_2}{f} \} \tag{5}$$

The summation is carried out over M size groupings.

If W_i represents the weight fraction in the size range i, then L_{s_1,s_2} represents the light energy falling on a ring bounded by s_1 and s_2, per unit mass of particles. The lumped constant, C'', depends on several factors such as incident light energy, but can be determined.

If the particles are not spherical, as assumed in the above theory, the size distribution is expressed in terms of an equivalent sphere based on the projected area of the particle.

The total light energy distribution is the sum of the products of the energy distribution for each size range and the weight fraction in that range. Expressed in a matrix equation this is:

$$L(j) = W(i)T(j,i) \tag{6}$$

where $L(j)$ is the light energy falling on ring j, $W(i)$ is the weight fraction in the size range i, and $T(j,i)$ defines the light energy distribution curve of each particle.

This equation is solved by assuming a weight distribution $W(i)$ and substituting in (6) to calculate the theoretical light energy distribution. The parameters of the weight distribution are then iteratively adjusted until the sum of the squared errors in $L(j)$ is a minimum. That is,

$$\Sigma \{ L(j) - W(i) . T(j,i) \}^2$$

is a minimum. Thus the relative weight distribution can be obtained from the diffraction pattern.

Anomalous Diffraction

Of great importance in diffraction is m, the ratio of the refractive index of the 'particle' (solid, liquid) to the suspending medium.

When m is close to one, the particle is not opaque and is also large ($\gg 1$ μm), it is possible to trace a light ray through the particle. The field behind the particle is changed in phase but not in amplitude.

The transmitted ray will, therefore, interfere with the diffracted light and produce so-called Anomalous diffraction patterns. Light absorption by a particle reduces the amplitude of the transmitted light and so the interference effect becomes smaller.

For Anomalous diffraction, the light energy distribution for a non-absorbing spherical particle may be calculated in a manner similar to that for Fraunhofer diffraction. The fraction of the light energy E falling on a particle of radius a which is contained within a circle of radius $\omega_o = \dfrac{S_o}{f}$ in the focal plane is

$$L(\omega_o) = \frac{1}{E} \int_0^\omega \int_0^{2\pi} I(\omega) \; \omega \, \delta\omega \, \delta\psi$$

$$= \frac{4I_o . 2\pi}{E} \int_0^\omega \{ (ReU)^2 + (ImU)^2 \} \, \omega\delta\omega$$

$$L(\omega_o) = 2k^2 a^2 \int_0^\omega \{ (ReU)^2 + (ImU)^2 \} \, \omega\delta\omega \tag{7}$$

where I_o = Light intensity at centre of diffraction pattern

$$\omega = \frac{S}{f} \, ,$$

ReU = Real part of the light intensity amplitude function

ImU = Imaginary part of the light intensity amplitude function.

The integration of Equation 7 is difficult because of the complexity of the real part of the amplitude function. However, the light energy in a ring in the focal plane may be approximated by dividing the ring into steps and summing the contributions by arithmetic integration. The light energy distribution for a group of particles may be predicted by summing the contributions made by each individual particle. The Anomalous diffraction light energy distribution for 100 particles spaced equally across the size range 39.0–50.2 μm is shown in the bottom half of Figure 2. The refractive index ratio used was m = 1.039. The corresponding Fraunhofer diffraction light energy distribution is shown in the upper half of the Figure.

The volume and weight of spherical and non-spherical particles

The undiffracted light which is focussed onto the 'centre spot' at the focal plane yields additional information which can be used to obtain the absolute weight distribution. The ratio of the light intensity measured at the centre spot before and after the sample is placed in the laser beam gives the fraction of light obscured by the particles. This obscuration is related, by the Beer–Lambert Law, to the total projected cross-sectional area of the particles.

The Beer–Lambert Law is:

$$\ln \left(\frac{I}{I_o} \right) = - \tau 1 \tag{8}$$

where I and I_o are the light intensities with and without the sample respectively and 1 is the optical path length. Now,

$$\tau = \sum_{i=1}^{15} N_i A_i Q_i \tag{9}$$

where N = the number of particles of size band i per unit volume, A_i = the cross-sectional area of the particles i and Q = extinction efficiency. Q is related to the diameter of the particles, to the refractive indices of the medium and particles and to the wavelength of the radiation. For Fraunhofer diffraction, Q = 2. For Anomalous diffraction, $Q \neq 2$. It is assumed that the particles are non-absorbing and that Q for extinction equals that for scattering.

Substituting the Fraunhofer value of 2 for Q into equation (9) gives:

$$\tau = 2 \sum_{i=1}^{15} N_i A_i \tag{10}$$

Now for a sphere,

$$\frac{\text{Volume}}{\text{Total surface Area}} = \frac{4\pi r^3/3}{4\pi r^2} = \frac{r}{3} = \frac{d}{6} \tag{11}$$

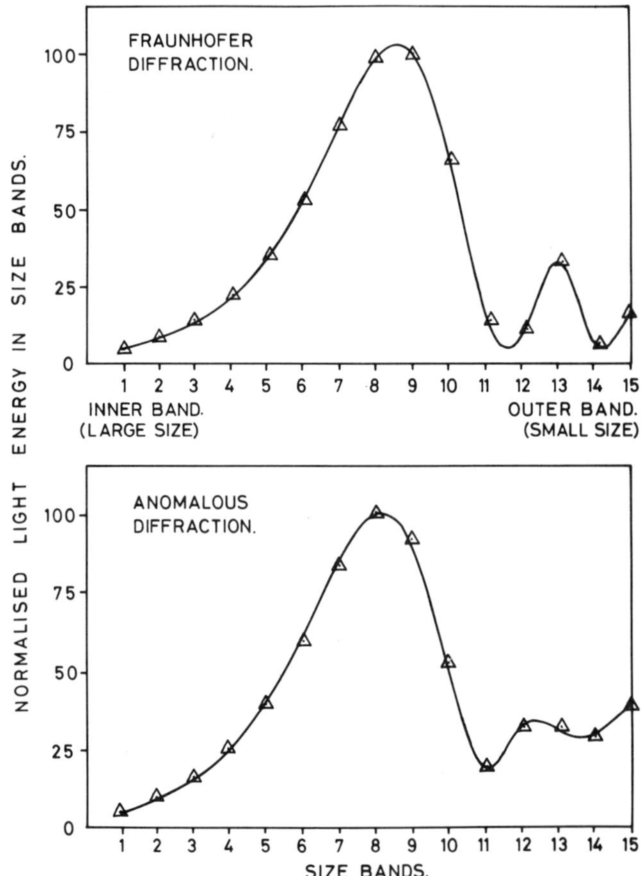

Figure 2.   Fraunhofer and Anomalous Light Diffraction for Glass Spheres of 39·0-50·2 Microns Diameter.

and the total surface area is four times the projected area

$$\frac{\text{Volume}}{\text{Projected Area}} = \frac{4\pi r^3/3}{\pi r^2} = \frac{4r}{3} = \frac{2d}{3} \tag{12}$$

Projected Area = Volume . 3/2d i.e.

$$N_i A_i = \frac{3V_i}{2d_i} \tag{13}$$

where V_i is the volume of particles in size range i, per unit volume. We measure for a beam volume of V_B which is the cross-sectional area times the path length. What we measure is the relative volume.

$$V_{mi} = \frac{\text{Volume of particles in size range i}}{\text{Total volume of all solid scattering particles in the beam, } V_S}$$

Hence,

$$V_{mi} = \frac{V_i V_B}{V_S} \quad \text{and} \quad V_i = V_{mi} \frac{V_S}{V_B} \tag{14}$$

Substituting for $N_i A_i$ and V_i in (10),

$$\tau = 2 \sum_{i=1}^{15} \frac{3V_{mi}}{2d_i} \frac{V_S}{V_B} = \frac{3V_S}{V_B} \sum_{i=1}^{15} \frac{V_{mi}}{d_i} \tag{15}$$

and substituting for τ in (8),

$$\ln\left(\frac{I}{I_o}\right) = \left(\frac{-3V_S}{V_B}\right) \sum_{i=1}^{15} \frac{V_{mi}}{d_i} 1 \tag{16}$$

Therefore, the absolute value of the total particle concentration is given by

$$\frac{V_S}{V_B} = (C)_v = \frac{\ln(I/I_o)}{-3\Sigma(V_{mi}/d_i)1} \tag{17}$$

If $Q_i \neq 2$, for Anomalous diffraction, then

$$(C)_v = \frac{V_S}{V_B} = \frac{2 \ln(I/I_o)}{-3\Sigma(V_{mi}Q_i/d_i)1} \tag{18}$$

$$V_S = \frac{2V_B \ln(I/I_o)}{-3\Sigma(V_{mi}Q_i/d_i)1} \tag{19}$$

The theory developed thus far means that as the light intensity I changes, the change in V_S may be followed directly and the concentration of the sample in the cell may be found by dividing V_S by the beam volume.

<u>Non-spherical particles</u>

As stated earlier, the size distribution for non-spherical particles is expressed in terms of an equivalent sphere based upon the projected area of

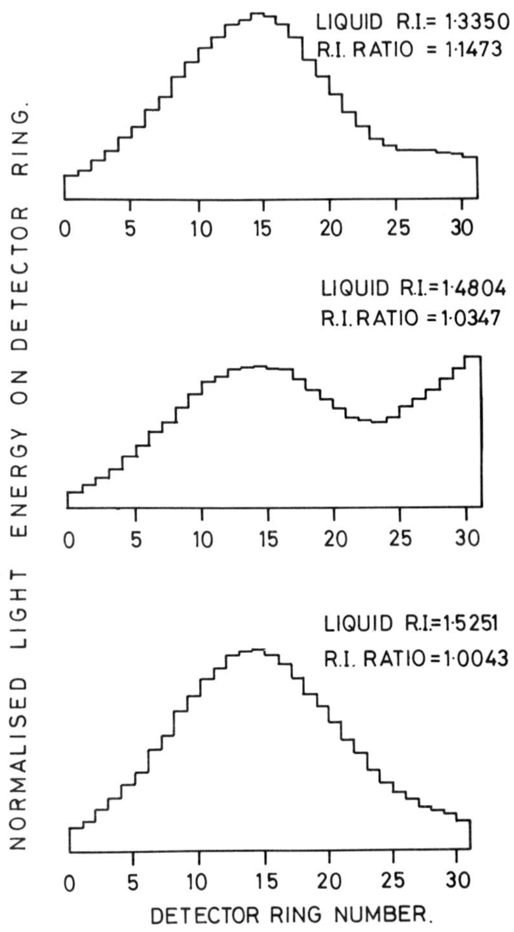

Figure 3. Change in LIght Energy Pattern with Change in Refractive Index Ratio.

the particle. Further calculation is necessary to obtain the actual volume
of non-spherical solid particles. The starting point is the theorem that
the average geometrical projected area of a convex particle with random
orientation is one quarter of the total surface area. This was first
demonstrated by Cauchy [1] and much later by van de Hulst [2] amongst
others. The procedure may be illustrated by taking the case of a cube.

If the projected area is taken to be that of a sphere of diameter d_s
instead of that of a cube of side d_c, then:

$$\frac{\pi d_s^2}{4} = 1.5 d_c^2$$

The volume of the equivalent sphere is thus:

$$\frac{\pi d_s^3}{6} = \frac{\pi}{6} \left(\frac{6}{\pi} \right)^{3/2} d_c^3 = 1.382 d_c^3$$

The volume recorded for the cube will be high, by a factor of 1.382 and
$d_c = 0.7236\, d_s$.

EXPERIMENTAL TECHNIQUE AND RESULTS

The apparatus consisted of either a 1mW or a 5 mW He/Ne laser fitted
with a spatial filter, a neutral density filter and a collimating lens so
as to provide a parallel beam of monochromatic coherent light. The
particles were placed in this beam, diameter 0.9 cm, and the diffracted
light was collected by a lens and brought to a focus on a special detector
consisting of thirty-one semi-circular photo-sensitive rings plus the
'centre spot.'

The particles were placed in the beam by putting a known weight of
them with a known volume of liquid in a small aluminium cell which had
circular windows of float glass, 14.3 mm apart, sealed into opposite sides.
The cell was so mounted that the windows were normal to the laser light
beam and adequate stirring to keep the crystals in suspension and give a
good circulation rate was provided by means of a magnetically-driven
stirrer mounted in the cell base.

True volume and weight measurements

In each experiment, the weight of particles used was in the range
2-20 mg, weighed to an accuracy of 0.1 mg; the volume of liquid used was
15 cm^3. Water was used for measurements on glass and silicon carbide. For
all other materials, the liquid used was propan-2-ol (density 780 kg m^{-3})
presaturated with the substance being measured.

Measurements were made on two different samples of glass spheres and
on potassium iodide, potassium aluminium sulphate dodecahydrate, sucrose,
copper sulphate pentahydrate and silicon carbide.

Details of the samples and the results obtained are shown in Table 1.

Anomalous diffraction experiments

The novel feature of the experimental technique and measurements,
described more fully elsewhere [3], was our use of two miscible liquids of
similar density but different refractive index. The liquids were 1-methyl

Table 1.   The ratio Measured Volume %/Predicted Volume %

Material	Formula and Density kg m^{-3}	Size Range Microns	Shape	Measured Volume % / Predicted Volume %	
				Calculated	Found
Glass (SRM 1003)	SiO_2 2380	5.0-39.0	Sphere	1	0.99
Glass (SRM 1004)	SiO_2 2460	30.3-261.6	Sphere	1	1.14
Potassium iodide	KI 3130	50.2-261.6	Cube	1.38	1.42
Potassium aluminium sulphate	$KAl(SO_4)_2 \cdot 12H_2O$ 1760	39.0-564.0	Octahedron	1.15-1.19	1.14
Sucrose	$C_{12}H_{22}O_{11}$ 1590	64.6-261.6	Monoclinic	1.49	1.46
Copper sulphate	$CuSO_4 \cdot 5H_2O$ 2280	39.6-564.0	(Irregular) Triclinic	-	1.25
Silicon carbide	SiC 3217	30.3-261.6	Irregular	-	1.34

naphthalene (Refractive index 1.617) and propan-2-ol (Refractive index
1.378). Using an addition-and-withdrawal technique, the liquid volume was
maintained constant whilst the refractive index of the liquid mixture
changed. Careful filtering of the liquids and cleaning of the optical
components ensured a high signal-to-noise ratio.

The solids used in the experiments were standard glass spheres and
sucrose crystals and the liquids were water, propan-2-ol, sucrose solution
and mixtures of propan-2-ol and 1-methyl naphthalene.

Measurements of light energy distribution are shown in Figure 3 for
glass spheres (Refractive index 1.5316) suspended in three liquid mixtures
of different refractive index and different values of m, the ratio of
refractive indices. The refractive indices were measured at 19°C for the
sodium D-line wavelength (λ = 0.5893 μm) and not for the He/Ne wavelength
(λ = 0.6328 μm). However, correction of refractive index for wavelength
and temperature gave values of m which were virtually identical with those
obtained using the sodium D-line data.

DISCUSSION

True volume and weight of particles

In Table 1, the ratios of measured to predicted volume are the slopes
of the lines obtained by a least-mean-squares analysis. For the glass
samples, the volume percentage predicted from the weight and density of the
spheres in a known volume of liquid was close to that measured for one

sample, confirming the applicability of the direct method without calibration. The poor results obtained with the coarser sample resulted from difficulties in obtaining a uniform suspension.

The correction for the shape produced a calculated volume ratio which was in good agreement with that found for potassium iodide, potassium aluminium sulphate and sucrose. For the commercial samples of copper sulphate and silicon carbide taken, the particle shapes were so irregular that the calculation of the correction factor was not made. For particles known to be irregular, it is less time-consuming to perform calibration experiments and then use the volume ratio so found to correct further experimental measurements rather than do a detailed analysis of photomicrographs.

It should be noted that the conditions for all of these experiments were such that Fraunhofer diffraction would be expected to occur, with $Q = 2$, since the value of m was large in each case.

Anomalous diffraction

The pattern of change in light energy distribution shown in Figure 3 is as would be expected from the theory presented previously and the results of calculations such as those plotted in Figure 2. At a high value of m there is the normal Fraunhofer diffraction. As m is reduced there is a relative movement of light energy to the outer detector rings (and smaller 'particle sizes'). When the value of m is close to unity, there is a loss of light energy in the outer rings corresponding to a reduction of the value of Q. This follows from the equation

$$Q = 2 - \frac{4}{\rho} \sin \rho + \frac{4}{\rho^2} (1 - \cos \rho), \tag{20}$$

where $\rho = \frac{4\pi a}{\lambda} |m-1|$

Each of the three histograms shown in Figure 3 came from measurements made on glass spheres with the same size distribution. Only in the case of the first histogram was the correct particle size distribution obtained.

The effect of correction for anomalous diffraction is illustrated by the data in Table 2 for sucrose crystals in sucrose solution.

The correction procedure removed spurious fine 'crystals' which had appeared to increase in quantity but not 'grow'; they were, in fact, an artefact of the measurement technique. There is great potential for the application to the sizing of other solid-in-liquid and liquid-in-liquid systems of some or all of the theory, techniques, results and conclusions described.

CONCLUSIONS

The small-angle laser diffraction sizing technique has been successfully extended from measurement of the size and relative concentration of spheres so that it may now be applied to:

1. The measurement of absolute concentration;

2. The measurement of size and concentration of convex solids;

3. The measurement of size and concentration of solids in liquids even when the values of their refractive indices are close together.

Table 2. Sucrose Crystal Size Distribution either Uncorrected or Corrected for Anomalous Diffraction

Equivalent sphere diameter Microns	Size Range Microns	Weight % Under		Weight % in Size Range	
		Uncorrected	Corrected	Uncorrected	Corrected
564.0		100.0	100.0		
	564.0 - 261.6			1.0	0.7
261.6		99.0	99.3		
	261.6 - 160.4			9.8	9.9
160.4		89.2	89.4		
	160.4 - 112.8			21.3	22.0
112.8		68.0	67.4		
	112.8 - 84.3			27.7	29.1
84.3		40.3	38.2		
	84.3 - 64.6			19.8	20.0
64.6		20.5	18.3		
	64.6 - 50.2			9.8	10.0
50.2		10.7	8.3		
	50.2 - 39.0			4.2	4.5
39.0		6.4	3.8		
	39.0 - 30.3			1.2	0.9
30.3		5.3	2.9		
	30.3 - 23.7			1.5	1.4
23.7		3.8	1.4		
	23.7 - 18.5			1.4	1.3
18.5		2.4	0.1		
	18.5 - 14.5			0.1	0.1
14.5		2.2	0.1		
	14.5 - 11.4			0	0
11.4		2.2	0		
	11.4 - 9.1			0	0
9.1		2.2	0		
	9.1 - 7.2			0.1	0
7.2		2.1	0		
	7.2 - 5.8			0.4	0
5.8		1.6	0		

ACKNOWLEDGMENT

The financial support of the Science and Engineering Research Council and British Sugar p.l.c. is gratefully acknowledged.

REFERENCES

1. A. Cauchy, 'Oevres Complètes d'Augustin Cauchy', 1er Sér, Vol II, 167, Gauthier-Villars, Paris (1908).
2. H. C. van de Hulst, 'Light scattering by small particles', 110, Dover, New York, 1981.
3. D. J. Brown, E. J. Weatherby and K. Alexander, 'An experimental study of anomalous diffraction'. To be published.

SCATTERED LIGHT PARTICLE SIZE COUNTING ANALYSIS:

INFLUENCE OF SHAPE AND STRUCTURE

Michael Bottlinger and Heinz Umhauer

Institut für Mechanische Verfahrenstechnik der
Universität Karlsruhe (TH)

CONCEPT

In various methods of the scattered light particle size
counting analysis the scattered light signals of <u>single</u> particles, which are crossing the measuring volume are analyzed
and registered as events (counted). The measured signal, resp.
the signal height, is the primarily determined property of
the particles. Generally the signal height is associated with
another dimension of dispersity, e.g. the diameter of the
particle.

For homogeneous, spherical particles this correlation
can be done with high accuracy mathematically (Mie-Theory)
as well as by measurements. If the light source and scattering
angle are suitable chosen, there is for spherical particles
of a specific material, a definite correlation between signal
height U and particle diameter x. For irregularly shaped
particles however and for particles with an internal structure,
the correlation is no longer unequivocal, because the orientation of the particle in the measuring volume influences
the signal height. The influence of the particle shape causes
a broadening of the measured distribution and a decrease in
resolution in comparison to methods insensitive to shape
effects.

To study the influence of the particle shape it is significant to record of individual particles of a specific
material, the representative spectrum of all possible signal
amplitudes reflecting all of the possible orientations of
the particle in the measuring volume (subsequently called
single particle impulse height spectrum h(U)'). This spectrum
gives a measure for the probability that a certain signal
height is measured when an irregularly shaped, randomly oriented particle crosses the measuring volume once. After having
determined the spectrum, the particle can be safely put aside.
It is now possible to measure the volume and surface of the
particle as well as characteristic parameters of the particle
shape. Thus it becomes possible to relate the value of a form

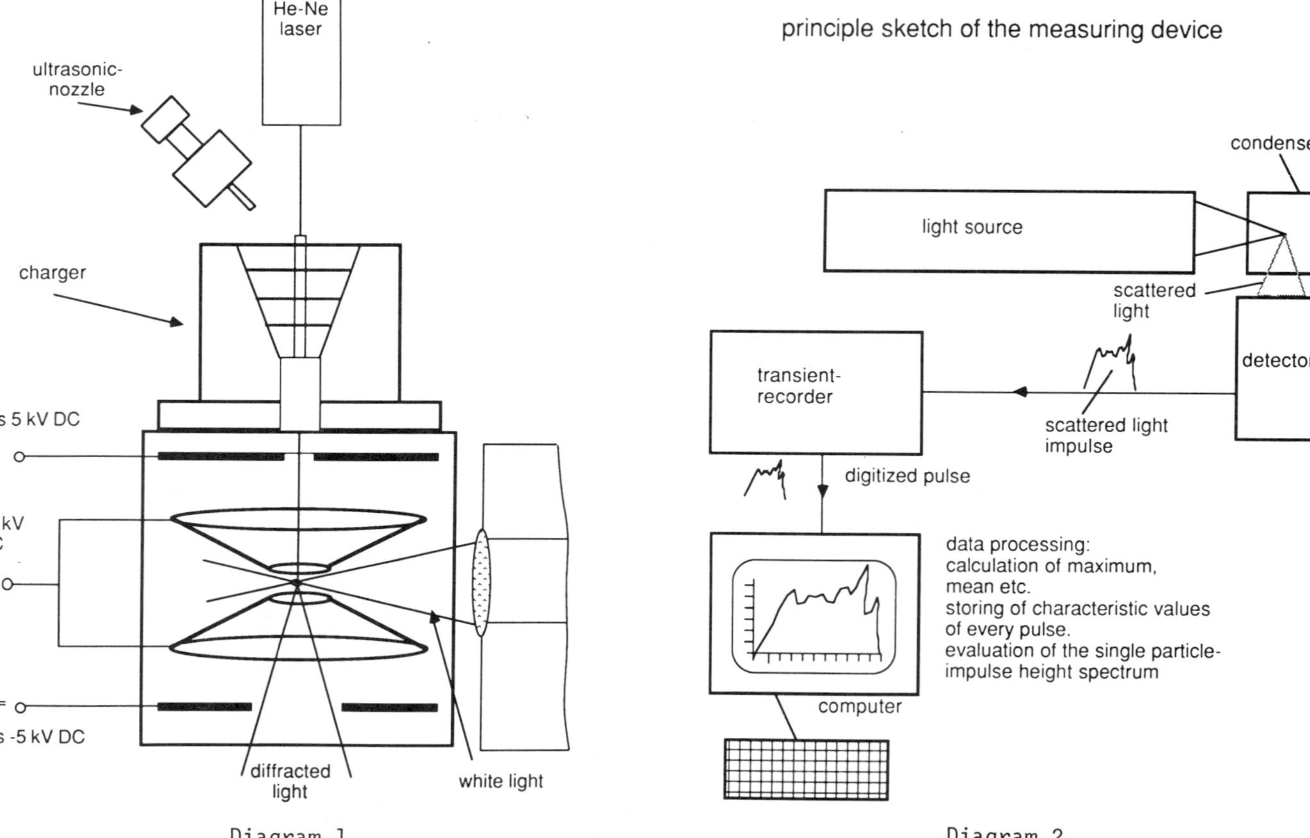

He-Ne laser
ultrasonic-nozzle
charger
0 bis 5 kV DC
=
0-10 kV AC
≈
=
0 bis -5 kV DC
diffracted light
white light
Diagram 1

principle sketch of the measuring device
light source
condenser
scattered light
detector
transient-recorder
scattered light impulse
digitized pulse
data processing:
calculation of maximum, mean etc.
storing of characteristic values of every pulse.
evaluation of the single particle-impulse height spectrum
computer
Diagram 2

independent property (e.g. the volume resp. the associated
equivalent diameter) to every impulse height spectrum.

The result of this correlation can be used as follows:

Choosing a characteristic value of the spectrum - for instance
the mean $\bar{U}$ - and relate it to the equivalent diameter x_V, a
calibration curve $x_V = K(U)$ for the investigated material is
obtained. Moreover the spectra which belong to particles in
the same size interval Δx_V can be combined, to form spectra
$h'(U)$, which are representative for the specific size intervals Δx_V. The calibration curve allows now the conversion of
every characteristic spectrum $h'(U)$ in a corresponding
spectrum $h^*(x_V')$.

Thereupon the result of a scattered light particle size
counting analysis of irregularly shaped particles can be expressed as a superposition of the representative spectra
$h^*(x_V')$, whereby every spectrum is weighted with the number of
particles in the corresponding size intervals Δx_V. The representative spectra again can be combined to form a transformation matrix T.

On the one hand it is now possible to simulate results of
measurements and to determine in this way quantitatively the
influence of the particle shape on the results of scattered
light measurements. If the transformation matrix can be inverted, it becomes on the other hand possible to correct subsequently results of real measurements and thus eliminate the
influence of the particle shape.

EXPERIMENTS

The measurements of the single particle impulse height
spectra were made by employing a scattered light counting
analyzer with a purely optically defined measuring volume |1|.
For electrodynamic suspending of single particles a modified
Millikan condensor |2| was used (diagr. 1). The measurements
were carried out as follows:

Particles of a specific material are fractionated by
sieving, dispersed by an ultrasonic-nozzle and charged
through contact with a potential of 10 kV. Particles reach
the inside of the condenser through the aperture of the upper
DC-electrode. A particle with suitable charge and mass is
captured and suspended by an inhomogeneous alternating field
produced by a special kind of ring-electrodes. If the AC-vol-
tage is suitably chosen then the particle is suspended
solely by the alternating field and is not oriented by an -
otherwise necessary - static field.

The particle swings with constant frequency and amplitude between the two ring-electrodes. The condenser is adjustable in all three dimensions, to be able to bring the
particle in the center of the measuring volume of the
scattered light particle size counting analyser. The particle
crosses the measuring volume during every oscillation with
random orientation and produces a scattered light signal
which is detected by a photomultiplier (scattering angle 90°).
Every signal is digitized by a transient recorder and

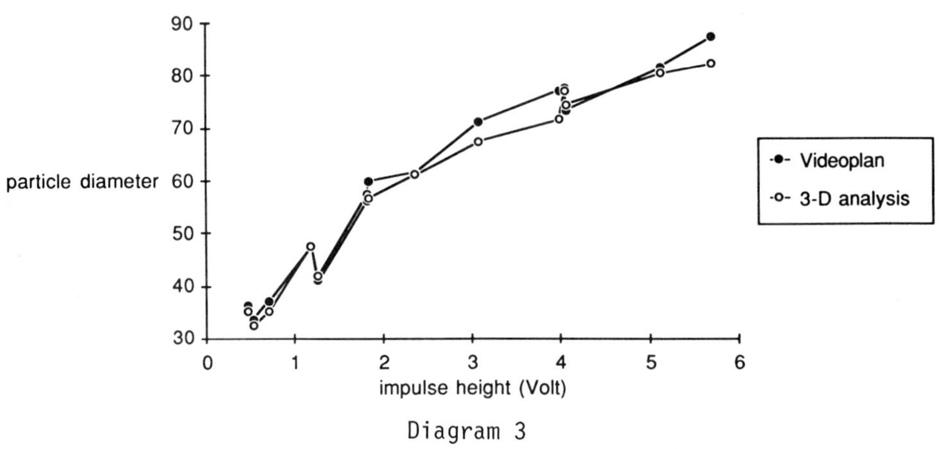

Diagram 3

calibration: limestone

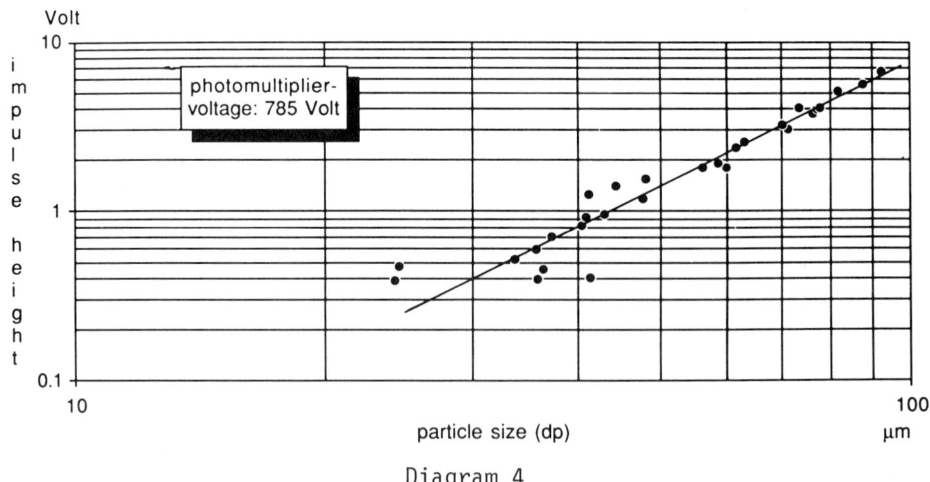

Diagram 4

transferred to a micro-computer (diagr. 2). The computer eva-
luates the signals and stores the characteristic values of
every pulse. Later the stored impulse heights (800-1500
values) are classified and the frequency distribution of the
scattered light-impulse height is generated (the formerly
mentioned single particle-impulse height spectrum).

After the measurement the particle is safely put aside
on an object slide. Pictures are taken from the particle from
three orthogonal directions. The areas of the three projected
images are determined by an image analyser (Kontron Video-
plan). From the mean of these values the diameter of the
sphere with the same projection area is computed (equivalent
diameter). Additionally the form of the particle is recorded
by a method for three dimensional form-analysis |3| and the
diameter of the sphere with same volume is calculated. The
results of the two methods are used as form independent
equivalent-diameters and corresponds almonst perfectly. Both
methods are suitable for determine form independent equivalent diameters (diagr. 3).

RESULTS

So far measurements with glass spheres, quartz and limestone particles have ben carried out. In this paper the results for limestone are presented.

Of this material particles with diameters between 20 and
100 µm were investigated. Diagram 4 shows the calibration
curve, obtained when the equivalent diameter x_V and the mean
$\bar{U}$ of the corresponding impulse height spectrum of every particle were related.

In diagram 5 the standard deviation over the mean of
every spectrum is plotted. Additionally the slant of every
spectrum was determined. The average of all slant values is
very near to zero.

From these results as well as from comparisons between
analytic distribution and measured spectra, it follows that
the impulse height spectra of limestone particles can be
described rather well by a normal distribution with a
standard deviation of approx. 12 % of the average.

SIMULATION AND INVERSION

As described above, it is possible to construct
representative spectra for every particle size class from the
measured single particle impulse height spectra when
characteristics attributes of the measured spectra are regarded. By combining these spectra to form a matrix $T(x_V', x_V)$
the result of a measurement, obtained from a conventional
scattered light counting analyzer can on principle be described by equation (1).

$$q_O^*(x_V') = \sum_{x_V min}^{x_V max} T(x_V', x_V) \cdot q_O(x_V) \cdot \Delta x \qquad (1)$$

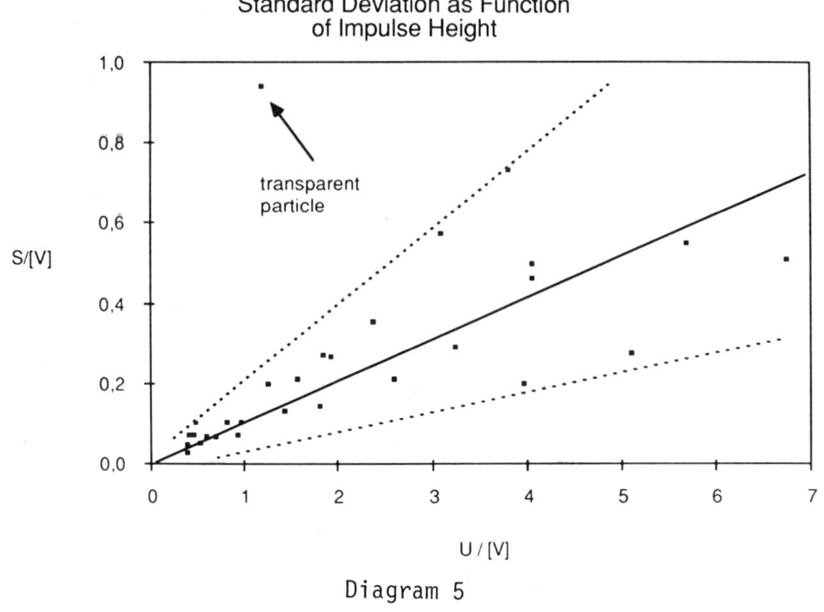

Diagram 5

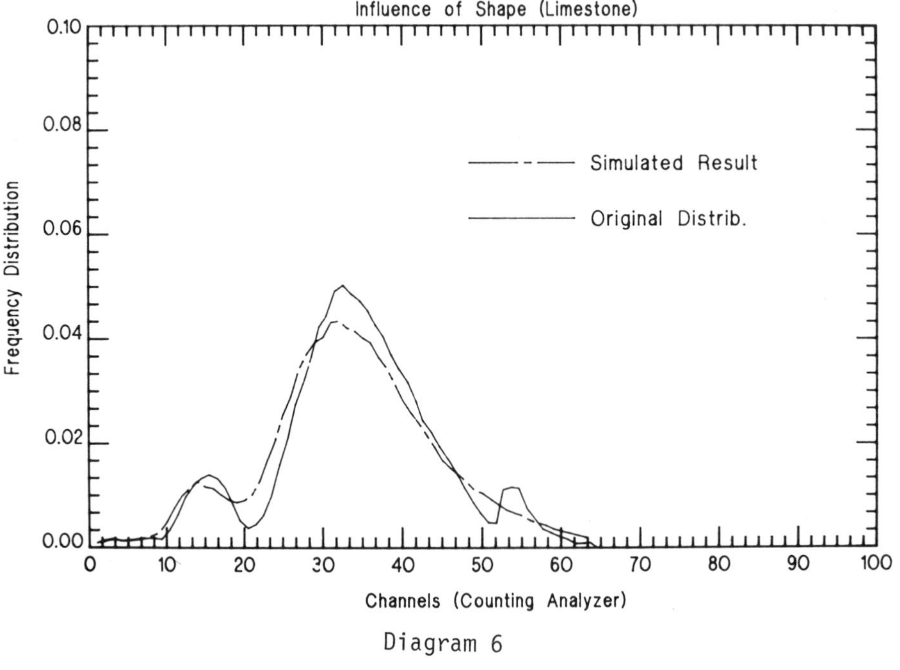

Diagram 6

Whereby $q_o^*(x_v')$ is the measured frequency distribution, which
is influenced by shape effects and $q_o(x_v)$ is the frequency
distribution, that could be determined by a method insensitive to shape effects. If q_o is given, the formula allows the
simulation of measurements to estimate the effect of the influence of particle form and structure. In the case of limestone, the transformation matrix can be calculated directly
from the calibration curve and the typical shape and width
of the determined single particle impulse height spectra.

If the calibration function is written as: $x = K(U)$ and
the inverse function as $U = K^{-1}(x)$ (U-impulse height in Volt,
x-particle diameter), it is valid that:

$$T(x', x_v) = K(Gauss\ (K^{-1}(x_v),\ K^{-1}(x'),\ K^{-1}(x') \cdot 0,12)) \quad (2)$$

$$1 \qquad\qquad 2 \qquad\qquad 3$$

Whereby the meaning of the function parameters is as follows
(from left to right):

notation commonly
used in literature

1. parameter x
2. average μ
3. standard deviation σ

With the equation (1) and (2) the influence of particle shape
can be simulated computationaly. Diagram 6 shows the effect
on a given frequency distribution of the equivalent diameter x_v. The calculated distribution is broader and the
maximum are less distinct.

In principle the inverse operation can be performed. It
is then possible to eliminate the influence of shape and
structure from the measurements. However the direct inversion
of the transformation matrix leads only in the case of ideal
representative measurements to the desired results. In fact,
other kinds of inversion techniques have to be used. In our
investigations, the <u>constrained linear inversion</u> by Twomey
has proven to be very useful.

REFERENCE

|1| Umhauer, H.   Particle Size Distribution Analysis
                            by Scattered Light Measurements
                            Using an Optically Defined Measuring
                            Volume
                            J. Aerosol Sci. 14 (1983) 6, 765/770

|2| Straubel, H.   Elektro-optische Messung von Aerosolen
                            Technisches Messen 48, Jahrg.,
                            1981, Heft 6

|3| Huller, D.   Quantitative Formanalyse von Partikeln
                            Dissertation Universität Karlsruhe (TH),
                            1984

VISIBLE INFRA-RED DOUBLE EXTINCTION MEASUREMENTS IN DENSELY
LADEN MEDIA, NEW PROGRESS.

G. Gouesbet, P. Gougeon, J.N. Le Toulouzan,
M. Thioye, and J.B. Guidt

Laboratoire d'Energétique des Systèmes et
Procédés,
INSA de Rouen, U.A. CNRS n°230, BP 08, 76130 Mont-
Saint-Aignan, France

INTRODUCTION

Optical techniques for measurements of particle sizes
and concentrations are usually limited to the investigation
of low optical thickness media. For instance, in LDA-based
systems, it is generally required, as a rule of thumb, that
the control volume should not contain simultaneously more
than one particle. Furthermore, the incoming beams to produce
the optical probe and the outgoing scattered light must not
be spoiled by the presence on the optical pathes of a too
large amount of refractive index inhomogeneities. Another
example is Malvern diffractometry (1,2,among many others).
Recent developments in this technique, including multiple
scattering corrections, permit to investigate media for which
the transmittance is as small as 2% (3,4), corresponding to
an optical thickness equal to 4. Although other progress are
likely, optical thicknesses limitations are expected to be
encountered in a near future (a dangerous statement indeed).
Unfortunately, industrial and laboratory applications
may require the use of optical diagnosis in densely laden
flows and systems at rest, for which multiple scattering
phenomena are significant. Examples are fluidized beds,
sedimentation devices, particle pneumatic transport, crystal
growth processes, coal Diesel engines, etc...
The aim of the present paper is to report new progress
in the development of a measurement technique (the so-called
Visible Infra-red Double Extinction (VIDE) method) well
suited for simultaneous measurements of particle sizes and
concentrations in densely laden media. Early states of
development can be found in refs 5,6,7. In the present stage,
measurements are possible for optical thicknesses up to 9,
corresponding to a transmittance of about 10^{-4}.

PRINCIPLE OF THE METHOD

We consider a slab of width L, embedding spherical, homogeneous, isotropic, non-magnetic scatterers of diameter d and complex refractive index m, with a number-density N. The particles are surrounded by a non-absorbing medium. Scattering by a discrete particle can be described by means of the Lorenz-Mie theory. Multiple scattering phenomena are accepted but dependent scattering, occuring when the distance between the particle is very small (8), can be neglected. We also assume a random distribution of the particles in the slab to dismiss interference effects. The theoretical analysis relies on a simple N-flux model described in refs 9,10.

The slab is illuminated by a (collimated) laser beam. We define the collimated-collimated transmittance τ_{cc} by the ratio of the transmitted collimated flux over the incident collimated flux. The medium surrounding the particles being the same as the one surrounding the slab, the transmittance τ_{cc} follows the Beer-Lambert law :

$$\tau_{cc} = \exp[-N \frac{\pi d^2}{4} Q_{ext} (d,\lambda,m) L] \tag{1}$$

in which λ is the light wave-length in the surrounding medium and Q_{ext} is the extinction efficiency factor depending on λ, d and m. Actually, d and λ are coupled to form a dimensionless group, the size parameter $\alpha=\pi d/\lambda$. The function Q_{ext} is given by the Lorenz-Mie theory and, when required, has been computed using the computer program Supermidi (11).

Relation (1) was originally designed in refs 9,10 for an incident collimated beam having an infinite lateral extension but it is also valid for the case of a laser beam, because the collimated flux does not gain (ideally) any energy from the diffuse radiation present in the slab.

In the VIDE-technique, we measure simultaneously the transmittance of the medium under study at two wave-lengths λ_1 et λ_2, leading to :

$$\frac{Ln \, \tau_{cc} (\lambda_1)}{Ln \, \tau_{cc} (\lambda_2)} = \frac{Q_{ext} (d,\lambda_1,m_1)}{Q_{ext} (d,\lambda_2,m_2)} = R(d) \tag{2}$$

in which m_1 and m_2 are the relative complex refractive indices of the particle material at the wave-lengths λ_1 and λ_2 respectively. When the right-hand-side R(d) is a sensitive and monotonic function of the diameter d, relation (2) enables us to determine a diameter d that we shall call d_{vide}. For λ_1, we choose a value for which the variation of Q_{ext} is important and monotonic on all the range of diameters to investigate. For λ_2, we choose a small wave-length (large size parameter) corresponding to a (nearly) constant value for Q_{ext}. For diameters ranging between 10 µm and 120 µm, a good choice is $\lambda_1 = 337$µm (HCN laser) and $\lambda_2 = 0.6328$µm (He-Ne laser).

To evaluate the variation of R(d) with respect to the diameter, we must know the nature of the particles and more precisely the optical properties at the two wave-lengths. There is usually no difficulty in the visible range because the literature is extensive enough. Furthermore, the size parameter being large, the value of Q_{ext} is nearly equal to 2, its asymptotic value, i.e it does not depend sensitively on m_2. The knowledge of m_1 is more important because it has a significant influence on R(d), particularly if we want to investigate particles which are transparent or nearly transparent.

Early measurements have been made on coal particles, with a specific application in mind (5,6,7). The determination of the optical properties of coal is discussed in these previous works. The present paper is devoted to the study of glass particles which are spherical and then approach more closely the assumptions behind the theory. The glass optical properties can be found in the literature for various mineral compositions, up to the far infra-red (12). After a chemical analysis of the used glass particles, we found that the experimental composition was almost similar to the one given in the reference (12), the difference lying on Na_2O and SiO_2 concentrations, and we retained the value $m_1 = 2.53 - 0.17i$. The computed ratio R(d) is shown in figure 1. From the experimental value R(d), the diameter d_{VIDE} is readily obtained. The particle number-density N_{VIDE} is afterward determined by the relation :

$$N_{VIDE} = -\frac{0.64\ \mathrm{Ln}\ \tau_{cc}\ (\lambda_2)}{L\ \ d_{VIDE}^2} \tag{3}$$

THE EXPERIMENTAL SET-UP

<u>Generalities</u>

The VIDE experimental set-up is shown in figure 2. It is basically a two wave-length (visible and far infra-red) photometer. Each channel is mainly constituted by a laser source (He-Ne or HCN), an optical system, and a detector associated with a lock-in amplifier. Signals from the lock-in amplifiers are fed to a two-channel recorder. All the optical apparatus, except the HCN laser, rests on a concrete block with anti-vibration supports. The medium under study is a suspension of glass particles created in a rotating and vibrating cylinder, similar to Ishihama and Enamoto's technique (13). It could also be a solid multiple scattering medium made of glass particles embedded in a polymer.

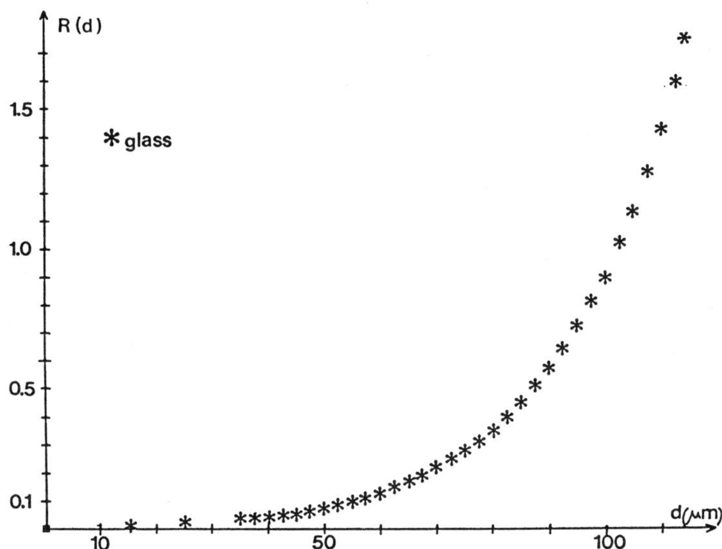

Fig. 1.   R(d) values versus particle diameters. Case of glass particles.

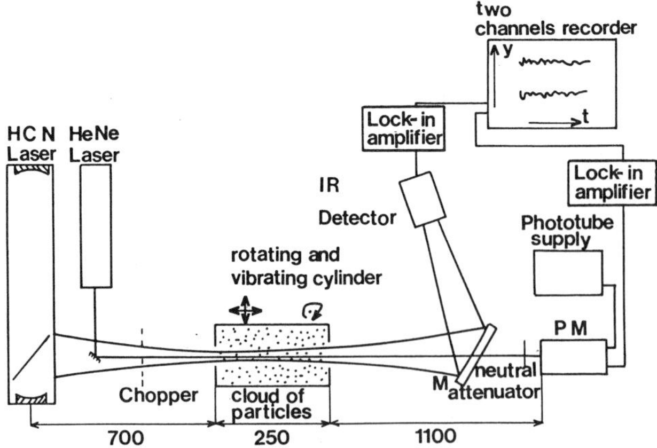

Fig. 2.   The VIDE experimental set-up. The medium represented is a
suspension of glass particles. Values are in mm.

<u>The infra-red channel</u>

The HCN laser source has been designed and assembled in the laboratory by Thenard (14) in collaboration with Veron (15). The wave-length of the radiation is 337 µm. Taking care of the composition of the gas mixture at the cavity input (CH_4, N_2, H_2) and for a good adjustment of the discharge current, the laser produces an output power equal to about 10mW in the TEM_{00}-mode, with a beam waist diameter of 21mm (at $1/e^2$-intensity) and a beam divergence equal to about 15 mrd.

The transmitted radiation is detected by a thermal detector after reflection of the beam by a concave mirror M having a 20 cm focal length and a 10 cm diameter. To measure very small transmittances (down to 10^{-4}), we used a pyroelectric detector (P4-45 Molectron) with a high sensitivity value (10^3V/W) at 337 µm. This value must be compared with the one for a Moll thermopile (10^{-1}V/W) used in our previous works (refs 5,6,7).

The beam is modulated at 75 Hz, this frequency being a good compromise between the requirement of linearity of the detector (controlled by means of an oscilloscope) and the necessity of a good noise equivalent power (we have for this value :NEP $=10^{-8}$W/(Hz)$^{1/2}$). The output of the detector is selectively amplified at the modulation frequency by means of a lock-in amplifier (Princeton Applied Research, 5207-model).

<u>The visible channel</u>

The visible source is a standard commercial He-Ne (Spectra-Physics) laser, working in the TEM_{00} mode with an output power of about 10 mW, a beam diameter of about 1 mm at $1/e^2$, and a 1 mrd divergence.

The two optical axes (infra-red and visible) are slightly misaligned for geometrical convenience.

The transmitted visible light is collected by a photomultiplier (RTC 2008), protected by an attenuator (MTO) with an optical density of about 4, to set the detector at a correct operating point.

The visible beam is also modulated with a frequency equal to 75 Hz. The transmitted light is amplified by a lock-in amplifier (Princeton Applied Research, 128-model).

<u>The collection angles</u>

For the visible light, the photomultiplier input pinhole is located at 1.225 m from the center of the medium under study and its diameter is about equal to 0.8mm in order to limit the collection of the diffuse radiation flowing in the same direction as the collimated transmitted laser beam. Then, the detector collection angle complies with a criterion given by Hodkinson (16). For single scattering, this criterion ensures that we effectively measure a true extinction, not spoiled by the amount of scattered radiation which also reaches the detector. This problem has been extensively discussed in the literature (see also refs 17,18). For multiple scattering problems, the criterion is expected to remain still sufficient because the scattering

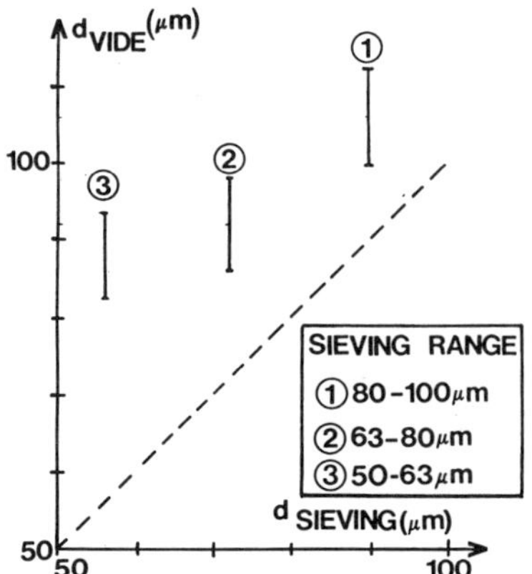

Fig. 3. d_{VIDE} diameters versus $d_{SIEVING}$ diameters for the three sieving ranges of glass particles.

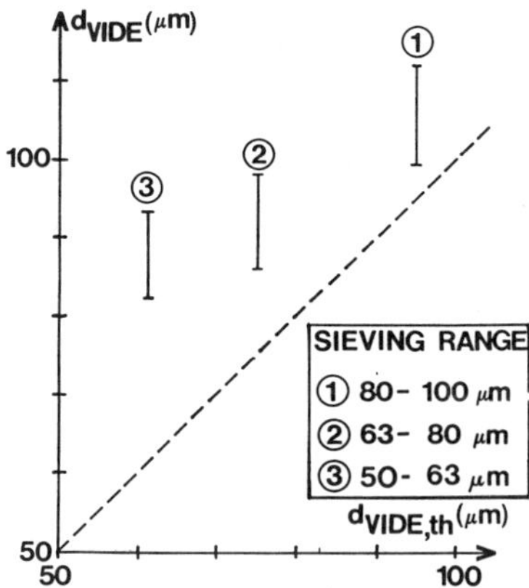

Fig. 4. d_{VIDE} diameters versus $d_{VIDE,th}$ diameters for the three sieving ranges of glass particles.

diagrams become more isotropic when the amount of multiple scattered light increases.

For the infra-red radiation, the collection angle is larger than for the visible light, but the problem is less stringent because the single scattering processes lead to scattering diagrams which approach isotropy, particularly for small diameters (small size parameters).

The media under study

The first kind of medium under study was a suspension of particles in air produced by the rotation (around an horizontal axis) and the vibration of a cylinder, as done similarly by Ishihama and Enamoto (13) for the study of dust explosions. At the beginning of the experiments, the particles are deposited as uniformly as possible at the bottom of the internal surface of the cylinder (having a diameter of 50 mm and a length L=250 mm). The cylinder is metallic and its internal surface is covered by a felt-like material to ensure an efficient dragging away of the particles when the system is rotating. This device is mechanically independent of the two wave-length photometer to prevent vibration disturbances of the measurements.

The two vertical limiting sides are window-free to avoid corrections due to specular reflections at the system boundaries, enabling us to use Relation (1). Rotation velocity and vibration amplitude were carefully adjusted, after several preliminary experiments, to obtain a satisfactory behaviour of the device.

The system meets the following requirements :

(i) A capacity for producing a medium of sufficiently high optical thicknesses. In the case of glass particles , the lowest value for the transmittance T_{cc} was 10^{-4} (optical thickness equal to about 9), for particles having a diameter equal to 100 µm.

(ii) Experiments which are easy to reproduce with reasonable accuracy.

(iii) Optical access.

(iv) Homogeneity. Longitudinal homogeneity (invariance of number-density along the cylinder axis) is not easy to check and has been assumed. Conversely, radial homogeneity has been checked by measuring visible transmittances for various values of the distance r to the axis of the cylinder, from r=0 (axis) to r=20 mm (geometrical limitation).

(v) Steadiness during a period of time long enough to perform the measurements. Reasonable steadiness of the transmittance is observed for more than one minute with fluctuations smaller than 20%.

The particles are carefully prepared using a standard procedure which is repeated in exactly the same way for each experimental run. To qualify the VIDE-technique, we used a monosized distribution of particles, as narrow as possible (within reasonable cost). To meet this requirement, the particles are sieved by means of an Alpine aspiration system (19) which combines the effect of an air jet cleaning the meshes and breaking up possible aggregates, and of an aspiration system for the removal of the finest particles. The sieves comply with the French standard AFNOR (NFX.11.501, Oct 1968) and allow the selection of particle samples, with

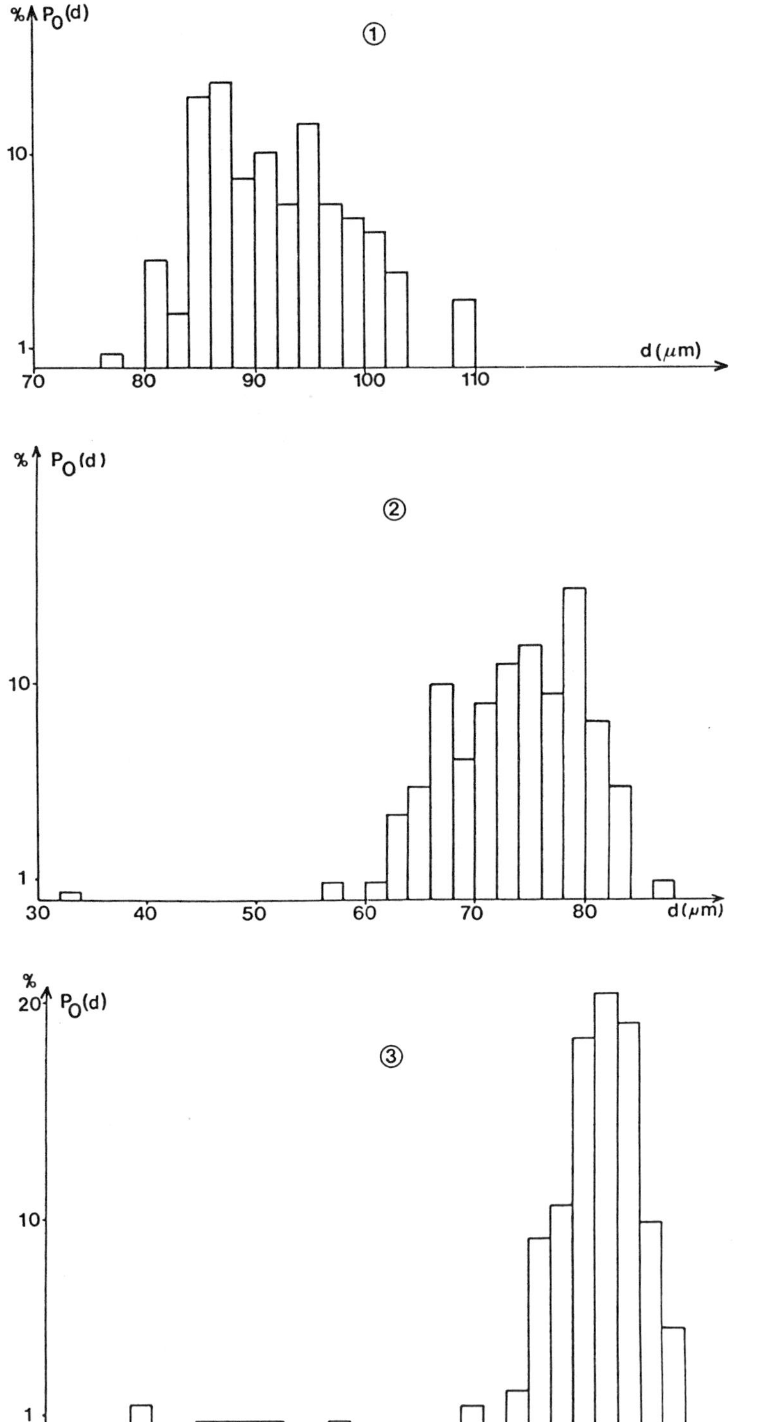

Fig. 5. Glass particles, size distribution given by optical microscopy

① Case of 80-100 μm sieved particles

② Case of 63- 80 μm sieved particles

③ Case of 50- 63 μm sieved particles

diameters in the ranges 20/32, 32/40, 40/50, 50/63, 63/80, and 80/100 µm. These ranges correspond to mean arithmetic nominal values, designated by $d_{sieving}$, equal to 26, 36, 45, 56, 72, and 90 µm respectively. This sieving operation is a very time-consuming one. Furthermore, after their preparation, the samples must be stocked under dry air conditions.

The second medium under study is a solid multiple scattering medium produced by embedding glass particles in a solid material. Such a medium is permanent and can be used to calibrate number-densities or sizes for various optical sizing instruments. Standard multiple scattering media can be produced by selecting the nature, the shape, the number-density and the size distribution of the particles.

For use with the VIDE-technique, the bulk material surrounding the particles must comply with the following requirements :

(i) Absorption coefficient at both wave-lengths small enough. To give a figure, we can for instance accept a transmittance equal to 80% for a sample having a geometrical thickness equal to 2 mm.

(ii) Mechanical and thermophysical properties permitting to homogeneously spread the particles in the bulk material, using a procedure to define, for diameters in the range 10-100µm, with volume fractions up to 50%

(iii) Chemical compatibility between the bulk and the material, and chemical stability of the samples.

Mineral materials such as quartz are widely used in optics, both in the visible and submillimetric ranges, but does not permit the embedding of particles. Conversely, embedding is possible with polymers whose plasticity can be temperature-controlled but the requirement of low absorption is difficult to meet. Actually, we find only one polymer meeting all the above requirements : the poly 4-methyl 1-pentene (TPX, seller : Aldrich).

RESULTS AND DISCUSSION

Glass particles in air

The (spherical) glass particles used in this work were easier to select by sieving than the (non-spherical) coal ones we previously used (5,6,7), and were expected to provide a more academic qualification of the VIDE-technique. Three sieving ranges (from 50 to 100 µm) have been selected, using the procedure described above. But, glass particles being denser than coal particles, the cloud of particles has been found more difficult to produce. A new definition of the cylinder operating condition (rotation velocity and vibration amplitude) has been required.

Results are shown in figure 3, exhibiting the d_{vide}-values versus the $d_{sieving}$-values. For each sieving range, various clouds with different number-densities, up to 3.10^9 m^{-3} for the sieving range 63/80 µm, have been studied. The d_{vide}-values does not significantly depend on the N_{vide}-values, as expected. The vertical bars designate the spreading of experimental results when the same experiments are repeated.

In all cases, we observe that d_{VIDE} is greater than the nominal $d_{sieving}$-values, by 18%,28% and 56% for the cases labelled 1,2,3 respectively. Similar results were obtained in our previous works on coal particles. At this time, the fact that the d_{VIDE}-values agreed very well with Malvern diffractometry (for low number-densities where both optical techniques, VIDE and Malvern, could work simultaneously) led us to the conclusion that the disagreement between optical and sieving diameters was not due to a defect of the optical system nor to an inadequation of the basic theory. This statement is not modified by the present work. Furthermore,we attributed the disagreement to the non-spherical character of the coal particles, producing for instance a systematic bias during the sieving process. Among other reasons, the present experiments on spherical glass particles were designed to test the above assumption. The figure 3 shows that the disagreement under discussion must be attributed to another cause.

Researching for such a cause, we examined the problem of departure of the particle size distribution from a monosized distribution. From an experimental point of view, the size distribution of the three sieving ranges have been independently measured by means of an image analysis system coupled with an automatic optical microscope (laboratory UA.CNRS.251, Caen,France). The results (fig. 5) show that the particles are far from being monosized, with the presence of particle diameters counted outside of the sieving range, particularly for the case 3 of the smallest sieving range. Consequently, for a more precise interpretation of the net experimental data, Relation (1) must be replaced by :

$$\tau_{cc} = \exp[-\frac{\pi.L}{4} . \int_0^\infty n(d) \, d^2 \, Q_{ext} \, (d,\lambda,m). \, d(d)] \tag{4}$$

in which $n(d)dd$ is the particle number-density between d and d+dd.

Using relation (4), not forgetting that the extinction efficiency factor is equal to its asymptotic value (equal to 2) for the visible light, and assuming a size distribution, the measured d_{VIDE}-value should ideally be equal to $d_{VIDE,th}$ given by :

$$\int_0^\infty n_o(d) \, d^2 \, Q_{ext} \, (d,\lambda_1,m_1) \, d(d) =$$
$$Q_{ext} \, (d_{VIDE,th},\lambda_1,m_1). \int_0^\infty n_o(d) \, d^2 \, d(d) \tag{5}$$

in which $n_o(d)$ is the normalized $n(d)$ according to :

$$\int_0^\infty n_o(d) \, d(d) = 1 \tag{6}$$

In figure 5, $P_o(d)$ is given by :

$$P_o(d) = n_o(d) \, \delta d \tag{7}$$

in which δd is the histogram width class (equal to 2 µm in the present case).

Clearly, a comparison between the VIDE-values d_{vide} and $d_{vide,th}$ is a better test to discuss the validity of the technique than a comparison between the values d_{vide} and $d_{sieving}$ which own a different character. The d_{vide}-values are still greater (by 17%, 23%, 44% for the cases 1,2,3, respectively) than the $d_{vide,th}$-values. Note that, for the cases 1 and 2, the difference is typically 20% and would certainly satisfy an amount of users, particularly if we remember the difficulty of such measurements in high optical thickness media.

The remaining disagreement might be explained by the sensitivity of the measurements with respect to the size distributions, and accordingly to the fact that the size distribution measured by optical microscopy after sieving is certainly not identical to the actual one in the cylinder.

First, the automatic optical microscope does distinguish close and even stuck particles which would be treated by the VIDE-technique as a single equivalent particle. Secondly, extra-aggregates are expected to be formed in the rotating cylinder, a phenomenon presumably more important for finer particles (see case 3). Note also (see figure 2) that the sensitivity of the VIDE-technique is not so good for fine particles (say below 70 μm). This limitation was less severe for coal particles (5,6,7) due to the values of the optical properties. In any case, a good sensitivity could be recovered by modifying the infra-red wavelength.

To evaluate the effect of possible particle clustering, we superpose to the distribution given in figure 5, case 3, an extra class of particles, with an uniformly repartited number-density equal to 5% of the total number-density, and with a diameter between 110 and 120 μm (corresponding to two original particles stuck together). Then, the value of $d_{vide,th}$ is modified from 61 μm (figure 4, case 3) to 83 μm which would agree with the d_{vide} value.

Experiments with independent mode Malvern diffractometry (on the cloud inside the cylinder), under low number-density conditions, confirm this interpretation.

They give us actual in situ size distributions which, when used with relation (5), may provide $d_{vide,\ th}$- values agreeing with d_{vide} -values.

Solid standard media

Much effort has been devoted to the preparation of solid standard media. The quality of the preparation of the polymer from raw material monomer delivered by Aldrich, before casting operation, is very important to ensure good optical properties of the final sample. In the present stage, the samples have the shape of a 50 mm-diameter disk with a thickness varying from 3 to 30 mm. The operating procedure must meet the following requirements:

(i) Prevent the inclusion of undesirable particles which would spoil the quality of the medium.

(ii) Prevent the oxidation of the chain of the polymer, leading to an yellow colour of the samples, with a decrease of the transmittance.

These requirements are met by using an inert gas
environment during preparation, and casting the polymer by
using a single block instead of smaller blocks of polymer.
When the chemical procedure to prepare the samples is carried
out, the two working faces of the disk must be polished to
ensure a satisfactory optical quality. The effort is
currently pursued to improve the fabrication process.

Infra-red transmittance preliminary measurements for
unloaded samples (pure polymer without particles) gave us a
transmittance equal to 55% for a sample of 10 mm-thickness,
to be compared with the expected theoretical value (70%).
Accordingly, the quality of the polymer preparation seems to
be fairly good . Conversely, the optical quality of the disk
faces in the visible range (polishing) is not good enough at
the present stage of development, leading to too much
speckles.
Also, preliminary tests concerning the homogeneity of the
spatial particle repartition show that the number-density is
smaller near the two disk faces than in the bulk. In the
bulk, however, the spatial repartition seems to be fairly
homogeneous.

CONCLUSION

A Visible Infra-red Double Extinction technique for
simultaneously measuring sizes and concentrations of
particles in densely laden media is described. The principle
is briefly recalled. The experimental set-up is described.
Results for glass particles in air are given and commented.
Taking size distributions into account, it is concluded that
the VIDE-technique is well suited for such measurements. At
the present time, data can be obtained for optical
thicknesses up to 9. The opportunity is taken to also discuss
the preparation of standard samples obtained by inclusion of
particles in a polymer bulk. Although the procedure of
preparation and the quality of the samples are not yet fully
satisfactory, the possibility of using such samples in future
work is very promising.

Acknowledgments

The authors wish to thank Mrs A.Gaudichet (Laboratoire
d'Etude des Particules Inhalées, DASES, Paris) for chemical
analysis of glass particles, the Laboratoire d'Analyse
d'Images, UA.CNRS.251, Caen, for automatic optical microscopy
of glass particles, and Mr.F.Chabot (Laboratoire des
Substances Macromoléculaires, UA.CNRS.500,ROUEN INSA) for aid
in polymer preparation.
This work has been supported by the Ministère de la
Recherche et de la Technologie (MRT), in collaboration with
BERTIN Company, Grant n° 85EO184, MRT Grant n° 05T1050, and
by the Centre National de la Recherche Scientifique, DVAR, n°
1303.

REFERENCES

1.P.G.Felton, A.A.Hamidi, and A.K.Aigal. Measurement of drop size distribution in dense sprays by laser diffraction, Proceedings of ICLASS'85 (Institute of Energy), London (1985).

2.D.Allano, M.Ledoux, and D.Lisiecki. Etude de quelques aspects de l'utilisation du granulometre ST 2200. 6th International Congress of the International Society for Aerosols in Medicine, Oct 2-4, Vichy, France (1986).

3.E.D.Hirleman. Modeling of multiple scattering effects in Fraunhofer diffraction particle size analysis. Spring Meeting, Western States Section, The Combustion Institute, April 27-29, Banff, Alberta, Canada, 1986.

4.A.A.Hamidi and J.Swithenbank. Treatment of multiple scattering of light in laser diffraction measurement technique in dense sprays and particle fields. Journal of the Institute of Energy, June, 101, 1986.

5.P.Gougeon, J.N.Le Toulouzan, C.Thenard, and G. Gouesbet. Simultaneous measurements of sizes and concentrations of coal particles in multiple scattering media by means of a double extinction technique. Proceedings of the Conference organized by the Particle Size Group of the Analytical Division of the Royal Society of Chemistry, Sept 16-19, Bradford, England, 1985.

6.G.Gouesbet, P.Gougeon, J.N.Le Toulouzan, B.Maheu and M.Thoye. New progress in optical diagnosis in densely laden flows. Proceedings of ICALEO'86, Nov. 10-13, Arlington, Virginia, Vol 58, Flow and Particle Diagnostics, 1986.

7.P.Gougeon, J.N.Le Toulouzan, G.Gouesbet, C.Thenard. Optical measurements of particle size and concentration in densely laden media using a visible infra-red double extinction technique, Journal of Scientific Instruments, To be published.

8.L.P.Bayvel and A.R.Jones. Electromagnetic scattering and its applications, Applied Science Publisher, London, 1981.

9.B.Maheu,J.N.Le Toulouzan, and G.Gouesbet. Four-flux models to solve the scattering transfer equation in terms of Lorenz-Mie parameters, Applied Optics, 23, 3353, 1984.

10.B.Maheu and G.Gouesbet. Four-flux models to solve the scattering transfer equation in terms of Lorenz-Mie parameters, special cases. Applied Optics, 25, 1122, 1985.

11.G.Grehan and G.Gouesbet. Mie theory calculations : new progress, with emphasis on particle sizing. Applied Optics. 18, 3489, 1979.

12.J.R.Birch, R.J.Cook, A.F.Harding, R.G.Jones and G.D.Price. The optical constants of ordinary glass from 0.29 to 4000 cm^{-1}.J.Physics D : Applied Physics, 8, 1353, 1975.

13.W.Ishihama and H. Enamoto. New experimental method for studies of dust explosions. Combustion and Flames, 21, 177, 1973.

14.C.Thenard. Plasma de xenon cree par onde de choc : etude de la cinetique de creation des electrons par interferometrie submillimetrique. These d'Etat, Rouen, France, 1980.

15.P.Belland and D.Veron. A compact cw HCN laser with high stability and power output. Opt. Commun. 9,146, 1973.

16.J.R.Hodkinson. Aerosol Science. Academic Press, New-York, 1966.

17.R.O.Gumprecht and C.M.Sliepcevich. Scattering of light by large spherical particles. J. Phys. Chem. 57,90,1953.

18.Z.G.Habib and P.Vervisch. The Mie theory and the nature of attenuation of thermal radiation in pulverized coal flames. NATO Workshop : Fundamentals of physical-chemistry of pulverized coal combustion. Les Arcs, France, July 1986.

19.Alpine. Service Instruction BV 390/15F. Alpine A 200LS, 1973.

OBSERVATIONS OF ELASTIC ANGULAR SCATTERING
FROM ORIENTED SINGLE CYLINDERS

J. D. Eversole*, H.-B. Lin, and A. J. Campillo

Optical Sciences Division, Naval Research Laboratory
Washington D.C. 20375
*Potomac Photonics Inc., College Park, MD 20742

Abstract

Elastic scattering measurements and observations have been made on
single cylindrical particles suspended in an electrodynamic trap. Under
certain circumstances the particles could be maintained in a stationary position
and orientation, and scattered light intensities from a He-Ne laser were obtained
as a function of angle. The particles were made by cutting and breaking glass
wool fibers which were then sorted by filtering. Suspended particles were
cylinders approximately 7 micrometers in diameter with aspect ratios from 2 to
20. Photomicrographs were taken to determine the particle size, quality of the
cylinder ends, and the extent of small dust or debri attached to the particle
surface. Angular scattering patterns of the longer cylinders resembled the
"infinite" cylinder case obtained by scattering from a macroscopic fiber. Minor
differences observed in the shortest cylinders included a "broadening" of the
cone of scattered radiation which increases with decreasing scattering angle,
and a higher rate of intensity fall-off as a function of scattering angle.

Introduction

The work described here is part of an on-going program at the Naval
Research Laboratory to determine optical properties of characterizable,
nonspherical particles.   While a substantial amount of prior information has
been obtained for spherical particles over wide ranges of size and composition,
relatively little work has been done with nonspherical particles. Furthermore,
much of the available nonspherical particle data consists of an average over
either size, shape, and/or orientation. Rigorous comparison of detailed
theoretical and computational models requires experimental data for which
composition, size, shape, and orientation are known. Our approach has been to
examine single particles, and obtain data at fixed orientations. This was
achieved by suspending charged particles in an electrodynamic trap which has
been modified to permit access to a wide range of scattering angles. Our main
interest has been in rod-shaped or cylindrical particles with large aspect ratios,
which may have applications as obscurants.   Several candidate sources of such
particles exist.   However, in this study we have chosen to examine particles
produced from ~ 7 µm diameter pyrex fibers which have a well-characterized
scattering pattern as described below. Particles in this size range are easily
imaged, so that size and shape information was taken directly by microscope
observation and microphotography.

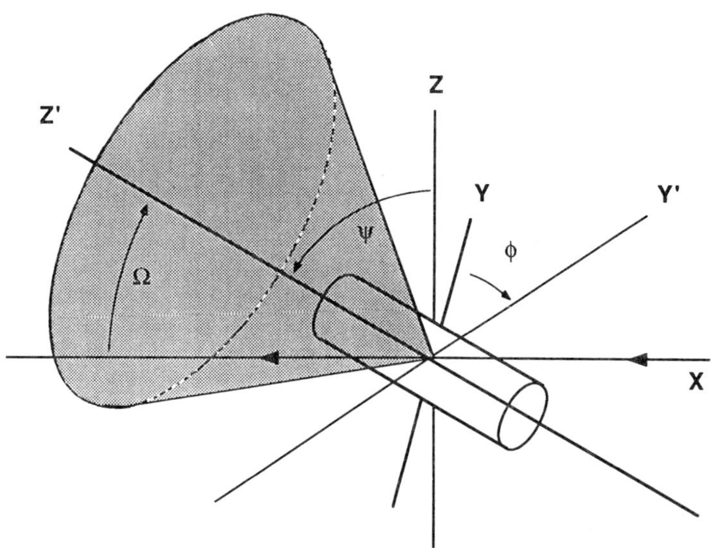

Figure 1. A schematic drawing of an arbitrarily oriented cylinder in the laboratory coordinate frame (unprimed). Radiation incident along the X axis is scattered outward along a conical surface defined by the half angle, Ω.

Exact light scattering solutions for the case of an infinite circular cylinder can be obtained as a series expansion of Bessel functions similar to Mie theory.[1-3] When the incident radiation and observation point are normal to the cylinder axis, the physical correspondence between a sphere and cylinder result in similar scattering patterns. The radiation pattern for the sphere is rotationally symmetric about the incident beam axis (for unpolarized radiation), and the scattering pattern is only a function of the spherical angle θ. For the infinite right cylinder there is no z dependence, and the scattering is only a function of the cylindrical angle θ. In either of these cases the angularly scattered light from a transparent material displays a familiar oscillatory structure in which the number of peaks per unit angle is indicative of the radius of the sphere or cylinder. More generally, the scattering pattern from an infinite cylinder with abitrary orientation can be described as a sheet of radiation whose wave front propagates outward from the cylinder along the surface of a cone[3]. Figure 1 defines this geometry for a cylinder whose coordinate frame (primed) is rotated with respect to the laboratory frame (unprimed) through an angle ψ about the Y axis, and then through an angle ϕ about the Z axis. The incident laser beam is coincident with the X axis.   The center line or symmetry axis of the cone is the same as the cylinder (Z'), and it has an apical half-angle of Ω, in such a way that the forward scattered beam is always included as a ray of the cone.   In the special case of normal incidence, when ψ goes to zero, the cone is flattened to the horizontal X-Y plane. It follows that when the scattered cone of radiation is intersected by a flat surfaced screen, the projected light forms a line which can be either hyperbolic, parabolic, elliptic, or straight depending on the angle of intersection of the cone and the screen. These features are easily demonstrated with any optical fiber and a low-power cw laser, and will be discussed later in the context of our observations of finite single cylinder scattering.

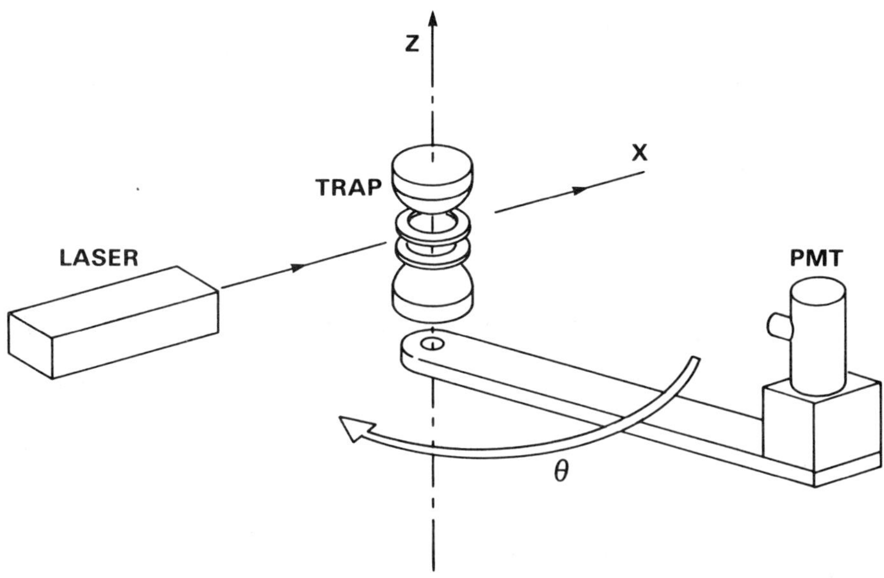

Figure 2. A schematic illustration of the basic experiment, showing the arrangement of the electrodynamic trap and light scattering apparatus.

Experimental Arrangement and Procedure

Figure 2 is a schematic diagram of the basic experimental configuration showing the trap electrodes in relation to the incident laser beam and the photomultiplier tube (PMT) for detecting scattered light. In practice, the maximum range of the scattering angle, θ, was physically constrained to ~120° due to the electrode support and trap enclosure. An enclosure was required to prevent laboratory air currents from disturbing the suspended particle, and collection of scattered light was obtained through a wide window. An unfocused 3 mW polarized He-Ne laser was used as a light source. The PMT (Hammamatsu R-985) was mounted on a metal beam projecting from a machinist's rotary table which was driven by an ac synchronous motor. The resulting scan speed was ~ 22°/min. Output of the PMT was amplified and plotted on a chart recorder. A narrow band filter blocked extraneous room light, and for all but the smallest particles, collection optics other than apertures to block stray light were not required. Angular resolution was determined simply by the solid angle of the PMT photocathode, and was ~0.3°.

Details of the trap electrodes are shown in Figure 3 which combines the electrical circuit diagram used with a scaled cross sectional drawing. The design of the electrodynamic trap was similar to that of previous investigators[4-8], but was distinguished from the classical bihyperboloidal trap by having the middle electrode split to permit optical access for angular scattering, and by the fact that the electrode surfaces were spherical rather than hyperboloidal. A detailed description of this arrangement and a comparison of its performance to other electrode geometries is available elsewhere[9], so that only a brief outline of its operation is given here. The ac trapping voltage applied to the center or ring

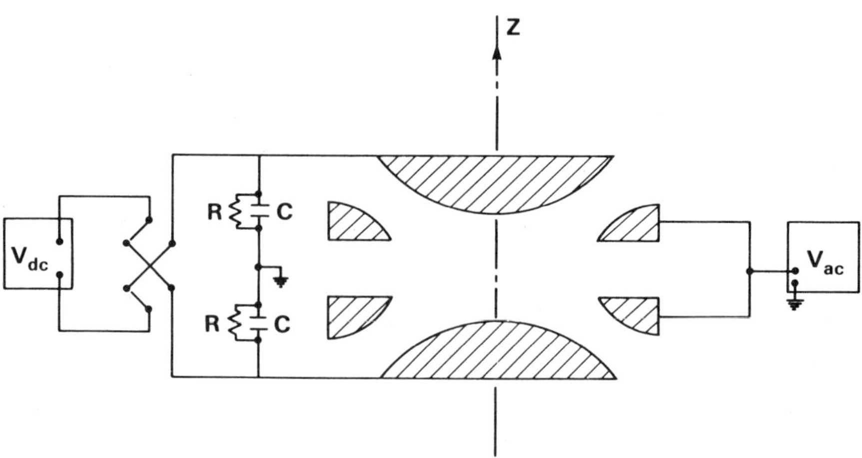

Figure 3. A schematic illustration of the circuit diagram for the electrodynamic trap, and a scaled cross sectional drawing of the electrodes.

electrodes was typically 600 V rms at ~100 - 200 Hz, while the dc balancing voltage was typically between 20 to 60 V. Particles were delivered to the trap by dragging a small polyethylene tube through a sample of the glass cylinders, by which some of them electrostatically attached to the tube. The tube was then lowered through a small hole in the upper end cap electrode and lightly tapped to allow some of the particles to fall through the trap. Application of the ac voltage at that point usually resulted in suspension of one or more of the particles. Multiple particles were eliminated from the trap by a process of momentarily reducing the ac voltage. With one particle suspended, the dc voltage was applied and increased until the null point was reached as determined by observation through a microscope.

As the particles were initially captured, they usually were oriented in the vertical direction. This natural tendency may be due, at least in part, to an uneven distribution of surface charge since the longer particles were observed more stably oriented than the shortest. It is interesting to note that such natural orientations were not dependent on the presence of the dc (balancing) field. In the absence of this field, the trapped particle's motion was oscillatory at the ac voltage frequency, and usually described an arc or curve. Nevertheless, in many cases it could be observed that a particular orientation was maintained throughout the oscillation period despite aerodynamic drag forces which would orient a rod-shaped particle in the direction of its motion.

Once the particle was stably suspended it was found that by decreasing the dc voltage, and decreasing the ac voltage frequency toward the "spring point" (frequency of unstable suspension) of the particle; that another stable orientation in the horizontal (X-Y) plane could be achieved. Sometimes a particle could be "flipped" from the horizontal back to the vertical orientation by similar manipulation of the suspending voltages. However it should be pointed out that some risk of losing the particle exists with each transition, so that flipping the same particle back and forth from vertical to horizontal several times was rare. Stable orientations at intermediate angles were observed only for irregular or odd-shaped particles. The angle of the cylinder with respect to the incident beam could be influenced with a charged electrode placed near the trap, and also by focussing of the laser beam.   Normally, once an orientation was obtained it would persist.   Microscopic observation of the particle's motion showed sporadic jumps in the particle position or orientation in which the particle would momentarily translate or rotate significantly and then return to its stable postion for a much longer period.   The forgoing description applies generally to all the observed particles.   For the particles with the shortest aspect ratios (less than 6:1), the orientational angles were less stable demonstrating both high frequency (wobble), and very low frequency (drift) oscillations in the particle attitude.

Results and Discussion

Nearly all of the scattering data was obtained with the laser beam horizontal and the cylindrical particles in the vertical orientation.   Under this circumstance, as discussed previously, the scattering is expected to be predominantly radiated in the horizontal (X-Y) plane.   Indeed, an intense line of light, modulated along its length, could be directly observed by holding a piece of thin white paper to the chamber window.   Since the chamber window was normal to the horizontal plane, the projected line of scattered radiation was straight.   For the infinite cylinder case, represented by a taut fiber, the width of this line was primarily determined by the diameter of the incident beam (~1 mm).   For the longer particles (aspect ratios greater than ~ 8) there were no differences discernable by eye in the appeerence of their scattered light pattern compared to the macroscopic fiber.   For a finite cylinder, some light scattering will occur in all directions due to the cylinder ends.   With normal incidence geometry this type of scattering was much less intense than the characteristic scattering pattern just described.   However, when a cylinder was oriented in the horizontal position, it was possible to align the incident beam along the cylinder axis.   In this circumstance the characteristic conical radiation pattern collapsed to a single ray coincident with the forward scattered beam, and scattering at all other angles, primarily due to the ends, greatly increased.   No quantitative results were recorded for this geometry since, as expected, the randomly rough ends of our samples did not produce any consistent pattern. For these cases, however, it is interesting to note that moving the incident beam slightly away (~10°) from the cylinder axis created an intense characteristic elliptical scattering pattern as projected onto the enclosure chamber interior.

Shorter particles were more problematic due to a combination of their greater tendency for rotational oscillation and the long data collection times. For particles which exhibited a wobble at the ac voltage frequency, the line of scattered light would appear to be broadened to as much as 10 mm at the chamber window.   For such particles, data collection was not usually satisfactory, and they were discarded.   On other occasions a low aspect ratio particle would drift slowly in its orientation angle, providing a narrow, well-defined line of scattered light which slowly moved around on the window-screen.   The shape of this line directly indicated the attitude of the particle.   When the particle was vertical, the line was straight.   If the particle tilted slightly about the X axis (defined by the incident beam) the projected line was a segment of a hyperbola in which the vertex was centered in the window. For small rotation about the Y axis, the line became a segment of one branch of the hyperbola.   Generally, any motion of the line was least at the edge of the window closest to the forward direction, since the cone of scattered radiation is constrained to include the forward scattered beam.   In order to obtain an angular spectrum of such a particle it was necessary to put two lenses in the collection path so that the vertical acceptance solid angle was increased to f 4. A slit was placed on the front lens to preserve the horizontal acceptance solid angle and maintain the same angular resolution.

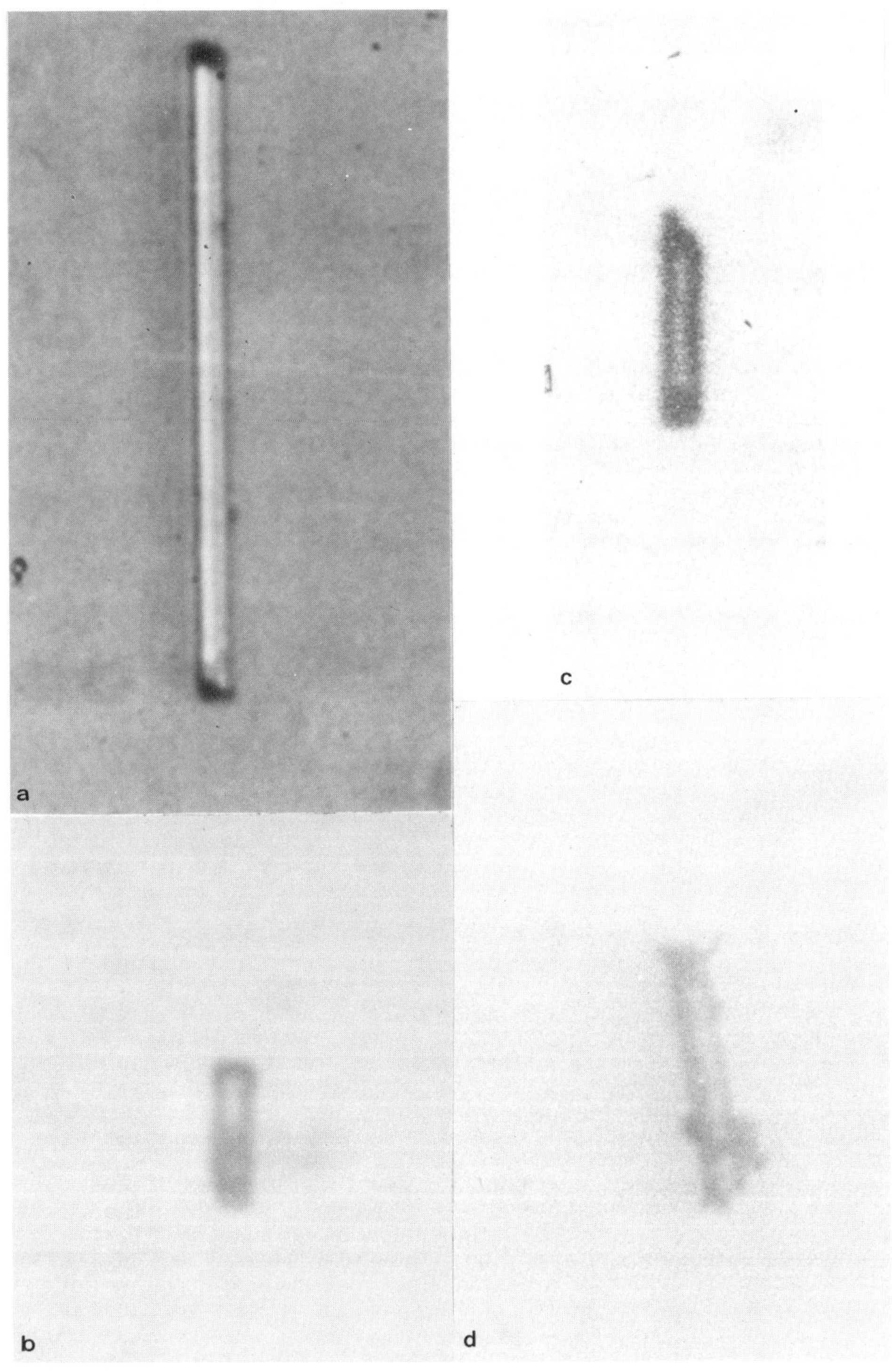

Figure 4. Four microphotographs of cylindrical particles of ~ 7 μm diameter and different aspect ratios as described in the text.

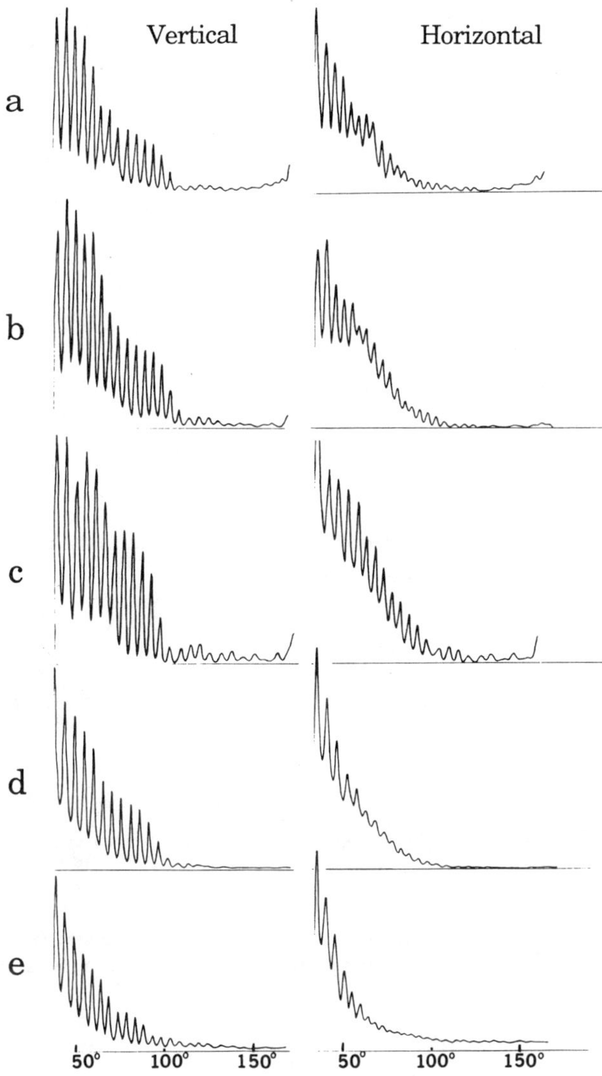

Figure 5. Scattergrams from various particles as described in the text. The left and right columns show vertical and horizontal incident polarizations respectively.

In two cases with the shortest particles (aspect ratios ~3 or less), a distinct broadening of the scattered radiation line was observed.   While the line remained narrow and sharp in the back scattered direction, it gradually became wider (~3X), and more fuzzy at the near-forward part of the window.   Curving the screen to keep the projected radiation equi-distant from the particle did not significantly alter this pattern.   While we have been unable to find any prediction or model for this observation, we believe it may be a characteristic of scattering from short aspect ratio cylinders.

Figure 4 shows photomicrographs of four selected particles all taken with a 47X magnification to the film, and subsequently enlarged the same amount. The longest cylinder, 4(a), has a measured diameter of ~6 μm, while the others are appoximately 7μm in diameter.Photos 4(a) and (b) are typical of the longest and shortest cylinders studied, with aspect ratios of 18.5 and 3.5 respectively. Generally the ends of the cylinders may be characterized as rough but flat, although there are occasional exceptions where an end may be angular as seen in Fig. 4(c). The sides of the suspended cylinders were typically uniform and clean, although again there were some clear cases of debri attached to the main particle as illustrated in Fig. 4(d). In this last example, the amount of excess material was sufficient to result in stable suspension at a noticible angle to the vertical. On two occasions particles composed of two distinct cylinders were stably suspended, but their orientation was not along either cylinder axis so that data could not be obtained. Observations of cylinders with smaller particles attached to their side showed that rotation about the cylinder axis also occurred.

Angular scattering in the horizontal plane was recorded from ~35° to 150°. Suspended cylinders oriented in the vertical direction provided well-defined and reproducible oscillatory patterns (scattergrams) over this range, with a rms signal to noise ratio for the longer particles was typically between 5 and 10%. Scattered light intensities from shorter particles were correspondingly weaker. In Figure 5 typical relative scattered light intensities are plotted versus angle for several different cases. Both vertical (TE) and horizontal (TM) incident beam polarizations were obtained as shown in the left and right columns respectively. The different cases are arranged by row from top to bottom: (a) ~8μm diameter sphere, (b) "infinite" fiber, (c) long finite cylinder (Fig. 4(a)), (d) intermediate length cylinder, and (e) short cylinder. Since the different particles were not precisely the same diameter, detail in the scattering patterns can not be compared. In general terms, however, the scattergram of the longest finite cylinder (c) resembles the patterns of the fiber (b) and the sphere (a) more closely than that of the shorter finite cylinders (d) and (e). From our observations, there is a general trend by which smaller aspect ratio cylinders were consistantly distinguished, which shows up best in scattering with a horizontally polarized beam. That is, the rate of decrease of the relative scattered light intensity as a function of angle is significantly higher for shorter cylinders than for either the longer cylinders, fiber, or sphere. As yet, no predictive calculations are available to compare with these observations, partly because of computational difficulty due to the large diameters of the cylinders compared to the incident light wavelength. We are currently attempting to extend this techique to cylinders with diameters of ~1 mm, and are optimistic that we wil l soon be able to compare experimental data with a finite cylinder scattering model.

References

1. H.C. van de Hulst, " Light Scattering by Small Particles ", Dover Publications, Inc.New York (1981), pp 297 - 328.
2. M. Kerker, " The Scattering of Light and Other Electromagnetic Radiation ", Academic Press, New York (1969), pp 255 - 310.
3. C. F. Bohren and D. R. Huffman, " Absorption and Scattering of Light by Small Particles ", Wiley-Interscience,New York (1983), pp 194 - 202.
4. E.J. Davis, Electrodynamic Balance Stability Characteristics and Application to the Study of Aerocolloidal Particles, Langmuir,1, 379 (1985).
5. S. Ataman and D.N. Hanson,Measurement of Charged Drops, Ind. Eng. Chem. Fundam., 8, 833, (1969), and J.W. Schweizer and D.N. Hanson, Stability Limit of Charged Drops,J. Colloid and Interface Sci., 35, 417 (1971).
6. T.G.O. Berg and T.A. Gaukler,Apparatus for the Study of Charged Particles and Droplets, AM. J. Phys., 37, 1013, (1969).
7. C.B. Richardson and J.F. Spann,Measurement of the Water Cycle in a Levitated Ammonium Sulfate Particle, J. Aerosol Sci., 15, 563, (1984).
8. R.F. Wuerker, H.M. Goldenberg, and R.V. Langmuir,Electrodynamic Containment of Charged Particles, J. Appl. Phys., 30, 441, (1959).
9. J. D. Eversole and H.-B. Lin, Effects of Design on Null Point Motion of Spheres Suspended in the Electrodynamic Balance, Rev. Sci. Inst. (1987), (in press).

EFFECT OF PARTICLE SHAPE ON THE RESPONSE OF

SINGLE PARTICLE OPTICAL COUNTERS

J. Gebhart and A. Anselm

Ges. f. Strahlen- u. Umweltforschung mbH
Paul-Ehrlich-Str. 20
D-6000 Frankfurt/Main

1. INTRODUCTION

An optical aerosol particle counter measures the size and number concentration of airborne particles in a limited size range by means of light
scattering on single particles. The amount of light which an individual
particle scatters is a function of its size, refractive index and shape.
Particle sizing based on this principle is known since more than 35 years.
Meanwhile the subject has been steadily developed and since about 25 years
optical particle counters using white light illumination are commercially
available. After the invention of the laser principle successful attempts
have been made to replace the white light illumination by coherent and
monochromatic laser light. Nowadays optical particle counters have found
useful applications in basic aerosol research, particle technology, air
pollution studies and clean room monitoring. Since in these fields of
research particles are mainly nonspherical it is of general interest how
optical particle counters respond to particles of irregular shape.

In the present work theoretical approximations valid for particle dia-
meters $d \ll \lambda$ and $d \gg \lambda$ (λ: wavelength of light) are used to derive some
general predictions about the influence of the particle shape on light
scattering. For $d \ll \lambda$ the particle oscillates like a dipol and the
scattering properties of a nonspherical particle can be expressed by its
polarizability. In the limiting case $d \gg \lambda$ the scattered light can be
considered as existing of three components which are due to the physical
effects of diffraction, reflection and refraction. The influence of the
particle shape on these components can then be considered separately.

The predictions derived from these approximations are compared with
experiments on optical particle counters using nonspherical particles.
Useful test aerosols are agglomerates of uniform polystyrene spheres and
mono-sized NaCl-particles before and after their conversion into saturated
NaCl-H$_2$O-droplets. The instruments include three light scattering spectro-
meters of the own laboratory and three commercial optical particle counters.
They differ in the kind of illumination (Laser or incandescent light),
the mean scattering angle and the receiver aperture. The experimental
findings are correlated with angular scattering patterns of nonspherical
particles obtained from microwave analog measurements (Zerull et al. 1977).

2. PARTICLES SMALLER THAN THE WAVELENGTH

2.1. Theoretical Considerations

When the particle size is much smaller than the wavelength the particle
is subjected to an almost uniform field. The particle then oscillates like
a dipole with polarization proportional to the electric field of the incident wave. The scattering properties of such a particle can be expressed
by its polarizability $\wp$ with is generally a tensor and reduces to a scalar
for a homogeneous sphere. In the dipol-approximation and for unpolarized
monochromatic light the power P of radiation scattered by a spherical
particle per unit solid angle in direction θ is given by:

$$P\,(\theta,\,d,\,\lambda,\,m)\; = \; I_o \cdot \frac{\lambda^2}{8\pi^2} \cdot \left(\frac{\pi\,d}{\lambda}\right)^6 \cdot \left|\frac{m^2-1}{m^2+2}\right|^2 \cdot \left(1+\cos^2\theta\right) \qquad (1)$$

where $m = n - ik$ is the refractive index of a particle and I_o the intensity
of the incident radiation. Equation (1) contains the polarizability
of a sphere which is independent of direction and turns out to be (van de
Hulst 1957):

$$\wp_s \; = \; 3 \cdot \left(\frac{m^2-1}{m^2+2}\right) \cdot V \qquad (2)$$

where V is the volume of the sphere. Substituting equation (2) into
equation (1) results in:

$$P\,(\theta,\,d,\,\lambda,\,m) = I_o \cdot \frac{\pi^2}{2\,\lambda^4} \cdot \wp_s^2 \cdot (1+\cos^2\theta) \qquad (3)$$

which means that the power of the scattered light is proportional to the
square of the polarizability. In the dipol approximation any anisotropy
of the particle can be described by three polarizabilities along the axes
of a cartesian coordinate system. In the case of a homogeneous ellipsoid
these axes coincide with those of the particle and the polarizabilities
along the three main axes of the ellipsoid can be calculated (van de Hulst
1957; Kerker 1969). In general an ellipsoid which oscillates along its
axis ν scatters light which is proportional to $\wp_\nu^2$.
The polarizability $\wp_\nu$ can be calculated by the general formula:

$$\wp_\nu \; = \; \frac{(m^2-1) \cdot V}{L_\nu\,(m^2-1)+1} \qquad (4)$$

whereby the factor L_ν depends on the ratio of the axes. Stimulation of
the dipol with the electric vector of the incident wave parallel to the
major axes of the ellipsoid produces more scattered light than an oscillation along its minor axis. For spherical particles $L_\nu = 1/3$, independently of direction, and $\wp_\nu$ gets identical with $\wp_s$ of equation (2). In
the case of a prolate spheroid with major axis a and minor axes b = c
L_a and $L_b = L_c$ can be calculated as function of the ratio a/b. Some
numerical results based on a formula of van de Hulst (1957) are listed
in Table 1.

Table 1. Light scattering diameter of prolate spheriods compared to the
diameter of a volume-equivalent sphere. Calculated for unpolarized light in dipol-approximation and refractive index of m=1.5.

ratio of semiaxes a/b	L_a	L_b	$\dfrac{S_a^2}{S_s^2}$	$\dfrac{S_b^2}{S_s^2}$	$\dfrac{S_a^2+S_b^2}{2\,S_s^2}$	$\left(\dfrac{S_a^2+S_b^2}{2\,S_s^2}\right)^{\frac{1}{6}}$
1.2	0.286	0.357	1.091	0.960	1.025	1.004
2	0.173	0.413	1.363	0.872	1.116	1.018
5	0.056	0.472	1.753	0.793	1.273	1.041

In accordance with equation (3) the power of scattered light of a
spheroid in relation to that of a sphere of equal volume and refractive
index is given by:

$$\frac{S_\nu^2}{S_s^2} = \left| \frac{m^2 + 2}{3 \cdot \left(L_\nu\,(m^2 - 1) + 1 \right)} \right|^2 \tag{5}$$

Corresponding light scattering ratios valid for prolate spheriods are also
contained in Table 1. For unpolarized light an average scattering ratio
$S_a^2 + S_b^2 / 2\,S_s^2$ has to be taken. As can be seen from the table a
value of a/b = 5 results in a variation of the scattering intensity
of a factor 1.273. But due to the d^6 – dependence of the scattered light
on the particle diameter, d, the light scattering diameter of such a
spheroid deviates only about 4% from that of a volume-equivalent sphere.

2.2. Experiments

For experimental investigations in the particle size range below the
wavelength of light agglomerates of uniform polystyrene spheres are useful objects to study the response of an optical particle counter to nonspherical particles. To analyse the scattering properties of agglomerates
of uniform spheres smaller than the wavelength of light optical particle
counters using laser light are advantageous because they combine high
sensitivity with good size resolution. In the following two such instruments
are presented together with their theoretical calibration curves. A laser
aerosol size spectrometer (LASS) developed in the own laboratory (Roth
and Gebhart, 1978) is shown in Fig. 1. The spectrometer is supplied with
either a helium-neon laser (λ = 0,63 µm) or an argon ion laser (λ =
0,514 µm). The laser light is focused by an astigmatic system of lenses
into the sensing volume. The light scattered by a single particle is
collected by a microscope objective under a mean scattering angle of 40°
and an aperture angle of 20° and is passed via a mirror to a photomultiplier. The aerosol nozzle is directed perpendicular to the laser beam and

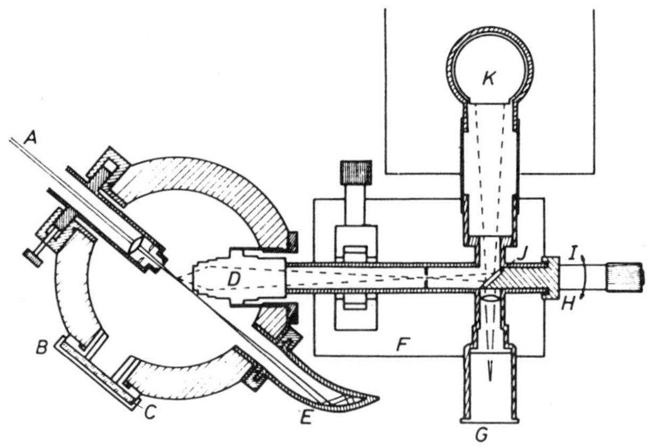

A: Laser light   D: Microscope objective   G: Eye piece   J: Mirror
B: Opaque material   E: Light trap   H: Observation   K: Photomultiplier
C: Glass window   F: Cross slide   I: Measurement

Fig. 1. Laser aerosol size spectrometer (LASS) for the submicron size range.

to the plane of observation. With the LASS particles between 0,08 µm and 0,6 µm in diameter can be classified with high size resolution. Some calculated response curves of the instrument are drawn in Fig. 2. They show the characteristic d^6 - dependence of dipol scattering up to particle diameters of about 2/3 of the wavelength. This means that for practical purposes particles with dimensions below 0,4 µm will behave like oscillating dipols.

A commercial laser particle counter is for instance the model Royco 236. Its optical sensor shown in Fig. 3 is manufactured by Particle Measuring Systems (PMS) and employs a high Q He-Ne-Laser cavity to achieve a high illumination intensity inside the active cavity. The primary collector of the scattered light is a parabolic reflector. Particles in the sample stream intersect the laser beam within the cavity at the focus of the parabolid which collimates the scattered light onto a 40° flat mirror. The light reflected from that mirror is refocused onto a photodetector by an aspheric lens. The whole system collects scattered light from 35° - 120° providing a 2.2 π steradian solid angle. The sensitivity depends on the sample flow rate and is 0,12 µm at 300 cm^3 min^{-1}. Theoretical calibration curves of the instrument valid for different refractive indices are drawn in Fig. 4. As can be seen the d^6-relationship holds up to particle sizes of about 1/2 of the wavelength, i.e. up to about 0.3 µm.

In conclusion it can be stated that response curves of optical particle counters can be approximated by a d^6-slope up to about d = 1/2 λ if the aperture includes rectangular scattering and up to about d = 2/3 λ for forward scattering (60° > θ > 20°). For low angle scattering (θ < 15°) this range can be even extended to about d = 3/4 λ. This means that for practical applications light scattering properties of particles below these size limits and with a refractive index m < 1.6 can be derived from the concept of an oscillating dipole and expressed in terms of polarizability.

Light scattering properties of agglomerates of uniform spheres may be characterized by a relative light scattering diameter as follows: Let $d_{sc,j}$ be the light scattering diameter of an agglomerate formed by j uniform spheres with diameter, d_1, and refractive index, m. Then a relative light scattering diameter, F_j, of the aggregates can be defined as:

$$F_j = \frac{d_{sc,j}}{d_1} \qquad (6)$$

In this definition $d_{sc,j}$ is the diameter of a sphere with refractive index m which scatters the same amount of light as an agglomerate of j uniform spheres of the same refractive index.

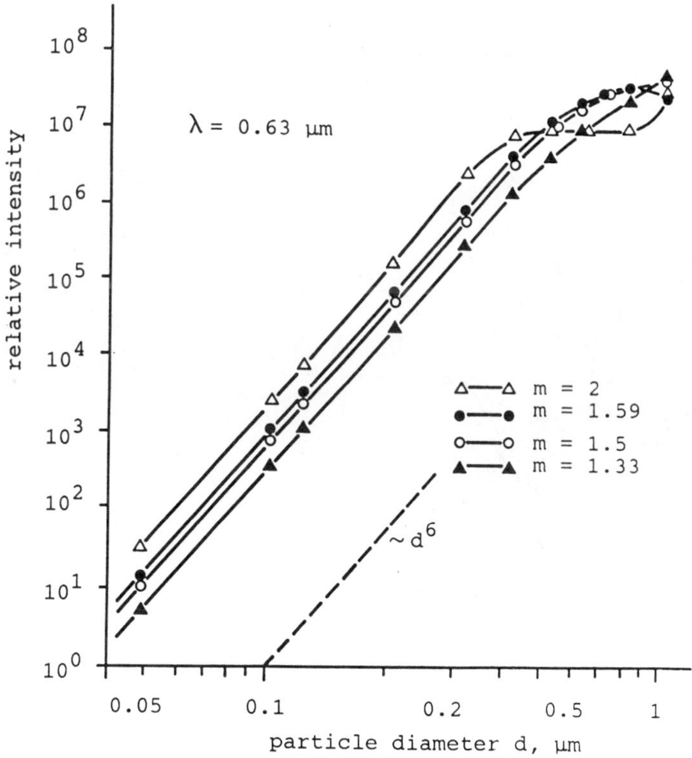

Fig. 2. Theoretical calibration curves of the Laser Aerosol size spectrometer (LASS).

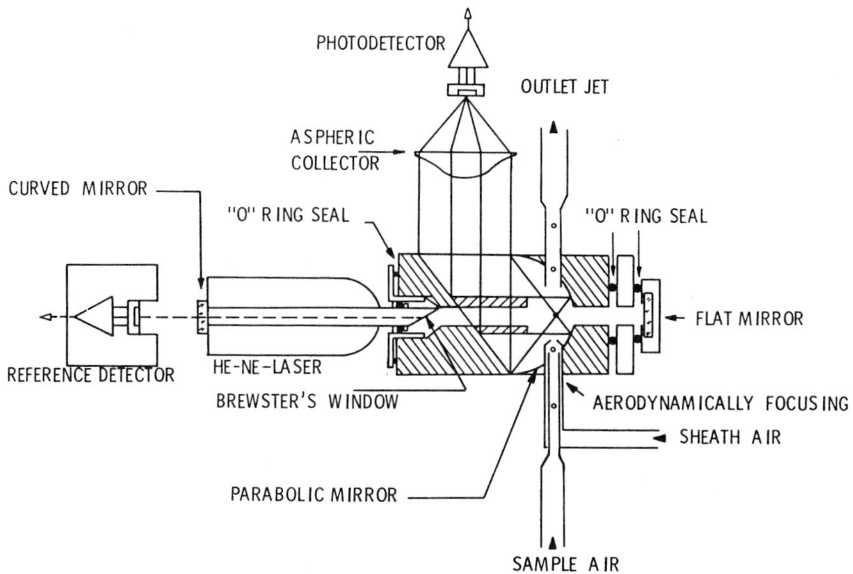

Fig. 3. PMS-Sensor of the Laser Particle Counter model Royco 236

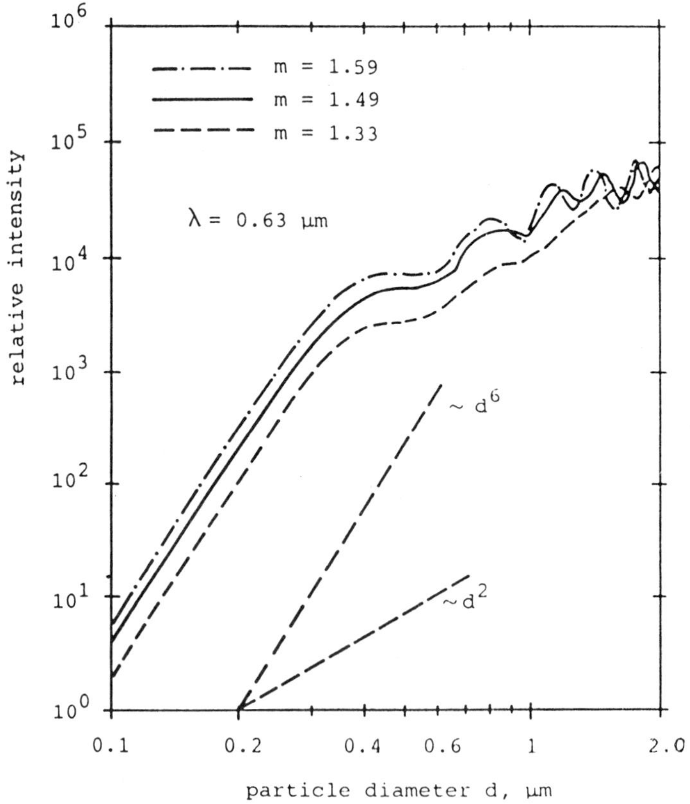

Fig. 4. Theoretical calibration curves of the Laser Particle Counter
Royco 236.

Some relative light scattering diameters measured with the LASS are summarized in Table 2. Due to the high resolution power of the instrument aggregates up to six spheres can be distinguished. As can be seen, $F_j = j^{1/3}$, so that within an experimental error of 3%, $d_{sc,j}$, is identical with the light scattering diameter of a sphere of equal volume. Chen et al. (1984) measured agglomerates of polystyrene spheres with the Royco 236 Laser counter and found also a volume-equivalent response for their submicron particles (Fig. 5).

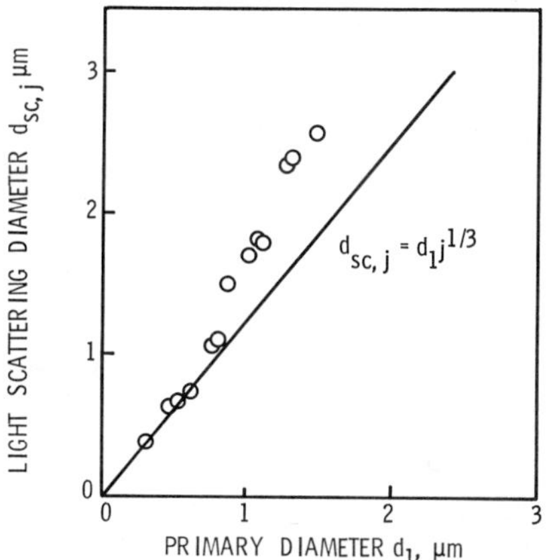

Fig. 5. Light scattering diameters $d_{sc,j}$ of doublet aggregates of polystyrene spheres measured with the Royco 236 Laser counter after Chen and Cheng (1984).

Microwave analog measurements on a cube comparable to the wavelength and on a cube larger than the wavelength (Zerull et al. 1977) are consistent with the results of Table 2 and Fig. 5. Whereas the mean scattering pattern of the smaller cube comes closely to the angular distribution of scattered light originating from a sphere of equal volume the mean scattering diagram of the larger cube deviates considerably from that of a sphere outside the range of low angle scattering (see Fig. 10).

Table 2. Relative light scattering diameters, F_j, of aggregates of j
         uniform polystyrene spheres measured with the LASS.

d_1 (µm)	F_2	F_3	F_4	F_5	F_6
0.100	1.254	1.446	1.610	1.708	--
0.151	1.285	1.457	1.616	1.702	1.841
0.206	1.233	1.417	1.588	1.718	1.830
0.318	1.267	1.478	--	--	--
average values	1.259	1.449	1.604	1.709	1.836
$F_j = j^{1/3}$	1.260	1.442	1.588	1.710	1.818

3. PARTICLES LARGER THAN THE WAVELENGTH

3.1. Theoretical Considerations

In the limiting case $d \gg \lambda$ the scattered light can be considered as
existing of three components which are due to the physical effects of
diffraction, reflection and refraction. For unpolarized and monochromatic
light the scattering coefficient Q can be expressed as the sum of its
components according to:

$$Q\,(\theta,\, d,\, \lambda,\, m,) = Q_o\,(\theta,\, d,\, \lambda) + Q_1\,(\theta,\, m) + Q_2\,(\theta,\, m) \qquad (7)$$

The scattering coefficient Q is the power P of radiation scattered per
unit solid angle in relation to the power $I_o \cdot \frac{\pi}{4}\cdot d^2$ striking the projected
area of the particle, i.e.:

$$Q\,(\theta,\, d,\, \lambda,\, m) = \frac{P\,(\theta,\, d,\, \lambda,\, m)}{I_o \cdot \frac{\pi}{4} \cdot d^2} \qquad (8)$$

$Q_o\,(\theta,\, d,\, \lambda)$ is the diffraction part of scattered light. It is independent
of the optical constants of the particle material, its angular distribution, however, depends on the size parameter:

$$\alpha = \frac{\pi \cdot d}{\lambda} \qquad (9)$$

$Q_1(\theta, m)$ is the fraction of light scattered by reflection on the surface

of the particle. Its angular distribution is independent of the particle
diameter but influenced by the optical constants of the material.
$Q_2(\theta,m)$ is the component scattered by two refractions on the particle
surface. Its angular distribution again depends on the optical constants
but not on the particle diameter. Higher order internal reflections (rainbows) are not considered in equation (7). Typical angular distributions
of the three components of the scattering coefficient valid for transparent spheres are shown in Fig. 6.

For low scattering angles diffraction dominates and can be calculated
separately via Huygen's principle in which at any point in the plane of
the obstacle (particle) elementary spherical waves are produced which
have to be summed up in the far field in accordance with their phase.
By this way the well-known Fraunhofer diffraction formula is obtained
which combines the cross-sectional area of the particle and its diffraction pattern by a Fourier integral. This means that the intensity distribution of the scattered light in the Fraunhofer plane and the light intensity in the object (particle) plane are Fourier-reziprocal and can be
converted into each other by a set of Fourier coefficients. In all cases
where the Fourier integral over the particle contour can be solved analytically or numerically the Fraunhofer diffraction pattern of the particle
and the corresponding Fourier coefficients are obtained. It turnes out
that the diffraction formula of any object consists of two parts: an intensity factor and an angular function. Whereas the intensity factor
$\left(\frac{F}{\lambda}\right)^2$ is a function of the projected area F of the particle only the
angular distribution changes with particle shape.

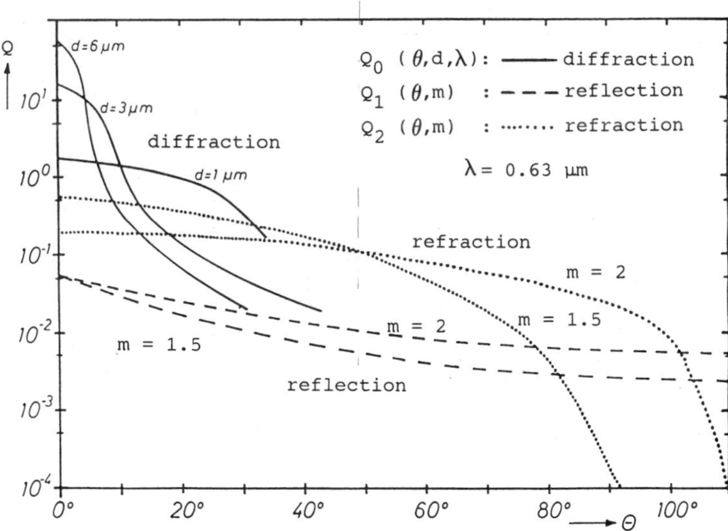

Fig. 6. Angular distribution of the three components of the scattering
coefficient

This means that the total amount of light scattered due to diffraction
is a function of the projected area F of a particle regardless of its
shape. What is modified, however, with the particle contour is the angular
distribution of the diffracted light. Since the diffraction pattern changes
with particle size different particle diameters d are contained in Fig.6.
The forward lobe of diffraction is limited to the angular range $\theta < \theta_{min}$,
where

$$\sin \theta_{min} = \frac{5\lambda}{4d} \tag{10}$$

Following Fourier's principle the limitation of the forward lobe of diffraction reflects the rough contour of a particle in a reciprocal manner
and thus contains information about its projected area. Details of the
contour, on the other hand, are mainly stored at angular ranges outside
the diffraction lobe resp. in Fourier coefficients of higher order. Thus
micro-structures of the particle contour like sharp edges or granules
tend to shift diffracted light to larger scattering angles.

While the component $Q_1(\theta, m)$ scattered by external reflection covers
the whole angular range: $0° < \theta < 180°$, the refracted part $Q_2(\theta, m)$ does
not exceed an upper angular limit θ_{max} which is given by:

$$\cos \frac{\theta_{max}}{2} = \frac{1}{m} \tag{11}$$

The smaller the refractive index m the more refracted light is concentrated in forward direction so that the particle acts like a focusing
lens. This focusing effect, however, is confined to ideal dielectric
spheres. The angular distributions of the light scattered by external
reflection $Q_1 (\theta, m)$ should be about equal for a random assembly of large
irregular particles and for large spheres, since there exists in the average
a simular probability for the angles of reflection. Considering light
refraction on the surfaces of randomly oriented nonspherical particles,
however, there is a greater possibility of high internal reflection angles
inside irregular particles which results in more total internal reflections
at the expense of scattering by refraction alone.

Concerning single particle observation light scattering by reflection
and refraction on an irregular surface is rather arbitrary. In other words,
outside the diffraction lobe specular reflections and internal reflexes
on an individual irregular particle can produce angular distributions
which deviate considerably from the scattering diagram of a sphere of
equal projected area. Consequently, the effect of the particle shape on
the response will be highest if light scattered outside the diffraction
part is collected through a relative small aperture.

For irregularly shaped particles with dimensions above the wavelength
of light the effect of the particle shape on light scattering is lowest
if the flux of scattered light is a function of the projected area of
the particle. Better conditions than to have a projected area response
cannot be realized for nonspherical particles above the wavelength of
light. There are two possibilities to build instruments with a projected
area response: either to collect only diffracted light (low angle scattering) or to collect almost all reflected and refracted light (4π-geo-
metry). The projected area of a nonspherical particle then, of course,
still depends on its orientation in the sensing volume.

3.2. Experiments

A low angle scattering instrument which collects only diffracted light
is shown in Fig. 7. It utilizes a mercury lamp, Q, and separates the

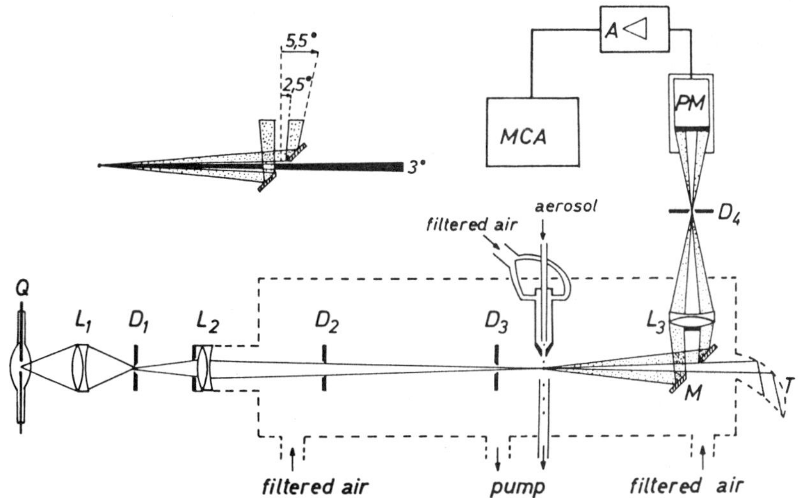

Figure 7.   Low angle scattering instrument for the size range above the
wavelength after Gebhart et al. (1976).

diffracted light from the primary beam by a hole in a flat mirror.
The scattered light is sampled by the lens, L3, which forms an image of
the particle onto the stop, D4. Behind this stop the scattered light
reaches the photomultiplier, PM, where the light flashes are converted
into electrical pulses which are then amplified and classified in a multi
channel analyser, MCA. The aperture of the concentric receiver optics
covers an angular range from 2.5° to 5.5°. The response of the low angle
scattering instrument to agglomerates of polystyrene spheres with diameters $d_1 > \lambda$ is summarized in Table 3.The results indicate, that the
relative light scattering diameter $F_j = j^{1/2}$, so that within an experimental error of about 3% d_{scj} is identical with the diameter of a sphere
which has the same projected area as the nonspherical particle. In other
words, as long as a compact particle larger than the wavelength is considered and light is collected within the forward lobe of diffraction the
response is a function of the projected area of a particle regardless
of its shape. It should be mentioned in this connection that the response of a projected area instrument to a nonsperical particle, of course,
depends on its orientation in the sensing volume. It can be shown, however,
by statistical considerations that even for randomly oriented agglomerates
of spheres those orientations are more frequent which exhibit projected
areas being multiples of the cross-section of a single sphere.

An instrument with a relatively large receiver aperture for the collection of reflected and refracted light is for instance the Climet CI 208
counter shown in Fig. 8. The Climet CI 208 utilizes an elliptical mirror
in its optical system. In the sensor, the particle sensing zone is
located at the primary focal point of the elliptical mirror.

High intensity light from a quartz halogen lamp is focused on the
sensing zone where it interacts with each traversing particle. Light
scattered from each particle is collected over an angular range from
15 to 105° and is directed to a photodetector located at the secondary
focal point of the ellipsoid. The response of this instrument to doublet
aggregates (j=2) of polystyrene spheres has been investigated by Chen
and Cheng (1984). Their results presented in Fig. 9 indicate that for

Table 3. Relative light scattering diameters, F_j, of aggregates of uniform polystyrene spheres measured with low angle scattering.

d_1 (μm)	F_2	F_3
1.158	1.39	1.69
1.83	1.41	1.75
average values	1.40	1.72
$F_j = j^{1/2}$	1.41	1.73

$d_1 > 0,6$ μm the light scattering diameter d_{scj} of such doublet aggregates is identical with the diameter of a sphere which has the same projected area as the cluster.

Results obtained from microwave analog measurements (Zerull et al. 1977) show that the diffraction patterns of a nonsphere and of a sphere of equal projected area agree quite well in the angular range where diffraction dominates. Outside the diffraction lobe, however, the intensity of light scattered by a nonspherical particle deviates considerably from that of a sphere of equal projected area. This is demonstrated in Fig. 10, where the scattering diagram of a cube obtained from microwave analog measurements is compared with Mie-theory calculations for a sphere of equal projected area. Outside the forward lobe of diffraction two characteristic features can be observed: (1) In the angular range:

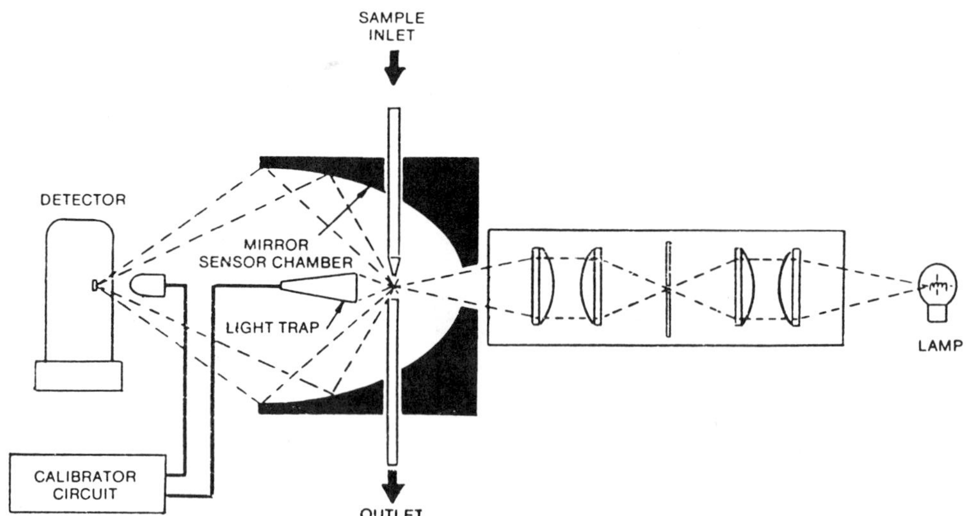

Figure 8. Optical sensor of the Climet CI 208 particle counter.

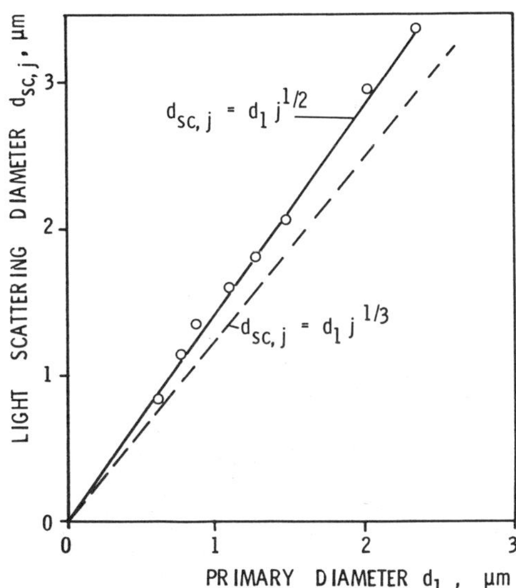

Figure 9. Light scattering diameters $d_{sc,j}$ of doublet aggregates of polystyrene spheres measured with the Climet CI 208 counter after Chen and Cheng (1984).

$60° < \theta < 160°$ the mean intensity of scattered light (average over all orientations) is higher for the cube than predicted from theory for an equivalent sphere. (2) The variation of the scattering intensity with particle orientation at a fixed scattering angle as expressed by the variance is lowest within the forward lobe of diffraction, increases with increasing scattering angle and is highest for rectangular scattering. Both characteristic phenomena can be recovered in corresponding responses of optical particle counters to irregular particles.

Büttner (1983) measured particles of quartz and limestone with the rectangular instrument HC 15 of POLYTEC (Fig. 11). The particle counter HC 15 forms its sensing volume by optical means only without using an aerosol nozzle. The system consists of two optical path-ways one for illumination and one for the collection of scattered light. The lenses I and II project miniaturized images of square masks (stop I and II) into the measuring chamber, where the two optical axes cross at right angles. The images of stop I and II which coincide with the crossing point of the two axes form the sensing volume. Particles passing this optically defined sensitive area are uniformly illuminated with incandescent light and the light scattered on individual particles is collected under a mean scattering angle of 90°. From the results in Fig. 12 it is obvious that the particles consisting of quartz and limestone scatter much more light in the 90°-range than spherical particles of polystyrene and glycerin of the same Stokes-diameter although the optical constants of the materials are comparable to each other. This may give evidence that light reflected and refracted on an irregular surface leads to angular distributions which differ from those of ideal spheres (see Fig.6). Angular distributions of scattered laser light measured recently by Coletti (1984) on collectives of isometric nonspherical particles agree with this general statements.

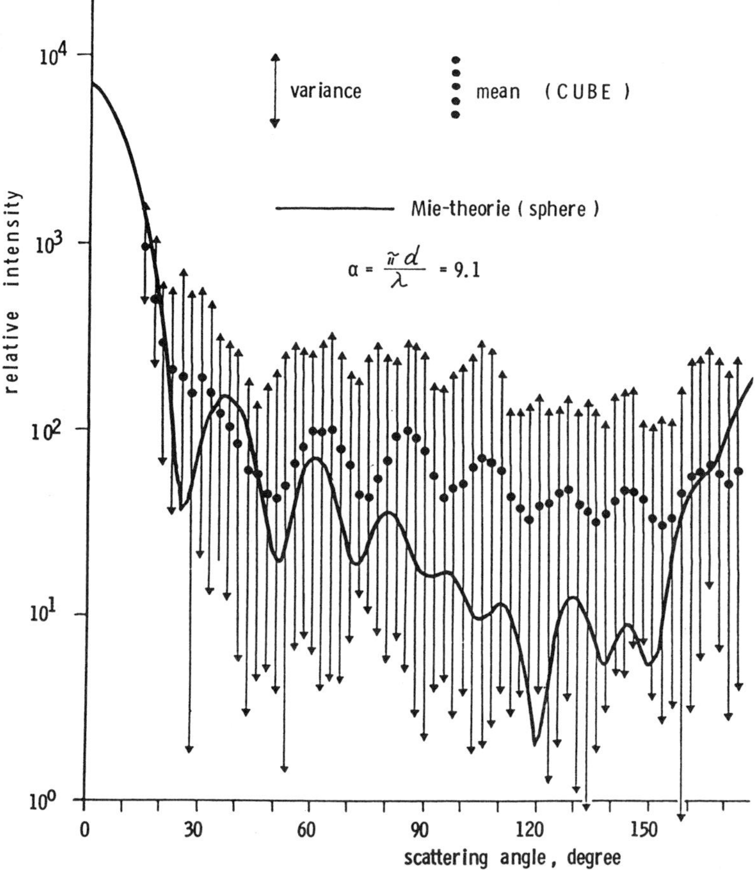

Figure 10. Scattering diagram of a cube obtained from microwave analog
measurements versus Mie-theory calculations for an equivalent
sphere after Zerull et al. (1977).
(refractive index m = 1.57 - 0.006i).

In a recent study mono-sized NaCl-crystals produced in a vibrating
orifice generator have been classified in the counting mode of a 2-Mode-
Laser-Aerosolphotometer developed for aerosol inhalation studies
(Gebhart et al.1980). The opto-mechanical set-up of the photometer is
shown in Fig. 13.The parallel beam of an argon ion laser (2W) traverses
a system of diaphragms and cylindrical lenses to form within the sensitive volume a sheet of light of 80 μm thickness and 15 mm height. The
sheet of light is inclined at an angle of 60° to the direction of the
aerosol flow and covers almost the whole cross section of the aerosol
channel. The receiver unit consists of a microscope objective which
collects scattered light under an angular range of 90 ± 12°, a field
stop and a field lens. A photomultiplier tube at the end of the receiver
channel converts the light flashes scattered from individual particles
crossing the light sheet into electrical pulses. Since the sheet of light
coincides with the object plane of the microscope only those particles
are illuminated which are imaged sharply into the field stop. Whereas
the Low Angle Scattering instrument of Fig. 7 indicates a monodisperse
NaCl-aerosol at the outlet of the generator, the 2-Mode-Photometer

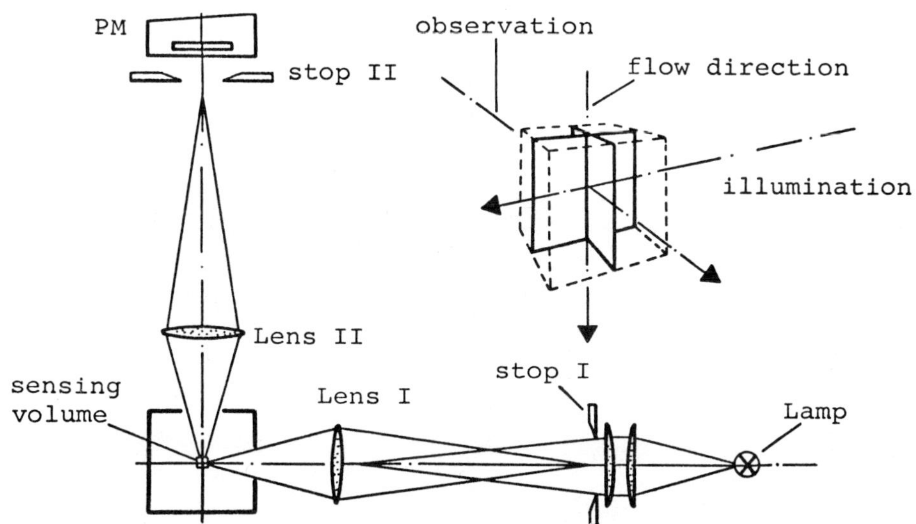

Figure 11. Optical system of the POLYTEC HC 15 counter.

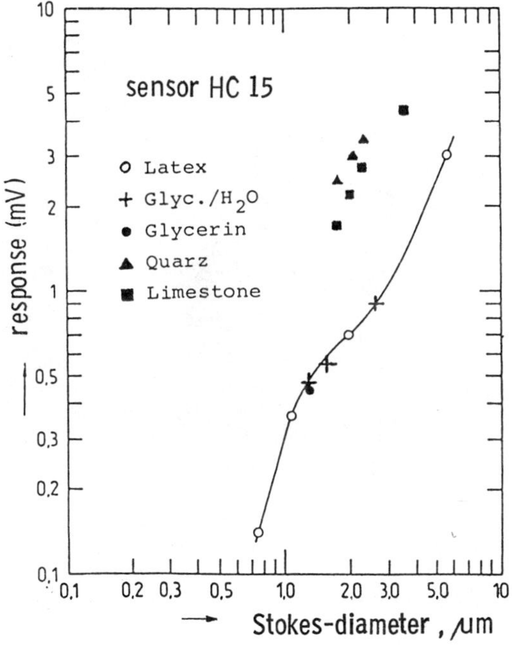

Figure 12: Response of the HC 15 rectangular particle counter to
spherical and irregular particles after Büttner (1983).

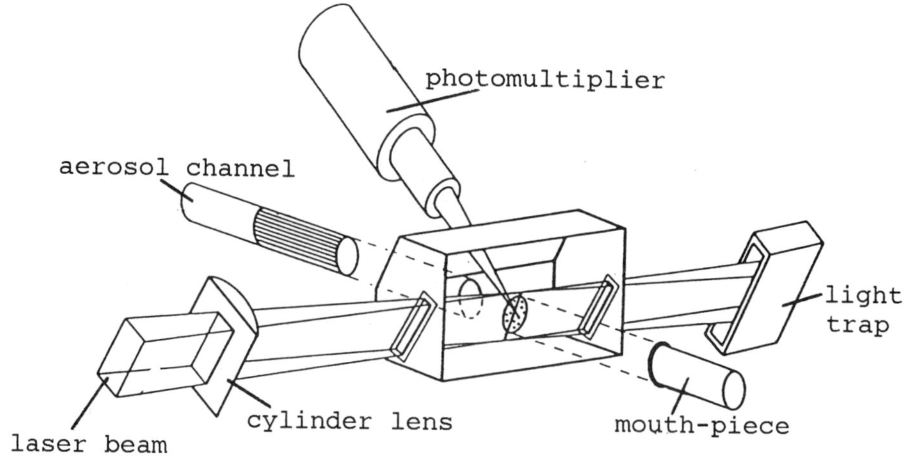

Figure 13. Schematic diagram of the rectangular 2-Mode-Laserphotometer
for aerosol inhalation studies after Gebhart et al. (1980).

classifies the NaCl-crystals as polydisperse (see Fig. 14). Converting
the same NaCl-crystals into saturated NaCl-H$_2$O-droplets, however, leads
to a size distribution in the Aerosolphotometer, which can be characterized
as monodisperse (Fig. 14).

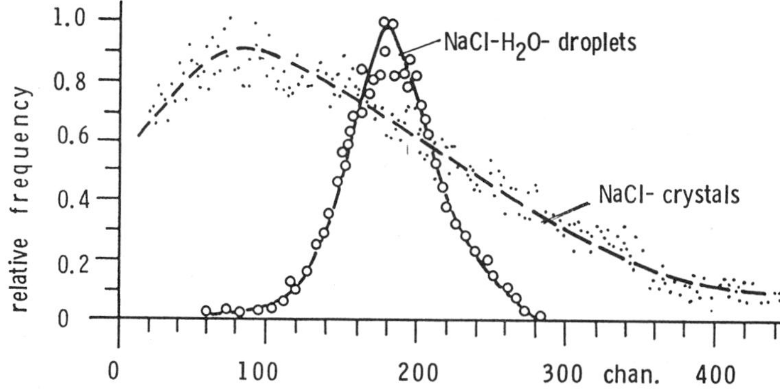

Figure 14. Size distribution of monodisperse NaCl-particles of 1.5 μm
volume-equivalent diameter and size distribution after conversion of the crystals into saturated NaCl-H$_2$O-droplets measured
with the rectangular 2-Mode-Laserphotometer.

REFERENCES

Büttner, H. (1983), Kalibrierung einer Streulichtmeßeinrichtung zur Partikelgrößenanalyse mit Impaktoren, <u>Chemie. Ing. Techn.</u>, 55:65.

Chen, B.T., Cheng, Y.S. (1984), Optical diameters of aggregate aerosols, <u>J. Aerosol Sci.</u>, 15:615.

Coletti, A. (1984), Light scattering by nonspherical particles: a laboratory study, <u>Aerosol Sci. and Techn.</u>, 3:39.

Gebhart, J., Heyder, J., Roth, C., Stahlhofen, W. (1976), Optical aerosol size spectrometry below and above the wavelength of light - a comparison, <u>in</u>: "Fine Particles", B.Y.H. Liu, ed., Academic Press, New York.

Gebhart, J., Heigwer, H., Heyder, J., Roth, C., Stahlhofen, W. (1980), A new device for aerosol and gas inhalation studies and its application in lung investigations, <u>J. aerosol Sci.</u>, 11:237.

van de Hulst, H.C. (1957), Light scattering by small particles, John Wiley, New York.

Kerker, M, (1969), The scattering of light and other electromagnetic radiation, Academic Press, New York, London.

Roth, C., Gebhart, J. (1978), Rapid particle size analysis with an ultramicroscope, Microscopica Acta, 81: 119.

Zerull, R.H., Giese, R.H., Weiss, K. (1977), Scattering functions of nonspherical dielectric and absorbing particles vs Mie-theory. <u>Applied Optics</u>, 16:777.

FORWARD SCATTERING SIGNATURE OF A SPHERICAL PARTICLE

CROSSING A LASER BEAM OUT OF THE BEAM WAIST

J-P. Chevaillier*, J. Fabre*, P. Hamelin*, and J-L. Lesne**

* Institut de Mécanique des Fluides de Toulouse
2, rue Charles Camichel - 31071 Toulouse Cedex, France
** Electricité de France. Direction des Etudes et
Recherches. 25, allée Privée - 94000 Saint Denis, France

INTRODUCTION

The development of optical methods for the measurement of particle size, presents a great interest in many different industrial and scientific areas concerned with dispersed two phase flows. For the simplest case of spherical particles, several methods [1,2,3,4] using the detection of light intensity scattered by particles crossing a laser beam have been realized to permit an "in situ" measurement of size and velocity simultaneously. However, solutions proposed to solve the problem of trajectory ambiguity induced by the gaussian intensity distribution of laser beams, lead generally to the design of expensive devices[1,2,3].

This paper is devoted to show that some scattering properties of particles situated in a laser beam can be used to develop a simple method for sizing. This method is founded on the analysis of the particle signature represented by the light attenuation signal occurring on the beam axis when a particle crosses the beam. This signature contains some information regarding the particle size and trajectory [5]. These scattering properties are pointed out in experimental records of glass sphere signatures. The results are confirmed by the numerical solutions obtained with a simplified model.

SIGNATURE OF A SPHERICAL PARTICLE CROSSING A FOCUSED LASER BEAM OUT OF THE FOCAL PLANE

When a spherical particle crosses a focused laser beam, the signature results in a light attenuation signal on the axis of the beam. Some typical examples corresponding to various particle trajectories are presented in figure 1. The output current of a photodiode, placed on the beam axis at 1 meter from the beam waist, is plotted versus time.

For these experiments, the glass sphere was electrostatically held in position on a vertical glass plate and a three dimensional micropositioner was used to translate the particle and to control the two main parameters of its trajectory :

- d : distance between the particle and the beam waist plane
- a : distance between the particle trajectory and the beam axis.

Each signature corresponds to different values of **a**, with d = 10 mm.
Two important comments may be made :

- The shape of the signature depends on the trajectory of the
particle in the beam. As long as the trajectory is close to the beam
axis, the signature displays a hump. The hump has a maximum amplitude
when the trajectory crosses the beam axis (a = 0), and disappears when
the distance **a** is greater than some given value (a_0).

- The attenuation **A** does not depend on **a** when the hump is present in
the signature ; it depends on the particle size and the distance **d**
between the trajectory and the beam waist plane.

Moreover, the particles crossing the beam waist give signatures
nearly gaussian whatever the distance **a** to the axis. Then, no specific
information is available to determine the particle trajectory so that the
size measurement cannot be obtained.

These qualitative results suggest a new method for sizing, based on
the shape analysis and the attenuation measurement. In order to confirm
the experimental results and to choose the geometrical parameters, namely
the distance **d**, a simple model of the light, forward scattered by a
sphere located anywhere in a gaussian beam, has been developed.

THEORETICAL MODEL OF FORWARD SCATTERED LIGHT

The exact solution of light scattered by a homogeneous spherical
particle placed in a uniform plane wave, is given by the well-known
Lorenz-Mie solution of the Maxwell equations. Some work have been devoted
to extend the solution for a gaussian beam [6,7] but calculations are still
laborious, and none of the formalisms enables one to compute scattered
intensities whatever the position of the particle in the beam.

A simplification of the problem is obtained if the difference
between the optical indices of refraction of the particle and the
surrounding medium is high enough : then the diffraction is predominant
[8,9] for small scattering angles ($\theta < 6°$) and particle diameters greater
than 10 μm.

Starting with the above assumption, a simplified model based on the
calculation of Fraunhofer integral was developed [10] in order to predict
the far field intensity forward scattered by a sphere smaller or equal to
the local beam diameter.

The relative intensity on the beam axis far from the particle is
expressed from the Fraunhofer integral by [8] :

$$\frac{I}{I_0} = \left| 1 - \int_0^{2\pi} \int_0^R e^{-h} \, r \, dr \, d\omega \middle/ \int_0^{2\pi} \int_0^\infty e^{-h} \, r \, dr \, d\omega \right|^2 \qquad (1)$$

in which h is :

$$h = \frac{r^2}{b^2} \left(1 - \frac{ikb^2}{2R_c}\right)\left(1 + \frac{2\delta}{r} \cos \omega\right) \qquad (2)$$

$$R_c = \frac{b^2}{2} \frac{d}{d_0} k \qquad (3)$$

where :

I_0 is the intensity when the particle is absent
R is the particle radius

412

δ   is the distance from the particle to the beam axis
b   is the beam radius at the distance **d**
R_c is the curvature radius of the wave
d_0 is the Rayleigh distance
k   is the wave number

Using **b** as a length scale to define the dimensionless variables R^+, δ^+, r^+, yields :

$$\frac{I}{I_0} = \left| 1 - \int_0^{2\pi} \int_0^{R^+} e^{-h} r^+ dr^+ d\omega \Big/ \int_0^{2\pi} \int_0^{\infty} e^{-h} r^+ dr^+ d\omega \right|^2 \qquad (4)$$

$$h = (1 - i\frac{d}{d_0})(r^{+2} + 2\delta^+ r^+ \cos \omega) \qquad (5)$$

Clearly the solution of the relative intensity depends on the following non dimensional parameters : the radius R^+, the distance of the particle to the axis δ^+ and the ratio d/d_0. It does not depend on the wave length of the laser :

$$\frac{I}{I_0} = F(\frac{d}{d_0}, \frac{R}{b}, \frac{\delta}{b}) \qquad (6)$$

For a rectilinear trajectory contained in a plane normal to the beam axis, the attenuation **A** is expected to depend on :

$$A = G(\frac{d}{d_0}, \frac{R}{b}, \frac{a}{b}) \qquad (7)$$

For some range of the distance d/d_0, the amplitude does not depend on **a**.

It turns out that the relevant parameter is the dimensionless distance to the beam waist $d* = d/d_0$.

NUMERICAL AND EXPERIMENTAL RESULTS

Experiments were carried out with particles of various trajectories and diameters (18 μm, 36 μm, 77 μm), crossing a He-Ne laser beam of 60 μm beam waist diameter [8]. The accuracy was improved recently and new results were obtained [11] : the figure 2 gives an example of a 50 μm particle crossing the beam axis at $d* = 2.6$. A good agreement is found between predicted and experimental signatures.

Using the model, an extensive study of particle signatures has been realized. It can be shown that :

- as long as $d*$ is less than unity, the shape of the signature is closely gaussian whatever the particle trajectory in the beam is ;

- for $d*$ greater than unity, a hump appears in the signature ; its amplitude increases with $d*$; the hump is present as long as the distance **a** is smaller than some value a_0 ($a_0/b \simeq 0.3$-0.5 from [8]) ; the attenuation is given by a simplified form of eq. 7 :

$$A = G(d*, R/b) \qquad [d > d_0 \text{ and } a < a_0] \qquad (8)$$

The relationship is plotted from numerical results in figure 3. For a given position of the trajectory plane ($d* = d/d_0$), a calibration curve is obtained.

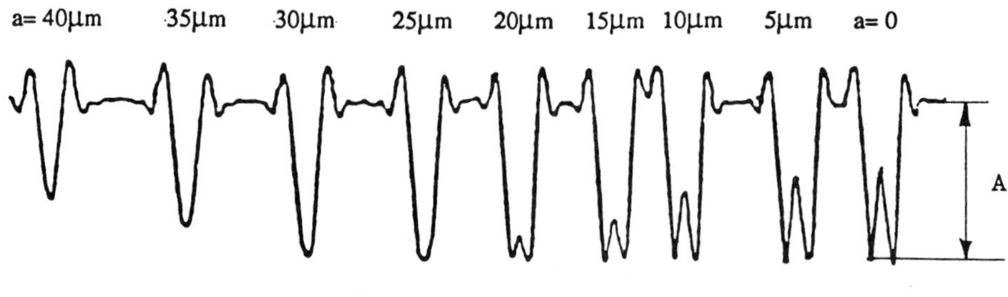

Figure 1 : Successive signatures of a particle crossing a
laser beam

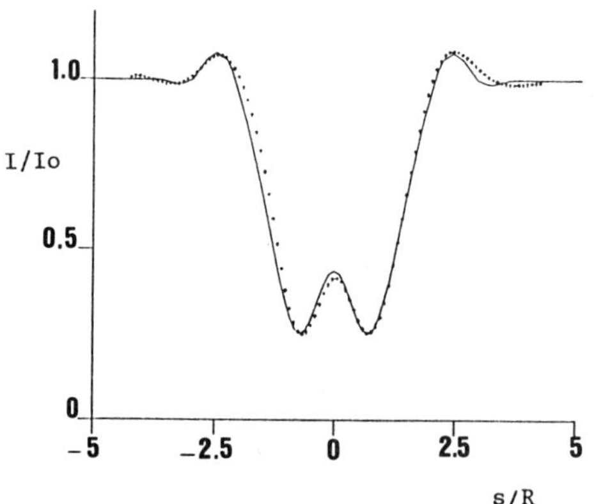

Figure 2 : Particle signature. Comparison between experimental
results (...) and numerical results (———)

INTERPRETATION OF SIGNATURE SHAPES

In order to explain the features of the shapes given in figure 1, let us examine the general solution of the diffracted field **E** in a plane located at a distance z from the diffracting plane. After some algebra, the Fraunhofer integral is written :

$$\left| E^+ \right| = (\pi b_0^+)^{-1} \exp(-\delta^{+2}) \left| \int_\Omega H(X^+,Y^+,U^+,V^+,z^+).\exp(-2X^+\delta^+)dX^+dY^+ \right| \qquad (9)$$

where :

x,y   are the coordinates in the diffracting plane
u,v   are the coordinates in the plane located at a distance z
δ,0   are the coordinates of the particle
X,Y   are the shifted coordinates defined by : $X = x-\delta$, $Y = y$
U,V   are the shifted coordinates defined by : $U = u-\delta z/R_c$, $V = v$
Ω is the external domain to the circle of radius R/b

The superscript $^+$ indicates a dimensionless variable ; the lengths are scaled by b.

$$E^+ = E(u,v,z)/E(0,0,z)$$

H   is a function which does not depend on δ, expressed by :

$$Ln(H) = -(X^{+2}+Y^{+2})(1-id/d_0)+ik^+(U^{+2}+V^{+2})/2z^+-ik^+(X^+U^++Y^+V^+)/z^+ \qquad (10)$$

For $\delta \ll b$, the equation 9 becomes :

$$\left| E^+ \right| = (\pi b_0^+)^{-1} \left| \int_\Omega H \, dX^+ \, dY^+ \right| \left| 1-2\delta^+\int_\Omega H \, X^+dX^+dY^+ / \int_\Omega H \, dX^+dY^+ \right| \qquad (11)$$

The first term of the r.h.s. corresponds to the field diffracted by a centered particle (δ=0) : it gives the well-known ring pattern in the z-plane. The second term of the r.h.s. represents the small modification of the pattern under the influence of a small displacement of the particle in the x direction.

The field $\left| E^+ \right|$ of an off-axis particle can be interpreted as the result of :

   - a deformation of the pattern corresponding to a centered particle
   - a displacement ($\Delta=\delta z/R_c$) of the pattern in the u-direction.

For a plane wave (particle crossing the beam waist), the pattern is motionless and the shape is only controlled by the deformation. For a spherical wave (particle crossing the beam out of the beam waist), the rings are displaced in the detector plane : the minimum value of the signal intensity (figure 2) corresponds to the presence of the first dark ring in front of the detector ; the maximum value corresponds to the presence of the first bright ring. This property could be used for measuring the particle velocity.

A useful representation of the signals diffracted by spherical gaussian waves is given in figure 4. The light intensity on the beam axis is plotted versus the x and y coordinates of the particle ; each line of the 3-D plotting corresponds to a linear trajectory.

The minimum value of light intensity occurs each time a particle is located at some distance of the beam axis a_0. Then the afore-mentioned properties of the signatures are clearly understood from the axis-symmetrical surface of figure 4.

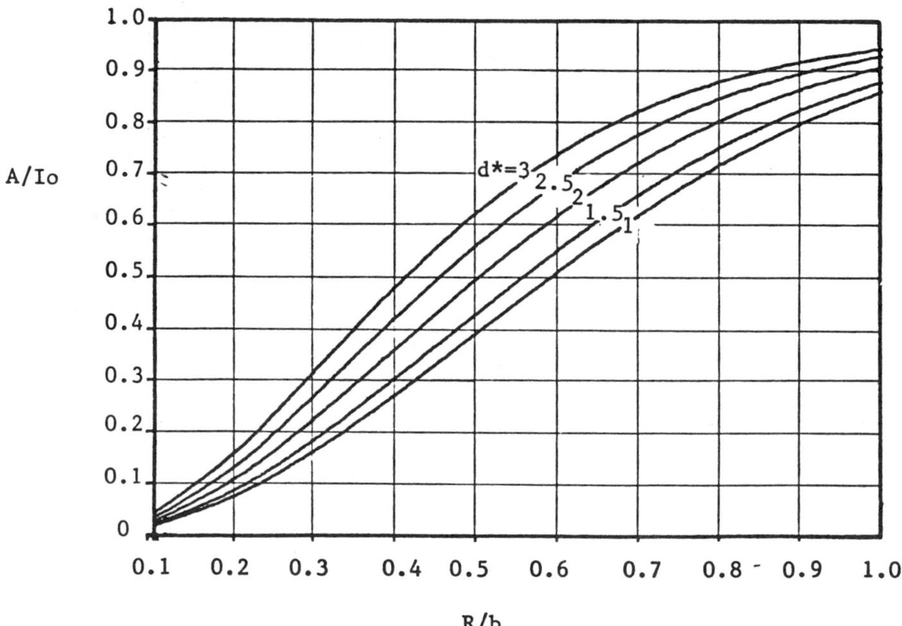

Figure 3 : Signal amplitude versus particle size

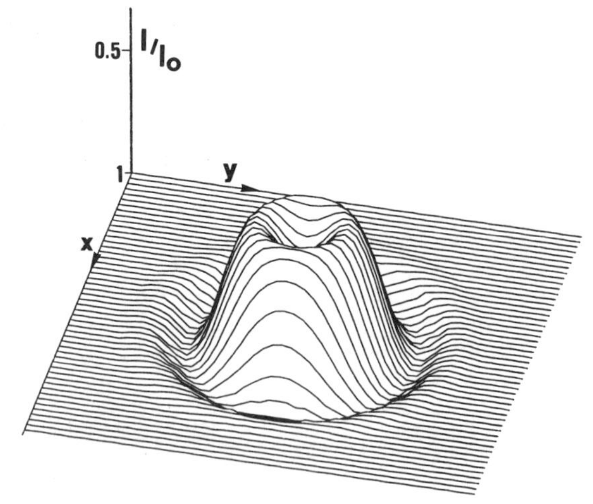

Figure 4 : 3-D display of particle signatures

APPLICATION FOR PARTICLE SIZING

The sizing method is based on the determination of the particle distance **d**. This can be achieved using two beams intersecting at a distance **d** from their beam waist. The size is obtained by checking the simultaneity of both attenuation signals and by analyzing the shape and the minimum value of the signatures. For this purpose, an apparatus combining a laser Doppler velocimeter and an optical sizing system has been developed. The electronic device for data acquisition and signal processing is able to perform simultaneous measurement of size and velocity of spherical particle in the range 10 μm – 500 μm and 1 cm/s – 5 m/s without any restriction of the particle optical index.

REFERENCES

1.   M.L. Yeoman, H.J. White, B.J. Azzopardi, C.J. Bates, P.J. Roberts, Optical development and application of two colors L.D.A. system for simultaneous measurement of particle size and velocity, Report AERE R 10458 Harwell, (1982)

2.   G. Gouesbet, G. Grehan, R. Kleine, Simultaneous optical measurement of velocity and size of individual particle in flow, Proceedings of the Int. Symp. on Application of Laser Doppler Anemometry to Fluid Mechanics, Lisbon, (July 2-5, 1984)

3.   K. Bauckhage, H.H. Floegel, Simultaneous measurement of droplet size and velocity in nozzle sprays. Proceeding of the Int. Symp. on Application of Laser Doppler Anemometry to Fluid Mechanics, Lisbon, (July 2-5, 1984)

4.   R. Semiat, A.E. Dukler, Simultaneous measurement of size and velocity of bubbles and drops : a new optical technique, AIChE Journal, vol.27, n°1, pp.148-159, (1981)

5.   J.N. Lecomte, In situ bubble sizing using a reference beam laser doppler anemometer, Proceedings of the Symposium in long range and short range optical velocity measurements, Saint-Louis, (September 15-18, 1980)

6.   G. Gouesbet, G. Grehan, Sur la généralisation de la théorie de Lorentz-Mie, J. Optics, Paris, V.13, n°97, (1982)

7.   W.G. Tam, R. Corriveau, Scattering of electromagnetic beams by spherical objects, J. Opt. Soc. Am., 68, 763, (1978)

8.   P. Hamelin, Application de la diffusion lumineuse à la métrologie des particules en écoulement diphasique dispersé, Thèse, Institut National Polytechnique de Toulouse, published in Bulletin de la Direction des Etudes et Recherches d'Electricité de France, série A, n°3/4 (1986)

9.   J.P. Chevaillier, J. Fabre, P. Hamelin, Scattering properties of spherical particles situated in a laser beam and application for sizing, Bradford, Particle Size Analysis Conference (September 1985)

10. J.P. Chevaillier, J. Fabre, P. Hamelin, Forward scattered light intensities by a sphere located anywhere in a gaussian beam, Applied Optics, vol.25, n°7, pp.1222-1225, (1986)

11. J.P. Chevaillier, Signature lumineuse de particules sphériques en faisceau gaussien, Rapport IMFT/EME n° 284, (Février 1987)

EFFECTS OF SHAPE AND ORIENTATION TO BE CONSIDERED

FOR OPTICAL PARTICLE SIZING

R.T. Killinger, R.H. Zerull

Ruhr Universität Bochum
Bereich Extraterrestrische Physik
4630 Bochum, FRG

INTRODUCTION

The light scattering features of particles are closely related to their size, material, shape, and structure. This is the reason why this phenomenon is frequently used for particle measurements (aerosol photometer, optical particles counters, etc.). The major advantage of this technique is, that the measurements can be carried out without disturbing the particle system significantly.

The main aspect for the selection of measurements is to illustrate, how the response of scattering devices can be affected by the shape of particles with special emphasis on orientation. Spheres are chosen as a base for comparison, not only because they are frequently used for calibration measurements but also because their scattering properties can be calculated relatively easy. The discrepancies to be expected with respect to spheres will be discussed for particles of different sizes (.1 to ≈ 50 μm), different materials (from transparent to opaque), and different shapes.

Concluding the presentation the influence of particle orientation for optical particle sizing will be illustrated. Due to perfect symmetry obviously only spheres scatter independently from orientation. Deviations due to orientation can easily reach more than one order in magnitude, in particular for scattering angles in the backward hemisphere. It will be shown, how these strong effects are smoothed out by integration over a large range of scattering angles and by use of a multicolour light source.

THE EQUIPMENT

The results presented in this paper are derived form single particle multicolour laser scattering experiments and from microwave analog studies.

Microwave Laboratory

The microwave analog technique takes advantage of the fact, that the electromagnetic laws are valid over a wide range of wavelengths, i.e. 4

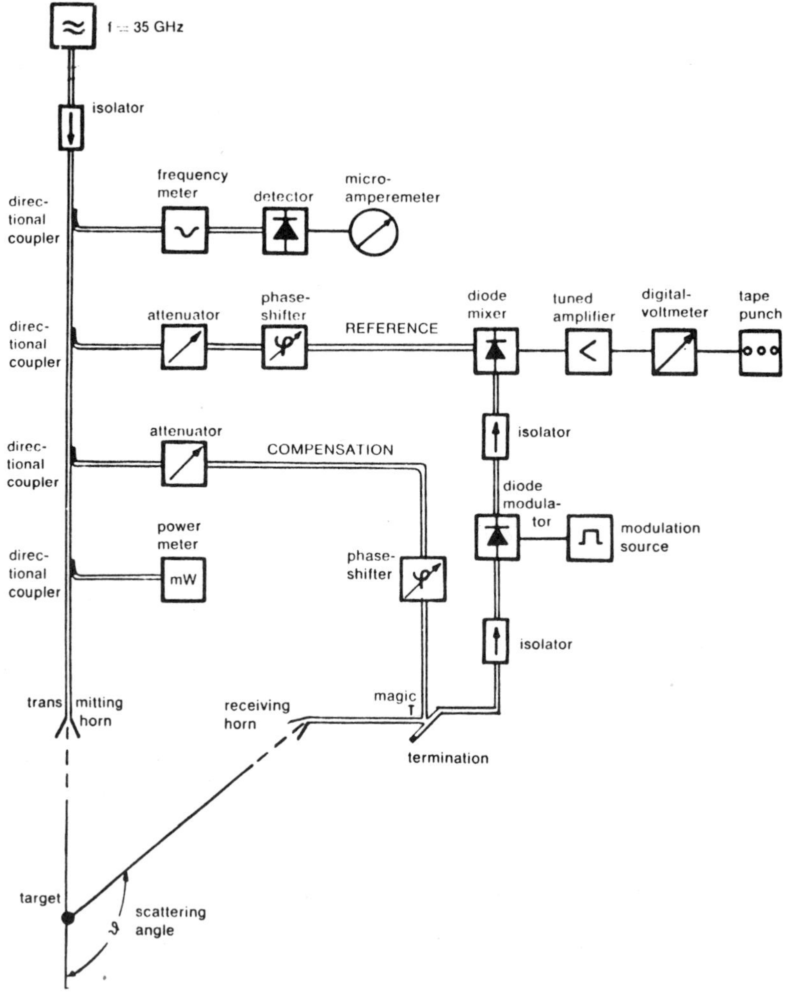

Fig. 1. Block diagram of the microwave analog experiment

orders of magnitude between the optical and the microwave region. Analog
conditions in both regions are provided for the same ratio of particle
size to wavelength and by choice of materials with the same refractive
index. Scattering results for mono- and polydisperse mixtures must be
composed of data for different orientations and sizes, which can become
quite laborious. The advantage of the microwave analog studies is, that
orientation, size, shape, structure, and refractive index of the
particles can be varied systematically and exactly.

Fig. 1 shows the schematic diagram of the microwave equipment at
Bochum as described in Zerull (1976). It operates at a frequency of 35
GHz and is placed into an unechoic chamber. The power of a reflex
klystron is fed to the transmitting horn antenna. The intensity scattered
by a target at varying scattering angle Θ is received by a similar
antenna and fed to a diode mixer which serves as receiver. The intermediate frequency of 1 kHz is generated by use of two auxiliary
oscillators. The rectified signal is proportional to the received signal

over 4 orders of magnitude power range. Various unwanted signals may be
superposed on the true scattering signal of the target, such as direct
signals via side lobes of the antennas and reflections from the walls and
parts of the set-up. These parasitic signals, which are also present, if
the target is removed, are eliminated by superposition of a signal of the
same amplitude but opposite phase via a compensation line and a hybrid T
junction. After this nulling procedure the scattering body, which is
suspended by nylon threads, is moved back to the scattering center for
measurement of the unblended scattering signal. For applications, where
specific orientations are of minor interest, the azimuthal average of the
scattering signal of the rotating target can directly be obtained by
means of an integrator. This mode keeps the measuring time within
economic limits. For investigation of the influence of orientation
another mode allows discrete measurements.

The major part of the measurements presented in the following
section is derived from microwave analog studies.

The Multi Colour Laser Light Scattering Device

As it is very difficult to simulate very complex particles for
measurement in the microwave laboratory, a scattering setup operating in
the optical region was developed. In this experiment single particles are
illuminated by a multicolour laser beam. The particles are charged and
electrodynamically suspended by a superposition of a Millikan field and
the alternating field of an double ring electrode (Straubel, 1960; Weiß-
Wrana, 1983). To stimulate rotation of the particle additionally six
electrodes around the scattering volume are subsequently fed by a high
voltage to produce a rotational field. The scheme of the levitation
chamber is shown in Figure 2.

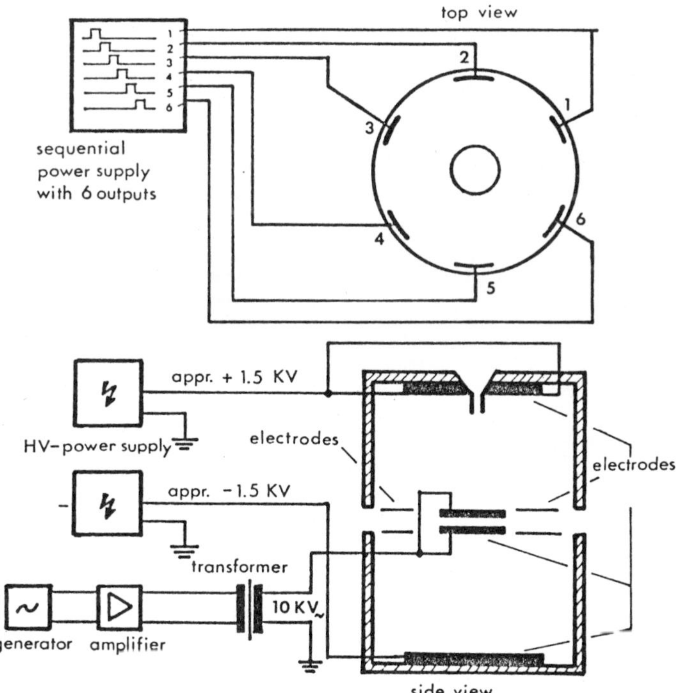

Fig. 2. Schematic diagram of the particle levitation chamber

The particles are illuminated by the beam of a Krypton Ion laser
with monochromatic linearly polarized light of the wavelengths 476.2,
520.8, 568.2 and 647.1 nm. The plane of the polarization of the incident
ray can be changed by means of a pockels cell. The scattered light passes
a polarizing beam splitter and is measured by two photomultipliers. The
range of scattering angles Θ is between 10° and 170° and varied by means
of a stepper motor.

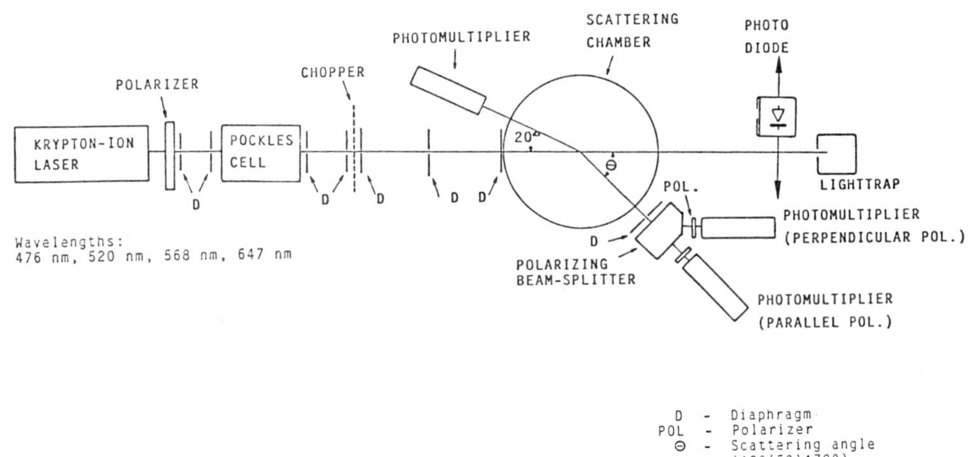

Fig. 3. Schematic diagram of the single particle multi colour light
scattering experiment

The plane of polarization is subsequently changed between the direction perpendicular and parallel to the scattering plane, respectively.
The use of a polarizing beam splitter allows simultaneous measurement of
both states of polarization. With these measurements, after calibration,
one gets the total scattering function i (van de Hulst, 1981), the degree
of linear polarization, the so called cross-polarization and the wavelength dependence of the scattered light. For this paper we will only
refer to effects, concerning the total scattered intensity.

THE INFLUENCE OF SHAPE

Most particle counters obtain their information about the size of a
particle by measuring the scattered intensity around Θ = 90°. In this
region, however, all nonspherical particles larger than about 1.5 μm
scatter significantly different from spheres. Figure 4. illustrates the
influence of moderate roughness (< 1/10 wavelength). The measured scattering diagram for a rough sphere (3 μm) does not deviate very much from
the Mie calculation.

The next example shown in Figure 5 is a rough and slightly elongated
glass sphere (D = 50 μm). It can be seen, that relatively small
deviations from spherical shape cause distinct differences concerning the
scattering properties.

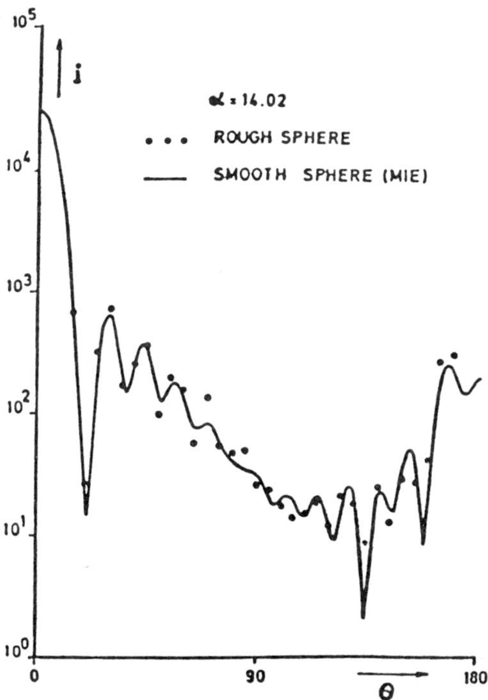

Fig. 4. Scattering function of a rough sphere (3 µm) compared to a smooth sphere.

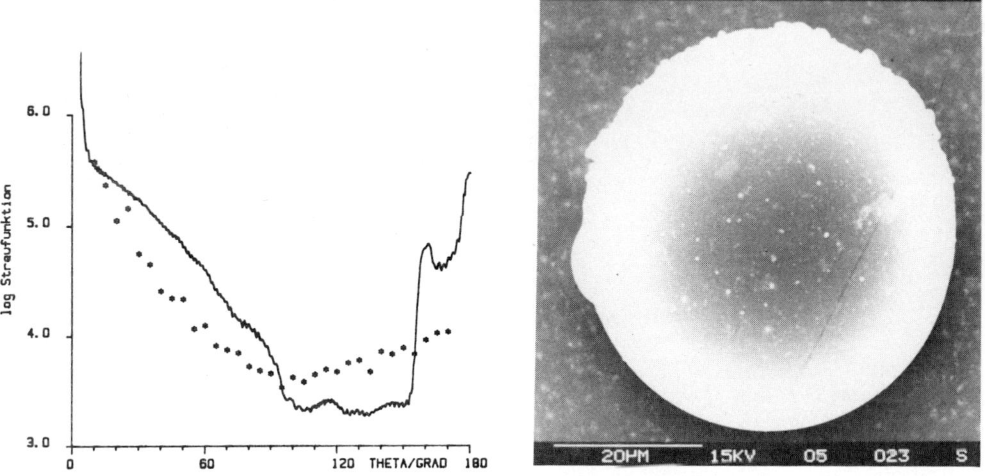

Fig. 5. Scattering function of a rough and slightly elongated glass sphere (50 µm). The line shows the calculated scattering function of an ideal sphere of the same size.

The next two examples (Fig. 6 a, b) are polydisperse mixtures of cubes. The scattering behaviour of the mixture of the smaller particles is relatively close to volume-equivalent spheres. For the larger cubes, however, similarity is restricted to the forward domain.

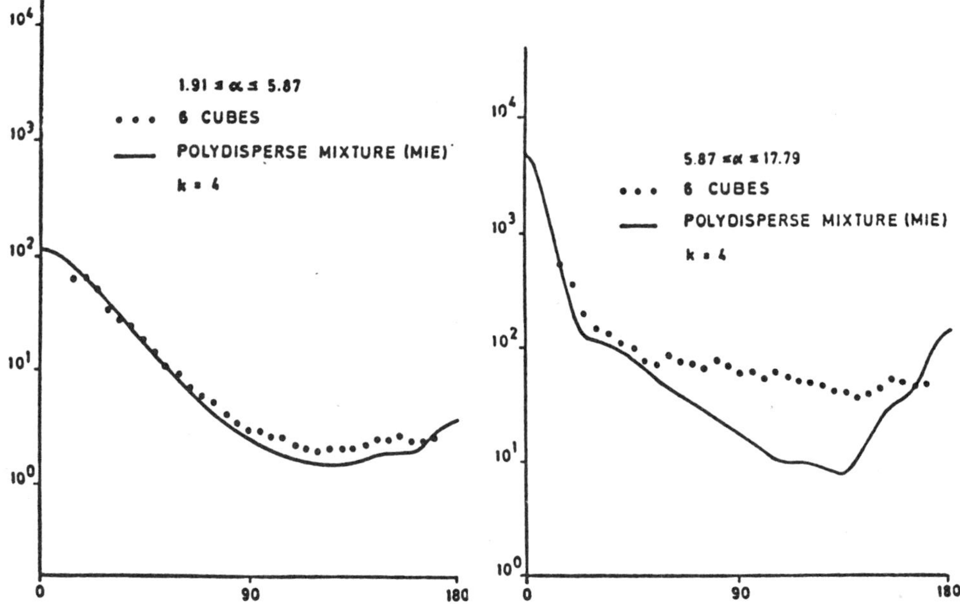

Fig. 6. Averaged scattering functions of two mixtures of single
measured cubes: a)   0.4 μm ≤ D ≤ 1.2 μm
b)   1.2 μm ≤ D ≤ 3.6 μm

Especially for scattering angles around $\Theta \approx 90°$ the values for the cubes are larger by a factor of five. A similar behaviour one can find for all nonspherical particles. The last example in this chapter (Figure 7.) shows the scattering function of a 'spherical' fly ash particle with even stronger deviation from Mie-theory.

THE INFLUENCE OF ORIENTATION

All measurements concerning the influence of orientation have been
made with the MICROWAVE-ANALOG-EXPERIMENT at the Bochum. The volume
equivalent sizes are given with respect to the optical wavelength of 555
nm.

Due to perfect symmetry obviously spheres scatter independently from
orientation. The influence of orientation for other shapes is illustrated
in the following figures. Figure 8. shows the scattering function of a
sphere with statistical roughness of a size parameter $\alpha = 8$,
corresponding to a diameter of 1.4 µm in the optical wavelength range.
The roughness, here defined as difference between minimum and maximum

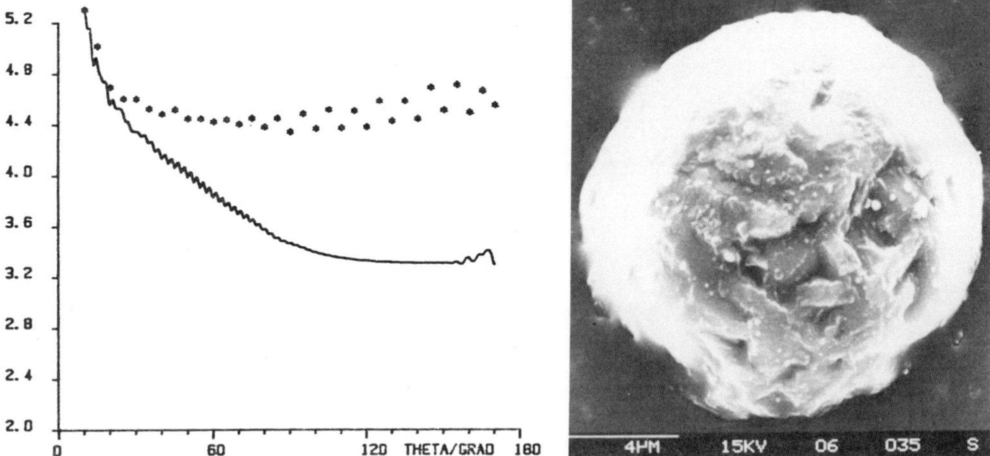

Fig. 7. Scattering function of a fly ash particle (m = 1.55-0.01i,
D ≈ 15 µm, α ≈ 85)

radius, is approximately 1/5 of the wavelength. For each measurement the
maximum, the average, and the minimum value of 36 different orientations
considered are plotted. The dots show the mean values. The bars are the
variance of the single measurement. It can be seen, that the average
value is relatively well described by the Mie calculations for a smooth
sphere of equal volume, which is also shown. However, a distinct
influence of orientation can be seen at all scattering angles including
the diffraction lobe. At medium scattering angles minimum and maximum
values differ by about one order in magnitude. For these measurements the
angular resolution Θ is less than 1°. Figure 8 b shows the same
measurement averaged over an interval of $\Theta = 20°$. The influence of
orientation clearly decreases, but there can still be a factor of 2
between the extreme values.

Figures 9. and 10. present analogous results for cubes with size parameter 3.6 and 9.1, respectively.

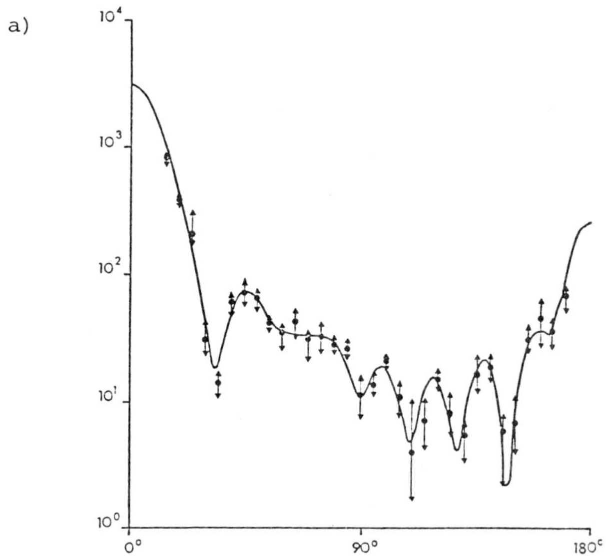

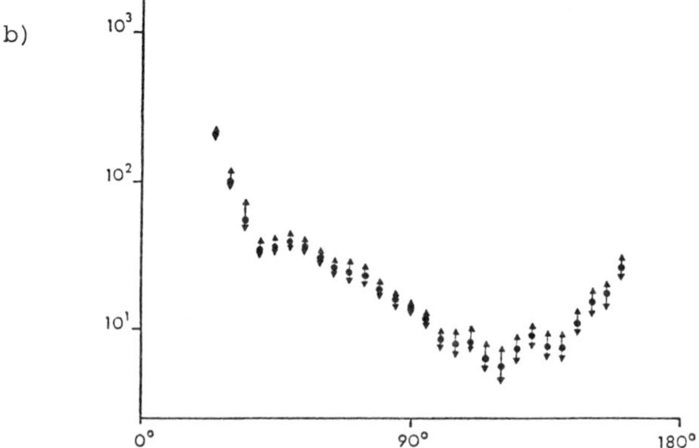

Fig. 8. Influence of orientation on the scattering function of a rough sphere with size parameter $\alpha = 8$ ($D \approx 1.4$ µm). Angular resolution: a) $\Theta < 1°$, b) $\Theta = 20°$.

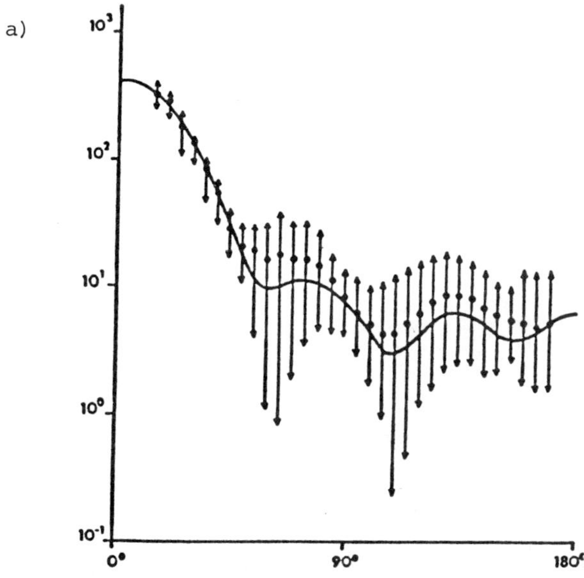

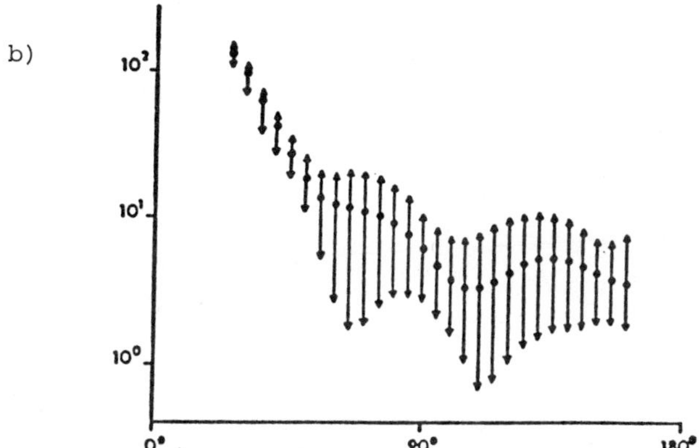

Fig. 9. Influence of orientation on the scattering function of a cube
with size parameter $\alpha = 3.6$ (D $\approx$ 0.65 µm). Angular resolution:
a) $\Theta < 1°$, b) $\Theta = 20°$.

It can be seen, that, apart from the diffraction lobe, Mie-
calculation for a sphere of equal volume provides a useful fit only for
the small cube. The influence of orientation is by far higher than for
the rough sphere. Extension of the viewing angle to $\Theta = 20°$ causes a
remarkable reduction of the difference between maximum and minimum
values, but there remain still nearly 2 orders of magnitude for the large
cube.

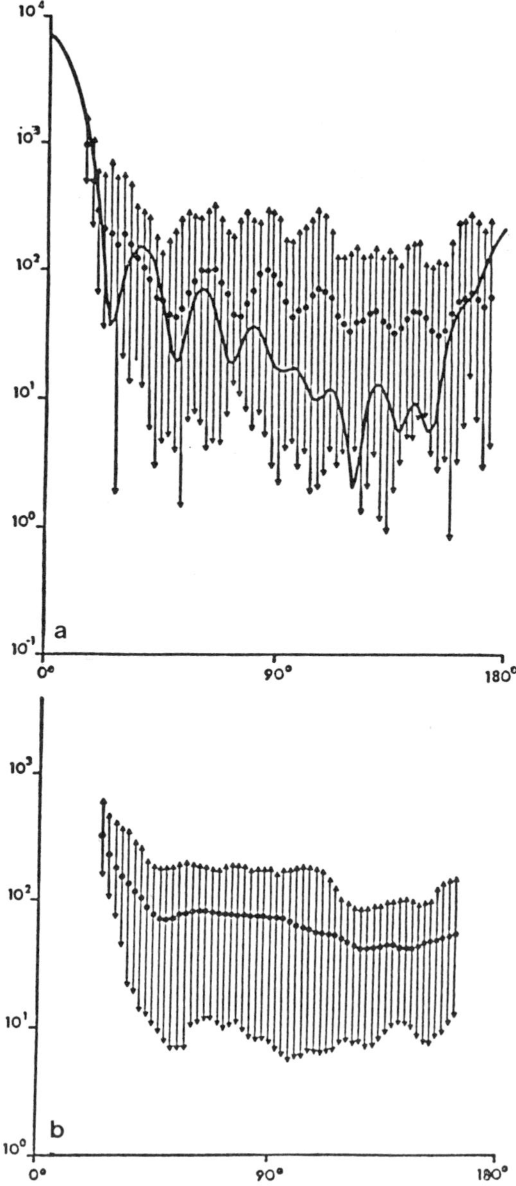

Fig. 10. Influence of orientation on the scattering function of a cube
with size parameter $\alpha = 9.1$ ($D \approx 1.6$ µm). Angular resolution:
a) $\Theta < 1°$, b) $\Theta = 20°$.

This leads to the conclusion, that the influence of orientation at different scattering angles does not change abruptly. Therefore it can not be expected, that it completely averages out even by integration over larger angular intervals. A smoothing effect, however, can be predicted by use of a multicolour light source. The size parameter α of a single particle is related to the wavelength (α = circumference / wavelength). Hence multicolour illumination of a single particle converts it to an artificial polydisperse mixture.

REFERENCES

Straubel, H., 1960, Die Stabilisierung elektrisch geladener Teilchen in Wechselfeldern, Acta Physica Austriaca, 8:265.
van de Hulst, H. C., 1981, "Light scattering by small particles," Dover, New York.
Weiß-Wrana, K., 1983, Optical properties of interplanetary dust: comparison with light scattering by larger meteoritic and terrestrial grains, Astron. Astrophys., 126:240.
Zerull, R. H., 1976, Scattering measurements of dielectric and absorbing nonspherical particles, Beiträge zur Physik der Atmosphäre, 49:168.

APPLICATION OF TOP-HAT LASER BEAM TO PARTICLE SIZING IN LDV SYSTEM

Masanobu Maeda and Kouichi Hishida

Department of Mechanical Engineering
Keio University
Kohoku-ku Yokohama, 223 Japan

INTRODUCTION

Efforts have been made to obtain information on mixing to relate with ignition in the peripheral region of spray jets with clouds dilute enough to be measured by an laser Doppler velocimeter system. Investigators have attempted simultaneous size measurements within the same optical system, or combined with another light source, by Mie-scatter intensity, Doppler visibility or Doppler phase difference [1-11]. The authors have investigated heat transfer in two-phase mist flows and analyzed flow behaviour and the variation of droplet size distribution in boundary layers with a two-beam LDV system [12,13]. The LDV system with particle sizing proposed previously [14,15] has worked enough in detecting the scatter intensity with discrimination of the particle paths in the measuring volume. However, the electronic circuits had to be more sophisticated, and the ambiguity of size estimation can not be avoided in highly turbulent flows.

To a further improvement with two-phase mist flows involving coarser particles, a monochromatic four-beam LDV system is being designed and tested to obtain a higher resolution in particle sizing. The measuring volume of the four-beam LDV is made up of two coaxial volumes with different diameters, and they are linearly polarized in the orthogonal direction. This equipment is being adapted for basic treatment with an advanced two-colour four-beam LDV system to be applied to complex practical flow fields such as a spray. On the basis of experimental studies of the monochromatic four-beam LDV [16,17], the authors extended the technique to a two-colour four-beam LDV system, which permits simultaneous measurements of particle size and two-component velocities with higher resolution. The larger measuring volume has a uniform top-hat intensity distribution to reduce the fluctuation in scatter intensity due to the variation of particle path in the measuring volume. The location of the particle path was confirmed by the presence of a Doppler burst signal from the smaller measuring volume. The present two-colour four-beam LDV system achieves higher resolution of particle sizing and better experimental results for atomizing spray indicated flow fields in gas and particle phases. The intensity method has essentially a disadvantage that signals are distorted, if dense particle cloud arround the measuring point obstacles the incident beam and scattering light to the photosensor. Meanwhile the advantage to permit to measure particles of high velocities and with high data sampling rate. Besides the mean flow schemes of droplets, more detailed local information is needed to analyzing two-phase mist flows. The add-on function of particle size discrimination makes it possible to estimate the time-dependent velocities in a cyclic spray jet for each class of droplet size.

OPTICAL SETUP

For opaque particles the Mie-scatter theory applies with sizes under 100 μm [18]. When the particles to be measured are limited to transparent and spherical liquid droplets, the measurements can be extended to larger sizes. It has been experimentally pointed out that when the collecting apertures (angle) of the receiving optics are large enough, the collected power is smoothly proportional to the square of the particle diameter. Figure 1 shows the relation between particle diameters and light scattering intensity measured in an uniform incidence beam at a constant off collection angle. Here the light scattering intensities are normalized by the pulse height value of the particles with the largest diameter (470 μm) used in the experiments. At a constant angle, the peak value of the light scatter signal is proportional to the square of the particle diameter, i.e. the relation $I_p = Kd^2$ holds. This confirms with the experimental results of [1],[2] and [5] for particles sizes less than 100 μm.

To avoid the dependence of scatter intensity on particle path, the measuring volume is made to an uniform intensity. The part of beam out of the uniform intensity is not necessary, namely the Gaussian distribution of the beam is reformed to a top-hat intensity distribution with a steep edge, where the coherency should preserve for applying to a LDV system. It is required that all particles to be measured have to pass the part of uniform intensity. Thus the measuring volume should have a larger diameter than the particle and not too large so that the multiparticle scattering may not occur. Beam diameter must be adjustable and be able to optimize to the particle size range.

The authors extended the system to a two-colour, four-beam LDV with a top-hat intensity distribution at the location of the measuring volume to obtain a higher resolution in particle sizing for larger diameters. The methods of producing top-hat laser beam are discussed in [19],[20]. The

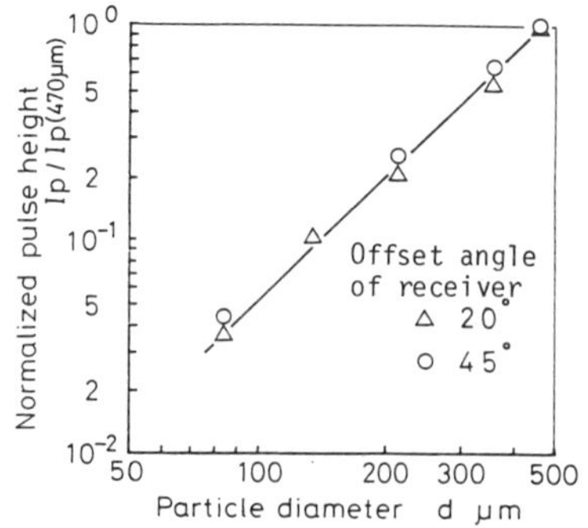

Fig. 1   Experimental relation between scattering intensity and
particle diameter by glass beads

432

sizing of droplets using a top-hat laser beam by a holographic filter was reported in [21]. First, a flat-intensity beam is made with a distributed density filter. The beam is filtered by a spatial filter to remove the noise,   and is magnified by a lens to a diameter of 1.3 mm. The beam passes through an anti-Gaussian density filter [22] made using a holographic plate (AGFA–GEVAERT 10E75), and is transformed into a uniform-intensity beam (Fig.2). However, the filter is damaged by higher power laser sources and the steepness of the top-hat edge is blurred, so an optical system with pinholes and lenses is used to obtain a top-hat beam. The optical setup of the system is illustrated in Fig.3. It can be employed with a powerful laser source and gives a top-hat intensity distribution with any beam diameter. The same idea to make a top-hat beam from a laser beam by a spatial filter was proposed by [23]. The laser beam diffracts and has coaxial fringes of higher order following a pinhole as shown in Fig.4. The obstruction of the higher order fringes causes the

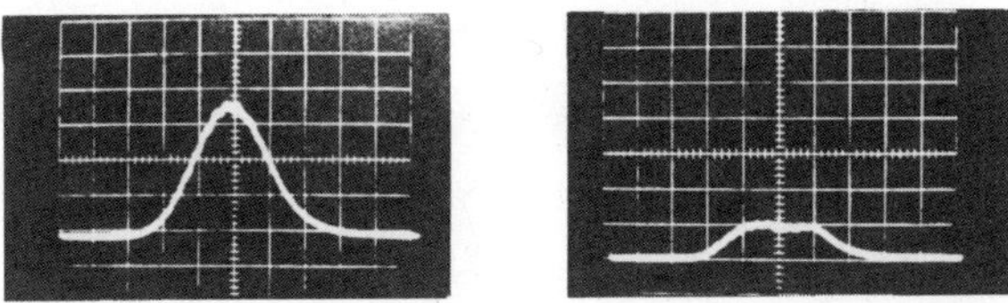

Fig. 2   Top-hat beam by Gaussian density filter using a
         holographic film

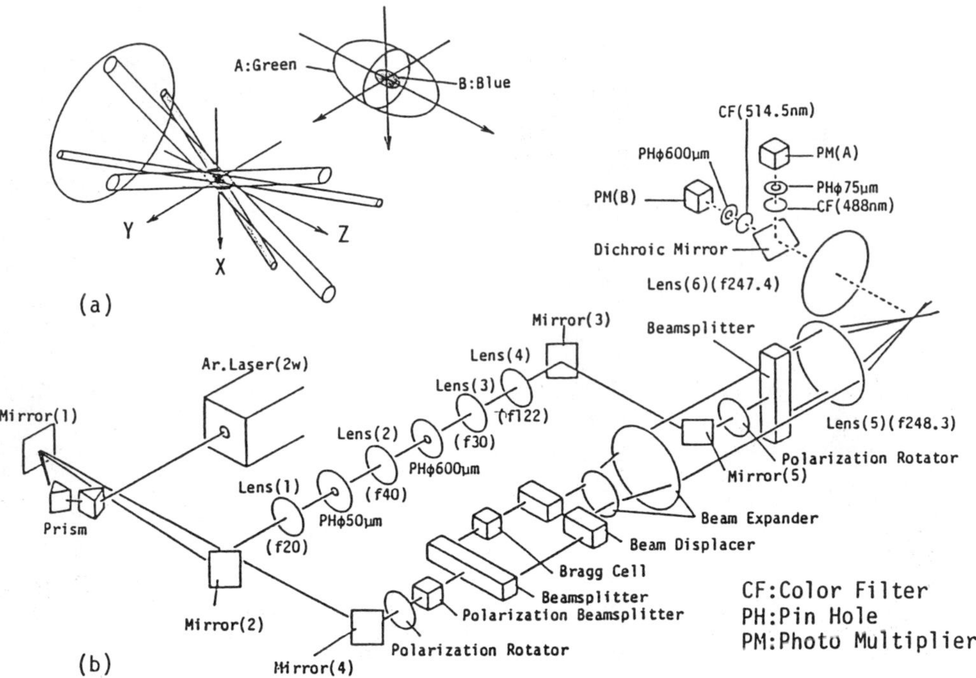

Fig. 3   Optical setup of two colour four beam LDV with coaxial
         measuring volumes of different diameters

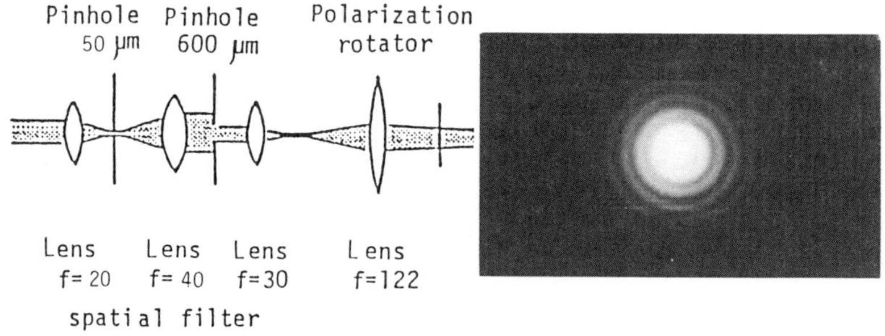

Fig. 4   Photograph of the beam following pinhole

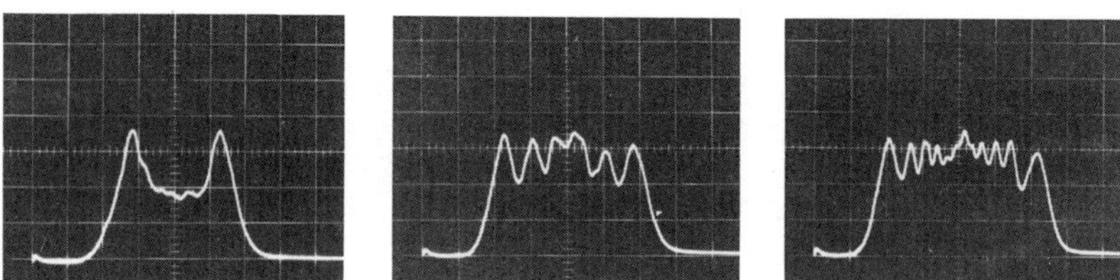

Fig. 5   By adjusting the system, the diffraction fringes of
beam become smaller and show steep edge

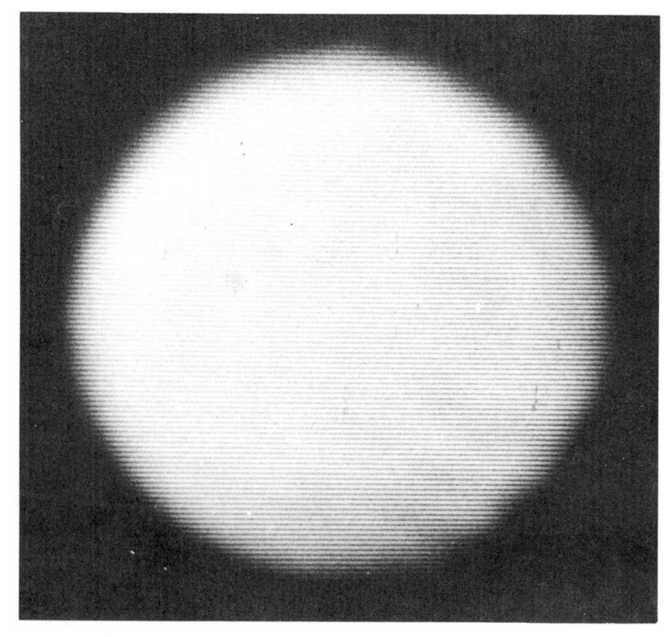

Fig. 6   Oscilloscope trace of a line CCD sensor output and the
projection of measuring volume

distortion of the beam form and the loss of steepness of the edge to produce the uniform intensity distribution at the measuring volume. When the image of the pinhole following the spatial filter is focused on the measuring point, the fluctuation of the intensity distribution caused by diffraction is smoothed giving a steep edge. At the beam crossing the intensity distribution is changed by adjusting the position of lenses. Observing the display of CCD line sensor as shown in Fig.5, the image of the spatial filter is focused onto the same position making the measuring volume of beam crossing. The measuring volume is easily adjusted to any diameter in from 100 to 600 µm. The blue beams were focused in a small diameter of 50-100 µm by employing a conventional beam expander. The basic optical construction consists of a TSI two-colour, four-beam system. Figure 6 shows an oscilloscope trace of a line sensor output by the top-hat green beam and its projection with fine fringes. It has a much steeper edge compared to the trace made by a density filter.

In the LDV system the frequency shifting is usually made by AOM, the diffraction efficiency is worse for a large diameter beam and this sort of beam with diffracted beam. To solve this one should set AOM before the pinhole. The incident beam splitted to the shifted and unshifted beams. Both beams are superposed again but their polarizations are set each other in a orthogonal direction e.g. [24]. The polarization of beam is rotated again after passing filter and separated to a pararel beam to be coherent by a polarization rotator (Fig. 7).

The receiving optic module is set in a sideward scattering mode, so the measuring volume can be more restricted by the receiving optics. This technique is applicable for measurements in a more concentrated suspension flow and in a practical flow field such as in sprays where the particles exist in a restricted region. The position of receiving optic must be perpendicular to the plane involving the crossing of two top-hat green beams, nevertheless the signals do not realize the top-hat trace due to the difference of the scattering intensity depend on the receiving angles. The characteristics of the photomultiplier are also significant in detecting light scattering intensity accurately. Particularly important is the linearity between quantity of light and the output signal voltage (or current). The first amplifiers after PM.(A) and PM.(B) have a frequency range from DC to 100 MHz and an amplification of 20 dB. The condition of the electronics for the photomultiplier, such as the supplied voltage between anode and cathode, is set to measure the signals for the largest particle without overloading. The photomultiplier as a sensor had a breeder circuit to provide enough current to guarantee a good S/N and a wide measuring range of linearity. The pinhole apertures of the receiving optics 75 µm and 250 µm in diameter are respectively fitted for the dimensions of each measuring volume diameter.

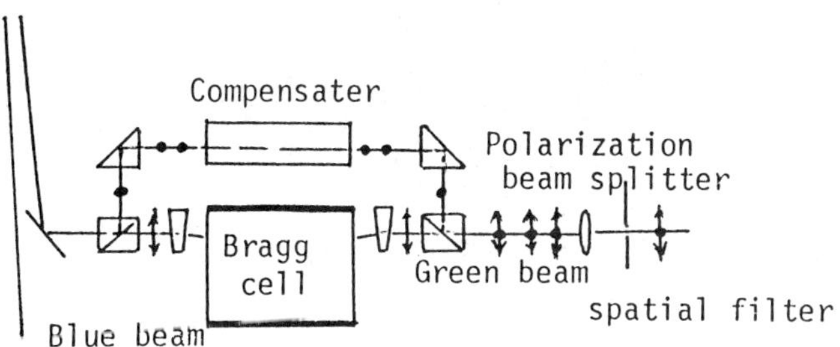

Fig. 7 Frequency shifting for top-hat beam

SIGNAL PROCESSING

The principle of signal processing is almost the same as in the previous report [12]. The light scattering intensity from a particle passing through the central core can be determined by the presence of a pedestal signal with the Doppler burst signal obtained from the smaller measuring volume at the middle of a longer duration pulse signals. Figure 8 shows a signal model of the output from a particle passing through the measuring volume. The signal from the larger volume has a longer duration time than that from the smaller one. When a particle passes through the central core of the measuring volume, both signals must be clearly distinguished. The signal from the smaller volume generated by the blue beams is located in the middle of the duration of the signal from the larger volume. The signal from a particle passing outside the center core of the measuring volume will have no signal from the blue beams.

Figure 9 presents an example of signals of a transparent 100 μm droplet generated by an ink jet printer head (piezoelectric capillary injector) working at a frequency of 2 kHz. The picture shows pulses of the droplets passing the central core of the top-hat beam (bottom), Doppler burst signals with high frequency components (middle), and Doppler burst signals from the blue beam (top). The figure confirms the top-hat intensity distribution of green beams and the existence of a blue beam signal in the

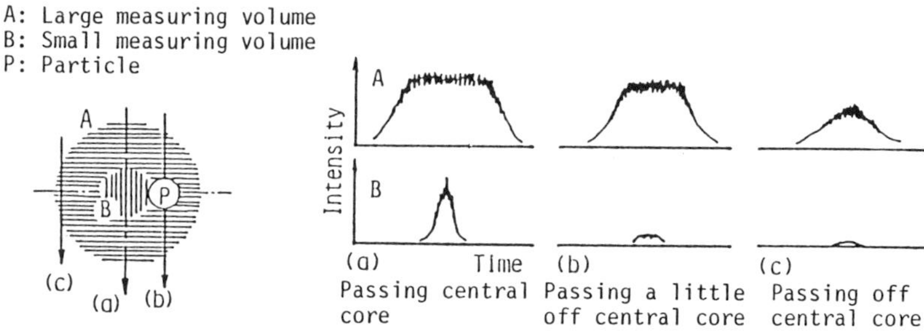

Fig. 8   Particle path in measuring volume and output signal from photomutipliers.

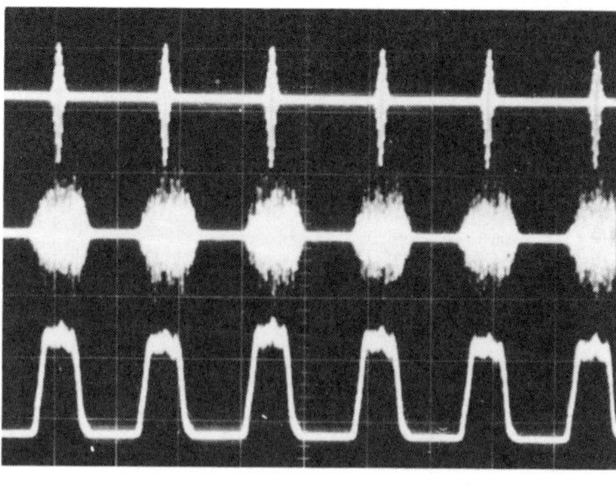

Fig. 9   Signals of monodispersed droplets by an ink jet; Doppler signals from blue and green beams and pulse signals from green beam.

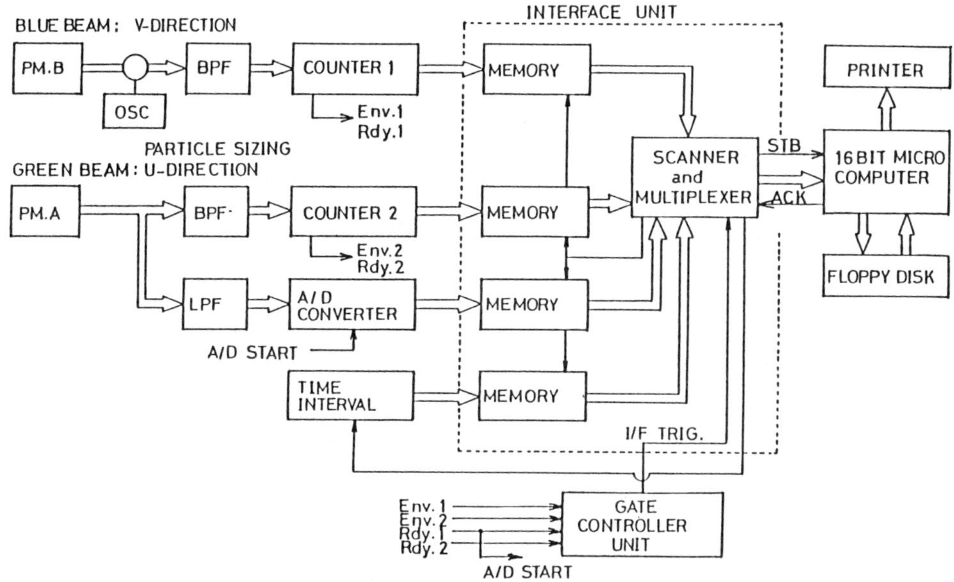

Fig. 10 Block diagram of electronic circuit of signal processor.

middle point. The decrease in steepness at the edge of the signals is
caused by the finite diameter of droplets passing through the measuring
volume. The processing is shown in a block circuit diagram in Fig.10.
After filtering the high frequency signals, the height of the pulses with a
low frequency component are digitized by a 10 bit A/D converter and
transferred to a computer (HP9000) with an interface trigger pulse (I/F
Trigg.). The high frequency Doppler burst signals from both the green and
the blue beams were processed by counter type processors. The processed
outputs of velocities were transferred, together with the particle size
information, by I/F trigger pulses generated by a coincidence of the gates
of the two Doppler burst envelopes.

APPLICATION TO A FLOW OF AN ATOMIZING NOZZLE

An air atomizing nozzle for water was adopted for an basic application
of the present measuring system. The utility of the system was discussed on
velocities concerning with their droplet sizes. The analysis was made by
5000 sampled data for one position measurements. Figure 11 illustrated an
air atomizing nozzle, the set up of optical system and the coordinate. The
air of outer annular jet can be heated to 80°C. The size distribution of
spray in room temperature was measured by a liquid immersion method and a
microscopic method and by LDA were shown in Fig.12. The probability of
number count to be captured by the center blue volume depends on the
droplet diameter passing through the measuring volume. Therefore, the
number counts are compensated by the cross-sectional area which is formed
by the view area for blue beam and the surrounded area bounded by the
distance of particle diameter. As the velocity and size of each droplet was
measured simultaneously, and velocities for each class of size were
obtained from the distribution of total sample droplets as shown in
Fig.13(a-c). The figure illustrated velocities of classified for below 10
μm, 10-20 μm, 20-30 μm and over 30 μm at measurement locations in the
distance from the nozzle X= 100, 200, and 300 mm, respectively. The fine
droplets below 10 μm acted as a tracer particles, since strong acceleration
in droplet flow unexisted in the present experiments. At X=200 mm, the

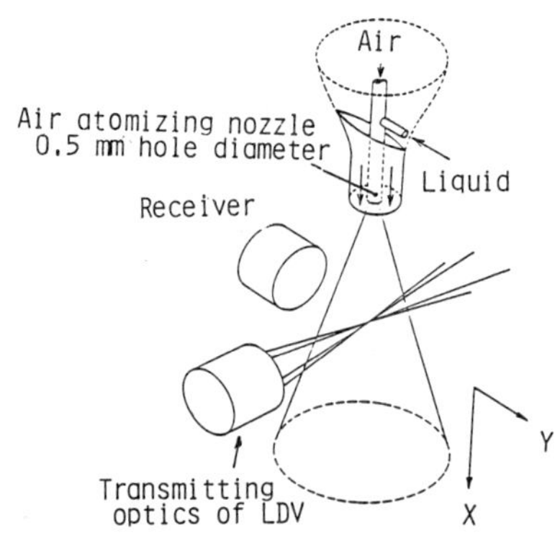

Fig. 11 Air atomizing nozzle and the setup of receiving optics

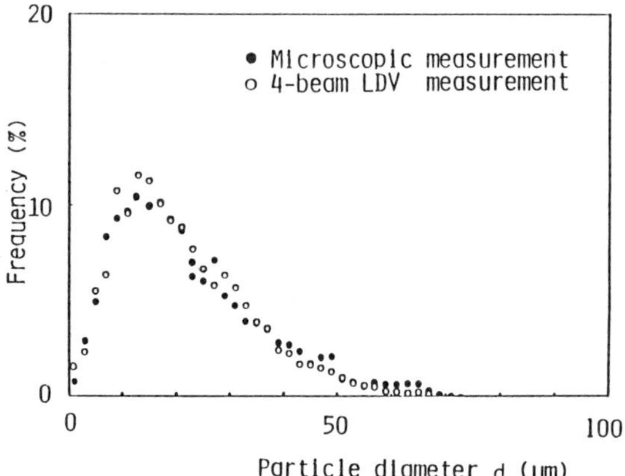

Fig. 12 Comparison of measureed droplet size distributions
between liquid immersion method and LDV

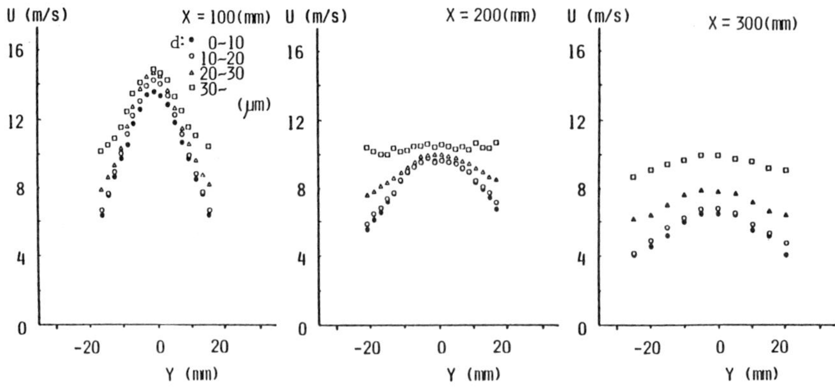

Fig. 13 Velocity distributions for each class of droplet sizes

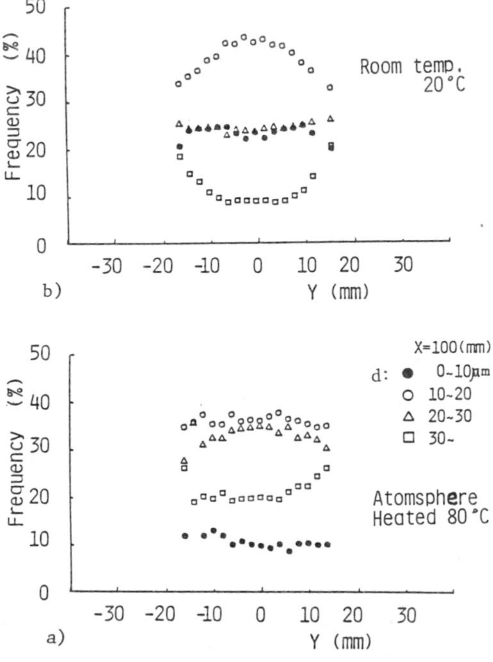

Fig. 14   Distributions of droplet number count per unit time
a) heated and b) unheated condition

velocities of coarser droplets in outer regions of the jet stream were
indicated as a larger value than fine droplets. The velocity difference due
to droplet sizes were remarkably observed at the downstream of X=300 mm.

Figures 14(a),(b) provide distributions of droplet number count in the
conditions of 80 C and room temperature at the location X=100 mm. They show
somewhat curious tendency that coarser droplets increase and small droplets
decrease. That is the reason why the small droplets evaporate fast and
coarse droplets also decrease their size, however, the decrease in diameter
due to evaporation is much faster in small droplets. As a consequence the
coarcer droplets increase in relative number.

CONCLUSIONS

The present paper demonstrated simultaneous measurements of droplet
size and two component velocities. The improved LDA system of the authors
employed co-axial double measuring volumes of two colours with different
diameters and a top-hat intensity distribution of green beam, so that the
data acquisition logic became much simpler and more reliable to apply in a
high turbulent flow. The adjustable beam diameter and the top-hat form in
the measuring volume would contribute to the signal quality for higher
concentration. The high resolution in measurement for individual droplet
would contribute to the analysis of the droplet motion in mist flows.

REFERENCES

[1]   Holve,D. and Self, S.A.,1979, Optical particle sizingfor in situ
      measurements part 1, Appl. Optics, 18-10, pp.1632-1645.
[2]   Kline,R., Gouesbet,G. and Ruck,B., 1982, Simultaneous, local
      measurements of particle velocity, particle size and particle
      concentration, Sensor 82 Transducer Technology and Temperature
      Measurement vol. 3, pp. 15-41.
[3]   Hadded,O., Bates,C.J., Yeoman,M.L. and White,H.J., 1984, Evaluation
      of drop size and two-component velocitymeasurement technique in
      annular two phase-flow, 2nd Int. Symp. on Appl. of LDA to Fluid
      Mech., pp. 16.2, 1984.
[4]   Yule,A.J., Chiger,N.A., Atakan,S. and Ungut,A., J. 1977, Particle
      size and velocity measurement by laser anemometry, Energy, 1, 4, pp.
      220-228.
[5]   Gouesbet,G and Ledoux,M., 1984, Supermicronic and submicronic optical
      sizing including a discussion of densely laden flow, Optical
      Engineering vol.23-5, pp. 631.
[6]   Ruck,B. and Pavlowski,B., 1984, Combined measurement of particle
      size amd velocity distribution in ducts, 2nd Int. Symp. on Appl. of
      Laser Anemometry to Fluid Mechanics, pp. 18.3.
[7]   Durst,F. and Umhauer,H., 1975, Local measurements of particle
      velocity size distribution and concentration with a combined laser
      Doppler particle sizing system, Proc. of the LDA-Symp. Copenhagen,
      pp. 430.
[8]   Durst,F. and Zare,M., 1975, Laser Doppler measurements in two-phase
      flow, Proc. of the LDA-Symp. Copenhagen, pp. 403.
[9]   Bauckhage, K and Schoene, A., 1985, Measurement of velocities and
      diameters of small particles by using fast digital circuits for the
      on-line evaluation of laser-Doppler signals, Part. Caract. 2, pp.
      113.
[10] Bachalo,W.D., Houser,M.J. and Smith,J.N., 1986, Evolutionary
      behaviour of sprays produced by pressure atomizers, AIAA 24th
      Aerospace Sciences Meeting, AIAA-86-0296.
[11] Saffman,M., Buchave, P. and Tanger,H., 1984, Simultaneous measurement
      of size, concentration and velocity of spherical particles by a laser
      Doppler method, Int. Symp. on Appl. of Laser Anemometry on Fluid
      Mech./Lisbon, 8.1.
[12] Hishida,K., Maeda,M. and Ikai,S., 1980, Heat transfer from a flat
      plate in two-component mist flow, Trans. ASME J. Heat Transfer
      vol. 102, pp. 513-518.
[13] Hishida,K., Maeda,M. and Ikai,S., 1982 Heat transfer in two-component
      mist flow; boundary layer structure on an isothermal plate, Proc. 7th
      IHTC, Munich/FRG vol. 5, pp. 301-306.
[14] Maeda,M., Hishida,K. and Furutani,T., 1980. Optical measurements of
      local gas and particle velocity in a upward flowing dilute gas-solids
      suspension, ASME Polyphase Flow and Transport Technology, BK#
      HOO158, pp. 211-216.
[15] Maeda,M., and Hishida,K., 1982, Velocity and turbulent intensity
      measurements of gas and spherical particles in two-phase flows by a
      modified LDA system, Sensor 82 Transducer Technology and
      Temperature Measurement vol. 3, pp. 43.
[16] Hishida,K., Maeda,M., Imaru,J., Hironaga,K. and Kano,H., 1982.
      Proc. Int. Symp. on Appl. of LDA to Fluid Mech., pp. 5.6, 1982.
[17] Hishida,K., Maeda,M., Tajima,K. and Kano,H., 1984, Measurements of
      two-phase turbulent flow by LDA with particle size discrimination,
      2nd Int. Symp. on Appl. of LDA to Fluid Mech., pp. 18.4.
[18] Maheu,B., Letoulouzan,J.N. and Gouesbet,G., 1984, Four flux models to
      solve the scattering transfer equation in terms of Lorenz-Mie
      parameters, Appl. Optics 23-19, pp. 3353.

[19] Belvaux,Y. and Virdi,S.P.S., 1975, A method for obtaining a uniform non-Gaussian laser illumination, Opt. Commun. 15-(2), pp. 193.

[20] Veldkampf, W.B. and Kastner, C.J., 1982, Appl. Opt. 21-(2), pp. 345.

[21] Allano,D., G.Gouesbet, Grehan,G. and Lisiecki,D., 1984, Droplet sizing using a top-hat laser beam technique, J. Phys. D: Appl. Phys., 17, pp. 43-58.

[22] Quintanilla,M. and de Frutos,A.M., 1981, Holographic filter that transforms a Gaussian into an uniform beam, Appl. Optics, 20-5, pp. 879.

[23] Ruck,B and Schmitt,F., 1982, Verfahren und Vorrichtung zum Erzeugen einer zylinderformigen Intensitatsverteilung eines Laserstrahl, Patentanmeldung/Deutsches Patentamt, P 33 01 232 6, 82/15369-Hf

[24] Knuhtsen, J., Olldag,E. and Buchhave,P., 1982, Fibre-optic laser Doppler anemometer with Bragg frequency shift utilizing polarization-preserving single-mode fibre, J. Phys. Sci. Instrum.,15 pp. 1188-1191.

SIZE AND VELOCITY MEASUREMENTS OF SPHERICAL PARTICLES IN MULTIPHASE

FLOWS AND THE PREDICTION OF ABSOLUTE PARTICLE CONCENTRATIONS

Klaus Bauckhage, Hans-H. Flögel, and Frank Schöne

Universität Bremen
Postfach 330 440
2800 Bremen 33, F.R. Germany

INTRODUCTION

For the important field of multiphase flows, where dispersed particles are immersed in a continuous phase there exists a strong demand for simultaneous and non-intrusive measurements of the size, velocity and concentration of the dispersed particles in order to establish the momentum, mass and energy balances of these processes.

Using the well known two-beam LDA for generating the fringe system and receiving the Doppler-bursts by means of two or more photodetectors, this phase-Doppler-difference method (LDVS) opens the possibility for simultaneous measurements of the sizes and velocities of individual spherical particles over an extremely wide size range, /1/ to /3/. While the technique for the velocity measurements is identical with the conventional LDA technique, that for the size correlations uses the phase shift of the Doppler-bursts. This permits an analysis of the local behaviour of the flowing particles in a multiphase-flow-situation with a high spatial resolution.

Very important preconditions for the successful use of the LDVS-method are the sphericity of the dispersed particles and the stationary multi-phase-flow situation. The latter condition has to be satisfied because of the counting character of the LDVS (of single events versus time). This might be a severe restriction to the applicability in spite of the multiphase systems. But in most two-phase flow situations, where fluid particles are dispersed in the continuous phase, these particles exist as spherical drops or bubbles, and very often also solid particles have a spherical shape with a surface of sufficient smoothness, and the flow is stationary.

MEASURING PRINCIPLE

During the motion of a spherical particle through the LDA interference fringe system the interferometric patterns reflected from or refracted by the particle are moving as a far-field scatter distribution with streak distances and a velocity corresponding to the size and velocity of the particle across the photodetectors.

The phase difference between adjacent photomultipliers responding to these signals from the fringe mode LDA is linearly proportional to the diameter of the spherical particle in the measuring volume. Thus the new instrument allows, with long distances between the optical devices and the measuring volume in the spray, non-intrusive measurements and from these the correlation of size and velocity of the individual droplet.

It is advantageous for the Doppler signal detection and especially for the phase difference measurement of the two Doppler-bursts received from transparent particles to choose photodetector positions where either reflected or refracted light is dominant.

As it has been shown earlier (see ref. /4/) for transparent particles under the laws of reflection or refraction there exist different types of linear dependencies of Φ verus d at off-axis angles φ between 40 and 90 degrees. Computations were carried out for non-absorbing particles, in which the only distinction made, was whether the dispersed phase of the transparent particle was the denser one (like water drops in air) or the less dense one (like air bubbles in water). The formulas for the particle diameter of transparent spheres under a dominant light scattering situation of refraction or reflection (as given in Table 1) have been received by the simplified equations of the geometric optics. These formulas have been proved by the numerical calculations of the complete Mie's equations and have been found in excellent agreement with these results.

The equation (2) may also be used for opaque or metallic particles which show a sufficient reflection /4/. In cases where the laser light intensity and the reflection result in good signals, the LDVS-instruments may be used as a backscattering set-up, which helps to save time by avoiding difficult adjustments as well as expensive preparations for the experimental plant.

The geometry of the LDVS-set-up, especially the arrangement of the different angles Θ, ψ and φ can be seen from Fig. 1, some specifications are given in Table 2.

As for all measuring techniques, the question of systematic or possible errors has to be discussed in spite of the LDVS. Thus in addition to the specifications of Table 2 there should be given the possible deviations, which may occur even if the adjustments have been very precisely executed.

Table 1. Equations for the particle diameter

	$d = \dfrac{1}{2b}\left\|\dfrac{\lambda_0}{\pi\,n_c}\right\|\Phi$	eq.(1)
$b_{reflect}$	$= \sqrt{2}\,[(1+\sin\frac{\Theta}{2}\sin\psi -\cos\frac{\Theta}{2}\cos\psi\cos\varphi)^{1/2}$ $-(1-\sin\frac{\Theta}{2}\sin\psi -\cos\frac{\Theta}{2}\cos\psi\cos\varphi)^{1/2}\,]$	eq.(2)
$b_{refract}$	$= 2\left\{[1+n'^2-\sqrt{2}n'(1+\sin\frac{\Theta}{2}\sin\psi +\cos\frac{\Theta}{2}\cos\psi\cos\varphi)^{1/2}]^{1/2}\right.$ $\left. -[1+n'^2-\sqrt{2}n'(1-\sin\frac{\Theta}{2}\sin\psi +\cos\frac{\Theta}{2}\cos\psi\cos\varphi)^{1/2}]^{1/2}\right\}$	eq.(3)
n'	$= \dfrac{n_d}{n_c}$	eq.(4)

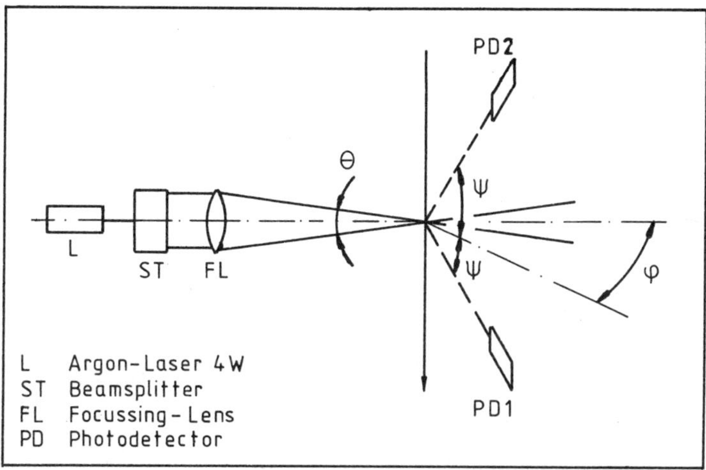

Fig.1. Experimental set-up for the LDVS equipment

Ignoring for instance, for a dynamic range from 5 to 300 µm, the theoretical source of error arising from differences between the ($\overline{x}$ ÷ d) dependencies of the simplified equations of the geometric optics in comparison with the numerical calculations of the complete Mie's equation, and omitting argumentations about the mistakes arising from incorrect allocation of the counts, we will attempt in the following discussion to quantify the different types of errors belonging to the LDVS-method.

Table 2. Data of LDVS-experimental set-up

laser	Ar-Ion	
power	4 W	
λ_0	0.5145	µm
intersecting angle Θ	2.7	deg.
elevator angle ψ	3.1	deg.
off-axis-angle φ	6 9	deg.
distances		
foc.lens – meas.volume	600	mm
meas.vol. – photodet.lens	600	mm
max.accepted counts	250	sec^{-1}
measuring volume		
δ_x	6.672	mm
δ_y	0.157	mm
δ_z	0.157	mm
δ_p	0.600	mm
dynamic measuring range		
particle velocity	0.7 to 21.8	msec^{-1}
particle diameter	5 to 190	µm

DISCUSSION OF TYPICAL LDVS-ERRORS

Though the LDVS-analysis results in a couple of absolute measured values (the size and velocity of the spherical particle) there exist, as in any other measuring problem, some errors and insufficiencies in the measurements, which in the case of particle size can be classified as follows:

i) the maximum possible error which emerges from using the formulas (1) to (4). This error can be explained in detail by three main factors:
   a) deviations of the geometrical adjustments for the optical parts of the set-up, i.e. in the case of Θ , ψ and φ ;
   b) deviations of the system-values n_d, n_c and λ_o;
   c) discrepancies in the measured value of ϖ , which are caused by the electronic data-processing;

ii) the insufficiencies, arising from the incoming light signals in spite of the validation criteria of the signal processing. These insufficiencies cause data corrections in order to establish further properties from the bivariate distributions;

iii) the insufficiencies of the particles caused by inhomogenities or rough surface structures.

The Maximum Possible Error

The first mentioned relative maximum possible error of the particle diameter is given by

$$\left(\frac{|\Delta d|}{d}\right)_{1;r} = \sum_{i=1}^{7} \left| \Delta x_i \; \frac{\partial d_{1;r}/\partial x_i}{d_{1:r}} \right| \tag{5}$$

with d_1 given by equation (1) and using the constituting equation (2), for reflection (index 1) and with d_r given by equation (1) and using the constituting equations (3) and (4) (i.e. the formulae for refraction, thus using index r) and with

$$x_1 = \Theta \; ; \; x_2 = \psi \; ; \; x_3 = \varphi \; ; \; x_4 = n_d \; ; \; x_5 = n_c \; ; \; x_6 = \lambda_o \; ; \; x_7 = \varpi \; ;$$

as given in Table 3.

Let us leave the discussion about the insufficiencies caused by differing light signals (i.e. part ii) for a moment and turn our attention to the maximum possible error. As described by Flögel /5/ this error shows in the regions of interest linear dependencies from the deviations $\Delta x_i \, \partial d/\partial x_i$ (here with i = 1; 2 and 3).

As the angles Θ and ψ are small compared with the off-axis-angle φ , the relative values of their deviations as given by equation (5) are about three times larger than that caused by $\Delta x_3 \, \partial d/\partial x_3$ in relation to the high value of 69 degrees. This means that in spite of the adjustments of the optical components the main attention should be paid to the angles Θ and ψ .

The deviations of the system values n_d, n_c and λ_o are comparatively small and can be set for the first approach equal to zero. Thus we can neglect the influence of error i/b). Only the last part i/c) caused by the electronic data processing has to be discussed in detail.

Table 3: relative maximum possible error including electronic components (of error of 1 %)

			Refl.	Refr.
$\Theta =$ 2.7°	$\Delta \Theta = \pm$ 0.05°	$\Delta \Theta\, \partial d / \partial \Theta$ =	3.7%	3.7 %
$\psi =$ 3.1°	$\Delta \psi = \pm$ 0.10°	$\Delta \psi\, \partial d / \partial \psi$ =	3.2%	3.2 %
$\varphi =$ 69 °	$\Delta \varphi = \pm$ 1.00°	$\Delta \varphi\, \partial d / \partial \varphi$ =	1.3%	0.5 %
$n_c =$ 1.0000	$\Delta n_c = \pm$ 0.0001	$\Delta n_c / n_c$ =	0.01%	0.01%
$n_d =$ 1.3334	$\Delta n_d = \pm$ 0.0001	$\Delta n_d / n_d$ =	0.01%	0.01%
$\lambda_0 =$ 514.5nm	$\Delta \lambda_0 = \pm$ 0.05nm	$\Delta \lambda_0 / \lambda_0$ =	0.01%	0.01%
$\bar{Q} = 2\pi \,\hat{=}\, 360°$	$\Delta \bar{Q} = \pm$ 7.2°	$\Delta \bar{Q} / \bar{Q}$ =	2 %	2 %
$\left(\dfrac{\lvert \Delta\, d \rvert}{d} \right)_{l;r}$ = after equation 5			=11.2 %;	10.4%

From the different ways of signal processing for establishing the values of velocity and size via the Doppler-bursts only the use of a) a multi channel transient recorder for a single data analysis and b) an electronic phase-frequency-counter for an analysis of high data rates can be discussed. At the outest it should be mentioned that the quantity of measuring error is clearly a question of investment, thus the shorter the temporal steps for instance of the transient recorder, the smaller its error, but the higher its price. Turning our attention to the alternatives a) and b) we assume that the amplification and filtering processes work in a strictly identical fashion and in the required manner. Thus the additional error arising from these parts of data processing should be of the same magnitude as that of the electronic phase counter.

The transient recorder digitalizes the incoming burst with a constant temporal time cycle. As in the following computer analysis the exact temporal position of the Doppler-amplitudes have to be found, and as this holds for both Doppler-bursts, the approach to the requested maximum of the amplitudes of the Doppler-bursts can be executed for each amplitude only to a distance of $\pm$ 0.5 cycles. From the comparison of the temporal positions of the identical amplitudes of both Doppler-bursts there results an error of t_c = 1.0 cycle per phase difference. As the temporal resolution is reciprocal to the Doppler-frequency $\nu_D^{-1} = \Delta z/v$ (i.e. the particle velocity v) the error rises linearly with the particle velocity and decreases with increasing clock rates. As shown by Flögel /5/ the relative error is given by the formula

$$\frac{\Delta \Phi}{2\pi} = v\, t_c \cdot \Delta z^{-1} = t_c\, \nu_D \qquad (6)$$

its magnitude is about 5.5% in the case of a Biomation 8100 (5 MHz) and a particle velocity of about 3 m/sec. This maximum possible phase difference error has to be devided by the number of amplitudes which have to be compared by the computer. Thus we can speak in general of a relative error of less than 2 % .

As the special phase-frequency-counter generally uses a cycle frequency which exceeds that of the transient recorder by a factor of 50 and as it often works with a cross correlation method to find out the phase differences of the bursts, there exists no other way of error detection than to use a burst generator with synthetic bursts. In our case this has been done with the result that the maximum possible (relative) error is less than 1 %, over the whole measuring range.

If we now use equation (5) for summing up the error contributions, which partly have been listed in Table 3, distinguishing between reflection and refraction, the total relative maximum possible error of the phase-Doppler-method can be specified as $(\Delta d / d)_r = \pm$ 10.4 % for refraction and as $(\Delta d / d)_l = \pm$ 11.2 % for reflection.

Corrections For Insufficient Light Signal Qualitites
<u>Corrections For Insufficient Light Signal Qualitites</u>

Now type ii) of error has to be discussed, arising from the attempt to use the bivariate distributions in order to receive further properties of the dispersed particles such as concentration data or kinetic energy distributions.

As already described by Glantschning and Chen /6/ or by Buchhave et al /2/ the scattered light intensity from the particles is proportional to their diameter squared. This holds for the geometrical optics conditions. Thus this scattered light causes electrical signals at the photodetectors, also $\sim d^2$ if the signals pass the trigger levels which have to filter out background noises. We receive filtered Doppler-bursts with amplitudes proportional to the square of the particle diameter under the basic requirement that the amplifications of the two photodetectors are exactly the same and linear between the level voltage value and that of the largest particle.

On the other hand the incident laser beams may be assumed to have Gaussian intensity cross sections, causing the so called Gaussian diameters of the measuring volume δ_x, δ_y and δ_z in the x-, y- and z-directions. This has the effect that only those particles will be received in the exact d^2-manner at the burst-counters which pass the measuring volume in its centre. Decentral passing particles lose a lot of their scattered light intensity. Thus their Doppler-bursts amplitudes may be recognized as being generated by a smaller particle *) or in the final consequence may even be held back by the trigger level. Missing the centreline of the measuring volume by the same small distance causes higher counting deficits for the smaller particles compared with the coarser ones.

It has to be kept in mind that each couple of Doppler-bursts which passes the trigger level will be counted and will result in the absolute values of measured velocity and size. But the smaller the particles the more often they fail to be counted and thus used for velocity and size calculations. The question whether the scattered light is sufficient in its intensity for generating a count is influenced by the refraction index of the continuous and the dispersed phase, the observation angles, the pinholes of the photodetectors, the laser power, the adjusted amplification and the trigger level.

----

*) which remains without any consequence because the phase-difference
   between the Doppler-bursts (not their amplitude or visibility) is
   used for size-calculations of the LDVS.

This leads to the gain correction formula, for which - see Buchhave et al /2/ - it is necessary to use calibration particles and by which the absolute values of the counted particles have to be corrected.

The correction formula quantifies the shifting in weight of a measured (real) distribution in favour of the finer particles. For a special particle diameter d_o the formula is set to the value 1, for finer particles, $d_{min,t} < d \lesssim d_o$, their counted number will be raised. The correction formula (using index t for tracer) for an off-axis angle φ of nearly 90° is given as

$$a_t = \frac{\pi}{8 A} \, \delta_x \, \delta_y \left\{ \ln\left(\frac{G d^2}{d_o{}^2}\right) - 2 \frac{N_o{}^2}{N_f{}^2} \right\} \qquad (7)$$

with

$$A = \frac{\delta_x \, \delta_y}{4} \pi \quad \text{and} \quad d_{min;t} = \frac{d_o}{\sqrt{G}} \exp\left(N_o{}^2/N_f{}^2\right), \qquad (8)$$

where δ_x and δ_y are the e^{-2}-Gaussian diameters of the measuring volume in x- and y-direction, δ_p is the projected image of the pin-hole diameter (of the photomultiplier), N_o is the number of fringes above a fixed trigger level required by the burst detector, N_f is the number of fringes within the Gaussian intensity profile in the z-direction and d_o is that particle diameter by which the correction formula has been fixed and the gainfactor G has been measured. This gainfactor G is given by the maximum voltage $U_{max}(d_o)$ of the particle under focus d_o and the trigger voltage U_t

$$G = U_{max}(d_o)/U_t$$

While a typical correction curve after equations (7) and (8) is given in /4/, the region below d_{min} is uncertain till now. Attention has to be paid to differences between the formulae (7) and (8) and those cited in /4/ and those in /2/ arising from different coordinate systems.

Another formula, established by Schöne /7/, does the correction procedure without the need of particles for calibration, if the absolute values of the size distribution of the particles (i.e. the number of particles n per diameter category d) show a maximum $n_{max}(d)$ in their uncorrected values. Schöne uses this diameter category as mean diameter value d_o for his calculations, thus avoiding time consuming calibration measurements. For this diameter category he can be sure that it contains the highest probability for central trajectories of particles through the measuring volume (i.e. x,y, = 0, and $- \delta_z/2 < z + \delta_z/2$). Then he uses, from a sufficient number of stored bursts (from this diameter category), the one which shows the highest number of burst-amplitudes N_1 above the trigger level. Thus he is allowed to assume, that this particle has exactly passed the centre line (z-axis) of the measuring volume. The correction formula after this period counting (using index p) is given as

$$a_p = \frac{\pi}{8 A} \, \delta_x \, \delta_y \left\{ \ln\left(\frac{d^2}{d_o{}^2}\right) + 2 \frac{N_1{}^2 - N_o{}^2}{N_f{}^2} \right\} \qquad (9)$$

with

$$d_{min,p} = d_o \exp\left\{ - (N^2{}_1 - N^2{}_o)/N^2{}_f \right\} \qquad (10)$$

If for large particles $d > d_t$ or d_p the sensitive area in the x-direction of the measuring volumes becomes larger than that which can be imaged by the receiver optics, the following formulae have to be used instead of the equations (7), (8) or (9) and (10).

$$a_t = \frac{\pi}{A\sqrt{8}} \; \delta_p \; \delta_y \left\{ \ln\left(\frac{Gd^2}{d_o^2}\right) - 2\frac{N_o^2}{N_f^2} \right\}^{1/2} \qquad (7a)$$

with

$$d_t = \frac{d_o}{\sqrt{G}} \; \exp\left\{ \frac{\delta_p^2}{\delta_x^2} + \frac{N_o^2}{N_f^2} \right\} \qquad (8a)$$

and

$$a_p = \frac{\pi}{A\sqrt{8}} \; \delta_p \; \delta_y \left\{ \ln\left(\frac{d^2}{d_o^2}\right) + 2\frac{N_1^2 - N_o^2}{N_f^2} \right\}^{1/2} \qquad (9a)$$

with

$$d_p = d_o \; \exp\left\{ \frac{\delta_p^2}{\delta_x^2} - \frac{N_1^2 - N_o^2}{N_f^2} \right\} \qquad (10a)$$

For off-axis-angles differing from those used for the LDVS-method till now, (i.e. 90 $>$ φ $<$ 60°), the equations (7) to (10) have to be changed in spite of the geometrical derivations. This holds especially for back-scattering positions, which may be chosen for instance for flow analyses of metallic particles in inert-gas streams. The equations have also to be changed if the particle trajectory is not strictly parallel to the z-axis. Though this is the more general case of LDVS-application this problem will not be solved before additional information can be used for instance in measuring the second or third component of the particle velocity. Here correction formulae are in derivation.

Application Of The Correction Formulae

In spite of the various reasons which may cause for the incoming light signals an insufficient quality compared with those required of the validation criteria we now have an important instrument by means of the equations (7) to (10a) for correcting the number of counts per particle size categories and getting closer to the truth.

Only under the assumption that the gates of the electronic data processors are open for every incoming signal, i.e. that the electronic equipment works faster than the events being researched occur in the multiphase flow system, we have a real chance to correct the particle size categories in order to receive from these data the particle flux or the particle concentration of the dispersed phase.

The correction procedure is scetched in Fig. 2. From the original size distribution n(d) at the top of Fig. 2 we receive, after correction by means of the factor a^{-1} (see the mid position), the corrected size distribution n'(d) (dotted line histogram at the top of Fig. 2).

Our intention is to receive a corrected bivariate distribution. Thus we have to check whether there exist a sufficient number of original events permitting the use of the correction formula also for the second piece of information: the particle velocity. As the basic requirement for the application of the LDVS-method was the stationary multiphase-flow-situation, we have to examine this by comparing different numbers of counts versus time or versus number of counts. This should have been done by means of grey scale diagrams (see ref. /8/) in which the normalized counts of - for example - 3,000, 4,000 or 10,000 can be compared, whether these diagrams show variations in the grey scales or remain unchanged.

If the grey scale diagrams remain unchanged, we can be sure that for this location under focus in the multi-phase-flow system and for this very special duration of measuring time, the flow behaves in a stationary manner. Thus we are allowed to use the correction formula for the particle size as well as for the particle velocity, but the latter only under the condition that the areas of the bivariate size and velocity diagrams have been covered with sufficient event figures, especially for the parcels of the smaller particles.

This must be mentioned explicitly as in the other case - if there exist only singular event figures per singular parcels as shown in the third diagram of Fig. 2 (bottom) the enlargement of these original values by the factor a^{-1} is unjustifiedly high and lacks any physical explanation. At the moment we have no better argument than to use the factor $a^{-1}(d_1)$ for all velocity values $v(d_1)$ in the same manner.

The LDVS-data n appear as counts per measuring area A and measuring time τ. This is equal to the local particle flux. In order to establish further properties of the dispersed phase from the correlated pair: size and velocity, we must for instance use the product

$$\dot{m}_{ij} = n_{ij} \frac{\pi d_i^3}{6} \rho$$

and divide $\dot{m}_{ij}$ by v_j thus receiving the specific local concentration of particles of diameter d_i and velocity v_j.

It's helpful to use the distribution curves and equations for diameter and velocity instead of histograms in order to work with the computed mean values or distinct distribution parameters. In this way it is easier to establish the momentum, mass or energy balances. This becomes evident if for instance more than one velocity component has to be used in the further correlations.

<u>Insufficiencies Of The Particles</u>

The second basic requirement for the application of the LDVS-method was the sphericity of the dispersed particles in the measuring volume. In answering this the main argument was, that in most multiphase-flow-systems the dispersed phase exists as bubbles or drops, which under the influence of surface tension have a spherical shape. The smaller the particles are the more will this argument be accomplished. Nevertheless there will be pulsatile flows and deformed bubbles or droplets in a prolongated or elongated form. Here a two-component measuring device as described from /3/ will in future be able to analyse these deviations from the sphericity and give relevant size informations from these particles.

By means of the LDVS-method also solid spherical particles can be analysed. Here not only the sphericity is in question but also the homogenity of the material and the structure of the particle surface.

As has been shown by Flögel /5/ especially for glass beads one has to pay attention for very small enclosures like air bubbles or dust particles. Also the surface structures may cause troubles and last but not least the glass-material itself very often is not fully transparent but exists of a partly absorbing component. Here severe deviations from the real particle diameter may be measured.

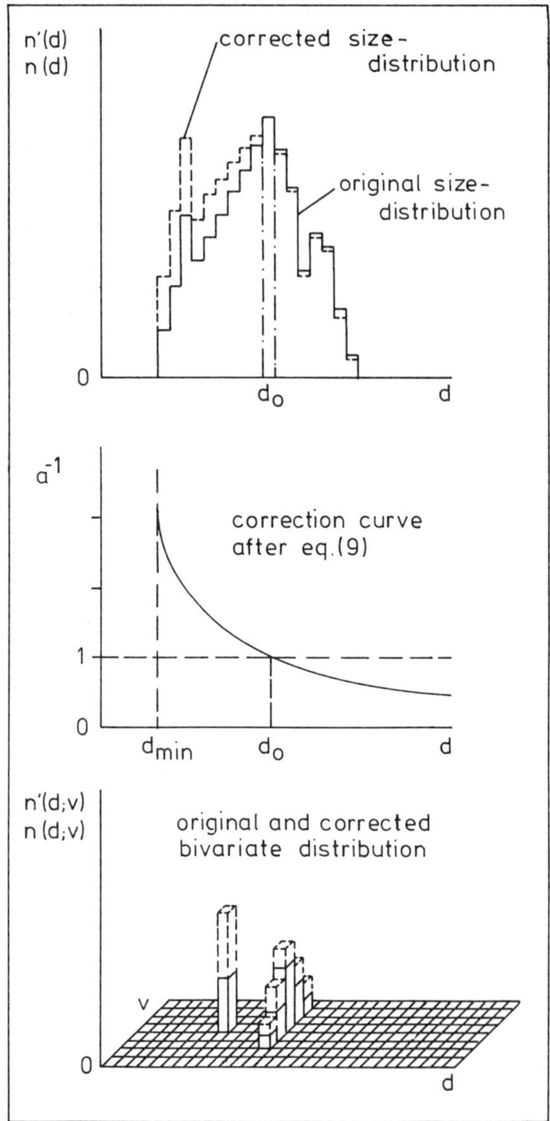

Fig. 2. Correction of original datas

Solid spherical particles which may have been generated by solidification processes from drops of molten metal only in some cases show smooth surfaces without shrinking zones, rift valleys or crater formations. Especially during rapid quenching processes the droplets from special crystal structures at the surface like cellular colonies or dendrites. These surface structures of the particles of most metal powders lead to diffuse light diffractions or reflections and affect the quality of the light signals at the photodetectors. The far field scatter distribution of the interferometric patterns looses the sharply seperated stripe structure.

It seems that the surface structures of solid spherical particles, if described by the roughness parameter s, could be classified in a similar way as the particle diamter in relation to the wavelength of the laser light, i.e. the Mie-parameter:

$$\alpha = \pi \frac{d}{\lambda_o} \quad ; \quad \beta = \pi \frac{s}{\lambda_o} \quad .$$

The smaller the roughness parameter s in relation to the particle diameter the stronger the differences between the governing rules of scattering.

CONCLUSIONS

The size and velocity measurements can be done by the LDVS with a certain maximum possible (relative) error. This error has been quantified for the particle size measurements. The concentration measurements have to be corrected by wighting factors. These factors have been discussed and also quantified. A correction formula has been given without the need of tracer particle measurements. Possible errors arising from the particle material and from surface structures have been described and discussed.

ACKNOWLEDGEMENT

The authors gratefully acknowledge the financial support for this work provided by the Deutsche Forschungsgemeinschaft, Bonn/Bad Godesberg.

REFERENCES

/1/ Bauckhage K., Flögel H.-H.; Simultaneous Measurements of Droplet Size and Velocity in Nozzle Sprays. 2nd Int. Symp. on Appl. of Laser Anemometry to Fluid Mechanics, 2.-4.6.1984, Lisbon, Proceed.

/2/ Buchhave P., Saffman M., Tanger H.; Simultaneous Measurement of Size, Concentration and Velocity of Spherical Particles by a Laser Doppler Method. ibidum.

/3/ Bachalo W.D., Houser M.H.; An Instrument for Two-Component Velocity and Particle Size Measurement. 3rd Int. Symp. on Appl. Laser Anemometry to Fluid Mechanics, 7.-9.6.1986, Lisbon, Proceed.

/4/ Bauckhage K., Flögel H.-H., Fritsching U., Hiller R., F. Schöne; Analysis of Particle sizes with the Aid of Laser-Doppler-Anemomtry: The Influence of Specific Geometrical and Optical Properties of Particles; 3rd Int. Symp. on Appl. Laser Anemometry to Fluid Mechanics, 7.-9.6.1986, Lisbon, Proceed.

/5/ Flögel H.-H.; Modifizierte Laser-Doppler-Anemometrie zur simultanen
     Bestimmung von Geschwindigkeit und Größe einzelner Partikeln, Diss.
     Universität Bremen (1987).

/6/ Glantsching W.S., Chen S.; Light Scattering from Water Drops in the
     Geometrical Optics Approximation. Appl. Optics; Vol. 20, No. 4,
     February 1982.

/7/ Schöne F.; Bestimmung der Geschwindigkeits- und Größenverteilung
     von Partikeln in Sprühkegeln mit verschiedenen Laser-Doppler-Ane-
     mometer-Aufbauten und Ableitung geeigneter Korrekturen. Diplomar-
     beit Universität Bremen (1986).

/8/ Bauckhage, K.; Size, Velocity and Flow Concentration Measurements
     in Sprays by Laser Doppler Anemometry. Int. Conf. on Laser Anemome-
     try-Advances and Application, 16.-18.12.1985, Manchester/UK,
     Proceed.

AN LDA GATING TECHNIQUE FOR BUBBLE MEASUREMENT IN DILUTE

THREE-PHASE SUSPENSION FLOWS

S. L. Lee, Y. Z. Cheng and Z. H. Yang

Department of Mechanical Engineering
State University of New York at Stony Brook
Stony Brook, N. Y. 11794, U.S.A.

ABSTRACT

A new method of measuring bubble size and velocity simultaneously in three-phase suspension flows by using a reference beam LDA system is developed in response to the need of the new, novel method of extracting additional underground oil by the introduction of compressed gas. Three photodetectors are used to obtain velocity, blocking period and other information of bubble characteristics. The operational principle, calculations, the block diagrams of discrimination and control logic circuits are described. Experimental investigations and sample results on bubble size and velocity measurements are also presented.

INTRODUCTION

For more than a decade, methods on the in-situ simultaneous measurements of the size and velocity of particles or bubbles in two-phase flows have been developed (see references) by making use of a laser-Doppler anemometer. For example, Durst and Zare (1975) and Martin et al (1981) developed a technique to measure the velocity and size of bubbles in a dilute air-water two-phase suspension flow for bubbles in the size range of 0.2-1.0mm. It is based on the Doppler signal created from a fringe pattern in space produced by two reflective or refracted light beams from a single moving sphere. The fringe pattern is of complex shape and changes its position in space as the sphere moves. Lee and Durst (1982) used this scheme to measure the movement of glass particles in a turbulent flow of a glass particle-air bubble two-phase suspension with uniform particles of a known size. It has been shown that the resulting signal frequency is independent of the detector location and is a function of the particle velocity. The scheme produces essentially identical information whether the object is a spherical bubble or a spherical solid particle.

Lee and Srinivasan (1982) developed an optical gating technique of particle measurement using an one-dimensional reference beam LDA system for particle larger than the smaller dimension of the optical measuring volume in a two-phase dilute suspension flow. Lee and Cho (1983) extended this particle measuring technique to measure the size and two velocity components of large particles in a two-phase suspension flow using a two-dimensional reference beam LDA system. The scheme combines the conventional technique of optical gating for particle sizing and reference beam laser-Doppler anemometry for velocity measurement. The optical gating is achieved by observing the blockage of one of the stationary beams by the moving particle. However, this scheme as it is can not be used to measure large bubbles for the obvious reason that the blocking of the stationary beam becomes terminated when it passes through the central protion of a moving bubble. Based on these observations, it may become feasible to measure bubbles suspended in a dilute gas-solid-liquid three-phase suspension flow if some additional information from a bubble can be used to distinguish it from a solid particle.

PRINCIPLE AND CONCEPTS

The optical discrimination scheme as shown in Fig. 1 gives distinguishable characteristics of the signals for a bubble or a particle. An off-axis photodiode D2 is adopted as a detector to receive the distinguishable optical signals from a bubble or a particle.

During the period when a solid particle intercepts the measuring volume, a block-off signal from the photodiode D1 is produced. Since the stationary scatter beam is blocked off by the solid particle of a size larger than the smaller dimension of the measuring volume, Lee and Srinivasan(1982), a characteristic signal from the photodiode D2 with only one pulse due to the reflection from the surface of the particle, and a Doppler burst from the photomultiplier at the end of the block-off period when properly aligned are obtained as shown in Fig. 2.

When a bubble intercepts the measuring volume, similar signals from D1 (except for the trough pulse in the middle) and the photomultiplier PM are obtained. However, the characteristic signal with two pulses from D2, as shown in Fig. 3, is different from that of particle which has only a single pulse. The first pulse from D2 is generated by beam reflection when the surface of a bubble or a particle first touches the stationary beam. The second pulse is generated by beam refraction when the bubble has blocked the scatter beam which is partially refracted twice by the surfaces of the bubble as shown in Fig.'s 4(a) and 4(b). In order to use the blocking signal from D1 to calculate bubble size, the trough in the middle of the blocking signal should be reduced and shaped out. It is easy to do this by utilizing a low-pass filter and a comparator with a reference voltage equal approximately to one half of the maximum amplitude of the blocking signal from D1, if the exposure time (Fig. 3), T_{exp}, is known. Apparently, the time constant, T_f, of the low-pass filter mentioned above should be at least twice as large as T_{exp}.:

$$T_f \geq 2T_{exp}.$$ [1]

On the other hand, T_{exp} can be calculated according to the geometrical parameters of the optical scheme as shown in Fig. 5.

$$T_{exp} = \frac{d_1 \, r}{2L(n_1 - n_2)V_B}$$ [2]

where $n_1 > n_2$, and

T_{exp} -- Exposure time,

d_1 -- Aperture of the photodiode D1,

L -- Distance between D1 and measuring volume,

r -- Radius of the bubble,

n_1 -- Refractive index of the surrounding medium (e.g. water)

n_2 -- Refractive index of the bubble medium (e.g. air),

V_B -- Velocity of the bubble.

Table 1 provides some sample values of the exposure time, T_{exp},

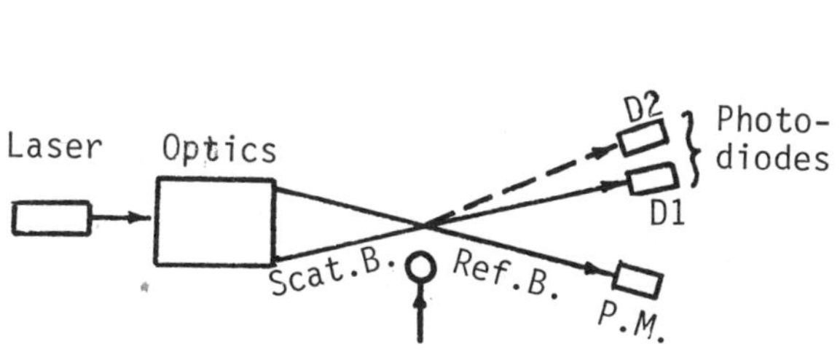

Fig. 1. Optical scheme for bubble measurement in three-phase flows.

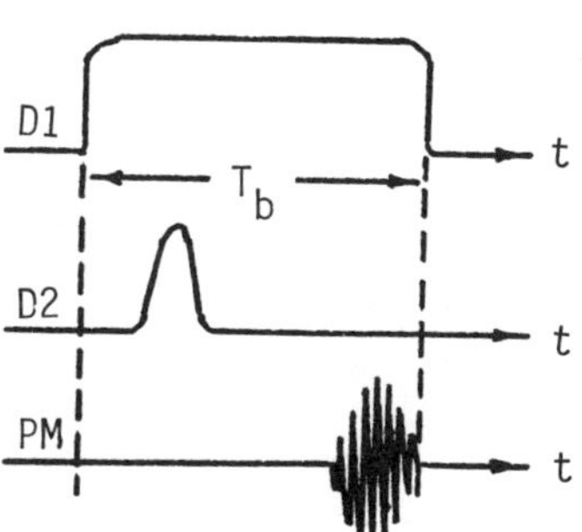

Fig. 2. Outputs of photodetectors from a particle.

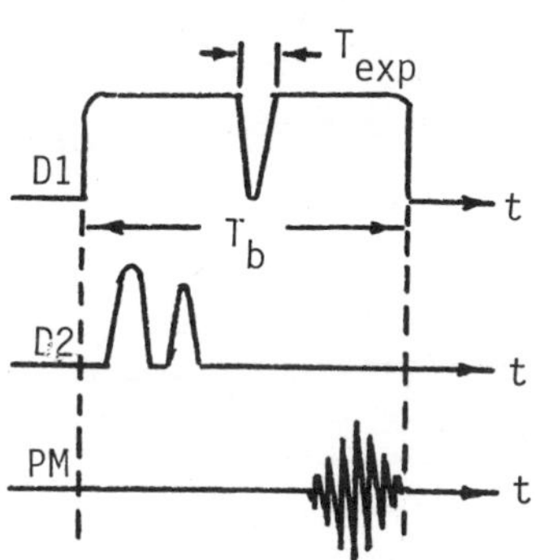

Fig. 3. Outputs of photodetectors from a bubble.

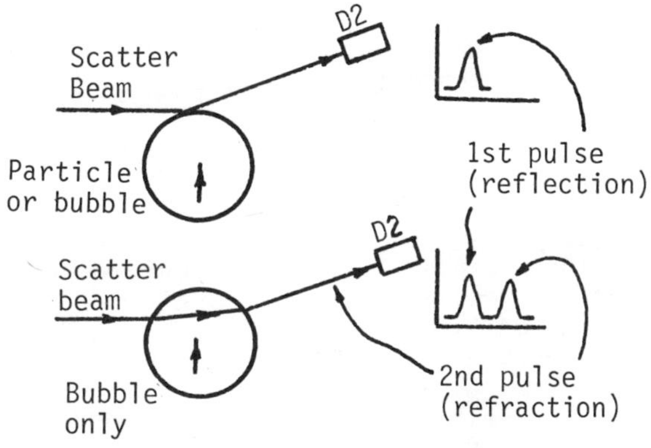

Fig. 4. Pulse signals from D2.

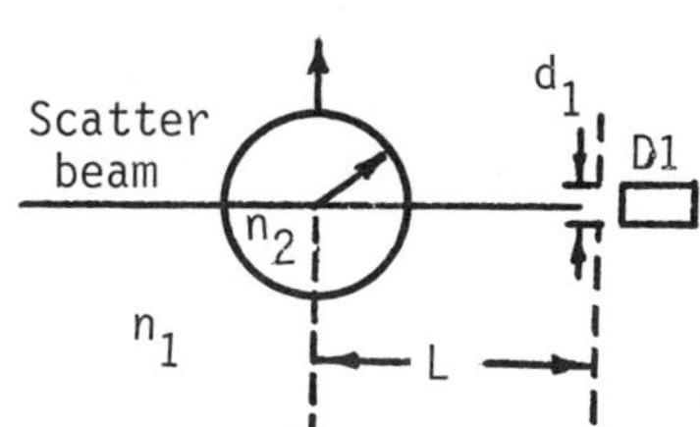

Fig. 5. Geometrical Parameters for calculating T_{exp}.

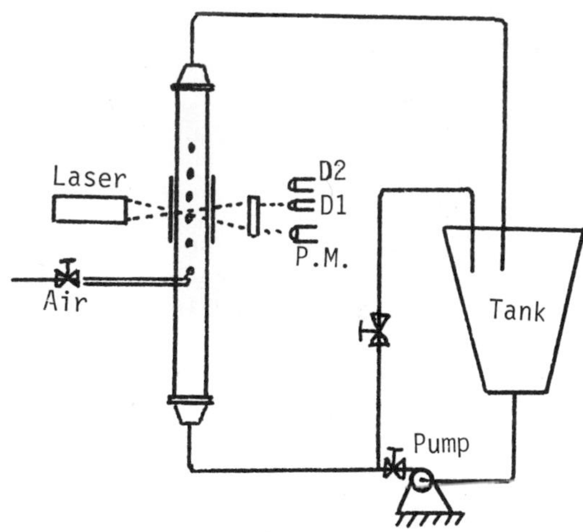

Fig. 6. Flow system

calculated when d_1 = 4mm, L = 250mm, n_1 = 1, n_2 = 1.33 are given.

The blocking time, T_b, which deals with the size of the bubble, can be obtained from the shaped blocking signal. The number of pulses from D2 can be used as the characteristic signal of a bubble or a particle. One pulse during the blocking period stands for a particle while two pulses stand for a bubble.

EXPERIMENTAL FLOW ARRANGEMENT

The arrangement of flow system for the bubble measurement in three-phase flows is shown in Fig. 6. The test channel with an intersection of 20mm x 15mm is vertically placed between the laser and the photodetectors. The air nozzle with a certain inner diameter (0.2mm or 0.7mm) is placed below the measuring volume to generate the bubbles required. The size range of the bubble could be varied by changing the nozzle.

ARRANGEMENT OF OPTICAL SYSTEM

Figure 7 shows the optical arrangement of the reference beam LDA system. The incoming beam from a 15mw He-Ne laser is split into three beams: two stronger beams with equal intensity but mutually orthogonal polarities to serve as scatter beams for the two-dimensional measurement and a weaker beam with a polarity angle of 45° to the stronger beams to serve as the reference beam. The beam angle between the reference beam and each of the two scatter beams is 8.02°, and the smaller dimension of the measuring volume is 240μm. Two photomultipliers are used for two-dimensional measurement to receive the reference beam and the scattered lights from the two scatter beams when the measuring volume is intercepted by the surface of the moving bubble or particle. The Doppler signals which are produced through optical heterodyning on the photocathodes of the photomultipliers are thus obtained. The photodiodes D1 and D2 are used to receive one of the scatter beams together with its reflected or refracted light respectively when a bubble or a particle intercepts it. Therefore, a block-off signal and a characteristic signal are produced.

Since the beam with the horizontal polarity is not used during the bubble measurement in the present flow system with a vertical test section, three signals are obtained: the Doppler signal from the photomultiplier standing for bubble or particle velocity, the blocking signal form D1 relating to the bubble or particle size when it is used in conjunction with velocity information obtained from the Doppler signal, and the characteristic signal of the bubble or particle from D2.

ELECTRONICS AND SIGNAL PROCESSING

A block diagram of electronic scheme for bubble measurement in three-phase flows is shown in Fig. 8. A counter and its control logic, and a logic discriminatior are designed as discriminating electronics for bubble measurement in three-phase flows. The counter counts the pulse form the photodiode D2 during the blocking period. If a particle intercepts the

Table 1. Sample of calculated T_{exp} (d_1=4mm, L=250mm, n_1=1, n_2=1.33)

r(mm)	V_B(m/s)	T_{exp}(ms)	r(mm)	V_B(m/s)	T_{exp}(ms)
2.0	0.2	0.121	1.0	1.0	0.024
2.0	0.5	0.097	0.5	0.2	0.031
2.0	1.0	0.048	0.5	0.5	0.025
1.0	0.2	0.061	0.5	1.0	0.012
1.0	0.5	0.049			

stationary scatter beam, the number count in the counter is one. If a
bubble intercepts the stationary scatter beam and the stationary scatter
beam is nearly coplanar with the center of the bubble, the number count
is two. Therefore, the gate is switched on only when the number count of
the pulses from D2 is two during the blocking period. This means that
only the Doppler signals from bubbles are allowed to pass through the gate
to be accepted by the counter processor during the bubble measurement.

   The blocking period is measured by a special electronic circuit, and
the velocity is measured synchronously by a commercial counter processor
during the blocking period if a Doppler burst appears at the input of the
processor. Both data are sent to a PDP 11/34 computer system via an
interface. The bubble size can then be calculated from

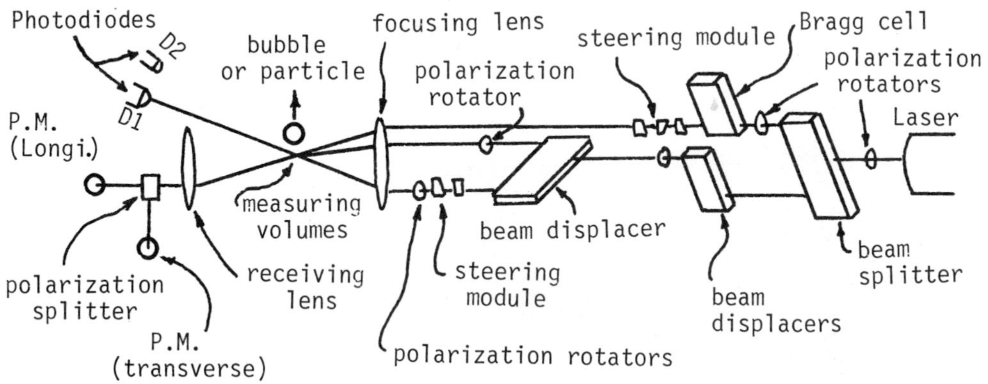

Fig. 7. Two-dimensional optical arrangement for bubble
measurement in three-phase flows.

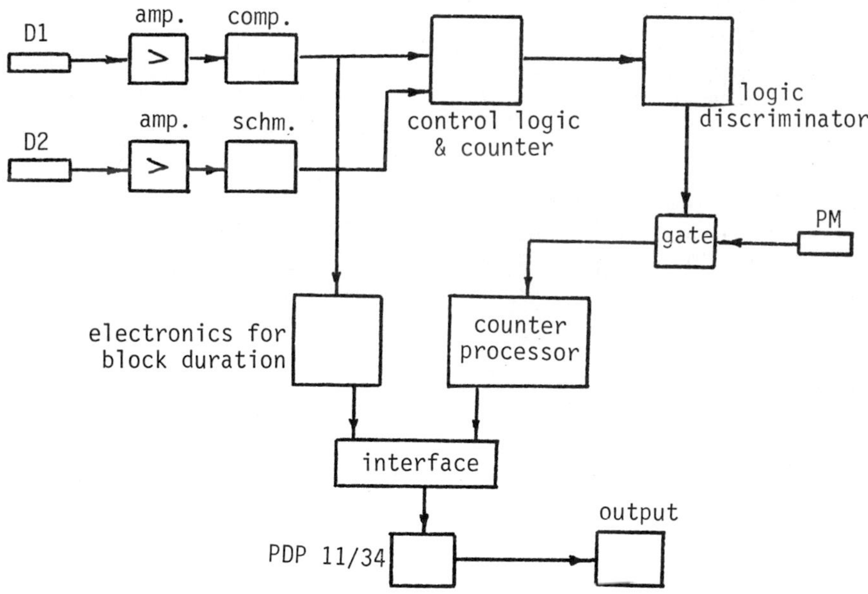

Fig. 8. Block diagram of electronics and signal processing
system for bubble measurement in three-phase flows.

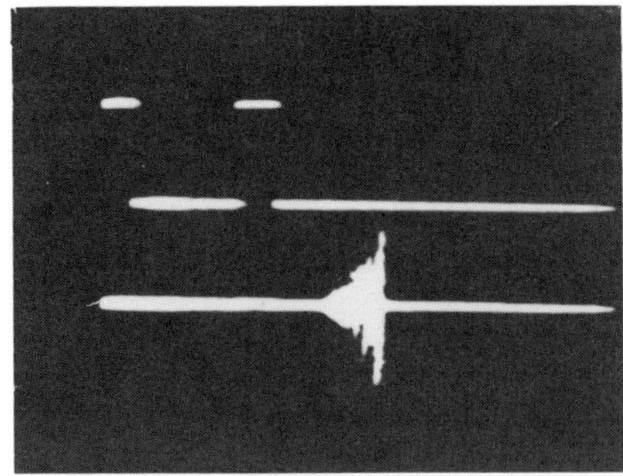

shaped characteristic
signal from D2

Doppler signal
from PM

Fig. 9. Signals from bubbles in a bubbly water flow.

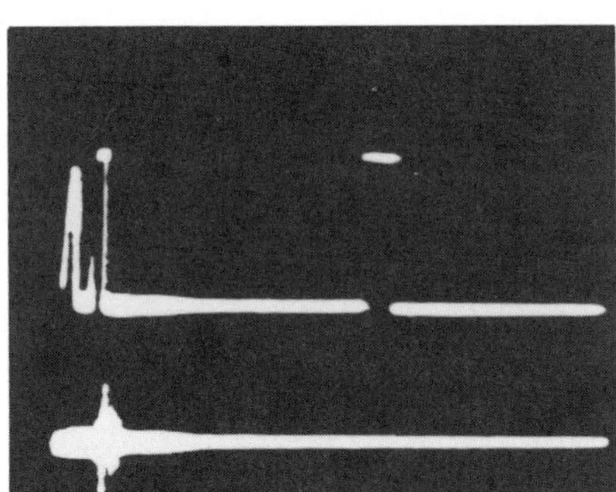

shaped characteristic
signal from D2

Doppler signal
from PM

Fig. 10. Bubble signals from a three-phase flow.

Table 2. Sample bubble size-velocity data in a three-phase flow.

Velocity(m/s)	Size(μm)	Velocity(m/s)	Size(μm)
0.41346E+00	0.23346E+04	0.45456E+00	0.22354E+04
0.50179E+00	0.24097E+04	0.37339E+00	0.16372E+04
0.37213E+00	0.15944E+04	0.37417E+00	0.14421E+04
0.41248E+00	0.19494E+04	0.38425E+00	0.18944E+04
0.35015E+00	0.17601E+04	0.39651E+00	0.18605E+04
0.36700E+00	0.16803E+04	0.38466E+00	0.20593E+04

$$d_B = T_B V_B \cos(\alpha/2) \qquad [3]$$

where $\quad T_B$ -- Blocking period when a bubble intercepts a
stationary scatter beam,
V_B -- Velocity of the bubble with negligible
transverse components,
α -- Beam angle,
d_B -- Bubble diameter.

Since a hard disk is used in the computer system, it is possible to collect as many data points as required under the software control.

EXPERIMENTAL INVESTIGATION

Figure 9 shows a sample bubble measurement in a bubbly water flow. The Doppler signal from the photomultiplier (lower) following the bubble characteristic signal (upper) from the photodiode D2 appears only when the number count of the pulses from D2 is two. Figure 10 shows a bubble measurement in a dilute air bubble-solid particle-water three-phase flow. The Doppler signal (lower) from the filtered output of a counter processor closely follows the two pulses (upper left) from D2 when a bubble intercepts the measuring volume. In contrast, when a solid particle intercepts the measuring volume only one pulse (upper right) from D2 appears.

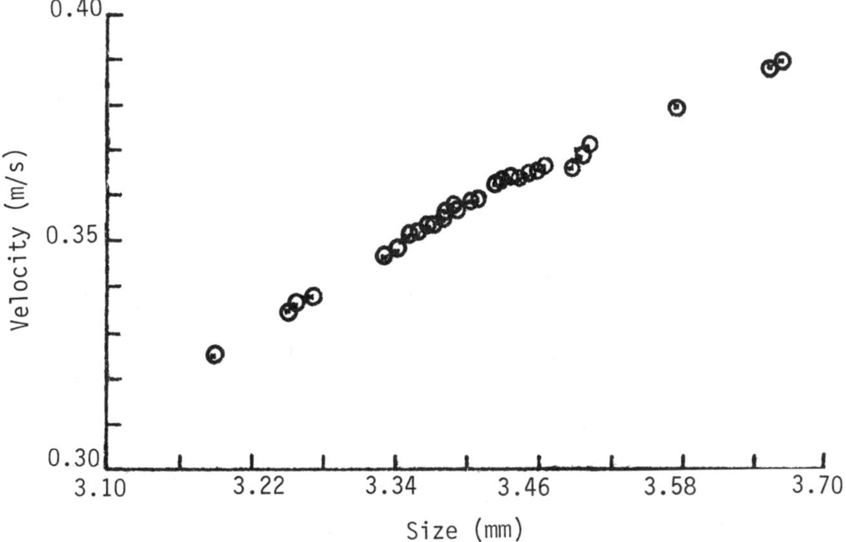

Fig. 11. Sample bubble slze-velocity data in quiescent
         water

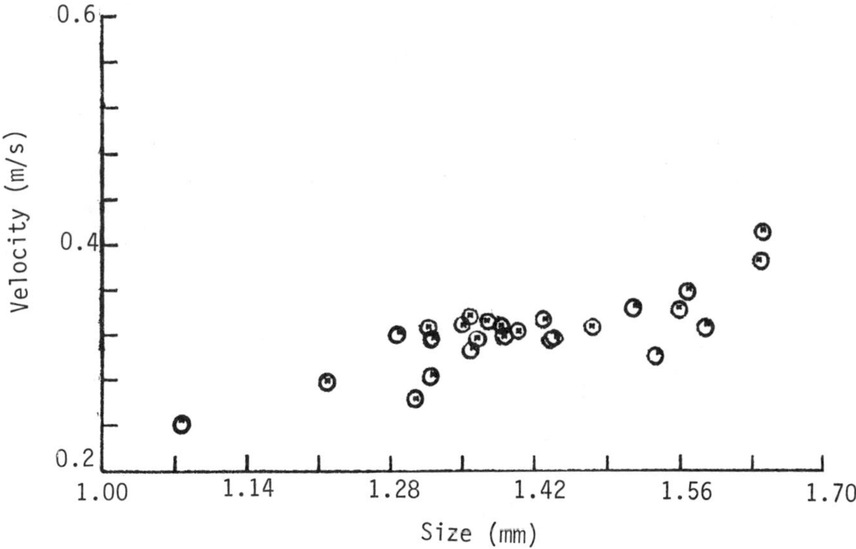

Fig. 12. Sample bubble size-velocity data in a bubbly
         water flow system.

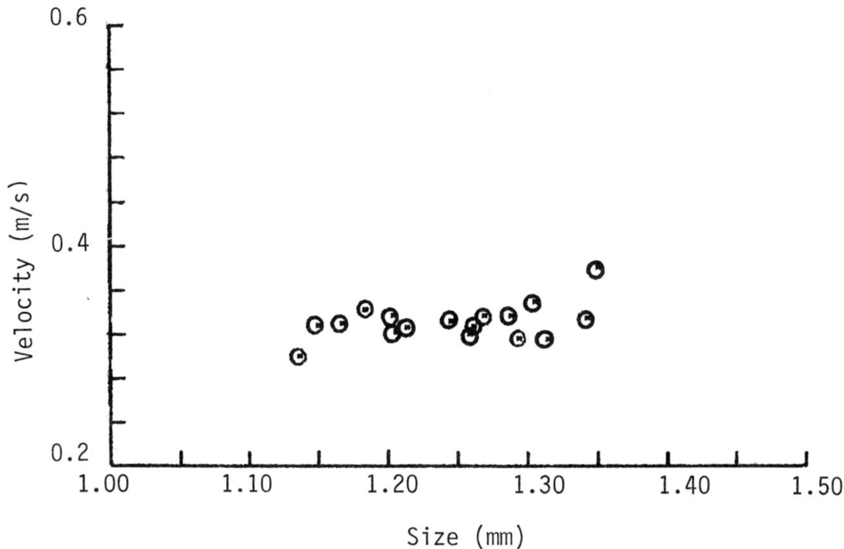

Fig. 13. Sample bubble size-velocity data in a
three-phase flow.

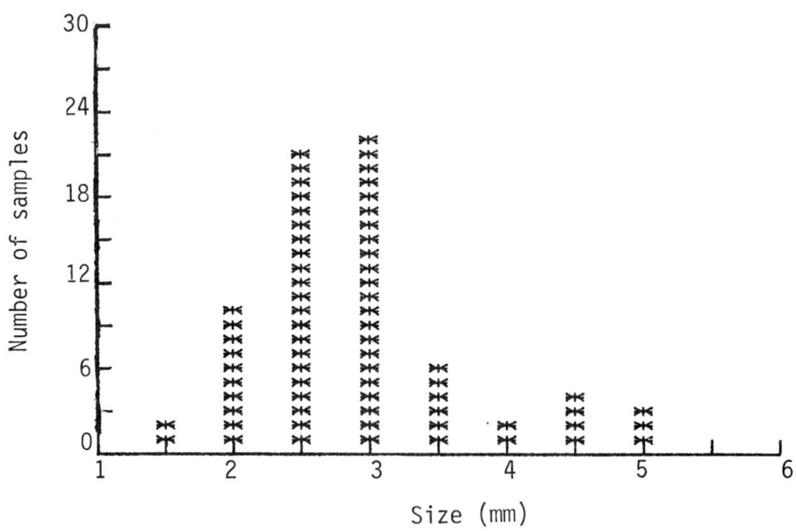

Fig. 14. Sample bubble size distribution in quiescent
water.

Therefore no Doppler signal follows it because the Doppler signal in this case is not allowed to pass through the electronic gate.

Figure 11 shows the velocity and size of bubbles in quiescent water with a large air nozzle of 0.7mm in diameter. Figure 12 shows the results in a tap water bubbly flow system with a small nozzle. By the use of a pumped air bubble-solid particle-water flow system with a small nozzle, similar results are obtained as shown in Fig. 13 and in the sample in Table 2. The increasing linearization between velocities and bubble sizes can be explained by the buoyancy and the resistance imposing on a moving bubble in the quiescent water.

The size distribution of the bubbles in a bubble measurement is shown in Fig. 14, which gives the sample number at different representative values of the bubble size. The predominant size range of the bubbles is from 2 to 3 mm.

CONCLUSION

Experimental results have shown that a special electronic control logic and discrimination circuits with a commercial LDA system can be used for the bubble measurement in dilute three-phase suspension flows. Based on this scheme a separate measurement for each of the two suspended phases in a three-phase flow has become feasible. It is clear that a two-dimensional LDA system for bubble measurement can be carried out, if all the three beams as shown in Fig. 7 are used to obtain the longitudianl and transverse velocity components.

REFERENCES

Durst, F. and Zare, M., 1975, Laser-Doppler measurements, in Two-Phase Flow, Proc. LDA-Symp. Copenhagen, 403.

Lee, S.L. and Cho, S. K., 1983, Simultaneous measurement of size and two-velocity components of large droplets in a two-phase flow by laser Doppler anemonetry, in:"Measuring Techniques in gas-liquid Two-Phase Flows", Springer-Varlag, 149.

Lee, S. L. and Durst, F., 1982, On the motion of particles in turbulent duct flows, Int. J. Multiphase Flows, 8 (No. 2): 125.

Lee, S. L. and Srinivasan, J., 1982, An LDA technique for in-situ simultaneous velocity and size measurement of large spherical particles in a two-phase suspension flow, Int. J. Multiphase Flows, 8 (No. 1): 47.

Martin, W. W., Adbelmessih, A. H., Liska, J. J. and Durst, F., 1981, Characteristics of Laser-Doppler signals from bubbles, Int. J. Multiphase Flow, 7: 439.

LIGHT SCATTERING BY DUST PARTICLES: EXPERIMENTAL RESULTS

B. Bliek[1,2] and P. Lamy[2]

Universite de Provence, Marseille, France[1]
Laboratoiré d'Astronomie Spatiale, Marseille, France[2]

1. EXPERIMENTAL PRINCIPLE AND SET-UP

1.1. Description

The principle retained for our experimental investigation
consists in generating a continuously flowing aerosol, a section of
which is illuminated in order to study its scattering properties over
the 0-180° interval of scattering angles. It can therefore be considered
as a nephelometer whose key part is a fluidized – bed generator which
receives a mixture of the dust to be studied (in the form of powder) and
of "large" glass spheres having diameters of 100 to 200 µm. An air flow
forces the mixture to "boil", the collisions between the glass spheres
desagglomerate the powder and the liberated
dust particles are transported by the air flow. The fluidized bed is fed
by an endless screw which continuously enrichs the mixture so as to
compensate for the loss of transported particles. It can be shown that
the operation of the generator is governed by a very simple equation
which relates the ponderal concentrations of the mixture and the flow
rates to the concentration of the aerosol. After a transitory phase of
approximately 10 mn, the stationary regime given by the equation is
reached and was experimentally shown to be remarkably stable. A nozzle
concentrates the aerosol in a column having a diameter of 5 mm; it is
then allowed to flow freely over a length of 20 mm before being taken
over by a suction device. The free part is of course used for the
optical measurements. It is illuminated by a collimated beam from a
quartz-iodine lamp combined with narrow-band filters and a polaroid
polarizor oriented either in the parallel or perpendicular directions.
The optical part of the detection system is based on classic principles
of photometry with two doublets and a field diaphragm (Fig. 1). The
scattered light is measured by a photomultiplier tube working in the
analog mode. Two identical optical detection systems have been set up,
one fixed to provide a reference so as to cancel any variation in the
aerosol or in the illuminating source and one mounted on a rotating
table moved by a stepping motor. The whole experiment is under computer
control. For a given scattering angle, we take the average of typically
3 measurements, each one having been previously divided by the
corresponding reference signal. Measurements are repeated for the set of
selected filters (see below). Complementary measurements are performed
on a MgO diffuser so as to normalize the incident light intensity for
the various filters and for the two directions of polarization.

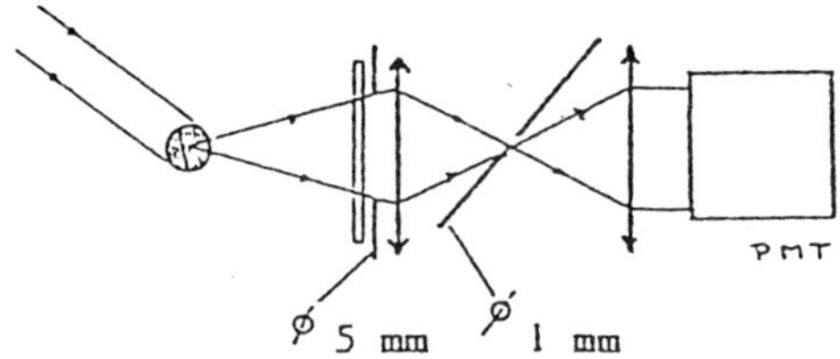

Fig. 1. Optical layout of the detection system

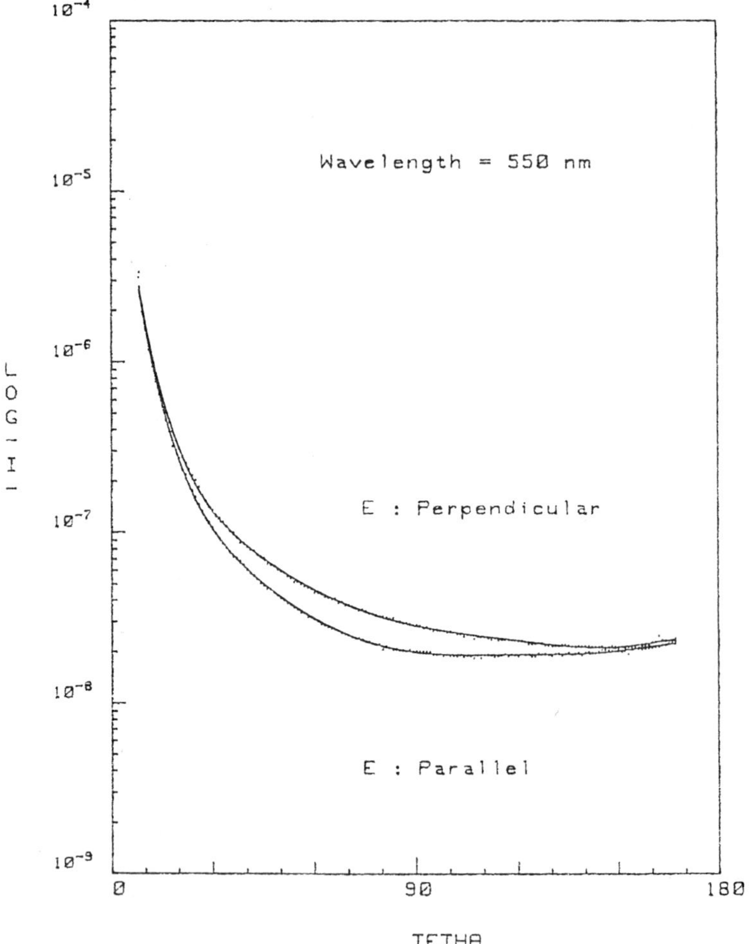

Fig. 2. Experimental results and fitting curves for SiO at 550 nm for the two directions of polarization. The points correspond to experimental measurements.

1.2. Characteristics

The size range of dust particles allowed by the fluidized-bed
generator depends upon the density of the dust but is typically 1 to 40
μm. The scattering volume has a diameter and an height of 5 mm and
contains approximately 100 to 1000 particles; a 250 W quartz iodine lamp
 is used as illuminating source with an optical system giving a
collimated beam having an internal angular dispersion of ± 2° and narrow
band filters centered at 447, 550, 645, 706, 744 and 824 nm (typical
bandpass ~ 45 nm).   The two detection systems have an angular aperture
of 2° and are equipped with Hamamatsu R 378 PMT whose sensitivity
extends over the interval 125-850 nm. Finally, the mechanical set-up has
been optimized to allow a range of scattering angles as large as
possible, 8 to 168°.

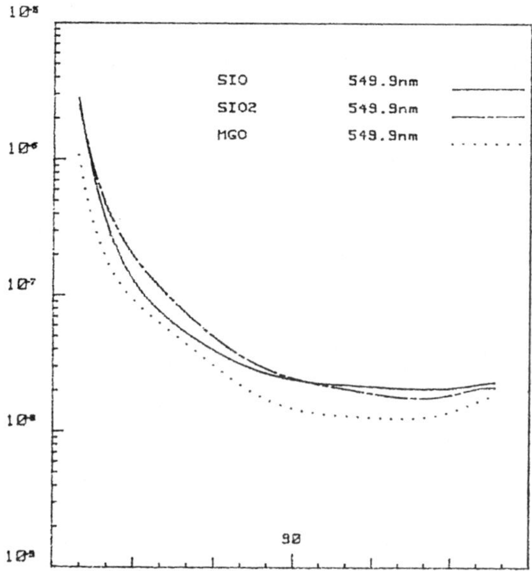

Fig. 3. Total volumes scattering functions versus scattering angle for
SiO, SiO$_2$ and MgO at 550 nm.

1.3. Main advantages

The presence of a large number of grains in the scattering
volume directly average the measurements over the various parameters
characterizing them (e.g., roughness). The flow being turbulent, the
average further extends over the randow orientations of the grains. The
scattering light is also sufficiently intense to allow spectral
measurements in a wide spectral domain with a source of reasonable
power. The system has the capability to directly measure the volume
scattering function for a dust population having a selected size
distribution. This function can be exactly determined independantly
using for instance a Coulter counter.

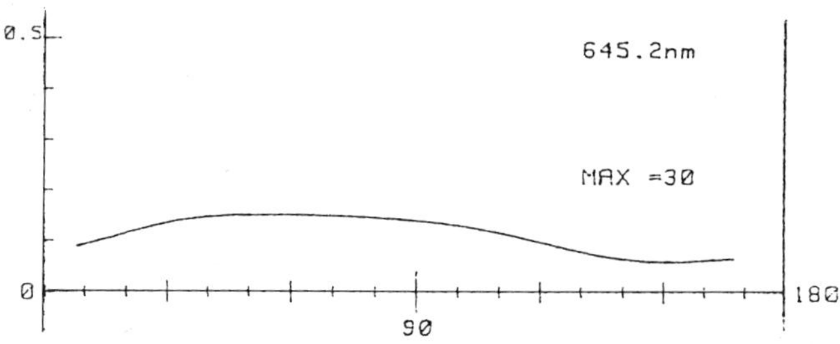

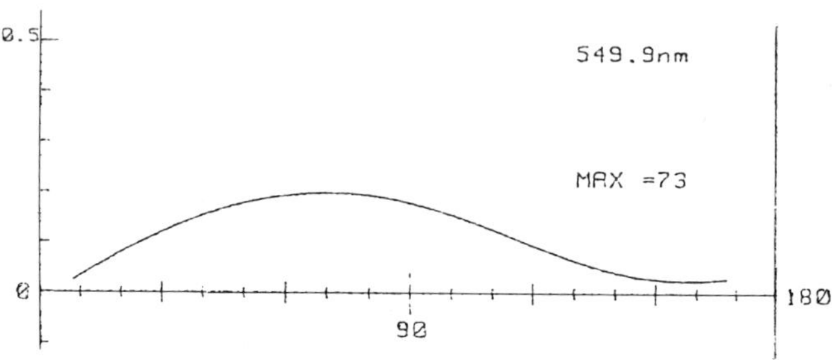

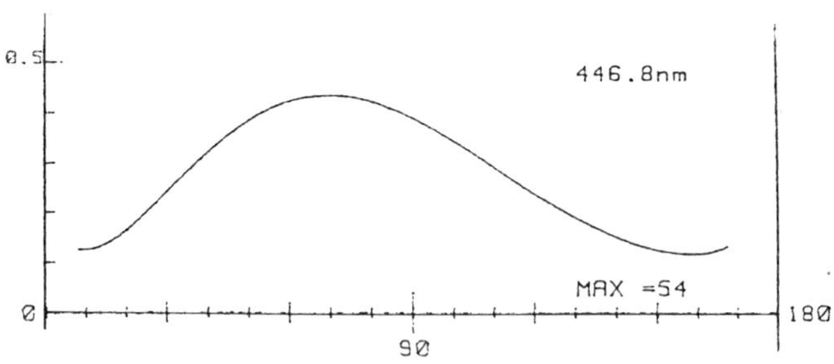

Fig. 4. Polarization versus scattering angle far SiO at 447, 550 and 645 nm.

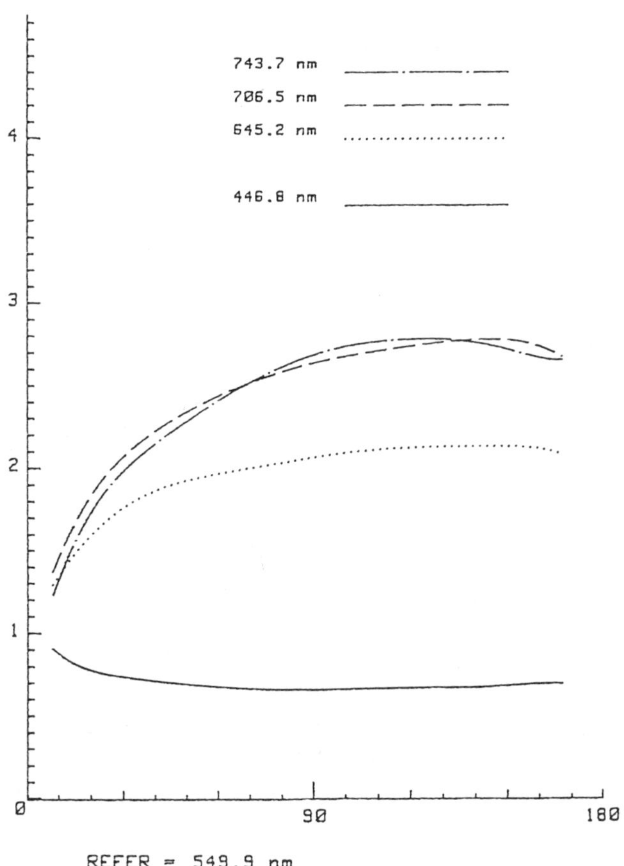

Fig. 5. Color effect for SiO: relative volume scattering function at 447, 645, 706 and 744 nm normalized to that at 550 nm versus scattering angle.

2. RESULTS

The results presented here have been obtained with commercially available powders of simple chemical composition and whose size distribution extends typically over a few microns. Fig. 1 displays the scattering functions for the two direction of polarization of silicon monoxide SiO at λ = 550 nm. The experimental data points are given to appreciate the error in the measurements as well as polynomical best fits representing the data. The total scattering functions of SiO, SiO_2 and MgO powders at λ = 550 nm are regrouped in Fig. 3. The polarization for the SiO sample powder at 3 wavelengths, λ = 447, 550 and 645 nm are shown in Fig. 4. These curves are polynomial fits to the data points. They do not appear totally satisfactory at this present stage as particularly evident for the forward and backward directions. Tests are presently under way to understand the problem. Another potential of these measurements is to study the color of the scattered light as function of the scattering angle. To illustrate this, the total scattering function at different wavelength is divided by that at λ = 550 nm used as a reference. The result for SiO is presented in Fig. 5 and indicates in fact an important color effect as a function of scattering angle.

OPTICAL PARTICLE SIZING : DIGITAL VIDEO

IMAGE PROCESSING APPLICATION

Gérard Lavergne, Yves Biscos, Francis Bismes,
and Patrick Hebrard

ONERA/CERT/DERMES
2 avenue Edouard Belin
B.P. 4025, 31055 Toulouse Cedex, France

INTRODUCTION

The development of microcomputers, when these one are associated with
visualization techniques using video or C.C.D, camera, represents a particu-
lary suited way to obtain, quite in real time, a lot of informations about
liquid drops. Such techniques have been developed at CERT/DERMES in which,
by digitizing the video picture from a magnified image of spray, one is able
to obtain, in non interfering way and without too much restrictive hypothesis about drops shape and size distribution, local values of drops size as
function of space and time.

After a description about the principle and calibration of this technique, the paper presents some applications of the method applied to combustion studies. The performances of different elementary injectors are
studied either as function of various parameters or when they are submitted to simulated aerodynamic conditions on isothermal models of turbojet
or ramjet engines . Other basic experiments dealing with vaporization and
influence of intense acoustic perturbations on drop size distribution are
also presented.

PRINCIPLE OF MEASUREMENT

For estimation of the evaporation rate of sprays, knowledge of the
mean drop size is inadequate and the distribution of drop sizes must be
taken into account. Also, the behaviour of these drops in the initial phase
and their repartition in the different zones of the combustor is mainly
function of their initial spatial distribution. So, we have tried to develop
a method allowing local size measurement of a great number of liquid
particles.

In this method described on figure 1, the flow to be studied is
illuminated with a thin sheet of coherent light. Generally, this sheet is
obtained by an optical arrangement from a 15 m watt He – Ne Laser but in
some experiments in which the instantaneous values of drop size is needed
this light is furnished by a laser diode which may be pulsed at the desired
frequency. Owing to the small size of the drops to be observed (from 5μm to
some 100μm), the scene has to be magnified. So, the light diffused by the
spray is observed at right angle by a video camera (Hitachi 230 x 240)

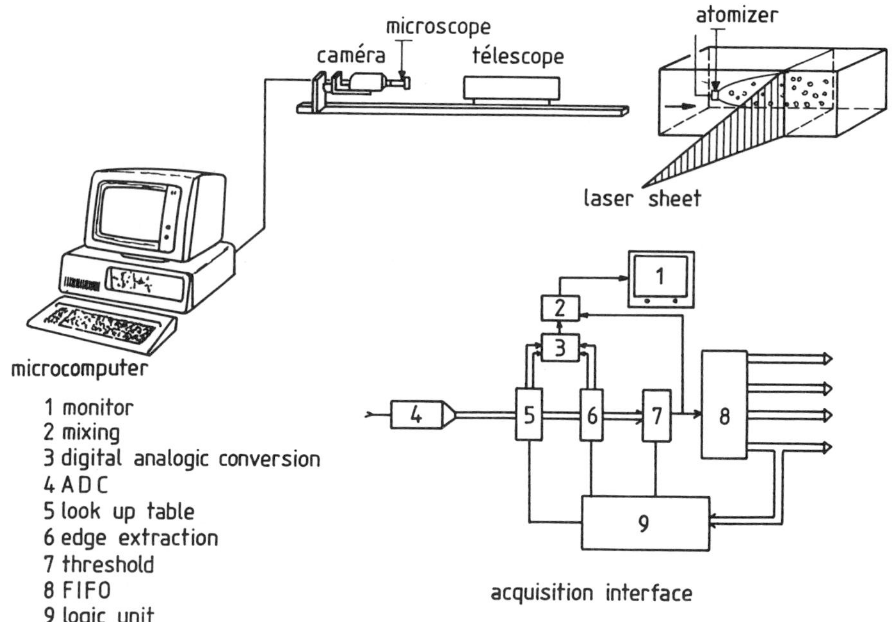

Fig. 1. Optical particle sizing principle

through an optical arrangement comprising a telescope (diameter 100mm, focus
600mm) and a microscope(X 8 or 4) in order to provide on the video or
computer monitor an image with sufficient magnification. With such an optical
arrangement the size of the analysed scene is some mm^3 with a depth of
field of about some $100\mu m$. At each frame, the image of the drops present in
the analysed volume appears on the monitor as elementary spots which may be
detected by one of the following image analysis method.

First method

In this first application, the Sauter mean diameter is measured at
each frame (every 16ms) from the edge detection of the magnified digitized
(1 byte) images of the drops present in the field and a grate furnished by
the matrix of the camera. This matrix surface is related to the real size of
the object field by two calibration coefficients Kx and Ky in order to take
into account the effects of optics, scattering, light power and digitizing
level. The principle consists in measuring at each frame, the number P_p of
points located inside of the drop images and the number P_L of points of the
intersection between the drop image periphery and the matrix. If so, the
Sauter mean diameter, S.M.D., defined as following[1]

$$S.M.D. = \frac{\sum nd^3}{\sum nd^2}$$

may be expressed as

$$S.M.D. = \frac{3P_p}{P_L}$$

To perform such a treatment, we use a microcomputer Apple II E
synchronised with the camera and an acquisition module composed of three
electronic boards : synchronization and digitizing, frame memorization and
frame analysis.

From the analysis of a sufficient number of frames (between 200 and
1000) the Sauter mean diameter is given in real time and a first approach of
drops size distribution may be achieved after a time of about two minutes.
As inconvenient, this system is limited by the fact that it is unable to
detect the position of each particle and also for the drop size for which
only mean value at each frame may be computed. So a second technique has
been developed in order to increase the possibilities of this method.

Second method

For the second method, the edge of each drop image is digitized in real
time by the Robert gradients method and the use of a delay line. In order to
be able to treat a great number of frames without using a too important capacity memory, we have developed a system in which only a "compressed" image is
memorized. The principle consists in saving the coordinates of points or
lines the brightness of which is greater than a fixed level ; so for each
one, only three values are to be transmitted (first point coordinates and
segment length) via a 32 bytes parallel bus and memorized. With this system,
developed with an I.B.M. P.C. microcomputer, the numerical processing allows
then to give at each frame the following informations : size of each drop
and distance between drops deduced from the coordinates of each center.

Here also, the analysis of a sufficient number of frames allows to
produce quite in real time, local information such as : drop size distribution, Sauter mean diameter and other mean diameters, drops distance distribution and value.

For the two methods, the possible measured drops sizes are between $5\mu m$
and some $100\mu m$ so, such a technique is particularly suitable to study the characteristics of injectors as used in turbojet or ramjet engines.

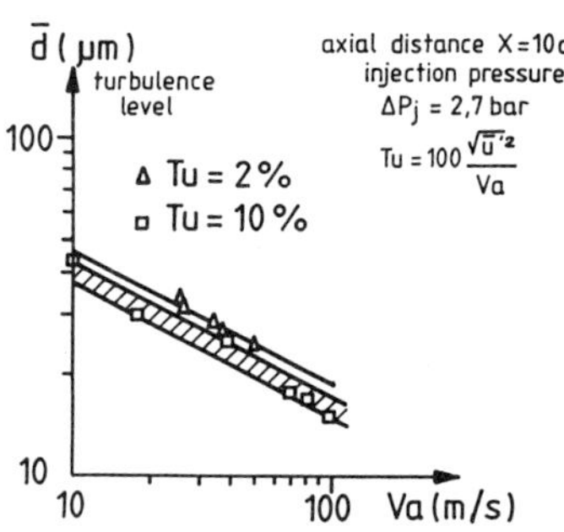

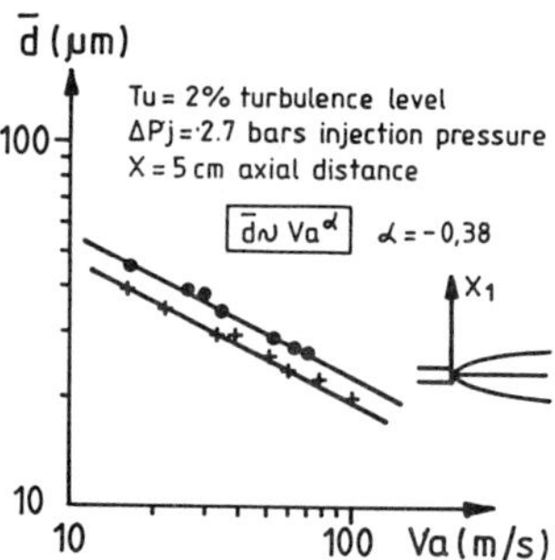

Fig. 2. Ramjet injector – Sauter mean diameter as funcion of air velocity and turbulence level.

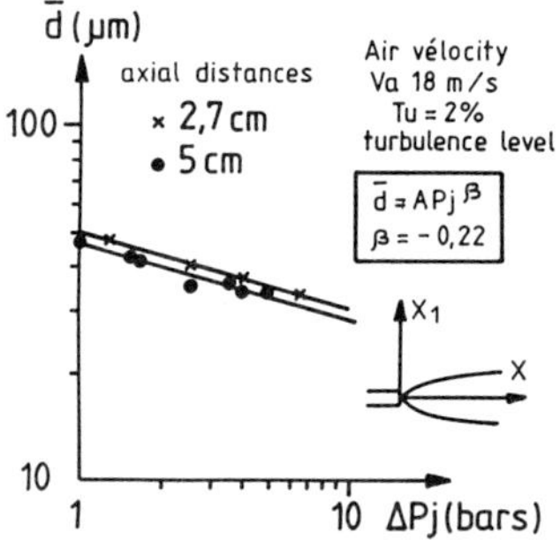

Fig. 3. Ramjet injector – Sauter mean diameter as function of injection pressure.

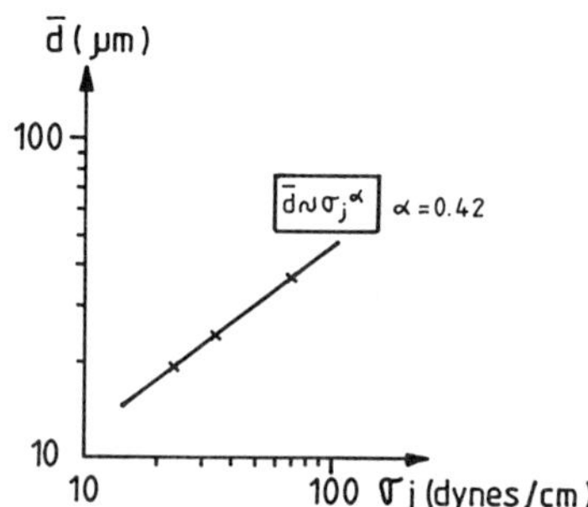

Fig. 4. Ramjet injector – Sauter mean diameter as function of surface tension.

<u>Other possible measurements</u>

In this paper we present only results dealing with drop size measurements but, as described elsewhere[2,3], we have used such techniques associating video camera, microcomputers and electronic devices to perform other kinds of measurement dealing with two phase flows. For example, the local value of drop concentration may be also deduced from the same image analysis or from the local video voltage, proportional to the total surface of drops present in the analysed window (if concentration is sufficiently small). Another possibility is also to measure drops velocities by analysing the successive positionsof the same drop illuminated with a stroboscopic light.

CALIBRATION

As said before, a calibration is necessary in order to link the size of the images seen by the camera and the real drop size. For a given light beam received by the drop (intensity Io and wave length λ) and angle between observation and illumination (θ) this calibration factor is only dependant on the magnification of optical devices (g), the digitizing level (S) and the drops refractive index (n). As we have no monodisperse spray generator, we generally use small balls of glass of similar refractive index and for different lighting condition and size. By comparison with the only geometrical calibration obtained from the observation of small displacement of such balls, we can say that the results are in good agreement with those deduced from the light scattering theory[4] in the following range.

$$\frac{\pi d}{\lambda} > 10$$

EXAMPLES OF APPLICATION

All following examples are dealing with studies linked with combustion chambers ; indead, CERT/DERMES is interested for many years by experimental and computational research about turbojet and ramjet engines. In these studies,for which the experimental part is performed on simulated isothermal similar or simplified chambers, the possibility to study the spray characteristics of different kinds of injectors as function of their operating condition is a particularly usefull tool. Also, for two phase flow modelling, it is often necessary to test elementary laws about pulverisation, vaporization and drops impact on hot surfaces. Also, basic studies about the influence of acoustic perturbations on the behaviour of elementary injectors is an element of the different research conducted about combustion instabilities.

<u>Ramjet injector performances</u>

In this first example, the characteristics of elementary injectors used in ramjet engines are studied as function of the operating conditions. Such results are obtained in a test ring shown on figure 1 in which air at different velocity (Va < 50 m/s) and turbulence level (2% < Tu < 15%) is moving in a squared (150 x 150) section of perspex. The elementary injector to be studied is placed at the center of the test section and the analysed plane defined by the laser sheet may be moved at different positions or orientationsfrom the injector. In order to see the influence of surface tension on the drop size, we have used either wateror mixing of water and alcohol.

Some results showing the variation of Sauter mean diameter as function of : air velocity, injection pressure, turbulence level, surface tension are presented on figures 2, 3, 4. All these results may be summarized in the following correlation which is in good agreement with litterature results[5]

$$S.M.D. = K \ Va^{-0,4} \ \nabla P_J^{-0\ 23} \ T_j^{0\ 42}$$

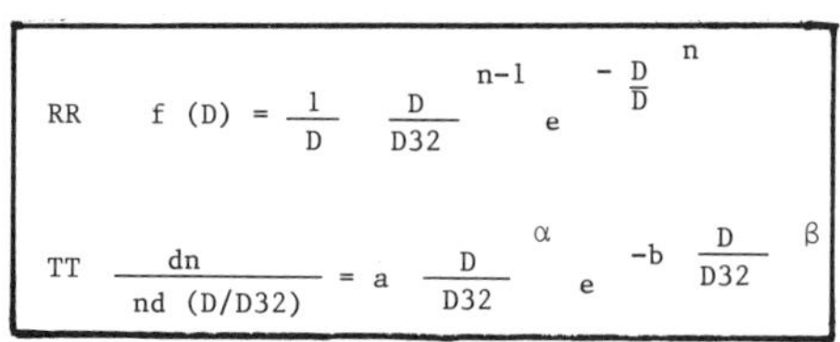

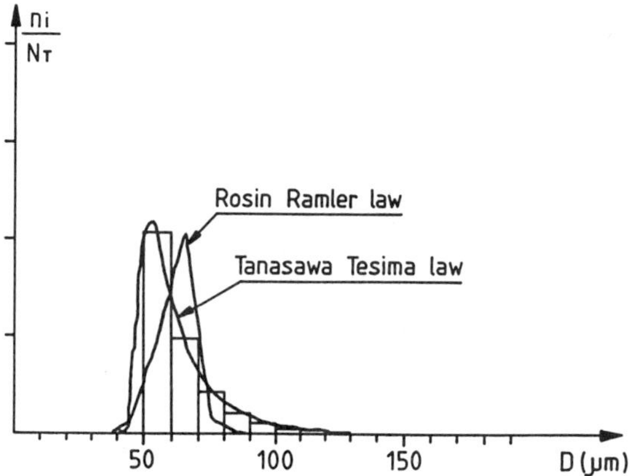

Fig. 6. Ramjet injector - Example of experimental drop size
distribution - Comparison with R.R. and T.T. laws.

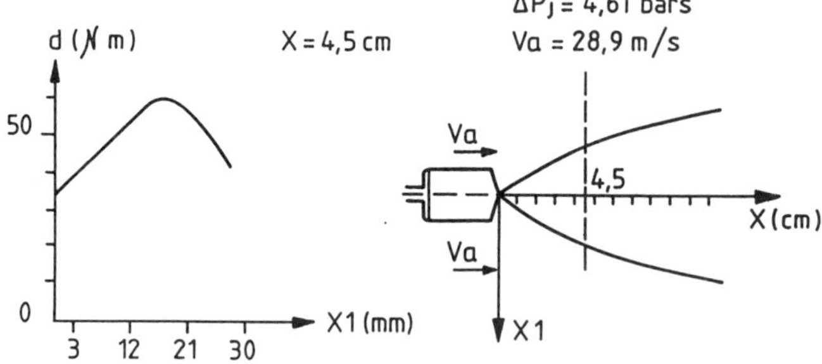

Fig. 5. Ramjet injector - Radial distribution of Sauter mean
diameter.

From such correlation, the benefic influence of air and liquid velocities on the pulverisation is quite obvious.

Another interesting result is shown on figure 5. As it can be seen, the radial distribution of Sauter mean diameter measured at some distance from the injector reveals important variations (from 35 μm to 60 μm in this example) which may be critical when giving the initial boundary conditions for the drops movement computation.

Always for such injector, figure 6 points out some examples of drop size distribution showing a better agreement with Tanasawa Tesima[6] law than with the Rosin Ramler one.

<u>Turbojet injector</u>

In order to see the influence of local aerodynamics on the spray characteristics issued from a double swirler airblast injector, a second set of experiments has been performed in a 2D isothermal model of turbojet combustion chamber operating at different flow conditions.

The radial distribution of Sauter mean diameter at different distances of the injector shows at figure 7 the following results :
- an important initial radial variation linked with the aerodynamics of jets in the formation region,
- a smooth distribution when going downstream,
- a diminution of S.M.D. mean value for the downstream distances. As noted in ref. 7, this result may be easilly explained by the behaviour of small particles which accelerate within an extremely short distance and then are underrepresented near the injector.

Such other results, not represented here, are also obtained when changing the operating conditions of the atomizer (ratio between internal and external air flow).

Always for the same atomizer, the figure 8 demonstrates that, with constant air flow rate and velocity, the S.M.D. remains quite constant when varying the liquid flowrate. Such a result could be completed by drop size distribution (not represented here) showing an influence of the modification of this function or on other mean diameters.

<u>Vaporizing studies</u>

Another example showing the possible applications of such a method is related to the vaporizing of small liquid drops when they are injected in hot air of different temperature and velocity. Such experiment is interesting in order to test the basic laws used in the different codes about liquid particles.

In this experiment represented on figure 9, hot air at variable temperature and flowrate can go inside of a pyrex tube (2m length and 60mm diameter) in which small water drops are introduced by an elementary injector built with a small orifice (0,4mm diameter) mounted on a can. The drop vaporization, which is the result of temperature and velocity differences between liquid and gas phases, is studied by analysing the drop size distribution at different downstream sections.

In order to make complete comparison, it would have been necessary to measure the local velocities of air and droplets at the different sections. Only the initial droplet velocity beeing known with the local air velocity, the local velocity has been estimated from the usual law representing the drops movement for every size class :

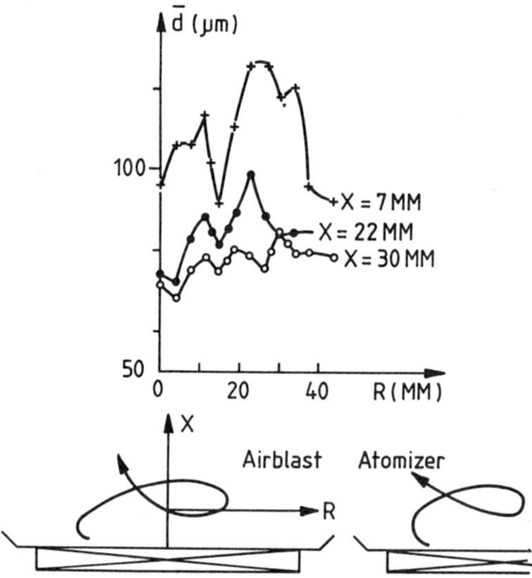

Fig. 7. Turbojet injector - Example of spatial distribution of Sauter mean diameter.

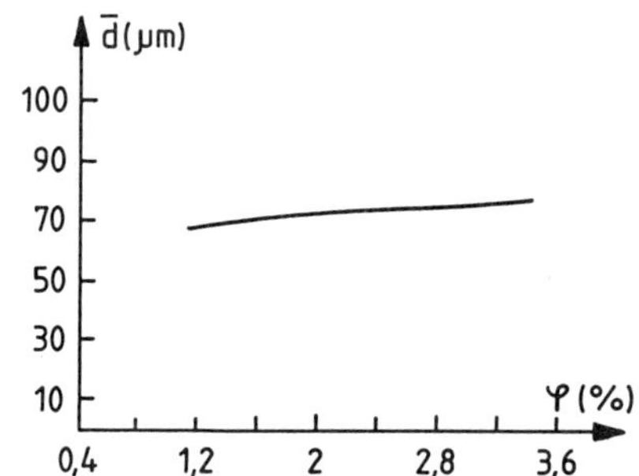

Fig. 8. Turbojet inejctor - Sauter
mean diameter as function of
simulated equivalence ratio.

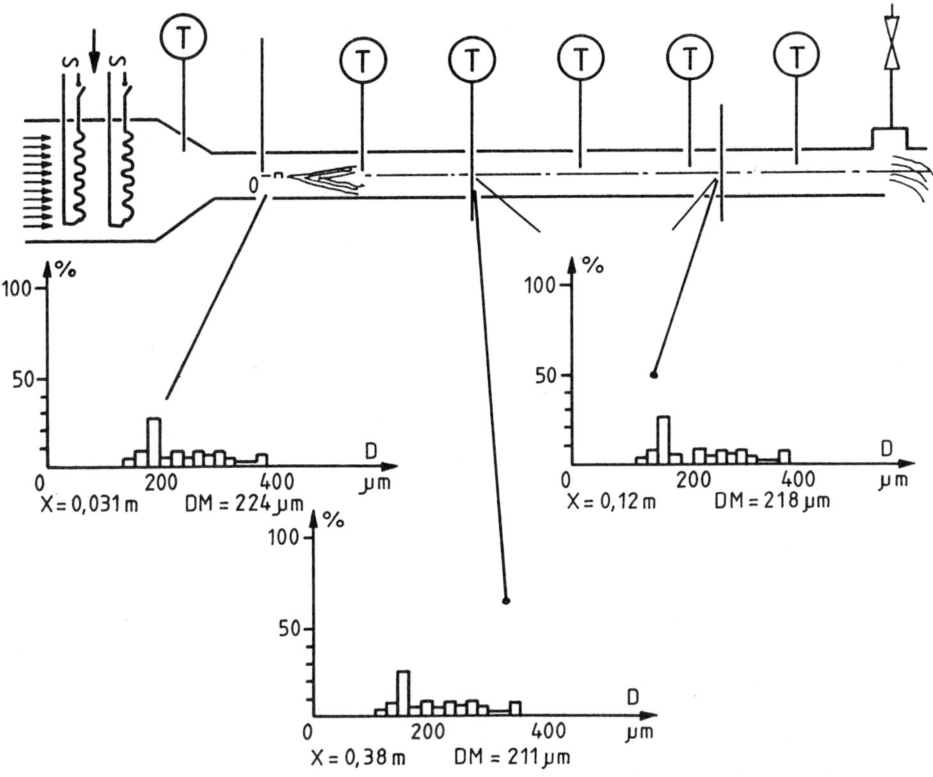

Fig. 9. Vaporizing studies – Experimental set up and evolution of drop size distribution.

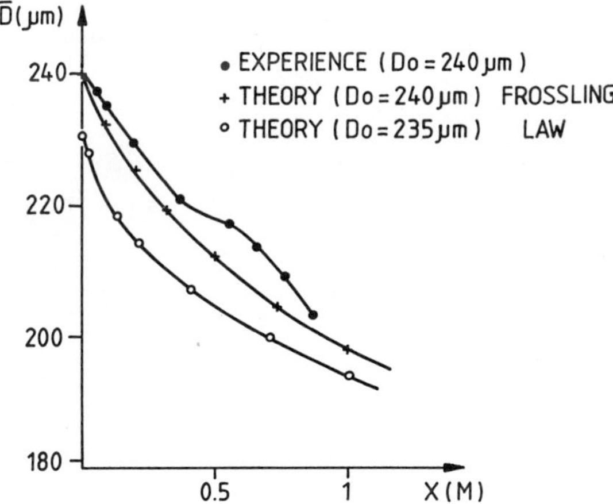

Fig. 10. Vaporizing studies – Comparison between experimental and computed values of Sauter mean diameter.

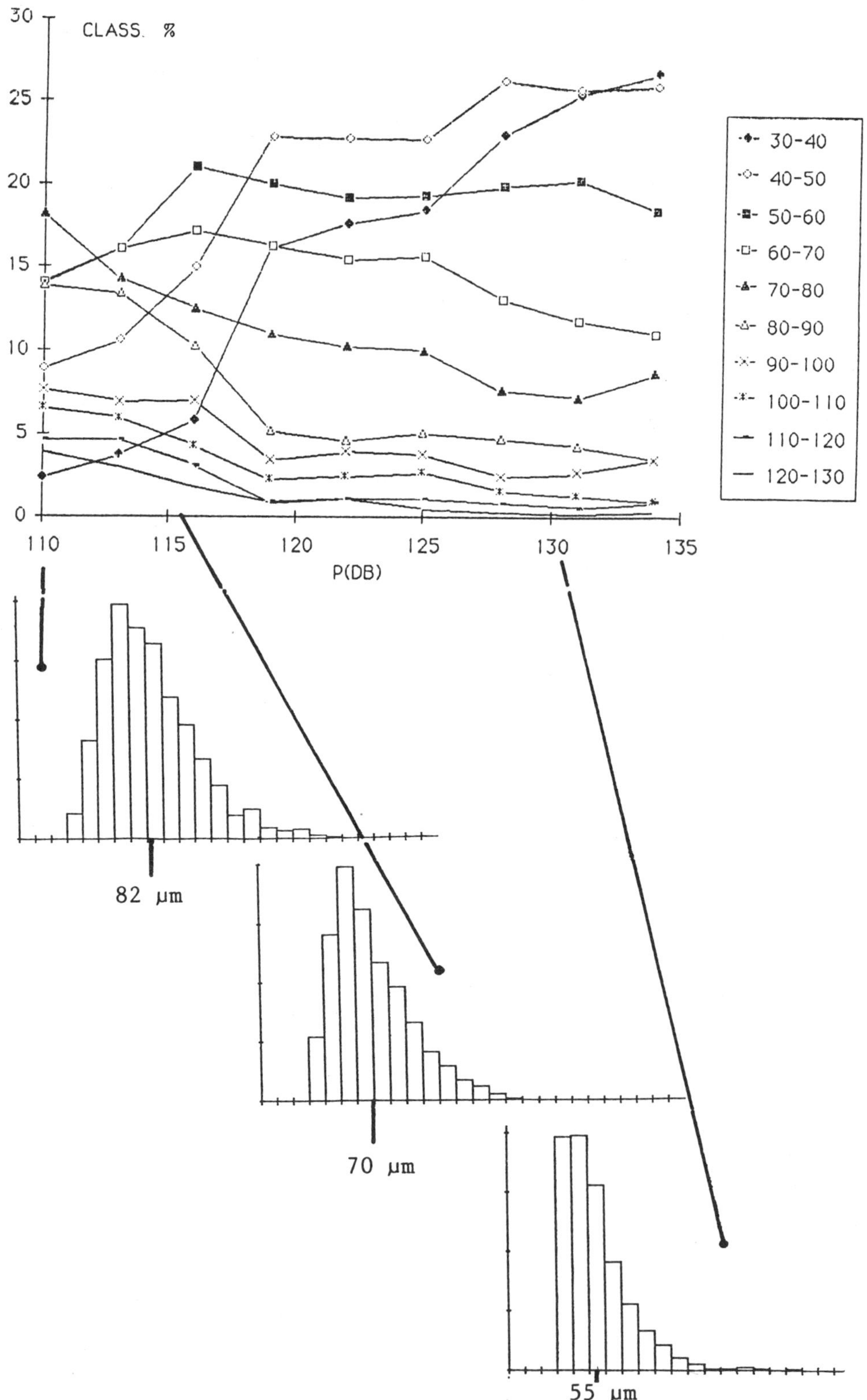

Fig. 11. Influence of acoustic excitation on the drop size distribution.

$$\frac{dV_R}{dt} = -\frac{3}{4} \; \frac{Cd}{D} \; \frac{\rho g}{\rho l} \; V^2_R$$

With $V_R = |V1 - Va|$
and

$$C_D = \frac{24}{Re} \quad \text{for Re} < 0,48$$

$$C_D = 27 \; Re^{-0,84} \quad \text{for } 0,48 < Re < 78 \qquad \text{from Dickerson and Shuman[8]}$$

$$C_D = 0,27 \; Re^{-0,21} \quad \text{for Re} > 78$$

with R_E, Reynolds number defined as

$$R_E = \frac{V_R D}{\nu}$$

In order to compare the measured results with computed results, a simple 1.D model[3] has been used in which the vaporization is computed from the Frosling law[9] .

$$M_F = M_o \; (1 + 0,244 \; Re^{0,5}) \qquad M_o = \pi \, d_L \, \frac{\lambda}{C_V} \, \text{Log} \, (1 + B) \; \text{avec} \; B = \frac{(T_\infty - T_L)}{L}$$

Figure 9 shows some example of drop size distribution from which the decrease of S.M.D. due to vaporization is quite obvious and represent a good comparison with the computed results as shown by figure 10.

Influence of acoustic perturbations

In the last example, the influence of acoustic perturbations of selected level and frequency on the behaviour of ramjet injectors has been studied. To do that, the facility used to study the ramjet injector performances has been modified in order to be able to superpose to the constant air velocity an acoustic perturbation coming from a pneumatic loud speaker which can produce noise till 135 db at frequencies in the range of 200 c/s to 5000c/s As seen by the photograph of figure 12 an optical arrangement with light, lenses and photodiode has been added to detect the characteristic frequency of drop formation by spectrum analysis of light received traversing the spray. This frequency, which is function of air and liquid velocities for a given injector vary from some 100 c/s to 2 ou 3 K c/s [10].

A lot of studies have been performed at CERT/DERMES showing that, when acoustic perturbations of sufficient level are superposed with an appropriate frequency related to the characteristic one, the drop formation mechanism may be completely modified[10]. This is particularly obvious on the figure 11 on which the following effects appear when the acoustic power is increased :
- narrowing of the distribution
- important decreasing of the Sauter mean diameter.

Presently these results are completed by experiments with a pulsed laser diode in order to detect possible pulsation in the drop formation.

CONCLUSION

This simple and obvious method is able to measure drop size within the range of 5 µm to some 100 µm quite in real time and in non interfering way. By simple analysis of successive video frames it can produce local measurements of drop size distribution, Sauter mean diameter and other moments, local concentration without any hypothesis about size distribution providing that a calibration can be executed in the same illumination conditions.

As seen by the presented examples, this method is particularly attractive in evaluating fuel atomizer performances and for basic research about liquid phase behaviour in combustion chambers studies : vaporizing, acoustic excitation.

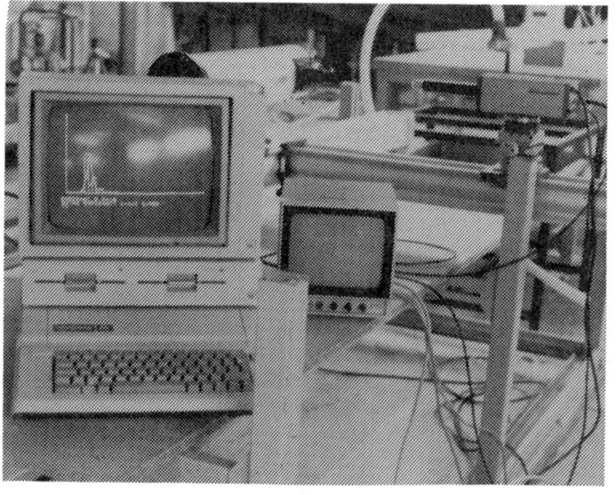

a)

b)

Fig. 12. Experimental set up for pul-
verization studies
a) optical sizing equipment
b) test rig for acoustic
excitation.

REFERENCES

1. J. M. Veteau, R. Charlot, Techniques de mesure des aires interfa-
    ciales dans les écoulements à bulles, Rapport CEA - R - 5075.
    (1983).
2. Y. Biscos, P. Hébrard, G. Lavergne, G. Toulouse, Digital video image
    processing application to drop size and concentration measure-
    ments, International symposium on visualization, Paris (1986).
3. P. Hébrard, G. Lavergne, Etude d'injection de carburant, Internal
    report,n° 2243/DERMES (1986).
4. H. C. Van der Hulst "light scattering by small particles", Chapman
    and Hall (1957).
5. A. H. Lefebvre, Rizkalla, the influence of air and liquid proper-
    ties on airblast atimization, ASME Journal of fluids engeneering
    (1975).
6. S. Nukizama, Y. Tanasawa, An experiment on the atomization of fluid
    Trans SME Japan 6, vol. 4, n° 14 (1938).
7. S. Witig, M. Aigner, K. H. Sakbaw and Th. Sattelmayer, Optical
    measurement of droplet size distribution : special considera-
    tions in the parameter definition for fuel atomizers, AGARD
    C.P. 353 (1983).
8. M. El Kotb, Fuel atomization for spray modelling. Prog., energy,
    combustion. Vol. 8 (1982).
9. N. Frossling, Geophysic 52, 170 (1938).
10. P. Hébrard, B. Platet, analyse du dynamique d'un injecteur soumis à
    une perturbation d'origine acoustique, Internal report n° 2/2249/
    DERMES (1986).

ACKNOWLEDGEMENTS

Thanks are due to G. Frager and B. Platet for their help in performing
the experiments and to C. Lempereur and J.M. Mathé for their contribution
in the development of image processing.

THE DIRECT COMPARISON OF THREE 'IN-FLIGHT' DROPLET SIZING TECHNIQUES FOR
PESTICIDE SPRAY RESEARCH

B. W. Young

Imperial Chemical Industries PLC
Plant Protection Division
Jealott's Hill Research Station
Bracknell Berkshire RG12 6EY
U.K.

W. D. Bachalo

Aerometrics Inc.
PO Box 308
Mountain View
California 94042
USA

INTRODUCTION

The successful atomization of a pesticide spray is generally
fundamental to the biological efficacy of the product; in that it is only
by thoroughly dispersing the product over the target surface that an
adequate response can be achieved. The interactions of the droplets with
the surface (for example a leaf) are complex and depend on such factors as
the physical properties of the liquid, the properties of the leaf, and the
kinetics of the impacting droplets. If the spray droplets do not remain
on the leaf surface after impaction, but bounce off again, then the
product may be wasted and the efficacy reduced. The production of
excessive quantities of very small droplets within the spray cloud can
lead to problems if they escape, are blown away in the wind, and perhaps
cause damage to neighbouring crops. For these, and other, reasons the
pesticide industry has always been aware of the potential importance of
droplet physics and has maintained an active involvement in the
measurement of these properties.

Until recently the traditional and laborious techniques of droplet
capture on artificial targets (such as soot layers, cards, oil films) or
photographic techniques were used to measure droplet size, with varying
levels of success; but the systemmatic collection of data was difficult.
Within the last decade a number of new optical techniques, that also
seemed appropriate to the needs of the pesticide industry, have been
developed and have become widely used for studying pesticide sprays. The
two techniques that have been most widely accepted have been the composite
diffraction technique from Malvern Instruments Ltd.[1] and the direct
shadowing technique from Particle Measuring Systems Inc.[2]. Both
instruments offered the 'new benefits' of direct in-flight measurements as
the droplets cut across a laser beam, and rapid data analysis and
presentation. More recently a new instrument based on the Doppler shift
principle has been introduced by Aerometrics Inc.[3] and this is beginning
to be used in pesticide spray research.

Unfortunately the enthusiastic response to these new instruments has
led to some problems. Much time has been spent by individual users in
collecting data for a wide range of typical hydraulic nozzles (and other
atomizers). However when attempts have been made to centrally collate the
data considerable discrepancies have been found. Members of both the
British Crop Protection Council (BCPC) and the American Society for
Testing and Materials have conducted 'user-surveys' with standard nozzles
in an attempt to develop acceptable testing protocols and data. A number
of comparative studies from other disciplines have also been reported[4,5].
In this paper an attempt has been made to systematically understand the
apparent discrepancies between these techniques. Their basic calibration
and testing using reticles and reference samples is discussed, using image
analysis as an independent check where appropriate. For some years freely
falling clouds of glass or plastic beads have been used to simulate
droplet clouds - in which case good agreement between the Malvern and PMS
instruments has been obtained. However as soon as dynamic droplet clouds
have been measured distinct differences in the data have been recorded.

With the introduction of the Aerometrics system a new series of
problems were envisaged and an unique series of tests using all three
instruments simultaneously to measure freely falling glass bead and
dynamic droplet clouds is reported. Using the findings from this 3-way
test the earlier data has been re-evaluated; and it has been shown that
close agreement is possible - provided the correct parameters are
considered.

CALIBRATION TECHNIQUES AND COMPARATIVE STUDIES FOR THE MALVERN AND PMS
SYSTEMS.

The design and operating principles of these instruments has been
published previously and will not be discussed here. Since its
introduction the design of the optical receiver for the Malvern has
improved significantly, eliminating much of the variability in early
instruments[6], and a standard test method has now been developed using a
specially designed reticle[7]. This consists of an accurately known
distribution of opaque spots on a glass mount. By positioning this
correctly in the laser beam the performance of any instrument can be
verified. Data for our Malvern 2600 has been compared with the
manufacturers data for one reticle (ref: RR-50-3.0-0.08-102-CF-115)[6] and
with image analysis data (using a Cambridge Instruments Ltd Quantimet
900[8]). This has shown close agreement between the measured and
theoretical data and supports previously published data on the use of such
reticles.

A similar technique can be used for the PMS system. The instrument
measures individual droplets (or in this case opaque discs) as they pass
through the laser beam, rather than the composite diffraction from a
multiplicity of droplets at an instant. A simple reticle has been
designed and produced consisting of a thin glass plate with seven opaque
discs which can be repeatedly passed through the laser beam[9]. This
technique is now used as a standard for our PMS model OAP-2D-GA2 and has
been tested by other users. It is possible to obtain single channel data
for each individual disc; and to assess the effects of depth of field
limitations etc. A more complex reticle has been produced as a circular
glass plate covered in an array of discs that is spun through the beam.
This has been used to extensively study the calibration of PMS model OAP-
260X probes[10].

Thus using these techniques as a starting point it can be shown that such instruments are capable of correctly measuring an artificial standard.

Planar reticles are of course not representative of three - dimensional droplets in a spray cloud and a more realistic reference system has been desirable. For this reason the use of freely falling clouds of glass beads has been adopted within ICI Plant Protection Division as a secondary technique. Such clouds have the advantage over sprays that they can be collected and re-used, they do not evaporate or change with time and they can be measured by other techniques - for example microscopy. Results for directly comparative measurements - where the samples have been passed through the beams of the Malvern and PMS instruments simultaneously - have been published previously[11], again showing that good agreement is possible. Generally the beads are of industrial quality (not expensive calibration standards) and their distributions improved by simple sieving.

Care is needed with each laser system in order to present an acceptable sample - particularly in terms of cloud density. The Malvern requires a relatively dense cloud in the beam in order to create an adequate diffraction signal, whilst the PMS measures individual droplets by shadowing and so a dilute cloud is preferred. In dense situations there is a danger of multiple imaging from overlapping shadows falling on the detector giving an oversizing effect. In some models there is only limited correction made for this, so it is important that operators understand the limitations of the instrumentation. The use of bead clouds does show that these instruments are intrinsically capable of giving comparative data for samples covering the size range typically encountered with agricultural hydraulic nozzles. It is of course important to ensure that appropriate lenses are used to adequately cover the size range in question. For the Malvern this is particularly important for sizes above 200 micrometres because the size increments determined by detector geometry become progressively wider. For the majority of our work with pesticide sprays the 1000mm lens is necessary.

During work for the comparative study for the British Crop Protection Council (BCPC) survey[12] a simultaneous sampling technique was used for the dynamic sprays from representative flat fan nozzles. The two instruments were positioned such that the PMS laser beam was immediately below (1cm) and aligned with the Malvern beam. The spray cloud was orientated across the beams and shielded such that the effective path length of the beam through the cloud was restricted to the 6cm beam length of the PMS. In this way the two instruments sampled the same portion of the cloud as closely as was physically possible. Data for a typical hydraulic nozzle (Lurmark Ltd) F110-01, sampled 30cm directly below the stationary nozzle, on the centreline, is shown for a range of pressures in Table 1.

Table 1. Volume Median Diameters (micrometres) obtained for F110-01 nozzle at a range of pressures; 30cm below nozzle, 0.1% ^{W}/v non-ionic surfactant.

	Spray pressure - bar.				
	5	4	3	2	1.5
Malvern 2600	81	88	97	117	130
PMS OAP. 2D.GA2	125	141	150	203	210

        This data clearly shows an immediate difference between the data from
the two instruments for a dynamic spray. Both are showing similar trends
-i.e. a coarsening with decreasing pressure - but with a significant
shift. Such a difference has been recorded on each occasion direct
comparisons have been made. In the pesticide industry at least it has
however been the cause of much controversy and criticism of the
instruments, and yet little attempt seems to have been made to resolve the
problem. The work of the BCPC has been directed at developing a nozzle
classification system for guiding users on the correct choice of nozzle
for a particular purpose.   For understandable reasons a relative
classification of Very Fine, Fine, Medium, Coarse, Very Coarse has been
initially adopted without giving any numerical values, but this is
suggesting a lack of confidence in the value of the data and/or the
techniques. At the other extreme widely differing numerical data for
nozzles measured by the two techniques have been published, with a
disclaimer regarding the validity of either technique.[13]

        It was against this background that the introduction of the
Aerometrics system to the UK in 1985 [14] was seen as potentially raising
even more problems. In this case the technique measures individual
droplets at the intersection region of two crossed laser beams,
essentially at a point. The results of comparisons between the Malvern
and Aerometrics have been published, for fine fuel atomisers, and
generally good agreement has been obtained.[4,5]. No data has been
available for the coarser sprays generally used for pesticide spraying
however; thus the idea of attempting to simultaneously sample with all
three systems was conceived.

EXPERIMENTAL METHOD

Instruments used

        A model 2600 Malvern with 1000 mm lens and Model Independent
programming, a PMS OAP.2D.GA2 with size range to 900 micrometres and the
Aerometrics P/DPA system were used. The Malvern beam was used as the
reference point. The PMS probe was fitted with a Dove Prism, allowing
rotation of the beam through 90°, such that the probe body was horizontal
but the beam was orientated to measure droplets falling through it. The
beam was positioned along the Malvern beam, 50cm from the receiving lens
and physically approximately 1cm above it. The Aerometrics was positioned
such that the transmitted beams intersected to give the sample volume
immediately below the centre line of the other two beams. In this way any
sample that passed through the PMS or the Aerometrics must have passed
through the Malvern beam - although obviously the converse is not the case
because of the greater width of the Malvern beam. However from a
practical point of view this was judged to be the closest possible to
direct simultaneous sampling. For each measurement all three systems were
started as nearly as practicable at the same instant.

Glass bead samples

        Six sieved samples of glass beads (Jencons Scientific Ltd. Leighton
Buzzard, U.K.) were used. These were nominally classified as indicated
in the results below. The beads were dispersed as a freely falling cloud
from a funnel with an adjustable flow, held above the three laser beams,
to achieve an 'hour-glass' effect.

<u>Hydraulic nozzles</u>

A selection of typical agricultural flat-fan hydraulic nozzles was
tested. In each case the nozzle was mounted stationary 30cm above the
three laser beams in such a way that the beams passed through the short
axis of the fan on the centreline. In this way the 'active' beam length
of the Malvern was minimised; for the other instruments the sampling
volume is physically determined. A 0.1% ^{W}/v nonionic surfactant solution
(Agral 90-I.C.I PLC) was sprayed at pressures of 2.5-3.0 bar.

RESULTS

Summary data for the glass bead samples are shown in Table 2, giving
the Volume Median Diameters as obtained from each instrument. Also
included are the corresponding values obtained by microscopy/image
anlaysis using the Quantimet. Summary data for the nozzles are given in
Table 3.

Table 2. Volume median diameters for glass beads from simultaneous
sampling.

Sample No.	Reference	Malvern	PMS	Aerometrics	Quantimet
1	53-75	59	67	57	56
2	75-106	73	79	79	74
		72	83	73	
3	106-150	128	139	133	125
4	212-425	251	269	245	282
		251	269	257	
5	425-600	493	523	459	526
		485	518	471	
6	102/1-2	145	144	134	-

Table 3. Volume median diameters for nozzles from simultaneous sampling.

Nozzle ref.	Malvern	PMS	Aerometrics
Lurmark Ltd	267	352	345
02-F80	273	354	337
Hardi Ltd	143	224	191
F4110-10	-	202	
Hardi Ltd	185	267	234
F4110-14			
Hardi Ltd	183	337	259
F4110-20	161	324	254
Hardi Ltd	220	232	212
2080-10	-	221	-

DISCUSSION

<u>Glass beads</u>

It is clear from Table 2 that the three laser systems have given
close agreement for all the samples - and that these agree well with the
image analysis data. The volume median diameters have been plotted in
figure 1 and this indicates two interesting trends:-

1)   for the fine samples the PMS tends to slightly oversize relative to
      the others, (for the calibration procedure used) as reported
      previously[9].
2)   for the coarser samples the PMS and Quantimet agree well, but the
      Malvern and Aerometrics begin to undersize by comparison.

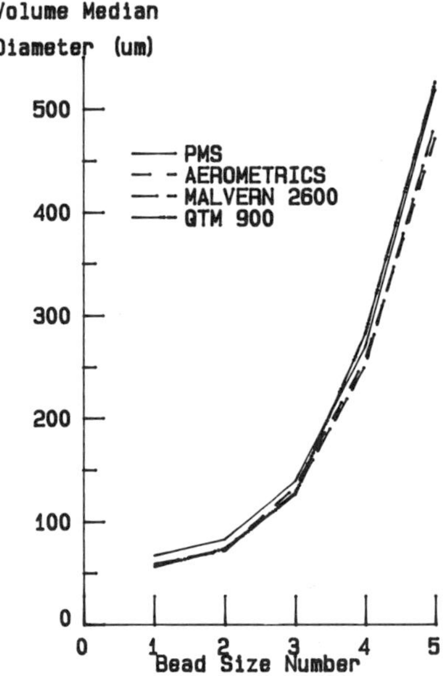

Figure 1: Comparative Volume Median Diameters for five glass bead
          samples.

The latter effect may be due to a fall off in response with
increasing size for the Malvern, due to the very small diffraction angle
from this size range, and the data falling within the first few channels.
A sampling and/or optical limitation with the Aerometrics (as used) may
have limited the acceptance of large beads through the sample volume.
Overall however this data supports the previous conclusions that the
different techniques can give comparative data.

488

<u>Hydraulic nozzles</u>

In contrast to the glass bead data, the data for the nozzles show some distinct differences between the instruments. The most striking effect is that the Malvern data is significantly finer than the other two – which show moderate to good agreement. (The exception is the 2080-10 nozzle, where close agreement is found – this is unexpected in the light of the rationale presented below and suggests an error in the data).

In considering possible explanations for the repeated difference in agreement when sampling sprays rather than the beads it was realised that a fundamental difference in the nature of the data was being overlooked. In the manner used the Malvern measures the composite diffraction pattern at an instant and therefore gives a spatial distribution; whilst the PMS and the Aerometrics accumulate data from individual samples over a time period—and thus give a temporal distribution. Because of the heterogenous nature of dynamic droplet clouds in terms of droplet size and velocity (especially at some distance from the nozzle as in this case) these distribution forms are bound to be different. Thus the data should not be expected to be comparable. The significance of these distribution forms has been published[15,16], but in the pesticide field at least seems to have been ignored when comparing data.

Table 4. Spatially and temporally resolved volume median diameters from the Aerometrics system.

	Spatial VMD	Temporal VMD
Glass beads		
Sample No. 1	57	57
" " 2	79	79
	73	73
" " 3	133	133
" " 4	245	245
	256	257
" " 5	458	459
	471	471
" " 6	133	134
Hydraulic nozzles		
02-F80	300	345
	290	357
F4110-10	130	191
F4110-14	190	234
F4110-20	187	259
	178	254
2080-10	165	212

With this in mind the data was re-examined, and here a particular
attribute of the Aerometrics system has been significant - because it can
provide both the spatially and temporally resolved data directly. (The
PMS system used measures both droplet size and velocity but the data
system has not been programmed to provide spatial data). In Table 4 the
spatial and temporal Volume Median Diameters from the Aerometrics are
given and these show a very significant fact. For the glass beads there
is no difference between the spatially and temporally resolved data, but
for the hydraulic nozzles there is.

What is more significant is that the spatially resolved data for the
nozzles is now in much closer agreement with the Malvern data (refer to
Table 3). Thus there seems to be a fundamental difference between
sampling the freely falling beads and the dynamic sprays; and a
significance of spatially or temporally resolving the data for the latter.
These effects are illustrated in figures 2 and 3. The former shows a
summary print-out from the Aerometrics for a glass bead sample (No. 3),
showing the virtually identical spatial and temporal data; the latter
shows equivalent data for one of the hydraulic nozzles (F4110-10) showing
a significant difference between the spatially and temporally resolved
data.

A difference between these two distribution forms would be expected
in a dynamic situation, and is dependent on the relative velocities of the
droplets in the cloud. Examination of the velocity data from the
Aerometrics gives clear support to this as the reason for the differing
responses. In figure 4 the velocity profile for the glass beads (No. 3)
is essentially horizontal, indicating (as would be expected in a free fall
situation) that there is little change in velocity with bead size over the
fall distance. In figure 5 however there is a clear variation with
droplet size as would be expected for a dynamic spray cloud (nozzle 01-
F80). The data is typical for this form of nozzle and shows close
agreement to equivalent data obtained from the PMS system.[9]

From this evidence a possible explanation for the differences
observed in dynamic situations with the Malvern and PMS becomes apparent;
and suggests that by conversion of the data to an equivalent format
agreement may be possible. A satisfactory approximation for the
conversion of temporal data to spatial data using the velocity profile has
been published and has been adopted here[17]. In outline this procedure
requires that the percentage number population of each size class of the
temporally resolved distribution is divided by the average velocity for
that size class and the new distribution calculated. In cases such as
these dynamic spray clouds the smaller droplets slow down relatively much
more rapidly as a function of distance from the nozzle, so the relative
proportion of the smaller droplets increases in the conversion from
temporal to spatial data. Thus a spatial distribution is always finer
than a temporal distribution. When describing agricultural sprays it is
common practice to use the volume distribution rather than the number
distribution and so new spatially equivalent volume distributions have
been calculated. An outline of this comparison has been published[18] and
the data shows that following this treatment the spatially resolved PMS
data moves to much closer agreement with the Malvern data. An example of
this for the F4110-10 nozzle is shown in figure 6. Treatment of the PMS
data for the other nozzles indicates a similar shift in the data although
as in Table 3 the agreement between the systems appears to worsen for the
coarser sprays.

The results for both the beads and the sprays indicate that the PMS
gives a coarser distribution for the large size ranges. It is possible
that in a heterogeneous spray cloud the relatively infrequent and weak

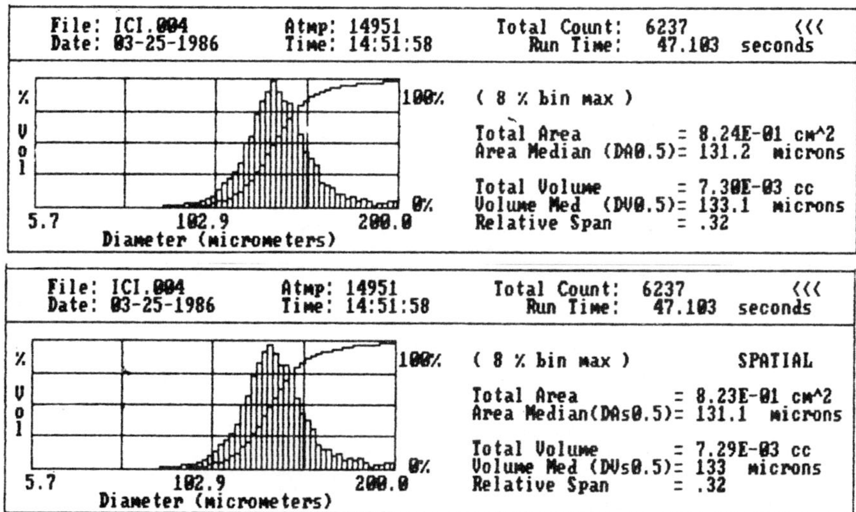

Figure 2: Data from the Aerometrics system for glass bead sample (No. 3) in free fall; showing equivalent temporal (upper) and spatial (lower) data

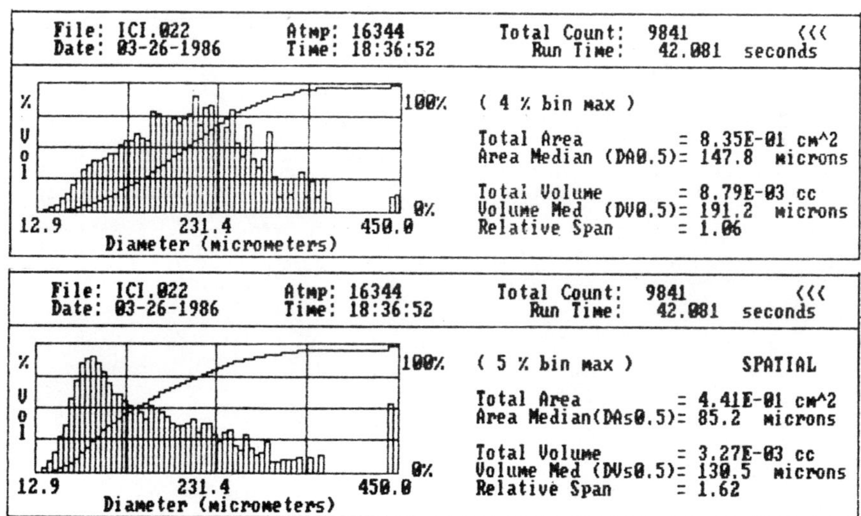

Figure 3: Comparison of spatial and temporal distributions for a dynamic spray cloud, showing the influence of droplet velocity on the volume distribution. (F4110-10 nozzle).

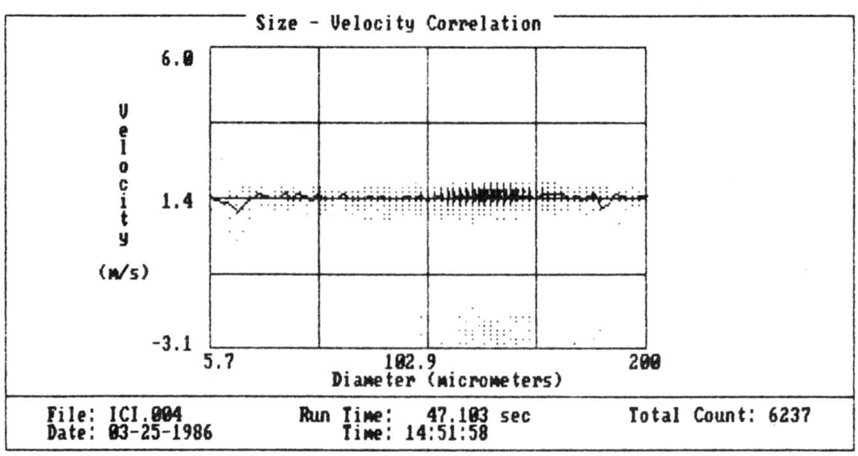

Figure 4: Velocity/size profile for freely falling glass beads (No 3).

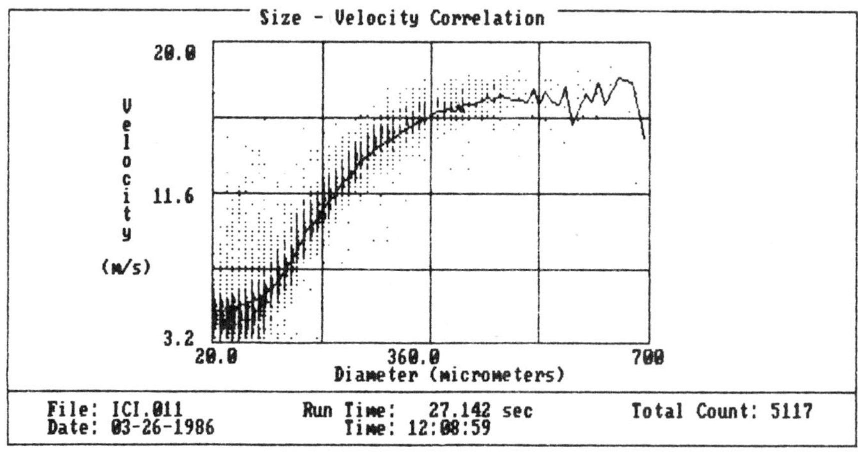

Figure 5: Velocity/size profile for the dynamic spray cloud from nozzle 02-F80.

diffraction signal from a large droplet is not adequately detected in the Malvern leading to an undersizing effect; and for the optical system used on the Aerometrics in these tests the low probability of successfully sampling the large droplets through the point source sample volume also reduced the sensitivity to these droplets. This issue will only be resolved by further careful comparisons with samples in this size range. These present tests have however indicated a new rationale for comparing the Malvern and PMS systems - which in hindsight of course should have been obvious.

REEVALUATION OF PREVIOUS COMPARATIVE DATA

Using the implications from this 3 way test, previous data from the Malvern and PMS has been reevaluated. The original work to characterise nozzle performance for the BCPC classification code highlighted the discrepancies between data from the Malvern and PMS. The nozzles have been classified as previously discussed. Data for the tests with the present instruments is given in Table 5. Once again this shows equivalent trends but with a definite shift in actual values. From this a 'Medium' nozzle would be expected to have a volume median diameter of approximately 335 or 215 micrometres depending on which instrument it was measured. This is clearly not a satisfactory situation.

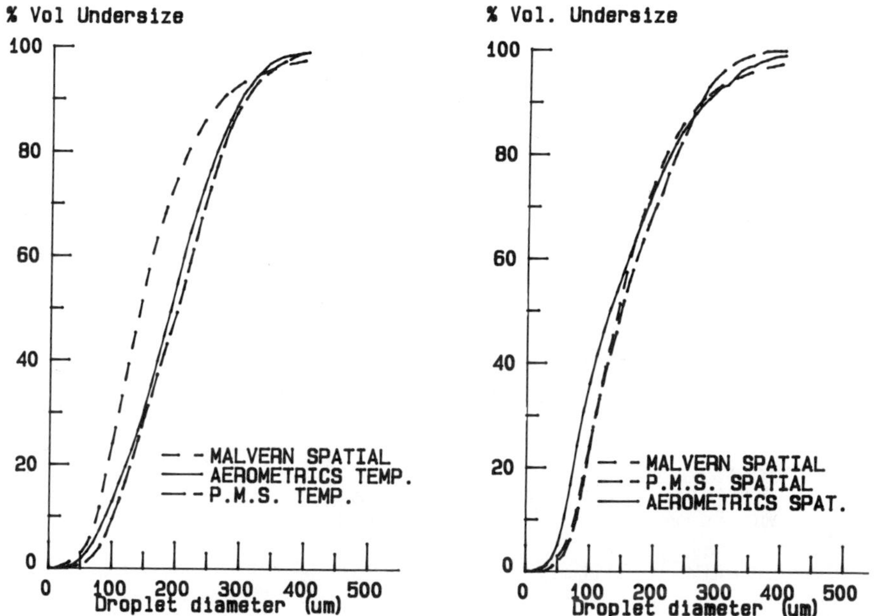

Figure 6: (a) distributions obtained directly from each laser, for F4110-10 nozzle.
(b) spatially converted data, showing improved agreement.

The data in Table 1 indicated similar trends in behaviour as a function of pressure. By using the velocity data the PMS spatially adjusted distributions can be calculated. Table 6 summarizes the comparison over the pressure range. This again indicates that the agreement appears to be a function of the range of the distribution itself and raises the question as to the relative sensitivities of the three instruments to broad spectrum, and by normal standards, coarse sprays.

Table 5. Volume Median diameters for BCPC nozzles using the Malvern and PMS lasers - sampled 30cm directly below the nozzle, at 3 bar pressure.

	Very Fine F110-01	Fine F110-02	Medium F110-04	Coarse F110-08
Malvern	97	140	215	350
PMS	150	250	335	450

Table 6. Spatial/temporal conversions for a range of pressures for F110-01 nozzle.

Spray pressure - bar

	5	4	3	2	1.5
Malvern	81	88	97	117	130
PMS temporal	125	141	150	203	210
PMS spatial	82	87	102	119	157

A SIMPLE SPATIAL/TEMPORAL MODEL

In comparison to previously published spatial/temporal data the differences observed here are much larger. This may be due to the coarser distributions being sampled further from the nozzle - allowing time for a greater change in the velocity profile. Previously published PMS data has shown how rapidly the profile changes with distance from the nozzle[9]. In order to illustrate the possible significance of this on the spatial distribution a simple model has been developed.

This takes a simplified case (based on typical experimental data) of a number distribution as shown in Table 7. In the ideal initial case all droplets are assumed to be travelling at 15m/s and the spatial and temporal forms are therefore the same. The equivalent volume distribution is shown in figure 7. At 20cm downstream a typical velocity profile has been assumed, as shown in Table 7. The corrected spatial volume distribution based on this is also shown and shows a decrease in median diameter from 240 to 215 micrometres. A third case, for typical 50cm conditions is also given; showing a further reduction to 180 micrometres. This magnitude of change (from 240 to 180 micrometres) is similar to that reported experimentally above. The temporal distribution would remain unchanged.

This simple model tends to confirm that any naturally spatially responsive instrument will be particularly sensitive to changes in the velocity profile of the spray cloud - such as caused by changing sampling distance. Ironically for the majority of work with these nozzles where comparisons between data have been made, the sampling distance used has probably maximised the difference between the spatial and temporal forms. This should be considered when developing future protocols.

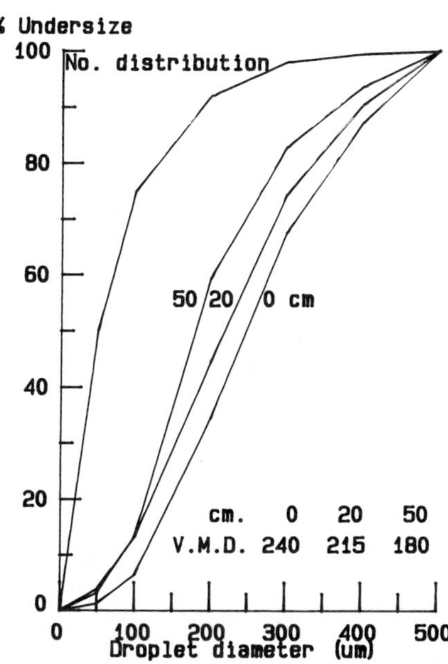

Figure 7: Model spatial distributions, showing the decrease in droplet size as a function of distance from the nozzle.

Table 7. Spatially resolved data for a simple model.

Diameter	No. %	Initial case Velocity m/s	Vol. %	20cm case Velocity m/s	Vol. %	50 cm case Velocity m/s	Vol. %
500	100	15	100	12	100	10	100
400	99.5	15	87.2	11	90.4	9	93.6
300	98	15	67.5	10	74.3	7	82.8
200	92	15	34.3	8	44.4	3	59.3
100	75	15	6.4	5	13.1	2.5	13.3
50	50	15	1.3	3	3.8	2	3.2

CONCLUSIONS

The performance of the Malvern and PMS laser systems has been
discussed. It has been shown that each can accurately respond to
artificial reference materials and that comparative data can be obtained
for freely falling clouds of glass beads.

In an unique 3 way test of these instruments with the new Aerometrics
system the significance of spatially and temporally resolving the data has
been demonstrated. This has shown that equivalent data can be obtained
between all three instruments when the results are correctly compared.
The agreement seems to deteriorate for coarse sprays and the relative
accuracy of each instrument for such situations must be questioned.

Further comparative testing would be of value in resolving this
problem. From the present data no judgement is made as to which
instrument is 'more correct', merely that there are relative differences.
It might be argued that data from such fundamentally different techniques
should not agree, but the experimental data does suggest that for
practical purposes useful comparisons can be made provided the limitations
of the techniques are appreciated.

ACKNOWLEDGEMENTS

We wish to thank Dr E Hislop, Long Ashton Research Station, Bristol, BS18
9AF, England for the loan of the Aerometrics Analyser for this work, and
Mr C A Hart, ICI Plant Protection Division for the image analysis data.

REFERENCES

1. Malvern Instruments Ltd, Spring Lane South, Malvern
      Worcestershire, WR14 1AQ. England.
2. Particle Measuring Systems Inc., 1855 South 57th Court, Boulder,
      Colorado. 80301 USA.
3. Aerometrics Inc., PO Box 308, Mountain View, California 94042,
      USA.
4. T. A. Jackson and G. S. Samuelsen. Performance comparison of two
      interferometric droplet sizing techniques. _S.P.I.E._ Vol 573
      Particle size and spray analysis. (1985).
5. V. G. McDonell, C. P. Wood and G. S. Samuelson. A comparison of
      spatially resolved drop size and drop velocity measurements in
      an Isothermal chamber and a swirl stabilized chamber. _21st
      Intl. Symposium on Combustion._ Munich. W Germany. August
      1986._
6. E. D. Hirleman and L. G. Dodge. Performance comparison of
      Malvern Instruments laser diffraction drop size analysers.
      3rd Intl. Conference on liquid atomisation and spray systems.
      Paper IVA/3. The Institute of Energy. London. (1985).
7. L. G. Dodge. Calibration of the Malvern particle sizer. _Applied
      Optics._ 23. 2415-2419 (1984).
8. Cambridge Instruments Ltd, Viking Way, Bar Hill, Cambridge. CB3
      8EL. England
9. B. W. Young. Practical applications of the PMS-2D Imaging
      spectrometer, _in:_ 'Pesticide formulations and Application
      Systems' volume 5. Spicer/Kaneko, eds., ASTM/STP 915. 128-
      133. (1986).
10. E. D. Hirleman. Arizona State University, Tempe. AZ 85287, USA.
      personal communication.

11. B. W. Young. The need for a greater understanding in the application of pesticides. Outlook on Agriculture, 15.(2) 80-87, (1986).

12. S. J. Doble, G. A. Mathews, I. Rutherford and E. S. E. Southcombe. A system for classifying hydraulic nozzles and other atomisers into categories of spray quality. British Crop Protection Council Conference - Weeds. 1125-1133. (1985).

13. The Exact Spraying System. Crop Control Products Ltd., 2/3 Clifton Street, Lincoln. LN5 8LQ. England.

14. W. D. Bachalo and M. J. Houser. Spray drop size and velocity measurements using the phase/Doppler particle analyser. 3rd Intl. Conference on liquid atomisation and spray systems. Paper VC/2. The Institute of Energy. London (1985).

15. A. R. Frost and J. R. Lake. The significance of drop velocity in the determination of drop size distributions in agricultural sprays J. Agric. Engng. Res. 26. 367-370 (1981).

16. R. W. Tate, Some problems associated with the accurate representation of droplet size distribution. 2nd Intl. Conference on liquid atomisation and spray systems. Paper 12-4. The University of Wisconsin. Madison. (1982).

17. W. D. Bachalo, M. J. Houser and J. N. Smith. Evolutionary behaviour of sprays produced by pressure atomizers. AIAA 24th Aerospace Science Meeting. Paper 86.0296. Reno, Nevada. (1986).

18. B. W. Young. Spray Application - more science, less art. Canadian Society of Agricultural Engineering. Summer Meeting. Paper 86-209. Saskatoon. Canada. (1986).

AN APPLICATION OF AN OPTICAL PARTICLE SIZING METHOD TO THE STUDY OF THE

EVAPORATION OF DROPLETS DIFFUSING IN A TURBULENT FLOW

H.Burnage and S.J.Yoon*

Institut de Mécanique des Fluides de Strasbourg
U.A. C.N.R.S. 854
2, Rue Boussingault, 67000 Strasbourg, France

I-INTRODUCTION

The physico-chemical process of evaporation of a cloud of
droplets moving in a turbulent gas flow is a very complex phenomenon of
major importance for many applications. Its difficulty stems from the
large number of interacting parameters on which it depends. When a droplet
is in motion, the flow around it modifies the boundary condition for
momentum, heat and mass transfert; the latter is directly related to the
velocity of the particle and to the partial pressure of vapour in the
neighbouring fluid. If these are sometimes known in the case of a single
drop displacing in a given environment, they are not when it follows a
random trajectory inside of a cloud of similar particles. It is then
necessary to elaborate mathematical models that represent the physical
phenomenon considered as a whole.
One of the most important variables is the diameter of the drops
as it determines, together with its velocity, the drag exerted by the gas
on the particle and it also influences strongly the rate of evaporation.
When the number of droplets undergoing the evaporation process is large
such as in a cloud, it is natural to regard it as a statistical ensemble,
and, rather than examining the evolution of a sample diameter, consider
its density probability function. Several models have been proposed to
describe the latter , in the present work, that of ROSIN and RAMMLER has
been prefered because, in spite of some analytical disadvantages, it is
defined by two parameters only, namely a length scale d and an exponent m
and is widely used in engineering applications. It states that the
probability of cumulating a volume V with the drops of which the diameter
is less than D, is given by:

$$P(V) = 1 - Exp[- (D/d)^m]$$

The aim of the present work is ,in a first stage to test the
applicability of a ROSIN-RAMMLER [17] model for sprays generated by the
type of air blast injector used here and ,eventually to observe if the
distribution of the particle diameters departs from it during the process
of evaporation.

------------+-----

(*) present address: Department of Mechanical Engineering, The Chonbuk
National University, 520 Chonju , Korea

The sensitivity of polydisperse flows to the intrusion of probes
that disturb both its mechanics and its thermodynamics called for an
optical method of particle sizing. The technique used here is based the
analysis of the light diffused by the drops. the basic measuring
instrument is a standard Laser Doppler Anemometer to which, as will be
detailed hereafter, minor modifications have been brought

II-EXPERIMENTAL SET-UP AND MEASURING METHODS

Experimental set-up

The experiment was performed in an open circuit wind-tunnel
(YOON [18]). The air is taken from the atmosphere cleaned from dust by a
filtration unit, then heated to the apropriate working section
temperature. A centrifugal compressor drives it through a diffuser and a
settling chamber equiped with seven stainless steel calibrated filters; it
is then accelerated by a convergent to the working section of which the
upstream part contains the injector. A turbulence grid is located at the
end of this section which is 50 cm. long .The working section has a
section of 20x20 cmm and is 1 m. long measured downstream of the grid. In
the present experiment the mean velocity in the undisturbed flow was kept
constant at 2 m/s .The two series of measurements hereafter refered to as
"series 1" and "series 2" were performed respectively at working section
temperatures of 4°C and 16°C.
The particles are droplets of dichlorotetrafluoroethane (Forane
114, the boiling point of which is 3.6°C at standard pressure and
temperature). They are generated by an axial air assist nozzle). Its inner
tube of 200 µm diameter carries the liquid and the air is forced under
pressure in a larger tube coaxial to the former. The rates of discharge of
both liquid and air have been adjusted so as to obtain a mean size of the
droplets of about 100 µm which offers the optimum compromise between the
performance of the opto-electronic device and the scientific interest of
the experiment. Therefore ,throughout the experiment, the rates of
discharge of air and liquid were respectively: 61 ml/s and 0.34 ml/s
whilst their exit velocities were: 92.5 m/s and 10.2 m/s.

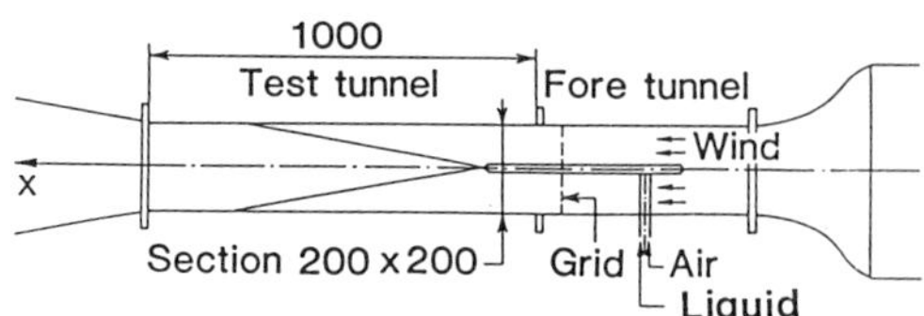

Figure 1 Experimental set-up. Test section

The injector has been set along the axis of the test section of
the wind-tunnel(Fig 1). It discharges downstream of the turbulence grid .
The average relative intensity of turbulence in the air stream is 4%.
Two series of measurements have been performed: first at an
external temperature of 4°C (refered to hereafter as "series 1") with
therefore a very weak rate of evaporation, then at 16°C (Series 2) which
enables a significant but low enough rate for the evolution to be
observed during the transit of the cloud of drops along the test section.
The first series may be considered as a reference datum

500

Measuring methods

        In order to determine the size distribution of the droplets and
its evolution during its migration and evaporation in the homogeneous
turbulent flow generated by the grid, it was preferable to use a non
intrusive optical method.
        Various such methods have been developed for this purpose from
L.D.A. techniques as described for instance by CHIGIER [5] and DURST [6].
The earlier ones make use of the visibility concept given by FARMER
[10,11,12] but they impose thet the drops diameters be of the same order
of magnitude as the interfringe. For larger diameters, YULE et al.[19]
have thought of using directly the pedestal information to estimate the
size distribution. It is then necessary to solve the so called trajectory
ambiguity problem due to the gaussian shape of the light intensity within
the laser beam. Two families of solution have been proposed: either the
light distribution is made uniform with the help of an electronic or
optical device as applied by ALLANO et al. [1] and EREAUT et al.[9] ,or
the probe volume is reduced, for instance by shortening the depth of focus
for on-axis observation or by getting out of axis, subsequently, the
residual ambiguity is removed by a

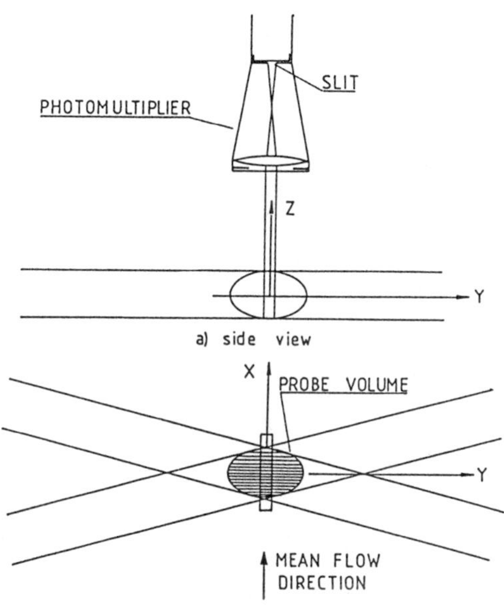

Figure 2 Optical set-up

suitable mathematical deconvolution usually based on the assumption that
the location of the particles in the measuring volume is equiprobable and
independent of their size as suggested by YULE et al [19],   HOLVE et al.
[15] and HOLVE [16].   BACHALO et al. [2] and BAUCKHAUGE [3] have proposed
a method based on phase difference which can be related to the diameter of
the particle when using two photomultipliers; the phase difference hence
measured is no longer sensitive to the refraction index variations and is
a linear function of the diameter. The technique used here belongs to the
"deconvolution" class of methods and has been developed by ECKEL [7],
ECKEL et al. [8] and GAIC. [13]. It has the advantage of making use of a
standard L.D.A. apparatus to which minor modifications have been
brought,ie: the photomultiplier has been mounted with a slit instead of a
pin-hole and the probe volume is observed at an angle of 90°.

The general lay-out of the optical system is given on fig 2,the laser beam (He-Ne 15mw of wave lenght 6328 A) which has a gaussian distribution of light intensity across its section is divided into two parallel beams that are focalised by a lens of focal length 600 mm to form the measuring volume. The superposition of the incident coherent beams is alternately constructive and destructive and produces a network of parallel interference fringes. The dimensions of this volume depend on the radius , the angle of the incident beams and on the wavelenght of the light. It is observed by a photomultiplier of which the optical axis is orthogonal to the plane of the incident beams . The pin-hole of the PM has been replaced by a slit so that only the central slice of the probe volume can be seen as shown on the figure. The overall response of the optical system was obtained by a direct calibration with glass balls as reported by ECKEL [7] and ECKEL et al. [8]. If I is the light intensity, Z a coordinate measured from the center of the volume and parallel to the direction of the mean flow as described on figure 5, D the diameter of the particle, K a constant depending on the geometry of the system and the refraction index of the particles and σ_z a length scale , it was found that for glass spherical particles, I , D, and Z were correlated by the relationship:

$$I(Z,D)=K.D^{1.73}.Exp(-0.4(Z/\sigma_z)^2) \qquad (1)$$

K, in the above equation can be computed so as to adapt the response of the system to the material of the particle. The computation, based on geometrical optics, has been made by GAIC [13] who obtained its value for dichlorotetrafluoroethane.
.   The deconvolution algorithm is based on the response curve I(D,Z) (relation 1) of the optical set up to the light diffused by spherical particles and has been described and discussed at length by ECKEL, ECKEL et al. and GAIC. However it will be briefly recalled here.
        As shown on figure 5, due to the replacement of its pin-hole by a slit, the photomultiplier only views the particles crossing the central slice of the measuring volume in the neighbourhood of the X,Z plane.The width of the slice beeing small as compared to the Y-scale length of the volume, it has been verified that the light intensity remains practically constant across the slit in the Y direction so that the ambiguity due to the Y location of the drop is eliminated. Moreover, since only the peak value of the pedestal voltage is retained, the effect of the X coordinate is also eliminated so that the sole remaining random coordinate is Z.
        Considering the case where the general relation I=f(D,Z) is not multiform in the domain of application,let p(I) be the probability density function of I and let p(D,Z) be the joint probality density of D and Z. The probability that I fall between two values I_0 and $I_0+\Delta I_0$ is given by/

$$P(I_0<I<I_0+\Delta I_0)=\iint_{I_0<f(D,Z)<I_0+\Delta I_0} p(D,Z)dD.dZ \qquad (2)$$

If the value I_1 corresponds to I_0 and I_2 corresponds to $I_0+\Delta I_0$, we have:

$$\int_{I_2}^{I_1} p(I)dI =\iint_{I_1<f(D,Z)<I_2} p(D,Z)dD.dZ$$

It is moreover reasonable to admit the the particle diameter D is statistically independant of the position Z where it crosses the equatorial plane and, furthermore, that this position is a uniformly distributed variable. We then have :

$$p(D,Z) = p(D).p(Z) = (1/l').p(D)$$

where l' is an equivalent length such that:

$$\int_{-\infty}^{+\infty} \frac{dZ}{l'} = 1 = \int_{-\infty}^{+\infty} p(Z).dZ$$

and $p(D)$ is the probability density of D.

It is also possible to represent $p(I)$ by a histogram of N classes with the classes boundaries chosen , for convenience, in decreasing values of I. The boundaries are denoted I_0, $I_1, I_2, \ldots I_j, \ldots I_N$, where I_0 is highest observed light intensity and I_N is the instrument resolution threshold.

Let n_i be the value of $p(I)$ for the ith class such that:

$$\int_{I_i}^{I_{i-1}} p(I)dI = n_i \Delta I_i$$

where: $\qquad\qquad I_{i-1} - I_i = \Delta I_i$

and: $\qquad\qquad \sum_{i=1}^{N} n_i \Delta I_i = 1$

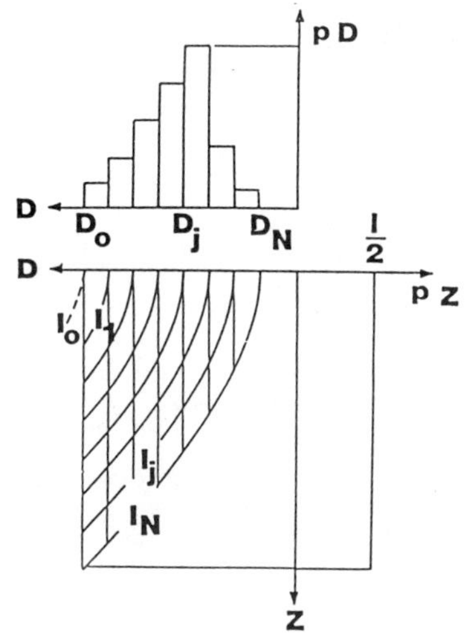

Figure 3 Deconvolution

The corresponding class boundaries of the diameter distribution can be determined from the calibration curves $I=f(Z,D)$ by taking I as a parameter. The jth boundary will be given by $I_j=f(0,D_j)$ since , if I_0 is the highest peak value observed, it can only occur when the largest particle of diameter D_0 passes near the center, or roughly Z=0. The first value I_1 may be due to particles of size D_1 passing through the center, or to larger particles passing farther away in the probe volume, and so forth. One can thus define a histogram of particles diameters such that:

$$\int_{D_j}^{D_{j-1}} p(D)dD = n'_j \Delta D_j$$

where: $\qquad D_{j-1} - D_j = \Delta D_j$

and: $\qquad \sum\limits_{j=1}^{N} n'_j \Delta I_j = 1$

Figure 3 shows the variation of D as a function of Z for several values of
I. as the curves are symmmetrical about the D axis, only the parts
corresponding to the positive values of Z are shown. The D histogram and
the uniform distribution of the Z's are displayed in regard. The integral
relation (2) can then be approximated in a discretized form giving ,for
instance:

$$P(I_5 < I < I_4) = n_5 \Delta I_5 = (1/2l')(a_{51}n'_1 + a_{52}n'_2 + a_{53}n'_3 + a_{54}n'_4 + a_{55}n'_5)$$

or, more generally:

$$n_i \Delta I_i = (1/2l')a_{ij}n'_j$$

n_i and n'_j are thus related by a system of linear equations of the form:

$$n_1 \Delta I_1 = (1/2l')(a_{11}n'_1)$$

$$n_2 \Delta I_2 = (1/2l')(a_{21}n'_1 + a_{22}n'_2)$$

$$n_3 \Delta I_3 = (1/2l')(a_{31}n'_1 + a_{32}n'_2 + a_{33}n'_3)$$

$$--- = -----------------------------------$$

$$n_N \Delta I_N = (1/2l')(a_{N1}n'_1 + a_{N2}n'_2 + a_{N3}n'_3 + ------ + a_{NN}n'_N)$$

The above expressions can be condensed in matrix form:

$$[n\Delta I] = (1/2l')[M][n']$$

Where $[n\Delta I]$ is a column matrix made up of elements calculated from
experimental data and $[M]$ is a triangular matrix of which the elements a_{ij}
are computed from the number of classes wanted and the calibration curve.
One obtains, therefore:

$$[n'] = 2l'[M]^{-1}[n\Delta I]$$

which enables to determine the drop size distribution when the
statistical distribution of the values of I is known.

The output voltage of the photomultiplier is fed in parallel to
two active filters with variable gain, band-pass and voltage offset
adjustments (Fig.4). Their purpose is to separate the low freqency
pedestal containing the size information (low pass filter) from the higher
freqency doppler signal (band-pass filter) which yields the velocity of
the particle . The two resulting voltages are then processed by an "ad
hoc" electronic device of which the operation is described by the synopsis
of figure 4. Basically, the pedestal peak value is measured by a "track
and hold" peak detector and the velocity is determined by counting the
time necessary for the particle to cross a predetermined number of
interference fringes. These two values are acquired and digitalised by a
two-channel GOULD Biomation 8000 when a validation pulse is delivered,
then transfered to a HP 217A desk computer for processing.

In order to avoid erratic trigerring by the noise, it has been
necessary to set threshold voltages both for the pedestal and doppler
sinals. They afford therefore two possibilities for triggering the

504

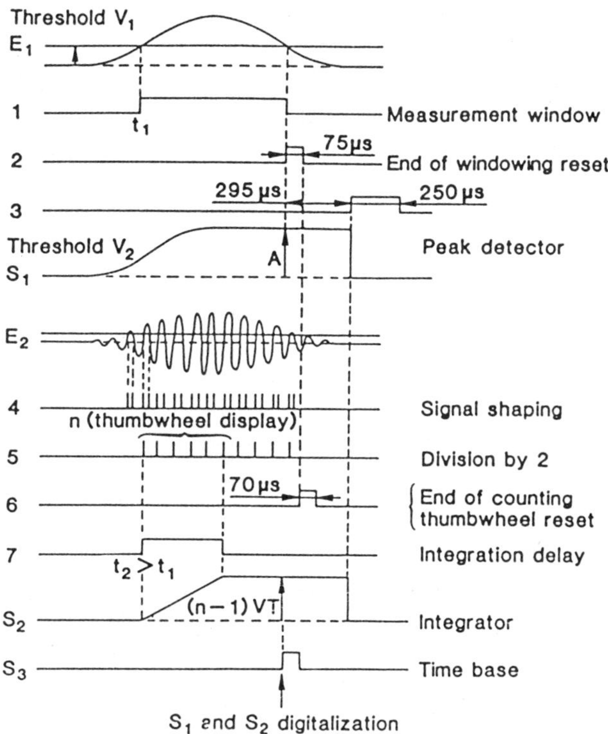

Figure 4 Signal conditioning

acquisition. When the first one is set higher, the small particles or
those passing on the edges of the measuring volume are eliminated, when it
is the doppler threshold which is higher, it the largest particles giving
the most powerfull doppler voltage that are retained, so that the overall
effect of both discriminators is the same. this seems to be an unavoidable
disadvantage of the method which can be minimised only by improving the
signal to noise ratio of the system. Besides, it very difficult to
estimate the magnitude of the error on the distribution of the sizes of
the particles and therefore on their mean values. However a direct
comparison of the results obtained by this method with water droplets and
a photographic observation has given a difference of less than 10% on the
values of the mean diameter

III-EXPERIMENTAL RESULTS

The above described method has enabled to observe the
evaporation process during the migration of volatile droplets in a grid
turbulence air flow.
Figure 5 shows the probability densities of the sizes of the
drops p(D) as a function of their diameter D. Y is the transverse
coordinate of the jet and X the longitudinal coordinate along its axis
,measured from the outlet section of the nozzle. For series 1, the rate of
evaporation is very low and the presence of small particles persists
throughout the jet, however, they seem to be in relative larger number
near the center and on the edge of the jet under the action of turbulent

diffusion since the smaller class of diameters falls in the neighbourhood
of the Kolmogorov scale.

When the external temperature is increased, the distributions
are strongly modified as shown by series 2 results, the small droplets
evaporate rapidly and hence, the distributions become flatter.

This becomes even more obvious if the above are represented by a
ROSIN-RAMMLER distribution [17] (Fig 6) as is common practice in spray
technology. This states that, for a given diameter D, the percentage
volume occupied by all the drops of which the diameter is less than D is
given by:

$$100(1-Exp(-(D/d)^m))$$

with the corresponding probability:

$$P(V)=(1-Exp(-(D/d)^m))$$

Where d is a scale diameter and m an exponent.

Although this model does not fit well the smaller diameters part
of the histograms, where, anyway, the experimental uncertainty is greater,
due to the noise, it gives an acceptable representation of the probability
of the cumulated volume P(V) as a function of the diameter of the drops.

For series 1, there is very little variation of the curves as
is generally observed when non- volatile liquids are used in the same
conditions (BURNAGE et al. [4] and GAIC et al. [14]).

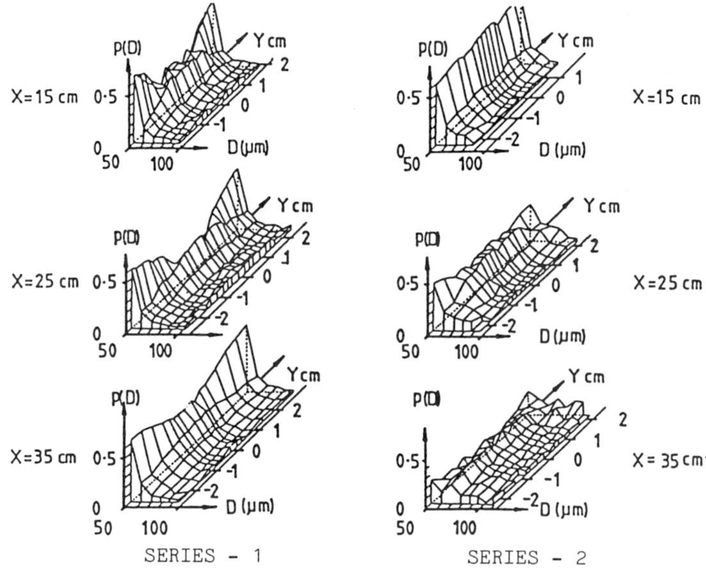

Figure 5 Particles size distributions

The situation is different for series 2 where evaporation
occurs. Then, for the most upstream section, the evolution of the ROSIN-
RAMMLER distribution shows a slight augmentation of the diameter scale d
when the point of observation moves towards the edge of the spray. Indeed,
one can observe that for a given value of the particle size D, the
cumulated volume probability P(V) diminishes. This corresponds to an
increase in the mean diameter of the distribution and is due to the fact

that the average time of migration for drops issued from the outlet
section of the nozzle is relatively larger if they have to move outwards.
The effect of evaporation is even more obvious when comparing
the above results to those obtained for the downstream station (X=35cm)
and one can estimate an overall increase of diameter of about 50%.

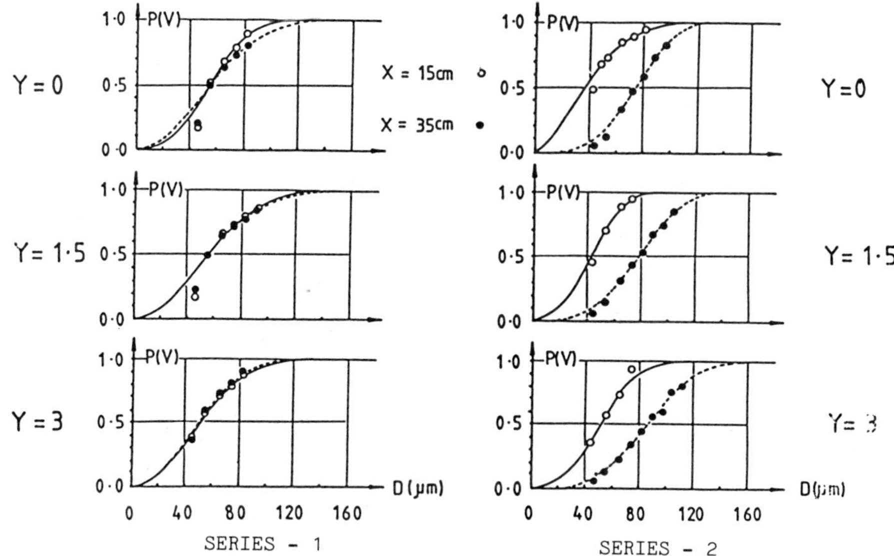

Figure 6 ROSIN-RAMMLER Distributions

IV- CONCLUSIONS

The evaporation of a spray in a turbulent air flow as been
investigated by means of an optical particle sizing method which had the
advantage is to be non intrusive and of making use of a standard L.D.A
set-up. However it operates in its optimal condition when concentration is
low enough so that the number of particles in the measuring volume is not
to high, if this is the case, too many samples are not validated which may
lead to prohibitive times for the acquisition of a sufficient number of
peak voltages to yield stable statistics On the other hand it is difficult to distinghish between the noise and the small pedestal voltage
values and consequently to asses the accuracy of the method. This is why
it is necessary to cross-check the results by comparing them to those
obtained by a different method. however it is believed that those presented here are significant.
Further improvements of the equipment are envisaged. The
electronic filters which separate the pedestal from the doppler signal
will have a sharper cut-off so as to increase the signal to noise ratio.
It is also possible to develop a quicker calibration technique by
analysing the response of the system to a known distribution of glass
balls instead of proceeding by separate investigation of its response to
diameter and to space location.
The application of this method to the present experiment has
shown that the size distributions of the droplets can be fairly well

represented by a ROSIN-RAMMLER distribution whatever the state of
evaporation. The parameters of the former law vary during the process in a
manner which is at least qualitatively consistent. Further experiments
will be carried out in order to attempt to link them to those on which the
flow conditions depend.

V-REFERENCES

1-ALLANO D., GOUESBET G., GREHAN G., LISIECKI D. Droplets sizing using a
          Top-Hot laser beam technique
          J.Phys. O: Appl.Physics 17 pp 43-58, (1984)
2-BACHALO W.W.,HOUSER M.J. Spray drop size and velocity measurements
          using the phase/doppler particle analyser
          Proc. of ICLASS-85 London (1985)
3-BAUCKHAGE H.Simultaneous measurement of velocity and size
          drops in free-flow fluid sprays
          Proc. of ICLASS-85 London (1985)
4-BURNAGE H., GAIC P., LOURME D., Size-velocity correlations measurements
          of large drops in a turbulent air flow.
          Proc. of ICLASS-85 London (1985)
5-CHIGIER N.A., Instrumentation techniques for studying heterogeneous
          combustion
          Proc. Energy Combustion Science 3, p.175 (1977)
6-DURST F.,Scattering phenomena and their application in optical
          anemometry
          Z.A.M.P. Vol.14,n°4, (1973)
7-ECKEL A., Contribution au développement de méthodes de mesure simultanée
          des dimensions, des vitesses et des concentrations de
          particules dans les écoulements diphasiques.
          Thèse de Docteur-Ingènieur,
          Université Louis Pasteur de Strasbourg (1983)
8-ECKEL A., BURNAGE H., LOURME D., Sur une méthode de mesure locale de la
          répartition statistique des diamètres de particules sphèriques
          diffusant dans un fluide.
          , La Recherche Aérospatiale, p.345, (1983)
9-EREAUT P.R., UNGUT A., YULE A.J., CHIGIER N.A., Measurement of drop size
          and velocity in vaporizing sprays.
          Proc of ICLASS-82 ,Madison Wisconsin (1982)
10-FARMER W.M., Measurement of particle size, number density and velocity
          using a laser interferometer.
          Appl. Optics Vol 11, p. 2603 (1972)
11-FARMER W.M., Observation of large particles with a laser
          interferometer.
          Appl. Optics Vol 13, p. 610 (1974)
12-FARMER W.M., Visibility of large spheres observed with a laser
          velocimeter: a simple model.
          Appl. Optics Vol 19, p. 3660 (1980)
13-GAIC P.Developpement et mise au point d'un système de mesures
          couplée des tailles et vitesses des particules au sein
          d'un écoulement diphasique disperse
          Thèse de Doctorat en Mécanique
          Université de Strasbourg I (1986)
14-GAIC P.,BURNAGE H., YOON S.J.,LOURME D. Distribution and mean
          concentration, dropsize and velocity, and size-velocity
          correlation in the spray of an air-assist nozzle.
          Int.J.Turbo and Jet Engines (to be published 1987)
15-HOLVE D., SELF S., Optical particle sizing for "in situ" measurements.
          Appl. Optics, Vol 18, N°10 (1979)

16-HOLVE D., "In situ" optical particle sizing technique.
        J. of Energy Vol 4, p.176 (1980)
17-ROSIN P., RAMMLER E. The lawws governing the fineness of powdered coal.
        J. of the Institute of Fuels. Vol.7 N°31 (1933)
18-YOON S.J., Contribution de la diffusion avec transfert de masse de
        gouttelettes dans un écoulement gazeux turbulent.
        Thèse de Doctorat en Mécanique
        Université de Strasbourg I,(1986)
19-YULE A.J., CHIGIER N., ATAKAN S.,UNGUT A. Particle sizing and velocity
        measurement by laser anemometry.
        J.Energy 1,220 (1977)

OPTICAL MEASUREMENT OF DROPLET EVAPORATION RATES

J. Timmler and P. Roth

Fachgebiet Verbrennung und Gasdynamik
Universitaet Duisburg
4100 Duisburg, W. Germany

INTRODUCTION

The evaporation of droplets suspended in a carrier gas is
an important physical process with many technical applications
such as spray combustion. Measuring evaporation rates of
liquid particles with different composition and physical
properties can show unknown details of the liquid material.
The gasdynamic heating of the gas/droplet mixture by a shock
wave can be used to initiate the process of droplet evaporation. Using laser light scattering it is possible to measure
the time dependent decrease in the droplet sizes. The
resulting scattered light fluxes measured under different
fixed angles can be interpreted by applying Mie-theory.

The purpose of this paper is to study the droplet evaporation rates of different liquid materials under well defined
conditions at temperature above 100°C. The process of mass
transfer between the liquid particles and the gas phase is
normally coupled with heat and momentum transfer and can
become very complicated, when particle group effects or
Knudsen number effects have to be taken into account[1]. In
order to neglect these effects, the particle volume fraction
of the initial gas/particle mixture has to be low and the mean
radius of the initial droplets have to be greater than 0.5μm.
These conditions and the requirement of a nearly monodispersed
aerosol were realized in this study by using a specially
constructed condensation-type aerosol generator.

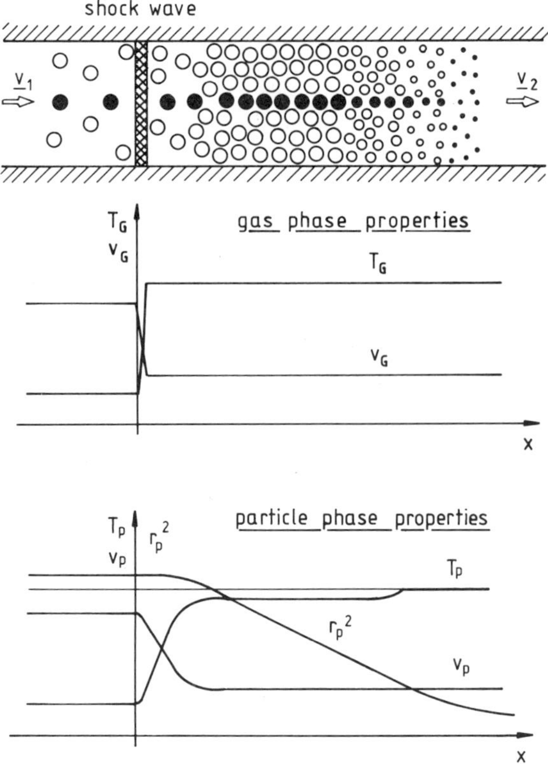

Fig.1.   Schematic of droplet evaporation behind a stationary
         shock wave.

A shock wave running through an aerosol disturbs the
quasi-equilibrium of the gas/particle mixture and initiates
the transfer processes. The situation of a stationary shock
wave in an aerosol is shown schematically in Fig.1. In our
experimental scale, the wave front can be assumed as a gas
phase discontinuity. Because of the very low volume fraction
of the suspended droplets, the gas properties are nearly
independent of the droplet processes. They can be assumed to
be constant behind the shock wave. The behaviour of the
droplet cloud going through the wave front is quite different
and more complex. The droplets are not directly influenced by
the gas phase discontinuity. Their velocity decreases because
of the momentum transfer with the carrier gas, the droplet
temperature rises because of heat transfer from the surrounding gas, and the droplet size decreases during the
subsequent evaporation process. In this model which describes
quite well our experimental situation, the particle behaviour
in the shock wave relaxation zone is determined by three

relaxing quantities, that is, the droplet velocity v_P, the droplet temperature T_P (assumed to be uniform for small Biot numbers), and the droplet radius r_P, all having different characteristic relaxation lengths. The experimental conditions are so chosen that the relaxation length (or time) for the evaporation process is much longer than for the other exchange processes. In this case, the measured time-dependent particle sizes behind shock waves can easily be interpreted in terms of evaporation rates.

THEORY OF DROPLET EVAPORATION

A general analytical description of shock wave inter-action in two-phase flow systems is very complicated. One way of approaching this problem is to model the ensemble of discrete particles as a pseudogas and to describe the fluid mechanical behaviour by the conservation equations of the two interacting continua. In our case, still a simpler approach can be used. Because of the very low particle volume concentration and the low particle number density, direct particle-particle interactions as well as the effects of the particle exchange processes on the gas phase properties can be neglected. Then the problem is reduced to one of a single droplet moving through a well-known gas continuum having different temperatures and velocities. For such a case, the analytical description of the exchange processes between the gas and the particles behind a shock wave is, for example, given by Roth and Fischer[1] and Reichelt et al.[2] They reduced the conservation equations of mass, momentum, and energy of a single droplet to a simple system of coupled differential equations. Assuming one-dimensional quasi-stationary conditions, computer simulations were done[2] using the well known empirical equations for the mass and heat fluxes and the drag force of a spherical droplet under continuum conditions. Their results of particle velocity and temperature show very short relaxation times. The particle size , represented by r_P^2, decreases linearly with time and can be described by the following equation[1,3]:

$$\frac{d\,r_P^2}{dt} = -\,D_{VG}\,p_V\,\frac{2\,M_V}{\mathfrak{s}_P\,R\,T_P} \tag{1}$$

where r_P is the droplet radius, D_{VG} the diffusion coefficient for vapour into gas, p_V the vapour pressure at the droplet surface temperature T_P, ρ_P the density of the liquid droplet, R the gas constant and M_V the molecular weight of the evaporating species. The evaporation rate dr_P^2/dt remains constant for isothermal conditions behind the shock wave.

The problem of droplet evaporation becomes more complicated if the particle ensemble consists of droplets of varying sizes. If the size distribution function is known, Eq.(1) can be applied to every radius group of the distribution function. It is well-known that aerosol particles generated by a condensation process have log-normal size distribution[4], described by the median particle radius $\bar{r}_P$ and the geometric standard deviation σ_g. An example of calculations done for an evaporating droplet ensemble having a log-normal size distribution is shown in Fig.2. At time t=0 the initial droplet size distribution function is characterized by a median particle radius $\bar{r}_P=1.57\,\mu m$ and a very small geometric standard deviation of $\sigma_g=1.13$. With increasing evaporation time the median droplet radius becomes smaller whereas the geometric standard deviation increases. At t=670μs the median particle radius and the geometric standard deviation have values of $\bar{r}_P=1.06$ and $\sigma_g=1.33$. When the smallest droplets of the ensemble are completely evaporated, the size distribution function of the droplets is no longer log-normal, as can be seen at time t=920μs. The geometric standard deviation σ_g is no longer defined.

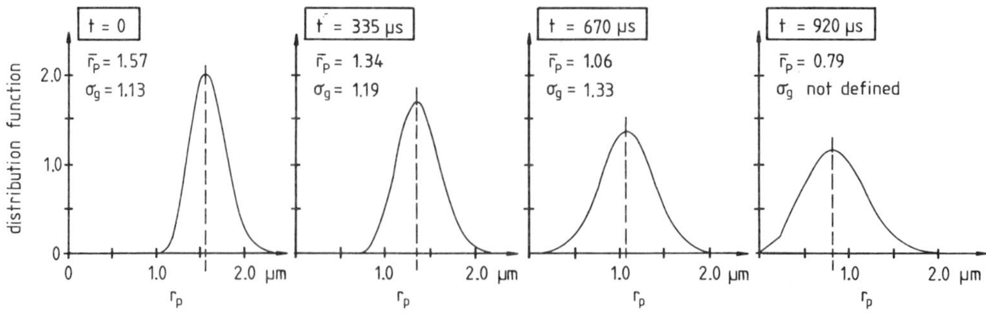

Fig.2. Calculated log-normal size distributions during evaporation of a DOP droplet ensemble at T=560K and p=4bar.

EXPERIMENTAL

The experiments on aerosol droplet evaporation were carried out in a conventional diaphragm-type shock tube with a low pressure section of 27 mm inner diameter and 3 m length. It is shown schematically in Fig.3. The initial pressure of the gas droplet mixture was measured by calibrated diaphragm-type pressure transducers. The shock velocity was determined by measuring the time interval between the pressure rise at the transducer and the change in scattered light signal at the optical measuring plane.

The particle size and the size distribution function are very important parameters in our evaporation experiments. Condensation aerosols are known to have very small log-normal quality size distribution[4]. Therefore the gas/droplet mixtures were generated by evaporation and subsequent condensation of liquids sprayed into an argon gas flow containing condensation nuclei. The liquids used in the present study are DOP($C_{24}H_{38}O_4$), silicon-oils PD5($C_{15}H_{32}O_3Si_4$), and PN200 ($C_{426}H_{758}O_{112}Si_{111}$). The initial particle size and the size distribution were measured by an optical particle counter (OPC). The geometric standard deviation of the particle size distribution was $\sigma_g \approx 1.15$, the particle number density about

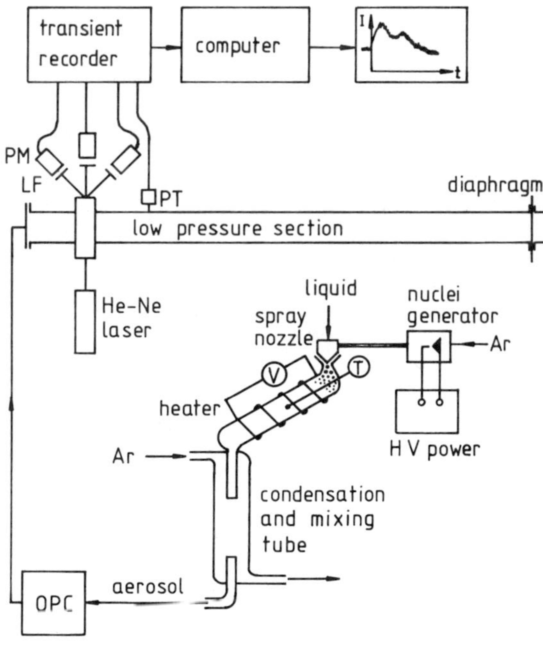

Fig.3. Shock tube setup with aerosol generator and particle diagnostics for aerosol studies.

10^6 cm^{-3}, and the mean initial particle radius between 0.6µm and 1.0µm.

Before each experiment, the aerosol was made to flow through the low pressure section of the shock tube for a while in order to get it homogeneously dispersed along the tube. The shock vave was started by rupturing the high pressure diaphragm. The time dependent change in the particle size during the droplet evaporation behind shock waves was measured by laser light scattering. The optical arrangement consisted of three light fiber probes, placed at fixed angles (Θ) of 20°, 30° and 45° relative to the outcoming laser beam having small opening angles of about 2°. The scattered light flux signals were led through laser line filters, transformed into electrical signals by photo multipliers and stored in a digital transient recorder. The recorded data were transfered to a microcomputer which plotted the scattered light intensities vs. time.

There is a possibility for the brake-up of the droplets into smaller parts caused by the strong momentum transfer in the shock wave. This process must be expected[5] at droplet Weber numbers (We)>12. For 12<We<25, the braking of the droplets has an insignificant effect on our measurements, because the larger droplets will split into a few equal sized parts resulting in a more monodisperse distribution function. When the droplet Weber number becomes very large, the initial droplets will burst into several submicron particles. This problem should be taken into account before carrying out the experiments and also in the interpretation of the measurements.

RESULTS

Typical results of directly measured scattered light intensities of two individual shock wave experiments are shown in Fig.4. In the upper part (4a), the light intensity signals at three different angles of DOP droplet evaporation are plotted against time. Fig.4(b) shows a similar measurement of the evaporation of PD5 droplets. The time scale is different for both experiments. Before the shock arrival, i.e. for time t<0, the scattered light signals are constant, characterizing the initial particle properties of the aerosol. The fast

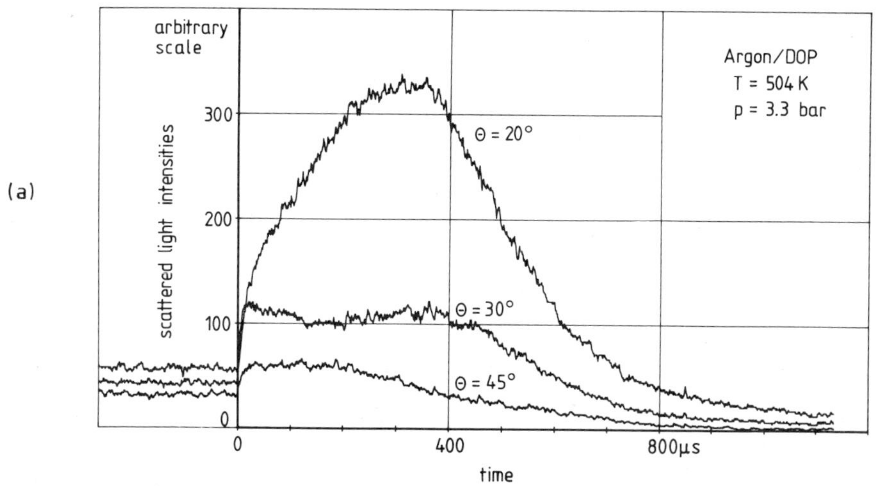

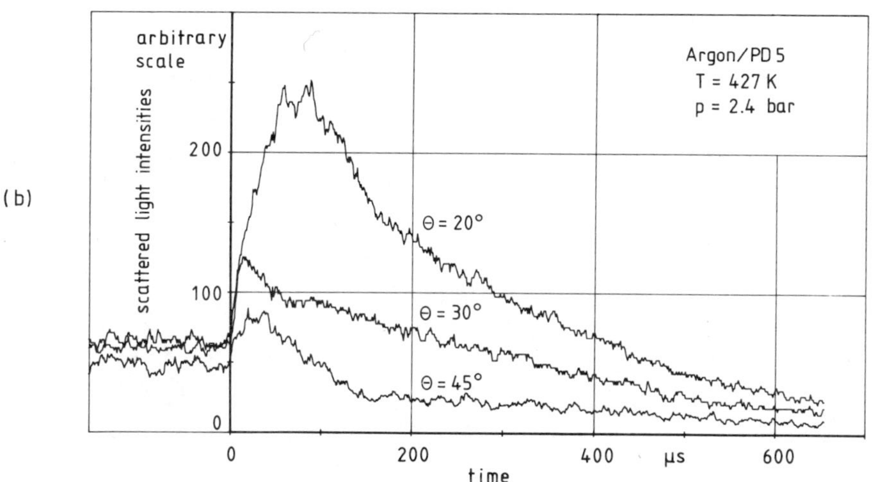

Fig.4. Examples of measured scattered light intensities behind incident shock waves.

signal increase at time t=0 after the shock arrival results mainly from the increase in particle number density during the particle velocity relaxation period. The coupled temperature increase causes a small change in the refractive index and the density of the particle material and thus increases the signal too. After this fast increase of the signals (roughly 30µsec), the relaxation of particle velocity and particle temperature are complete and the evaporation of the particles takes place, resulting in variations in the intensity signals for the remaining period of evaporation time.

The method of analysing the scattered light flux data is to compare them with intensities calculated from the well-known Mie-theory. The spherical particles are assumed to be homogeneously dispersed in the scattering volume and the size distribution of the particle ensemble is assumed to be log-normal. The Mie-intensity function has to be integrated over a log-normally ditributed particle ensemble resulting in an intensity which depends on the median particle radius $\bar{r}_p$ and the geometric standard diviation σ_g. In Fig.5 such a calculation is shown, where scattered laser light intensities for three different angles are plotted against $\bar{r}_p$. The value of σ_g is held to be constant in this case. The refractive index which also influences the Mie-intensity function depends mainly on the droplet temperature and is nearly constant for each experiment.

Both the measured signals (Fig.4) and the computed intensities (Fig.5) show maxima and minima. By comparing these typical maxima and minima, one can obtain the particle size distribution parameters $\bar{r}_p$ and σ_g at those times and the calibration factors relating the measured scattered light fluxes to the Mie-intensities. The calibration factors contain particle number density, optical response and geometries of the measuring facility. By using a set of computed intensities with varying $\bar{r}_p$ and σ_g at constant refractive indexes, one can determine the time dependent values of $\bar{r}_p$ and σ_g for the eva-porating droplets. Such a result is shown in Fig.6, where $\bar{r}_p^2$

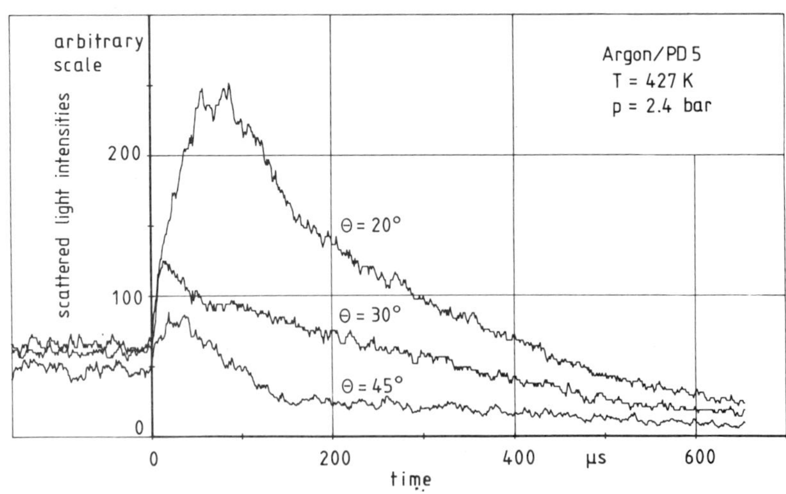

Fig.5.   Scattered laser light intensity calculated using the Mie-Theory assuming constant σ_g and refractive index $\hat{m}$.

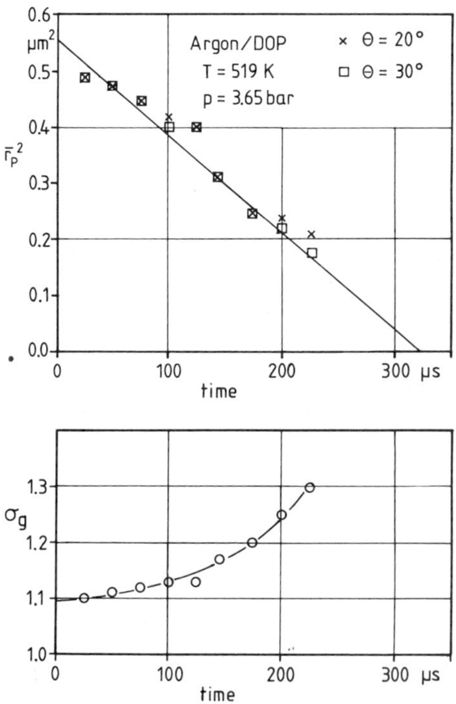

Fig.6. Evaluation of time dependent median particle sizes $\bar{r}_P^2$ and geometric standard diviations σ_g during DOP droplet evaporation.

and σ_g are plotted against time in the upper and lower parts, respectively. As can be seen, $\bar{r}_P^2$ decreases linearly with time, whereas the geometric standard deviation σ_g increases with time, as expected by the theory. At times $t > 200\mu s$ the values of $\bar{r}_P^2$ determined from the intensity signals measured at $\Theta = 20^o$ and $\Theta = 30^o$ differ more and more by using a single σ_g value. This indicates that the size distribution of the evaporating droplet ensemble is no longer log-normal.

Typical evaporation behaviours of Ar/DOP, Ar/PD5 and Ar/PN200 mixtures are shown in Fig.7. For the DOP droplets the evaporation rate $(d\bar{r}_P^2/dt)$ is nearly constant except at longer evaporation times. A completely different evaporation behaviour was observed by PD5 droplets. They show a fast evaporation rate upto about $130\mu s$ followed by a much slower rate, which again remains constant. Another time behaviour of evaporation was found for PN200 droplets. They first shows a fast linear decrease of $\bar{r}_P^2$ which deviates at longer evaporation times.

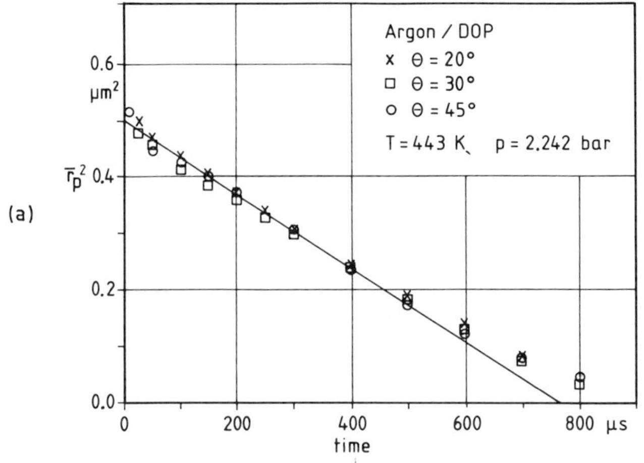

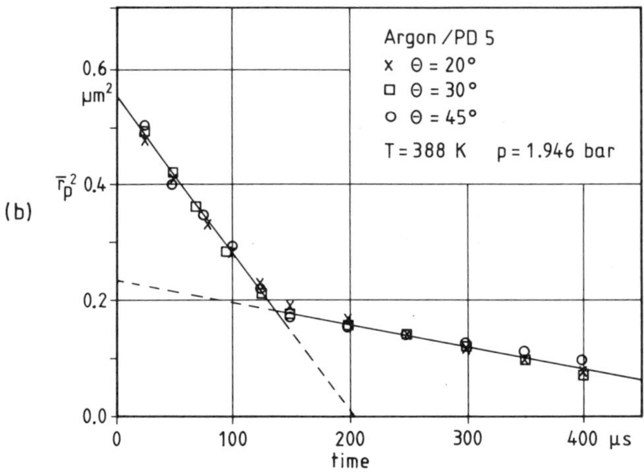

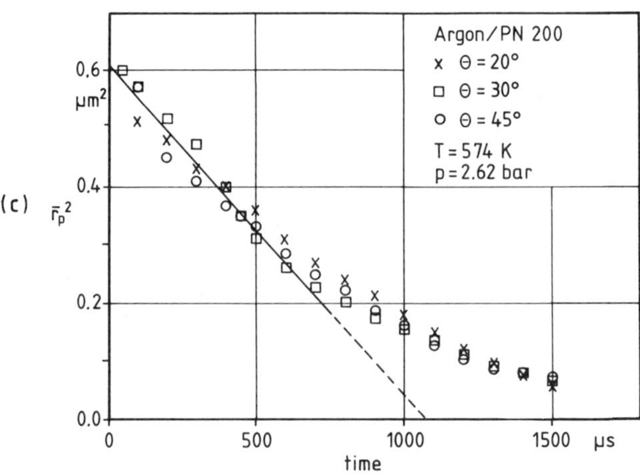

Fig. 7. Examples of experimentally determined droplet evaporation rates of DOP, PD5, and PN200.

DISCUSSION

An understanding of the experimentally determined droplet evaporation behaviour is possible by comparing it with the evaporation model. According to Eq.(1), the evaporation rate $(d\bar{r}_p^2/dt)$ is constant for our experimental conditions, if the droplets contain only one chemical compound and evaporate under isothermal conditions. It is obvious that the measurements of the DOP droplets show the same behaviour. The time dependent decrease can be represented by a straight line, as given in the upper parts of Figs.6 and 7. From the slope of this line, the product of the vapour pressure p_V and the diffusion coefficient D_{VG} can be determined, if the other properties are known. For low vapour concentrations, the diffusion coefficient D_{VG} of the droplet vapour into a monatomic gas can be described by the kinetic theory[6], if the Lennard-Jones parameters are known. We computed the diffusion coefficient D_{VG} for DOP using the collision diameter $(\sigma_V=10.5\text{Å})$ and the collision integral $(\Omega_V^{(1,1)}\approx1.2)$, given by Davis and Ray[3]. Based on this value of D_{VG}, the vapour pressure of DOP could be determined from the present evaporation rate experiments. Results are shown in Fig.8(a). The points obtained are close to the given solid line, which is the mean value of the DOP vapour pressures given by the manufacturer[7].

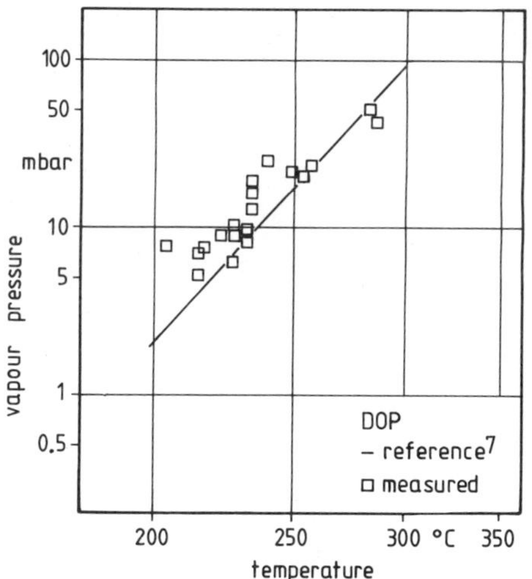
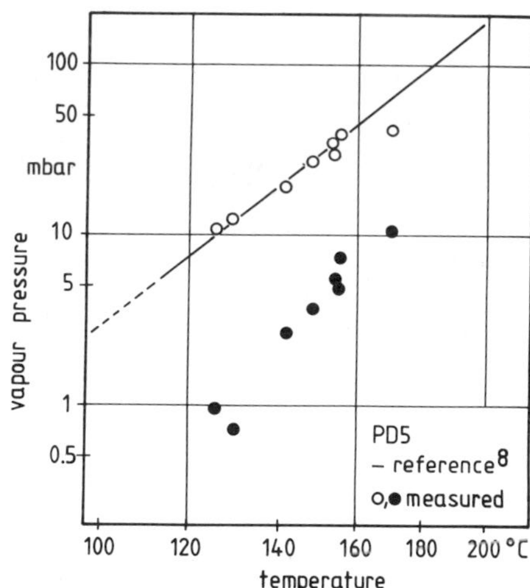

Fig.8. Experimentally determined vapour pressures of DOP and PD5, using estimated diffusion coefficients.

The evaporation of the silicon-oil PD5 droplets shown in Fig.7(b) can be approximated by two straight lines having different slopes indicating the presence of two compounds in the droplets. The two slopes of the evaporation rate are related to different vapour pressures and diffusion coefficients. Assuming $\sigma_V = 10\text{Å}$ and $\Omega_V^{(1,1)} \approx 1.2$ for PD5 droplets, a single diffusion coefficient for the particular temperature of the experiment is obtained. Based on this value of diffusion coefficient, two different vapour pressures (Fig.8(b)) can be determined from the two measured slopes of $\bar{r}_P^{\,2}$. The higher vapour pressure values agree well with the data given by the manufacturer[8]. The lower values of vapour pressures should correspond to a second component of PD5, which has not been measured earlier.

The silicon-oil PN200 seem to consist of several different compounds. The highest slope of evaporation rate at the beginning as seen in Fig.7(c), should correspond to the highly volatile component. By further evaluation of such experiments with varying evaporation rates it seems possible to estimate the volume fraction of the components.

REFERENCES

1. P.Roth and R.Fischer,
   Phys. Fluids, 28: 1665 (1985).

2. B.Beichelt, P.Roth, and L.Wang,
   Wärme und Stoffübertragung, 19: 101 (1985).

3. E.J.Davis and A.K.Ray,
   J. Chem. Phys., 67: 414 (1979).

4. O.G.Raabe,
   J. Aerosol Sci., 11: 289 (1971).

5. W.G.Reinecke and G.D.Waldman,
   "5th Int. Conference on Erosion by Liquid and Solid Impact",
   Newnham College, Cambridge (1979).

6. J.O.Hirschfelder, C.F.Curtis, and R.B.Bird,
   "Molecular Theory of Gases and Liquids", New York (1954).

7. Chemische Werke Hüls AG,
   "Weichmacher in der Kunststoffindustrie", W.Germany (1976).

8. Bayer AG, "Baysilone-Öle P.",
   Bayer-Leverkusen AG, W.Germany (1968).

STUDY OF CONDENSATIONAL GROWTH OF WATER DROPLETS BY

OPTICAL MIE SCATTERING SPECTROSCOPY

K.H.Fung and I.N.Tang

Environmental Chemistry Division
Department of Applied Science
Brookhaven National Laboratory
Upton, New York 11973

SUMMARY

Optical Mie scattering technique is applied in the study of
the growth of a single saline solution droplet. A quadrupole
particle chamber is used to suspend a charged sodium chloride
solution droplet in a water vapor environment. The droplet-vapor
equilibrium is momentarily shifted by the heating effect of an
infrared laser and the subsequent condensational growth of this
droplet is monitored by means of right angle Mie scattering.
Simultaneous mass and heat transfer processes are considered, and
the mass accommodation coefficient for water is deduced.

INTRODUCTION

Liquid droplet is a pertinent form of water present in the
atmosphere. The formation of such droplets is initiated by a
heterogeneous nucleation center which can be any airborne
particles such as sea salts, combustion exhaust or other air
pollutants. When the critical criteria are met, a phase
transition from dry particle to a sizable water droplet will
occur.[1] A collection of these water droplets can lead to clouds
and fog formation. In addition to the meteorological effects of
these droplets, they are also carriers for the airborne
pollutants such as NO_x and SO_x. Hence a sound knowledge of the
transformation of water droplets is vital to the modeling of
dispersion and precipitation of air pollutants.

Mie scattering is very suitable for studying and measuring
the size and the index of refraction of submicron to micron size
spherical droplets.[2,3] The presence of Mie resonance scattering
peaks makes extremely accurate optical measurement possible. Each
Mie resonance peak has its own unique characteristic profile
which is determined by the scattering angle, refractive index and
the droplet radius. Chylek *et al.*[4] have demonstrated that the
slope of a Mie resonance profile can be used to determine the
physical properties of optically levitated liquid droplets. Using
this method, the refractive index and the diameter of a silicone

oil droplet have been deduced with relative errors of 5×10^{-5}. The degree of accuracy depends on a precise knowledge of the wavelength of the incident radiation, since the Mie scattering intensity is a function of the size parameter $x = 2\pi r/\lambda$, where r is the radius of the droplet and λ is the wavelength of the incident radiation. A cw narrow bandwidth dye laser which has a typical linewidth of 0.01 A (1 GHz) would imply an ultimate resolution of $\Delta\lambda/\lambda = 2 \times 10^{-6}$.

In this report, we are presenting an optical technique using Mie scattering to study the condensational growth of a saline solution droplet. The size uncertainty in this method is $\approx 0.5\%$. The results show that the condensational coefficient α_M of water is found to be unity.

EXPERIMENTAL

The experimental setup is shown schematically in Fig. 1. The particle levitation cell[5] (Front View) consists of two endcap electrodes formed from stainless steel mesh and a central stainless steel ring electrode. This particle containment trap is mounted and centered in a stainless steel six-way cross. Salt particles are introduced at the top port of the chamber. After a particle is caught, this port is closed with a vacuum flange equipped with a capacitance manometer to monitor the chamber

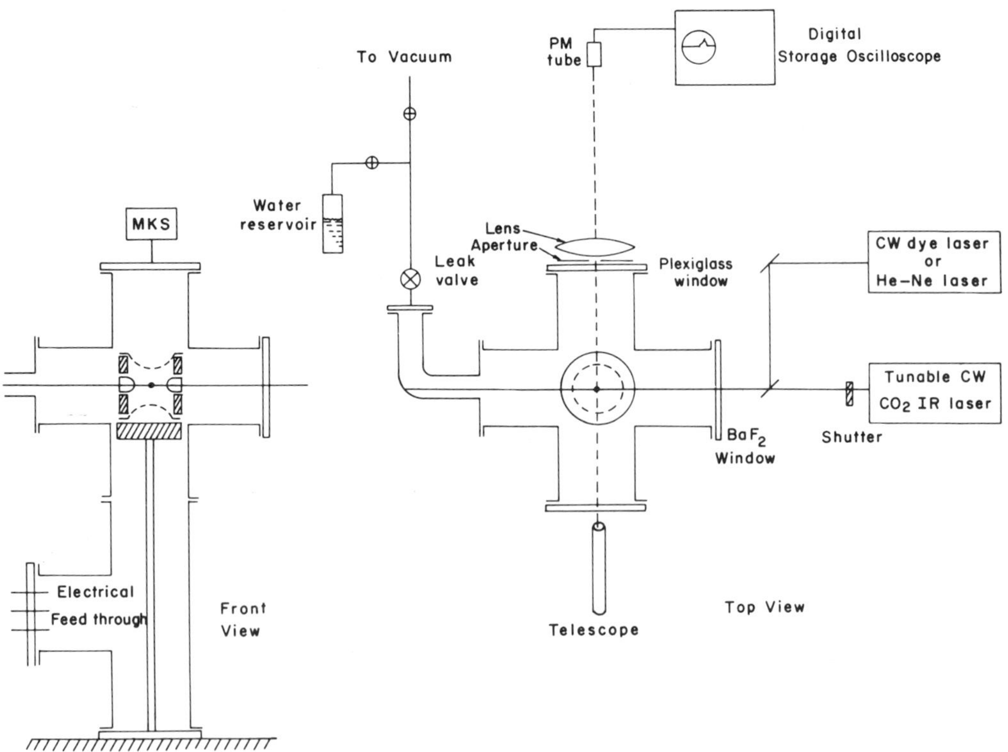

Figure 1: Schematic diagram of the experimental setup.

pressure. The size of the trapped particle can be measured by the
"spring-point voltage" technique.[6]

The infrared heating laser and the probe laser are aligned
to overlap at a germanium beam splitter and enter the chamber
through a BaF_2 window. The probe laser beam is perpendicularly
polarized with respect to the scattering plane. The front port is
fitted with a sight-glass flange to view and align the suspended
particle with a telescope. The back port is covered by a 1/2 inch
thick plexiglass flange. The scattered light is collected by a
40-mm asymmetric plano-convex lens and is focussed onto a
photomultiplier. A rectangular spatial filter in front of the
lens gives the total acceptance angle of 0.05 degrees for the
collecting optics. The photomultiplier signal is directly
processed and displayed on a digital storage oscilloscope.

After the size of the dry salt particle is determined, the
chamber is evacuated to a base pressure of 10^{-4} to 10^{-5} torr
through a precision leak valve. The vacuum system is then
isolated from the chamber. Water vapor is admitted slowly to the
suspended particle until it deliqueces. The size of the saline
solution droplet can be controlled by lowering or raising the
water vapor pressure, or the temperature of the system. The
levitation chamber is kept at room temperature and is measured by
means of a calibrated chromel-alumel thermocouple. No temperature
stabilization is necessary because the total time duration of
each measurement is in the order of a few seconds.

RESULTS

The equilibrium of a suspended single solution droplet can
be perturbed by the heating effects of an infrared laser.[3] When
this perturbation is removed, the droplet will return to its
equilibrium size by means of condensational growth. Mie
scattering can be used to monitor the droplet size during the
entire growth process. In Fig. 2, the waveform of such process is
shown. In order to model this physical phenomenon, both the heat
and mass tranport equations need to be considered, which are the
following:

$$\frac{dW}{dt} = 4\pi\, r^2\, \alpha_M \left(\frac{M_w}{2\pi\, RT}\right)^{1/2} (P_\infty - P) \, , \qquad (1)$$

and

$$C_T W \frac{dT}{dt} = L\frac{dW}{dt} - 4\pi\, r\, k\, \beta_T (T - T_\infty) \, , \qquad (2)$$

where α_M is the condensation coefficient, r is the droplet
radius, W is the droplet mass, M_W is the molecular weight of
water, R is the gas law constant, P_∞ is the system pressure at
temperature T_∞ as measured, P is the droplet vapor pressure at
the interface temperature T, C_T is the heat capacity of the
droplet, L is the latent heat of water , k is the thermal
conductivity of water vapor and β_T is a correction factor as
given by Fukuta and Walter.[7] When the equilibrium vapor pressure
P is linearized using a Taylor series expansion of W and T, the
coupled equations (1) and (2) can be rewritten in the vector
form,

$$\frac{d}{dt}\binom{W}{T} = \begin{bmatrix} B & D \\ F & G \end{bmatrix} \binom{W}{T} + \binom{A}{E} \ , \qquad\qquad (3)$$

and the analytical solution to Equation (3) is given by,

$$\binom{W}{T} = C_1 e^{\lambda_1 t} \binom{X_{11}}{X_{12}} + C_2 e^{\lambda_2 t} \binom{X_{21}}{X_{22}} + \binom{W_p}{T_p} \qquad (4)$$

This analytical expression gives the growth trajectory in a local growth region about which the Taylor series expansion is valid. Using the Equation (4), the growth waveform can be therefore calculated and compared with the experimental result.

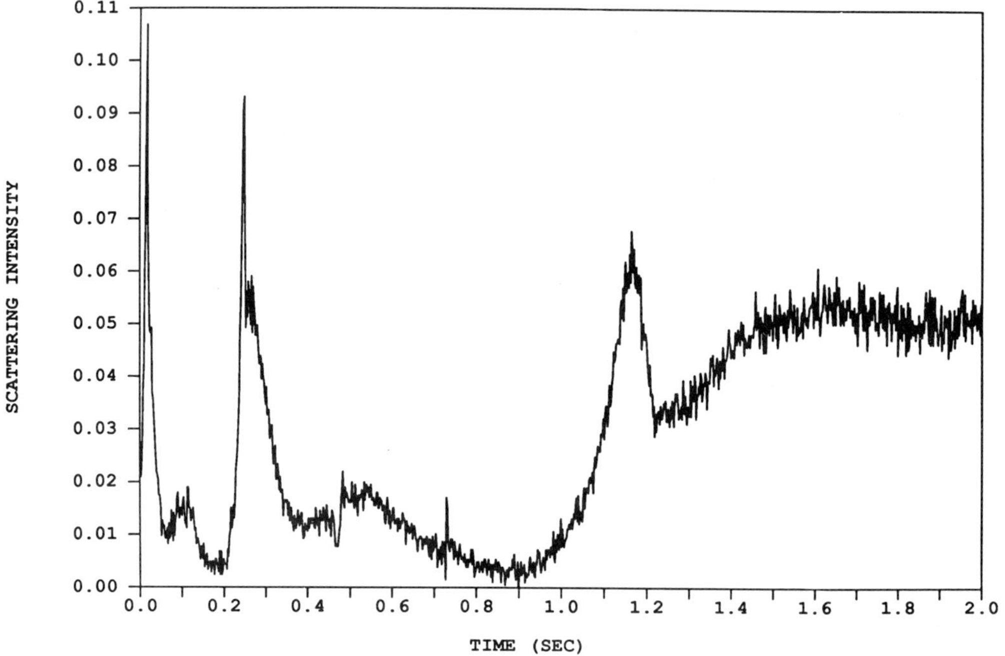

Figure 2: Experimental growth curve for a saline solution droplet. the radius of the dry salt is 4.76 μm, the water vapor pressure is 14.94 torr, and the temperature is 24.8 C.

In Fig. 3, a simulation is made based on the experimental conditions. The condensation coefficient α_M of water used in this calculation is 1.0.

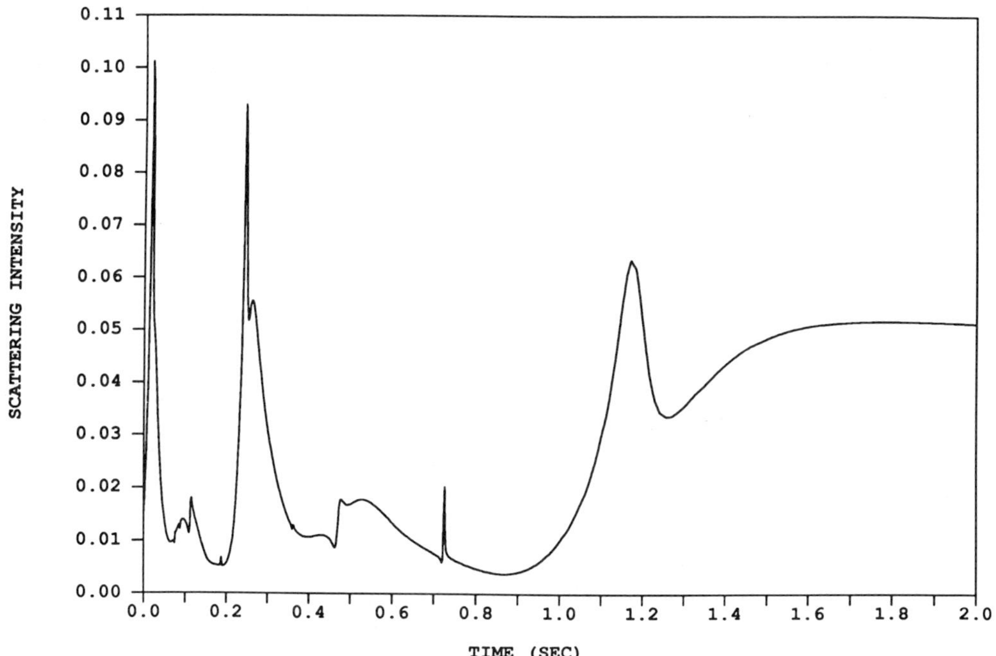

Figure 3: Calculated growth curve for a saline solution droplet which has a dry salt radius of 4.76 μm. The accommodation coefficient α_M of water used is 1.0.

ACKNOWLEDGMENT

This work was supported under the auspices of the United States Department of Energy under Contract No. DE-AC02-76CH00016.

REFERENCES

1. I. N. Tang, H. R. Munkelwitz, and J. G. Davis, Aerosol Growth Studies-IV. Phase Transformation of Mixed Salt Aerosols in a Moist Atmosphere, J. Aerosol Sci. 9:505 (1978).
2. M. Kerker, "The Scattering of Light and other Electromagnetic Radiation," Academic Press, New York, (1969).
3. C. B. Richardson, H.-B. Lin, R. McGraw, and I. N. Tang, Growth Rate Measurements for Single Suspended Droplets Using the Optical Resonance Method, Aerosol Sci. Tech. 5:103 (1986).
4. P. Chylek, V. Ramaswamy, A. Ashkin, and J. M. Dziedzic, Simultaneous Determination of refractive Index and Size of Spherical Dielectric Particles from Light Scattering Data, Appl. Opt. 22:2302 (1983).
5. R. F. Wuerker, H. Shelton, and R. V. Langmuir, Electrodynamic Containment of Charged Particles, J. Appl. Phys. 30:342 (1959).
6. R. H. Frickel, R. H. Shaffer, and J. B. Stamatoff, Chamber for the Electrodynamic Containment of Charged Aerosol Particles,

AD/A 056 236, National Technical Information Service, U.S.
Department of Commerce, Springfield, VA 22161 (1978).

7. N. Fukuta, and L. A. Walter, Kinetics of Hydrometeor Growth
from a Vapor-Spherical Model, _J. Atmos.Sci_. 27:1160 (1970).

OPTICAL SIZING OF COMETARY DUST :

TENTATIVE LESSONS FROM COMET HALLEY DATA

Jean François Crifo

Laboratoire de Physique Stellaire et Planétaire
B.P. 10 - F 91371 Verrières le Buisson - France

INTRODUCTION

The purpose of this work is to try to evaluate the level of
confidence that can be placed in the present techniques available for
sizing the cometary dust (and, at the same time, deriving all its other
characteristics). We use for this the new possibilities opened by the
recent in-situ sampling of Comet Halley dust. We look whether data from
this origin are compatible with visible and I.R. spectra of this comet
acquired from the Earth, and we also examine whether they are compatible
with the set of observations acquired on another comet considered
similar to Halley, i.e. Comet Kohoutek "1973f". We find that, as happens
frequently when remote sensing and local sampling techniques data can be
compared, none of these techniques can claim to be fully satisfactory.
In the case of cometary dust, current estimates based on optical sizing
only may be much less accurate than precedingly assumed. In any case,
improvements in the observational approaches appear needed, as well as
dedicated laboratory investigations and theoretical developments.

This text is organized as follows : an initial section introduces
the non-specialized reader with the objectives of cometary dust sizing ;
after it, the experimental data selected here for analysis are presented ; the principles of their analysis are then given, followed by the
results and by a discussion.

THE PHYSICAL SIGNIFICANCE OF COMETARY DUST SIZING

Dust particle sizing is at the very heart of cometary physics, even
though the motivations underlying this investigation have evolved
following improvements in our understanding of comets. The most
spectacular characteristics of comets, which gave rise to their
denomination, is the display of one or several bright, extended, and
roughly antisolar tails. Their dimensions can reach 10^7 to 10^8 kilometers. They originate from a bright, diffuse, and roughly spherical
region, the "head" or "coma", with typical dimension 10^5 km.

Early in the XIXth Century Bessel initiated celestial mechanics studies of cometary tails. He postulated that tails consist in "matter" subjected to an "antisolar force" varying with distance to the Sun according to an inverse square law. Its work was pursued mostly at Russian Observatories for about a century. This school, and in particular Bredikhin, established that there are different types of tails characteristized by the strength of the repulsive force. We retain from their work the labels "Type I" for tails which are almost straight and exactly antisolar revealing repulsive forces much stronger than the solar attraction, and "Type II" for tails which are curved and submitted to forces smaller than the solar attraction. Spectroscopy of the comets initiated by Donati in 1864 quickly revealed that type I tails were mostly molecular ions, while type II tails reflected the solar spectrum as large grains or gravels. In 1902 Lebedew evidenced radiation pressure at the laboratory and immediately proposed it as the source of the repulsive forces. But this idea was not really fully accepted until the years 1950 when Biermann predicted the Solar Wind as a the source of the strong repulsive force in type I tails, leaving the radiation pressure for type II tails. At the same epoch F. Whipple proposed a model now widely accepted for the "operation" of comets (see Whipple, 1976) ; his suggestion that cometary activity is controlled by the sublimation of dusty ice seems to have been confirmed, in particular, by the recent discovery that water vapour represents about 90% of the gas emitted by Comet Halley (Krankowsky et al., 1986 ; Mumma et al., 1986). The first quantitative model of the formation of a type II (dust) tail by sublimation of ice, acceleration of the dust by the vapour, and deflexion of this dust by radiation pressure was developed by Finson and Probstein (1968) and used to analyze the brightness contours of the tail of Comet Arend Roland "1956 h". This method is not, strictly speaking, an optical sizing : it is based on the fact that different grains assume different trajectories in an external field of force. Since the particle acceleration is proportional to its area and inversely proportional to its mass, the sizing is done in terms of the product $a\rho$ of the grain radius a by the grain density ρ. The size interval accessible from their analysis was

$$3.10^{-4} \leqslant a\ \rho\ (gcm^{-2}) \leqslant 10^{-2} \tag{1}$$

and a steep differential size distribution, proportional to $(a\rho)^{-3}$ to $(a\rho)^{-5}$ was found over this interval. The ratio of the rates of ejection of dust and of gas was deduced to be ~ 1.2 (by mass) —to an uncertainly of x 8 or $\div$ 5. The authors warned carefully that this was possibly only part of the total dust output ; in fact the comet revealed transiently a sunward oriented abnormal tail formed of grains much larger than those contained in the preceding interval.

The main importance of cometary dust is thus that it carries a large fraction, perhaps a dominant fraction, of the comet mass loss. Presumably, it carries away most of the heavy elements from the comet. It is believed that due to its small size the interior of a comet "nucleus" is not subject to high pressures and temperatures as planetary interiors. Therefore the remote past of the minerals has probably not been obliterated there, as in the planets. Comets may be "Rosetta stones" of the Solar System. As regards the observable grain size distribution it is not known whether it is genuine or whether it results from fragmentation during the nucleus sublimation. Theoretical models of the formation of the Solar System are not sophisticated enough at this time to predict the size distribution of dust formed in the process of condensation of the pre-solar nebula. But when progress is made on these points the size distribution will perhaps acquire a high significance.

There are less fundamental, but nonetheless important reasons to investigate the properties of cometary dust, including their size spectrum :

1. Comets can contribute a fraction or eventually all of the tenuous cloud of "zodiacal" particles which fill interplanetary space. It is observed that streams of large grains travel in bunches of orbits in the vicinity of a certain number of periodic comets, including comet Halley. When the Earth crosses the plane of these orbits it is submitted to a meteor shower. Thus it is possible that grains of cometary origin contribute a fraction of the tiny extraterrestrial grains that are routinely collected in the stratosphere with high altitude aircrafts.

2. The grains play a role in the formation of the cometary atmosphere (the comae) via energy and perhaps mass exchanges with the gas.

3. Our knowledge of cosmic dust (interstellar, circumstellar, extragalatic) comes from interpretation of observations similar to those that we will consider here, except that they are in general less favorable (the dust is distributed in an apriori unknown way on the line of sight, and the radiation field to which it is submitted is poorly known). This visit of comet Halley has provided the first opportunity to compare deductions based on remote sensing of the dust with in situ sampling of this dust. The conclusions one can draw from this comparison are of great significance to the credibility of all conclusions presently accepted regarding all other populations of extraterrestrial dust.

OPTICAL EMISSIONS FROM COMETS HALLEY AND KOHOUTEK

The observations to be considered here do not concern type II tails, but the comae of these comets. This is the region, centered on the nucleus where the effect of radiation pressure is not yet evident, and where dust grains outflow radially from the "nucleus". One can represent the nucleus by a point source of dust, since its dimensions (a few kilometers across) are much smaller than typical coma sizes (10^4 to 10^5 kilometers). Information on the dust in the coma comes from spectroscopy and polarimetry of its emission continuum. Data are available from 0.2 μm to 150 μm in wavelengths, in the most favorable cases, but in general the instrumental resolution is low ($\lambda/\Delta\lambda \sim 5$ to 50). This wavelength interval is of course not covered by a single instrument but only by generally non-overlapping discrete measurements. Intercalibration problems exist. Besides, corrections are needed for terrestrial absorption and emission. Finally, superimposed on the dust continuum are the cometary molecular emission, which are dominant in many regions of the visible spectrum.

At visible wavelengths, the dust continuum is due to scattering of solar light (the sun radiates approximately as a 5770K diluted blackbody). At long wavelengths, the emission is due to thermal radiation from the grains. The shape of the spectrum changes with comet position, due to changes in the heliocentric distance of the comet and to fluctuations in the properties of the emission. Here, we will neglect the latter.

Figure 1 shows a set of spectra of comet Kohoutek obtained by Ney (1974). The similarity with comet Halley is evident from Figure 6 where the measured spectrum is a composite of observations by Tokunaga et al. (1986), Campins et al. (1986), Herter et al. (1986) and Glaccum et al.

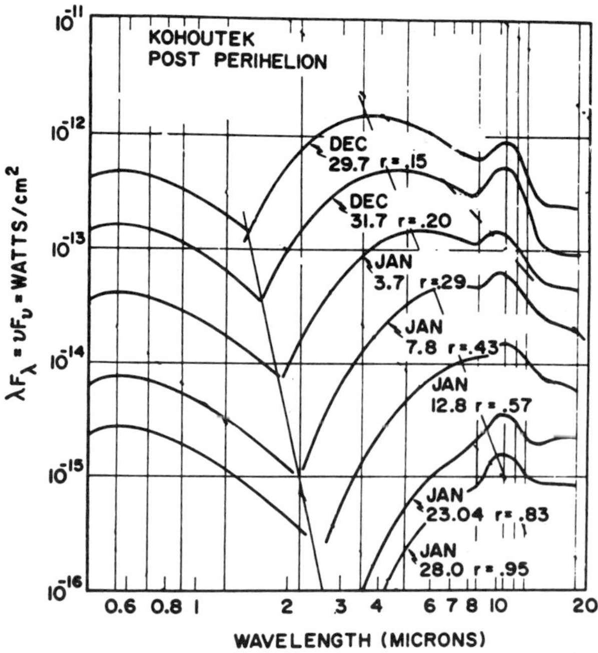

Figure 1. Comet Kohoutek spectra r is the heliocentric distance in A.U. Measurements were made by Ney (1974).

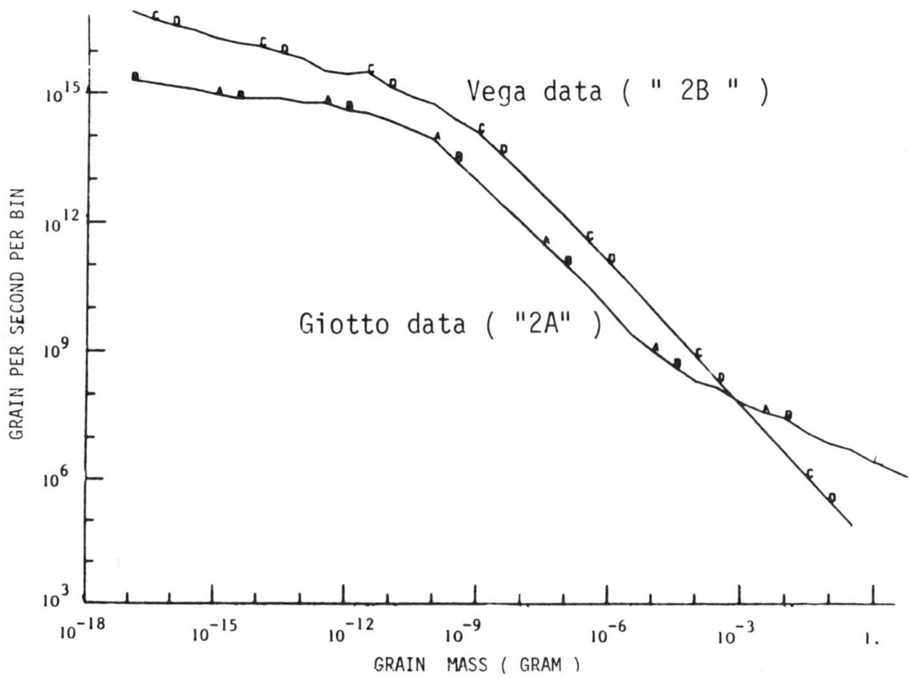

Figure 2. Total dust production rate deduced from the in-situ data. The method is described in Crifo (1987).1 bin = $[m, m\sqrt{10}]$/

(1986). One notices spectral features near $\lambda = 10$ µm and $\lambda = 20$ µm which are currently attributed to silicates although precise attribution to a precise mineral has not been achieved. The features recall emission bands from amorphous minerals rather than cristalline phases. They are sometimes absent from the spectrum. Other features have been discovered recently, in particular near $\lambda = 3$ µm, which will not be discussed here, as their origin is still a matter of speculations. Polarimetric observations are available on the comets considered here, but will not be discussed here. Conclusions of this work will be compared to the polarimetric data in the future.

Prior to the recent return of Comet Halley, fits to parts of the cometary spectra such as those of Figure 1 were made using single size grains or a size distribution inspired from Finson and Probstein results and assuming either absorbing material or silicate material or both (see an excellent review by Hanner, 1980). A fit on the complete spectral range of Kohoutek data was made by Crifo (1982) revealing some difficulties but no global disagreement. Thus the warnings by Finson and Probstein concerning the limited character of their results were superseded by the erection of their results in almost a physical law for cometary dust.

IN SITU SAMPLING OF COMET HALLEY DUST

Three space probes swept through Comet Halley dust coma during March 1986 (for a comprehensive account of the results see the proceedings of the 20th ESLAB Symposium on the Exploration of Halley's Comet, ESA report SP-250). They carried a variety of dust impact counters and dust elemental composition analyzers. Here, we will make use of the preliminary summary data presented by Mc Donnell et al. (1986) for the "Giotto" probe and by Mazets et al. (1986) for the "Vega 1" and "Vega 2" probes. They consist in "fluences", i.e., plots of the number of impacts per unit cross section of detector, F(a), versus grain mass (the grain size is not a priori known).

From the fluences one can deduce rates of emission of grains from the comet nucleus (see next sections). The result is shown on Figure 2. Both ends of both curves are extrapolations. In particular, the largest mass at which the curves must be terminated is unknown. The mass spectra derived from "Giotto" data and from "Vega" data do not agree with each other. This is probably not an instrumental problem, but an indication that the distribution of dust in the coma is not given by simple models. One notes that the range of sizes involved is about 6 decades, as compared to less than 2 decades precedingly inferred from dust tail analysis. There is a large population of particles much smaller than the lower limit in (1) -presumably extending below the lower limit of the in-situ dust detectors. On the other hand, most of the mass in the Giotto spectrum lies in grains much larger than the upper limit in (1). Nonetheless, Comet Halley dust emission spectra appear "normal". Thus the question of the reliability of past analysis of cometary dust is abruptly raised. These disappointing facts prompted the present study which tries to interpret consistently in-situ and remote observations of the dust.

COMETARY DUST SIZING

<u>Principles of the optical method</u>

The tenuous dust distribution (which can be easily computed to be optically thin) is illuminated by a well calibrated polychromatic source at infinity -the Sun-. This illumination heats the dust, and both the thermal emission of the dust and the scattered (polychromatic) light in a well given direction are observed. The absolute intensity of the source is changed over up to two decades (by virtue of orbital motion) and at the same time the scattering angle is changing (however, near forward scattering cannot be observed). From the set of continuum spectra observed, all properties of the dust (not only the size distribution) must be inferred.

In principle, inversion of the optical data alone could be attempted ; however, since information are available on the grain size distribution coming from observations of the dust tails, and from the recent Halley flyby results, an hybrid method is preferable, which consists in looking for a best fit to <u>both</u> sets of data.

<u>Simplified theory of the emission continuum</u>

Since the nucleus is 5000–10000 times smaller than the coma, we may take it as a point source emitting isotropically (on the average) grains with a normalized size distribution η (a) cm^{-1}. Calling Q_d the <u>total</u> rate of grain emission (sec^{-1}) when the nucleus is at distance r_o from the Sun, we have for the total number of grains of radius a within a given sphere of radius R_{max} (defined by the field of view of an observing instrument) :

$$N\ (a,\ r_o,\ R_{max}) = \frac{Q_d\ (r_o)\ \eta(a)\ R_{max}}{V_d\ (a,\ r_o)} \qquad (2)$$

where we have introduced the radial outflow velocities of the grains V_d, resulting from gas drag.

The flux received at Earth (above the atmosphere) from this sphere of dust is then :

$$F_\lambda\ (r_o,\ \Delta,\ R_{max}) = \frac{1}{\Delta^2} \int_0^\infty N\ (a,\ r_o,\ R_{max})\,\pi a^2\ \{f_o(r_o,\lambda)\ Q_s(a,\lambda)\ p(a,\lambda)$$

$$+\ Q_A\ (a,\lambda)\ B_\lambda\ [T_d\ (a,\ r_o)\} \ da \qquad (3)$$

in which Δ is the comet–Earth distance, Q_s and Q_A the grain scattering and absorption efficiencies, p the phase function for scattering evaluated in the direction of the Earth (steradian^{-1}), f_o the solar flux at the comet position, B_λ Planck's function, and T_d the grain equilibrium temperature.

$T_d\ (r_o,$ a) is given by the grain energy budget equation. For fast spinning spherical grains :

$$\int f_o\ (r_o,\ \lambda)\ Q_A\ (a,\ \lambda)\ d\lambda = 4 \int Q_A\ (a,\ \lambda)\ B_\lambda\ [T_d\ (a,r_o)]\ d\lambda \qquad (4)$$

Finally, V_d must be evaluated from an hydrodynamic model of the gas–dust interaction (Finson and Probstein, 1968).

Rate of grains impact on a flying-by spacecraft

Under the same assumptions as above, one computes easily that the total number of impacts on a unit surface element of a spacecraft detector passing by the comet at a minimum distance Rmin is given by :

$$F(a) = \frac{Q_r \, \eta(a)}{V_d(a) \, Rmin} \tag{5}$$

The connection between grain radius and grain mass (given by the detectors) implies an assumption about the density ρ of the grains. Once this is done, the set of equations (2)-(5) make it possible to relate in-situ and remote sensing observations of the comet, at least in a first-order approximate way.

Hybrid dust sizing concept

The following method suggests itself : (1) start from F (a) given by either Giotto or Vega results ; (2) assume a distribution of grain shape, density, and chemical composition throughout the mass spectrum ; (3) compute the continuum emission and compare with observations ; (4) iterate for best fit.

Inspection of formula (2), (3) and (5) reveals that, if optical data are considered at a distance r^*, the unknown ratio $q(r_o, r_o^*) = Q_d(r_o)/Q_d(r_o^*)$ appears in the computation of the optical flux. It represents the change in nucleus rate of emission of dust with distance to the Sun. One can either derive q from the fits, or use theoretical predictions based on a model of the sublimation of the nucleus. Here we use the simple assumption that the rate of emission of dust is proportional to the solar energy flux, thus :

$$q(r_o, r_o^*) = (r_o^*/r_o)^2 \tag{6}$$

In general, different comets reveal different production rates at a given distance to the Sun probably because the size of their nuclei are different. Here data from Comet Halley were transferred to Comet Kohoutek assuming (empirically) at each distance r_o :

$$Q \text{ (Kohoutek)}/Q \text{ (Halley)} = 0.5 \tag{7}$$

Finally, there is no a priori reason why $\eta(a)$ should be independent of r_o or constant from comet-to-comet. We make this assumption as a starting one, in accordance with the principle of simplest possible approach.

PHYSICAL ASSUMPTIONS CONCERNING THE GRAINS

Grain shapes

There is no evidence at all concerning the shape of cometary grains. In the solar system, nearly spherical grains are found (e.g. on the Moon) which formed from liquid droplets or gas recondensation following hypervelocity impacts. It is not likely that such processes are important in comets. Small extraterrestrial grains are currently gathered in the stratosphere, which possibly may originate in comets. They are extremely complex agregates of tiny constituants, with anything but spherical external shapes. However, following the gas/dust interaction that expels the grains, and following the effect of solar

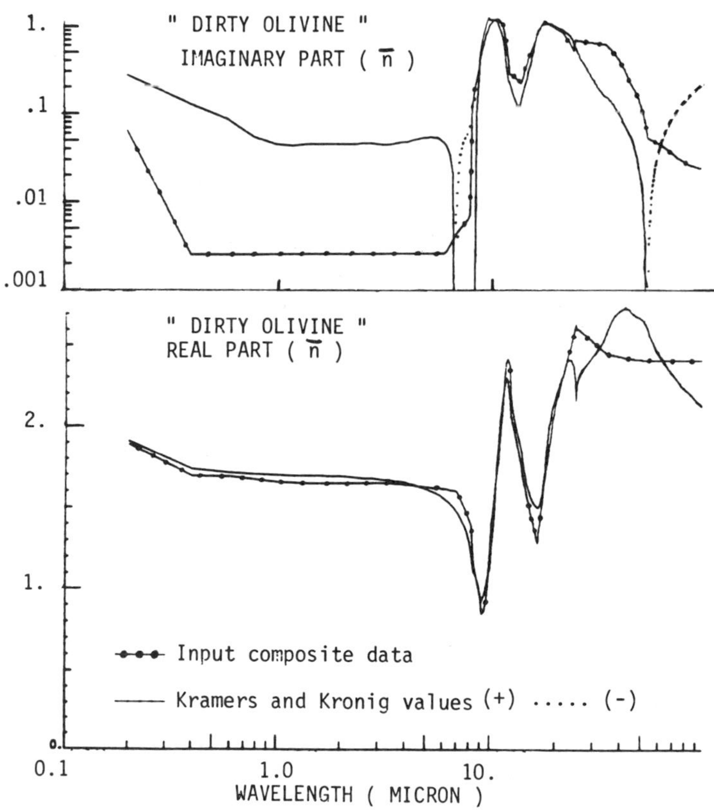

Figure 3. Model optical constants for Olivine. The sources are given in Table I. Also shown are values computed from the Kramers-Kronig relations. Discrepancies and, in particular, negative k values reveal imperfections in the inputs.

radiation torques, all grains are expected to be spinning very fast. This provides some physical justification for the use of spherical scattering theory to represent their properties, in the sense that all cross sections of interest will be averaged over all orientations of the grains in the absence of large scale spin coherence. To put numbers on the accuracy of this assumption seems beyond reach, since relevant studies (e.g. Chylek and Ramaswamy, 1982 ; Wang et al., 1979 ; Mugnai and Wiscombe, 1980 ; Roessler et al., 1983) are for much simpler sizes than involved here, and for a polydispersion of <u>similar</u> shapes.

We also assume spherical shapes when computing the hydrodynamic properties of the grains. Due to their real shapes, the real grain properties will be spread around the values computed here, by an amount yet to evaluate.

<u>Grain specific mass</u>

If one interprets the radius of the spherical grains occuring in (2 to 5) as an "effective equivalent sphere radius", it follows that the apparent grain specific mass will be :

$$\rho = \frac{m_d}{(4/3 \ \pi \ a)^3} \neq \frac{m_d}{\Omega_d} = \rho_d$$

where Ω_d is the real grain volume.

In general, $\rho < <_d \rho$ because one deals with loose agregates. We have followed the universal practice of computing optical properties from bulk material optical constants but with fancy bulk densities. This in fact is another formulation for the practice of computing optical properties from an "equivalent sphere radius" not coïncident with any of the real dimensions of the body.

<u>Grain chemical composition</u>

Based on (1) the fact that cometary dust survives for many hours at temperatures eventually above 1000 K, (2) independant indices coming from the analysis of extraterrestrial dust and meteorites collected on Earth, and (3) theoretical predictions concerning the formation of Solar System grains, one generally assumes that cometary dust includes both an absorbing component and a dielectric component, the latter giving rise to the emission spectral features. For the former, one often adopts the iron oxide magnetite, and for the latter the silicate olivine. However, ·the data gathered during this visit of Comet Halley indicate that a fraction of the dust is organic-like (rich in elements C, H, O, N). Based on this evidence, we prefer to use amorphous carbon instead of magnetite, as a representative of absorbing matter. We have verified that this does not change very much the emission spectra, and in any case we do not think likely that grains of pure amorphous carbon or pure magnetite exist at all in the coma of any comet.

<u>Input optical constants</u>

Most of the energy absorbed by the grains is at visible wavelengths since the Sun radiates approximately as a 5770 K black body. On the other hand, most of the radiated energy is at wavelengths extending from a few μm to ∼ 100 μm, considering the grain temperatures (see Figure 4). Observations of the grain emission are available from ∼ 0.2 μm to ∼ 150 μm. This spectral range lies within the range of measurements

available at a synchrotron facility. Unfortunately, studies of materials
of astrophysical interest have not in general been performed there. One
exception (noted in this work for the first time) is amorphous carbon,
another strong reason of our choice here. But for silicate materials,
one has to make a "patchwork" of independant measurements made on
different samples. Here, we have followed an earlier practice consisting
in selecting "amorphous olivine" as a representative dielectric mineral,
because the "silicate signature" of cometary dust is a structureless
feature. Data for this material are available only near the reststrahlen
bands at 10-20 μm, so we had to supplement them by data taken from
cristalline olivine, as indicated on Table 1. We show on Figures 3A and
B the result $\underline{\mathrm{of}}$ a Kramers-Kronig test conducted on our "composite"
complex index $\bar{n}$. The moderate agreement between "input" and "output"
values is in part due to the arbitrary extrapolation of our input data
beyond λ = 100 μm, and in part to the imperfect compatibility of the
input data themselves.

The heterogeneous grain problem

There is obviously not much sense in postulating that extraterrestrial
dust grains are chemically homogeneous. The imaginary part k of $\bar{n}$ plays
an essential role in determining the equilibrium temperature of the
grain. It is generally admitted that exceedingly small values of k do
not occur in natural minerals due to absorbing impurities. Thus, again
in accordance with earlier practice, we have contaminated the olivine,
by imposing the condition :

$$k = - \mathrm{Im}\ (\bar{n}) > k_v$$

where the suffix "v" is meant for visible even though, in practice, the
preceeding relation is effective at visible to middle IR wavelengths
(see Figure 3).

This manipulation of $\bar{n}$ is certainly arbitrary. But we are unaware of
better possibilities. The Mie-Güttler algorithms are used by some
authors, but we think that the special geometry implied there does not
represent well the very complex inhomogeneous agregates to be dealt
with. Recently, Mukai and Mukai (1984) have revived an earlier
suggestion of Bohren and Wickramasinghe (1977) to use an effective
medium theory algorithm to represent impure grain properties. However,
these algorithms are derived for media with dimensions much greater than
the wavelength, and for impurities of dimensions much smaller than it.
Both assumptions are in general violated in cometary dust, and therefore
these algorithms are not usable in this context. Indeed Bohren (1986)
recently pointed out that this practice can lead to unphysical results.

THEORETICAL GRAIN TEMPERATURES AND VELOCITIES

Once a physical grain model has been defined, the temperatures and
velocities of the grains are computed using the multicomponent radiation
hydrodynamic model of Crifo (1987). T_d and V_d vary with distance to the
nucleus, but, for the present work, one can neglect this effect and use
the asymptotic values reached in fact not far from the nucleus (see
Crifo, 1986 and Crifo, 1987).

<u>Hydrodynamic algorithms</u>

They are given in concise form in Crifo (1986) and will not be described
here. We just wish to mention that they incorporate —unavoidably—
assumptions concerning the properties of the nucleus of the comet. The
less founded one is probably that it distributes dust, on the average,
in a spherically symmetric manner. Departures from this assumption will
probably not change much the emission spectral shape, but will possibly
change the absolute value of the computed fluxes.

<u>Electromagnetic scattering algorithms</u>

The wide range of sizes and of wavelengths involved here implies Mie
size parameters x varying over the interval $3.10^{-3} - 2.10^6$ approximately. The absorption efficiency Q_A only is needed most of the time (to
compute grain temperatures). The scattering efficiency Q_S and phase
function p are needed only at a few wavelengths and scattering angles
for the interpretation of visible and near-infrared emissions. To manage
reasonable computer times, we were forced to optimize the optical
computations as follows :

(a)   for $|\bar{n}|.x < .1$ Rayleigh formulae are used as given in Van de Hulst
      (1957) ;

(b)   for $x < 70$ Mie formulae are used. Our program is an adaptation of
      Grehan and Gousbet (1979) "supermidi" code which makes use of the
      Lenz algorithm for Bessel computations. The number of terms
      retained in the series summation is that recommended by Wiscombe
      (1980).

(c)   for $x > 70$ and $n > 1$ the Complex Angular Momentum Theory
      formulation is used (Nussenzveig and Wiscombe, 1980). An
      unfortunate limitation of this formalism, however, is that it is
      not applicable when $n < 1$, i.e., near some of the reststrahlen
      bands. In this case, we use the geometrical optics formulae in the
      form given by Van de Hulst (1957), when applicable, i.e. for
      $x > 400$ and $kx > 2$. When one of the latter conditions is not met,
      we go back to the Mie formalism, at the expense of a lot of
      computer time. This limitation of the Complex Angular Momentum
      theory to $n > 1$ will be circumvented in the future (Fiedler-Ferrari
      and Nussenzveig, 1987).

(d)   whenever the phase function is needed, we use Mie formalism, or
      Geometrical Optics formulae supplemented by Fraunhofer diffraction
      formulae because Complex Angular Momentum formulae for the phase
      are not yet published.

<u>Results</u>

Figure 4 shows T_d (a, r_o) for amorphous carbon and amorphous olivine
with k_v = 0,03. The general shape of the curve is in agreement with
earlier results on similar materials, when available, e.g. Hanner
(1980), Röser and Staude (1978).

Figure 5 presents V_d (a, r_o) appropriate for Comet Kohoutek (Comet
Halley has somewhat higher velocities). The curves apply to grains with
an assumed density of 0.3 gcm^{-3} ; the difference between olivine and
carbon grain is negligible and not indicated on the figure.

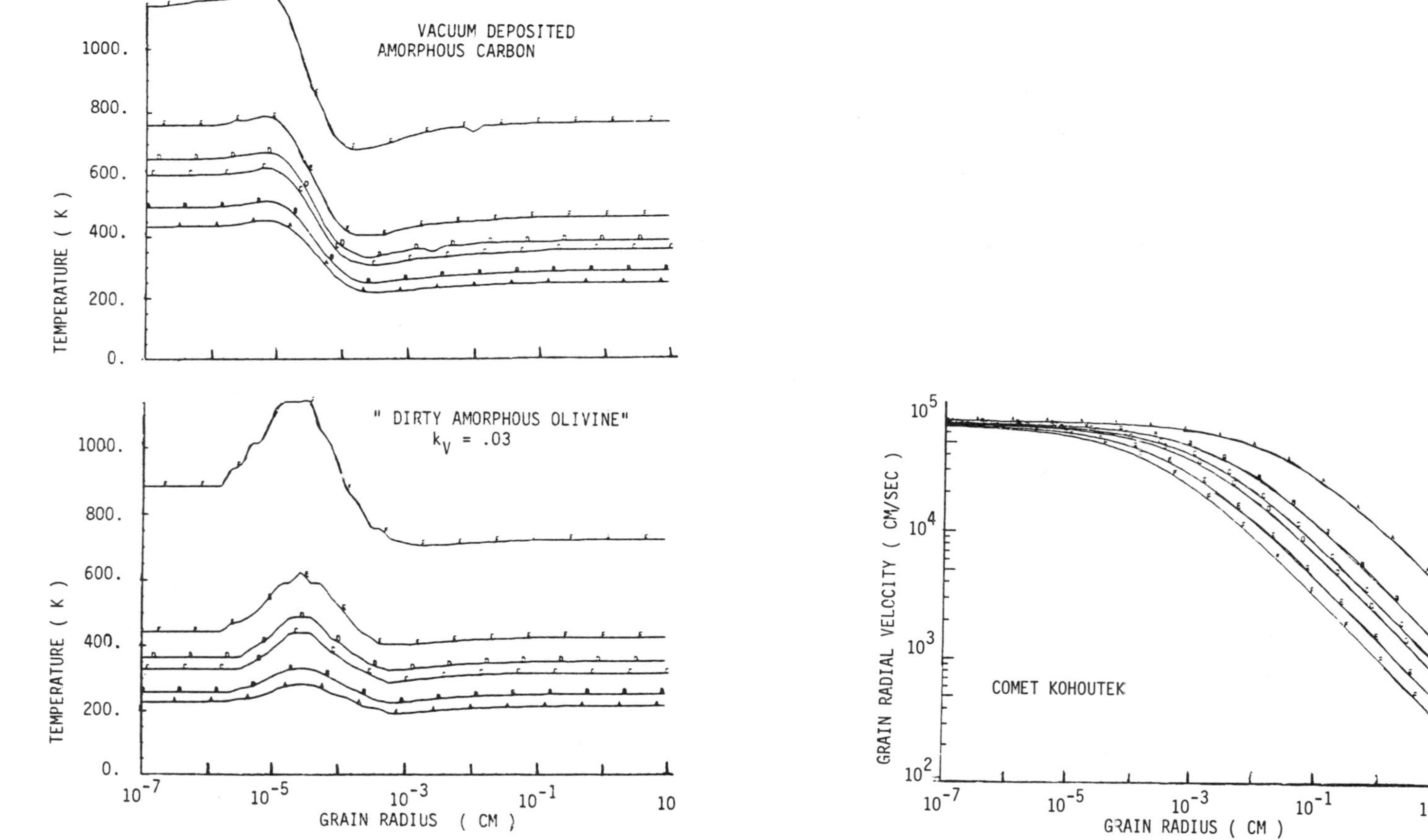

Figure 4. Grain temperatures. Each curve is for a given heliocentric distance; from top to bottom: 0.15, 0.44, 0.65, 0.8, 1.28, 1.77 A.U. The ripple in the curves reveals the limits in computation accuracy.

Figure 5. COMPUTED GRAIN RADIAL VELOCITIES. Each curve is for a given heliocentric distance. From top to bottom: 0.15, 0.44, 0.65, 0.8, 1.28, 1.77 A.U. Details of the model are given in Crifo (1987).

COMET HALLEY DUST

The full exploitation of what we called precedingly an "hybrid dust sizing concept" is far for being completed at this time. We have presently explored several tentative laws expressing (1) the variation of grain density with grain size (2) the variation of the ratio of carbon to silicate content. We also used three possible inputs from in-situ data, assuming (1) that the curve 2A is representative of the average conditions in the coma, or (2) that on the contrary it is curve 2B ; or (3) that it is a curve constructed from whichever segment 2A or 2B is greater at each given mass. Let us call these latter assumptions "SA", "SB" and "SAB" for conciseness. The best fits obtained presently using SA, SB or SAB are shown on Figure 6. They correspond to an equipartition of the dust between carbon and olivine, at each mass, to a constant apparent specific mass of 0.3 gcm^{-3}, and to a truncation of the mass spectrum at 3.6 gram. We believe that SA is excluded, based on the fact no silicate signature is visible (the emission from the large grains is overwhelmingly dominant), and also on disagreements on the shape of the continuum. The distinction between SB and SAB is perhaps less convincing, but is in favour of SAB. In any case the very fact that we may hesitate between SB and SAB is of great significance : it indicates clearly the limits of the information that can be extracted from this type of optical dust sizing, pretty much like Figure 2 showed the limits in the significance of in-situ cometary dust samplings. The difference in the rates of mass ejection resulting from (SB) and (SAB) is a factor 6 ! On the other hand, from the data of Figure 6 one can say that the mass spectrum "SA" and "SAB" given by Giotto must terminate near 3.6 gram, otherwise no silicate features can be seen and the radiated fluxes would be two high. Therefore, the factor 6 cited above represents the allowed range of uncertainties, with the highest probability in favour of the upper limit of this range.

COMET KOHOUTEK DUST

The rationale for looking whether Comet Halley fluences are compatible with Comet Kohoutek emissions is (1) to test the simple concept of a "universal size distribution" of cometary dust, suggested already by analysis of their dust tails, (2) to test the assumed invariance of η (a) against changes of heliocentric distance, since data for Comet Kohoutek are available from 0.15 AU to 1.77 AU (*), and (3) to test earlier interpretations of Comet Kohoutek emissions.

A word of caution is here needed. The experimental data that will appear on the next figure were obtained with multiband photometers featuring only $\sim$ 12 discrete bands at a resolution $\lambda/\Delta\lambda \sim 10$. In particular, there are only two bands inside the "silicate" emission features near $\lambda = 10$ μm. The "measured spectra" result from Black body fits to the measurements, except for the 10 μm feature which is hand-drawn through the two data points. An exact fit to the data would then be circumstancial, if not suspicious ! Figure 7 presents computed spectra from SB and SAB for the extremes in heliocentric distances. Data at 1.77 AU are from Rieke and Lee (1974) and at 0.15 AU from Ney (1974). Again, the dust is assumed to have $\rho_d = 0.3$ g cm^{-3} and to be equally partitioned between carbon and olivine. Maximum mass is at 1.3 gram. Inspection of the figures reveals rather good agreement between data and model, except near the 10 μm feature at $r_o = 0.15$ AU. Here the model

(*) 1 AU = 1 Astronomical Unit = 1 Mean Sun-Earth distance = 1.5 x 10^8 Km

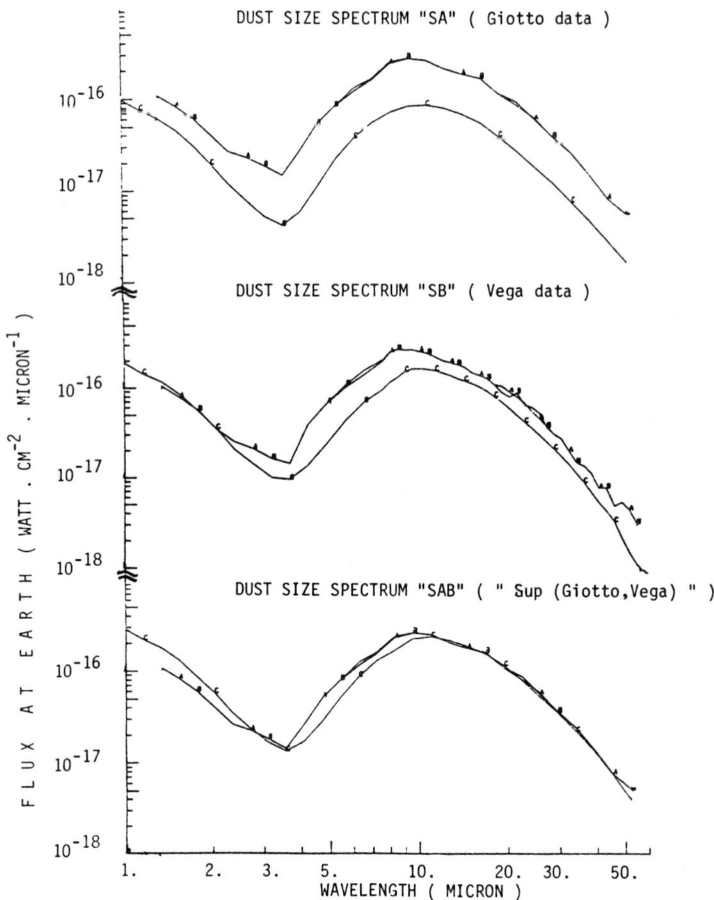

Figure 6. Theoretical fits (—c—) to comet Halley composite spectrum (a b). References for the composite are given in the text. The comet is at about 1.25 A.U. from the Sun, pre-perihelion. The three fits correspond to an assumed grain density of 0.3 g.cm^{-3}, to equipartition between amorphous carbon and dirty amorphous olivine, and to the dust size spectrum indicated on the figure.

gives a feature much fainter than the data, and possibly at too short wavelengths. It is not easy to distinguish between SB and SAB fits at 1.77 AU, and, at 0.15 AU, none of these distributions is satisfactory due to the 10 μm problem. To obtain a sufficiently strong emission feature, it is necessary to place the maximum mass near 10^{-9} gram (radius $\sim$ 10 μm), and to increase the computed Q_d by a factor $\sim$ 3. In other words : η (a) derived from Comet Halley (near r_o = 0.8 AU) are not applicable to Comet Kohoutek near 0.15 AU. On the other hand, an η (a) truncated at 10^{-9} gram (and Q_d multiplied by 3) would produce a silicate signature at 1.77 AU, while none is observed. This difficulty was known already before this study.

One possible lesson from these Kohoutek fits is that η (a) varies with heliocentric distance but is the same in the two comets. This cannot be checked before Halley spectra are available over a large range of heliocentric distances. On the other hand we cannot say for sure that η (a) changes with r_o : it may be possible to play with the ratio carbon/olivine to reduce sufficiently the 10 μm signature at large heliocentric distances. At the present time, this has not been achieved.

In conclusion, we cannot say at this moment whether two different distance-independant η (a) apply, one to Halley, the other to Kohoutek, or whether the same (but distance-dependent) η (a) apply to both with eventually the ratio of dust components being dependent upon heliocentric distance.

CONCLUSIONS

Detailed comparison of Comet Halley emission spectra and in-situ dust fluences reveals that none of these techniques can claim to provide exhaustive information on the dust. Each one reveals the limitations of the other. It is probable that Comet Halley average dust size distribution extends over a very broad spectrum, a conclusion to which optical sizing alone could not have led. In consequence, the dust production rate in this comet is probably $\sim$ 6 times greater than commonly estimated. The assumption of invariance of the size distribution η (a) with distance to the Sun cannot significantly be tested with the present set of data which were gathered on a restricted range of heliocentric distances.

Comet Kohoutek emission spectra at large distance from the Sun are compatible with η (a) derived in Comet Halley. However, at smaller heliocentric distances there is incompatibility, due to the presence of strong emission feature near λ = 10 μm. It is not possible to say at this time whether this means that there is a common η (a) for the two comets, but changing substantially with heliocentric distances, or if there are definitely different η (a) for the two comets, which could be, eventually but not necessarily, independent of heliocentric distances.

The reliability of the preceding conclusions is limited by the following problems :

1. Comparison between models and data is restricted by the insufficient completeness of the data. In particular, one would like to have at hand good continuum spectra extending from 0.25 μm to 7-8 μm without gaps, at low resolution, and high resolution spectra between 8 and 25 μm to clarify the origin of the silicate feature. Perhaps this would require a dedicated facility operating from outside the Earth atmosphere.

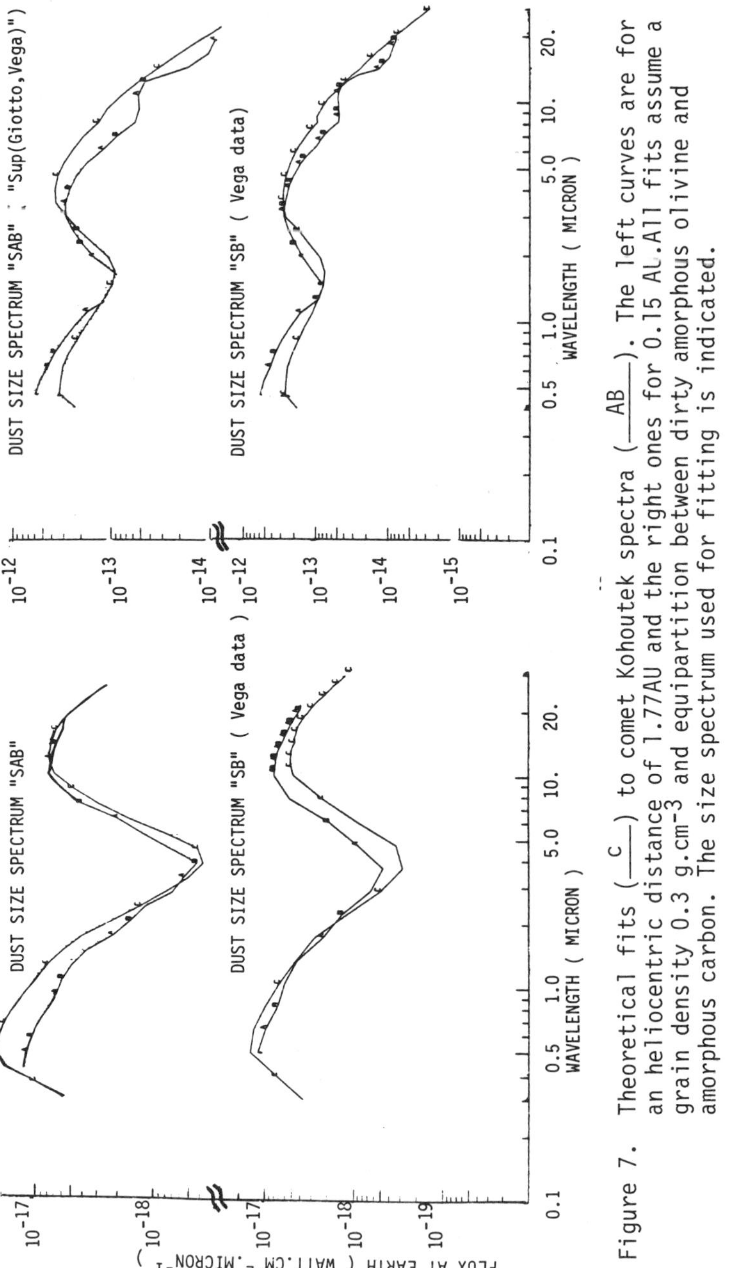

Figure 7. Theoretical fits ($\frac{C}{\quad}$) to comet Kohoutek spectra ($\frac{AB}{\quad}$). The left curves are for an heliocentric distance of 1.77AU and the right ones for 0.15 AL.All fits assume a grain density 0.3 g.cm^{-3} and equipartition between dirty amorphous olivine and amorphous carbon. The size spectrum used for fitting is indicated.

2.   The bank of optical constants available for fits to observed
      emission is very limited, calling for observational programs at
      polychromatic facilities.

3.   The optical computation algorithms available presently are not yet
      sufficiently powerful. In particular one would benefit very much
      from finalizations of the Complex Angular Momentum theory and from
      the development of algorithms suitable for inhomogenous grain
      scattering.

Table I. Origin of the optical constants adopted for Pure Olivine

Wavelengths (micron)	n	k
0.0775 - 0.2	Huffman and Stapp (1973)	Röser and Staude (1978)
0.2 - 3.0	"	Huffman (1975)
3.0 - 8.33	Interpolation between Huffman and Stapp (1973) and Krätschmar and Huffman (1979)	"
8.33 - 25.	Krätschmar and Huffman (1979)	Krätschmar and Huffman (1979)
25. - 100.	Hanner (1983)	Hanner (1983)

The data below 8.33 μm are for cristalline, and above 8.33 μm for
amorphous olivine. In the first case, the cristal orientation is not
always known.

ACKNOWLEDGEMENTS

This work was supported by C.N.R.S. and C.N.E.S. grants. Thanks are due
to Drs A. Hauchecorne and B. Gondet for assistance in some of the
computations, to Mr F. Bougnet, Mss. F. Marchand, G. Bargot, S. Ardoin
and M.T. Peyroux for their technical contribution. We are grateful to
Drs G. Grehan, G. Gouesbet and W.J. Wiscombe for supplying some of their
optical computation algorithms.

REFERENCES

Bohren, C.F. (1986), J. Atmos. Sci. 43, 5, 468

Bohren, C.F. and Wickramasinghe, N.C. (1977), Astrophys. Space Sci. 50,
461

Campins, H., Bregman, J.D., Witteborn, E.F., Wooden, D.H., Rank, D.M., Allamandola, L.J., Cohen, M. and Tielens, A.G.G.M. (1986), in 20th ESLAB Symposium on the Exploration of Halley's Comet, ESA report SP 250, Vol. 2, p. 121

Chylek, P. and Ramaswamy, V. (1982), Appl. Opt. 21, 23, 4339

Crifo, J.F. (1982) in Proceedings of the International Conference on Cometary Exploration, ed. by T.I. Gombosi, Budapest Academy of Sciences Press, Vol. 2, p. 167

Crifo, J.F. (1986), in Rarefied Gas Dynamics XV, edited by C. Cercignani and V. Boffi, Tuebner ed., Vol. 2, p. 229

Crifo, J.F. (1987) Optical and Hydrodynamic implications of Comet Halley dust size distribution : in Proceedings of the Symposium on the Diversity and Similarity of Comets, Bruxelles, April 1987, ESA SP 278 (in press)

Fiedler-Ferrari, N., and Nussenzveig, M. (1987), These proceedings

Finson, M.L. and Probstein, R.F. (1968), Ap. J. 154, 327 and 353

Glaccum, W., Moseley, S.H., Campins, H. and Loewenstein, R.F. (1986), in 20th ESLAB Symposium on the Exploration of Halley's Comet, ESA Report SP 250, Vol. 2, p. 111

Grehan, G. and Gousbet, G. (1979), Appl. Opt. 18, 20, 3489

Hageman, H.J., Gudat, W. and Kunz, C. (1974) Deutsches Elektronen Synthrotron report DESY SR-74/7 (Hamburg)

Hanner, M.S. (1980), in Solid Particles in the Solar System, ed. by I. Halliday and B.A. Mc Intosh, p. 223

Hanner, M.S. (1983), Private communication

Herter, T., Gull, G.E. and Campins, H. (1986), in 20th ESLAB Symposium on the Exploration of Halley's Comet, ESA Report SP 250, Vol. 2, p. 117

Huffman, D.R. and Stapp, J.L. (1973), in Interstellar Dust and Related Topics, ed. by J.M. Greenberg and H.C. Van de Hulst, D. Reidel, p. 297

Huffman, D.R. (1975), in Solid State Astrophys., edited by N.C. Wickramasinghe, Astrophysics and Space Science Library n° 55, D. Reidel, p. 191

Krankowsky, D., Lämmerzahl, P., Herrwerth, I., Woweries, J., Eberhardt, P., Dolder, U., Hermann, U., Schulte, W., Berthelier, J.J., Illiano, J.M., Hodges, R.R. and Hoffman, J.H. (1986), Nature 231, 326

Krätschmer, W. and Huffman, D.R. (1979), Astrophys. Space Sci. 61, 195

Mc Donnell, J.A.M., Kissel, J., Grün, E., Grard, R.J.L., Langevin, Y., Olearczyk, R.E., Perry, C.H. and Zarnecki, J.C. (1986), in 20th ESLAB Symposium on the Exploration of Halley's Comet, ESA Report SP 250, Vol. 2, p. 25

Mazets, E.P. and 14 co-workers (1986), in 20th ESLAB Symposium on the Exploration of Halley's Comet, ESA Report SP 250, Vol. 2, p. 3

Mugnai, A. and Wiscombe, W.J. (1980), J. Atmos. Sci. 37, 1291

Mukai, T. and Mukai, S. (1984), Adv. Space Res. 4, 9, 207

Mumma, M.J., Weaver, H.A., Larson, H.P., Scott Davis, D. and Williams, M. (1986), Science 232, 1523

Ney, E.P. (1974), Icarus, 23, 551

Nussenzveig, H.M. and Wiscombe, W.J. (1980) Phys. Rev. Lett., 45, 18, 1490

Röser, S. and Staude, H.J. (1978), Astron. Astrophys. 67, 382

Rieke, G.H. and Lee, T.A. (1974), Nature 248, 737

Roessler, D.S., Wang, D.S. and Kerker, M. (1983), Appl. Opt. 22, 22, 3648

Tokunaga, A.T., Golisch, W.F., Griep, D.M., Kaminski, C.D. and Hanner, M.S. (1986), Astron. J. 92, 5, 1183

Van de Hulst, H.C. (1957), Light scattering by small particles, J. Wiley editor, New York

Wang, D.S., Chen, H.C.H., Barber, P.W. and Wyatt, P.J. (1979), Appl. Opt. 18, 15, 2672

Whipple, F.L. (1976), Nature 263, 15

Wiscombe, W.J. (1980), Appl. Opt. 19, 9, 1505

MEASUREMENTS OF ABSOLUTE CONCENTRATION AND SIZE DISTRIBUTION

OF PARTICLES BY LASER SMALL ANGLE SCATTERING

Shigeru Hayashi

Aircraft Emission Research Group
National Aerospace Laboratory
Chofu Tokyo, Japan

INTRODUCTION

The accurate and in-situ measurement of size distribution and concentration of particles is a subject of interest in a variety of indus-
rial applications including spray drying, spray combustion and milling
machinery. A major concern is with particles within the size range of one
to several hundred microns. Among the various scattering techniques of
particle size measurement, methods based on the scattered light intensity
profile in the forward small angles have made a great progress in recent
years.

Chin and co-workers (1955) and Shifrin (1956) independently investigated the possibility of size determination of particles larger than a few
microns in diameter from the measurements of the intensities of light
scattered in the forward small angles.

Dobbins and co-workers (1963) have shown by a computer simulation
that the scattered light intensity versus reduced size parameter based on
surface mean diameter or Sauter mean diameter (SMD) is virtually independent of the parameters of the distribution function. This simple correlation allows the determination of surface mean diameter from the ratio of
the intensities at two angles or from the width of the scattered light
intensity profiles alone. A scanning minute photomultiplier was used to
obtain the intensity profiles (Chin, 1955b; Hayashi et al., 1981), though
a scanning slit in a rotating disk was used to measure the widths of the
scattered intensity profiles (Buchele, 1976). More recently, Hayashi
(1985) developed an ease-to-use apparatus for the determination of SMD of
transient sprays, which is based on the ratio of intensities measured by a
sensor comprised of two concentric full ring-shaped silicone photodetectors. The system displays the value of SMD at intervals of one second.

During the last fifteen years some great improvements were made in
the detection system. Cornillault (1972) used a rotating screen containing rectangular windows to measure the angular variation of the scattered
light intensity by a photocell mechanically positioned behind the screen.
Swithenbank et al. (1976) used a semicircular monolithic photodetector
array which was connected to a digital voltmeter by a scanner. Semi-
circular detector arrays similar to that and a liner detector arrays were
used by Ruscello and Hirleman (1983) and Hayashi and co-workers (1982).

Even in the diffraction based analysis of the scattering data, it is usual that a model of size distribution is assumed to solve a set of linear equations by the least squared method. This approach is easier than the classical one using the mathematical inversion, and can avoid physicaly meaningless solutions. The use of the laser and multichannel photodetector array has lead to a great progress in the measurements of scattered light in the forward small angles. Some systems that use sample-and-hold circuits can make snap shot measurements of the scattered intensity profiles, and is applicable to transient sprays such as Diesel injections. Data sampling by these circuits improves the quality of measurements even in steady sprays or particle suspentions.

There are some cases where not only the size distribution but also concentration is desired to be determined. It is well known that the extinction measurements of a collimated beam in suspended particles can determine the line-of-sight averaged concentration as far as the size distribution is separately determined. But its application is limited to particles smaller than the wave length used, typically 0.5 μm in diameter. Using a diffraction based spray analyzer, Yule and co-workers (1981) measured the variation of relative concentration of droplets in a spray from an air blast atomizer by assuming axisymmetry. Hayashi et al. (1985a, b) showed that absolute concentrations of particles as well as size distributions can be simultaneously determined.

This paper describes a laser scattering technique for the simultaneous and instantaneous measurements of absolute concentration and size distribution of spherical particles. Some examples of applications of the technique are also presented.

THEORETICAL

We consider scattering from homogeneous spherical particles that are shone by a collimated laser beam. It is assumed that the particle cloud is dilute. Since multiple scattering is negligible, the angular variation of the scattered light intensity $I(\theta)$ for the particle cloud can be related to the intensity profile $i(\alpha,\theta)$ of each single particles in the beam by the following equation

$$I(\theta) = c\,AL\int_{0}^{\infty} i(\pi D/\lambda,\ \theta)\,n(D)\,dD \tag{1}$$

where $n(D)dD$ is the number of particles with diameter between D to $D + dD$ in the unit spatial volume, A is the effective cross sectional area of the beam, L the optical path length, and α is the size parameter defined as $\pi D/\lambda$, λ being the wave length of the beam. In the optical Fourier transform arrangement, $I(\theta)$ can be determined by measuring the radial variation of intensity of scattered light at the focal plane of the lens.

By introducing a normalized volume size distribution function, $W(D)$, and a parameter, $X1$, representing volumetric concentration, Eq.(1) reduces to

$$I(\theta) = c'ALX1\int_{0}^{\infty} i(\pi D/\lambda,\ \theta)\,W(D)/D^3\,dD \tag{2}$$

where C' is a new constant which depends on laser power and the optical efficiency. Since $i(\alpha,\theta)$ is a known function, given by Mie scattering theory or geometrical optics, determination of particle size distribution from the measured intensity profile, $I(\theta)$, is, in a mathematical sense, equivalent to solving the integral equation.

Though the algorithm proposed by Lentz (1976) removes the upper bound

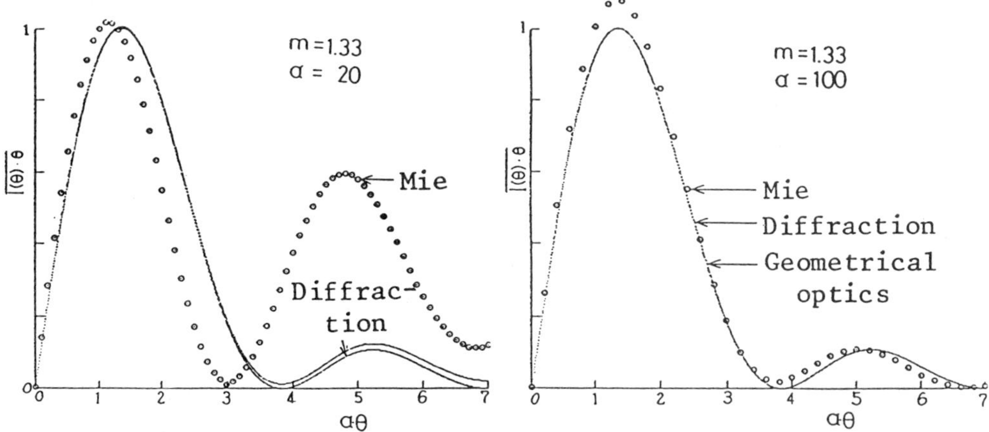

Fig. 1. Scattering intensity profiles by Mie theory, geomertical optics and diffraction approximations for small and large particles.

of size parameter in the calculation of the Mie scattering functions by using mainframe computers (Grehan and Gouesbet,1979), it is practical to calculate $i(\alpha,\theta)$ by the equations based on the classical geometrical optics for particles several times larger than the wave length used, and by the rigorous Mie scattering functions for particles of smaller sizes. Figure 1 compares the profiles of $I(\theta)\theta$ calculated by the Mie scattering functions with those by using the geometrical approximations for large and small particles for a refractive index m =1.33. For α =100, a fairly good agreement is observed even outside the Airy disk. The deviation decreases with increasing α and is indistinguishable at α =400. For α =20, the approximated $I(\theta)\theta$ peaks at a larger angle as compared to that based on the Mie equations and the deviation is significant at larger angles. The diameter of a particle with the size parameter of 20 is approximately 4 microns for He-Ne laser beams.

In the cases where scattering from each particles is approximated by the diffraction, Eq.(2) can be inverted analytically in a closed form to yield $W(D)$ independent of any prior assumption of modeling of the distribution on the basis of the Titchmarsh transform. The final expression is

$$W(D) = -C \int_0^\infty \partial/\partial\theta \, (I(\theta)\,\theta^3) \, J_1(\alpha\theta) \, Y_1(\alpha\theta) \, \theta \, \alpha^2 d\theta \quad (3)$$

The evaluation of the right hand side of of this equation requires very accurate measurements of the intensity, otherwise differentiation of the measured intensity $I(\theta)$ becomes a source of additional errors. It is usual that the number of particles in the beam as well as the size distribution changes with time. Therefore the technique based on Eq.(3) has only a limited number of practical applications.

For annular detector arrays, it is more convenient to introduce photoenergy instead of intensity. The photoenergy received by the i th element is given by integrating Eq.(1) over inner and outer scattering angles, $\theta i+, \theta i-$.

$$E_i = cALX1 \int_0^\infty W(D)/D^3 \int_{\theta i-}^{\theta i+} \theta i \,(\pi D/\lambda, \theta) \, d\theta dD \quad (4)$$

Assuming the Rosin-Rammler distribution function for $W(D)$, Swithenbank and co-workers (1976) solved a set of liner equations by the least squared method and determined the two parameters characterizing the size distribution. A similar, but more rigorous approach was employed in the analysis of the energy distributions (Hayashi et al., 1980). One of the familiar particle size distribution models having two characterizing parameters such as the Rosin-Rammler, Nukiyama-Tanasawa, Gaussian and lognormal functions is assumed as $W(D)$ to solve Eq.(3) by the nonlinear best fitting technique. The two parameters of the distribution as well as the concentration, $X1$, are so determined by an iterative procedure that the sum of the squared deviations between the observed energy distribution and theoretically predicted one may be minimized. The measured data presented in this paper were obtained by analyzing the scattering data on the basis of the Rosin-Rammler distribution function.

APPARATUS

One of the particle sizing apparatus developed at National Aerospace Laboratory is shown schematically in Fig. 2. A laser beam from a He-Ne laser tube is spatially filtered and expanded by using a beam expander to form a collimated beam of 8 mm in diameter. The scattered light in the

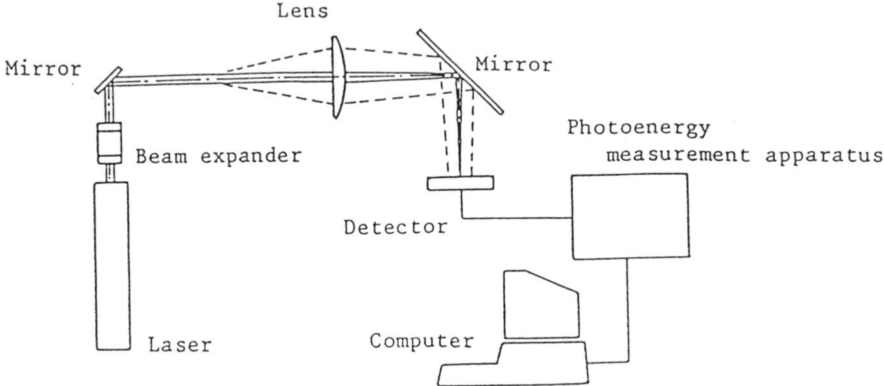

Fig. 2. Optical arragement and data processing unit of particle sizer.

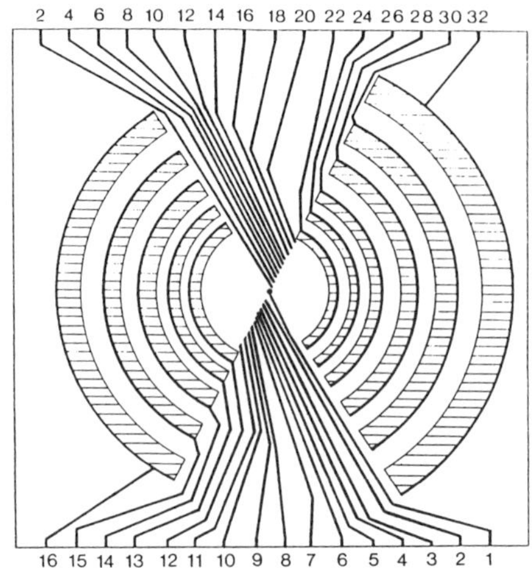

Fig. 3. Schematic drawing of semicircular silicone photodetector array.

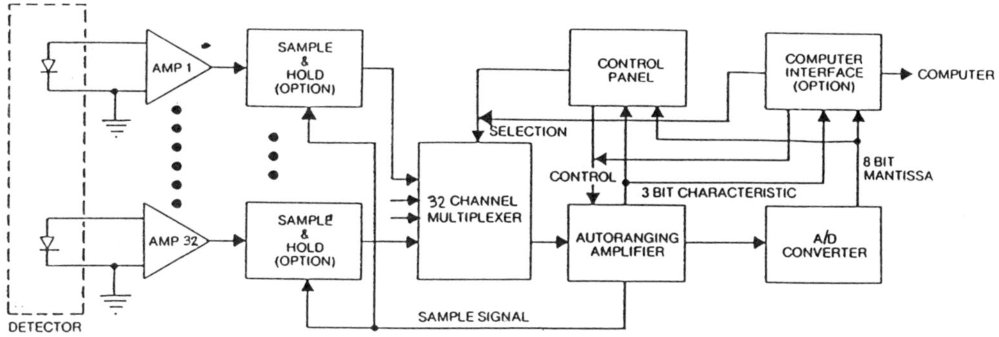

Fig. 4. Blockdiagram of photoenergy measurement system.

forward small angles is received by a lens and the angular variation of
the scattered intensity is detected by a scanning photomultiplier or an
semiannular detector array. The focal length of the standard lens is 760
mm for the scanning photomultiplier system (Hayashi et al., 1981) and 300
mm for the array, respectively. Figure 3 schematically shows the arrangement of the semicircular silicone photodetector elements on the array
fabricated at NAL by the present author. One of the features of the array
is that the outer circular arc of the i th element and the inner circular
arc of the i+1 th element are on the same circumference. Another is that
an aluminum electrode is attached to each element along the outer circular
arc to improve responsibility.

Figure 4 shows a block diagram of the photoenergy measuring apparatus. The photodetector amplifier outputs are multiplexed into an auto-
ranging amplifier. A sample and hold circuit is attached to each photodetector to facilitate snap shot measurements of photoenergy distributions. The measured scattering data are analyzed by a 16-bit microcomputer, interfaced with the apparatus to reduce the size distribution
and line-of-sight averaged volumetric concentration of particles.

EXPERIMENTAL

A series of the measurements were made to check the validity of the
computer code developed for the analysis of the particle size distribution
and volumetric or mass concentration. Monosized polystyrene latex particles (Duke Scientific Co.) of known diameters between 5 and 200 μm and
polydisperse glass beads were used. Measurements were made for the
particles suspended in ethanol in an optical cell. A magnetic stirrer was
used to prevent the settling of the particles and make the suspensions
homogeneous. A laser beam was passed through the optical cell and the
scattered light intensity profile was measured by the apparatus described
in the preceding chapter.

Figure 5 shows an example of the measured scattered light energy for
polystyrene uniform latex particles of 43.9 μm nominal diameter. The best
fitted theoretical prediction of energy profile is compared with the measured. The fitting inside the first minimum is quite satisfactory, though
a slight deviation in the energy levels exist at the outer detectors. The
oscillatory behavior is characteristic to the scattering of uniformly
sized particles.

The results of the particle size measurements of the polystyrene
latex particles and the glass beads are summarized in Fig. 6. Comparisons
of the data are made on the basis of surface mean diameter. In this

553

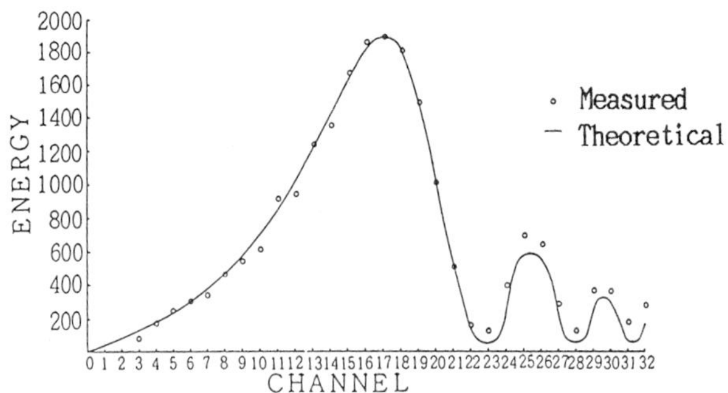

Fig. 5. An example of measured scattered light energy distribution.
Polystyrene latex particles, Nominal diameter 43.9 um.

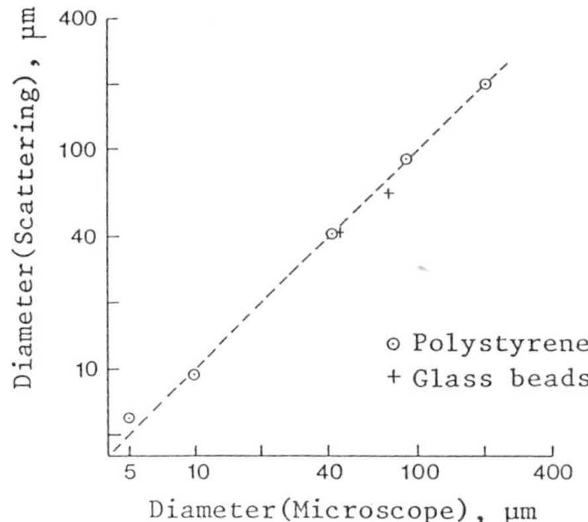

Fig. 6. Comparison of mean diameters by small angle scattering and
by microscopy for solid particles.

connection, it should be noted that, since the standard deviation is
smaller than 6% for the polystyrene latex particles used, the values of
SMD is very close to the count median diameters. The agreement for polystyrene latex particles is complete except for the 5 micron particles.
The scattering data for the 5 micron particles were analyzed based on the
Mie scattering theory by using a computer code developed for a mainframe
computer. The refraction of the scattered light at the inter-face between
liquid and air was taken into consideration. This analysis showed that
the deviation for the 5 μm particles in Fig. 6 was due to the approximation used. As can be seen from the left in Fig. 1, diffraction approximation gives a diameter somewhat larger than the actual. The accuracy of
size determination for the polydisperse particles is better than 5 %.

Calibration for Concentration Measurements

The optical constant C, being necessary for the determination of
beads, much less expensive than latex particles. The glass beads, being
sampled on a sheet of powder paper, were weighted by a precision
microbalance and put into an optical cell filled with ethanol, and then
the scattered light intensity profile was measured. This procedure was
repeated several times. The amount of the particles left on the paper was
taken into consideration each time. The scattered light energy data were
analyzed to reduce the droplet size distributions and concentrations.

554

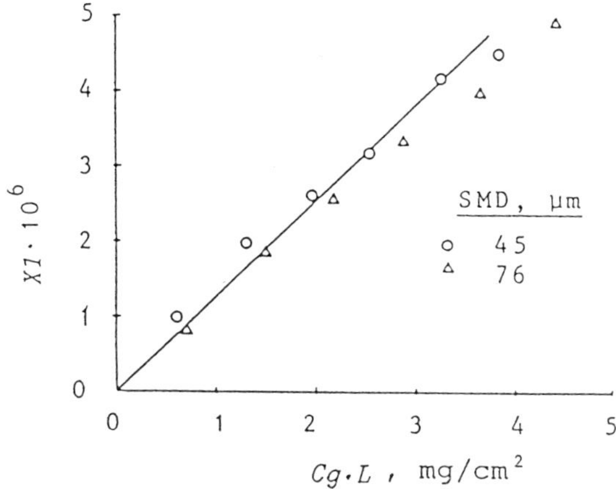

Fig. 7. Correlation between $X1$ and absolute mass concentration Cg.

Figure 7 shows the relation between $X1$ and the product of the mass of particles in unit volume Cg and the optical path length L. At the higher values of $Cg.L$, $X1$ is slightly lower than it should be. This departure from the linear relation was mainly due to the loss of particles adhered on the wall of the optical cell that was actually observed, and partly due to the effect of multiple scattering. It can be said that a better linearity would hold over the range of $Cg.L$ tested. The accuracy in the determination of mass concentration is moderately estimated as 10%.

APPLICATIONS

Solid Particle Size Measurements

The developed system can be used for the size measurements of solid particulate such as toner for xerography, cement powder, and reference solid particles specified by the Japanese Industrial Standards. Figure 8 compares the cumulative volume size distribution of toners measured by a prototype model developed at NAL and a Coulter counter type TA 11. The agreement is very satisfactory. The 50% diameters differs only 10% for both the fine and coarse samples.

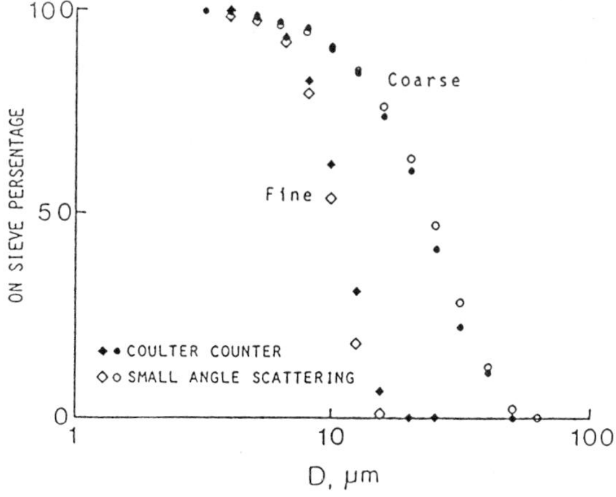

Fig. 8. Comparisons of cumulative size distributions of solid particles by Coulter counter and by small angle scattering.

Another important application of the developed system is measurements
of sprays. Since liquid particles are easily evaporate, coagulate into
larger droplet, or disintegrate into smaller droplets, the size distribution obtained by intrusive sampling may be quite different from the
actual.

An example of the size and absolute concentration measurements of
liquid droplets is presented. Water sprays from a swirl atomizer were
diagnosed. The sprays are assumed to be axisymmetrical. The scattering
data at different radial position reduces the radial variation of droplet
size distribution and volumetric concentration by using the onion peering

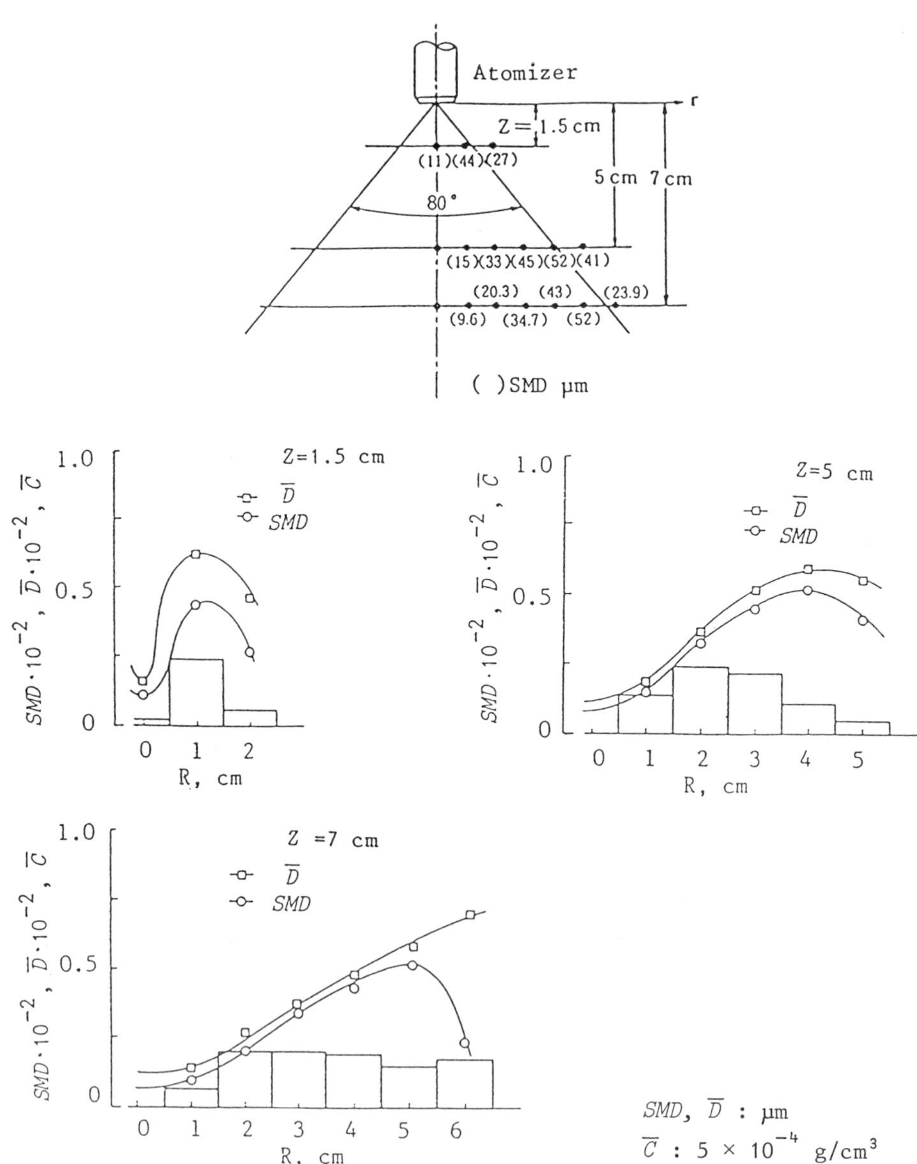

Fig. 9. Axial and radial variations of mean droplet diameters
concentration in a hollow cone spray.

technique. Measurements were conducted at 1.0 cm intervals on the plane 1.5, 3.0, 5.0, 7.5 cm below the nozzle orifice.

The radial variations of mass concentration and mean diameters at three axial distances are presented in Fig. 9, where $\bar{D}$ is the mean diameter in the Rosin-Rammler distribution function (63% under sieve diameter) and SMD is the Sauter mean or surface mean diameter. The feature of the development of hollow cone spray from a swirl atomizer is well detected. The mean diameters are maximum at around the spray sheath and minimum at the spray axis. Mass concentration, having a peak on the spray cone angle not far from the nozzle, becomes uniform as the distance from the nozzle increases. The entrained air stream into the spray sheath explains the observation that droplet size is smaller in the central region and larger on and around the sheath.

CONCLUSIONS

Volumetric concentrations and size distributions of particles can be determined simultaneously from the instantaneously measured scattered light intensity profiles in forward small angles. The accuracy of particle sizing is estimated to be better than 5 % over a dynamic range of about a hundred, and that for concentration is better than 10 %. It is pointed out that for accurate sizing of small particles about 5 μm in diameter, a rigorous analysis based on Mie scattering theory is required.

The technique have been successfully applied to measuring the solid particles suspended in liquids and liquid particles in air. The structure of a follow cone axisymmetric spray is clarified. Mass concentration, being maximum at around the spray sheath near the nozzle, becomes uniform as the axial distance increases. Mean droplet size has a maximum at around the sheath and smaller in the central region.

ACKNOWLEDGMENT

The author is grateful to Mr. M. Yoshitake and Mr. J. Ikeda for support in making the measurements. Financial support was given by Agency of Environments Japan for FY 1975-1978.

REFERENCES

Buchele, D. R. (1976). Scanning radiometer for measurement of forward-scattered light to determine mean diameter of spray particles, NASA TM X-3454.

Buchele, D. R. (1983). Particle sizing by measurement of forward scattered light at two angles, NASA TP 2156.

Cornillault, J.(1972). Particle size analyzer, Appl. Optics, 11, 265-268.

Chin, J. H., C. M. Sliepcevich, and M. Tribus (1955a). Particle size distributions from angular variation of intensity of forward scattered light at very small angles, J. Chem. Phys, 59, 841-844.

Chin, J. H., C. M. Sliepcevich, and M. Tribus (1955b). Determination of particle size distributions in polydispersed systems by means of measurements of angular variation of intensity of forward scattered light at very small angles, J. Chem. Phys, 59, 845-848.

Dobbins, R. A., L. Crocco, and I. Glassman (1963). Measurement of mean particle sizes of sprays from diffractively scattered light, AIAA J., Vol.1, No.1, 1882-1886.

Grehan, G. and G. Gouesbet (1979). The computer program "SUPERMIDI" for Mie theory calculations, without "practical" size nor refractive index limitations, Internal Report TTI/GG/79/03/20.

Hayashi S., T. Saitoh, S. Horiuchi, H. Yamada, and A. Meguriki (1980). National Aerospace Laboratory Technical Report, TR-614.

Hayashi, S., S. Horiuchi and T. Saitoh (1981), Development of an apparatus measuring droplet size distributions of sprays based on intensity profiles of scattered light in small forward angles, National Aerospace Laboratory Technical Memorandum, TM-454.

Hayashi, S. (1985a). Simultaneous measurements of size distribution and concentration of sprays, The 13-th Conference on Liquid Atomization and Spray Systems in Japan, pp.29-34.

Hayashi, S. (1985b). Simultaneous and instantaneous measurements of concentration and droplet size distribution in two-phase flows, Fluid Control & Measurement, Proc. of an Int. Symposium, Pergamon Press, 825-830.

Lascello, L. V. and E. D. Hirleman (1981). Determining droplet size distributions of sprays with a photodiode array, Paper WWS/CI-81-49, 1981 Fall Meeting Western State Section, The Combustion Institute.

Lentz, W. J. (1976). Generating Bessel functions in Mie scattering calculations using continued fractions, Appl. Optics, 15, 668-671.

Roberts, J. H. and Webb, M. J.(1964) Measurement of drop size for wide range particle size distributions, AIAA J., Vol.2, No.3, 583-585.

Shifrin, K. S. (1956). Trudy Wsesoyunogo sacochnogo lesotechnicheskogo instituta, No. 2.

Swithenbank, J., J. M. beer, D.S. Taylor, D. Abbot, and G. C. Mc Greath (1977). In R. T. Zinn (Ed.), <u>Progress in Astronautics and Aeronautics, Vol. 53, AIAA.</u>

Yule, A. J., C. Ah Seng, P. G. Felton, A. Ungat and N. A. Chigier (1981). A laser tomographic investigation of liquid fuel sprays, Eighteenth Symposium (International) on Combust., The Combustion Institute, 1501-1510.

SOME ASPECTS OF UTILIZATION OF

MALVERN DIFFRACTION GRANULOMETER

D. Lisiecki, D. Allano, and M. Ledoux

U.A. CNRS 230/CORIA - Place Emile Blondel
B.P. 67   76130 Mont Saint Aigan, France

Two extreme opinions can be heard about the Malvern granulometer :
1) expressing a quasi religous confidence in this apparatus ; 2) rejecting it as a whole.

The authors had to use the Malvern in many different situations (oil
burners, pharmaceutical or perfumery sprays, diesel ...) and found eviden-
tly both of these attitudes irrelevants. As many others |1||2| it was
recognized that a good knowledge of measurement and processing is necessary as well as modifications and improvement of procedures.

In a first, essential features of the structure of the granulometer
and of processing will be recalled in order to locate the points where an
intervention was necessary.

Different aspects of the utilization of the diffraction granulometer
will be considered here :

i)   some improvements of procedures : elimination of some accicental
(background noise) or systematical errors (calibration, starting point of
calculation in the (X, N) plane), instationary sprays.
ii) some modifications in model independant procedure.
iii) a study of the effects of high obscuration values.

The granulometer concerned in this paper is of Malvern ST 2200 type.
It is to be emphasized that some remarks made further could be irrelevant
for more recent models (as the need of calibration). Nevertheless main
features of this study can be extended to never types of the granulometer.

1 - Structure of the granulometer and processing

The Malvern ST 2200 granulometer can be divided into :

- an optical section which makes a Fourier Transform of the spray, considered as an array of circle apertures. This transform is created on
  a detector (Fig. 1).

- a detector, composed of 31 semi circular diodes (only 30 are used -
  Why ?) plus a central triple diode (Fig. 2). This detector provides 30
  useful signals S(J). (J is the number of diode, 1 to 30 from center

to periphery). The diffracted pattern is then analyzed through 30 zones of various width, each of them between incident angles θ_{J1} and θ_{J2}.

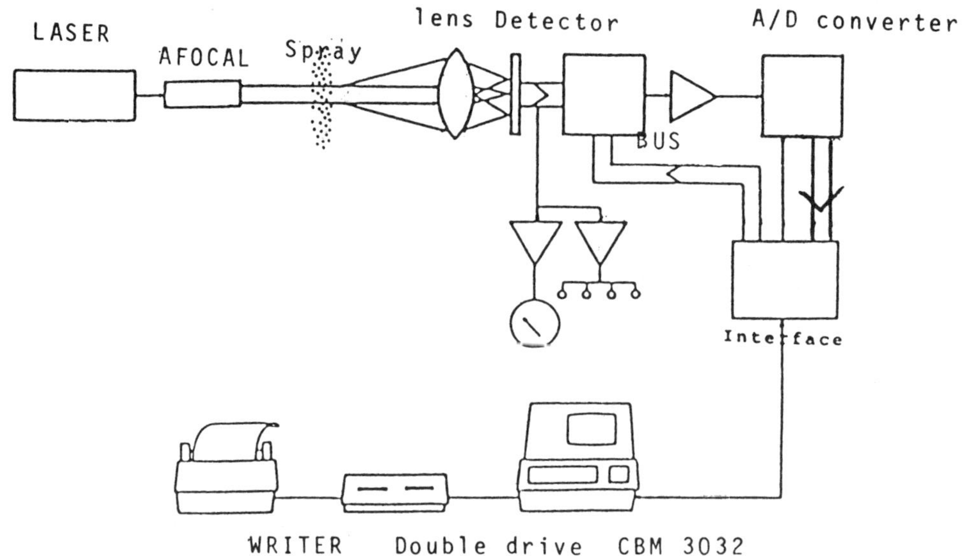

FIG. 1.   Structure of the MALVERN ST 2200

The central diode gives a double information : i) about the best setting of the optical axis (this triple diode disapeared from further version) ; ii) about obscuration, ie absorption. This term of obscuration is a "commercial" term and may be criticized. Perhaps extinction should be better. Nevertheless "obscuration" will be used in this paper. This obscuration, Ob is defined as $1-I/I_O$ where I is the intensity received on central diode, and I_O is the intensity emitted by the laser (received in absence of spray).

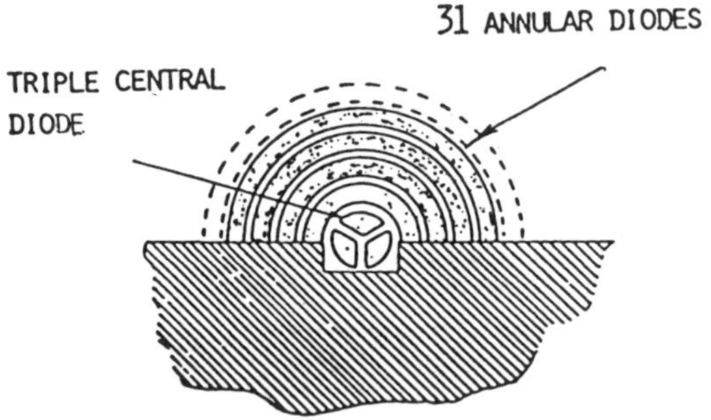

FIG. 2.   Sketch of the detector

From the 30 physical signals created by the detector, a set of datas is created. The commercial system works as follows (Fig. 3).

1.1 : Building of the datas

Each physical signal is digitalized. The exploration of the detector is sequential in the ST 2200 (This is not the case in further versions of the granulometer).

Taking 30 datas will be called a "sweep".

- First, a number N of sweeps of the 30 diodes is made without spray
  (measurement of background noise) the mean values of these N sweeps for
  each diode I is named B(I).

- Then the spray is examined. A number N of sweeps is made, and gives
  birth to 30 signals S(I) obtained, for each diode, from a mean on N
  experiments. The duration of a sweeps is 20ms, the processing time
  needed is 280ms. So a set of datas need 300ms to be obtained.

- It can be seen at this point that 30 informations can be obtained for
  the diffraction pattern.

In fact the processing reduces that information.

For each diode, a data D(J) is created through :

$$D(J) = K(J) \left[S(J) - B(J) \right] \qquad (1.1)$$

where K(J) is a calibration coefficient.

At this point datas are grouped two by two and only 15 resulting
datas are used in the calculation.

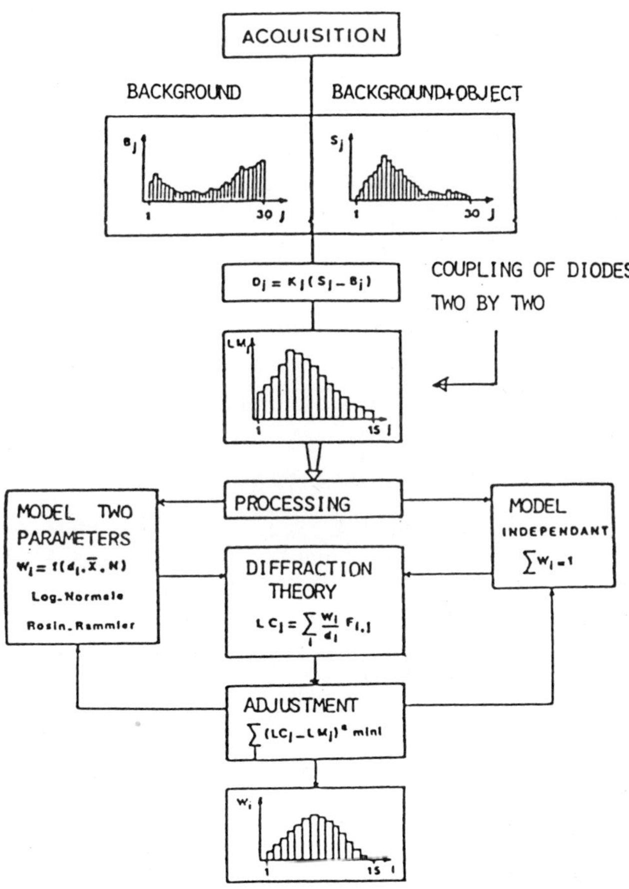

FIG. 3.   Acquisition, datas and processing

<u>1.2 : Processing</u>

The result obtained from the granulometer is a <u>relative</u> granulometry. Results can be given under two forms :

i) two parameters laws (Rosin-Rammler, Lognormal, normal law, not considered here)
ii) 15 classes histograms, is named model-independant.

The processing is based upon the Fraunhoffer diffraction theory. Each drop is treated as a disk. If W(I) is the relative weight fraction occupied by drops in a classe I centered on a diameter d(I), then, the theoretical (calculated light received by the diode (J) between angles θ_{J1} and θ_{J2} is $|3|$.

$$LC(J) = \sum_I \frac{W(I)}{d(I)} \, F(I, J) \qquad (1.2)$$

where F(I, J) is given, through theory :

$$F(I, J) = (J_O^2 + J_1^2)_{d(I),\theta_{I1}} - (J_O^2 + J_1^2)_{d(I),\theta_{I2}} \qquad (1.3)$$

So vector LC(J) can be related to the vector W(I) through a matrix T(I, J).

$$|LC(J)| = |T(I, J)||W(I)| \qquad (1.4)$$

Under commercial version T(I, J) is a given, built-in matrix. The inversion of a matrix like T(I, J) (of a great dynamic), is not useful. So, a trial and error procedure is used. A 15 classes W(I) histogram is "guessed", theoretical corresponding LC(J) are calculated, and then compared with measured LM(J). Comparison is made through a least square expression, named Log error LOGER.

$$LOG\ ER = Ln \sum_J \left[LC^2(J) - LM^2(J) \right] \qquad (1.5)$$

Two two-parameter laws are of interest :

Rosin Rammler (parameters α, N) and Log normal (X, σ). In these two cases a research is made in the (X, N) or (X, σ) plane of the minimum LOGER. The starting point of the research can be given by the apparatus of choosen by the operator. From a couple (X, N) or (X, σ) an 15-class histogram is built to construct the LOGER. Two parameters have to be found. In the case of model independant, 15 parameters have co be searched, but the building of LOGER is evident.

<u>2 - Some improvements in procedure</u>

<u>2.1 : Systematical error : calibration of diodes</u>

K(J) coefficients in (1.1) contain information on efficiency of each diodes. So they result mainly from a calibration. For the commercial apparatus the authors used, these K(J) were obtained apparently from a sampling among a certain number of apparatus.

A recalibration was made by the authors which showed substantiel differences with built in values. Effects on the results will be seen in § 3 on Fig. 7.

562

<u>2.2 : Accidental error</u> : effect of background fluctuation

A complete study of background noise was made. This study was led by
escaping the commercial system, and recording independently on the
diskette each value of B(J) obtained for each sweep (without time-mean
operation). For the situation tested, two main frequencies of variation
of background were observed : a "high" frequency (approx.some Hz) and
a great amplitude low frequency (approx. 2mn in period) probably due to
mechanical problems. So a compromise has to be found between duration of
experiment and number of sweeps.

It is evident that this study has to be made for each different
experimental situation. Complete results have been given in |4|.

Only another feature of this problem will be pointed out here. The
commercial system takes accout of the background only with <u>initial</u> values
of B(J). If any accident occurs during spray existence (e.g. wetting of
lens ...) it cannot be detected. So, we propose here another procedure
(Fig. 4) where a comparison is made between initial background BI(J),
before experiment, and final background BF(J), after experiment.

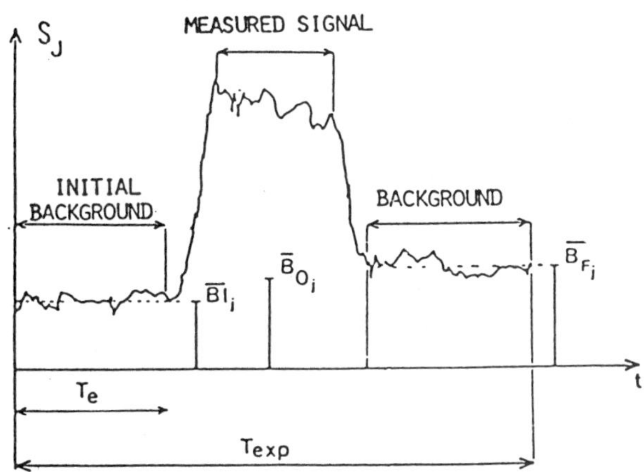

FIG. 4. A test based on the evolution of the background

In practice, the authors use the following criterium, if
$$\sum_J \frac{BF(J) - BI(J)}{BI(J)} > 5 \%,$$ the measurement is rejected. This criterium is
severe, as shown on the Fig. 5 (case of a wetting of optical surface
leading to a 100% variation in background).

<u>2.3 : Effect of the starting point in the (X, N) plane</u>

As related in § 1, the research of the (X, N) or (X, σ) point in the
(X, N) plane is made from an initial value. The search of a minimum LOGER
is dependant of the ΔX and ΔN, steps of X and N. In order to understand
the procedure followed by the ST 2200, values of LC(J) where calculated
for a given distribution (Rosin-Rammler X = 220µm, N = 8) and introduced
as measured LM(J) values in the computer. Using an order "CP" on the
computer, initial values of the iteraction are given by the software
$(X_i > X_o , N = 5)$. The results of the different iteraction are shown on
Fig. 6. It can be seen that X is reached from upper values and N through
lower values. The result is : X = 223.2 ; N = 7.4 with a LOGER of 3.5
which can be satisfactory. If better accuracy is needed, another method

can be employed : it should be better to fix lower values of ΔX and ΔN
but the duration of the operation would be too long. So, a first estimation of (X, N) can be made through a "CP" order, and then used as the
starting point for the research of a better (X, N) with smaller ΔX and ΔN.

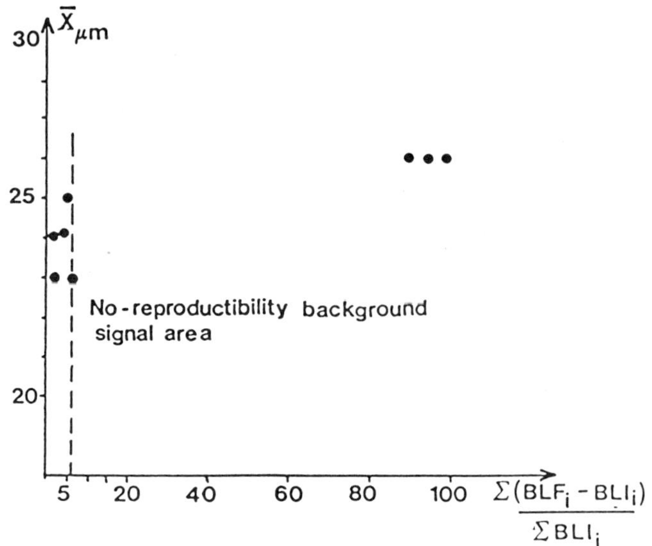

FIG. 5.   5% test on background

On curve (Fig. 6) appear curves of iso-values of LOGER. It can be
seen that research of final (X, N) is made through a "steepest slope"
procedure.

2.4 : Short duration sprays

For intermittent sprays, it is necessary to synchronise measurement
and apparition of the mist in the laser beam. The device described
here can only be applied to quasi-stationary, reproducible sprays (time
evoluting spray has been studied by Watson |8|).

Two methods have been employed to obtain this synchronisation :

- the start of measurement is given by the signal given by a photodiode placed on the side of the apparatus and collecting scattered light
by the spray. This method is now used by Malvern instruments under commercial form.

- The start can be synchronized by the central diode : a gating of obscuration signal gives the order of measurement.

3 - Modification of "model independant" procedure

The built-in T(I,J) matrix, defined in (1.4) leads to a fixed definition of the classes of histogram (related to the focal length of the
lens). These limits of classes are not always convenient, especially for
lightly disperse sprays. This is usually the case for pharmaceutical sprays
where particles can be limited to a narrow range (e.g. 50-100μm).

From (1.3) and (1.4) it can be seen that other repartitions of histogram classes can be obtained by changing the T(I,J) matrix. Another point

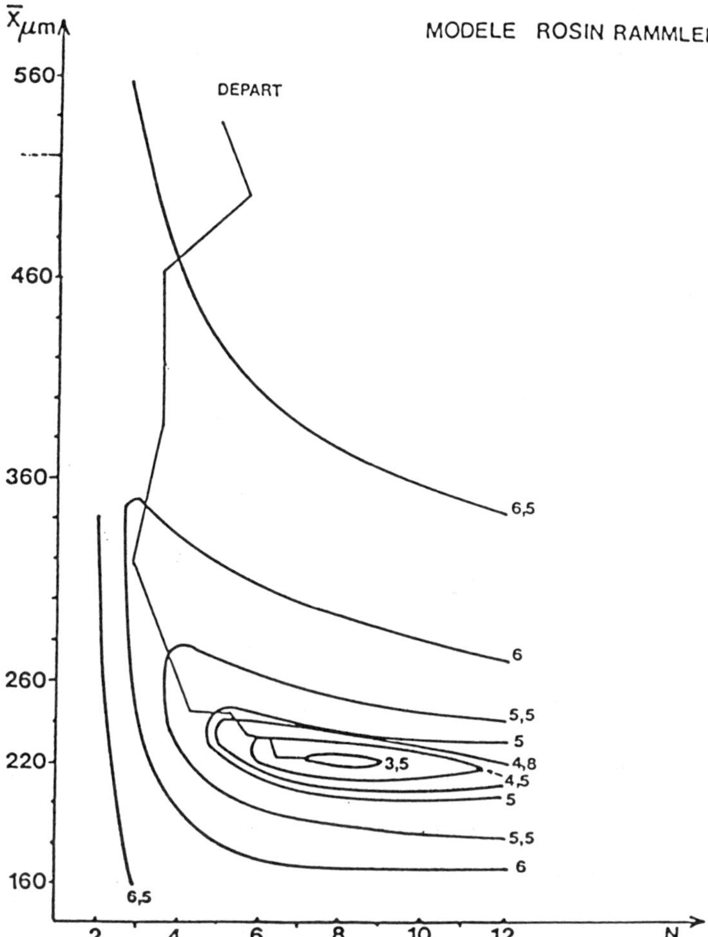

FIG. 6.   Research fo X,N - ISO-LOGER values

is that, in fact 30 informations (datas) D(J) are available. So an appropriate choice of D(J)|J=1 to 30| used and d(I) can theoretically be made. Some care is then to be taken :

i) the examination of the background noise on each diode (J) may lead to an elimination of the more "noisy" ones, detected as explained in §2.1.
ii) the built-in T(I,J) matrix is bounded by two considerations :

- the size of the computer : the limitation of T(I,J) to a 15 x 15 matrix seems to be due to the 30K of the computer. Two by two coupling is then arbitrary.

-the sensitivity of F(I,J) to d(J). As shown in |3| the choice of classes is related to the maximum slope of $J_0^2 + J_1^2$ function.

At first right, T(I,J) could be any n x m (n = 1 to 30, m = 1 to 30) matrix. In fact the remark above led to take care to the numerical

sensitivity of the method. The authors used a method of "visualisation"
of the matrix (not explained here) to obtain an ideal compromise between
sensitivity and choice of the classes of the histogram.

As a summary, this procedure allows :

- an optimal choice of datas D(J) used
- a model independant presentation more adaptated to a given spray.

Fig. 7 gives an illustration of the modifications presented in § 2
and 3. This Figure shows results of measurements made on a glass bead distribution. This distribution has been measured first through photograph,
manual and image processing (upon 600 beads). Results of the "time" distribution are shown on Fig. 7 as "standard distribution"($— \cdot \cdot —$)
under cumulative form.

By using the Malvern ST 2200 without any care (reading the book and
pressing the buttons) (i.e. using "commercial" K(J) values) the straight
line ($\_\_\_$) is obtained.

Recalibrating the K(J) leads to the "calibrated" ($— \cdot —$) curve.
The result is dramatic.

With a recalculated (10 x 10) matrix T(I,J) using classes 10µm in
width, the "new matrix" curve ($- - - -$) is obtained.

The advantage of the method is evident.

4 - Influence of strong obscuration

Long ago the possibility of errors due to multiple scattering has
be recognized. This multiple scattering is bounded to the present of a
strong ($\sim$ 1) value of obscuration parameter Ob.

Negus and Azzopardi |10| following Felton inferred that no measure-
ment was mossible for OB>.5.

Some authors proposed correction formulas to be applied to the
results.

From theoretical calculations, Felton |7| propose (e.g. for the
Rosin Rammler distribution) the formulas :

$$\frac{X}{X_{app}} = 1 + (0.036 + 0.4947 \, Ob^{8.997}) N_{app}^{(1.9-3.43706)} \qquad (4.1)$$

$$\frac{N}{N_{app}} = 1 + (0.035 + 0.1099 \, Ob^{8.65}) \, N_{app}^{(0.35 + 1.45 \, Ob)} \qquad (4.2)$$

where (X,N) and (X_{app}, N_{app}) are respectively real and measured (X,N).

These formulas are proposed for $65\% < Ob < 98\%$ and $1.2 < N_{app} < 3.8$.

Brault and Lourme |2| from experiments determined

$$\frac{X_{app}}{X} = 0.03 \cos\{ \frac{\pi}{4} (Ob^5 + Ob^{8.5})^2 \} + 0.7 \qquad (4.3)$$

$$N = N_{app} \; ; \; Ob < 0.8 - 0.1 \, N \qquad (4.4 \text{ a})$$

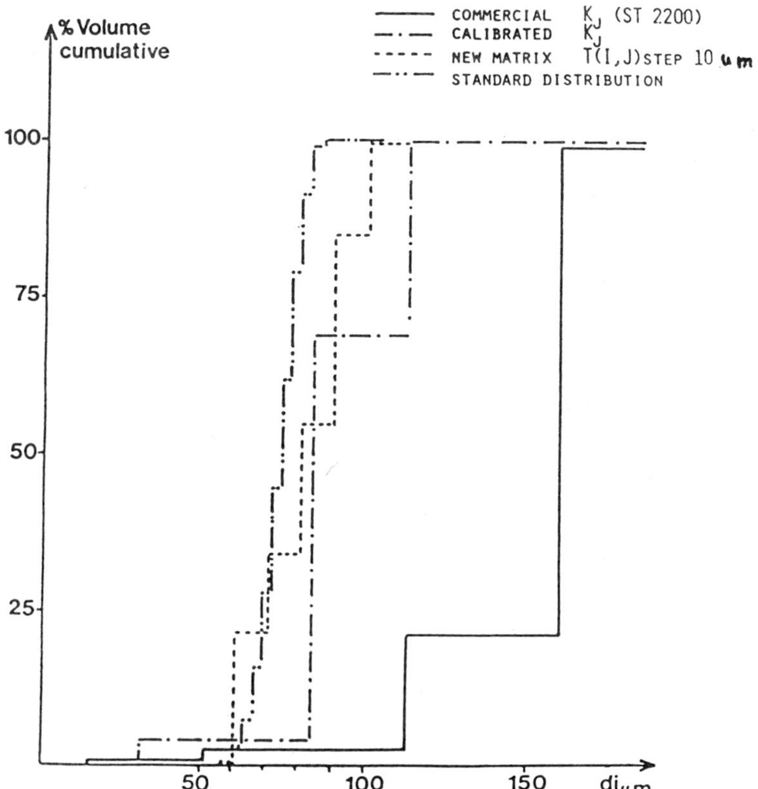

FIG. 7.   Different results obtained with a ST 2200
for a 63-80µm glass beads distribution.
Effect of K(J) and new matrix procedure.

$$N = 0.8 \; \frac{0.8 + N_{app} - Ob}{1.68 - 0.1 \, N_{app} - Ob} \quad ; \quad Ob > 0.8 - 0.1 \, n \qquad (4.4 \text{ b})$$

which agrees well with Felton's results.

Hirleman |9| is now carrying calculations following the same aim.

The fact is that the same obscuration can be obtained for different
length of spray traversy l_F, and volumetric ratios C_v

$$(C_v = \frac{\text{volume of particles}}{\text{volume of carrying flow}})$$

In order to get more information about the relative effects of C_v
and l_F, the authors compared the results given by the Malvern for different polymer beads suspensions of unique sizes (100µm, 200µm and 300µm
in dia) contained in glass cells of different lengthes l_F (Fig. 8)
shows the experimental situation.

Some results for 220 and 100µm polymer beads are shown in Fig. 9 and
10.

For a same Ob, the lowest l_F corresponds to the highest C_v (C_v from 1.4 to 5.3^{-3} were used). It can be seen that the global effect of Ob is to reduce X apparent and give a wider distribution. For Rosin Rammler X_{app} decreases, N diminishes ; For Log. normal X decreases, σ increases.

It can also be seen that, at the same Ob, better results are obtained for higher l_F, that is to say, the smaller C_v. In other words the multiple scattering effect is less for the greatest mean distances δ between particles. In these experiments values of δ are varying from 1mm to 500µm far beyond the wave length of the laser. So a situation of independant scattering can be easily assumed |10|.

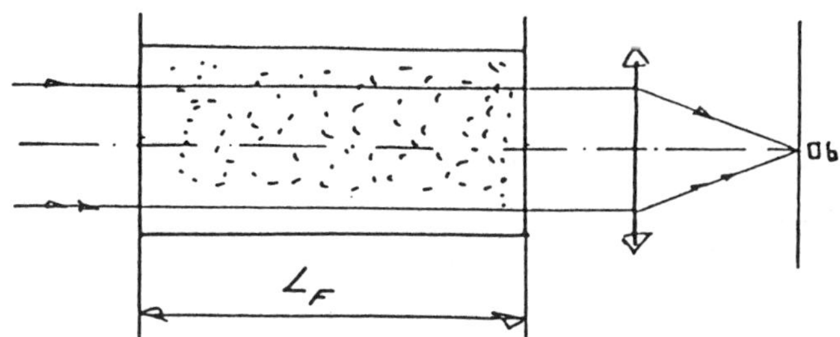

FIG. 8. Sketch of experiments for strong obscurations

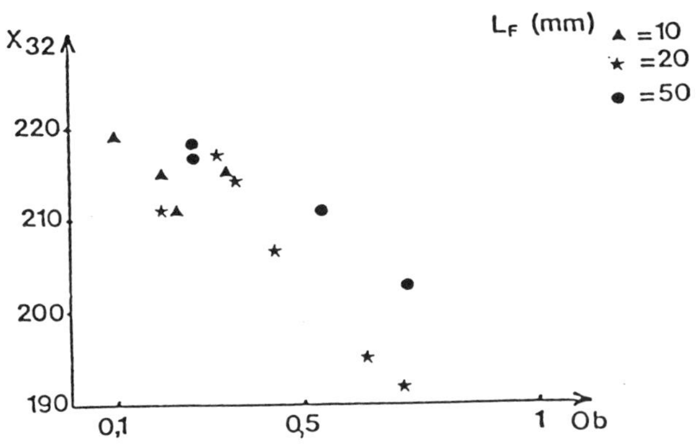

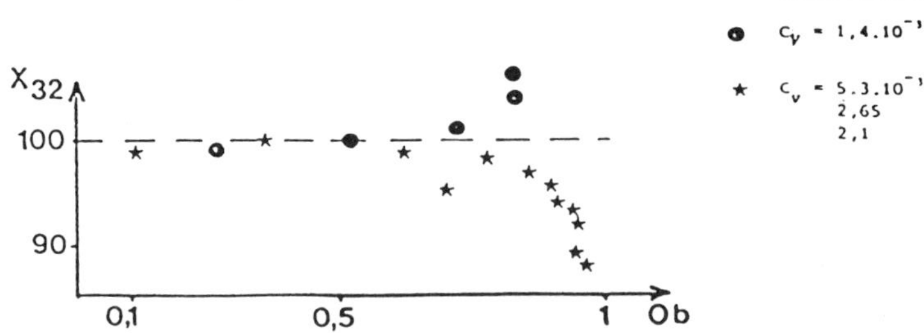

FIG. 9. Influence of Ob and l_F on measured sauter mean diameter

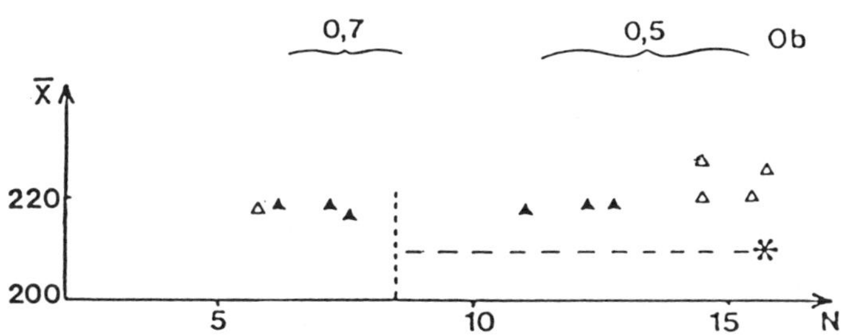

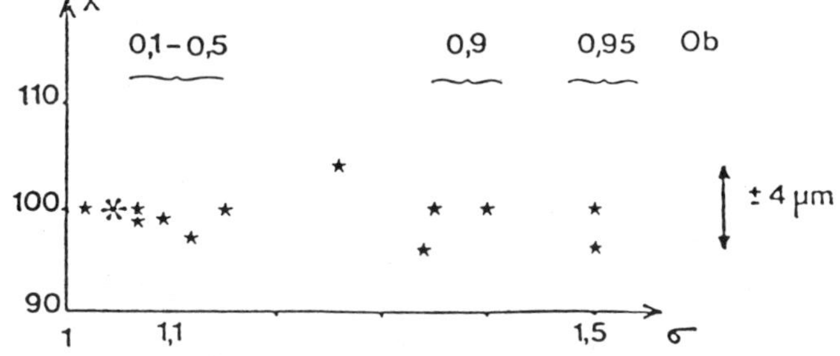

FIG. 10.   Influence of Ob and l_F on N and σ

In order to get more information on the influence of multiple scattering a plot of C_v/d against the extinction coefficient K_{ext} (deduced from Ob) is made in Fig. 11.

K_{ext} is related to Ob through :

$$K_{ext} = - \frac{1}{l_F} \, Ln \, (1 - Ob) \qquad (4.5)$$

for very light sprays (single scattering) C_v should be related to K_{ext} through :

$$C_v = \frac{2}{3k} \, d. \, K_{ext} \qquad (4.6)$$

A plot of C_v/d against K_{ext} allowt to check if k is equal or different from 2.

Fig. 11 shows that for values of obscuration Ob higher than 0.7 even taking care of limited accuracy (due to the use of central diodes) a value of K = 2 cannot be accepted.

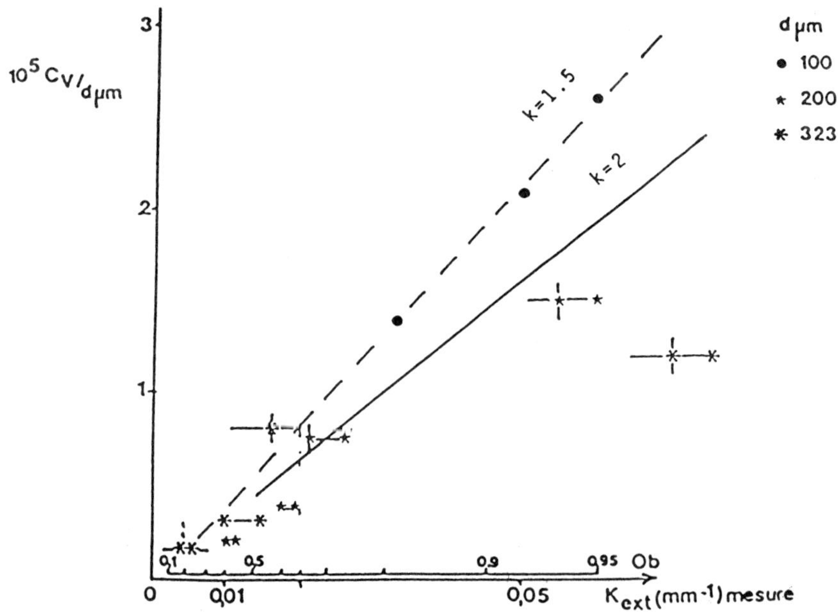

FIG. 11.   Variation of C_v/d against K_{ext} value of K

It can also be seen in Fig. 9 that good results can be obtained in
certain cases (high l_F) for obscuration up to O.8. The criterium of
Ob O.5 is then to be applied with some prudence.

CONCLUSION

Some suggestion are made in this paper to modify some of the procedures of utilization of the Malvern ST 2200.

A better control of background influence can be obtained from a
previous investigation of its frequency range. Also problems due to
wetting of optical surfaces can be avoided. The research of (X, N) values
seems to be satisfactory but, when required, better accuracy can be
obtained without waste of time.

A method for obtaining home made histograms has been tested. Its
interest appears to the authors as lying in the fact that width of classes can be choosen at the level of the calculation from the datas, and
not by a recalculation from the "commercial" histogram, which would
cumulate the errors.

At last it has be shown that good measurements can be obtained for
high ($\sim$ O.8) obscurations. But, to legitimate a result in such circumstances, the value of C_V has to be known. A better knowledge of the relationship between $C_V(l_F$ or $\delta)$ and k would be necessary to go further and
to determine concentration from obscuration measurements. At this
point an amount of calculations under multiple scattering conditions is
needed. It is to be remarked that correction formulas like 4.1 to 4.4
give a bijective dependance between (X, N) and (X$_{app}$, N$_{APP}$). Present
experiments show clearly that it is not usually the case.

570

BIBLIOGRAPHY

1   C. Negus, B.T. Azzopardi. A.E.R.E. Harwell Owfordshire. Report AERE
    H 9075, Dec. 1979.

2   F. Brault, D. Lourme. ICLASS 2985, London Paper IIC/4/1.

3   J. Swithenbank, J.M. Beer, D.S. Taylor, D. Abbot, G.C. Mc Greath.
    Paper 76-69. A.I.A.A. 14th Aerospace Science Meeting. January 1976.
    Dept of Chemical Engineering and Fuel Technology, University of
    Sheffield. Internal Report HIC 245.

4   D. Lisiecki, D. Allano, M. Ledoux. 4ème Congrès de Chimie Analytique.
    3ème journée d'Etudes sur les aérosols. Paris 8-12 Déc. 1986.

5   D. Lisiecki, D. Allano, M. Ledoux. 6th International Congress of
    International Society for Aerosols in medicine. Vichy, 2-4 Oct.1986.

6   L.G. Dodge. Applied Optics, Vol. 23, n° 14, July 1984.

7   P.G. Felton, A.A. Hamidi, A.K. Aigal. ICLASS 1985, London, Paper
    IVA/4.

8   D.J. Watson. ICLASS 1985, London, Paper VC/2/1.

9   E.D. Hirleman. The Combustion Institute Spring Meeting, Western
    State Section - April 27-29 (1986) Alberta Canada.

10 M.R. Brewster, C.L. Tien. Journ. of Heat Transfer, vol. 104,
    p. 573, Nov. 1982.

OPTICAL METHODS IN SUBMICRONIC AEROSOL MEASUREMENT

Denis Boulaud and Guy Madelaine

Laboratoire de Physique et Métrologie des Aérosols
IPSN/DPT/SPIN/LEPA, Commissariat à l'Energie Atomique
BP 6, 92265 Fontenay-Aux-Roses Cedex, France

INTRODUCTION

Industrial requirements for the detection of submicronic aerosols
are increasing : combustion processes, nuclear fuel cycles, microelectronics, space research, new materials, pollution prevention and control.

For each application, the equipment must be chosen on the basis of
two factors :

1) the parameters characteristic of the aerosol to be stutied ;
2) the conditions for measurement and the resultant constraints
imposed on the equipment.
The parameters characteristics of the aerosol may be qualitative or
quantitative : nature, morphology, numerical or mass concentration,
specific surface area and size distribution. It is necessary in all
cases to specify the properties to be quantified (inertial, diffusional,
optical, electrical properties).

There are three basic categories of conditions for measurements
relating to :

- the aerosol itself (size, maximum and minimum concentrations,
time-dependent changes) ;
- thermalhydraulic conditions (temperature, pressure, composition
and flow rate of fluid vector) ;
- economic considerations (frequency of measurement, data
acquisition speed, cost).
These three categories impose constraints on the equipment and
define the basic characteristics of the instruments according to the
intended use.

A short description is given of conventional solutions to the
problems of aerosol characterization in different conditions for
measurement and of the methods available for submicronic aerosol
measurements :

- inertial methods (impactor, cyclone, centrifuge),
- diffusional methods (diffusion batteries),
- electrical methods (electrical aerosol analyzer, differential
mobility analyzer).

The limits of application of these methods is given as well as a
description of their advantages and drawbacks.
The role of optical methods in this field of measurement is examined.
Their specific contributions, applications, advantages and drawbacks are
compared with those of conventional methods, through the use of two
examples.

WHICH MEASUREMENTS ? WHICH INTRUMENTS ?

The choice of equipment for aerosol measurement must be made
according to :

- the properties and parameters required, and,
- the conditions for measurement.

Properties and parameters characteristic of an aerosol

<u>Qualitative</u> : qualitative parameters include phase (solid or liquid),
electrical state, chemical form, morphology (sphere, cube, chain).

<u>Quantitative</u> : the most common quantitative parameters are numerical and
mass concentration, specific surface area, mass per unit volume,
refractive index, size distribution and electric charge.

Characterization of parameters requires definition of the
properties and behavior specified by these parameters :

- aerodynamic (inertia, sedimentation, diffusion), characteristic
of transport mechanisms,
- optical, characteristic of the interaction with electromagnetic
radiation,
- electrodynamic, characteristic of the transport of charged
particles.

Conditions for measurement

The conditions for measurement can be classified into three main
categories relating to :

- the aerosol itself,
- the thermalhydraulic conditions of the fluid,
- the operational aspects of the system of measurement.

These three factors impose constraints on the equipment and define
the essential features of the different instruments according to their
intended use. In order to illustrate this, the following discussion,
with the help of three tables (1 ; 2 ; 3), deals with the relations
between some measurement conditions and the constraints they impose on
measuring devices. The tables are not exhaustive, but they do illustrate
the main questions that must be answered before aerosol measurement.

Solutions to the problems related to aerosol equipment will be
defined by the properties of the aerosol in question and the conditions
for measurement. The following discussion first deals with conventional
methods and then with new optical methods and their place among existing
instruments.

Table 1. Relation between the aerosol to be studied and the
constraints on the measuring device

Conditions of measurement : aerosol		Constraints on the device
Size range	broad ->	broad measurement dynamic
	restricted ->	high power of resolution
Numerical or mass concentration	high ->	necessity or otherwise for dilution
	low ->	increase in sample volume
State of aerosol	solid ->	risk of rebounce, reentrainment
	liquid ->	risk of migration
	electrically charged ->	risk of deposition on the wall
Chemical nature	->	risk of interaction with materials constituting the apparatus
Time course	->	optimization : - number of samples - time of sampling - duration of sampling

Table 2. Relation between thermalhydraulic conditions of the fluid
vector and constraints on the measuring device.

Thermalhydraulic conditions of the fluid vector	Constraints on the measuring device
High or low pressure High or low temperature	technological aspects : - nature of materials - airtightness - telecommand calibration - simulated conditions - real conditions
Flow rate and velocity of fluid vector	in situ measurement sampling : optimization of sampling rate in order to respect the isokinetics of the velocities and to limit deposition in the sampling lines

Table 3. Relation between measurement characteristics and
constraints on the measuring device

Measurement characteristics		Constraints on device
Frequency of measurements	->	The time needed to reset the device between two must be less than the interval between consecutive readings
Rapidity of data acquisition	->	Detection and treatment of data in real time, or otherwise
Use	->	Function of the degree of complexity of the various steps necessary for operation of the system and their degree of automation

CONVENTIONAL METHODS

Conventional methods based on the dynamic or electrical mobility of submicronic aerosol particles can be divided into three groups :
- inertial,
- diffusional,
- electrical.

It is also possible to combine two or more methods in parallel or in series. Methods by collection and examination in optical or electron microscopes will not be dealt with here since they are often limited to qualitative parameters (nature, morphology).

Inertial methods

When a particle is sufficiently large, it is theoretically possible to define and predict changes in its movement as a function of the different forces acting on it. This movement may be rectilinear, i.e. in a straight line, with or without acceleration, or curvilinear, due to the effects of different forces acting on the particle.

Movement can be predicted by solving the force equilibrium equation, which gives the acceleration, velocity and position of the particle as a function of time :

$$m \, d\vec{v}/dt = \Sigma \vec{F} \tag{1}$$

where m is the mass of the particle, $\vec{v}$ its velocity vector and $\Sigma\vec{F}$ the sum of the forces acting on the particle. The forces of resistance or friction ($\vec{F}_r$) of the fluid acting on the moving particle can be incorporated into this expression :

$$m \, d\vec{v}/dt = - \vec{F}_r + \Sigma\vec{F}_{ext} \tag{2}$$

where $\Sigma\vec{F}_{ext}$ represents the sum of the external forces acting on the particle. The force $\vec{F}_R$ is a function of the relative velocity of the particle with respect to the fluid and may be complex.

$$\vec{F}_R = (3\pi\eta d \, (\vec{v}-\vec{U})/C) \, f \, (Re) \tag{3}$$

where $\vec{U}$ represents the velocity vector of the fluid, d the particle
diameter, η the dynamic viscosity of the fluid, C a correction factor
related to the ratio of particle diameter and the mean free path length
of the fluid molecules and f(Re), which is a function of the Reynolds
number of the particle.

$$Re = |\vec{v} - \vec{U}| \, d/\nu \qquad (4)$$

where ν is the kinematic viscosity of the fluid.

Equation (2) therefore becomes :

$$d\vec{v}/dt = \frac{(\vec{v}-\vec{U})}{mB} f(Re) + \frac{\Sigma \vec{F}ext}{m} \qquad (5)$$

where B is the dynamic mobility, $B = C/3\pi\eta d$. The product mB corresponds
to the relaxation time τ of the particles :

$$\tau = mB = d^2\rho_p C/18\eta \qquad (6)$$

where ρ_p represent the mass per unit volume of the particles

From equation (5) it can be seen that the equation of movement
relates the acceleration of the particle to its dynamic mobility, or its
relaxation time and to the external forces acting on it.

Equation (5) forms the basis of the various inertial and electrical
methods. The basic idea is the same : acceleration ($d\vec{v}/dt \neq 0$) or

constant drift velocity is applied to the particle so that $\vec{v} \neq \vec{U}$. The
trajectory of the particles will therefore differ from that of the fluid.

The value of these methods clearly relates to the characterization
of the aerodynamic behavior of the aerosol by its relaxation time.

Two types of devices based on equation (5) can be described : [1], [2],
[3] :

- those using an external force :
    * gravity field (sedimentation chamber),
    * centrifugal force (centrifuge),
    * electrical force, the use of which is dealt with in detail below,
- those imparting an acceleration to the particle by variation of the

velocity vector ($\vec{U}$) of the fluid. This variation may be :
    * in the direction : cyclone or impactor,
    * in amplitude (Aerodynamic Particle Sizer From TSI Inc)[3].

In general, size distribution of the aerosol through the use of
these devices is achieved by the following operations :

- <u>selection of particle</u> size, by application either of an external
force or by a sudden change in velocity vector of the fluid. This stage
is often carried out with simple devices (impactor, cyclone,
sedimentation chamber) but the solution of equation (5) is not always
easy, and this sometimes results in approximative resolution of particle
trajectory thus leading to uncertainty in the determination of particle
relaxation time.

- <u>Detection of particles</u> on the basis of size, either after
collection on a support (chemical or gravimetric analysis) or by optical
detectors, the aerosol still being in suspension in the fluid (case of
APS).

- <u>Reduction of data</u> collected after selection and detection in
order to restore size distribution.

In the most commonly used devices these operations are rarely
performed simultaneously and in real time (with the exception of APS),
and measurements are therefore relatively time-consuming. The various
devices available will not be described here, but on the basis of the
questions raised in the introduction table 4 provides a summary of
commonly used devices and their characteristics.

<u>Electrical methods</u>

All aerosols, natural and artificial, may be electrically charged (positive or negative). The forces exerted on charged particles in an electric field $\vec{E}$ may be much greater than those due to gravity, diffusion or heat. This property can, therefore, be used to apply an external force ($\vec{F}ex = pe\vec{E}$) and hence obtain the dynamic (B) or electrical (Z) mobility of the particles from the expression :

$$Z = peB \tag{7}$$

where p is the number of charges and e the elementary electric charge.

Four successive steps are usually employed with electrical methods[4] :

- <u>labeling of particles</u> by applying a known number of electric charges. Three techniques are generally used, by placing the particles in contact with :
  * a cloud of bipolar ions,
  * a cloud of unipolar ions,
  * a cloud of unipolar ions with the superposition of an electric field.

- <u>selection of particles</u> by application of an electric field in an electrostatic precipitator the power of resolution of which is classified in three categories :
  * order 0,
  * order 1 : Electrical analyzer of aerosols (EAA),
  * order 2 : Differential mobility analyzer (DMA).

- <u>detection of selected particles</u> either by electrical measurements or by the use of other detectors (optical counters, condensation nuclei counters),

- <u>reduction of data</u> collected after selection an detection, allowing establishment of size distribution.

Table 4 gives details of the most commonly used devices.

<u>Diffusional methods</u>[2]

Small particles suspended in a gas are subject to random motion (Brownian notion) consequent upon their collisions with gaz molecules. Due to this Brownian motion, when the particles are uniformly distributed their position changes but their distribution in space remains identical. Following perturbation, this uniform distribution is reestablished by Brownian motion, which results in movement of particles from areas of high particle concentration to areas of lower particle concentration. This process is known as particle diffusion and it obeys the laws of molecular diffusion known as Fick's Laws. Fick's first law relates particle flux (J) per unit area per unit time to the concentration gradient (dn/dx) by the use of a constant known as the diffusion coefficient (D).

$$J = - Ddn/dx \tag{8}$$

This diffusion coefficient is related to the dynamic mobility of the particle B, and hence to its diameter (d), by the Stokes-Einstein relation :

$$D = k\ TB = kTC/3\pi\eta d \tag{9}$$

where k is Boltzmann's constant and T the temperature.

<u>Deposition by diffusion</u> Gas molecules subject to Brownian motion rebound when they collide with a wall. In contrast, the particles of an aerosol adhere to the wall surface upon collision. This means that particle concentration in the vicinity of the wall is nil and a diffusional flow a particles to the wall therefore results.

<u>Diffusion batteries</u> These devices use the phenomenon of deposition described above to separate particles on the basis of their diffusion

coefficient D. The selection principle generally operates by circulating
the aerosol through a system subject to diffusion. This system may be a
bundle of cylindrical tubes or parallel plates, a screen or a granular
bed. Concentrations upstream and downstream from the measuring device
must be determined.

Data reduction is often complex because selectivity of the
diffusion mechanism is weak due to its dependence on aerosol size. The
characteristics of these devices are listed in table 4.

Combination of several methods

Several methods can be combined to cover a broader range of
particle sizes. We have been able to cover a 2000-Fold range of size by
combining a cascade impactor and a diffusion battery built with a
granular bed (SDI 2000)[5]. This system is still in the prototype stage
but its value is clear since it can be used to characterize the aerosol
by its true aerodynamic behavior over the whole dimension range :
inertial behavior from 15 to 0.3 µm, diffusional behavior from 0.3 to
0.0075 µm.

Summary

Although not exhaustive, table 4 outlines the conditions for
application of the conventional devices described in this section. It
can be seen that these methods are versatile but do not cover all
potential applications, notably :
- detection in real time of very low concentrations of aerosols
less than one micrometer in diameter,
- detection in real time of submicronic aerosols at high
temperature and pressure.

There are doubtless more cases not covered by conventional methods,
and the following section is devoted to the contribution of optical
methods to the whole range of applications, whether or not conventional
methods are adequate.

OPTICAL METHODS

Light scattering is an extremely sensitive tool for the measurement
of the concentration and dimension of aerosols. A single 0.1 µm particle
can produce a detectable signal of scattered light. Techniques based on
scattered light have the advantage that the aerosol and fluid vector are
relatively unperturbed and that an instantaneous response is obtained
allowing real-time monitoring of aerosol characteristics. One drawback is
that light scattering is sensitive to small changes in refractive index,
scattering angle, and particle size and morphology, which may confound
interpretation of experimental results.

The methods fall into three broad categories :
- characterization of the extinction or scattering of light by
aerosol clouds,
- characterization of light scattered by a single particle,
- use of light as a vector of non-specific information regarding
the optical properties of the particles (counting, measurement of
velocity, Brownian motion).

These different techniques have been dealt with in detail during
this symposium and fall outside the scope of the present article. We
shall, therefore, limit the discussion to their use for measurement of
submicronic aerosols and will consider two examples of applications
where conventional methods prove inadequate :
- detection of very low concentrations at normal or high
temperature and pressure,
- detection of small particles ($\leq$ 0.1 µm) at high temperature and
pressure.

Table 4. Comparison of different methods and conventional devices

Methods	Devices	Aerosols				Thermalhydraulic	Operation	
		Size range (μm)	Resolving power	Acceptable concentration or mass		Adaptation to variable conditions	Detection in real time	Use
				Max*	Min			
Inertial	Sedimentation	1 to 50	Average	100 g	b	awkward[e]	no	awkward
	Centrifuge	0.05 to 5	Very good	1 g/m^3	b	awkward[e]	no	awkward
	Impaction	0.05 (a) 0.3 to 15	Good	0.1 to 1 g	b	possible	no(f)	simple
	Cyclone	0.5 to 20	Average	50 to 100g	b	possible	no	simple
	APS	0.8 to 15	Very good	10^3 p/cc	c	difficult	yes	simple
Electrical	AEA	0.01 to 1	Good	1 to 10 mg/m^3	b	difficult	yes	simple
	DMA	0.008 to 0.5	Very good	100 mg/m^3	b	difficult	yes	simple
Diffusional	Diffusion Battery	0.005 to 0.3	Low	d	b	possible	no	awkward[g]
Inertial + diffusional	SDI 2000	0.0075 to 15	Good Low	d	b	possible	no	awkward[g]

a case of low-pressure impactors
b function of means of detection and analysis
c very low for large sample volumes
d function of diffusion medium
e preservation of laminar flow of fluid
f use of quartz balance
g data redution requires complex algorithms
* typical values

<u>Detection of low concentrations</u>

There is an increasing need in the microelectronics and space industries for reduction of particulate contamination to extremely low levels (a few particles per m^3). For this it is necessary to detect particles one-by-one, and only optical methods are able to achieve this. Optical methods have the additional advantage that they can be adapted to high temperatures and pressures. Currently available laser particle counters have a lower detection limit for particle diameter of approximately 0.1 µm, with a sampling rate of 2.8 1/min (Pacific Instruments, Particle Measuring Systems).

Two additional requirements are apparent :
- a tenfold increase or more in sampling rate in order to deal with very low concentrations in acceptable sampling times,
- a tenfold reduction in the lower detection limit in order to account for virtually all particles present in a clean room.

Tenfold increases in sampling rate are now a real possibility and laser counters operating at a rate of 28.3 1/min, with a lower detection limit of 0.1 µm, should soon be commercially available.

A tenfold reduction in the lower detection limit poses more complex problems in this size range (0.01 to 0.1 µm), since Rayleigh scattering dominates when a laser is used in the visible range and varies as d^6. This means that there will be a decrease in the amount of light scattered at 0.01 µm by a factor of 10^6 compared with the present limit.

This problem can be overcome by the expedient of increasing particle size, in supersaturated vapor, until the particles are detectable by conventional optical means. This method has already been employed in condensation nuclei counters (CNC)[6], which can be classified into three groups :
- expansion CNCs, in which supersaturation is achieved by pseudo-adiabatic expansion of an air-vapor mixture,
- continuous flow CNCs, in which supersaturation results from passage of the sample, previously saturated in condensated vapor, through a refrigerated tube,
- a third type of CNC in which supersaturation is obtained by continuous mixing of vapor and cold air containing the particles in question.
CNCs have the advantage of covering a wide range of particle sizes. However, they have the drawback that information on particle size is lost.

Devices are also available which only increase particle size ; detection is effected downstream by an optical counter. We have developped a continuous flow device of this type in which the condensable liquide is glycerol. This device has enabled us artificially to lower the detection limit of optical counters to slightly below 0.01 µm[7]. In this system the saturator is maintained close to 75 °C and the condensor at approximately room temperature. The present sampling rate is 2.8 1/min, but studies are under way to increase flow tenfold in order to reach extremely low concentrations ($\cong 100/m^3$) of particles of sizes ranging between 0.01 and 10 µm, with reasonable sampling times (between 10 and 60 minutes).

<u>Detection of small particles at high temperature and pressure</u>

The increasing importance of polymers and the production of very fine powders in industrial and biological applications call for control of combustion processes and various chemical reactions in order to carry out manufacturing controls. This need has led to efforts aimed at developing real-time detection methods for very small particles (0.005 to 1 µm) produced during various operations, frequently at extremes temperature and pressure.

Table 5. Optical methods for submicronic aerosols

Methods	Devices	Aerosols					Thermalhydraulic	Operation	
		Size range (μm)	Resolving power	Acceptable concentration or mass			Adaptation to variable conditions	Detection in real time	Use
				Max[*]	Min				
	White light	0.3 to several tens of μm	Very Good	10^5/cc	c		possible	yes	simple
Single particle counters	Laser	0.1 to several tens of μm	Very Good	10^5/cc	c		possible	yes	simple
	Enlarging device + Laser	0.01 to several tens of μm	None	10^5/cc	c		awkward	yes	simple
PCS		0.01 to 1	Good if aerosol monodisperse	h	h		possible	yes	awkward[g]

* typical values
c very low for large sample volumes
g data reduction complex
h function of particle size

It can be seen from table 4 that no conventional method completely
fulfills these requirements. Electrical methods are adversely affected
by high temperatures and conventional diffusional methods do not provide
real-time monitoring of parameters.

Optical methods, notably photon correlation spectroscopy (PCS), can
detect particles "in situ" and in real time, even at extreme
temperatures and pressures.

PCS requires coherent laser light and has only come into common use
with the development of high-resolution photocorrelators and fast
Fourier transform spectral analyzers. This method gives the coefficient
for Brownian diffusion from the measurement of the broadening of the
incident ray caused by the Doppler effect due to particle movement. This
method is commonly applied to liquid suspensions in order to determine
diffusion coefficients and molecular weights of macromolecules.
Experimental results have already been obtained in flames[8] and this
method should develop rapidly because of the numerous advantages it
presents over conventional methods.

PCS is absolute if it is assumed that the aerosol is monodisperse.
This important limitation should be overcome in the future since it is
already possible to obtain weak polydispersities[9],[10].

In applications where sampling is possible, PCS could be compared
with conventional diffusional methods, which give diffusion coefficient
distributions for the particles, whether or not the aerosol is
polydisperse.

Table 5 indicates the characteristics of optical methods in the
measurement of submicronic aerosols. The comparison is limited to single
particule counters, CNCs and PCS.

CONCLUSIONS

We have examined the value of optical methods in the measurement of
submicronic aerosols, bearing in mind the conditions for measurement and
the resultant constraints imposed on measuring devices. It has thus been
possible to evaluate their specific contributions to the measurement of
very low particle concentrations or of very fine aerosols produced at a
range of temperatures and pressures.

REFERENCES
/ 1 /   P.C. REIST (1984) "Introduction to aerosol science", Macmillan
          Publishing Compagny, New York.
/ 2 /   W.C. HINDS (1982) "Aerosol Technology", Wiley-Interscience
          Publication, New York.
/ 3 /   G.J. SEM (1984) "Aerodynamic Particle size : why is it
          important", TSI quarterly, 10, 3.
/ 4 /   K.T. WHITBY (1976) "Electrical Measurement of Aerosols", in :
          "Fine Particles" B.Y.H. LIU ed, Academic Press, New York.
/ 5 /   M. DIOURI, D. BOULAUD, C. FRAMBOURT and G. MADELAINE (1986)
          "Nouveau Spectromètre Diffusionnel et Inertiel", 3ème Journée
          d'Etude sur les Aérosols, GAMS-COFERA, Paris.
/ 6 /   J. BRICARD, P. DELATTRE, G. MADELAINE and M. POURPRIX (1976)
          "Detection of ultrafine particles by means of a continuous
          flux condensation nuclei counter", in : "Fine Particles"
          B.Y.H. LIU ed, Academic Press, New York.
/ 7 /   M. ASSA, D. BOULAUD and Y. METAYER (1986) "Ultrafine particle
          detection by vapour condensation", in : "Aerosols, formation
          and Reactivity", G. ISRAEL ed, Pergamon Press, Oxford.
/ 8 /   P. FLAMENT (1982) "Etude de la croissance des aéries dans les
          flammes laminaires par spectroscopie homodyne Doppler, Thèse
          3ème cycle, Université de Rouen.

/ 9 /   M. BERTERO, C. DE MOL and E.R. PIKE (1987) "Extraction of
        polydispersity information in photon correlation spectroscopy"
        in "Optical particle sizing : theory and pratice" G. GOUESBET
        ed, Plenum publishing company.
/ 10 / P.N. PUSEY and N. VAN MEGEN (1987) "Measurement of small
        polydispersities by dynamic light scattering", in : "Optical
        particle sizing : theory and practice", G. GOUESBET ed, plenum
        publishing company.

THE USE OF CALIBRATION TECHNIQUES FOR THE DEVELOPMENT AND APPLICATION OF

OPTICAL PARTICLE SIZING INSTRUMENTS

D.J. Hemsley and M.L. Yeoman

Materials Physics & Metallurgy Division,
Harwell Laboratory, Oxfordshire, OX11 ORA, UK

C.J. Bates and O. Hadded

University College, Cardiff, Dept. of Mech. Eng. & Energy
Studies, Newport Road, Cardiff, UK

INTRODUCTION

The development of optical instruments to infer the size of
transparent or opaque particles from their scattering characteristics is
currently an active research area. A number of techniques are being used
based on either the Doppler signal visibility[1-8], signal amplitude[9-11],
the spatial phase difference between Doppler signals[12-15], or a
combination of these parameters. Each one of these methods requires the
experimentally measured parameters to be converted into particle size
through theoretically computed relationships, using either Mie scattering
theory, geometrical optics computations, or Fraunhofer diffraction
theory.

Particle size distribution measurements require that the instrument
employed should measure accurately both the size of the particle and the
relative number of particles present in each size class covered by the
instrument. Often, the experimental processes being characterised require
correlations between particle velocity, size and number density. This
requires that these parameters are measured with sufficient resolution and
accuracy. In reality all particle monitoring instruments are subject to
inherent biases and inaccuracies which prejudice these measurements even
for spherical particles in simple flows. Moreover, when the particle
properties are unknown, with a range of shapes, surface texture and
refractive indices then the response of the instrument to each particle
type becomes more uncertain. Two approaches are currently employed to
minimise the errors in the experimentally measured particle size, velocity
and concentration distributions, namely (a) to reduce the number of
parameters which influence the response of the instrument through
optimised optical design and signal processing procedures and (b) to
calibrate the instrument with known particles following controlled
trajectories.

Optical particle sizing instruments which operate within the 1 1000
micron range either monitor the angular distribution of the scattered
light from many particles passing through a laser beam or monitor the
light scattered from single particles passing through a measurement region
defined by the intersection of two focused beams. Both techniques are

affected by particle properties as well as optical response variations due
to components drift, imperfections or simple misalignment of the various
components. Calibration procedures for the two types of instruments are
necessarily different since a distributed source of calibration particles
is required for ensemble averaging instruments whilst individual particles
can be used to calibrate counting instruments. The Malvern particle
size analyser uses a series of concentric photodiodes which are positioned
in the Fourier transform plane of a lens centred on the optical axis of
the incident laser beam. The light energy diffracted into the photodiode
array are converted into particle size distributions using inversion
techniques[16]. Although these instruments have been used extensively by
numerous workers it has only recently become apparent that calibration is
in fact necessary if errors are to be avoided[17,18]. By using calibration
reticles containing a two-dimensional array of opaque circular discs on a
glass substrate, to simulate the forward scattering characteristics of a
Rosin-Rammler droplet size distribution, the above workers reported
discrepancies between the measured and the predicted data of up to 15%.
These variations were attributed to a combination of detector calibration
errors and non-ideal lens effects. Thus, although the scattering function
for the particle ensemble is well understood, the instrument still
requires initial calibration.

The same is true for single particle counting instruments, but even
more so, because of the added effects of particle concentration (multiple
occupancy), particle trajectory (Gaussian probe volume intensity profile)
and the definition of the collection geometry, especially for off-axis
apertures. The signal processing requirements of this type of instruments
can be another source of error (or inherent bias) by imposing a
preferential acceptance of particular size classes by virtue of their
Doppler signals being more easily validated.

This presentation outlines the calibration procedures adopted for
single particle counting instruments based on the two-colour probe volume
configuration first described by Yeoman et al[19] and used in the LISATEK
series of instruments. Calibration methods are described for instruments
employing either forward diffraction, used mainly for on-line measurements
of absorbing particles in industrial plant, or near forward off-axis
scattering which is mainly used for transparent particle and drop size
measurements in various spray systems and two-phase flows.

TWO-COLOUR-LDA PROBE VOLUME

In the following discussion of calibration techniques it will be
assumed that the signals are generated from a two-colour probe volume as
proposed by many authors[6,7,8]. The two-colour probe volume is essentially
a normal LDA probe volume, formed at the waist and intersection of two
focused coherent laser beams. When such a probe volume is used to measure
the size as well as the velocity of a single particle passing
perpendicular to the fringe planes, only signals which pass very near to
the centre of the probe volume should be analysed. These particles are
identified by forming a smaller probe volume of a different colour at the
centre of the outer probe volume. Typically the diameter ratio is chosen
to be 5:1 and the size of the inner reference probe volume is made larger
than the largest particle to be measured.

When modelling the response of a single particle counter using
Fraunhofer diffraction theory it is assumed that:-

  i) The particles are large compared with the laser wavelength.
  ii) The particles are uniformly illuminated when they pass through

the centre of the probe volume, independent of their
trajectory.
- iii) The probe volume fringes are perfectly modulated.
- iv) Particles are spherical and the dominant mechanism for scattering
is diffraction.
- v) All of the light scattered into the collection aperture enters
the photodetectors.

Modelling the instrument response using Mie theory releases assumption i)
and iv) so scattering problems can be treated quite generally for
spherical particles.

PROCESSOR REQUIREMENTS

For any useful LDA based particle sizing instrument the processor
must be able to measure signal over an amplitude range of about 100:1 and
a frequency range greater than 10:1 with a processing speed of at least
1000 signals per second. This fast processing rate is usually required
for any aerosol system where:-

- i) large quantities of liquid are being atomised.
- ii) where the system is prone to drift and instabilities, or is a
pulsed system.
- iii) Where window fouling is experienced.

The processor must also have the facility for on-line validation of
good signals without bias. To effectively validate without bias has been
found to be a very demanding requirement. In any processor which
measures particle size from single Doppler signals it is usual to make
sure that:

- i) The signal has a minimum number of cycles.
- ii) The signal is symmetrical about its maximum.
- iii) The signal amplitude does not reduce either side of its maximum
faster than that expected of a gaussian profile.
- iv) There is only a small (10%) frequency deviation along the signal
to reject signals produced from two particles travelling with
different velocities (or different phases).
- v) The signal is above threshold and below saturation of the
electronics.

When the above checks are implemented on signals with a noise
component, over a large amplitude and velocity range, it is found that the
precise tolerances on all the above checks greatly influence the response
of the instrument. The processing systems which have been found to suffer
least from bias are those in which the validation checks are reduced to a
minimum. This can have the unfortunate consequence of reducing the
resolution of the measurements that can be taken. The most ideal
compromise seems to be to limit the validation to:

- i) Ratio test on amplitude between the REF and SIG signals.
- ii) Velocity (or frequency validation check).
- iii) Make sure that there are at least 5 peaks centred around a
maximum.
- iv) Do curve fitting to peaks and troughs.

Resolution of the instrument should be increased using optical signal
validation rather than electronic signal validation. In practice this
means that ideally a photomultiplier must be placed at 90° to the probe
volume to ensure that only particles which pass through its centre are
analysed.

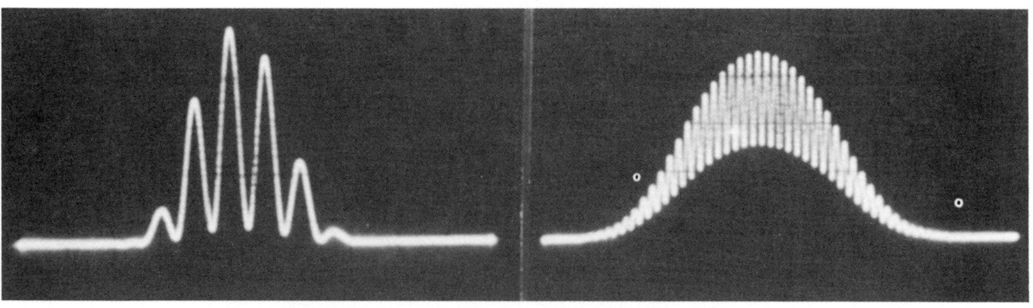

Figure 1   Collection signals generated by a Doppler signal simulator.

PROCESSOR OR SOFTWARE TESTING

It has been shown that the signal processing system will require careful testing whether the algorithm for Doppler signal validation and measurement are "hard wired" into dedicated units, or are analysed by computer after analogue to digital conversion.

Digital and analogue signal processor can be checked for uniformity of response with simulated Doppler bursts produced by a signal generator. One such generator produced at Harwell (R Wallace Sims 1985) is either triggered internally (400 Hz) or externally. Delayed trigger pulses fire an adjustable width triangular generator. Triangular pulses are shaped into Gaussian pulses which are used to modulate the amplitude of the output of a variable-frequency sinewave generator. The amplitude of the sinewave can be adjusted before it is added to the Gaussian. The summed signal therefore represents a Doppler burst with variable visibility.

Signals in the amplitude range 0 to 5V of duration 2.5-30μs and from 20-300μs can be continuously varied to simulate Doppler bursts over a visibility range and from zero to unity, frequency range from 0.05 to 0.5 MHz and 0.5-5 MHz. Fig. 1 illustrates a range of simulated Doppler signals. Two of these signal generators, triggered externally, can be used to simulate signals from a composite two-colour sample volume and the quality of the signal-processor response is evaluated over the range of particle size and velocities in the instrument specification.

CALIBRATION PROCEDURES FOR INSTRUMENTS BASED ON DIFFRACTION THEORY

Before any calibration experiments can be attempted it is necessary to have theoretical computation of the Doppler signal visibility, intensity or phase as a function of particle size for the exact optical geometry and particle properties in question. For initial instrument characterisation, experimentation is greatly simplified by selecting either optical geometries, or particle properties, which can be modelled using Fraunhofer diffraction theory[20]. The optical geometries dominated by diffracted light are, of course, those which are centred on the optical axis of the illuminating laser beams. Many commercial LDA instruments make use of the advantages of off-axis light collection. It has been shown[20] that the response of these geometries can still be predicted using Fraunhofer diffraction theory if suitable particles are selected. Methods of calibrating LDA instruments employing both on and off-axis collection will now be discussed.

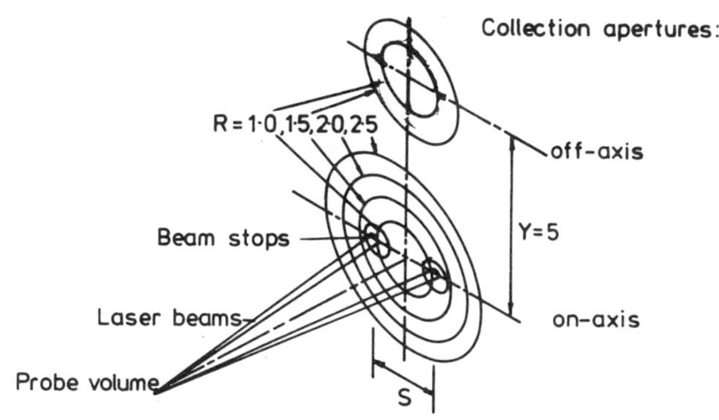

Figure 2   Collection geometries modelled using Fraunhofer diffraction
theory.

Precision pinholes

A rapid assessment of the probe volume quality, scattered light
collection optics and signal amplification and processing behaviour can be
made in simple experiments using a scanning pinhole unit. This consists
of a disc with precision pinhole of various sizes which span the sizing
range of the instrument mounted on precisely the same diameter. When the
disc is placed such that the pinholes pass consecutively through the probe
volume when it is rotated the pinholes diffract light into the collection
optics.

If a synchronous motor is used to rotate the disc then its precise
speed will facilitate the accurate measurement of the fringe spacing. A
variable speed DC motor will provide instant assessments of the dynamic
frequency range of the signal processing electronics. With the
combination of different pinhole sizes and motor frequencies the total
dynamic response of the system can be readily assessed. The use of
pinholes which are very much smaller than the fringe spacing should
produce signals with a visibility of unity. Imperfect probe volume
formation optics very often lead to the fringes being poorly modulated.
This can reduce the accuracy of LDA measurements and invalidates size
measurement using Doppler signal visibility.

Forward diffraction techniques presuppose that the laser beams
emerging from the particle aerosol are prevented from entering the
photodetector. Beam stops placed in the collection aperture to remove the
laser beam will also obscure a proportion of the diffracted light. The
code used to predict the behaviour of the various forward, and near
forward, collection geometries subtracted the light diffracted into the
beam stops where appropriate. The detailed effects of beam stops have
been treated elsewhere[4,21,20] so it is sufficient to say that their effect
tends to reduce the relative amount of light collected from larger
particles. This is because as the angular width of the diffraction cone
decreases with increased particle size, a greater proportion is stopped by
a beam stop placed on the optic axis of the illuminating laser beam. This
effect will be evident on the following visibility and intensity curves
calibrated using precision pinholes.

Figure 2 shows the collection geometries modelled using Fraunhofer
diffraction theory. In the computer code it was convenient to use the
following dimensionless variables, all measured at the collection plane:

    S = laser beam separation
    X = horizontal displacement of the collection aperture/(S/2)
    Y = verticle displacement of the collection aperture/(S/2)
    R = radius of the collection aperture/(S/2)
    R_b= radius of the beams stops/(S/2)
    D = particle diameter/fringe spacing.

Since diffraction theory scales with the geometric arrangement of the
collection optics, the solutions present in terms of the above parameters
are independent of beam crossing angles (provided the particle diameter is
such that the diffraction theory is valid).

Calibrated theoretical curves for on-axis collection geometries are
given in Fig. 3. The Doppler signal visibility is an accurate measure of
particle size over the steep portion of the curve but becomes ambiguous
above about 18 microns. Note that the visibilities produced by 3 micron
diameter pinholes is very near to unity indicating fully modulated
fringes.

The amplitude curves are a plot of the square root of the Doppler
signal mean amplitude as a function of pinhole diameter/fringe spacing.
Hence for "square law" scattering, the correlation should be linear. The
deviation of the curves from linear at the higher particle sizes is due to
the effect of beam stops. It is evident that for these optical geometries
the use of scattered intensity as the size dependent parameter would not
significantly increase the upper sizing range of the instrument.

An important feature of the visibility curves, produced when placing
the collection aperture at a distance Y = 5 from the probe volume, is that
the sizing range on visibility is extended to many times the fringe
spacing.[6] This allows the use of smaller fringe spacing for a given
particle size, and therefore a reduction in the probe volume size for a
given number of fringes. The probe volume size is also reduced because it
is being viewed at an angle rather than along its axis. These effects
increase the particle concentration range that the instruments can cover
without running into particle coincidence problems.

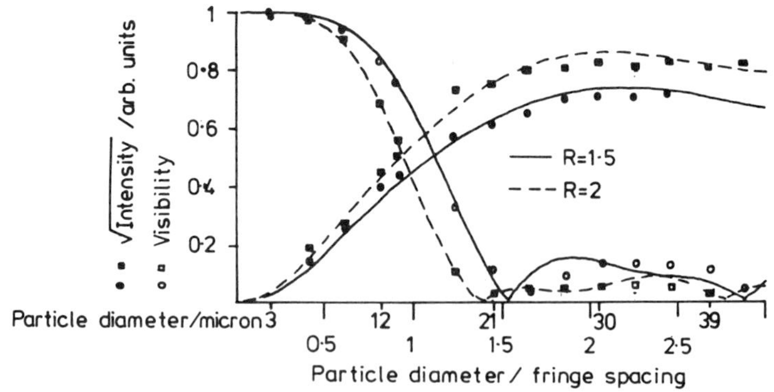

Figure 3   Calibrated on-axis collection geometries.

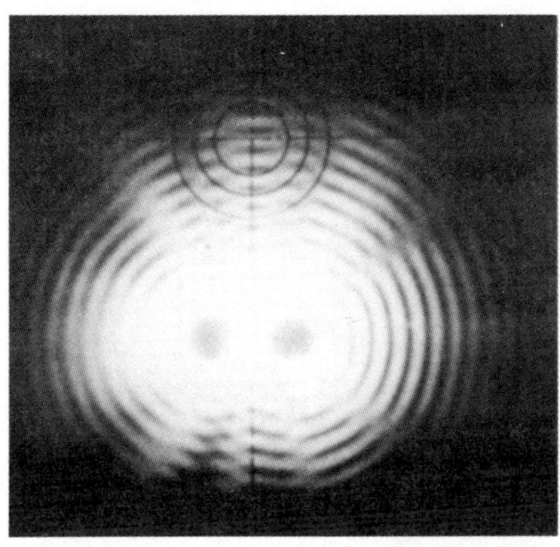

Figure 4   Position of off-axis collection apertures X=0, Y=5, R=1, 1.5,
            2.0 with respect to the diffraction pattern produced by a 60
            micron pinhole.

     The position of the collection apertures Y = 5, R = 1, 1.5 and 2.0
with respect to the beam stops and diffraction pattern produced by a 60
micron pinhole are shown in Fig. 4. The corresponding visibility curves
are shown in Fig. 5 it can be seen that visibility becomes more sensitive
to changes in particle diameter as the collection aperture radius
increases. The variation of visibility sensitivity with collection
aperture radius has been used in Fig. 6 to extend the dynamic range of an
instrument based on the visibility function. The scattered light was
simultaneously monitored using two different collection apertures with
radius parameters of R = 1.5 and 3.5 both placed at X = 0 and Y = 5.
These were chosen because they have an overlapping range at R = 3.5. This
geometry was implemented using a 50% beam splitter after the R = 3.5
collection aperture then inserting a further aperture of radius R = 1 in
one of the components. To do this effectively it is necessary to also
split the laser beams so that the smaller collection aperture can be
accurately positioned with respect to the beams. The data shown in figure
6 shows a considerable advantage in using the R = 3.5 aperture for sizing
particles below ∿ 22 microns.

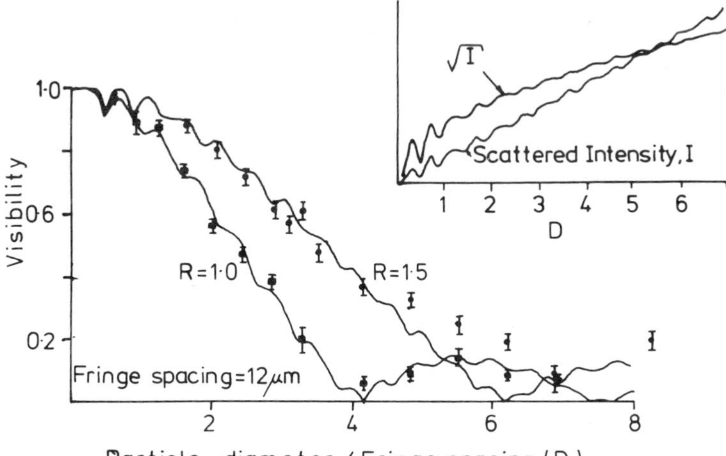

Figure 5   Calibrated off-axis visibility curves X=0, Y=5, R= (1, 1.5).
            The inset shows that the scattered intensity is proportional to
            particle diameter.

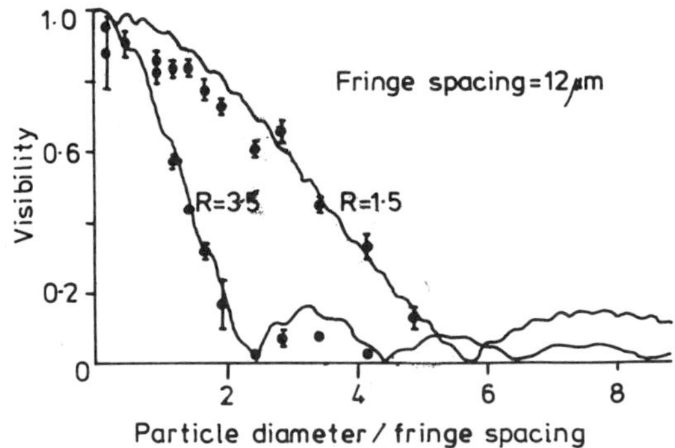

Figure 6   Extension of dynamic sizing range using the visibility function
            produced from two collection apertures of different diameters
            placed at the same scattering angle.

An additional feature of the diffraction theory response curves for
the off-axis collection apertures is that the scattered intensity is
proportional to "diameter", rather than the more normal "square of
diameter". This is shown in the inset of Fig. 5 where curves show a
linear relationship between the Doppler signal mean amplitude and
diameter. This has significant advantages when an increased dynamic
sizing range is required.

Figure 7 shows the data obtained when four pinholes were
consecutively passed through the probe volume and the 8 bit transient
recorder was optimised to monitor the intensities of pinholes between 20
and 70 microns. The scatter diagram shows the signal amplitude versus
diameter, which was computed from the $X = 0$, $Y = 5$ and $R = 1$ theoretical
visibility curve. The data shows that although the Doppler signal
amplitude is linear and tightly grouped over the whole sizing range, the
digitisation of the 8 bit analogue to digital converter effects the size
resolutions visibility for the 20 micron pinhole.

When the diameter histogram based on visibility is plotted, Fig. 8,
the lack of resolution in the measurement of the pinhole diameter is
clearly evident. It is for these, and other, reasons that it has been
made normal practice to particle size using the intensity parameter, but
to correlate the signal intensity with diameter using the Doppler signal
visibility. This correlation is obviously weighted against visibilities
measured on low amplitude signals.

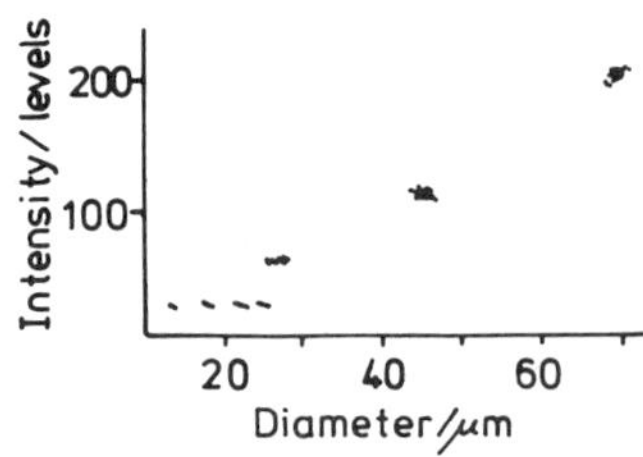

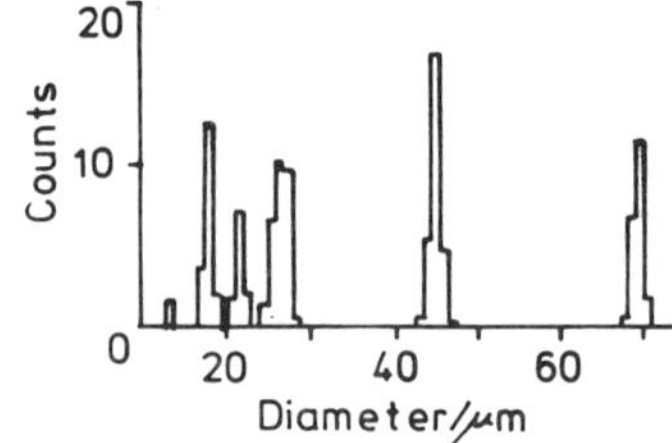

Figure 7   Scatter plot showing a   Figure 8   Particle size histogram
            linear correlation.   based on the data of
                                                       figure 7.

<u>Calibration plate</u>

Assessment of an on-axis collection geometry using the scanning
pinhole method has its limitations because the illuminating laser beams do
not reach the beam stops. Much of the system optimisation is in reducing
the beam stops as much as possible to increase the sizing range on
amplitude, whilst maintaining adequate signal-to-noise ratios to reliably
process the smallest signals.

As a consequence, a calibration plate has been designed and
manufactured. Chrome discs, with diameters in the range 5-200 micron,
were sputtered onto a glass plate of 100 mm diameter in five rings. The
innermost ring, with radius 30 mm, has ten discs starting at 5 microns and
increasing in 2 micron steps to 23 micron. The second ring on a radius of
35 mm starts at 20 micron and increases in 10 micron steps to 200 micron.
The third ring covers the size range 20 - 100 micron in 10 micron steps,
and the last ring at a radius of 45 mm has 20 discs starting at 5 microns
and increasing in 5 micron intervals to 100 microns in diameter. All
spots are positioned on the specified diameters to within +/-5 micron. By
spinning the disc about its central axis the spots are made to
consecutively passing through the probe volume a known size distribution
of "particles" are simulated.

When the spot or "particle" size becomes comparable to the
cross-section of the sample volume then beam obscuration effects begin to
affect the instrument response. In this situation the calibration plate
can be used to investigate the extent of beam obscuration effects due to
large particles.

A similar calibration slide has been developed independently in the
USA and has recently been described[18]. Photo-etched circular discs of
chrome film were used to simulate a Rosin-Ramler distribution of particles
with diameters from 80 micron to 455 micron on a glass substrate to
calibrate the response of the Malvern particle size analyser.

Recent experience with the Harwell calibration plate has shown the
need to gate the observation of the chrome discs in order to avoid
monitoring dust particles which settle on the surface. This can be
achieved with either spatial or, as recently suggested[22] with temporal
gating.

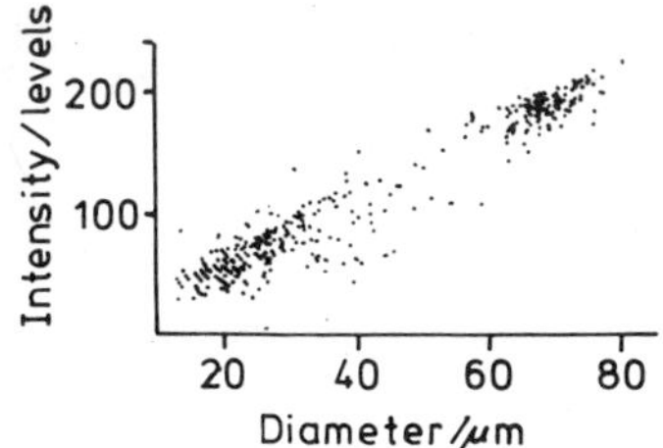
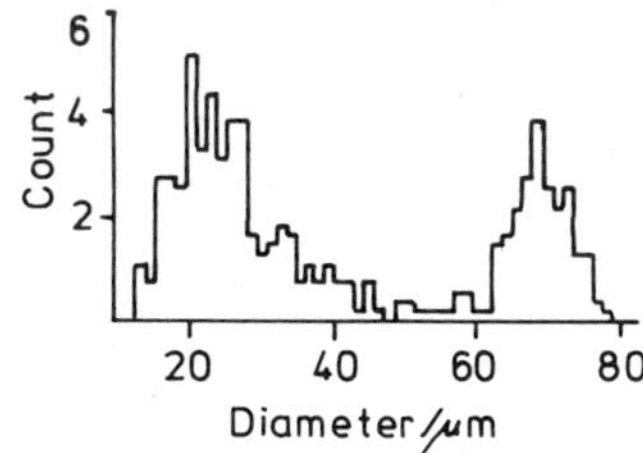

Figure 9   (a)Doppler signal amplitude versus particle size correlations
            for a bimodal distribution of bronze particles.
            (b)Probability histogram of the bimodal distribution of the
            bronze particles.

<u>Particle aerosols which predominately diffract</u>

Once the reliability of the optics and processor have been
demonstrated using the previous techniques, it is necessary to assess the
effect of random particle trajectories on the results produced by the
processor. The size resolution will depend on the optical elimination or
processor rejection of signals which do not pass through the centre of the
probe volume. There is normally a trade-off between resolution and
dynamic sizing range, and resolution and uniform instrument response with
respect to particle size. Assessment of the instrument performance can be
readily assessed using bronze powders.

Bronze powders are relatively inexpensive and have many of the
properties required for calibration purposes. They can be prepared as
spheres in narrow size ranges, they do not agglomerate and are therefore
readily dispersed into an aerosol, and can be recycled and stored for
regular use. The surface of the spheres has an "orange peel" texture so
the reflected component is diffuse rather than specular, and of course,
there is no refracted component. These properties make bronze powders
ideal for diffraction theory and instrument calibration purposes for
on and off axis visibility and intensity functions, and as it has recently
been found, for off-axis phase Doppler measurements. In order to test the
instrument performance with a size distribution of airborne particles,
which have a range of trajectories, a bimodal distribution of bronze
particles was prepared by sieving. The bronze powder was dispersed by an
aerosol nebulizer, and the aerosol was monitored using the laser
instrument employing the off-axis scattered light collection geometry,
X=0, Y=5, R=1. Fig. 9(a), 9(b) illustrates the data obtained for a powder
consisting of sieve cuts of - 45 microns and 63 - 75 micron. Good
agreement was observed between the sieved size cuts, the laser visibility
measurements, and the results of a Quantimet analysis. Fig. 10a shows a
strong correlation between the intensity of light scattered by the bronze
particles and the particle size. This demonstrates that when using the
two-colour probe volume the intensity function can be used as a measure of
the particles provided a correlation similar to the quality shown in Fig.
9(a) is observed.

CALIBRATION PROCEDURES FOR INSTRUMENTS BASED ON MIE SCATTERING THEORY

For instrument which employ off-axis collection optics which
monitor the light scattered from reflective and/or transparent spheres the
instrument response is modelled using full Mie scattering theory, or the
geometrical optics approximation to the full theory because the scattering
is dominated by reflected and refracted light.

It is suggested that these instruments should initially be assessed
in terms of frequency response and validation bias using the previously
described techniques and then be calibrated using one of the following
techniques.

<u>Glass spheres mounted on carbon fibre stalks</u>

Stalk-mounted particles can be used to determine the instrument
response for individual particles of known size and refractive index.
The difficulty in the past has been to produce stalks which are so narrow
that their projected area (length and diameter) is negligible compared
with that of the particle. Recent advances in the mounting of glass
microballoons for laser fusion studies have resulted in the use of 6
micron diameter stalks of carbon fibre. Fig. 10(a) illustrates how a
single coal particle was mounted on a carbon fibre which is fused into a
glass rod for mounting purposes. The diffraction pattern produced by this

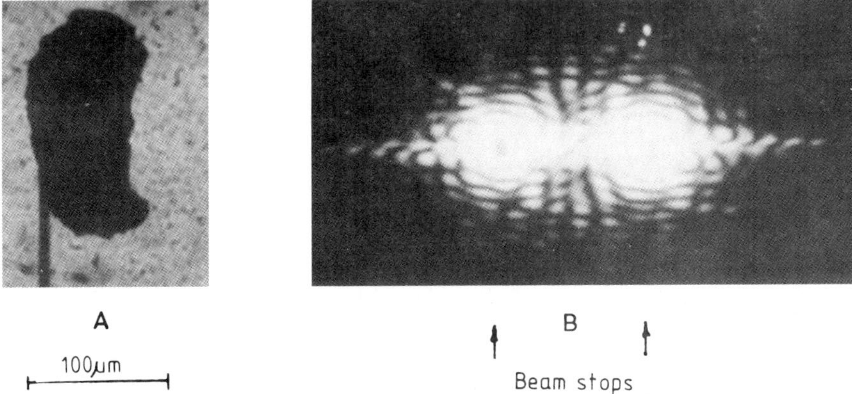

Figure 10   (a)   Coal particle mounted on a carbon fibre stalk.
(b)   Diffraction pattern produced when (a) was positioned in an LDA probe volume.

stalk-mounted particle is shown in Fig. 10(b). It is evident that the far-field diffraction pattern of the particle as well as the stalk can be significant but this depends on the actual position of the collection aperture. However, in most circumstances stalk-mounted particles can be used to calibrate the dynamic range of the instrument by using a range of particle diameters which have identical stalk dimensions.

The glass rods holding the stalks were mounted on a synchronous motor and traversed at controlled trajectories through the centre of the probe volume. Five different glass spheres were used to obtain the calibration points shown on the theoretical visibility curve in Fig. 11. Mie theory shows that large differences exist between the visibility curves for water and glass due to the refractive index change. In particular, the fine scale oscillations are much larger for glass than they are for water. The histogram obtained[23] using both 182 microns and 280 microns glass spheres is shown in Fig.12.

Stalk mounted particles have also been found useful in determining the variation of probe volume size with particle diameter for particle concentration measurements.

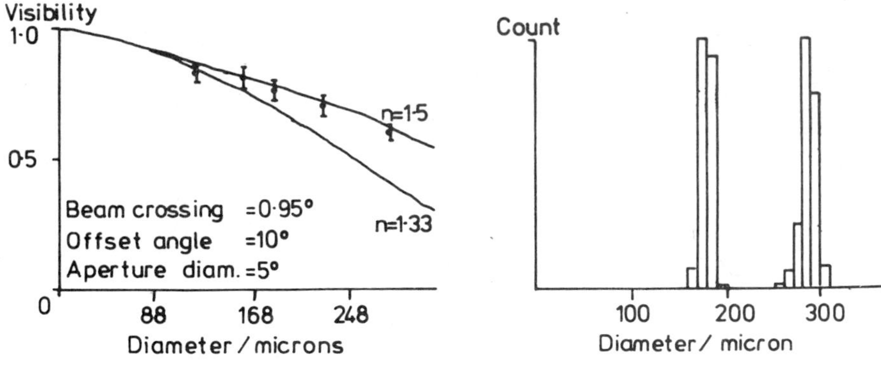

Figure 11   Wide size range visibility curves and data point obtained using stalk mounted glass spheres.

Figure 12   Size distributions obtained by rotating 182 and 280 micron glass spheres mounted on stalks through the probe volume.

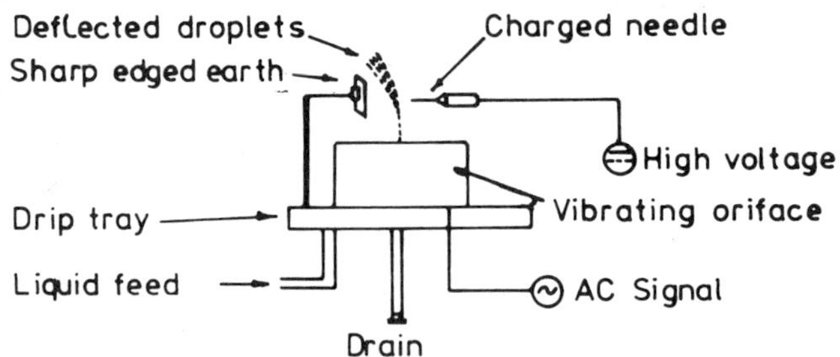

Figure 13   Schematic of the dispersion effect produced by the electric
field applied to the Berglund-Liu.

Berglund-Liu Monodisperse Droplet Generator

This instrument is by far the most widely used method of calibrating
LDA particle sizing instruments. It is invaluable for determining or
verifying the precise instrument response to known size particles.

Very briefly, the principle of operation is that a cylindrical liquid
jet, is naturally unstable and tends to break up into droplets. When left
uncontrolled, the breakup process produces non-uniform droplets. However
by applying a periodic disturbance via a piezo-electric crystal of an
appropriate frequency on the liquid jet, the breakup process can be
controlled to produce droplets of precise diameter. Furthermore, as one
droplet is produced per cycle of disturbance, the volume of a droplet can
be precisely calculated from the liquid feed rate and the frequency of the
disturbance.

In order to ensure single particle occupancy of the two-colour probe
volume a droplet dispersion mechanism was designed. This consisted of an
electrically earthed knife edge set at approximately 10 mm from a sharp
needle supplied with a high voltage (typically 2-3 kV) to create an
electric field. The stream of droplets which passed through this field
acquired an electrostatic charge proportional to the particle surface
area. This resulted in a 'fan' of droplets being produced as they
migrated towards the earthed knife edge, as indicated in Fig. 13. The
size and shape of this fan could be varied by the position, shape and
voltage supplied to the needle. It was found that voltages between 2.4
and 2.9 kV were most suitable for our application. An additional problem
with the earlier Berglund-Lui aerosol generator was that the pump
mechanism tended not to give a constant liquid feed rate. As a
consequence plunger driver mechanism on the syringes was replaced by an
carbon gas pressure system (W. Dalzell, A.E.R.E., Harwell). To determine
the correct piezo electric driving frequency and air pressure combination
the optical system shown schematically in Fig. 14 was used. A laser beam,

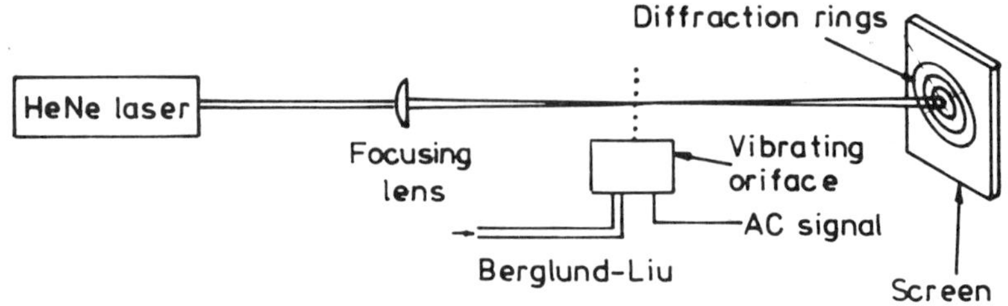

Figure 14   Simple optical configuration for assessing the generation of
monosized droplets from the Berglund-Liu.

from a He-Ne source, was focussed on the vertical stream of droplets
generated by the orifice. The oscillation frequency was then varied and
the diffraction ring pattern observed on a screen positioned one focal
length away from the vertical jet. Outside the range of frequencies where
'regular' sized droplets are generated, the diffraction ring pattern is
very ill-defined. As the droplets generated become more and more similar
in diameter, the pattern becomes clearer and the rings are almost
stationary when very narrow-sized droplet distributions are generated.

To finally determine the exact frequencies which determine true
monosize droplets the method suggested by Drain and Martin[24] was used.
This method is based on the simple assumption that true mono-disperse
particles must be equally spaced and so have a diffraction grating effect.
Their production will then be accompanied by horizontal fringes
superimposed upon the scattering rings produced by the particle. The
distance between the horizontal lines could be related theoretically to
the velocity of the droplet and the excitation frequency. Careful
operation and observation allowed such lines to be observed and thus
determine those frequencies at which monosize droplets were being
generated. By varying the frequency and the orifice sizes, it was
possible to obtain the experimental calibrations of the theoretical
visibility curves shown in Fig. 15. It is seen that despite the
fluctuations around the mean values, the agreement is extremely good for
visibilities in the range 0.4 to 0.9.

Latex spheres in a recirculating cell

In many instances it is required to determine the actual sizing range
of an instrument, and to assess the effect that particles at either end of
the sizing range have on the processor, as a function of particle
velocity. Such experiments are not possible using the Berglund-Liu.
Also, calibration measurement of LDA systems designed for measuring very
small (< 10 micron) solid particles in liquids are made tedious by
contamination of the pure liquid phase and by the production of air
bubbles in the recirculating system.

The system shown in Fig.16 overcomes the above problems by making the
flow cell as simple as possible whilst maintaining a unidirectional flow
perpendicular to the probe volume fringes. The system consists of a
circular glass beaker with a hemispherical bottom which ensures a smooth
fluid flow. This stands in an parallel sided glass container

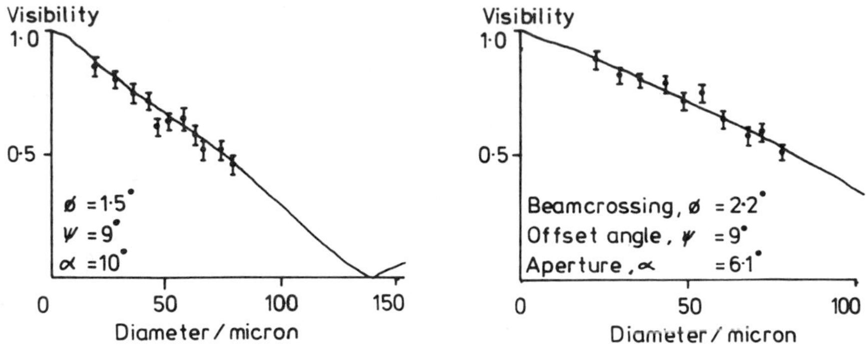

Figure 15   Experimental calibration of the visibility curves using
monosize droplets.

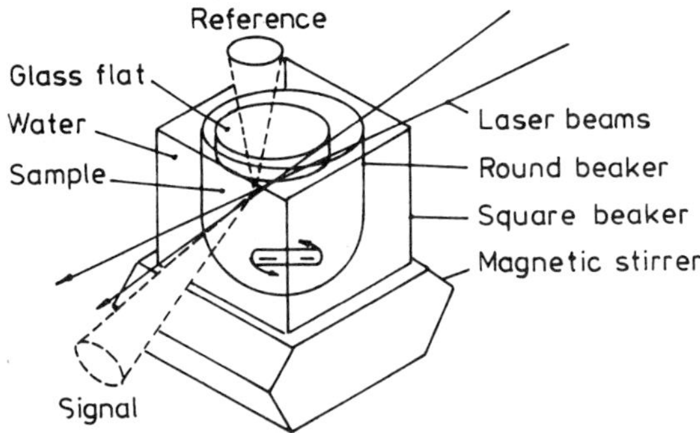

Figure 16   Experimental arrangement for calibrated studies using latex spheres in water.

(manufactured from microscope slide glass).   When the beaker is filled with pure liquid (ISOTON), and the outer container is filled with water, the "index matching" effect between the beaker and the water serves to practically eliminate the effect of the curved beaker walls.

The liquid in the beaker is stirred using a magnetic stirrer and a uniform flow is created through the probe volume. To increase the resolution of the optical particle size measurements a 90° photomultiplier can be placed above the probe volume if a glass disc is placed on top of the beaker after it has been fitted with liquid. The beaker is then a closed system which can be seeded with latex spheres for detailed calibration studies over extended periods of time.

Figure 17 show the particle sizing data obtained using the intensity of light scattered at 30° to the forward direction from 2.83 and 12.9 micron latex spheres in water.

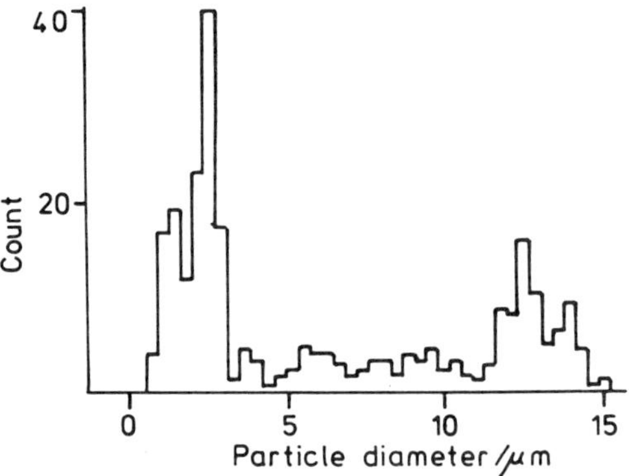

Figure 17   Data obtained from latex spheres in water.

DISCUSSION AND CONCLUSIONS

A number of calibration techniques have been presented which can be
used to characterise both the optical geometry and the signal processing
system of single particle counters. Particular emphasis has been placed
on the calibration of the visibility curves for on, and off-axis,
collection apertures predicted from Fraunhofer diffraction theory as well
as the full Mie scattering theory. The calibration techniques discussed
provide an insight to the complexities of using single-particle counters
to measure size distributions. The initial design of the instrument
should reduce the parameters which will effect the size distribution. A
primary calibration is required to verify the theory, and implementation
of the optical geometry modelled theoretically. The method of scanning
four apertures of different sizes through the probe volume highlights
three points, namely:

    (i) The accuracy with which size can be determined using the
        visibility function, over the whole size range with signal
        amplitudes which cover the amplitude range of the processor.

   (ii) The precision of measurement at the lower sizing range. It has
        been found that when using a 8 bit A/D converter measurements on
        signals with amplitude below 20 levels can be quite imprecise.

  (iii) Preferential rejection of signals at either end of the sizing
        range, due to the stringent validation checks, used to avoid the
        analysis of signals produced by more than one particle, and those
        particles which do not pass through the centre of the probe
        volume.

The pinhole calibration method has been extended by the production of
circular glass plates, with chrome spots of known sizes sputtered on fixed
diameters, with solid transparent particles mounted on stalks.

A second stage of calibration is recommended using solid particle
aerosols in narrow size cuts. Bronze spheres have proved very useful
because they sieve easily and are found to be good diffractors of light,
even for off-axis collection apertures. An alternative method for
transparent particles is to use latex sphere suspended in a recirculating
beaker of liquid. These methods have been summarised in Fig. 18.

This calibration stage is needed to assess and minimise the effects
of random particle trajectories through the probe volume, possible
multiple occupancy effects and particle obscuration of the beams.

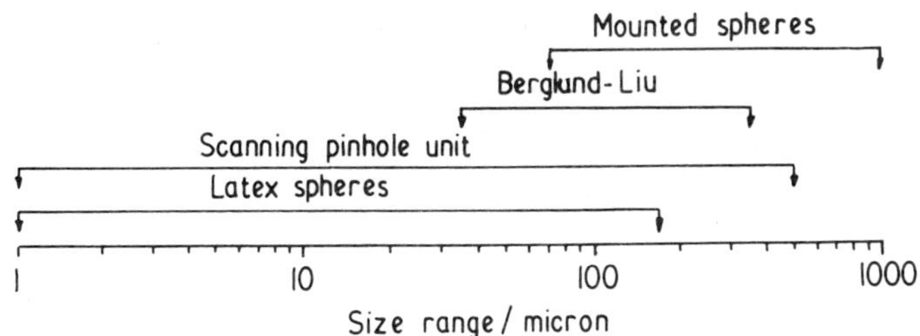

Figure 18 Calibration methods used for instrument response experiments.

A good test of a correctly set up instrument is for it to accurately
measure a bimodal distribution with peaks near both ends of the sizing
range. A useful indication of the optical quality of the instrument is
provided by observing the correlation between particle size (or
visibility) and Doppler signal amplitude.

Acknowledgements

The recirculating cell data was taken with the assistance of Mr. R.
Gillespie. Mr. S.R. Martin has made significant modifications to the Mie
scattering software. Thanks are also due to Mr. W. Dalzell and Mr. R.
Wallace-Sims for useful discussions.

References

1.   W.M. Farmer, Measurement of particle size, number density and
     velocity using a laser interferometer. Appl. Opt., 11: 2603 (1972).

2.   A.R. Jones. Light scattering from a sphere situated in an
     interference pattern, with reference to fringe anemometry and
     particle sizing. J. Phys. D (London). Appl. Phys., 7: 1369 (1974).

3.   F. Durst and B. Eliasson. Properties of Laser-Doppler signals and
     their exploitation for particle size measurement. Proc. of LDA Symp.
     Copenhagen: 457 (1975).

4.   D.W. Roberts. Particle sizing using laser Interferometry. Appl.
     Opt., 16, 7:1861 (1977).

5.   W.P. Chu and D.M. Robinson. Scattering from a moving spherical
     particle by two crossed coherent plane waves. Appl. Opt. Vol. 16, 3:
     619 (1977).

6.   W.D. Bachalo. Method for measuring the size and velocity of spheres
     by dual-beam light scattering interferometry. Appl. Opt. 14, 3: 363
     (1980).

7.   C.J. Bates, M.L. Yeoman, P.J. Roberts and W.D. Dalzell. Development
     of a laser system for the simultaneous measurement of particle size
     and particle velocity. AERE-R10210. H.M.S.O. (1981).

8.   C.F. Hess. Non-intrusive optical single particle counter for
     measuring the size and velocity of droplets in a spray. Appl. Opt.
     23, 23:4375 (1984).

9.   E.D. Hirleman. Laser technique for simultaneous particle size and
     velocity measurement. Opt. Letts. 3, 1:19 (1978).

10. N.A. Chigier, A. Ungut and A.J. Yule. Particle sizing in flames with
     laser velocimeters. "Laser Velocimetry and Particle sizing" (Ed.
     Thompson H.D. and Stevenson, W.H.):416 (1978).

11. G. Gouesbet, G. Grehan, R. Kleine. Simultaneous optical measurement
     of velocity and size of individual particles in flows in "The 2nd
     Int. Symp. laser anemometry to F. mech." Portugal (1984).

12. F. Durst and M. Zare. Doppler measurements in two phase flows.
     Proc. LDA Symp. Copenhagen: 403 (1975).

13. W.D. Bachalo and M.J. Houser. Phase Doppler spray analyser for
    simultaneous measurements of drop size and velocity measurements.
    Opt. Eng. 23,5: 583 (1984).

14. M. Saffman, P. Buchhave and H. Tanger. Simultaneous measurements of
    size, concentration and velocity of spherical particles by a laser
    Doppler method. Int. Symp. on Appl. LDA to F. Mech. Lisbon '84.
    (1984).

15. K. Bauckhage and H.H. Floegel. Simultaneous measurement of Droplet
    size and velocity in nozzle sprays. (1984) (reference 14).

16. J. Swithenbank, J.M. Beer, D.S. Taylor, D.S. Abbot and C.G. McCreath.
    A laser diagnostic technique for the measurement of droplet and
    particle size distribution.
    Progress in Astronautics and Aeronautics, 53: 421 (1977).

17. E.D. Hirleman, V. Oechsle and N.A. Chigier. J. Optical Eng. 23,
    5:610 (1984).

18. E.A. Hovenac, E.D. Hirleman, R.F. Ide. 2nd Int. Conf. on Liquid
    Atom. and Spray Sys. London (1985).

19. M.L. Yeoman, H.J. White, D.J. Azzopardi, C.J. Bates, P.J. Roberts.
    Optical development and application of a two colour LDA system for
    the simultaneous measurement of particle size and velocity.
    ASME Winter Annual Meeting, Pheonix, Arizona, USA (1982).

20. D.J. Hemsley. A study of Solid fine particle aerosols using a
    two-colour laser Doppler system. PH.D. Thesis. University of Wales
    (1985).

21. C.R. Negus and L.E. Drain. Mie calculations of the scattered light
    from a spherical particle traversing a fringe pattern produced by two
    intersecting laser beams. J. Phys. D. (London). Appl. Phys. 15:375
    (1982).

22. E.D. Hirleman, Private communication, 1985.

23. O. Hadded. A packaged instrument for drop size and velocity
    measurement in two-phase flow. Ph.D. Thesis, University of Wales
    (1986).

24. L.E. Drain and S.R. Martin. Experimental measurements of the
    visibility of water droplets crossing a fringe system produced by
    intersecting laser beams. AERE-R 11668 H.M.S.O. (1985).

OPTICAL PARTICLE SIZING AND PARTICLE CHARACTERIZATION

BASED ON POLARIZATION MEASUREMENTS

R.H. Zerull, R.T. Killinger, and R.H. Giese

Ruhr-Universität Bochum
Bereich Extraterrestrische Physik
4630 Bochum, F.R.G.

INTRODUCTION

Commercially available optical particle analyzers generally provide
satisfactory results after adequate calibration with the particles in
question. It is the purpose of this paper to point out, how the
performance of optical particle sizers can be improved by polarization
measurements.

The results presented in this paper are derived from multicolour
laser experiments providing electrodynamic suspension of single dust
particles. The size range of the particles is between 20 and 150 μm. For
details of the equipment see Killinger et al. (1987, this volume).

The light scattering features of particles are closely related to
their physical properties. Features addressed in this paper are the run
of intensity, linear polarization and cross polarization vs scattering
angle Θ.

If the scattering signature is used for **size discrimination**, the
most common application, it must be taken into account, that the
scattering properties strongly depend on **particle shape** and **composition**,
as well.

Basic connections between observed scattering features and physical
properties of the particles are schematically summarized in the
following Table 1. These connections will be illustrated by typical
measuring results. It will further be shown, how polarization effects
can be used for correct interpretation of particle counter and aerosol
photometer measurements as well as for particle characterization and
discrimination.

Table 1.

QUANTITY MEASURED	TYPICAL FEATURES	CONNECTION WITH PARTICLE PROPERTIES
INTENSITY	ENHANCEMENT OF FORWARD SCATTERING	SIZE
	ENHANCEMENT OF BACKWARD SCATTERING	SHAPE MATERIAL
LINEAR POLARIZATION	POSITIVE, NEUTRAL, OR NEGATIVE VALUES AT MEDIUM SCATTERING ANGLES	MATERIAL SHAPE STRUCTURE SIZE
CROSS POLARIZATION	RELATIVE MAGNITUDE OF CROSS POLARIZING ELEMENTS OF THE MUELLER MATRIX	MATERIAL SHAPE SIZE

FORMALISM

The scattering problem can be described proceeding from a linear transformation (see van de Hulst, 1957):

$$\begin{bmatrix} I_{s1} \\ I_{s2} \\ U_s \\ V_s \end{bmatrix} = \frac{\lambda^2}{4\pi^2\, r^2} \left| A_{ij} \right| \begin{bmatrix} I_{o1} \\ I_{o2} \\ U_o \\ V_o \end{bmatrix}$$

Index o of the Stokes vector characterizes the incident, index s the scattered radiation, respectively; indices 1 and 2 characterize the components polarized parallel or perpendicular to the scattering plane; λ is the wavelength; r is the distance between the scattering particle and the observer. The complete information of light scattering is contained in the 16 elelents of the 4 x 4 Mueller matrix, each of which depends on the scattering angle Θ. However, for a single particle in fixed orientation there are only seven independent matrix elements, which reduce to six for particles with symmetry plane in random orientation (Bohren and Huffman, 1983).

Although other elements also contain information about the particle (Perry et al., 1978, Bickel and Stafford, 1980, Bottiger et al., 1982), consideration of the elements A_{11}, A_{22}, A_{21}, and A_{22} is sufficient for the treatment of effects concerning total intensity, linear and cross polarization.

Based on these coefficients, the total scattered intensity is proportional to $(A_{11} + A_{12} + A_{21} + A_{22}) = i$ (total scattering function). The degree of linear polarization can be defined as

$$P = \frac{(A_{22} + A_{21}) - (A_{11} + A_{12})}{A_{11} + A_{12} + A_{21} + A_{22}}$$

Cross polarization is characterized by the elements A_{12} and A_{21}. Their magnitude ($A_{12} \equiv A_{21}$ for random orientation) relative to A_{11} and A_{22} provides a measure for the strength of cross polarizing effects.

THE IMPORTANCE OF POLARIZATION MEASUREMENTS
FOR PARTICLE CHARACTERIZATION

Particle sizing devices evaluating the sideward scattered intensity
are not able to eliminate effects of shape and material. For example,
particles of nonspherical shape scatter much more intensity in that
angular domain than spheres of the same size (factor between 2 and
about 6). Furthermore, "bright" particles cause the same signal than
much larger "dark" particles if only intensity measurements are
considered. It is the purpose of this chapter to illustrate, how these
ambiguities can be reduced by polarization measurements.

Linear Polarization

As listed in Table 1, the degree of linear polarization at medium
scattering angles dominantly depends on material and shape of the
scattering particles, but it is also influenced by their inner structure
and size. In the sense of geometrical optics the scattered intensity may
be roughly devided into 3 parts:

1. **Diffraction** (here of minor interest, because unpolarized and
   effective only to the forward scattered intensity).

2. **Reflection**, reaching the observer after one or higher order **external**
   reflections at the boundary of the particle.
   This part is **strongly polarized** at medium scattering angles
   (Fresnel's formulae).

3. **Refraction**, scattered energy having entered the particle and
   reaching the observer after one ore even more **internal** reflections.
   Due to multiple reflections, this part of the scattered radiation is
   unpolarized for the first approach (except for particles with high
   symmetry, for instance spheres).

For a non-absorbing particle typically more than about 90% of the
scattered radiation at medium scattering angles reaches the observer
after internal reflections, i.e. the strong polarization effects of
externally reflected rays are blurred out. For an absorbing particle, on
the other hand, rays entering the particle are absorbed and hence mainly
the reflected part is observed. This gives just a few to ten percent of
the incident energy, but strongly polarized with a maximum up to 100%
located in the angular interval between $60 \lesssim \Theta \lesssim 90°$.

These connections are the main idea of the "Polarization-Albedo-
Rule", familiar in the physics of planetary surfaces (Wolff, 1980). The
Albedo A is a measure of the energy scattered by a surface or particle
relative to the total energy intercepted (for details see Bohren and
Huffman, 1983, Hanner et al., 1981).

The Polarization-Albedo-Rule predicts a monotonic connection
between the albedo and the degree of linear polarization observed. This
relation can be evaluated to determine the size of small bodies in the
solar system (Asteroids) from polarization measurements. Although the
scattering conditions are not completely equivalent, the Polarization-
Albedo-Rule can to some extent be found again in the scattering
properties of small grains. In general the Albedo is a measure for the
overall scattering efficiency, and by this it is, for instance, also
related to the intensity scattered around 90° and therefore of interest
for particle sizing.

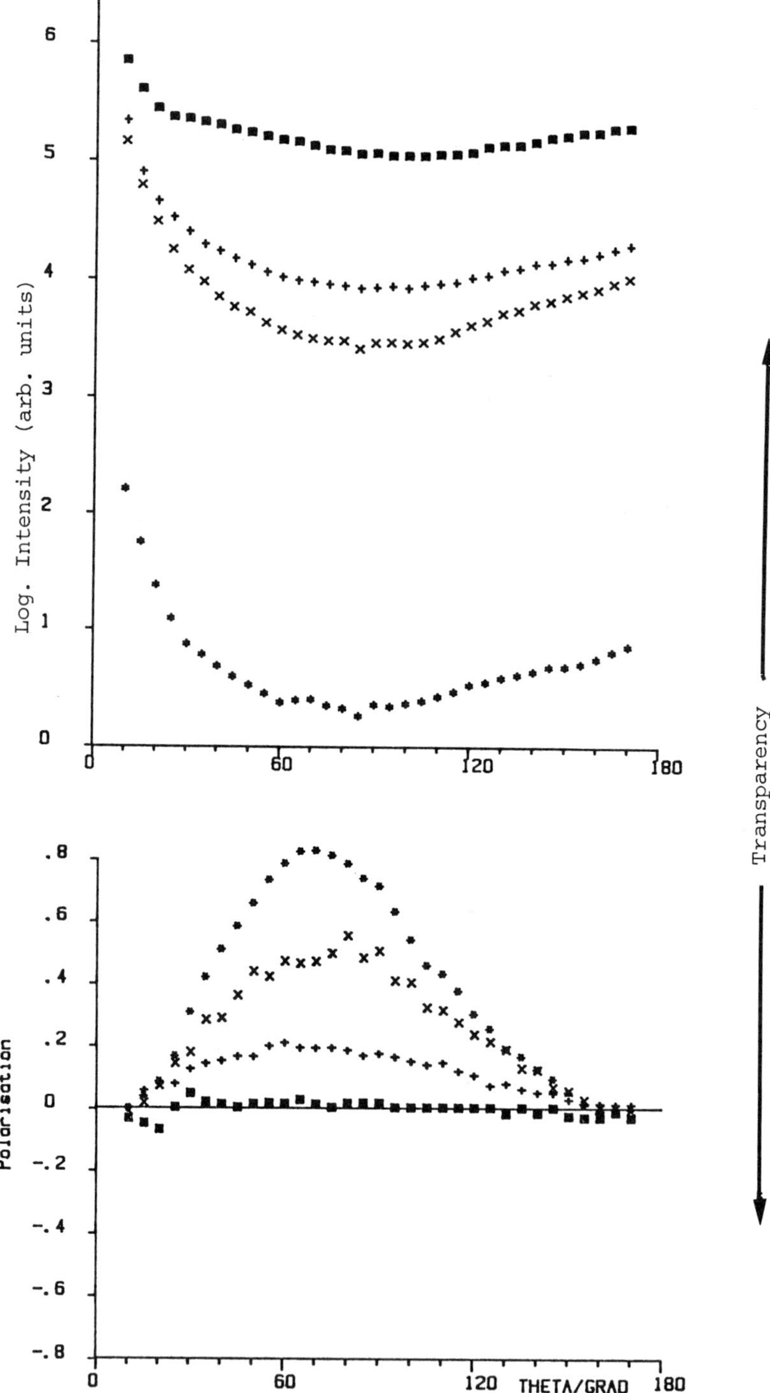

Fig. 1. Comparison of total intensity and polarization
vs. scattering angle for dust sumples with
decreasing transparency (increasing absorption).

For similar particles differing only concerning the imaginary part of the index of refraction qualitatively the following relation holds:

$$
\begin{array}{ccc}
0 & \leq m'' \leq & m''_{max} \\
1 & \leq A \leq & A_{min} \\
P_{min} & \leq P \leq & P_{max}
\end{array}
$$

The range of validity of this monotonic relation is restricted to $m'' \lesssim 0.1$; larger values characterize the region of semimetallic and metallic reflection, where the albedo increases again and the polarization decreases.

The relations mentioned above are illustrated in Fig.1, juxtaposing the scattering properties of 4 different kinds of terrestrial and extraterrestrial samples. The main aspect for the selection of the measurements was to clearify the definite relationship between transparency, scattering efficiency, and the degree of observed polarization. Each curve represents the average of several particles of the same kind in random orientation.

The mean size of the samples is $\sim$ 40 µm. The tendency observed is, of course, not only a feature of the specific samples but is also valid for all kinds of aerosols.

Cross Polarization
<u>
</u>

Cross polarization is characterized by the elements A_{12} and A_{21} of the Mueller matrix. Contrary to spheres, irregular particles can produce cross polarization even considering a single scattering process. The contribution of this part of radiation to the scattered intensity has been taken into account in the diagrams of the previous section.

After Fresnel's formulae a single reflection does not cause cross polarization. Hence, higher order reflections at the boundary or inside the particle must be considered. For symmetry reasons single spheres do not cause cross polarization (Bohren and Huffman, 1983).

As already stated in Table 1, the degree of cross polarized intensity is related to the material, shape, and size of the particles. Size- and shape dependence are discussed elsewhere (Zerull et al., 1976, 1986). In this paper the connection between transparency of the particle material, scattering efficiency, and cross polarization will be illustrated by 3 typical examples. In Fig. 2-4 the cross polarizing 1lements $(A_{12} + A_{21})$ are compared to the elements $(A_{11} + A_{22})$ of the Mueller matrix. For easier comparison the logarithmic scale (arbitrary units) is normalized to the geometric cross section of the particles.

Fig.4 shows the measurement of a carbon particle. As in this case refracted rays are nearly totally absorbed, the only mechanism for cross polarization is multiple reflection at the rough surface. The resulting reflection coefficient is very low, consequently the elements $(A_{12} + A_{21})$ are nearly two orders of magnitude below $(A_{11} + A_{22})$.

For a moderately absorbing flue ash particle, shown in Fig. 3, the difference between the curves is distinctly reduced to less than one order of magnitude.

The measurement of a very bright magnesium oxide particle, shown in Fig.4, illustrates, that $(A_{12} + A_{21})$ can be of same order of magnitude than $(A_{11} + A_{22})$ over a wide range of scattering angles.

607

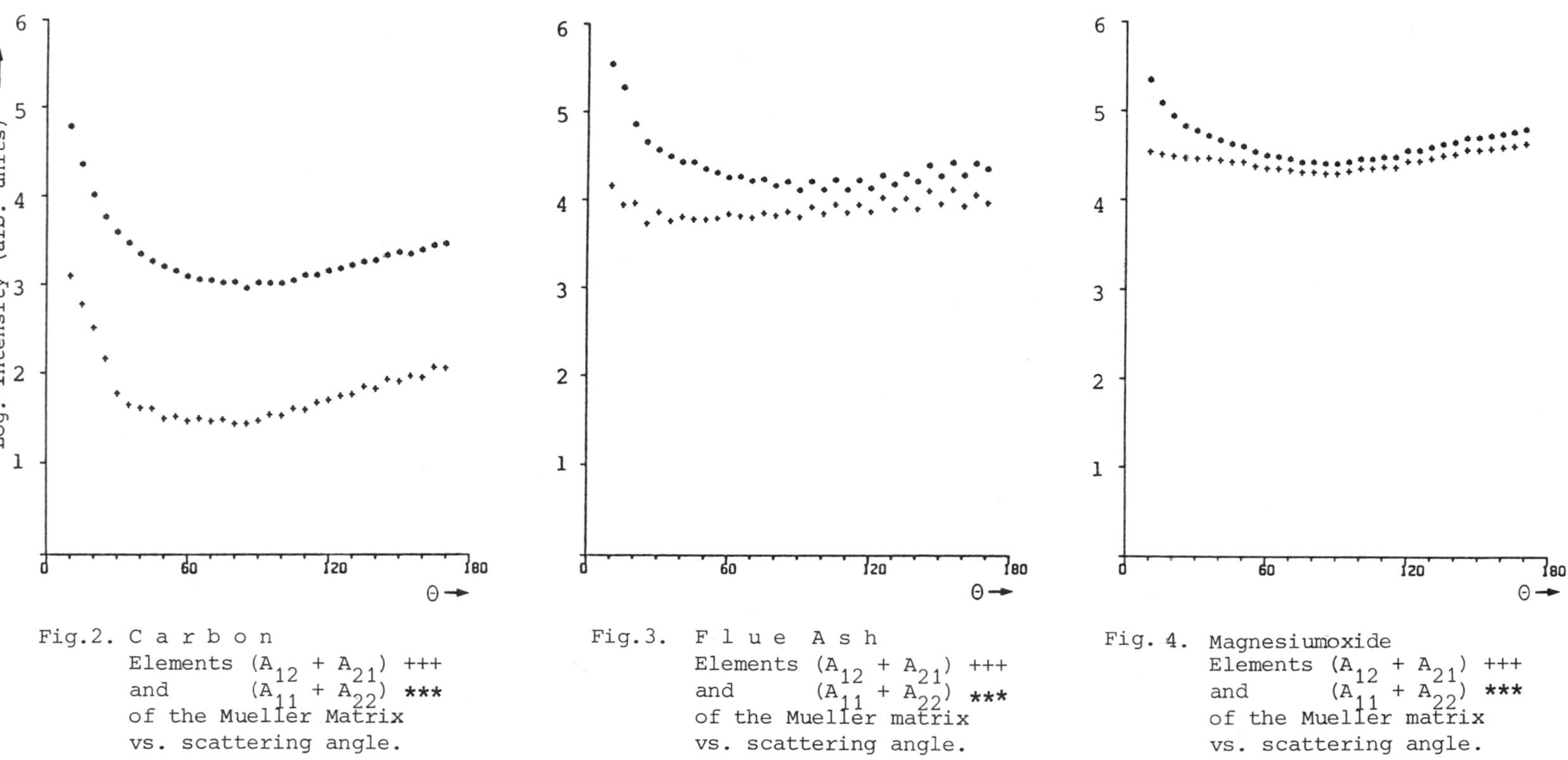

Fig.2. C a r b o n
Elements $(A_{12} + A_{21})$ +++
and $(A_{11} + A_{22})$ ***
of the Mueller Matrix
vs. scattering angle.

Fig.3. F l u e A s h
Elements $(A_{12} + A_{21})$ +++
and $(A_{11} + A_{22})$ ***
of the Mueller matrix
vs. scattering angle.

Fig. 4. Magnesiumoxide
Elements $(A_{12} + A_{21})$ +++
and $(A_{11} + A_{22})$ ***
of the Mueller matrix
vs. scattering angle.

CONCLUSIONS

In the last sections monotonic relations between the scattering
efficiencies of particles and the degree of the observed linear and
cross polarization have been demonstrated. These effects can be
evaluated to characterize and discriminate particles. They also provide
a useful tool for particle sizing, because different kinds of particles
may scatter the same amount of intensity within the angular interval of
interest, but will hardly have the same polarization properties.

REFERENCES

Bickel, W.S., Stafford, M.E., 1980, in: "Light Scattering by Irregularly
        Shaped Particles" (D. Schuerman, ed.), p.299, Plenum Press, New
        York.
Bohren, D.F., Huffman, D.R., 1983, "Absorption and Scattering of Light
        by Small Particles", John Wiley & Sons, New York, Chichester,
        Brisbane, Toronto, Singapore.
Bottiger, J.R., Voss, K.J., Fry, E.S., 1982, Aerosol Science and
        Technology, 1:317.
Hanner, M.S., Giese, R.H., Weiss, K., and Zerull, R., 1981, Astron.
        Astrophys., 104:42.
Hodkinson, J.R., 1963, in: "Electromagnetic Scattering" (M.Kerker, ed.),
        p.287, Pergamon Press.
Kerker, M., 1969, "The Scattering of Light", Academic Press, New York
        and London.
Perry, R.J., Hunt, A.J., Huffman, D.R., 1978, Appl. Opt., 17:2700.
van de Hulst, H.C., 1957, "Light Scattering by Small Particles", Wiley,
        New York.
Wolff, M., 1980, Icarus, 33:780.
Zerull, R.H., 1976, Beiträge zur Physik der Atmosphäre, 49:168.
Zerull, R.H., Killinger, R.T., Weiss-Wrana, K., 1986, in: "Proceedings
        1. World Congress Particle Technology Part I" (K. Leschonski,
        ed.) Nürnberger Messe- und Ausstellungs-GmbH., Nürnberg.

OPTICAL CALIBRATION OF ACCURATE PARTICLE SIZING STANDARDS

AT THE U.S. NATIONAL BUREAU OF STANDARDS

Thomas R. Lettieri

National Bureau of Standards
Gaithersburg, MD 20899   USA

ABSTRACT

This paper discusses the need for accurate calibration artifacts in the optical particle sizing community, and details work at the U.S. National Bureau of Standards toward the development of such materials using optical techniques for certification.

INTRODUCTION: THE NEED FOR PARTICLE SIZING STANDARDS

The importance of accurate reference standards is generally recognized, but often given short shrift by members of the optical particle sizing community. All too often, one faithfully accepts the numbers which come pouring out of one's optical sizing instruments, without stopping to think about what the numbers really mean. Many times, the particle size is taken to be a "given" in an experiment, so that all the time is spent trying to find ways to accurately measure the other parameters. However, a bit of reflection should convince one that the particle size information is usually just as important, and that this information is often not as well known as one might think.

Several questions should come to mind when using any type of optical instrument for particle sizing purposes: How was the instrument calibrated? Was it calibrated with the same type of particles you are trying to measure? Do the particles you are trying to measure satisfy the assumptions made about the sizing technique used in the instrument? If there is any doubt about the answer to any of these questions, then it is likely that the size information given by the instrument is not entirely accurate.

One step toward the goal of accurate measurements is the use of well characterized Standard Reference Materials (SRM's). SRM's are homogeneous, stable materials or simple artifacts for which a specific chemical or physical property has been measured and certified by the U.S. National Bureau of Standards (NBS). SRM's can be broadly grouped into three categories: chemical composition, physical properties, and engineering materials [1]. Examples of the first category include SRM's for the chemical analysis of low and high alloy steels, glasses, spectrometric solutions, biological materials, industrial hygiene samples, lubricating materials, cements, and nuclear materials. The second category includes SRM's for the measurement of ion activity, electron microscope calibration, coating thickness, viscosity, elasticity, density, microhardness, molecular weights of polymers, differential thermal analysis, thermodynamic fixed points, magnetic and optical properties, radioactivity, corrosion, and electrical resistivity. Finally, the third category includes SRM's for: rubber compounding, surface roughness, microscope resolution, color, magnetic computer storage, and fire research. The 900-plus SRM's available from NBS are used for quality control and other purposes in many industries and technologies: medicine, environment, metals, plastics, glasses, rubber, electronics, nuclear power, automobiles, and computers, among others.

The particular SRM's discussed in this paper have been developed at NBS to promote uniformity in size measurement of microparticles from about 0.1 um to about 100 um in diameter. These SRM's have several important uses: (a) to calibrate particle sizing instruments (both optical and nonoptical); (b) to help develop new test methods for measuring particle size; (c) to facilitate commerce between buyers and sellers of particle products and sizing instruments; and (d) to assure the long term integrity of the particle sizing measurement process.

(a) In all applications of optical particle sizing instruments, from basic science experiments to online particle production monitoring, it is imperative that accurate standard artifacts be used to calibrate the instruments. In online monitoring of particle properties, SRM's promote quality control and uniformity of product. In basic science experiments, the use of SRM's can lead to the discovery of new physical principles that were previously overlooked due to inaccuracies in particle sizing. As the measurements of the other physical parameters in an experiment (temperature, pressure, mass, etc.) become more refined, it is imperative that measurements of the particle dimensions and optical properties become equally refined.

(b) SRM's also assist in the development of new ways to measure particle properties. By using an accurate particle sizing SRM to test a new technique, the researcher can be confident that any discrepancy he measures must be due to something other than the particle dimensions.

(c) Standard Reference Materials also promote equity in trade and commerce in particle products and particle sizing instrumentation between companies and between countries. In some industries, small differences in particle properties from one batch to another can result in millions of dollars of lost sales from unusable products.

(d) Finally, SRM's promote long term measurement assurance in testing laboratories throughout private industry, and throughout local, state, and federal governments in the United States and elsewhere around the world.

From the above, it is clear that there is a critical need for standard calibration artifacts (i.e., SRM's) in the optical particle sizing community. For purposes of instrument calibration, such artifacts should ideally have three characteristics: (i) they should be certified for the same physical property the instrument measures (for example, mean particle volume), (ii) they should have the correct optical properties for the instrument (e.g., opaque particles for light blockage instruments), and (iii) they should be about the same size as the particles to be measured. Because of the wide variety of optical instruments available which measure many different types of particles (including single particles, agglomerates, distributions, spheres, aspheres, fibers, flakes, transparent particles, opaque particles, and liquid droplets), it is apparent that no single type or size of SRM will satisfy all needs. Thus, some compromises must be made when developing particle sizing SRM's.

NBS/ASTM PARTICLE SIZING STANDARD REFERENCE MATERIALS

Since it would be almost impossible to anticipate all present and future needs for particle sizing SRM's, the approach taken at NBS has been to develop monosized "benchmark" particle SRM's at discrete diameters covering several orders of magnitude. These primary benchmarks can then be used in the development of secondary standards tailored to specific particle sizing needs.

The NBS program is a joint effort with the ASTM S-21 Coordinating Committee for Standard Reference Materials for Particle Metrology, which is composed of representatives from over 20 other ASTM committees including: B-09 on Metal Powders, C-01 on Cement, D-22 on Sampling and Analysis of Atmospheres, E-29 on Particle Size Measurement, and F-21 on Filtration. The diameters to be certified through the program, 0.1, 0.3, 1, 3, 10, 30, and 100 um, cover a wide range of sizes in order to meet the requirements of most users. All of these SRM's will have well defined and well characterized properties. First, they will be as monosized as current technology permits. Second, their physical dimensions will be accurately measured by two or more definitive techniques having small systematic uncertainties. Third, the optical properties of the particles, although not certified, will be homogeneous and uniform. Fourth, their physical and optical properties will be stable, since particles which disintegrate or agglomerate with time are clearly undesirable.

The NBS particle sizing SRM's currently available, with nominal diameters of 0.3, 1, 10, and 30 um, are sold as 5 ml plastic vials containing a liquid suspension of polystyrene microspheres (Figure 1). The volume concentration of particles is somewhat different for each SRM, but is typically about 0.5%. To prevent the growth of biological

Fig 1. A typical vial of particle sizing SRM and a certificate.

matter, 50 ppm of sodium azide is added to each batch of particles before packaging. Each SRM vial comes with a certificate containing information about the certified mean particle diameter, the size distribution width, the approximate number of outliers, the certification techniques and procedures used, proper use of the SRM, and measurement results from cooperating laboratories. An example of such a certificate is shown in Figure 2.

For each of these SRM's several different techniques were used in the certification process, some of them optical in nature. The operating principles of these optical calibration methods and the results obtained with them are the focus of the remainder of the article.

OPTICAL TECHNIQUES USED IN SRM CERTIFICATION

The techniques chosen to certify a specific SRM were determined primarily by the size of the particles, although the accuracy achievable with the method was also an important consideration. In each case, optical and nonoptical methods were used to calibrate the SRM's in order to check for systematic errors in measurement, and to better characterize the dimensional properties of the material. Most of the nonoptical calibration methods involved some type of electron microscopy, in particular transmission electron microscopy (TEM) and a variation of scanning electron microscopy called metrology electron microscopy (MEM). These will not be discussed here, but details of the techniques and their use in SRM calibration are given elsewhere [2]. The other nonoptical method, also not covered here, used flow through, electrical sensing zone instrumentation.

The optical techniques which were used to certify the NBS SRM's can be classified into two categories: light scattering and optical microscopy.

<u>Light Scattering</u>

Several different light scattering methods were employed, depending on the size of the particles.

<u>Quasielastic Light Scattering</u> For the smallest SRM particles, of 0.3 um nominal diameter, quasielastic light scattering (QELS) was chosen as the preferred technique. In this well known method, the average lifetime of the Brownian motion of the particles in water suspension was measured as a function of scattering angle. This gave a diffusion coefficient which was used in the Stokes-Einstein equation to get the mean hydrodynamic particle diameter, independent of the particles' optical properties [3]. Although various methods have been developed to recover the size distribution width from the QELS spectra, none of them was tried at NBS, since only the mean particle diameter was of interest.

<u>Angular Intensity Light Scattering</u> For the next largest SRM, of 0.9 um nominal diameter, two light scattering methods were selected: Mie angular intensity light scattering (AILS) from single levitated particles and Mie AILS from a liquid suspension of particles. In the single particle experiments, an SRM microsphere was electrostatically suspended in a laser beam and the angular scattering pattern measured for two polarizations of light. The experimental AILS patterns were then computer fit with calculated Mie patterns to get the size of the microsphere. The total uncertainty of this technique, including random and systematic errors, was $\pm$ 0.011 um for the 0.9 um diameter

U. S. Department of Commerce
Malcolm Baldrige
Secretary

National Bureau of Standards
Ernest Ambler, Director

National Bureau of Standards

Certificate

Standard Reference Material 1690

Nominal One-μm Polystyrene Spheres

(In cooperation with the American Society for Testing and Materials)

This Standard Reference Material (SRM) is intended for use as a primary particle size reference standard for the calibration of particle size measuring instruments including microscopes. The SRM is a suspension of polystyrene spheres in water at a weight concentration of about 0.5%.

The number average particle diameter was determined by measuring the light scattered by the polystyrene spheres suspended in water. The value used for the refractive index of polystyrene was $n(\lambda_{vac} = 632.99) = 1.588$. The diameter was determined from the best fit of Mie light scattering theory to the measured intensity versus angle.

Number Average Diameter, μm	Uncertainty, μm
0.895	± 0.008

The uncertainty includes both random and systematic errors.

The sample-to-sample variability (standard deviation) of the number average diameter, as determined on single drops taken from 20 vials (light scattering measurements of water suspensions and electrical sensing zone counter measurements), was found to be 0.0008 μm.

The value certified for the number average diameter was confirmed by two other measurement techniques. The first of these was by measuring the light scattered by individual spheres (8 were measured) levitated in air. In this technique both the diameter and refractive index are determined by the best fit to light scattering theory. In the second technique the average diameter was determined by optically measuring the row length of particles in two dimensional arrays formed by air drying. Scattering by individual particles: (0.900 ± 0.011 μm). Optical array sizing: (0.900 ± 0.015 μm).

The particle size distribution of the polystyrene spheres (as determined by measurements with a transmission electron microscope) is narrow with a standard deviation of about 0.009 μm, excluding small particles with diameters less than 0.6 μm (about 0.5%) and large single particles with diameters in the range of 2-6 μm (about 0.1%). A discordancy test based on the sample kurtosis was used at the 5% level for rejecting these particles [V. Barnett and T. Lewis, Outliers in Statistical Data, (Wiley, 1978) p. 101]. The particles are spherical with an average deviation from sphericity, $(D_{max} - D_{min})/D_{ave}$, of about 0.006. Measurements with an electrical sensing zone counter and by optical microscopy indicated that about 1.5% of the particles are agglomerated doublets.

The material is expected to have a four year shelf life when stored at room temperature provided the cap on the vial is not removed. Care should be exercised once the cap has been removed to prevent contamination.

Before taking a sample by squeezing a drop from the vial, manually shake and/or expose to ultrasonics until the spheres are uniformly distributed. Use filtered (0.2-μm pore size filter) distilled water for dilution. When electrolytes are used for electrical sensing zone counter measurements, first dilute the sample with water to prevent agglomeration.

The technical direction and physical measurements leading to certification were provided by G. Mulholland, T. Lettieri, G. Hembree, A. Hartman, and E. Marx of the Mechanical Production Metrology Division, with guidance on statistical analysis provided by K. Eberhardt of the Statistical Engineering Division.

The overall coordination of the measurments by the cooperating laboratories was performed under the direction of R. Obbink, Research Associate, ASTM-NBS Research Associate Program.

The technical and support aspects involved in the preparation, certification, and issuance of this Standard Reference Material were coordinated through the Office of Standard Reference Materials by R.K. Kirby.

Washington, D.C. 20234
December 22, 1982

George A. Uriano, Chief
Office of Standard Reference Materials

Fig. 2. An example of a particle sizing SRM certificate.

microspheres. The other AILS experiments were similar, but instead of
using single microspheres, the laser beam was scattered from a liquid
suspension of SRM microspheres. From the angular light scattering
patterns (Figure 3), both the mean diameter (d) and the size
distribution width(s) could be obtained with a total uncertainty of
$\pm$ 0.008 um on the certified value of d. Further details of the
experiments, calculations, and errors for both AILS methods are given in
Reference 4.

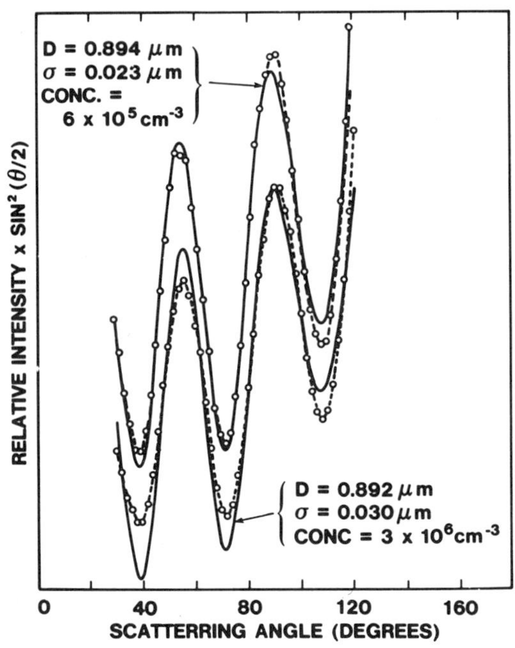

Fig. 3. Mie AILS pattern for 0.895 um spheres in liquid suspension.

<u>Resonance</u> <u>Light</u> <u>Scattering</u>   A fourth light scattering technique,
resonance light scattering (RLS), was used in the certification of the
10 um SRM. RLS makes use of the extremely sharp resonances which occur
in the intensity-versus-wavelength patterns of light scattered from
dielectric spheres [5]. The wavelengths of the resonances are a
sensitive function of the size of the particle. In principle, RLS can
be used to size a single microsphere to 10 ppm or better, although
uncertain knowledge of the other parameters, primarily the refractive
index, limits the accuracy achievable. To get the mean diameter from a
distribution of SRM microspheres, RLS spectra were obtained from a
liquid suspension of the particles (Figure 4), and the experimental peak
wavelengths were least-square fit with peak wavelengths calculated using
Mie theory [6]. Note that the size distribution of the particles
broadens the resonances, making the peak finding procedure less precise.
Nonetheless, the total uncertainty in mean diameter using this procedure
was better than $\pm$ 0.03 um for the 10 um diameter microspheres. Details
of the RLS technique and its errors when used to certify the 10 um SRM
will be presented in an upcoming publication [7].

<u>Optical Microscopy</u>

The optical microscopic techniques were array sizing and a
variation of this, center distance finding (CDF).

616

<u>Array Sizing</u> In conventional array sizing [8], a hexagonally ordered
array of microspheres is formed on a microscope slide, and the center-
to-center distance is measured along a number (say 10) of inline

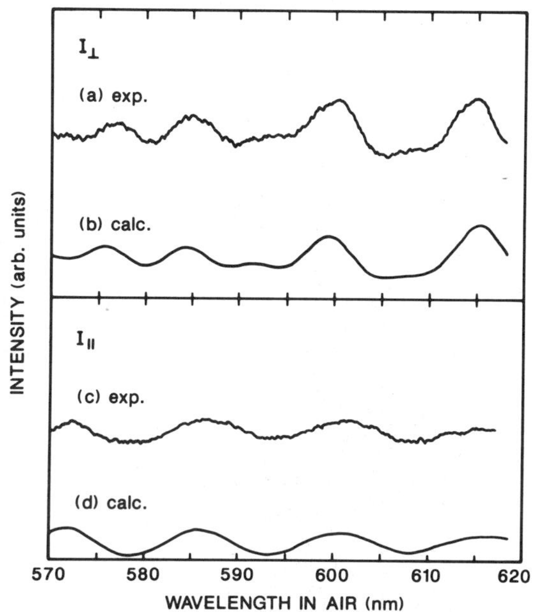

Fig. 4. RLS spectra for 9.89 um spheres in liquid suspension.

particles using either a calibrated eyepiece reticle or photographic
film.   By measuring across N microspheres, the random component of the
measurement error is reduced by $\sqrt{N}$. At NBS, careful measurements of the
distances on photographic film, along with accurate calibration of the
microscope magnification, image distortion, and film shrinkage, resulted
in total measurement uncertainties of $\pm$ 0.015 um on 0.9 um particles
[4].   For the 0.3 um SRM, it was possible to form hexagonal arrays and
resolve them with off-axis illumination, but the particles were somewhat
too small to get an accurate mean diameter with optical array sizing.

<u>Center Distance Finding</u> One of the problems with conventional array
sizing is that the hexagonal arrays always contain air gaps and voids.
This makes it difficult to get an accurate mean diameter for the
microspheres, and tends to broaden the measured size distribution [8,9].
In addition, it is not easy to precisely pinpoint the center of a
microsphere image with a reticle or on photographic film.   For this
reason, an optical microscopic technique called center distance finding
(CDF) was developed at NBS [9]. With CDF, the particles are illuminated
by parallel light;   each microsphere acts like a positive lens which
focusses the light into a small spot in a plane just above the
microspheres.   If the microscope objective is then focussed on this
plane, the resultant photographic image will have a pattern of well
defined dots which pinpoint the microsphere centers (Fig. 5).   The
distances between the dots contain the particle diameter information.
Also, to avoid the problem of air gaps, hexagonal arrays are not used.
Instead, the center-to-center distances are measured between two

contacting microspheres along unordered chains of spheres, which do not
have air gaps or voids (Fig. 6). This gives a much more accurate
measurement of the mean diameter, and also makes it possible to get an
accurate measurement of the diameter distribution. The resolution of
the CDF technique for metrology purposes can approach that of an
electron microscope, about ± 0.01 um, and with a well calibrated optical
train (microscope plus photographic film readout), an accuracy of ± 0.04
um was realized at NBS [7]. CDF was used to certify both the 10 um and
the 30 um SRM's.

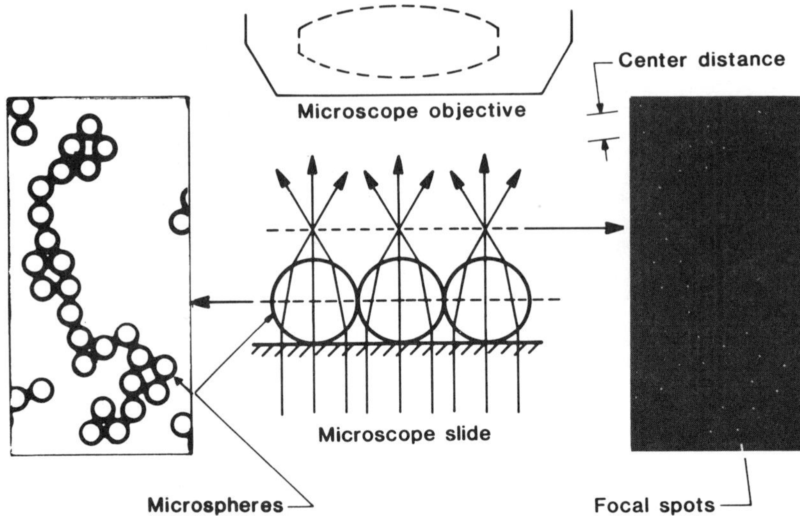

Fig. 5. The center distance finding technique.

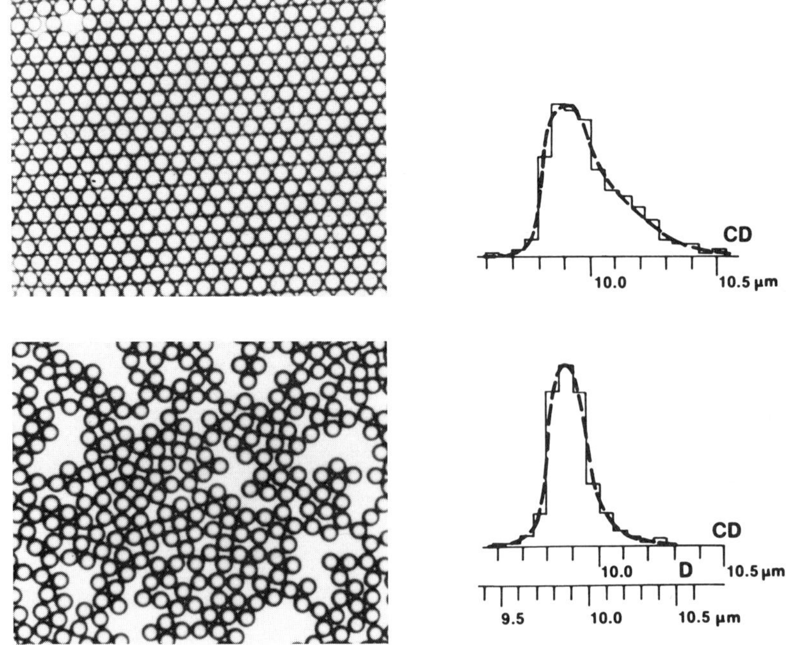

Fig. 6. The effect of air gaps on center distance (CD) measurements.

CERTIFICATION OF THE SRM'S

<u>SRM 1691</u>: <u>Nominal 0.3 um Spheres</u>

As noted, QELS was the optical technique utilized to certify the nominal 0.3 um SRM 1691. The result of the QELS measurements on five different samples of SRM 1691 was a mean diameter of 0.276 ± 0.007 um, somewhat higher than the dry diameter of 0.269 ± 0.007 um measured by TEM (Figure 7). The coefficient of variation (CV = standard deviation/mean diameter) was undetectable with QELS, but other methods showed it to be less than 1%. One other optical method, Mie AILS from a water suspension of the microspheres, was tried, but gave a featureless pattern from which it was difficult to get both a mean diameter and a CV. However, when the CV was fixed at 1%, the resultant mean diameter of 0.270 um obtained from Mie AILS was close to the other measured values, although this result was not used in the certification process. Applications for this submicron SRM include: integrated circuit and computer disc contamination monitoring, and the calibration of QELS instruments, electron microscopes, and atmospheric sampling equipment.

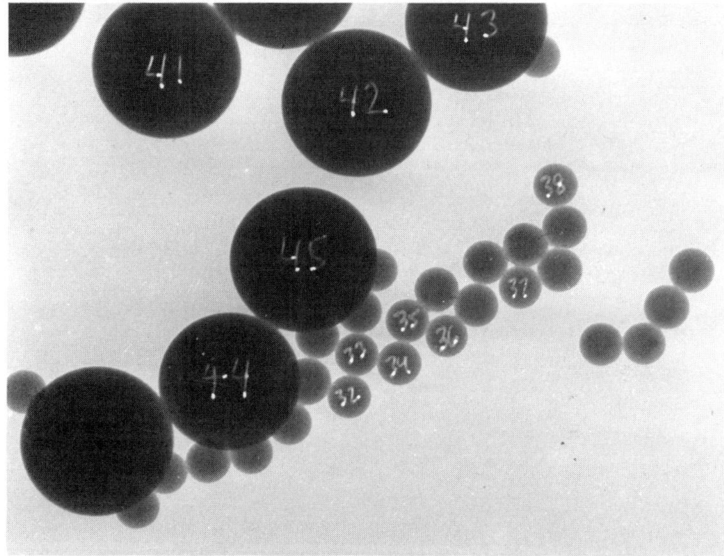

Fig. 7. TEM photograph of 0.269 and 0.895 um SRM spheres.

<u>SRM 1690</u>: <u>Nominal 0.9 um Spheres</u>

For the certification of these particles, three optical techniques were employed: Mie AILS from particles in liquid suspension, Mie AILS from single particles levitated in air, and optical microscope array sizing [4]. Careful measurements and detailed characterization of the errors in each technique resulted in mean diameters which agreed very well, with the first method giving 0.895 ± 0.008 um, the second 0.900 ± 0.011 um, and the third 0.900 ± 0.015 um. The most accurate value, 0.895 um, was chosen as the certified mean diameter for this SRM. The CV, although not certified, was measured to be about 1%. This SRM has about 0.5% undersized outliers (<0.6 um in diameter) and about 0.1% oversized outliers (>2 um in diameter). Applications are similar to those for the 0.3 um particles, with the addition of optical microscope calibration.

<u>SRM 1960</u>: <u>Nominal 10 um Spheres</u>

For the nominal 10 um polystyrene spheres (SRM 1960), two optical sizing methods were employed: CDF and RLS [7]. A unique feature of these calibration particles is that they were grown in microgravity aboard a NASA space shuttle and, as a result, are highly spherical [10]. The high degree of sphericity, together with their size, make them nearly ideal particles for calibration by optical means. Figure 1 shows a vial of SRM 1960, and Figure 6 shows hexagonal arrays and unordered chains of the microspheres. Center distance finding yielded a certified mean diameter of 9.89 $\pm$ 0.04 um and a CV of 0.9% for this SRM, while the RLS technique gave a mean diameter of 9.91 um, well within the uncertainty of the other measurement. The result of 9.886 $\pm$ 0.029 um from a third (nonoptical) method, metrology electron metrology, was also in excellent agreement with the optical results. This SRM 1960 has recently been considered by the U.S. Pharmacopeia for use as a national standard in the pharmaceutical industry. Other applications include blood cell counting, optical microscope calibration, and atmospheric sampling, especially with regard to the new PM-10 standard of the U.S. Environmental Protection Agency.

The material is also available as hexagonal arrays and unordered chains of microspheres formed on a microscope slide (SRM 1965 Microsphere Slide).

<u>SRM 1961: Nominal 30 um Spheres</u>

Only one optical method, CDF, was used in the certification of the 30 um SRM 1961, also made in microgravity aboard a NASA space shuttle. In this case, the mean diameter from CDF was 29.64 $\pm$ 0.06 um, with a CV of 0.8%. Again, there was excellent agreement between the optical method and the electron microscopy results, with the MEM technique yielding a mean diameter of 29.68 $\pm$ 0.05 um. The diameter distribution for this SRM was measured by CDF to be normal from 3% to 97% of the distribution, and the numbers of undersized and oversized outliers are both less than 1%. Uses for this material include online monitoring of particle production, as well as general instrument calibration.

<u>Future SRM's</u>

For the certification of the future SRM's (0.1, 3, and 100 um), other optical and nonoptical methods will be employed as needed. As before, the choice will depend on the particular size to be certified and the accuracies of the techniques. In particular, optical microscope-based techniques, such as CDF, will be used on the 3 and 100 um SRM's, but not on the 0.1-um-diameter SRM. In the latter case, QELS would be the best optical technique for certification, and MEM the best nonoptical technique.

SUMMARY

The results of the above discussion are summarized in the Table.

CONCLUSION

As evidenced here, Standard Reference Materials have many important uses in all areas of research, technology, and engineering involving the measurement of particle size, shape, and optical properties. With the advent of new, high accuracy optical techniques for particle sizing, it is now possible to calibrate the dimensions of such SRM's to $\pm$ 0.01 um or better in accuracy. With careful measurement methodology, and by accounting for all systematic errors, these new optical techniques can

Table 1

NBS Standard Reference Materials Calibrated by Optical Techniques

SRM Number	Certified Diameter	Uncertainty	CV	Optical Techniques	Comments
1691	0.269 um	± 0.007 um	< 2%	* QELS	
1690	0.895	± 0.008	1%	* AILS from particles in suspension * AILS from single particles * array sizing	
1960	9.89	± 0.04	0.9%	* center distance finding	grown in microgravity
1961	29.64	± 0.06	0.8%	* center distance finding	grown in microgravity

give dimensional information with resolution and accuracy comparable to that from electron microscopy. More importantly, with proper modelling of the measurement process and its errors, optical measurements of particle dimensions have always agreed with those from the nonoptical techniques, even though each technique uses a completely different physical principle to get the size information. Through the use of accurate particle sizing SRM's, the coming years will no doubt see more uniformity in measurements within laboratories, better agreement in measurements between laboratories, and, possibly, the discovery of new physical principles involving microparticles.

ACKNOWLEDGMENTS

The author wishes to acknowledge the contributions of Arie Hartman, Gary Hembree, Egon Marx, and George Mulholland in the SRM calibrations.

REFERENCES

1. "NBS Standard Reference Materials Catalog 1986-87," NBS SP260, R.W. Steward, Ed. (NBS OSRM, Gaithersburg, MD, 1986).

2. D. Swyt, T. Lettieri, A. Hartman, and S. Jensen, "Techniques for the Calibration of Microscopic Particle Size Standards," in *Particulate Systems: Technology and Fundamentals*, J.K. Beddow, Ed. (Hemisphere, Washington, DC, 1983).

3. See for example: B. Chu, "Laser Light Scattering," (Academic Press, New York, 1974).

4. G. Mulholland, A. Hartman, G. Hembree, E. Marx, and T. Lettieri, "Development of a One-Micrometer-Diameter Particle Size Standard Reference Material," J. Res. Nat. Bur. Stand. (U.S.) 90, 3 (1985).

5. See for example: P. Chylek, V. Ramaswamy, A. Ashkin, and J. Dziedzic, "Simultaneous Determination of Refractive Index and Size of Spherical Particles from Light Scattering Data," Appl. Opt. 22, 2302 (1983).

6. T. Lettieri and E. Marx, "Resonance Light Scattering from a Liquid Suspension of Microspheres," Appl. Opt. 25, 4325 (1986).

7. T. Lettieri, G. Hembree, E. Marx, and A. Hartman, Nat. Bur. Stand. (U.S.), Spec. Publ. 260 (to be published).

8. H.E. Kubitschek, "Optical Calibration of Some Monodisperse Polystyrene Latexes in Arrays," Nature 192, 1148 (1961); A. Hartman, "Investigations in Array Sizing Part 1. Accuracy of the Sizing Process," Powder Techol. 39, 49 (1984).

9. A. Hartman, "Investigations in Array Sizing 3. The Center Distance Finding Technique," Powder Technol. 46, 109 (1986).

10. G. Mulholland, G. Hembree, and A. Hartman, "Sizing of Polystyrene Spheres Produced in Microgravity," Nat. Bur. Stand. (U.S.), Internal Report #84-2914 (1985); D. Kornfeld, "Monodisperse Latex Reactor (MLR): A Materials Processing Space Shuttle Mid-Deck Payload, Nat. Aero. Space. Admin. (U.S.), Marshall Space Flight Center, Technical Memorandum #86487 (1985).

CONCLUSION AND PERSPECTIVE

A.R. Jones

Department of Chemical Engineering
Imperial College
London SW7 2BY

1. INTRODUCTION

As was eloquently shown by Professor Kerker in his plenary lecture,
light scattering by particles has a long history. The fundamentals were
laid down in the late nineteenth and early twentieth centuries. However,
interest in the subject has accelerated rapidly over the last decade, and
the conference came at an opportune time to review progress.

The practical applications of light scattering arise in many disciplines. In consequence, there are many meetings in which light scattering
arises, but very rarely are there cross-disciplinary meetings on the subject in its own right. Thus, an important aspect of this conference was
to bring together people with a common purpose who might not otherwise
meet. Table 1, culled crudely from addresses, indicates the wide range of
disciplines represented by the speakers.

Understandably, the meeting concentrated on particle sizing, where
there is currently most demand. An indication of the areas of interest is
given by Table 2. The most popular instrument areas were those employing
Fraunhofer diffraction and laser Doppler velocimetry. There was surpri-
singly little on photon correlation spectroscopy and particle counters.
Analytically, the major concerns were the influence of non-sphericity and
its possible measurement and the recovery of particle size distribution by
direct inversion.

This summary roughly follows the headings of Table 2.

2. DIRECT INVERSION

In all light scattering methods the problem is to infer properties of
the particles from some set of measurements. There are essentially two
ways of tackling this. In one it is assumed that the properties are described in some way and the theoretical light scattering data is generated
for a range of parameters. A "best fit" is then obtained to the measured
data. The most common example is in particle sizing where a two parameter
function is assumed, such as modal size and standard deviation.

In the second method an attempt is made to retrieve the particle
properties without making prior assumptions. This is the direct inversion

TABLE 1

Aerospace	1
Astronomony/Astrophysics	4
Chemical Engineering	4
Chemistry	5
Electrical Engineering	2
Energy/Combustion	5
Industry	7
Mechanical Engineering	8
Meteorology	1
Physics	9

TABLE 2

Fraunhofer Diffraction	11
Non-Spherical Particles	11
LDV	9
Direct Inversion	8
PCS	5
Particle Counters	4
Gaussian Beams	3
Multiple Scattering	3
Spectral Extinction	3
Miscellaneous	7

problem, of which there was much discussion during the meeting. Since it
reflects upon various techniques I will cover this subject first.

It is worth recalling the basic statement of the problem at this
stage. The mathematics of inversion is involved and a simple statement of
the problem is often lost in the detail.

The direct inversion problem is to recover the function g(y) from the
measured function f(x). They are related through

$$f(x) = \int K(x,y)g(y)dy + \varepsilon(x)$$

where K(x,y) is the kernel and $\varepsilon(x)$ is due to error of measurement and
noise. In matrix form

$$F_i = K_{ij}G_j + E_i$$

In principle when there is little or no error ($E_i \rightarrow 0$) this matrix
equation may be inverted directly. However, the kernel K_{ij} is ill-conditioned and sensitive both to error and numerical rounding. The art,
therefore, is to provide methods of inversion which can deal with noise.

There are three specific cases of relevance to this meeting:-

(a) <u>Fraunhofer diffraction</u>

$$f(x) = I_{sca}(\theta)$$

$$K(x,y) = a^4\left[\frac{2J_1(ka\theta)}{ka\theta}\right]^2$$

$$g(y) = n(a)$$

(b) <u>Spectral extinction</u>

$$f(x) = K_{ext}(\lambda)$$

$$K(x,y) = \pi a^2 Q_{ext}(a/\lambda)$$

$$g(y) = n(a)$$

(c) <u>Photon correlation spectroscopy</u>

$$f(x) = f(\tau)$$

$$K(x,y) = \exp\left[-2K^2 D_f\tau\right]$$

$$\left\{K = \frac{4\pi}{\lambda}\sin\frac{\theta}{2} \; ; \; D_f = \frac{KT}{6\pi\mu a} - \text{diffusion coefficient}\right\}$$

$$g(y) = n(D_f)$$

In the latter case we note that the inversion takes the form of a Laplace
transform.

An obvious approach is to minimise the influence of the error term by
the method of least squares. There are various methods based on this,
perhaps the most successful being that due to Philips (1962) and Twomey
(1963). This has been applied, for example, with some success to the

spectral extinction method by Walters (1980). One disadvantage of the
method is the requirement that the operator adjusts an unknown Lagrangian
multiplier to smooth the output to a desired amount. Much is thus left to
personal judgement.

Direct inversion schemes for all three cases were discussed in detail
by Bertero, Mol and Pike. They pointed out that least squares methods can
be very time consuming. They described a number of computationally efficient schemes making use of various filter functions or sampling methods,
combined with some prior knowledge of the size distribution - such as its
finite width.

In discussion, Pike implied that without such additional knowledge
an infinite number of solutions existed and a "correct solution" was not
possible. Nonetheless, as was demonstrated by several papers in the conference, for practical purposes the size distributions yielded by the
various instruments were very satisfactory.

3. FRAUNHOFER DIFFRACTION

This technique arose out of the suggestion by Hodkinson (1966) that
the shape of the forward scattered lobe is insensitivie to refractive
index and particle shape. Furthermore, near forward scattering for sufficiently large particles can be adequately described by Fraunhofer diffraction yielding particularly simple analysis. Since that time these points
have been confirmed by numerous authors, and they have led to diffraction
becoming the basis of very popular and powerful techniques for sizing
particles greater than a few microns in diameter. This popularity was
reflected in the number of papers on the subject at this meeting.

The most commercially successful machines based on this technique are
those from Malvern Instruments, and there were a number of papers devoted
to aspects of these. In particular, Hirleman considered the redesign of
the ring detectors and new ways of choosing the size classifications that
would optimise performance. A consequence of this is that the data matrix
becomes Toeplitz in form, requiring the storage of less information.
Also, the matrix becomes diagonally dominant easing the inversion and
making it less sensitive to measurement error.

A limiting feature to sizing by diffraction has been multiple scattering. Errors in particle size become signficiant when the transmissi-
vity falls below 50%. Hamidi and Swithenbank and Hirleman have tackled
this problem, and claim that corrections can now be made to provide size
at transmissivities down to less than 10%. Hirleman claimed that his
method is more general and more accurate at higher ring numbers, but
Hamidi and Swithenbank actually provide correlation functions for direct
use.

While the shape of the diffraction yields particle size, the extinction and magnitude of the scattered light provide evidence of the particle
concentration. Hayashi and Brown and Weatherby were able to obtain concentration from diffraction measurements.

As a consequence of the simplicity of diffraction theory an analytical inversion is known to exist. Bayvel showed the results of experiments
comparing this with the model independent software supplied by Malvern.
He claimed that the analytical inversion was superior. Similarly, Kouzelis et al found that the conjugate gradient method and analytical inversions were the best.

However, Bachalo and Young in an important paper compared the performance of a Malvern diffractometer with both a Particle Measuring Systems shadowing technique and an Aerometric Phase Doppler particle sizer. They recognised that the diffractometer provides a spatial average, whereas the others provide a temporal average. When they corrected for this difference the three instruments agreed well. We must conclude, therefore, that the software provided by Malvern is adequate. Nonetheless, there may be advantages of simplicity to the analytical inversion scheme.

Fraunhofer diffraction is proving to be a powerful tool for sizing particles between 3-4μm and 500μm or greater, for extinctions lying between about 10% and 50%. Major advances have been reported at this conference which extend the range to extinctions greater than 90%. Taken together with recommendations for redesign of the instrument which should require less data storage and ease the matrix inversion the applications of the method can only increase.

4. LASER DOPPLER VELOCIMETRY

Fraunhofer diffraction relies on the presence of sufficient numbers of particles to provide a constant size distribution in time. This requirement can be relaxed by averaging over longer periods, but this has its limitations. In tenuous systems, an alternative is particle counting in which a signal due to one particle at a time is recorded.

A second aspect of diffraction is that it provides a spatial average over the beam diameter and along the beam path. It provides poor spatial resolution along the path, and transverse to this if the beam is expanded.

In particle counting methods a temporal average is obtained since particles are observed crossing a small volume as a function of time. This provides superior spatial resolution, but if a spatial average is required the particle velocity must also be recorded so that a correction may be made.

Around the mid-seventies various authors (e.g. Farmer, 1974) pointed to the potential use of laser Doppler velocimeters to size particles, and thus combine both operations. The size could be determined in principle from either the AC or DC components of the Doppler burst, or their ratio - the visibility. An alternative possibility was to measure the velocity or acceleration in a diverging flow field. However, this does require knowledge of the flow divergence and is restricted to sampling methods or flow in the vicinity of a probe.

Crossed-beam methods provide excellent spatial resolution, as the test volume is very small. Against this there is a requirement that only one particle at a time can be present in the test space which places an upper limit on concentration. The value of this upper limit is the subject of debate, but may be of the order of $10^{12}m^{-3}$. Further, since the test volume is so small it can take considerable time to build up a picture of any spatial variation of properties.

Most of the early work was on the visibility method. A problem with this was the limited dynamic range. The particles had to be sufficiently large to yield a good signal and a visibility significantly less than unity for any reasonable fringe spacing and signal frequency. This limited the minimum size to about 1μm. At the same time, the particles could not be so large that the visibility fell beyond the first zero in the visibility-size curve. Extensive calculations (e.g. by Negus and Drain,

1982) suggested that with appropriate apertures and angles the upper limit
could be extended to a few hundred micrometers.

Another problem arose because for Gaussian beams the fringe contrast
and intensity vary throughout the cross-over volume. The fringe contrast
is unity only in a very small region near the centre (e.g. Hong and Jones,
1978). There are two possible ways around this difficulty. A beam with a
flat, or "top hat", profile can be generated. The alternative is to
define the central region in some way, usually by using a second beam of
orthogonal polarisation or different wavelength. Signals are then only
used that arise from both volumes. However, such methods invariably throw
away much of the available intensity.

The use and calibration of an instrument employing the visibity
method was described in the paper by Hadded et al.

For spheres the signal can be calculated from Mie theory. However,
when the particles are large this can be time consuming. Further, a problem arises when the particle size becomes a significant fraction of the
Guassian beam waste. The particle is then not uniformly illuminated. In
such cases it is simpler to turn to geometrical optics. Al-Chalabi et al
showed that there is reasonable agreement with Mie theory even for particles as small as 1μm. Börner and Zhan used geometrical optics and included a Guassian profile. They calculated the scattered intensity as a
function of time and showed that there are three peaks. They obtained the
size from the transit time between the two side peaks which correspond to
the edges of the particle. This method is similar to that of Birch (1968)
which used schlieren photographs of droplets.

Other techniques for treating large particles and bubbles were described by Lee, Chang and Zheng, and Ohba and Isoda. In the latter case an
interesting conclusion was that the fringe spacing depends on the local
radius of curvature. This has interesting potential for observing nonspherical particles.

The visibility method has the advantage of being relative and insensitive to incident intensity. Unfortunately, it does have the requirement
that the particle must pass through the very small control volume, and
must be smaller than this volume. Also, in an extensive system the visibility may be degraded by beam blocking. One, or both, beams may be influenced by particles along their path thus reducing the fringe contrast.
Density gradients which cause relative motion of the beams can also cause
degradation.

Intensity measurement is not relative and can suffer due to dirt on
windows and lenses. Both the incident and scattered light can be
affected. It is also subject to beam blocking, or extinction along the
incident beam path.

A relatively new technique which should not be subject to the same
problems is phase Doppler. This measures the phase difference in the AC
signals measured by two or more separate but closely spaced detectors.
Under appropriate conditions the phase difference is found to be a linear
function of size over a wide range. A number of commercial instruments
are now available.

Bachalo has done much to develop this method and in this meeting he
reviewed the theoretical basis, confirming the validity of geometrical
optics. He also discussed the problems that arise due to the Gaussian
nature of the beams. Bauckhage et al, discussed the practical use of a
phase Doppler system in a spray, and draw attention to the errors which
arise if the particles are non-spherical or inhomogeneous.

Laser Doppler velocimetry for particle sizing is a rapidly developing
area. At this meeting we witnessed interesting developments especially in
relation to the use of geometrical optics to describe the scattering process, and the influence of the Gaussian beam structure. There was also
much useful discussion of the use of these techniques in practical
systems.

5. PARTICLE COUNTERS

A number of commercial particle counters are available which measure
size from intensity scattered from a single incident beam. These usually
integrate over a wide angular range. Gebhart reviewed such instruments.

Minimum measurable particle sizes are approaching 0.1µm, though Heintzen-
berg, Gärdneus and Ogren have achieved 0.05µm. The ultimate limit is due
to noise from gas molecules, though if even smaller particles become of
concern in clean rooms the molecules themselves will be the contaminants.

The influence of particle shape on the response of particle counters
has been of some concern. Killinger, Zerull and Giese demonstrated that
the scattered signal may vary by an order of magnitude from an irregular
particle depending upon orientation. This variability can be reduced by
using multi-wavelength sources and large apertures, but some variation
remains.

An elegant means of calibrating for the influence of particle nonsphericity was described by Bottlinger, Umhauer and Loffler. They suspended single particles and imposed an AC field. The particle then oscil-
lated and made several traverses of a laser beam, each time with random
orientation. The particle was examined microscopically to obtain the
equivalent diameter. A plot of the average intensity over all orientations against equivalent diameter yielded the calibration. They suggested
a matrix relationship between the measured form dependent intensity and
the required form independent intensity which could easily be inverted.

It is known that particle counters are sensitive to refractive index
and shape. Because the calculations may be performed for spheres, numerous studies have been made on the influence of refractive index (e.g.
Makynen, 1983). The effects of shape have been dramatically demonstrated,
for example, by Seville et al (1984), but this has not been explored so
deeply. The papers at this meeting have therefore provided a very useful
contribution.

In view of the earlier discussion under LDV there is an obvious need
to know how particle counters deal with the trajectory problem in Gaussian
beams. The manufacturers' literature does not always discuss this. It
would also be useful to know whether they can measure velocity in order to
convert to a spatial distribution. More recent particle counters, such as
that described by Holve and Davis (1985), are beginning to consider these
problems.

6. PHOTON CORRELATION SPECTROSCOPY

A small particle suspended in a fluid is subject to molecular bom-
bardment, and will exhibit random Brownian motion. Light scattered by a
moving particle experiences a Doppler shift, and since Brownian motion is
random this results in a broadening of the line width. Smaller particles
have higher Brownian velocities and this results in greater broadening.
Thus the width of the scattered line is inversely dependent on the particle size.

The width of the line can, in principle, be measured using a high
resolution spectrometer. However, it may be very narrow for particles of
reasonable size and it becomes more practical to measure the auto-correlation function. This is obtained by comparing the signal with itself as a
function of increasing delay time. The auto-correlation function is exponential, as discussed in Section 2. For a size distribution the inversion
problem reduces to that of solving a Laplace transform. Bertero, Mol and
Pike, Ross, and Schmitz and Yu discussed various aspects of this inversion
problem.

A special difficulty arises when the distribution is almost monodisperse. In this case the inversion techniques are not sufficiently sensitive and problems arise due to the ill-conditioned nature of the Laplace
transform. To obtain extra information Pusey and van Megan exploited the
angular variation of the correlation function. They measured the initial
slope as a function of angle and were able to characterise very narrow
polydispersities.

Again complications arise if the particles are not spherical.
Rarity described an interesting technique to measure the axial ratio of
ellipsoids. The elongation produces an extra degree of freedom and in
addition to translational motion there is rotation. Rarity used two
detectors and calculated their mutual coherence function. The two terms
due to translation and rotation decay at very different rates and can
easily be distinguished. By comparing his measurements with theory based
on Rayleigh-Gans-Debye theory he was able to measure both axes of the
particles.

Photon correlation spectroscopy is a useful technique for sizing
small particles suspended in a fluid. Since it is mechanical in nature it
is not sensitive to refractive index, though it is sensitive to shape. A
major problem has been the recovery of the size distribution from the
correlation function, and considerable progress on this was reported.

If the fluid is not stagnant, but moves itself, there is an additional problem. The motions of the fluid and the particles are additive and
influence the spectrum. This is a limiting factor to the more general
application of the method especially in gas flows. There has been some
work in this, for example by Chang and Penner (1981), who propose ways of
separating out the motion of the gas and particles, and the Doppler ambiguity which arises from the transit time of the particles in the laser
beam. In a laminar flow this latter uncertainty can be minimised by
making the transit time large, as for example in the work of Scrivner et
al (1986). This will be a fruitful area for further analysis.

7. SPECTRAL EXTINCTION

For sufficiently small particles the extinction efficiency varies
rapidly and monotonically with the ratio of size to wavelength. Thus
measurement of extinction coefficient against wavelength can yield size
uniquely. This is the spectral extinction method, the inversion problem
for which was outlined in Section 2.

The principles of the method are straightforward and well established. However, there are two problem areas which were covered at this
meeting; non-spherical particles and dense systems. In the first case,
Quinten and Kreibig explored the extinction spectra of aggregates using
dipolar coupling. They found that the resonances seen for spheres split
into several peaks for aggregates. They were able to confirm these experimentally.

At high densities multiple scattering disrupts the simple interpretation of data. This problem has been tackled by Gouesbet, Gougeon, Le
Toulouzan and Thioye using a four flux model; forward and backward diffuse
fluxes, and reflected and transmitted collimated beams obeying Beer's law.
They measured the ratio of extinctions at wavelengths of 337μm and
0.6328μm for monodisperse particles in the size range 10-100μm. They were
able to obtain sizes for concentrations up to $3\times10^9\mathrm{m}^{-3}$. They also found
that particle irregularity results in oversizing since extinction is
always increased by non-sphericity.

Spectral extinction is a ueful method that is simple to use. However, it is very sensitive to refractive index, which is a limitation if
this parameter is not known.

8. NON-SPHERICAL PARTICLES

Most studies on light scattering have assumed the particles to be
spherical. This is because of the availability of the Mie theory. However, most particles in nature are not spherical and there is the need to
explore their scattering properties, the influence on particle sizing and
whether characteristics of the structure can be measured directly.

The effect in sizing has been investigated in a number of papers at
this meeting. These have been reviewed in the sections devoted to specific techniques.

Before the structure of non-spherical particles can be measured it is
necessary to have an adequate theory to describe how they scatter. There
are two classes: (a) non-spherical but regular (e.g. cylinders, ellipsoids, cubes, Tchebyshev particles) and (b) irregular where the surface
cannot be predicted from point to point or particle to particle.

The first class can be treated by a variety of techniques employing
either approximations (e.g. RGD, anomalous diffraction, geometrical
optics), or boundary condition methods. To solve the boundary condition
problem numerical techniques (e.g. T-matrix method) are usually employed,
though in a few cases analytical results are available. The solution for
the infinite circular cylinder is well known, and there is a solution for
the ellipsoid due to Asano and Yamamoto (1975). In the second class the
shape of the surface cannot be predicted. This implies that a statistical
approach is required averaging over many particles. A good review of work
in both classes is provided in the book edited by Schuerman (1980).

Non-spherical particles alter the state of polarisation between the
incident and scattered light. This can be described via the Stokes' vectors.

$$
\begin{pmatrix} I_s \\ Q_s \\ U_s \\ V_s \end{pmatrix}
= \frac{1}{k^2 r^2}
\begin{pmatrix}
S_{11} & S_{12} & S_{13} & S_{14} \\
S_{21} & S_{22} & S_{23} & S_{24} \\
S_{31} & S_{32} & S_{33} & S_{34} \\
S_{41} & S_{42} & S_{43} & S_{44}
\end{pmatrix}
\begin{pmatrix} I_i \\ Q_i \\ U_i \\ V_i \end{pmatrix}
$$

The definitions of I, Q, U and V are given, for example, by van de Hulst
(1957), and properties of the matrix were discussed by Turner (1973). For
randomly oriented particles the matrix becomes:

$$
\begin{pmatrix} I_s \\ Q_s \\ U_s \\ V_s \end{pmatrix} = \frac{1}{k^2 r^2} \begin{pmatrix} S_{11} & S_{12} & 0 & 0 \\ S_{12} & S_{22} & 0 & 0 \\ 0 & 0 & S_{33} & S_{34} \\ 0 & 0 & -S_{34} & S_{44} \end{pmatrix} \begin{pmatrix} I_i \\ Q_i \\ U_i \\ V_i \end{pmatrix}
$$

Further, for spheres $S_{11} = S_{22}$ and $S_{33} = S_{44}$.

Barber and Hill considered which elements of the matrix it may be most profitable to measure to find various properties of non-spherical particles. They compared calculations for spheres from Mie theory with randomly oriented ellipsoids using the T-matrix method. They proposed that $1-S_{22}/S_{11}$, which is zero for a sphere, is a useful measure of non-sphericity. They also observed that for an ellipsoid at nose on incidence there are eight zero elements, whereas at other orientations all sixteen elements are non-zero. This may be used to indicate alignment. Further, they noted that by observing the reduction in scattering resonances at wavelength is varied axial elongation as low as 1:1.05 may be measurable.

Detailed measurements on non-spherical particles were reported by Zerull, Killinger and Giese, and Eversole, Lin and Campillo. In particular, Zerull et al pointed to a relationship between albedo and degree of polarisation. By incorporating polarisers into particle counters this could be used to provide extra information on the particles present.

From the point of view of calculating the scattering properties of non-spherical particles, Barber and Hill's description of the T-matrix method was very informative. The three dimensional representations of the matrix were fascinating.

There have been some very useful contributions to this meeting concerning the study of non-spherical particles and the influence they have on various particle sizing methods. However, the emphasis is still on particles which can be easily described, such as ellipsoids. For genuinely irregular particles a statistical approach was suggested by Jones, but this was restricted to scalar diffraction.

9. GAUSSIAN BEAMS

The pre-eminent light source for scattering studies is the laser, usually operating in the TEM_{00} mode. It is known that this mode has a Gaussian intensity profile while Mie theory assumes an infinite plane wave. There is no problem provided the particles are small compared to the beamwidth. This can be achieved by beam expansion but results in a low irradience.

When the particle is large or the laser beam is focused there may be considerable variation of field strength across the diameter. Then Mie theory requires modification to allow for the true nature of the beam. The work of Maheu, Gréhan and Gouesbet is valuable because it provides a solution which is very similar to the Mie equations with a deceptively simple correction factor. Unfortunately, in the general case this function is in reality a complicated integral which can involve considerable computer time. However, in the case of a simple Gaussian with the particle on the beam axis a very simple approximation to the correction function has been obtained by associating each term in the Mie series with

a geometrical light ray - the so-called "localisation principle". This
extremely elegant result enables very simple calculation of Gaussian beam
scattering, and was found to give good agreement with calculations based
on Mie theory and the RGD approximation.

Another paper by the same authors looked at scattering when the particle was off-axis in a laser beam. They pointed out that the intensity
is higher on one side of the particle than the other creating a gradient
in radiation pressure which pushes the particle sideways. Also, Chevaillier et al examined particles crossing a focused laser beam at various
distances from the beam waist plane. They found that a hump developed in
the signal depending on the distance, so that information was available on
the trajectory as well as the size.

The development of suitable readily computable theories of Gaussian
beam scattering will mean that the full intensity can be brought to bear
on a particle, by focusing if necessary. Detailed calculations should
provide evidence of the nature of the scattered signal in terms of position in the beam so that problems due to trajectory may be eliminated, or
taken into account.

10. MISCELLANEOUS

This section covers a few papers which did not easily fit into the
categories above.

Blieck and Lamy considered the generation of particles for study in
light scattering instruments. They used a fluidised bed in which the
required small particles were mixed with much larger glass particles.
Collisions with the glass spheres broke up any agglomerates in the fine
powder, and the small particles were elutriated from the bed. An apparatus was also described employing a white light source with filters and
polarisers. Various aspects of the particles may be studied.

Mie scattering at a number of angles was used by Timmler and Roth to
study the evaporation of droplets behind a shock wave. They used fibre
optics to collect the scattered light. They were able to confirm that the
rate of change of area was constant during evaporation.

Particles which are so small that they fall in the Rayleigh region
present special problems because the variation of scattering with angle
and wavelength is independent of size. A classical technique described by
van de Hulst (1957) is to measure the ratio of the scattering and extinction cross-sections. A common substance falling in this régime is soot,
and Bockhorn et al studied its formation in flames using this technique
when the particles were small. When they had grown larger Mie scattering
was employed. By sampling they obtained the standard deviation of an
assumed log-normal distribution and claimed that they were then able to
obtain the size and complex refractive index. However, they assumed the
particles to be spherical and the variations in apparent refractive index
they observed were probably in reality due to change in shape caused by
agglomeration.

The need for standard particles for the calibration of light scattering instruments was emphasised by Lettieri. He described how the National
Bureau of Standards are producing microspheres and certifying them. They
are monodisperse, homogeneous and stable in time, and cover sizes in the
range 0.1µm to 100µm.

On a very different theme Crifo discussed the optical properties of
cometary dust. This may be observed both by the scattering of sunlight

and infrared thermal emission. There are certain constraints on the particle size due to knowledge of their motion, so that the scattering properties can be predicted. The author showed that the dust size distribution usually assumed is compatible with observation.

Finally, a very interesting theory describing the scattering by large
bubbles was presented by Fiedler-Ferrari and Nussenzveig. Here the refractive index of the paticles is less that than of the surrounds and total
external reflection can occur. The authors noted that this produces a
sharp cut off of intensity equivalent to an edge. The influence of this
can then be calculated from diffraction theory.

11. CONCLUSIONS

This conference has demonstrated the lively status of light scattering, especially its use for particle sizing and characterisation. A wide
range of papers has been presented illustrating not only the variety of
techniques available, but also the expanse of applications.

We have witnessed significant developments in a number of general
areas. Consideration has been given to instrument design, calibration and
standard materials. There has also been discussion on the extension of
existing methods into new areas. Thus we found studies of dense media and
multiple scattering, the influence of Gaussian beam profile and non- spherical particles. We have also seen interesting new progress in direct
inversion methods.

In addition, the conference was complemented by three excellent plenary lectures. Kerker provided a fascinating insight into the early history of the subject and the properties of small colloidal particles. On a
very practical level Durst discussed the design of two beam scattering
systems with emphasis on noise and bandwidth, and the methods of choosing
correct components. Finally, the meeting was brought to reality by Galant
who described some of the problems which arise in industrial applications,
especially in terms of scale, cost and ease of implementation.

Looking to the future we may expect further developments on all
fronts. Experimentally new methods should become available to measure not
only the size of non-spherical particles but also characteristics of their
shape and structure, as well as refractive index. As computational power
increases and becomes cheaper we should see powerful new numerical techniques for predicting scattering properties of a wide range of non-spherical, irregular and inhomogeneous particles.

REFERENCES

Asano, S. and Yamamoto, G., App.Optics, 14, 29-49 (1975).
Birch, K.G., Optica Acta, 15, 113 (1968).
Chang, P.H.P. and Penner, S.S.,J.Quant.Spect.Rad.Trans.,25,97-110 (1981).
Farmer, W.M., App.Optics, 13, 610-622 (1974).
Hodkinson, J.R., App.Optics, 5, 839-844 (1966).
Holve, D.J. and Davis, G.W., App.Optics, 24, 998-1005 (1985).
Hong, N.S. and Jones, A.R., J.Phys.D., 11, 1963-1967 (1978).
Makynen, J., J.Aerosol.Sci., 14, 361-363 (1983).
Negus, C.R. and Drain, L.E., J.Phys.D., 15, 375-402 (1982).
Phillips, D.L., J.Assoc.Comp.Mach., 9, 84-97 (1962).
Schuerman, D.W., Light Scattering by Irregularly Shaped Particles (Plenum
     Press, 1980).
Scrivner, S.M., Taylor, T.W., Sorensen, C.M. and Merklin, J.F., App.
     Optics, 25, 291-296 (1986).

Seville, J., Coury, J., Ghadiri, M. and Clift, R., 3rd Europ.Symp.Particle
    Characterisation, 305-322 (1984).
Turner, L., App.Optics, 12, 1085-1090 (1973).
Twomey, S., J.Assoc.Comp.Mach., 10, 97-101 (1963).
Walters, P.T., App.Optics, 19, 2365 (1980).
van de Hulst, H.C., Light Scattering by Small Particles, (Chapman and
    Hall, 1957).

EXHIBITORS LIST

        The following organisations
assisted in the equipment exhibition
held   simultaneously   with   the
Conference

Dantec Electronique   Scop
22 A. de la Baltique   27-33 Rue d'Anthony
91540 Les Ullis - France   94533 Rungis - France

Elsevier Science Publisher   Sympathec
PO Box 1527   Burgstcsetter Strasse 6
1000 BM Amsterdam - The Netherlands   3392 Clausthal - R.F.A.

Insitec   TSI France
2110 Omega suite F   68 Rue de Paris
San Ramon LA 94583 - U.S.A.   93804 Epinay sur Seine - France

Malvern   Wild Leitz
Worcestershire WR 40 1A - England   86 A. du 18 Juin 1940
                                             92506 Rueil Malmaison - France
Pacific Scientifique
91120 Palaiseau - France

Palas Gmbh
7500 Karlsruhe 1 - R.F.A.

Pergamon Books
Oxford OX 3 OBW - England

Photonetics
52 A. de l'Europe - BP 39
78160 Marly le Roi - France

Polytec
7517 Waldbrown - R.F.A.

Retsch Gmbh
Rheinische Strasse 36
5557 Haan 1 - R.F.A.